ASPARTIC PROTEINASES

Structure, Function, Biology, and Biomedical Implications

ADVANCES IN EXPERIMENTAL MEDICINE AND BIOLOGY

Recent Volumes in this Series

Volume 357
LACTOFERRIN: Structure and Function
Edited by T. William Hutchens, Sylvia V. Rumball, and Bo Lönnerdal

Volume 358
ACTIN: Biophysics, Biochemistry, and Cell Biology
Edited by James E. Estes and Paul J. Higgins

Volume 359
TAURINE IN HEALTH AND DISEASE
Edited by Ryan J. Huxtable and Dietrich Michalk

Volume 360
ARTERIAL CHEMORECEPTORS: Cell to System
Edited by Ronan G. O'Regan, Philip Nolan, Daniel S. McQueen, and David J. Paterson

Volume 361
OXYGEN TRANSPORT TO TISSUE XVI
Edited by Michael C. Hogan, Odile Mathieu-Costello, David C. Poole, and Peter D. Wagner

Volume 362
ASPARTIC PROTEINASES: Structure, Function, Biology, and Biomedical Implications
Edited by Kenji Takahashi

Volume 363
NEUROCHEMISTRY IN CLINICAL APPLICATION
Edited by Lily C. Tang and Steven J. Tang

Volume 364
DIET AND BREAST CANCER
Edited under the auspices of the American Institute for Cancer Research; Scientific Editor: Elizabeth K. Weisburger

Volume 365
MECHANISMS OF LYMPHOCYTE ACTIVATION AND IMMUNE REGULATION V: Molecular Basis of Signal Transduction
Edited by Sudhir Gupta, William E. Paul, Anthony DeFranco, and Roger Perlmutter

ASPARTIC PROTEINASES

Structure, Function, Biology, and Biomedical Implications

Edited by

Kenji Takahashi

The University of Tokyo
Tokyo, Japan

SPRINGER SCIENCE+BUSINESS MEDIA, LLC

Library of Congress Cataloging-in-Publication Data

Aspartic proteinases : structure, function, biology, and biomedical implications / edited by Kenji Takahashi.
p. cm. -- (Advances in experimental medicine and biology ; v. 362)
Proceedings of the 5t International Conference on Aspartic Proteinases, held Sept. 19-24, 1993 at the Naito Museum of Pharmaceutical Science and Industry, in Kawashima-cho, Gifu Prefecture, Japan.
Includes bibliographical references and index.
ISBN 978-0-306-44830-0 ISBN 978-1-4615-1871-6 (eBook)
DOI 10.1007/978-1-4615-1871-6
1. Aspartic proteinases--Congresses. I. Takahashi, Kenji. II. International Conference on Aspartic Proteinases (5th : 1993 : Naito Museum of Pharmaceutical Science and Industry) III. Series.
[DNLM: 1. Aspartic Proteinases--physiology--congresses. W1 AD559 v. 362 1994 / QU 136 A8376 1994]
QP609.A86A88 1994
612'.01516--dc20
DNLM/DLC
for Library of Congress 94-43762
CIP

Proceedings of the Fifth International Conference on Aspartic Proteinases, held September 19–24, 1993, in Gifu, Japan

ISBN 978-0-306-44830-0

10 9 8 7 6 5 4 3 2 1

This work is dedicated to the loving memory of
Tomoko and Matsue Takahashi

PREFACE

The 5th International Conference on Aspartic Proteinases was held on September 19 through 24, 1993, at Naito Museum of Pharmaceutical Science and Industry, Kawashima-cho, Gifu Prefecture, Japan, about 15 miles northwest of Nagoya City. About 100 scientists attended the conference, including 52 from 14 countries outside Japan, and 32 papers were presented by invited speakers, and 58 papers as posters. The purpose of this conference was to present and discuss new information on the structure, function, and biology, and related topics, including biomedical implications, of aspartic proteinases, and this book is a collection of nearly all the papers presented at the meeting.

Aspartic proteinases belong to one of the four major classes of proteinases, the others being serine, cysteine, and metalloproteinases, and are so called since they have two catalytic aspartic acid residues in common in their active sites. Most of them are optimally active at acidic pH, hence the long-used name "acid proteinases," which, indeed, was the major title of the first conference of this series. However, some of them are active at around neutral pH, indicating their physiological roles in a wider range of pH than hitherto considered.

The previous history of this series of international conferences and related additional meetings was summarized in some detail in the preceding book by Ben Dunn (3). The first conference was organized by Jordan Tang and held at Norman, Oklahoma, in 1976. The second conference was organized 8 years later by Vladimir Kostka and held in Prague, Czechoslovakia in 1984. After 4 years, the third conference, organized by Bent Foltmann, was held at Elsinore, Copenhagen, Denmark in 1988. This was followed 2 years later by the fourth conference, organized by Ben Dunn and held at Sonoma County, California, in 1990. The proceedings of the first, the second, and the fourth conferences were published (1-3). In addition, two small meetings (workshops) were held. One was organized by Tom Blundell and John Kay and held at the Birkbeck College, London, England, in 1982, and the other was organized by Michael Samloff, sponsored by Chugai Pharmaceutical Company of Japan, and held in Tokyo, Japan in 1985. Thus, the fifth conference was indeed the first major meeting of this series organized and held in Asia, which further demonstrated that the conference was really international.

On the occasion of the Sonoma conference, several people suggested that Japan might be the candidate for the site of the next conference. Through occasional communications and discussions with Ben Dunn and Jordan Tang during 1991 to 1992, I finally decided to organize the fifth conference and to hold it somewhere in Japan in 1993. Although the previous conferences had not been numbered, I dared to use the term "The 5th" for this conference just to demonstrate the continuity of this important series of conferences.

In this book, the papers are arranged, except for some papers, in the order as they were presented at the conference, essentially following the style of the previous volume (3). The presentations were grouped into four categories, *i.e.*, gastric, non-gastric, retroviral, and microbial aspartic proteinases. The grouping is not always very strict, and the latter two groups are treated separately from the non-gastric proteinases group. Each group is devided into two sections, *i.e.*, plenary lectures and poster reports, corresponding to oral and poster presentations, respectively. As can be seen from the book, significant new advances have been made these few years in various aspects of aspartic proteinases. They include X-ray structural studies on progastricsin, renin, cathepsin D, etc.; molecular dynamics; protein engineering by site-directed mutagenesis; and design and chemical syntheses of new inhibitors and even new enzymes. Further, several new aspartic proteinases have been found and characterized, and notably a greater number of papers than before were presented this time on intracellular aspartic proteinases, including cathepsins D and E, novel processing proteinases, several plant proteinases, plasmodium proteinases, etc., thus expanding further the realm of aspartic proteinases and simultaneously demonstrating their wider range of biological importance. Much attention was also paid from the biomedical point of view to several reports on aspartic proteinases related to diseases, such as renin, HIV and related retrovirus proteinases, *Candida* proteinases, and malaria parasite proteinase, and on the clinical application of serum pepsinogens. Thus, aspartic proteinases appear to become increasingly important in the biomedical area.

I deeply acknowledge Eisai Company, Ltd., for generously providing us with the beautiful and comfortable site and facilities for the conference and lodging, and thank people in the Eisai Industrial Park and Kawashima Co., Ltd., for their kind and warm accomodations. This also made it possible for all participants to enjoy more intimate academic discussions and communications until late every evening after the sessions. I am especially grateful to Professor Ben Dunn for helpful advice and suggestion about the organization of the conference, selection of speakers and editing of this book, and assistance in raising funds. My sincere thanks also go to Drs. Jordan Tang, John Kay, and Alexander Wlodawer for helpful suggestions and advices about selection of speakers, and Drs. Shin-ichi Ishii, Teruhiko Beppu, and Kazumasa Miki for help in raising funds. I am thankful to all my colleagues, including Drs. Takashi Kageyama, Akihiko Moriyama, Masaru Tanokura, Hideshi Inoue, Tomonari Muramatsu, Masao Ichinose, Shinko Tsukada, and others and to the 13 graduate students in my laboratory, for assistance in the organization and operation of the conference. Drs. H. Inoue and T. Muramatsu especially devoted much time to this work. Further, thanks are due to Mrs. Hisako Kamioka for skillful assistance as a secretary for the conference and for preparation of this book, and Mrs. Masako Taguchi for help with several aspects of the secretary work. I also express my thanks to Mss. Mary Safford and Eileen Bermingham and Messrs. Thomas L. Flood and Brian P. Halm of Plenum Publishing Corporation for helpful suggestions and cooperation for the preparation of this book.

Finally, I express my sincere thanks to International Science Foundation for cooperation as the coorganizer of this conference and to many groups that have provided finantial support for the conference, without which this successful meeting would not have been realized. Their names are listed in a preceding page.

It is hoped that this book will serve as one of the milestones for further advances in studies on this unique class of enzymes.

Kenji Takahashi
Tokyo, Japan

REFERENCES

1. "Acid Proteases : Structure, Function, and Biology," Jordan Tang, ed., Advances in Experimental Medicine and Biology, Vol. 95, Plenum Press, New York (1977).

2. "Aspartic Proteinases and Their Inhibitors : Proceedings of the FEBS Advanced Course No. 84/07," Vladimir Kostka, ed., Walter de Gruyter, Berlin (1985).
3. "Structure and Function of the Aspartic Proteinases. Genetics, Structures, and Mechanisms," Ben M. Dunn ed., Advances in Experimental Medicine and Biology, Vol. 306, Plenum Press, New York (1991).

ACKNOWLEDGMENTS

Financial Support for the Conference was Provided by the Following Sponsors:

Ajinomoto Co., Inc.
Applied Biosystems Japan, Inc.
DAINABOT Co., Ltd.
Eisai Co., Ltd.
Funakoshi Co., Ltd.
Glaxo Group Research, Ltd.
Japan Energy Corporation
Kato Memorial Bioscience Foundation
Kikkoman Corporation
Kirin Brewery Co., Ltd.
Meito Sangyo Co., Ltd.
NEC Corporation
Nippon Chemiphar Co., Ltd.
Osaka Pharmaceutical Manufacturers Association
Roche Products, Ltd.
Seikagaku Corporation, Tokyo Research Institute
SmithKlein Beecham Seiyaku K. K.
Suntory Institute for Biomedical Research
Terumo Life Science Foundation
The Asahi Glass Foundation
The Kajima Foundation
The Mitsubishi Foundation
The Naito Foundation
The Pharmaceutical Manufacturers' Association of Tokyo
The Tokyo Biochemical Research Foundation
Werner-Lambert / Parke-Davis Pharmaceutical Research
Yakult Honsha Co., Ltd.

CONTENTS

Gastric Aspartic Proteinases: Plenary Lectures

Gastric Aspartic Proteinases: Poster Reports

Non-Gastric Aspartic Proteinases: Poster Reports

Retroviral Aspartic Proteinases: Plenary Lectures

Retroviral Aspartic Proteinases: Poster Reports

Microbial Aspartic Proteinases: Plenary Lectures

Microbial Aspartic Proteinases: Poster Reports

COLOR PLATES

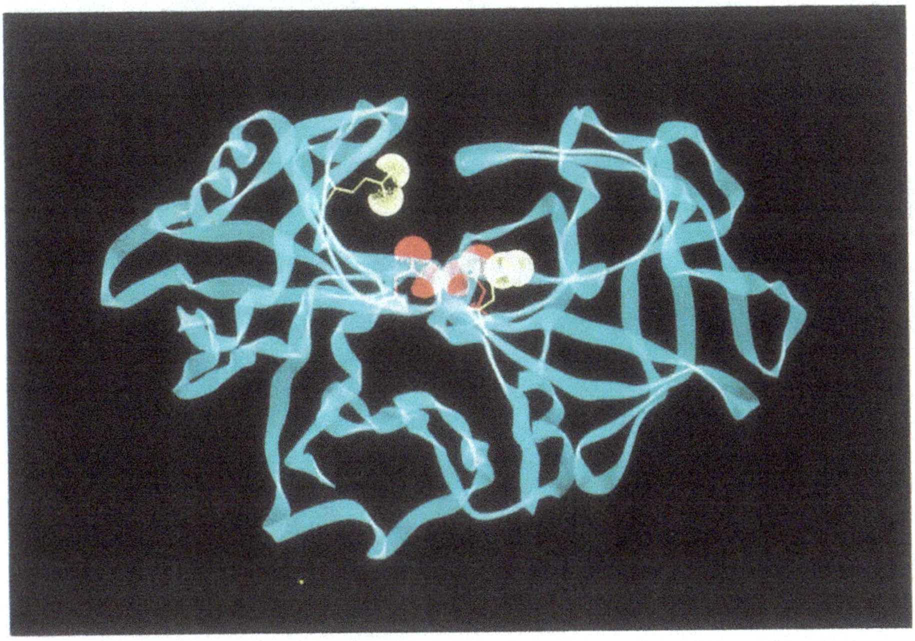

Chapter 1, Dunn et al.: Figure 1. Ribbon diagram representation of the structure of an archetypal aspartic proteinase. The N-terminal domain is on the right in this view. The side chains of the two catalytic residues are represented by red stick structures, and the two oxygen atoms are indicated as van der Waals surfaces, seen as red dots. The positions of two important Glu residues in porcine pepsin (13 and 287) are also given by white stick structures, with the oxygen atoms of these indicated as white van der Waals surfaces.

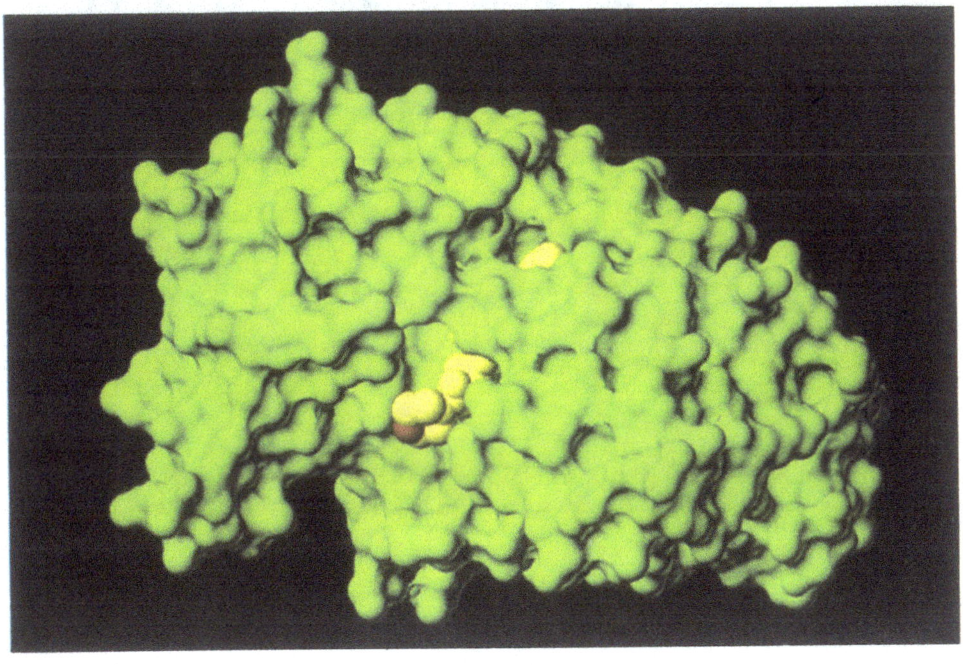

Chapter 21, Hoover et al.: Figure 2. Predicted binding of a morpholine-containing inhibitor to human renin. Inhibitor CP-81,282 (Mor-CO-Phe-Nle-DifluoroC-StatoneNHMe, endothiapepsin-bound X-ray coordinates [30]), modelled into the X-ray coordinates of recombinant human renin (from the renin/CP-85,339 crystal structure [31]. View shows the solvent-accessible (Connoly) surface of renin (green) and the solvent-exposed N-terminus of the inhibitor (yellow, with morpholine oxygen in red). The N-terminus of the inhibitor is solvent-exposed.

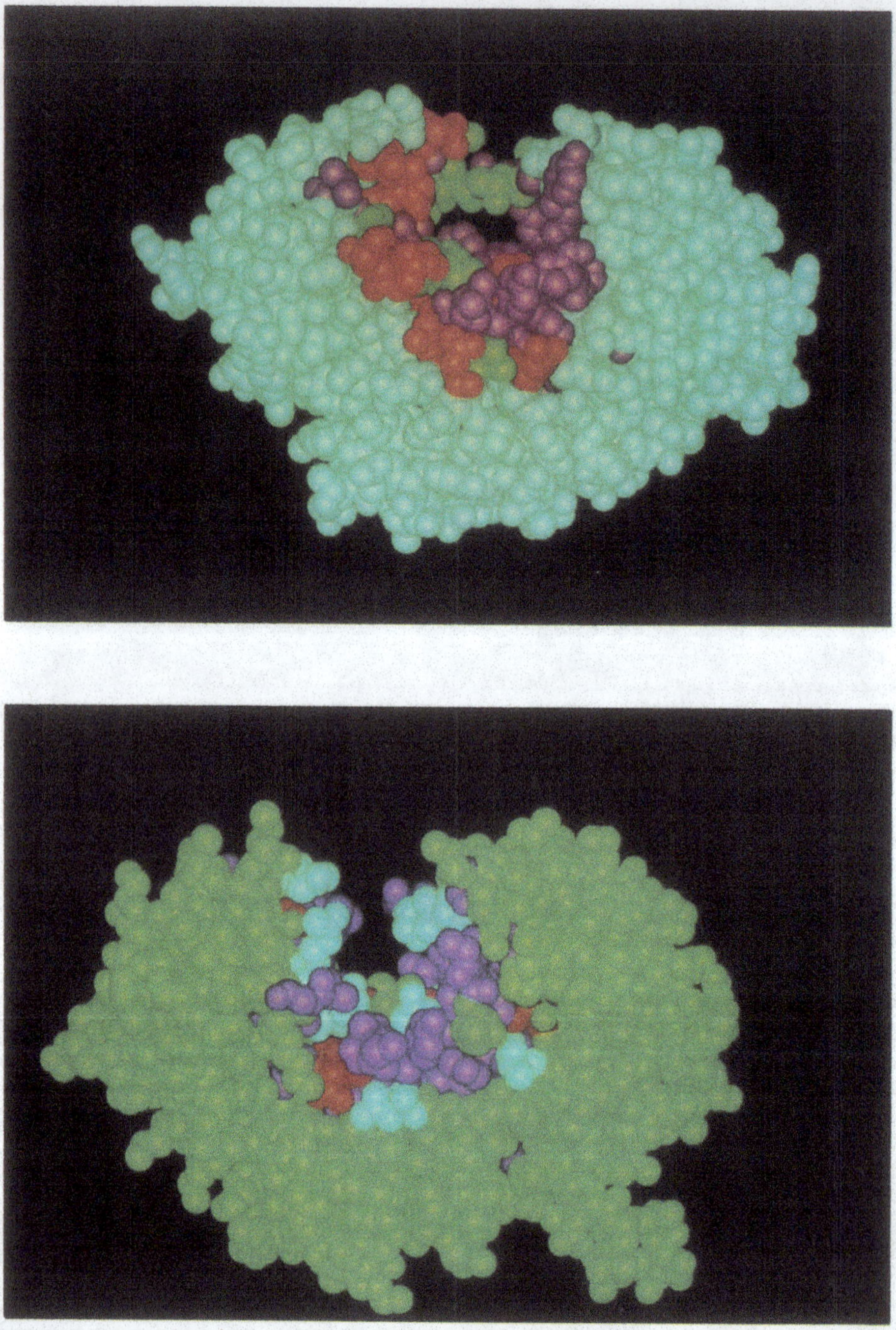

Chapter 1, Dunn et al.: Figure 4. Representation of the acidic (Asp/Glu) and hydrophobic residue distribution within the active site clefts of (A) porcine pepsin and (B) human cathepsin D. Asp and Glu side chains are colored red in these space-filling diagrams. Hydrophobic residues are colored purple, and hydrophobic residues (e.g., Ser, Thr, Asn, Gln) are colored green. Residues outside the active site cleft are shown as cyan spheres. Both structures are oriented at the same angle.

COMPARISON OF THE ACTIVE SITE SPECIFICITY OF THE ASPARTIC PROTEINASES BASED ON A SYSTEMATIC SERIES OF PEPTIDE SUBSTRATES

Ben M. Dunn, Paula E. Scarborough, W. Todd Lowther, and Chetana Rao-Naik

Department of Biochemistry and Molecular Biology
University of Florida College of Medicine
Gainesville, FL 32610-0245

INTRODUCTION

The first crystal structures of members of the aspartic proteinase family were discussed at the 1976 Conference in Norman, Oklahoma [1]. Since that time, an increasingly rapid stream of structural information has issued from the crystallography groups [2]. At this point it is clear that members of this family are closely related structurally, with a central core composed of two domains, each of which has a two-layered β-sheet arrangement (see Figure 1 for an overall representation of a typical aspartic proteinase).

At the interface of these two domains in all the enzymes is a clearly-defined active site cleft where substrate and inhibitor binding takes place. The dimensions of this cleft are longer (20-30 Angstroms) than wide, so that a peptide substrate binds in an extended fashion. From structural studies of bound inhibitor molecules at least seven amino acids can be accommodated within the cleft in an extended β-strand conformation. This agrees with early specificity studies [3,4] where demonstrable preferences for specific amino acids were observed in analyses of protein cleavage by porcine pepsin. Based on these reports, we designed and synthesized [5] a peptide substrate which proved of value in the kinetic analysis of enzymes in the aspartic proteinase class.

In addition, work initially described by Fruton [6] and more recently pursued by Hofmann and his colleagues [7,8] has clearly demonstrated that extending a peptide substrate through at least five amino acids leads to significant increases in k_{cat} when compared to shorter peptides. Thus, it is obvious that the extended cleft of the enzymes in the aspartic proteinase class has evolved to provide multiple points of interaction between substrate and enzyme in order to maximize efficiency of cleavage.

The increase in rate of cleavage is usually ascribed to added hydrogen bonds between enzyme residues and the backbone -NH- groups at P_3 and P_2' or P_3'. [Positions of a substrate or inhibitor are designated according to Schechter and Berger, BBRC (1967), 27:157]. Indeed,

Aspartic Proteinases: Structure, Function, Biology, and Biomedical Implications
Edited by Kenji Takahashi, Plenum Press, New York, 1995

Figure 1. Ribbon diagram representation of the structure of an archetypal aspartic proteinase. The N-terminal domain is on the right in this view. The side chains of the two catalytic residues are represented by red stick structures, and the two oxygen atoms are indicated as van der Waals surfaces, seen as red dots. The positions of two important Glu residues in porcine pepsin (13 and 287) are also given by white stick structures, with the oxygen atoms of these indicated as white van der Waals surfaces. (See color plate preceding page 1.)

examination of all enzyme/inhibitor complexes has shown that almost all -NH- and C=O groups of a bound peptide or peptidomimetic are involved in hydrogen bonding. It is easy to imagine a picture where a bound peptide is "tied down" at multiple points by this hydrogen bonding. This would maximize residence time of a peptide within the active site cleft.

In recent years, we have turned our attention to a second question in these enzymes: How can we account for differences in specificity of cleavage between various members of this family? While it is clear that the extensive backbone hydrogen bonding mentioned above is one critical factor in the rate of cleavage, this cannot account for specificity, as all peptide sequences have identical -NH- and C=O groups.

To account for differences in the specific points of cleavage it is necessary to consider the second feature of a bound peptide or inhibitor: the side chain groups. While hydrogen bonding "ties down" a bound peptide, and may contribute to the energy of binding, the side chains of the peptide are able to interact with specific regions within the cleft, potentially with either a positive or a negative effect upon the energy of binding.

A schematic diagram illustrating the extended mode of binding of a peptide within the active site cleft of an aspartic proteinase as well as the concept that individual side chains of the bound peptide interact with defined regions ("subsites") of the cleft is presented in Figure 2.

We can then hypothesize that differences in the interactions at specific positions, when comparing two enzymes in this family interacting with one peptide substrate, can account for observed differences in the rates of cleavage of the peptide sequence. This hypothesis is based upon knowledge of the structures of these enzymes (and their complexes) which reveals that the overall features of the active site clefts are similar, while the specific

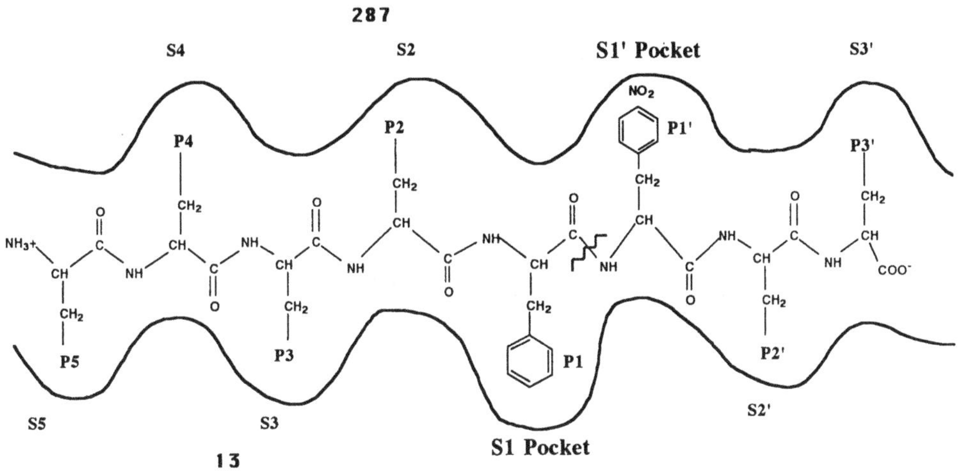

Figure 2. Schematic representation of the binding of a substrate in the active site cleft of an aspartic proteinase. Note the extended β-strand conformation of the substrate, in agreement with the pattern seen in inhibitor complexes. The scissile peptide bond is juxtaposed with the catalytic apparatus. Each amino acid side chain of the substrate interacts with a specific region of the active site, or "subsite". The involvement of residues 13 and 287 in the S_3 or S_2 regions, respectively, are indicated by those numbers in the appropriate

amino acid residues of the enzyme surface show some variation in positions that we believe to be critical to specificity.

In order to explore the differences in specificity between members of this family, and to test the hypothesis we have developed above, a set of peptide substrates was constructed. This set is based upon previous results with porcine pepsin [9], with the parent peptide having the sequence:

Lys-Pro-Ala-Lys-Phe*Nph-Arg-Leu

where the asterisk (*) indicates the point of cleavage, and Nph is the three-letter code for *p*-NO_2phenylalanine. The two aromatic residues, Phe in the P_1 position and Nph in the P_1' position, fit the primary specificity of most members of the aspartic proteinase family well. In the series given in Figure 3, we have kept the two primary specificity determining residues (-Phe*Nph-) constant, but have varied the residues in the other six positions to include the small, hydrophobic Ala, the small, hydrogen-bonding Ser, the small, acidic Asp, the larger basic residue Arg, and the larger, hydrophobic Leu. In this series, we have avoided the aromatic residues to avoid creating new cleavage points and the sulfur-containing amino acids in favor of long-term stability.

This report will compare selected results with porcine pepsin, human cathepsin D, and the aspartic proteinase from *Rhizopus chinensis*. The three-dimensional structures of those proteins are known at this point, facilitating interpretation of the differences observed.

MATERIALS AND METHODS

All peptides used in this study were synthesized by solid-phase methodology, utilizing *t*-BOC chemistry. Following cleavage from the synthetic resin by anhydrous HF, the purity of the peptides was assessed by amino acid analysis, HPLC (reversed-phase C-18

```
                Peptide XVI Series
               "Ben's Bookshelf"

LYS-PRO-ALA-LYS-PHE*NPH-ARG-LEU        XVI

ALA- - - - - - - - - * - - - - - -     XVIP5A
ASP- - - - - - - - - * - - - - - -     XVIP5D
SER- - - - - - - - - * - - - - - -     XVIP5S
LEU- - - - - - - - - * - - - - - -     XVIP5L
ARG- - - - - - - - - * - - - - - -     XVIP5R

LYS-PRO-ALA-LYS-PHE*NPH-ARG-LEU        XVI
 - -ALA- - - - - - - * - - - - - -     XVIP4A
 - -ASP- - - - - - - * - - - - - -     XVIP4D
 - -SER- - - - - - - * - - - - - -     XVIP4S
 - -LEU- - - - - - - * - - - - - -     XVIP4L
 - -ARG- - - - - - - * - - - - - -     XVIP4R

LYS-PRO-ALA-LYS-PHE*NPH-ARG-LEU        XVI
 - - - -ASP- - - - - * - - - - - -     XVIP3D
 - - - -SER- - - - - * - - - - - -     XVIP3S
 - - - -LEU- - - - - * - - - - - -     XVIP3L
 - - - -ARG- - - - - * - - - - - -     XVIP3R

LYS-PRO-ALA-LYS-PHE*NPH-ARG-LEU        XVI
 - - - - - - -ALA- - * - - - - - -     XVIP2A
 - - - - - - -ASP- - * - - - - - -     XVIP2D
 - - - - - - -SER- - * - - - - - -     XVIP2S
 - - - - - - -LEU- - * - - - - - -     XVIP2L
 - - - - - - -ARG- - * - - - - - -     XVIP2R

LYS-PRO-ALA-LYS-PHE*NPH-ARG-LEU        XVI
 - - - - - - - - - - * - -ALA- -       XVIP2'A
 - - - - - - - - - - * - -ASP- -       XVIP2'D
 - - - - - - - - - - * - -SER- -       XVIP2'S
 - - - - - - - - - - * - -LEU- -       XVIP2'L

LYS-PRO-ALA-LYS-PHE*NPH-ARG-LEU        XVI
 - - - - - - - - - - * - - - -ALA      XVIP3'A
 - - - - - - - - - - * - - - -ASP      XVIP3'D
 - - - - - - - - - - * - - - -SER      XVIP3'S
 - - - - - - - - - - * - - - -ARG      XVIP3'R
```

Figure 3. Comparison of the sequences of peptides created as an internally consistent set of substrates. Use of this set of peptides provides the opportunity to investigate the specificity of a newly discovered enzyme. Cleavage between Phe and Nph results in a chromogenic shift that can be quantitated.

with CH_3CN as the mobile phase), and, in some cases, capillary electrophoresis and mass spectroscopy. The typical preparation contained 90% of the desired peptide, with small amounts of peptides containing residual blocking groups as the most common contaminant.

The enzymes utilized herein have been described in previous publications from this laboratory[2].

Active site titration with the tightest binding inhibitor available for each enzyme was used to quantitate the concentration of enzyme in the assays.

In all cases, the assay depends on the shift in absorption due to cleavage of the -Phe*Nph peptide bond, and the resulting effect of the new -NH_3+ group on the *p*-nitrophenyl chromophore of Nph. This shift, from an absorbance maximum of 280 nm to one of 272 nm, has been documented in previous reports. The procedures utilized in this laboratory are described in full in a recent chapter [10]; briefly, a Hewlett-Packard diode-array spectrophotometer is utilized to collect multiple absorbance points in the range of 284-324 nm. We take advantage of the averaging capacity of the computer to provide data smoothing as well as error estimates of the measurements. All experiments were conducted at 37°C.

Analysis of the structural relationships was carried out using the SYBYL program of Tripos Associates (St. Louis, MO, USA) and the coordinate sets 5PEP, 1LYA, and 2APR.

RESULTS AND DISCUSSION

Comparisons of kinetic data between different enzymes are dangerous, as it is possible that the mechanism of the systems might be different. In our studies, we have attempted to make all our evaluations within sets of closely related peptides. In addition, by our extensive variation of structure we have managed to identify at least some substrates that yield comparable optimal levels of activity for each enzyme, as will be detailed below. This indicates that the *intrinsic* mechanism and catalytic efficiency of each enzyme is similar, thus validating our analysis.

In the following discussion we will focus on the parameter k_{cat}/K_m, which is denoted the *specificity constant* for a given enzyme-substrate pair. This is a fundamental kinetic constant, and is preferable for the illustrations to be described. The use of the K_m parameter for comparing results between enzymes should be avoided, unless the true meaning of that constant has been defined for the enzymes in question.

The specificity constants for the three enzymes are listed in Table 1 for ten peptide substrates with changes in the P_3 and P_2 positions. The larger set of data, including all the peptides listed in Figure 3 and the kinetic parameters k_{cat} and K_m, will be reported separately (P. E. S., W. T. L., and C. R.-N., manuscripts in preparation).

Table 1. Specificity Constants for the Enzyme-Catalyzed Cleavage of Peptide Substrates of the Form Lys-Pro-*Xaa*-*Yaa*-Phe*Nph-Arg-Leu

	PORCINE PEPSIN	HUMAN CATHEPSIN D	*RHIZOPUS CHINENSIS* PROTEINASE
P_5 - P_4 - P_3 - P_2 - P_1 * P_1' - P_2' - P_3'		k_{cat}/K_m $(M^{-1}s^{-1}) \times 10^{-6}$	
Lys-Pro-*Ala*-*Lys*-Phe*Nph-Arg-Leu	1.1	< 0.02	1.3
Lys-Pro-*Asp*-*Lys*-Phe*Nph-Arg-Leu	0.1	< 0.02	0.7
Lys-Pro-*Ser*-*Lys*-Phe*Nph-Arg-Leu	0.3	< 0.02	0.6
Lys-Pro-*Leu*-*Lys*-Phe*Nph-Arg-Leu	0.4	< 0.02	1.4
Lys-Pro-*Arg*-*Lys*-Phe*Nph-Arg-Leu	< 0.02	< 0.02	1.4
Lys-Pro-*Ala*-*Ala*-Phe*Nph-Arg-Leu	0.8	0.2	2.4
Lys-Pro-*Ala*-*Asp*-Phe*Nph-Arg-Leu	0.5	0.3	0.8
Lys-Pro-*Ala*-*Ser*-Phe*Nph-Arg-Leu	0.5	0.1	1.0
Lys-Pro-*Ala*-*Leu*-Phe*Nph-Arg-Leu	0.7	1.0	2.3
Lys-Pro-*Ala*-*Arg*-Phe*Nph-Arg-Leu	0.5	< 0.02	2.2

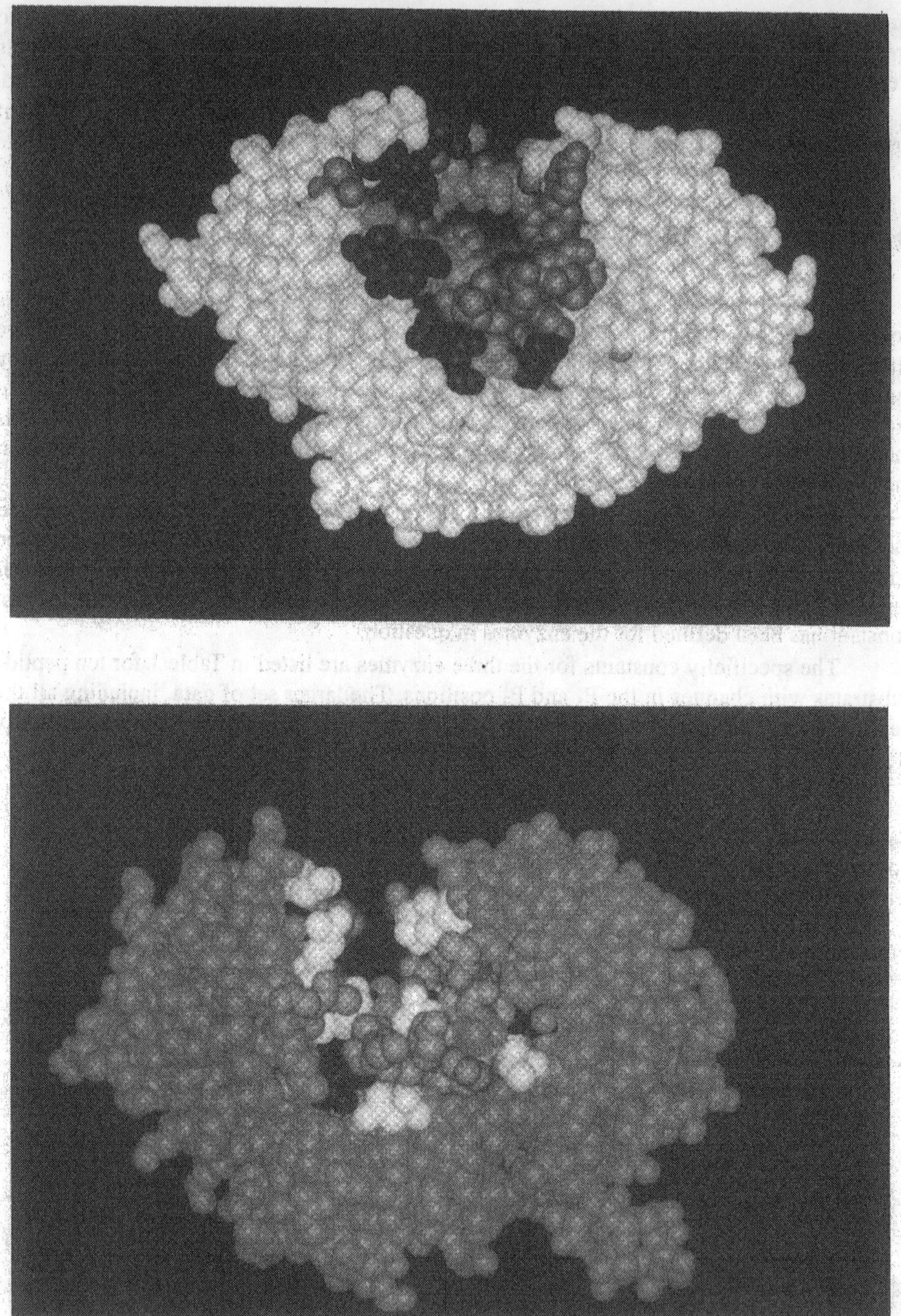

Figure 4. Representation of the acidic (Asp/Glu) and hydrophobic residue distribution within the active site clefts of (A) porcine pepsin and (B) human cathepsin D. Asp and Glu side chains are colored red in these space-filling diagrams. Hydrophobic residues are colored purple, and hydrophobic residues (e.g., Ser, Thr, Asn, Gln) are colored green. Residues outside the active site cleft are shown as cyan spheres. Both structures are oriented at the same angle. (See color plate preceding page 1.)

In examining these data, we will begin with several generalizations. First, it can be seen in the upper section (P_3 substitutions) that the three enzymes show significant differences in their specificity toward the five substrates with Lys in the P_2 position. Pepsin cleaves the peptides with Ala, Asp, Ser, and Leu in P_3 at roughly equivalent rates, but does not cleave the peptide containing Arg in the P_3 position; cathepsin D cleaves none of the five peptides at measurable rates; while *Rhizopus chinensis* proteinase cleaves all five peptides at roughly similar rates, and is the most efficient enzyme.

In the lower section it can be seen that all three enzymes cleave most of the peptides at nearly equivalent efficiencies; the *significant* exception is that cathepsin D does not cleave the peptide with Arg in the P_2 position. This result can be combined with the topmost peptide, with Ala-Lys in P_3-P_2, confirming that cathepsin D does not tolerate a positively-charged residue in the P_2 position.

In comparing all the information given in Table 1, we can see that there are two contrasting themes. On the one hand it is clear that some substitutions within the peptide substrate sequence are without (major) consequence; this helps to confirm that this *series* of peptides is interacting with the various enzymes in a common fashion. As long as the side chain substituted into the peptide can be accommodated within the corresponding enzyme subsite without penalty, the resulting catalytic efficiency is approximately equivalent. For example, all three enzymes cleave the peptide of sequence, Lys-Pro-Ala-Leu-Phe*Nph-Arg-Leu, with nearly equal k_{cat}/K_m values ($0.7 \times 10^6\ M^{-1}s^{-1}$, $1.0 \times 10^6\ M^{-1}s^{-1}$, and $2.3 \times 10^6\ M^{-1}s^{-1}$, for pepsin, cathepsin D, and *Rhizopus chinensis* proteinase, respectively).

On the other hand, when a substitution is made that leads to an energetically unfavorable interaction with the enzyme surface at that particular point in the active site cleft, the catalytic reaction is dramatically altered.

The source of the energetic penalty in both cases shown in Table 1 is essentially the same; placing a positively charged residue in a position where it is forced to interact with a hydrophobic subsite of the enzyme surface.

In the case of porcine pepsin, the only substrate that shows reduced cleavage is the one in which an Arg is present in the P_3 position. Examination of the structure of the S_3 region of the active site cleft of this enzyme reveals a preponderance of hydrophobic residues (Tyr_{75}, Phe_{111}, Ala_{115}). Thus, it is reasonable that the positively charged side chain of Arg would not interact favorably with this environment. The only residue in this vicinity that could offer an energetically beneficial interaction is the side chain of Glu_{13}; however, at the pH of the assay reported in Table 1 (pH 3.5), the side chain of this residue would be neutral. When the pH of the assay is increased, the rate of cleavage of the peptide containing Arg in the P_3 position increases, presumably due to loss of a proton from Glu_{13} and the consequent formation of an ion-pair interaction (C. R.-N. and B. M. D., unpublished observations). Site-directed mutagenesis at position 13 is now being employed to test this hypothesis.

For human cathepsin D, all substrates with a positively-charged residue in the P_2 position are cleaved at rates significantly below substrates with Ala, Asp, Ser, or Leu in that position. Examination of a molecular model of human cathepsin D [11] revealed that the S_2 region of that enzyme is more hydrophobic, due to the presence of a Met residue at position 287, than that of porcine pepsin, which has a Glu in the same position. These observations were later confirmed upon the publication of the three-dimensional structure of human cathepsin D by two groups [12,13]. Site-directed mutagenesis has been utilized to alter the subsite preferences of human cathepsin D by changing the Met at position 287 into a Glu [14]. This mutant enzyme cleaves the peptides containing Lys or Arg in the P_2 position of substrate at readily measurable rates [14].

Overall, the active site cleft of porcine pepsin is significantly more acidic than that of human cathepsin D, as illustrated in Figure 4, panels A and B. The large number of Asp or Glu residues within the active site cleft of porcine pepsin accounts for the rapid cleavage of peptides such as those shown in Figure 3 and Table 1, where the net charge on each peptide varies from + 1-4 in the acidic pH range under consideration. We describe this phenomenum as a "general electrostatic effect", whereas the effect observed on titration of Glu_{13} in porcine pepsin would be described as a "specific electrostatic effect". We differentiate these effects based on the results generated with specific substitutions, such as those listed in Figure 3.

Finally, it can be seen in Table 1 that the proteinase from *Rhizopus chinensis* is the most versatile enzyme of the three compared, as it is able to cleave all the peptides listed at roughly comparable rates. In particular, it is of great interest that the *Rhizopus* enzyme is the only one that can cleave the peptide with Arg in the P_3 position at pH 3.5. Comparing the active site structures of the three enzymes, it is clear that *Rhizopus chinensis* aspartic proteinase is unique in the presence of two *additional* acidic residues: Asp_{30} and Asp_{77} (using the pepsin numbering system). Asp_{30} is located at the interface of the S_1 and S_3 subsites, on the N-terminal domain, while Asp_{77} is located on the mobile "flap" that hangs over the active site cleft. Asp_{77} can interact with residues in several positions, including at least P_1 and P_2. In addition, *Rhizopus chinensis* aspartic proteinase has Glu at position 13, while the residue corresponding to 287 is a Gly. Thus, the *Rhizopus* enzyme is identical to pepsin at position 13, but is more like human cathepsin D at position 287. The lack of an acidic residue at 287 is compensated for by the additional acidic side chains at 30 and 77. This leads to the *Rhizopus* enzyme demonstrating the most flexibility in accommodation of substrates, cleaving all the 10 peptides offered to it in this study.

Again, the unique residues in the *Rhizopus chinensis* aspartic proteinase are the targets of site-directed mutagenesis (T. L. and B. M. D., unpublished observations) and will be the subject of future publications from this laboratory.

In summary, the comparison presented here illustrates the dominant role of a few residues in determining the "secondary specificity" of the aspartic proteinases [5]. While the overall conformation of the three enzymes are extremely similar, the specific amino acid side chains lining the pockets of the various subsites of the enzymes are variable. The residues can provide either an attractive or a repulsive force when interacting with a complementary side chain of a ligand.

The differences in specificity revealed by the comparison of cleavage rates of the sets of peptides presented in Figure 3 and Table 1 provide information that could be of value in the design of selective inhibitors for each enzyme. Conversely, substituents could be chosen, based on this information, to design a compound that would *avoid* interaction with a specific enzyme. It will be of interest to determine the patterns of specificity of other members of this class. Finally, the data obtained by analysis of the set of peptides given in Figure 3 provides an important database of information that is essential to the analysis of the effects of site-directed mutagenesis of various members of the aspartic proteinase class.

REFERENCES

1. *Acid Proteases: Structure, Function, and Biology* (J. Tang, Ed.), Advances in Experimental Medicine and Biology, Volume 95, Plenum Press, New York, (1977).
2. *Aspartic Proteinases and Their Inhibitors* (V. Kostka, Ed.), Walter de Gruyter, New York, (1985); *Structure and Function of the Aspartic Proteinases* (B. M. Dunn, Ed.), Advances in Experimental Medicine and Biology, Volume 306, Plenum Press, New York, (1991).
3. J. C. Powers, A. D. Harley, and D. V. Myers, *in: Acid Proteases: Structure, Function, and Biology* (J. Tang, Ed.), Advances in Experimental Medicine and Biology, Volume 95, Plenum Press, New York, pp. 141-157 (1977).
4. V. K. Antonov, *in: Acid Proteases: Structure, Function, and Biology* (J. Tang, Ed.), Advances in Experimental Medicine and Biology, Volume 95, Plenum Press, New York, pp. 179-198 (1977).
5. B.M. Dunn, B. Kammermann, and K.R. McCurry, Anal. Biochem., 138: 68-73 (1984).

6. J. S. Fruton, Adv. Enzymol. 44:1-36 (1976).
7. T. Hofmann, B. Allen, M. Bendiner, M. Blum, and A. Cunningham, Biochem., 27:1140-1146 (1988).
8. M. Balbaa, A. Cunningham, and T. Hofmann, Archives B. B., 306:297-303 (1993).
9. J. Pohl and B.M. Dunn, Biochem., 27: 4827-4834 (1988).
10. B.M. Dunn, P.E. Scarborough, and W. Swietnicki, *in: Peptide Synthesis and Purification Protocol*, (B.M. Dunn and M.W. Pennington, Eds.), Humana Press, Clifton, N.J., in press (1993).
11. P.E. Scarborough, K. Guruprasad, T.L. Blundell, G.R. Richo, and G.E. Conner, Protein Science, 2: 264-276 (1993).
12. P. Metcalf and M. Fusek, EMBO J. 12:1293-1302.
13. E. T. Baldwin, T. N. Bhat, S. Gulnik, M. Hosur, R. C. Sowder II, R. E. Cachau, J. Collins, A. M. Silva, and J. W. Erickson, Proc. Natl. Acad. Sci. (USA) 90:6796-6800.
14. P.E. Scarborough and B.M. Dunn, Protein Engineering, in press (1994).

THE MOLECULAR STRUCTURE OF HUMAN PROGASTRICSIN AND ITS COMPARISON WITH THAT OF PORCINE PEPSINOGEN

Michael James, Stanley Moore, Anita Sielecki, Maia Chernaia, and Nadezhda Tarasova[1]

Medical Research Council of Canada
Group in Protein Structure and Function
Department of Biochemistry
University of Alberta
Edmonton, Alberta T6G 2H7, Canada
[1]ABL-BRP
National Cancer Institute
FCRDC, P.O. Box B, Frederick, MD 21702

INTRODUCTION

Mammalian aspartic proteinases are synthesized as inactive precursors or zymogens. Stomach zymogens undergo a conversion to the active enzyme form autocatalytically at pH < 5.0 (1). The human gastric juice has two major groups of aspartic proteinases, the pepsins (EC3.4.23.1) and the gastricsins (EC3.4.23.3). Progastricsin or pepsinogen C (PGC) is converted to gastricsin by removal of the 43 amino-terminal residues of the prosegment. The resulting mature gastricsin has 329 amino acid residues. The sequence of human PGC has been determined independently in two laboratories by nucleotide sequencing of the gene (2) and of cDNA clones (3).

The most thoroughly studied conversion of zymogen to active enzyme is that of porcine pepsinogen to pepsin. Not only is there a wealth of biochemical and biophysical data for the system (4-11), but also the molecular structures of porcine pepsinogen (12-14) and of porcine pepsin (15-17) are known at high resolution. Reasonably detailed pathways of this molecular conversion have been elaborated (9,18,8).

Progastricsin is activated through a pathway of conformational change and partial proteolysis that is different from that of pepsinogen (19,20). In order to contribute to a more complete understanding of the activation process of progastricsin we have determined the molecular structure of the human zymogen at 1.6 Å resolution. Comparison of this structure with that of porcine pepsinogen shows many similarities but some unexpected and remarkable differences.

Table 1. Crystallographic Refinement Statistics for Human Progastricsin

Resolution range (Å)	20.0 - 1.62
R factor*	0.180
Number of reflections	50353
Completeness of data	96%
Total no. of protein atoms	2863
Total no. of solvent molecules	243
rms deviations from ideal stereochemistry:	
bond distances	0.016 Å
bond angles	2.7°
planar groups	0.016 Å
ω of peptide bonds	2.9°

* R-factor = $\Sigma||F_o|-|F_c||/\Sigma|F_o|$ where $|F_o|$ and $|F_c|$ are observed and calculated structure factor amplitudes, respectively.

MATERIALS AND METHODS

The isolation of human progastricsin (hPGC), its purification and crystallization have been reported (21). Shortly following this report, a collaboration between the Moscow groups and our laboratory in Edmonton was established. The crystals used in the structure determination were grown under slightly different conditions than those reported previously. The hanging drop method was used with solutions of protein (conc. = 10 mg/ml) equilibrated against 1.8 M sodium formate with trace amounts (3-4% v/v) of ethanol at pH = 6.0. The

Table 2. Disordered Regions in the Structure of Human Progastricsin

Residue Range	B-factors (Å^2)†	Comments
Glu25p-Lys31p	49-71	Breaks in main-chain density, poor side-chain density.
Arg39p-Leu43p	48-71	Breaks in main-chain density, poor side-chain density.
Tyr75-Ser79	41-56	Poor side chain density (tip of flap).
Ala127-Thr134	41-82	Breaks in main-chain density, poor side-chain density. Region displaced by prosegment C-terminus.
Asn157-Ser161	52-73	Breaks in main-chain density, poor side-chain density.
Trp206-Glu210	42-46	Poor side chain density in region of S-S bridge.
Glu243-Tyr244	41-43	Poor side chain density.
Asn280-Gly282	44-53	Poor side chain density, small breaks in main chain.

† Average B-factor ranges for the main-chain atoms of the residues involved.

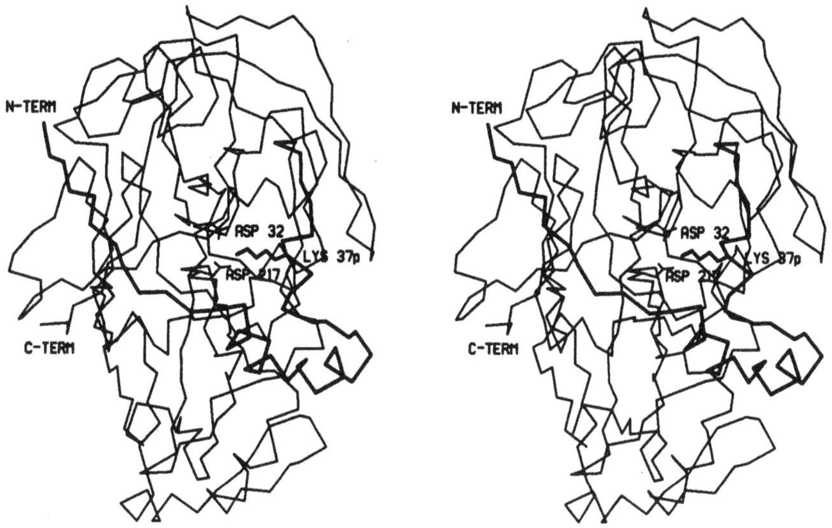

Figure 1. Stereoview of the C^{α} tracing of hPGC. The 43 residues of the prosegment are represented by a thick line whereas the 329 residues of gastricsin are drawn with thinner lines. The N-terminal and C-terminal residues are labeled and the side chains of the active site aspartate residues Asp32 and Asp217, as well as of Lys37p, are shown.

crystals are tetragonal, space group $P4_22_12$, with unit cell dimensions of $a = b = 105.1$ Å and $c = 70.3$ Å. Intensity data were collected to a resolution of 1.62 Å with the San Diego Multiwire Area Detector (22) installed in our laboratory in Edmonton. The processing of these data followed standard procedures (23). The molecular structure of hPGC was determined by a combination of *m*ultiple *i*somorphous *r*eplacement (mir) and molecular replacement phasing. Four heavy atom derivatives were used for the mir component of the process. The template for the molecular replacement contribution to the phasing was a model based on the refined structure of porcine pepsinogen (13). The resulting electron density map was reinterpreted by fitting the sequence of hPGC into the positive electron-density contours. Extensive restrained-parameter least-squares refinement (24) of this initial model has resulted in an excellent molecular structure of human progastricsin.

RESULTS AND DISCUSSION

Table 1 summarizes the refinement results obtained for the current model. Table 2 lists the regions in the molecule that have poor associated electron density and for which this molecular model should be considered tentative. These regions of disorder mostly relate to regions of conformational flexibility and many are in common with highly mobile regions in other aspartic proteinase structures that have been determined.

The overall molecular structure of hPGC is similar to that of porcine pepsinogen (Fig. 1). The conformation adopted by the first 38 residues of the prosegment is very similar to the one exhibited by the equivalent residues of the porcine zymogen (Fig. 2). The general fold of the gastricsin portion of the structure clearly resembles that of the other aspartic proteinases. The molecule is formed by two structurally similar domains that are related by an approximate molecular 2-fold axis that runs through the active site between the two

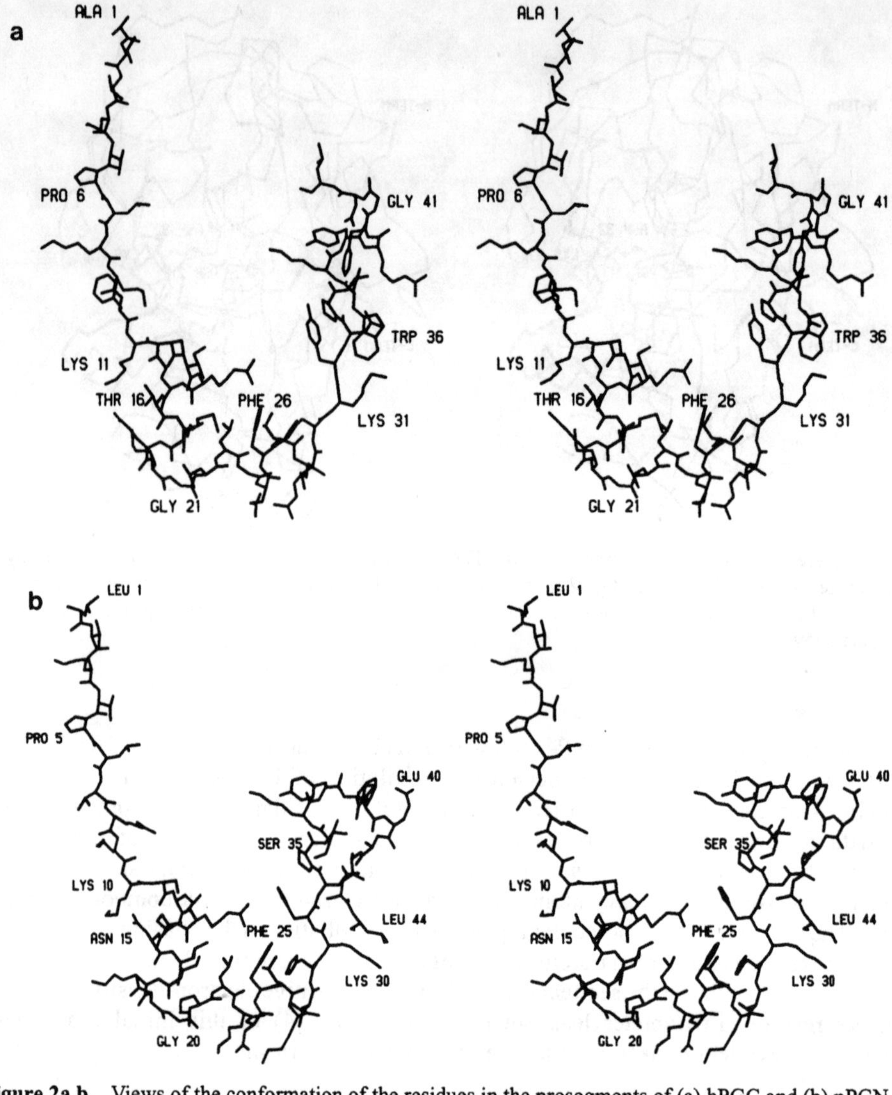

Figure 2a,b. Views of the conformation of the residues in the prosegments of (a) hPGC and (b) pPGN. The two prosegments are represented in the same (arbitrary) orientation to emphasize the similarity of conformation. Every 5th residue is labeled for identification.

catalytic aspartic acid side chains (Fig. 3). It is clear that in both the porcine and the human zymogens, the corresponding enzymes (pepsin and gastricsin) have their active sites in a close-to-active conformation. A superposition of the various residues in the active sites of pPGN and porcine pepsin indicates a faithful correspondence. The only conformational change observed is a rotation of -120° around the C^{α}-C^{β} bond of Thr218 (side chain torsional angles χ^1), disrupting the hydrogen bond Thr218 O^{γ}....Asp215 O^{δ} observed in the active pepsin. Although the structure of gastricsin is not known, a similar conformational change of the side chain of the equivalent residue Thr220 is also likely to occur. In this way, the active sites of these enzymes would attain their active conformations upon removal of the critical prosegment residue Lys37p (Lys36p)* from its observed position in the zymogens.

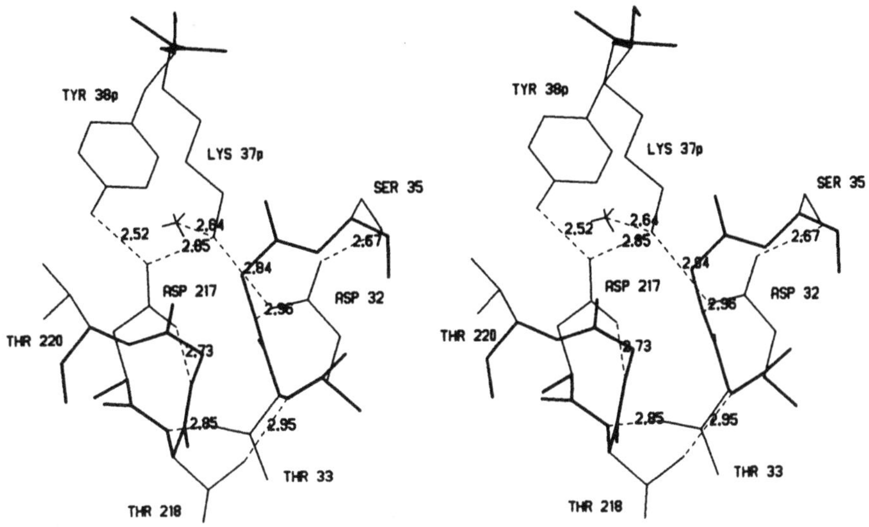

Figure 3. Stereoview of the catalytic site of hPGC showing the approximate 2-fold symmetry relating the two active site peptides. Asp32 to Ser35 and Asp217 to Thr220. The position and hydrogen-bonding interactions involving Lys37p and Tyr38p of the prosegment are included.

The amino-acid sequences of the prosegments of hPGC, of porcine pepsinogen (pPGN) and of human pepsinogen A (hPGA) are highly similar (Fig. 4); the hPGC prosegment is 44% identical to pPGN. In both structures residues 1p to 11p (10p) are in an extended conformation (Fig. 2a,b) and the first α-helical segment is initiated by the highly conserved Ser12p (Ser). The second α-helical segment runs from Leu22p (Lys21p) to Thr29p (Thr28p). In hPGC this section of the molecule is rather disordered and many of the side-chains of the external residues are not seen on the electron density map (Table 2). The region around the highly conserved Lys37p-Tyr38p (Lys36p-Tyr37p) adopts a conformation that is virtually identical in both zymogens and the N^{ζ} atom of Lys37p interacts with the active site aspartate residues as shown in Fig. 3. This is clearly one of the important electrostatic interactions stabilizing the zymogen conformation at high pH. Upon dropping the pH, these and other aspartate and glutamate carboxyl groups will become protonated, thereby disrupting the electrostatic attractions between N^{ζ} of Lys37p and the Asp32 and Asp217 carboxylates.

In spite of the high sequence similarity throughout the molecules, structural comparison of hPGC with pPGN reveals several remarkable conformational differences. Most of these differences occur in the N-terminal domain and are associated with loop regions in contact with residues of the prosegment. The polypeptide chain following the conserved Lys37p-Tyr38p (Lys36p-Tyr37p) takes a very different path in the two zymogens (Fig. 2a,b). In pPGN, the stretch from Pro39p to Leu44p forms a loop that winds around the tip of the flap, from residues Tyr75 to Ser79. In contrast, the equivalent segment in hPGC, Phe40p to Leu43p, points away from the flap and causes a major displacement of the gastricsin loop from Tyr125 to Thr134 relative to its position in pPGN (Fig. 5a,b). Because of these differences in conformation of the C-terminal residues of the prosegments, the conformation of the first 5-6 residues of the enzymes are also different. In hPGC the residues Ser1 to Pro6 are in an extended conformation running almost parallel to the strand Leu73-Gly76 of the flap (Fig. 5a). This alters the position of the flap so that Tyr75 O^{η} no longer forms a hydrogen bond

```
           p1      p5        p10       p15       p20
           |       |         |         |         |
                 * * * *         * *   *               *   *
hPGC       A V V K V P L K K F K S I R E T M K E K G L L G
hPGA3      I M Y K V P L I R K K S L R R T L S E R G L L K
pPGNA        L V K V P L V R K K S L R Q N L I K D G K L K
                 * * * *         * *   *           |   *   *
             |       |         |         |   |         |
             p1      p5        p10       p15           p20


        p25  |    p30       p35       p40                     1
         |   |     |         |         |                      |
           * *     *         * *   * *                        *
hPGC     I F L R T H K Y D P A W K Y R F G D L                S V T Y E P
hPGA3    D F L K K H N L N P A R K Y F P Q W E A P T L          V D E Q P
pPGNA    D F L K T H K H N P A S K Y F P E A A A L              I G D E P
           * *     *         * *   * *                                *
           |       |         |         |                        |
           p25     p30       p35       p40                      1
```

Figure 4. Comparison of the amino-acid sequences of the prosegments of human progastricsin (hPGC), human pepsinogen A_3 (hPGA3), and porcine pepsinogen P (pPGNA). The sequence of hPGC is in the top row (with its residue numbering above); that of pPGN is in the bottom row (pPGN numbering below). Thick letters have been used to denote identical residues in all 3 sequences to emphasize the highly conserved nature of the substitutions among the zymogens. Taken from reference #25.

with Trp39 N^{ε} as in most other aspartic proteinases. In contrast, the first five residues of pepsin in the pPGN structure (Fig. 5b), Ile1-Pro5, form a 3_{10} helix that penetrates beneath the flap near residues Thr77-Gly78 thereby shifting the position of the flap in the opposite direction to that seen in hPGC. There are other smaller differences in conformation in regions bordering the residues of the prosegments (of ~ 1 to 2 Å in magnitude); they are much less impressive than the 10 to 15 Å differences detailed above.

Conversion of Progastricsin to Gastricsin

Even though the tertiary structure of mature gastricsin is not available, the present structure of progastricsin contributes to a more in depth understanding of the activation process. Foltmann and Jensen (1982) and Kageyama *et al.* (1989) have studied this conversion biochemically and have demonstrated that there is a stable intermediate in which the N-terminal residues Ala1p to Phe26p remain associated with gastricsin following the autolytic cleavage of the Phe26p-Leu27p bond. The peptide Ala1p-Phe26p remains bound even after the peptide Leu27p-Leu43p is removed from the N-terminus of gastricsin (i.e. after the Leu43p-Ser1 peptide bond is hydrolyzed). The first scissile peptide bond Phe26p-Leu27p is ~ 15 Å distant from the two aspartic acids of the active site of progastricsin, Asp32 and Asp217 (Fig. 1 and Fig. 5a). In addition, before this peptide bond can be hydrolyzed in the conversion process (19,20) major conformational changes to move other parts of the prosegment out of the active site region will have to take place.

It is not clear how the N-terminal peptide of the activation intermediate of gastricsin remains bound to gastricsin in the complex. It is most likely that the removal of Leu27p-Leu43p is accomplished by an intermolecular cleavage reaction by another active gastricsin intermediate. The intermediates are stable at pH > 5.4 (19,20). In order to elucidate the nature

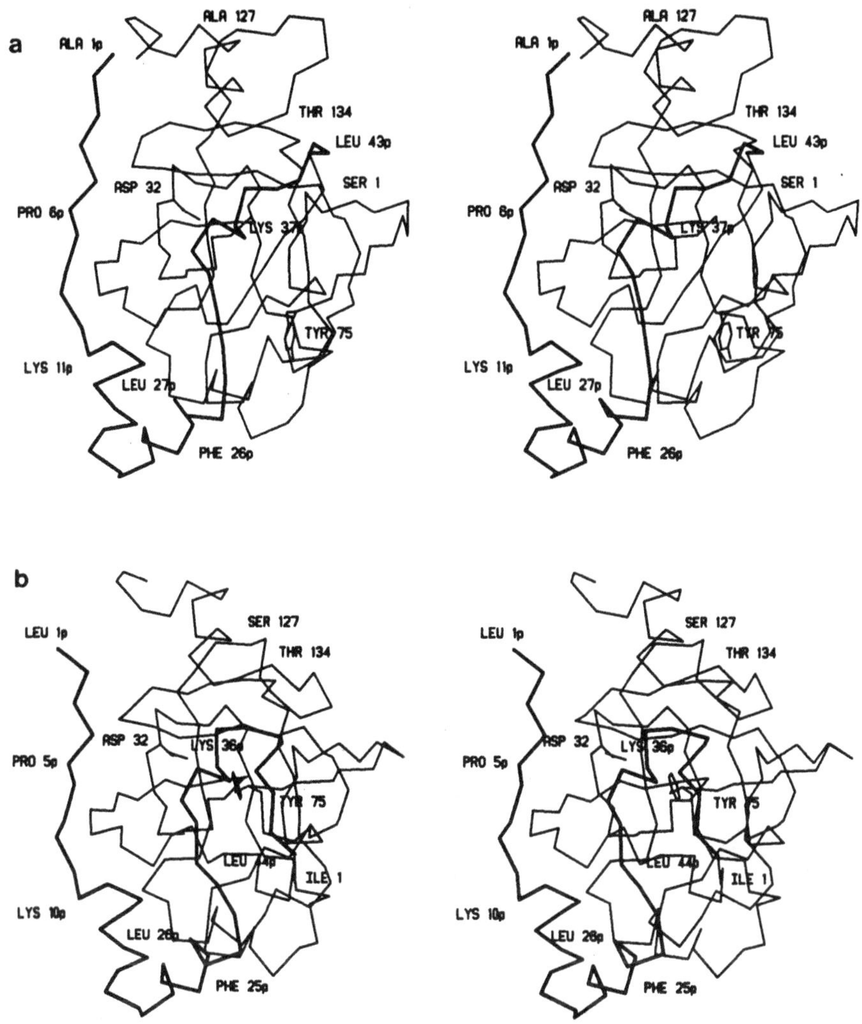

Figure 5. Stereoviews of the C$^{\alpha}$ tracing of the prosegment and the N-terminal domains of (a) hPGC and (b) pPGN. The two domains have been oriented similarly with the flap regions of both molecules (residues 71-82) shown closest to the observer. The residues of the prosegments have been represented by thicker lines in each rendering. Residues at the junctions of key regions are labeled to give the viewer the correct relative orientation. The side chains of residues Asn32 in the active side and Tyr75 in the flap are also shown.

of this complex we have recently crystallized the intermediate with the view to determining its molecular structure.

ACKNOWLEDGMENTS

We are indebted to Dr. Natalia Andreeva for bringing the first crystalline samples of hPGC to Edmonton, thereby initiating this collaboration. The Medical Research Council of Canada has generously supported our research on aspartic proteinases over the years. The research was sponsored in part by the National Cancer Institute, DHHS, under contract No. NCI-CO-74101 with ABL. S.M. was holder of a Studentship from the Alberta Heritage Foundation for Medical Research during part of his study. Mae Wylie has been most patient in transcribing the first draft of this manuscript into a wonderfully clear typed version.

REFERENCES

1. B. Foltmann, *Essays in Biochem.* 17:52-84 (1981).
2. T. Hayano, K. Sogawa, Y. Ichihara, Y. Fujii-Kuriyama and K. Takahashi, *J. Biol. Chem.* 263:1382-1385 (1988).
3. R.T. Taggart, L.G. Cass, T.K. Mohandas, P. Derby, P.J. Barr, G. Pals and G. Bell, *J. Biol. Chem.* 264:375-379 (1989).
4. R.M. Herriott, *J. Gen. Physiol.* 22:65-78 (1939).
5. G.E. Perlmann, *J. Mol. Biol.* 6:452-464 (1963).
6. P. McPhie, *J. Biol. Chem.* 247:4277-4281 (1972).
7. M. Bustin and A. Conway-Jacobs, *J. Biol. Chem.* 246:615-620 (1971).
8. J. Al-Janabi, J.A. Hartsuck and J. Tang, *J. Biol. Chem.* 247:4628-4632 (1972).
9. M. Funatsu, Y. Harada, K. Hayashi and T. Kaneda, *Agr. Biol. Chem.* 36:305-312 (1972).
10. V.B. Pedersen and B. Foltmann, *FEBS Lett.* 35:255-256 (1973).
11. D.M. Glick, Y. Shalitin and C.R. Hitt, *Biochemistry* 28:2626-2630 (1989).
12. M.N.G. James and A.R. Sielecki, *Nature* 319:33-38 (1986).
13. A.R. Sielecki, M. Fujinaga, R.J. Read and M.N.G. James, *J. Mol. Biol.* 219:671-692 (1991).
14. J.A. Hartsuck, G. Koelsch and S.J. Remington, *Proteins: Struct. Funct., Genet.* 13:1-25 (1992).
15. A.R. Sielecki, A.A. Fedorov, A. Boodhoo, N.S. Andreeva and M.N.G. James, *J. Mol. Biol.* 214:143-170 (1990).
16. J.B. Cooper, G. Khan, G. Taylor, I.J. Tickle and T.L. Blundell, *J. Mol. Biol.* 214:199-222 (1990).
17. C. Abad-Zapatero, T.J. Rydel and J. Erickson, *Proteins: Struct., Funct. Genet.* 8:62-81 (1990).
18. J.A. Hartsuck, J. Marciniszyn, J.S. Huang and J. Tang, *in:* Acid Proteases, Structure, Function and Biology, J. Tang, ed., pp. 85-102, Plenum, New York (1977).
19. B. Foltmann and A.L. Jensen, *Eur. J. Biochem.* 128:63-70 (1982).
20. T. Kageyama, M. Ichinose, K. Miki, S.B. Athauda, M. Tanji and K. Takahashi, *J. Biochem.* 105:15-22 (1989).
21. P.K. Ivanov, M.M. Chernaya, A.E. Gustchina, I.V. Pechik, S.V. Nikonov and N.I. Tarasova, *Biochim. Biophys. Acta* 1040:308-310 (1990).
22. A.J. Howard, C. Nielsen and N.H. Xuong, *Methods in Enzymology* 114:452-472 (1985).
23. R. Hamlin, *Methods Enzymol.* 114:416-452 (1985).
24. W.A. Hendrickson and J.H. Konnert, *in:* Computing in Crystallography, R. Diamond, S. Ramaseshan and K. Venkatesan, eds., pp. 13.01-13.23, Indian Institute of Sciences, International Union of Crystallography, Bangalore, India (1980).
25. B. Foltmann, *in:* Proceedings of the 18th Linderstrøm-Lang Conference on Aspartic Proteinases, B. Foltmann, ed., pp. 7-20, Elsinore, Denmark (1988).

A NEW WAY OF LOOKING AT ASPARTIC PROTEINASE STRUCTURES: A COMPARISON OF PEPSIN STRUCTURE TO OTHER ASPARTIC PROTEINASES IN THE NEAR ACTIVE SITE REGION

Natalia S. Andreeva, Alexey Bochkarev, and Igor Pechik

V. Enhelhardt Institute of Molecular Biology
Russian Academy of Sciences
Moscow, Russia

INTRODUCTION

A detailed inspection of the highly refined three-dimensional structure of porcine pepsin (1) offers the explanation of important properties of this enzyme. One of the first results of these studies was the detection of charged groups responsible for unusual electrophoretic behavior of pepsin and its abnormal stability in extreme acidic conditions (1,2). However, very important steps of these investigations have to explain the structural grounds for catalytic properties of pepsin. This problem is closely connected with the analogous research on other aspartic proteinases which share the common type of the active site structure with that of pepsin. The extensive X-ray crystallographic studies of their complexes with various inhibitors have culminated in description of the generally accepted mechanism of their catalytic activity (3,4,5). Three-dimensional structure of all aspartic proteinases gives a clear explanation of the preference for hydrophobic residues flanking the scissile bond of their substrates. However, evidence which would help in understanding the prominent difference between a rather wide specificity of pepsin and a strict requirement for the renin substrate, as well as for specificity of chymosin towards κ-casein, is not in hand. This phenomenon cannot be explained solely on the basis of differences of their substrate binding pockets.

Another very important property of pepsin and other aspartic proteinases which awaits for a convincing structural explanation is a marked dependence of the enzyme kinetics on the length of peptide substrates, described for pepsin in the seminal works of Fruton (6). A prominent jump of K_{cat} is observed, up to three orders of a magnitude, when a peptide substrate of pepsin is extended by the addition of only one P3 residue. This occurs for various substrates with different amino acid residues at the P3 position, even with glycine. Therefore,

Aspartic Proteinases: Structure, Function, Biology, and Biomedical Implications
Edited by Kenji Takahashi, Plenum Press, New York, 1995

it is hard to suppose that atomic displacements in the S3 binding pocket are main determinants of this feature.

Detailed comparisons of the structure of pepsin to its liganded forms and to the structures of other aspartic proteinases offer the needed structural grounds which can explain their common properties, as well as some unique features of this enzyme. The whole work on the comparison of pepsin molecules to the other enzymes of this class is too large to be presented in one paper. Here, a part of this work concerning some groups and water molecules in the vicinity of the active site region is described. It includes mainly the so-called 'flap' with residue numbers 70-85*, which is located in the vicinity of the active site, protects it from solvent and participates in substrate binding.

The comparison reveals a possible role of intramolecular motions in the catalytic properties of pepsin. The role of the flap movement in specificity of chymosin, suggested in the work performed in collaboration with Gilliland (7) is also discussed. A new approach to compare three-dimensional protein structures has been used, and it provides additional information for these studies.

MATERIALS AND METHODS

X-ray crystallographic studies of monoclinic porcine pepsin crystals have been completed with the highly refined structure at 1.8 Å resolution (1), and the atomic coordinates have been deposited with the Brookhaven Protein Data Bank (8) as the entry 4PEP. The list of other coordinate files used in these studies is presented in Table 1. It contains files for the two native (monoclinic (1) and hexagonal crystals (9)) and two liganded forms of pepsin (10); two sets of coordinates for porcine pepsinogen (11-13); two files for the native bovine chymosin - cloned (14) and isolated from natural sources (15); one file for the native (16) and for four liganded forms of rhizopuspepsin (3,36); files for the native (17) and for seven liganded forms of penicillopepsin (4,18,30); files for the native (19) and for ten liganded forms of endothiapepsin (20-24,37); and files for the native HIV protease (25) and its complex with a hydroxyethylene-based inhibitor (26).

Various superposition procedures are widely employed to compare protein three-dimensional structures, and one of them (27) has been used in these studies. However, important results have been obtained with the use of a special approach which describes differences in three-dimensional structures in rather simple terms (7). This approach implies the use of the common intramolecular coordinate (I.C.) system for all compounds to be studied. The I.C. system used in this work has been based on the three reference intramolecular points corresponding to the positions of three C_α atoms in the most conserved and the best refined region of aspartic proteinase molecules. This so-called template region of the N-terminal domain (28) includes the active site loop (residues 28-41) and the arms of the loop related to the active one by the intradomain axis of symmetry (residues 99-104 and 118-123). The arms of the two loops form the central β-sheet - the B-layer (1,29) located at the middle of a pile of the three sheets in the N-terminal domain with approximately orthogonal direction of chains (Fig. 2). The bottom A-layer is formed from the two antiparallel β-hairpin loops, while the upper layer which does not belong to the template region is formed from the flap and the adjacent strand.

Various C_α positions in the B-layer have been tested to create I.C. systems. This work has been mainly done with the use of I.C. system based on the positions of C_α of Gly-34 (conserved in all aspartic proteinases), Phe-101 (conserved as a hydrophobic residue) and

* Pepsin numbering of residues is used here for all discussed structures.

Table 1. Coordinate Files Used in These Studies

Native structures:	PDB files	References
1. Monoclinic porcine pepsin	4PEP	(1)
2. Hexagonal porcine pepsin	3PEP	(9)
3. Porcine pepsinogen A	2PSG	(11,12)
4. Porcine pepsinogen A	1PSG	(13)
5. Bovine chymosin	1CMS	(14)
8. Bovine chymosin	2CMS	(15)
6. Rhizopus pepsin	2APR	(16)
7. Penicillopepsin	3APP	(17)
8. Endothia pepsin	4APE	(19)
9. HIV1-protease	3HVP	(25)
Liganded forms:		
10. 4 liganded forms of Rhizopus pepsin	3APR	(3)
	4APR	(3)
	5APR	(36)
	6APR	(36)
11. 6 liganded forms of penicillopepsin	1APV	(4)
	1APW	(4)
	1PPL	(18)
	1PPM	(18)
	1APT	(30)
	1APV	(30)
12. 10 liganded forms of Endothia pepsin	2ER6	(37)
	2ER7	(23)
	2ER9	(20)
	3ER3	(24)
	3ER5	(24)
	4ER1	(21)
	4ER2	(22)
	4ER4	(20)
	5ER1	(24)
	5ER2	(24)
13. HIV-1 protease+inhibitor	4HVP	(26)

Gly-119 (conserved Gly in all non-viral enzymes). All these atoms have very low values of the temperature B-factors. The position of the active site groups is consistent with these reference points, it means that observed differences of some atomic positions in the studied structures do correspond to their differences in relation to the active site. The validity of this approach which depends on the accuracy of reference point coordinates can be checked by the number of atoms in compared structures which have internal coordinate differences of less than 0.2-0.3 Å. Usually this includes more than a hundred points for the domains of aspartic proteinases. There were 236 main chain points in the pepsin part of pepsinogen and in pepsin itself which have internal coordinate differences close to the limit of their

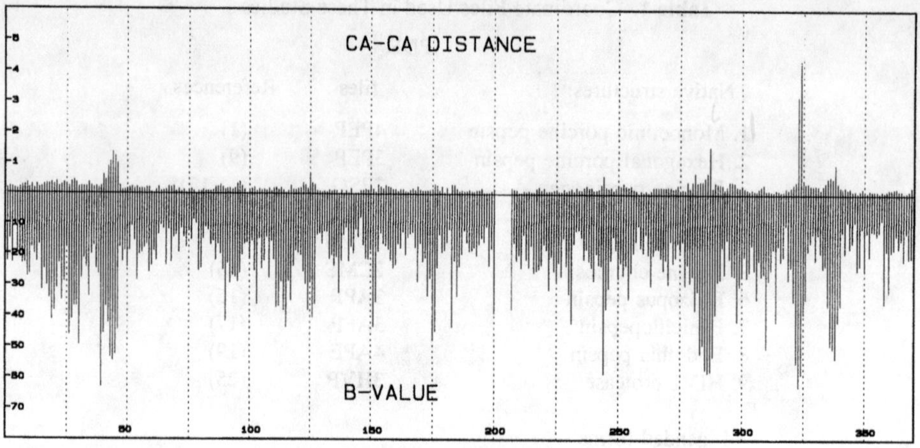

Figure 1. The upper part of the plot: the differences in positions of C_α atoms in the two files of atomic coordinates of porcine pepsinogen A - 1PSG (13) and 2PSG (11-12) converted into the internal coordinate system based on the positions of C_α Gly-34 (the origin), C_α Phe-101 and C_α Gly-119. The scale in Å is shown at the upper part of the vertical axis. Numbers of C_α atoms are shown at the bottom horizontal line of the plot. The lower part of the plot: the temperature B-value of C_α atoms. One can see an obvious dependence between the discrepancies of C_α positions and their B-values. The average difference of C_α positions is 0.21 Å for B values less than 15, it is 0.25 Å for B values between 15 and 25; for B-values between 25 and 35 it is equal to 0.27 Å, for B-values between 35 and 45 it is 0.40 Å.

experimental accuracy. Their root-mean-square (r.m.s.) deviation is equal to 0.17547 Å, while the superposition of the same points gives r.m.s value 0.17454 Å. This means that the approach is equivalent to inspection of structures from the system of points which have correlated positions. Aspartic proteinase molecules are known to be very flexible, which manifests itself in the I.C. system by displacements corresponding to changing of atomic positions in relation to the reference points.

Fig. 1 shows the results of a comparison of the two porcine pepsinogen coordinate files - 1PSG (11) and 2PSG (12), obtained in different laboratories and referred to in different unit cell axes settings. This Figure demonstrates the validity of the comparison with the use of the proposed I.C. system and a high quality of both coordinate files.

During the comparison of the flap region in pepsin to that in other aspartic proteinases, as well as in pepsinogen and in complexes of pepsin with inhibitors, the view approximately perpendicular to the B-layer has been chosen to follow prominent conformational changes at the level of the main chain of the enzymes (see Fig. 2). Computer graphics studies have been performed utilizing the Tektronics graphics station of the Institute of Molecular Biology in Moscow. Studies of the structural changes at the level of side chains and water molecules have been performed by the analysis of coordinate differences.

RESULTS

The inspected region near the active site contains the upper surface of the B-layer in the N-terminal domain and the flap. As is well known, this surface is formed with hydrophobic side chains and serves as the first binding site S1 for the P1 residue side chain before

Table 2. Internal Coordinates of Some Conserved Bound Water Molecules in Aspartic Proteinases

1. W-334 (pepsin numbering)- present in all native and liganded forms of non-viral aspartic proteinases

	X	Y	Z
Mean internal coordinates for 29 compounds (in Å)	7.276	2.122	1.808
R.M.S. deviation of internal coordinates	0.109	0.195	0.233

This water molecule forms hydrogen bonds with the following groups:

a. Hydroxyl Ser-35 OG in all 29 compounds
b. Carbonyl oxygen of a residue at position 37 in all 29 compounds
c. Hydroxyl of Tyr-75 OH in all compounds except native chymosin and nativ penicillopepsin

2. W 328 (pepsin numbering) - present in all native and liganded forms of non-viral aspartic proteinases

	X	Y	Z
Mean internal coordinates for 29 compounds	-2.506	-1.753	-2.434
R.M.S. deviation of internal coordinates	0.132	0.12	0.212

This molecule forms in all 29 compounds hydrogen bonds with the following groups:

a. Carbonyl oxygen of Thr-33
b. Carbonyl oxygen of a residue at position 214
c. NH group of Trp-190

the scissile bond of a substrate. Ile-30, Ile-120, and Trp-39 outline this site in pepsin, while the flap residue Tyr-75 forms another wall of the pocket. It is highly conserved in all non-viral enzymes. The inspected region is larger than the binding pocket. Several water molecules are located there, and most of them are displaced during substrate/inhibitor binding. Now one can put the question: what does it mean the conserved position in the three-dimensional structure. The definition can be done with the use of the I.C. system. An atomic group has a conserved position in the three-dimensional structure if it has similar internal coordinates, i.e. if it has internal coordinates with the low r.m.s. deviation from the average position. A low value of r.m.s. deviation from the average position has been found for several groups of 28 non-viral proteinases compared here. Table 2 demonstrates how two water molecules (334 and 328 in pepsin numbering) near the active the site are conserved. They are found in all native and liganded forms of aspartic proteinases and in zymogens. The conservative character of these water molecules has not yet been discussed, and it requires the explanation. Figure 3 shows that water molecule 334 forms a conserved network of hydrogen bonds with O of Ser-35, peptide carbonyl oxygen of Asn-37, and OH group of Tyr-75. Table 3 shows the close similarity of the internal coordinates for this hydroxyl group in all liganded forms of aspartic proteinases. One can think that this network of hydrogen bonds is essential to preserving the position of the Tyr-75 ring in the wall of the binding pocket, however this does not always occur.

Table 3. Internal Coordinates of Some Residue Atoms Located in the N-Terminal Domain Flap of Aspartic Proteinases

a. OH-group of Tyr-75 (pepsin numbering)

	X	Y	Z
1. Native porcine pepsin	7.553	4.158	3.022
2. Porcine pepsinogen A	7.554	3.877	3.400
3. Pepsin+inhibitor 63218	7.348	3.954	3.347
4. Pepsin+inhibitor 66702	7.356	3.792	3.340
5. Mean internal coordinates for all native and liganded forms of fungi enzymes with exception of native penicillo-pepsin (23 compounds)	6.913	4.222	3.872
R.M.S. deviations of internal coordinates	0.153	0.145	0.124
Exceptions:			
6. Native penicillopepsin	5.866	5.530	4.846
5. Native bovine chymosin	2.067	1.954	10.723

The results of the comparison of the flap position and its conformation in pepsin and other aspartic proteinases are presented in Fig. 2. Fig. 2a shows a view of the pepsin main chain if one looks along the direction perpendicular to the B-layer (1,28). In Fig. 2b, the same view is drawn for the complex of pepsin with a peptidomimetic inhibitor (10). As one can see, the flap in the complex is displaced down towards the upper surface of the B-layer, which binds leucine side chain of the inhibitor P1 residue. This kind of movement has been observed in many studies of aspartic proteinase complexes with various inhibitors. At the same time, a prominent decrease of the B-value has been detected for the flap atoms; it shows the flap becomes more rigid after a ligand binding. Descriptions of this phenomenon have been presented previously (3,4,10,18,20-24,30,31). We present here data on internal atomic coordinates of the flap atoms participating in substrate binding (Table 4). One can see that they are rather different in different native compounds; however the position of these groups becomes rather similar for all of them after a ligand binding. Of course, one should keep in mind that these positions are relative to the chosen I.C. system. The exception is the oxygen in the side chain of the residue 77 (O Thr-77 in pepsin, and OD Asp-77 in the fungal enzymes).

A quite different conformation of the flap is observed in pepsinogen (11-13). Figure 2c shows that pepsinogen flap is very high up in relation to that in pepsin, that means that during activation process it moves down towards the B-layer, and even lower during substrate binding. The movement is not a simple one since the tip of the flap changes its position in relation to the body of this loop. In other words, the movement of the flap is not the movement of a rigid body. The displacement of C_α-77 is a little less than 5 Å, of C_α-76 and C_α-78 slightly more than 4 Å, and it is about 1 Å for C_α-75. It looks rather unexpected that in spite of such large movement the hydroxyl group of the flap residue Tyr-75 is displaced less than 0.5 Å, and the conserved network of hydrogen bonds which it forms is preserved (see Fig. 3). This network forces Tyr-75 to play a role of an anchor, which permits only special conformational changes of the flap.

The position of the flap in chymosin is rather unusual. Figure 2d shows that the flap in chymosin is displaced as a whole in relation to that in pepsin, and the displacement looks

Table 4. Internal Coordinates of Flap Atoms Participating in Formation of Hydrogen Bonds with Substrates

a. Carbonyl oxygen at position 74

	X	Y	Z
1. Native porcine pepsin	3.626	9.534	1.327
2. Porcine pepsinogen A	5.412	8.661	.856
3. Native calf chymosin	4.167	7.093	3.914
4. Native penicillopepsin	2.880	11.329	2.640
5. Native rhizopus pepsin	3.927	9.999	1.858
6. Native rndothia pepsin	2.648	8.912	2.288
7. Mean for 21 liganded forms of fungi enzymes	2.862	9.166	2.119
R.M.S. deviation of internal coordinates for these liganded forms	0.175	0.211	0.163
8. Pepsin+inhibitor 63218	3.444	8.595	1.632
9. Pepsin+inhibitor 66702	3.419	8.478	1.695

b. NH group of a residue at position 76

	X	Y	Z
1. Native porcine pepsin	.451	8.612	4.564
2. Porcine pepsinogen A	2.431	10.714	2.509
3. Native calf chymosin	1.603	7.513	7.707
4. Native penicillopepsin	.783	11.164	5.281
5. Native rhizopus pepsin	.050	8.543	3.833
6. Native endothia pepsin	.595	7.849	4.840
7. Mean for 21 liganded forms of fungi enzymes	.565	7.850	4.694
R.M.S. deviation of internal coordinates for these liganded forms	0.176	0.193	0.300
8. Pepsin+inhibitor 63218	.134	7.717	4.705
9. Pepsin+inhibitor 66702	.070	7.566	4.816

c. NH-group of a residue at position 77

like a turn around the axis perpendicular to the B-layer with a sliding of the flap in a plane parallel to this layer. Moreover, the observed conformation of the flap is rather different from those found for other aspartic proteinases (7,14,15). The flap is displaced, and Tyr-75 cannot play a role of an anchor like in the other aspartic proteinases. The tyrosine ring occupies the S1 and part of the S3 substrate binding pockets instead. The use of the common I.C. system for all compounds makes it possible to see in chymosin a water molecule at the position of Tyr-75 hydroxyl in other aspartic proteinases, at the same time the whole network of hydrogen bonds shown in Figure 3 is preserved. There was a suggestion that crystal packing forces induce conformational changes of the chymosin flap and the observed structural

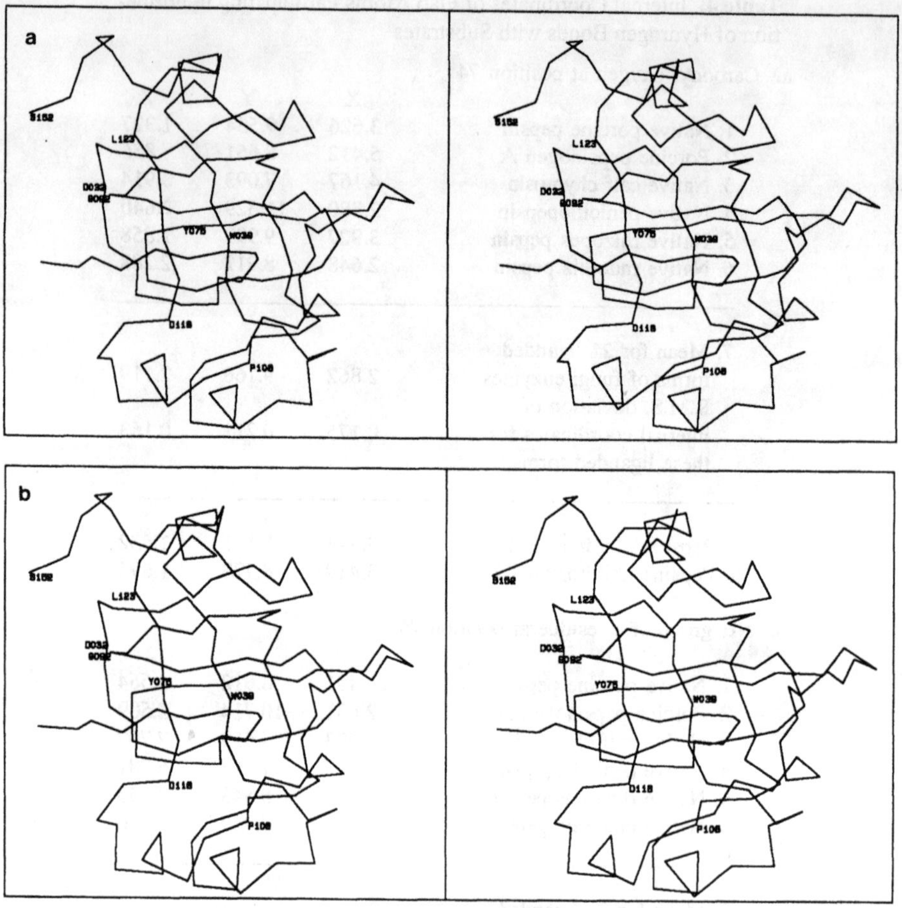

Figure 2. The position of the N-terminal domain flap in relation to the B-layer in different aspartic proteinases and their complexes with inhibitors: a) the native porcine pepsin main chain (1,28); b) the complex of porcine pepsin with peptidomimetic inhibitor (10); the flap is displaced down towards the surface of the B-layer.

properties have nothing in common with the free enzyme in solution (15). A detailed analysis of the intermolecular contacts in chymosin crystals has shown that it is not the case (7). It was shown that the most movable part of the chymosin flap (residues 74-76) is surrounded by water molecules, and that there are many intermolecular contacts on the other side of the flap (residues 77-82). However, they have to push this loop in the direction opposite to that found in crystals. Therefore, it looks probable that the observed structure of the chymosin flap with an unusual orientation of the Tyr-75 residue corresponds to one of its native states, and the enzyme can adopt a self-inhibited form (7).

The view of the HIV-1 protease molecule (25) perpendicular to the B-layer is presented in Fig. 4a. Homologous positions to pepsin Gly-34 (Gly-27), Phe-101 (Val-75) and Gly-119 (Asn-83) have been chosen for the transformation of this enzyme atomic coordinates to the common I.C. system. As can be seen in Figure 4, the enzyme three-dimensional structure of this enzyme is not very similar to that of non-viral proteinases, although the A- and the B-layers (the template region) look homologous. The main difference concerns the flap position and the structure of the connecting segments. In the native HIV-1 protease the flap does not form the orthogonal sheet with the B-layer, and the angle between

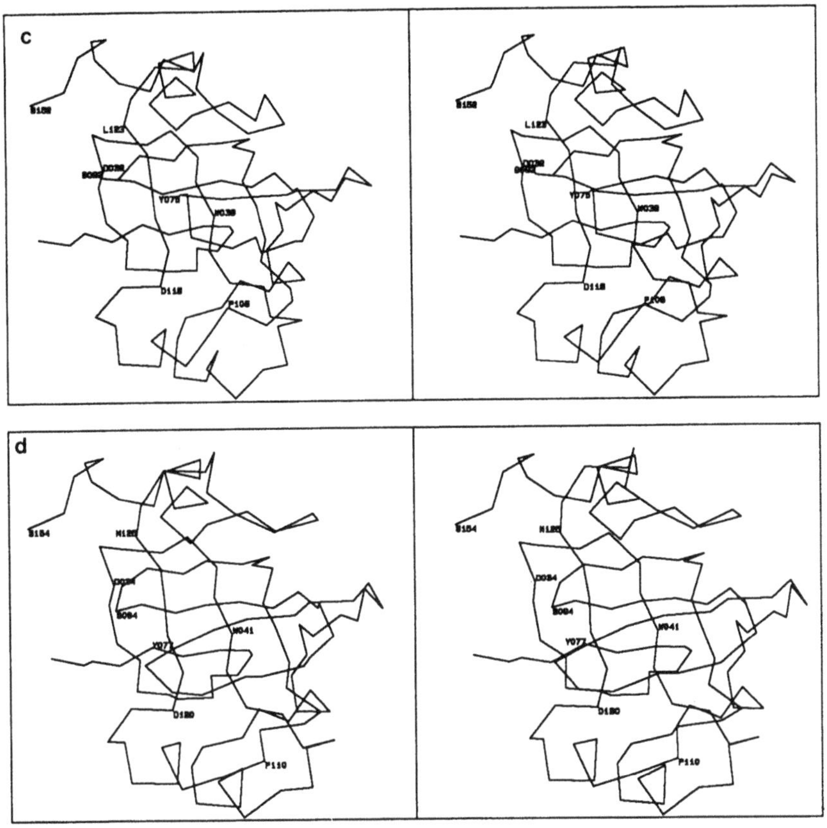

Figure 2. c) Porcine pepsinogen (11-13); the flap is very high up in relation to the B-layer; d) chymosin (14,15), the flap is displaced in relation to that in pepsin by sliding in a plane parallel to the B-layer.

the direction of chains in the flap and in the layer is closer to that for piles of β-sheets of the second type (32). This structural property of HIV-1 protease has not been discussed before. However, in the complex of HIV-1 protease with a hydroxyethylene inhibitor (26) the flap takes up another position with chain direction close to that in non-viral aspartic proteinases, i.e. like in orthogonal sheets (Fig. 4b). Of course, HIV-1 protease cannot have any anchor similar to Tyr-75, since its movement has to be too extensive. As shown before (33), the peptide inhibitor binds to HIV-1 protease in a different way comparatively to that for non-viral enzymes, and a special water molecule connects the flap and the peptide chain of the inhibitor.

DISCUSSION

These considerations show that the flap is a very mobile element of the three-dimensional structure of aspartic proteinases. It moves as a whole, at the same time it changes its own chain conformation, especially in the region of the turn. This movement can be more or less extensive, depending on the particular enzyme. The flap is less flexible in rhizopuspepsin and is extremely mobile in chymosin and HIV-1 protease. The comparison of the pepsin and pepsinogen structures shows that it is also rather mobile in pepsin.

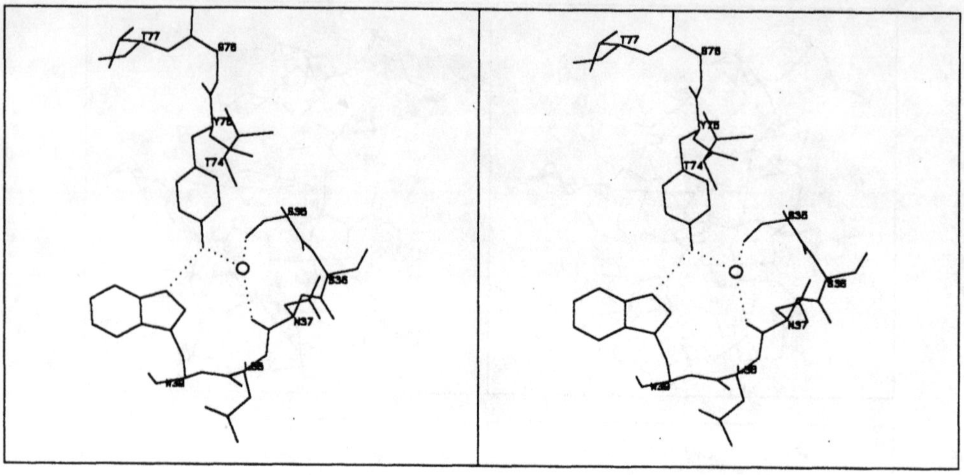

Figure 3. The network of hydrogen bonds which forms the hydroxyl group of Tyr-75 and hydrogen bonds of the water molecule which is bound to Tyr-75 hydroxyl. The position of this water molecule is completely conserved in all non-viral aspartic proteinases and their complexes with inhibitors. The position of Tyr-75 hydroxyl group is also highly conserved, the exceptions being chymosin and native penicillopepsin.

One can predict what kind of flap movement has to occur during substrate binding to chymosin, if the flap takes in the enzyme-substrate complex the standard position like in all complexes of aspartic proteinases with inhibitors. One can expect that the main component of the chymosin flap movement must be the sliding parallel to the B-layer with the simultaneous reorientation of Tyr-75 ring. Thus structural transformations in chymosin have to be more complicated than in the other aspartic proteinases, therefore they require a special interaction with a substrate. It was a suggestion that chymosin substrate specificity may include the ability of the chymosin substrate (κ-casein) to expel the tyrosine ring from the binding pocket (7). This hypothesis explains equally well the low general proteolytic activity of the enzyme, as the required special interaction to transform the flap cannot occur equally well with any peptide substrate. The fixation of the flap in a standard conformation and position are prerequisites for the development of the catalytic reaction. For all these studies experiments on the site specific mutagenesis of the flap residue Tyr-75 in chymosin and residues close to it are extremely relevant (34,35). It is important to emphasize that the major part of hydrogen bonds between aspartic proteinase molecule and its substrate is formed with the flap, and most of them are with the groups located at the very movable tip of this flap. It is clear that only definite fixed position and definite conformation of the flap can provide unique productive binding. Let us consider how the position and the required conformation of the flap can be fixed. For this purpose, any well-refined data for the complex of an aspartic proteinase with a peptide inhibitor can give an answer. Figure 5 shows a part of the three-dimensional structure of penicillopepsin with one of its inhibitors. This inhibitor Ival-Val-Val-diflourostatone contains only P1, P2, P3 residues and difluorostatone group which mimics transition state interactions in the active site (4).

As can be seen, the inhibitor is bound in the extended conformation and it is antiparallel to the 217-219 fragment of the enzyme molecule (215-217 in penicillopepsin numbering) with the formation of the three hydrogen bonds standard for all complexes of aspartic proteinases with peptide inhibitors. It forms also the three hydrogen bonds with the groups located at the tip of the flap, and these bonds are also standard. Thus, the inhibitor is held between the two blocks of the protein molecule and simultaneously clamps them to

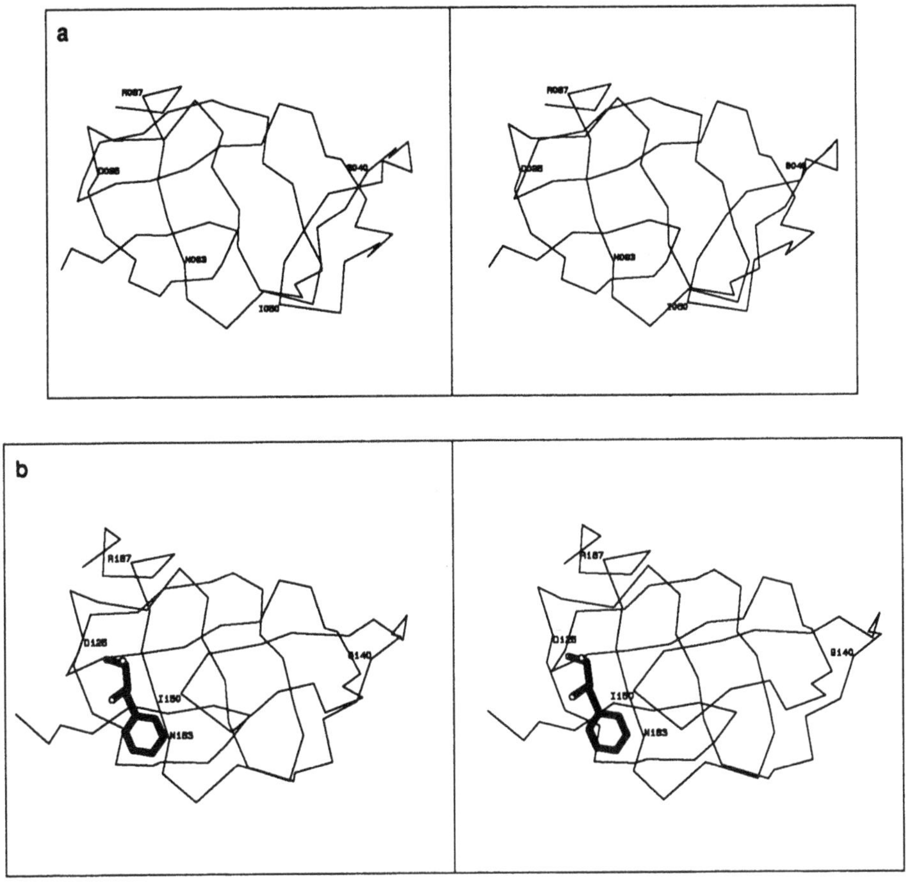

Figure 4. a) The view of HIV-1 protease molecule (25) perpendicular to the B-layer. One can see that the flap takes up quite different position from that in pepsin, and it does not form orthogonal sheet with the B-layer; b) - the same view for the complex of HIV-1 protease with a hydroxyethylene inhibitor (26). The flap does form orthogonal sheet with the B-layer. The phenolic ring of the inhibitor is shown.

each other. The flap in such a complex cannot move. Let us suppose that a substrate has only P2, P1 and P1' residues. In this case all three bonds with the flap will be retained, while only one bond will exist with the less movable segment 217-219 (with CO-217). The substrate will be fixed by this less movable segment only at one point, therefore it can follow the flap movement turning to some extent around the bond NH of P1 with CO 217. The productive mode of binding will be one of many possible, and the catalytic reaction cannot be efficient. A quite different event will be observed if one adds a P3 residue to this substrate (Figure 5). The main chain groups of this residue (which form hydrogen bonds with the O and NH of the residue at the 219 position) make up the clip which fastens the flap in a definite conformation via P2-P3 peptide bond, providing a unique mode of a substrate-main chain binding. In this case, the catalytic reaction must be much more efficient. That is the proposed explanation of the prominent increase of K_{cat} after an addition of a P3 residue to a short substrates observed in the works of Fruton (6) on pepsin. At the same time, if the P2 position of a substrate is taken by proline residue, the fastening of the flap cannot be so efficient as the P2-P3 peptide group does not form the requisite hydrogen bond. That is the only explanation of the role of proline residues at the P2 position of pepsin peptide substrates

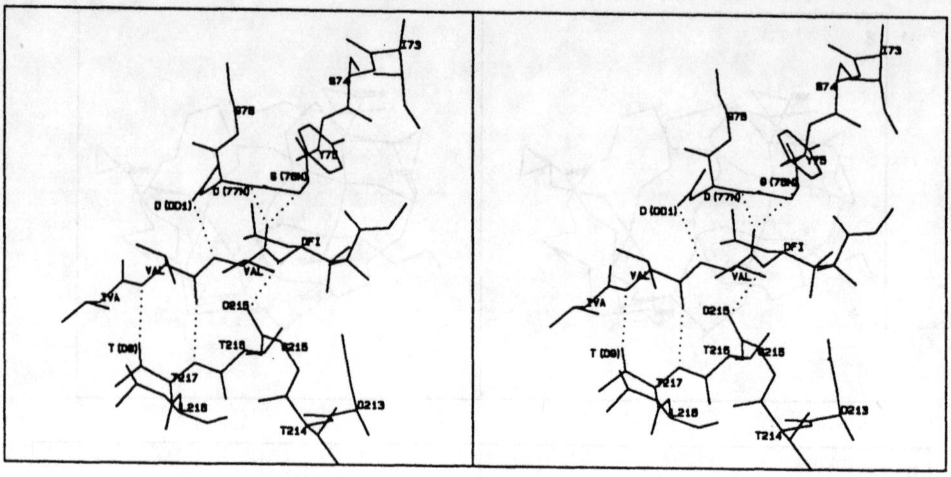

Figure 5. The standard hydrogen bonding of the N-terminal peptide portion of any aspartic proteinase inhibitor with P1, P2, P3 and P4 residues, which represents the binding of a real substrate peptide fragment. Data on the complex of penicillopepsin with fluorine containing inhibitor are used (4). One can see that the N-terminal portion of substrates forms three hydrogen bonds with the groups located at the most movable tip of the flap and with the more rigid 217-219 enzyme segment. These bonds fix the flap in a definite conformation. The presence of a P3 residue is critical for the fixation as it occurs via P2-P3 peptide group of a substrate.

observed in (6) as well as in other studies. We think that investigations of the intramolecular movements in pepsin and other aspartic proteinases can give an essential information on their main properties.

ACKNOWLEDGMENTS

This work has been supported with the grant provided by the Academy of Sciences of Russia. We are grateful to Michael James, Anita Sielecki, John Erickson, and Alexander Wlodawer for sending us data before they became available from the Protein Data Bank. Part of this work has been performed with a Tektronics computer graphics workstation, and we are grateful to Vladimir Tumanjan for enabling us to use it and to Sergey Korolev for his help in its utilization.

REFERENCES

1. A. Sielecki, A. Fedorov, A. Boodhoo, N. Andreeva, M. James, Molecular and crystal structure of monoclinic porcine pepsin refined at 1.8 Å resolution, *J. Mol. Biol.* **214**: 143-170 (1990).
2. N.S. Andreeva, M.N.G. James, Why does pepsin have a negative charge at very low pH? An analysis of conserved charged residues in aspartic proteinases, *In*: Structure and Function of the Aspartic Proteinases, B.M. Dunn, ed., Plenum Press, New York, London, 39-45 (1991).
3. K. Suguna, E.A. Padlan, C.W. Smith, W.D. Carlson, D.R. Davies, Binding of a reduced peptide inhibitor to the aspartic proteinase from *Rhizopus chinensis*: Implications for a mechanism of action, *Proc. Natl. Acad. of Sci.* **84**:7009-7013 (1987).
4. M.N.C. James, A.R. Sielecki, K. Hayakawa, M.H. Gelb, Crystallographic analysis of transition state mimics bound to penicillopepsin: difluorostatine- and difluorostatone-containing peptides, *Biochem.* **31**:3872-3886 (1992).

5. B. Veerapandian, J. Cooper, A. Sali, T.L. Blundell, R.L. Rosati, B.W. Dominy, D.B. Damon, D.J. Hoover, Direct observation by X-ray analysis of the tetrahedral 'intermediate' of aspartic proteinases, *Protein Science* **1**:322-328 (1992).
6. J.S. Fruton, The mechanism of the catalytic action of pepsin and related acid proteases, *Adv. Enzymol. Relat. Areas Mol. Biol.* **44**:1-36 (1976).
7. N.S. Andreeva, J. Dill, G.L. Gilliland, Can enzymes adopt a self-inhibited form. Results of X-ray crystallographic studies of chymosin, *Biochem. Biophys. Res. Comm.* **184**:1074-1081 (1992).
8. F. Bernstein, T. Koetzle, G.J.B. Williams, E.F. Meyer, M.D. Brice, J.R. Rogers, O. Kennard, T. Shimanouchi, M. Tasumi, The protein data bank: A computer-based archival file for macromolecular structures, *J. Mol. Biol.* **112**:535-542 (1977).
9. J.B. Cooper, G. Khan, G. Taylor, I.J. Tickle, T.L. Blundell, X-ray analysis of aspartic proteinases, ii. three-dimensional structure of the hexagonal crystal form of porcine pepsin at 2.3 Å resolution, *J. Mol. Biol.* **214**:199-222 (1990).
10. C. Abad-Zapatero, T.J. Rydel, D.J. Neidhart, J. Luly, J.W. Erickson, Inhibitor binding induces structural changes in porcine pepsin, *In*: Structure and Function of the Aspartic Proteinases, B.M. Dunn, ed., Plenum Press, New York, London, 9-21 (1991).
11. M.N.G. James, A.R. Sielecki, Molecular structure of an aspartyl proteinase zymogen, porcine pepsinogen, at 1.8 Å resolution, *Nature* (London) **319**:33-35 (1986).
12. A.R. Sielecki, M. Fujinaga, R.J. Read, M.N.G. James, Refined structure of porcine pepsinogen at 1.8 Å resolution, *J. Mol. Biol.* **219**:671-692 (1991).
13. J.A. Hartsuck, G. Koelsh, S.J. Remington, The high-resolution crystal structure of porcine pepsinogen, *Proteins: Structure, Function and Genetics* **13**:1-25 (1992).
14. G.L. Gilliland, E.L. Winborn, J. Nachman, A. Wlodawer, The three-dimensional structure of recombinant bovine chymosin at 2.3 Å resolution, *Proteins: Structure, Function and Genetics* **8**:82-101 (1990).
15. M. Newman, M. Safro, C. Frazao, G. Khan, A. Zdanov, I.J. Tickle, T.L. Blundell, N. Andreeva, Structure and refinement at 2.2 Å resolution of bovine chymosin, *J. Mol. Biol.* **221**:1295-1309 (1991).
16. K. Suguna, R.R. Bott, E.A. Padlan, E. Subramanian, S. Sheriff, G.N. Cohen, D.R. Davies, Structure and refinement at 1.8 Å resolution of the aspartic proteinase from *Rhizopus chinensis*, *J. Mol. Biol.* **196**:877-900, (1987).
17. M.N.G. James, A.R. Sielecki, Structure and refinement of penicillopepsin at 1.8 Å resolution, *J. Mol. Biol.* **163**:299-361 (1983).
18. M. Fraser, N.C.J. Strydnatka, P.A. Bartlett, J.E. Hanson, M.N.G. James, Crystallographic analysis of transition-state mimics bound to penicillopepsin phosphorous-containing peptide analogue, *Biochem.* **31**:5201-5214 (1992).
19. T.L. Blundell, J.A. Jenkins, B.T. Sewell, L.H. Pearl, J.B. Cooper, S.P. Wood, B. Veerapandian, The three-dimensional structure at 2.1 Å resolution of endothiapepsin, *J. Mol. Biol.* **211**:919-941 (1990).
20. T.L. Blundell, J.B. Cooper, S.I. Foundling, B. Atrash, M. Szelke, On the rational design of renin inhibitors: X-ray studies of aspartic proteinases complexed with transition-state analogues, *Biochem.* **26**:5585-5590 (1987).
21. J.B. Copper, S.I. Foundling, A.M. Hemmings, T.L. Blundell, D.M. Jones, A. Hallet, M. Szelke, The structure of synthetic pepsin inhibitor complexed with endothiapepsin, *Eur. J. Biochem.* **169**:215-221 (1987).
22. J.B. Cooper, S.I. Foundling, T.L. Blundell, J. Boger, R.A. Jupp, J. Kay, X-ray studies of aspartic proteinase-statine inhibitor complexes, *Biochem.* **28**:8596-8603 (1989).
23. B. Veerapandian, J.B. Cooper, A. Sali, T.L. Blundell, The three-dimensional structure of endothiapepsin complexed with a transition state isostere inhibitor of renin at 1.6 Å resolution, *J. Mol. Biol.* **216**:1017-1029 (1990).
24. A. Sali, B. Veerapandian, J.B. Cooper, S.I. Foundling, D.J. Hoover, T.L. Blundell, High-resolution X-ray diffraction study of the complex between endothiapepsin and oligopeptide inhibitor: the analysis of inhibitor binding and description of the rigid body shift in the enzyme, *EMBO J.* **8**:2179-2188 (1989).
25. A. Wlodawer, M. Miller, M. Jaskolski, B.K. Sathyanarayana, E. Baldwin, I.T. Weber, L.M. Selk, L. Clawson, J. Schneider, S.B.H. Kent, Conserved folding in retroviral proteases: crystal structure of a synthetic HIV-1 protease, *Science* **245**:616-621 (1989).
26. A.L. Swain, M.M. Miller, J. Green, D.H. Rich, J. Schneider, S.B.H. Kent, A. Wlodawer, X-ray crystallographic structure of a complex between a synthetic protease of human immunodeficiency virus I and a substrate-based hydroxyethylamine inhibitor, *Proc. Nat. Acad. Sci., U.S.A.* **87**:8805 (1990).

27. Y. Satow, G.H. Cohen, E.A. Padlan, D.R. Davies, Phosphocholine binding immunoglobulin FabMcPC603. An X-ray diffraction study at 2.7 Å resolution, *J. Mol. Biol.* **190**:593 (1986).
28. N.S. Andreeva, A consensus template for the aspartic proteinase fold, *In*: Structure and Function of the Aspartic Proteinases, B.M. Dunn, ed., Plenum Press, New York, London, 559-572 (1991).
29. N.S. Andreeva, A.S. Zdanov, A.E. Gustchina, A.A. Fedorov, Structure of ethanol-inhibited porcine pepsin at 2 Å resolution and binding methyl ester of phenylalanyl-diiodotyrosine to the enzyme, *J. Biol. Chem.* **259**:11353-11365 (1984).
30. M.N.G. James, A.R. Sielecki, F. Salituro, D.H. Rich, T. Hofmann, Conformational flexibility in the active sites of aspartyl proteinases revealed by pepstatin fragment binding to penicillopepsin, *Proc. Natl. Acad. Sci., USA*, **79**:6137-6141 (1982).
31. V. Dhanaraj, C.G. Dealwis, C. Frazao, M. Badasso, B.L. Sibanda, I.J. Tickle, J.B. Cooper, H.P.C. Driessen, M. Newman, C. Aguilar, S.P. Wood, T.L. Blundell, P.M. Hobart, K.F. Geoghegan, M.J. Ammirati, D.E. Danley, B.A. O'Connor, D.J. Hoover, X-ray analyses of peptide-inhibitor complexes define the structural basis of specificity for human and mouse renins, *Nature (London)* **357**:466-472 (1992).
32. C. Chotia, D. Janin, Orthogonal packing of beta-pleated sheets, *Biochem.*, **21**:3955-3965 (1982).
33. A.E. Gustchina, I.T. Weber, Comparison of inhibitor binding in HIV-1 protease and in non-viral aspartic proteases: the role of the flap, *FEBS Lettr.* **269**:269-272 (1990).
34. J. Suzuki, K. Sasaki, Y. Sasao, A. Hamu, H. Kawasaki, M. Nishiyama, S. Horinouchi, T. Beppu, Alteration of catalytic properties of chymosin by site-directed mutagenesis, *Protein Eng.* **2**:563-569 (1989).
35. P. Strop, J. Sedlaczek, J. Stys, Z. Kaderabkova, I. Blaha, L. Pavlickova, J. Pohl, M. Fabry, V. Kostka, M. Newman, C. Frazao, A. Shearer, I.J. Tickle and T.L. Blundell, Engineering enzyme subsite specificity: preparation, kinetic characterization, and X-ray analysis at 2 Å resolution of Val 111-Phe site mutated calf chymosin, *Biochem.* **29**:9863-9871 (1990).
36. K.D. Parris, D.J. Hoover, D.R. Davies, Crystal structures of rhizopuspepsin/inhibitor complexes, *In*: Structure and Function of the Aspartic Proteinases,.B.M. Dunn,ed., Plenum Press, New York, London, 217-231 (1991).
37. S.I. Foundling, J. Cooper, F.E. Watson, A. Cleasby, L.H. Pearl, B.L. Sibanda, A. Hemmings, S.P.Wood, T.L. Blundell, M.J. Valler, C.G. Norey, J. Kay, J. Boger, B.M. Dunn, B.J. Leckie, D.M. Jones, B. Attrash, A. Hallett, M. Szelke, High resolution X-ray analysis of renin inhibitor-aspartic proteinase complexes, *Nature (London)* **327**:349-352 (1987).

REARRANGING PEPSINOGEN AND PEPSIN BY PROTEIN ENGINEERING

Xinli Lin and Jordan Tang

Protein Studies Program
Oklahoma Medical Research Foundation
University of Oklahoma Health Science Center
Oklahoma City, Oklahoma

Eukaryotic aspartic proteases, such as pepsin, are synthesized as a single-chain zymogen. After activating into single-chain proteases, they contain bilobal structures. The three-dimensional structures of the N- and C-terminal lobes are homologous. It has been hypothesized that the origin of this internal similarity in eukaryotic aspartic proteases is derived from an evolutionary process of gene duplication and fusion (1). A primordial form of aspartic protease is probably a homodimer. The retroviral aspartic proteases are homodimers and structurally related to the eukaryotic aspartic proteases (2). Thus, the retroviral proteases may represent the primordial forms of eukaryotic enzymes. In this scenario, the gene of a homodimeric retroviral protease is integrated in the host genome of a primordial eukaryotic cell. The protease gene is retained by mutational processes which provided survival advantages to the host in evolution. It ultimately evolved through gene duplication and fusion to a single-chain, internally symmetrical eukaryotic enzyme.

This hypothetical evolutionary scheme suggests that many intermediate forms of aspartic proteases may have existed but are now extinct with their hosts. Therefore, we were interested in "reversing" the evolutionary process by protein engineering, i.e. to make a two-chain aspartic protease from a single-chain eukaryotic aspartic protease. Such mutagenesis studies, it is thought, may lead to insights on the structure and function relationships of these two different types of aspartic proteases. This study focused on three aspects: the study of heterodimer and homodimer pepsin, the use of pepsin lobes as probe for structural denaturations, and structural rearrangement of pepsinogen. These results are described below.

TWO-CHAIN PEPSIN AND ITS N-TERMINAL LOBE HOMODIMER

We decided that the first step in 'reverse evolution' is to engineer a two-chain pepsinogen and a two-chain pepsin. Pepsinogen and pepsin were chosen because they had been cloned and expressed for mutagenesis studies in our laboratory (3,4). The strategy, as summarized in Fig. 1, is to separately express two chains of pepsinogen in *E. coli* and

Aspartic Proteinases: Structure, Function, Biology, and Biomedical Implications
Edited by Kenji Takahashi, Plenum Press, New York, 1995

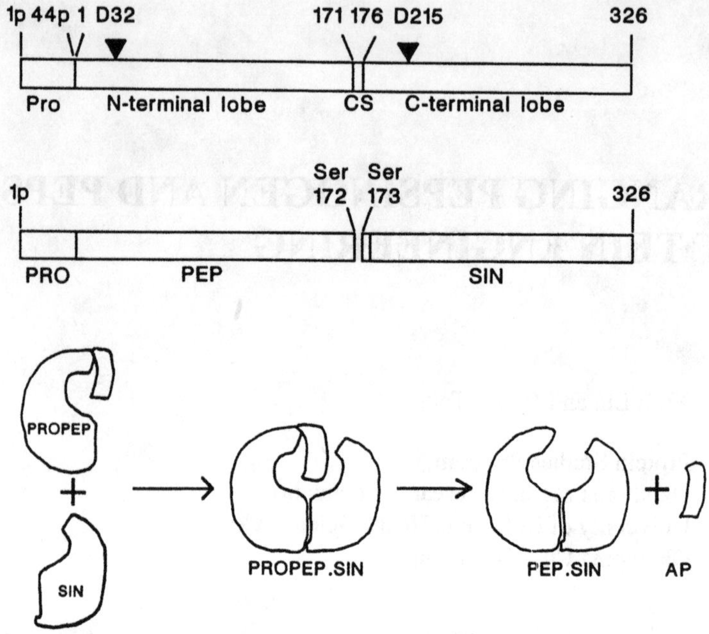

Figure 1. Strategy for making two-chain pepsinogen and two-chain pepsin. Top line shows native pepsinogen cDNA encoding, from the N-terminal side, the *pro* region (residues 1p to 44p), N-terminal lobe (*pep*, residues 1 to 170), connecting strand (CS, residues 171 to 176), and the C-terminal lobe (*sin*, residues 177-326). The locations of the active-site Asp^{32} and Asp^{215} are shown in solid triangles. The second line illustrates the constructed genes for the expression of *propep* (residues 1p to 172) and *sin* (residues 173 to 326). The bottom line shows the expressed fragments propep and sin are recombined to generate a two-chain pepsinogen, *propep.sin*, which is activated to form a two-chain pepsin, *pep.sin*, with the removal of the activation peptide (AP = *pro*).

recombine them into a two-chain zymogen. The zymogen is then activated into two-chain pepsin. The designed pepsinogen chains are: the N-terminal lobe, propep[1] (residues Leu^{1p}-Ser^{172}), and the C-terminal lobe, sin (residues Ser^{173}-Ala^{326}). The chain separation point is between two serines (residues 172-173) which are located in the strand connecting two lobes and are on the surface of the tertiary structure. Thus the cutting of this peptide bond will create minimum structural perturbations.

The genes for the chains were constructed from porcine pepsinogen cDNA using recombinant DNA techniques (5) and used to direct the synthesis of the chains separately in *E. coli*. The pepsinogen chains, which were synthesized as insoluble 'inclusion bodies', were recovered, dissolved in 8 M urea, and refolded by dilution (5). By recombining the refolded chains, we observed the generation of acid proteolytic activity, suggesting the formation of a two-chain pepsinogen, *propep.sin*. We confirmed this by the isolation of two-chain pepsinogen and its activation product, two-chain pepsin. Somewhat surprisingly, we also observed the activity of presumably homodimer of *propep* and *pep*. These observations (5) are summarized below:

(a) *Properties of two-chain pepsinogen, propep.sin*: When refolded *propep* and *sin* are mixed, the activity of *propep.sin* is generated. This observation demonstrated that both *propep* and *sin* are capable of independent folding. However, the refolding characteristics of these two chains are different. Gel filtration of refolded *sin* on a column of Sephacryl S-300 revealed that about 25% to 30% of total *sin* are in a single peak (5) which can combine

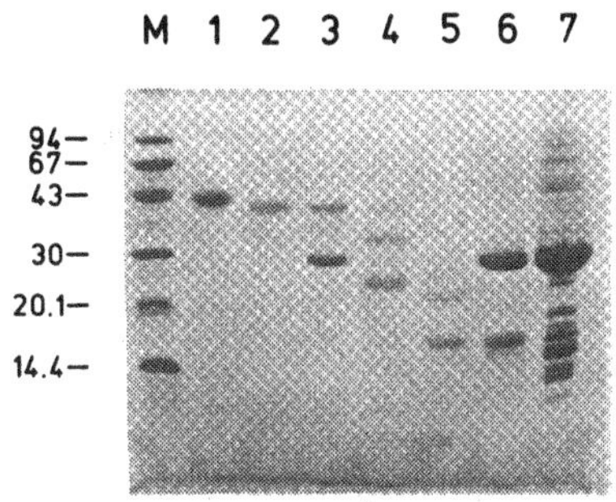

Figure 2. SDS-polyacrylamide gel electrophoresis of pepsinogen fragments, recombined zymogens and activation products. Lane 1, pepsinogen; lane 2, pepsin; lane 3, purified *propep.pepsin$_a$*; lane 4, acid-activated *propep.pepsin$_a$*; lane 5, acid-activated *propep.sin* (resulting in *pep.sin*); lane 6, purified *propep.sin* (not acidified); lane 7, recombinant *propep* in *E. coli* cell lysate. M = molecular weight markers.

with propep into active two-chain pepsinogen ('potential activity'). Refolded *propep* chromatographed under the same conditions produced a broad peak with at least two potential activity peaks, suggesting that *propep* is refolded into several slightly different 'potentially active' conformations. *Propep* and *sin* are optimally refolded at pH 6 and pH 7-8 respectively. However, to make *propep.sin*, it is more efficient to refold *propep* and *sin* together. *Propep.sin* was purified from Sephacryl S-300 and monoQ columns to apparent purity as judged by two bands in polyacrylamide electrophoresis corresponding to the 30-kDa *propep* and 18-kDa *sin* (Fig. 2). Like pepsinogen, propep.sin is stable at pH 8.0.

(b) *Conversion of two-chain pepsinogen to two-chain pepsin*: *Propep.sin* converts spontaneously to *pep.sin* in acidic solutions, as does native pepsinogen, with a reduction in the size of the *propep* band in gel electrophoresis (Fig. 2). Both the first-order and second-order activation rate constants (k_1 and k_2) for *propep.sin* (0.02 min^{-1} and 0.2 $min.mg/ml^{-1}$ respectively) are larger than those for pepsinogen (0.016 min^{-1} and 1.3 $min.mg/ml^{-1}$ respectively). Since both activation mechanisms depend on the catalytic site of pepsin (6), the slower activation of two-chain pepsinogen is apparently due to a decreased catalytic efficiency of pep.sin as compared to pepsin. *Pep.sin* and pepsin have similar K_m (0.046 mM and 0.032 mM respectively) but more than 6-fold difference in k_{cat} (59 sec^{-1} and 343 sec^{-1} respectively). The reason for this loss of catalytic efficiency by *pep.sin* is intriguing, because the point of chain break is on the opposite side of the molecule in 3-D structure. We speculate that the chain separation may result in a less rigid catalytic apparatus, hence lower activity, due to increased motion of the chains.

(c) *Propep.propep and pep.pep homodimers*: Refolded *propep* contains residual proteolytic activity. In order to confirm this activity, we chromatographed refolded *propep* on a gel filtration column. As shown in Fig. 3, proteolytic active peak (*propep**) is eluted at fraction 37, clearly separated from the potential activity of *propep*. The yield of *propep** activity is very low but its presence can be reproducibly demonstrated by a highly sensitive radio-labelled hemoglobin assay. The activity of *propep** is also less stable at pH 8.0 or to heat as compared to *propep.sin* or *pep.sin*. For these reasons further purification of this activity was unsuccessful.

Fragment *propep* contains only a single active site residue in Asp^{32}, it is unlikely that the folded chain by itself can catalyze proteolysis. It is most probable that *propep** activity is generated from the homodimer of this fragment. Since the *pro* segment does not have a definitive conformation in acid, the activity must have resulted from the homodimer of the *pep* lobes. To confirm this, we expressed and refolded *pep* and observed residual activity

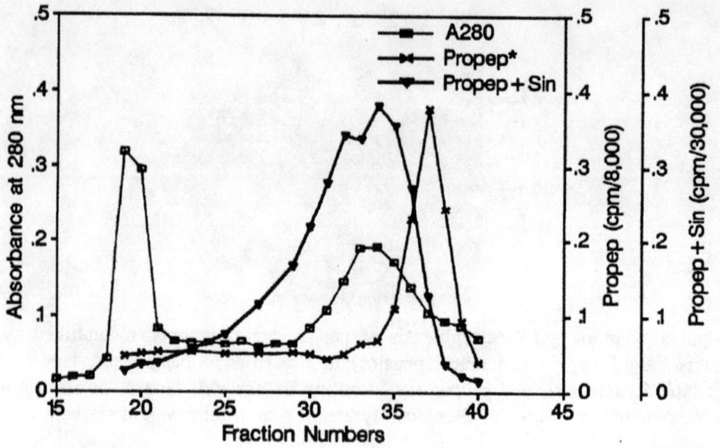

Figure 3. Gel filtration chromatography on Sephacryl S-300 of refolded recombinant *propep* and the location of proteolytic and potential activities. The potential activity was measured for each fraction after the addition of refolded *sin*. Proteolytic activity (*propep**) was assayed using radio-labeled hemoglobin as substrate.

similar to that produced by *propep** (X. Lin and J. Tang, unpublished observations). A graphic model of *pep.pep* homodimer was built by keeping the spacial relationships of two active-site Asp^{32} as in native pepsin. Interesting features of this model (Fig. 4) are as follows: (a) The interface of the homodimer model fits remarkedly well with relatively few serious spacial violations from each other. This is in contrast to the homodimer model of *sin.sin* which reveals rather poor surface contacts (results not shown). This model is consistent with the observation that refolded sin does not process detectable proteolytic activity. (b) The substrate binding cleft appears well preserved by two flaps which cover as in the case of retroviral aspartic proteases (2). In this respect and in overall appearance, the *pep.pep* model shown in Fig. 4 is remarkably similar to the structures of retroviral aspartic proteases.

The conformation of the *pep* monomers in the homodimer may be slightly different to its native conformation in pepsin. As discussed above, the shapes of the activity peaks in chromatogram of Fig. 3 suggests that *propep* is refolded into several slightly different conformations that can recombine with *sin* to form *propep.sin*. The minor fraction that

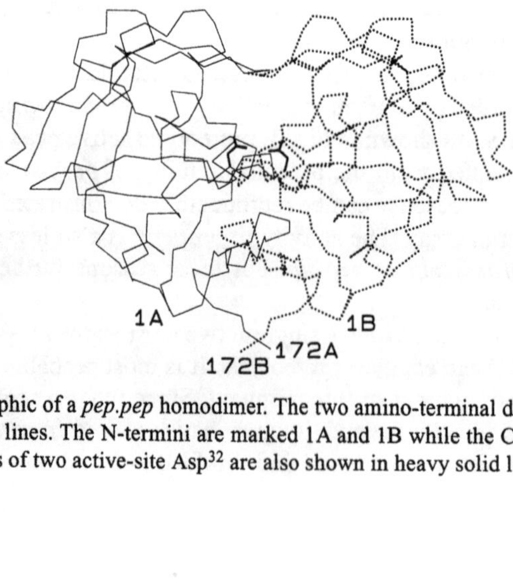

Figure 4. A computer graphic of a *pep.pep* homodimer. The two amino-terminal domains of pepsin, *pep*, are shown in solid and dotted lines. The N-termini are marked 1A and 1B while the C-termini are marked 172A and 172B. The side chains of two active-site Asp^{32} are also shown in heavy solid lines.

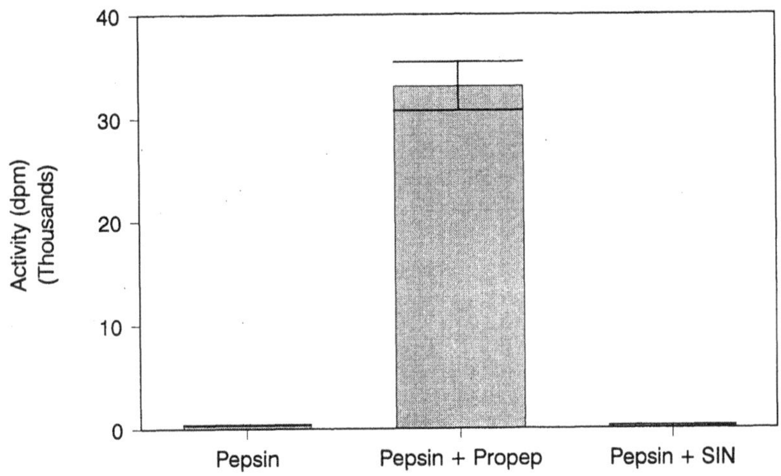

Figure 5. Regeneration of proteolytic activity of alkaline denatured pepsin by the addition of refolded N-terminal lobe (*propep*) and C-terminal lobe (*sin*). Left column, alkaline denatured pepsin shows no activity. Central column, the addition of *propep* regenerates the proteolytic activity. Right column, the addition of *sin* does not regenerate activity.

manifests activity (propep*) perhaps has a conformation that favors the interfacial fitting and hence the homodimer formation. In spite of this conformational uncertainties, it is remarkable that a pepsin lobe retains some structural information to form an active homodimer. It suggests that the interfacial structure is well conserved in evolution. This conservation is reasonable. Any mutation on one face requires a correctional mutation on the opposing face in order to keep an effective interfacial relationship of the monomers. These points suggest that the ability to form homodimers by lobes from eukaryotic aspartic proteases may not be an isolated event.

PROBING FOR ALKALINE DENATURATION OF PEPSIN

It has long been known that pepsin spontaneously denatures in neutral and alkaline solutions. Since *pep* and *sin* can be recombined to generate activity, they were used as probes for alkaline denaturation of pepsin. The addition of refolded *propep* to alkaline denatured pepsin, $pepsin_a$, regenerated protease activity (Fig. 5). *Sin*, however, had no effect. These observations suggest that in neutral and alkaline solutions, the *pep* lobe in pepsin is specifically denatured. The *sin* lobe in pepsin must have been stable in alkaline solution thus is capable of forming a complex with added *propep* to produce activity, as illustrated in Fig. 6. This explanation is confirmed by the isolation of $propep.pepsin_a$ complex from a mixture where the activity of alkaline denatured pepsin had been regenerated by the addition of *propep* (7). The isolated complex gives rise to two bands in SDS-gel electrophoresis with mobilities corresponding to *propep* and pepsin (Fig. 2).

The structure of $propep.pepsin_a$ complex is thought to be similar to that of two-chain pepsinogen (see Fig. 6) except that the denatured *pep* is still covalently attached. Acidification of the complex results in the activation of the complex and peptide cleavages in both *propep* and $pepsin_a$. Unlike the activation of pepsinogen and the two-chain zymogen, in which the entire pro peptide is rapidly removed, the activation of the complex $propep.pepsin_a$ gives rise to a gel electrophoretic band larger than *pep* (Fig. 2), indicating a cleavage near

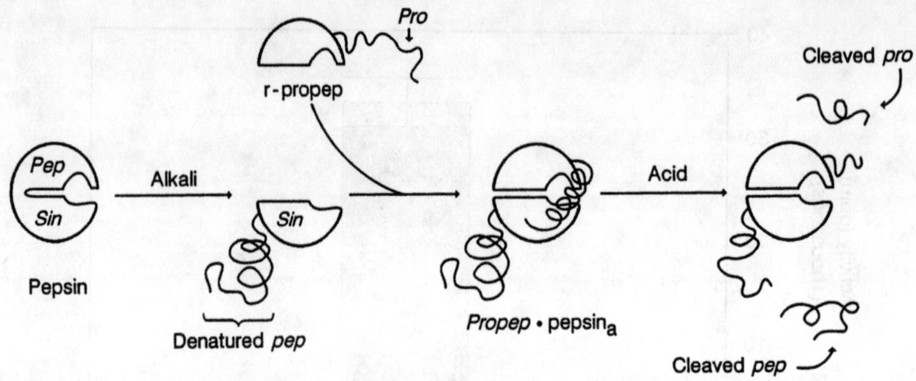

Figure 6. Schematic presentation of the formation of *propep.pepsin*$_a$ complex and proteolytic cleavage during activation. An alkaline solution denatures the *pep* lobe of pepsin to cause an inactivation. The addition of recombinant (r-) *propep* to the alkaline-denatured enzyme (*pepsin*$_a$) results in the formation of complex *propep.pepsin*$_a$, which resembles a two-chain pepsinogen (5). When the complex is in an acidic solution, the activation is accompanied by the cleavage of the *pro* peptide and the denatured *pep* domain of pepsin.

the middle of the pro sequence. The size reduction of *pepsin*$_a$ must be due to partial proteolysis of denatured *propep* moiety during the activation. These phenomena are illustrated also in Fig. 6. We have measured the activation kinetics of *propep.pepsin*$_a$ complex. The apparent first-order activation constant is at least 100 times slower than that of the native pepsinogen (8). This is in part due to the competition of denatured *pep* with the activation cleavage of *pro* peptide, a phenomenon similar to the competitive inhibition of activation by a pepsin substrate (6).

The most plausible reason why *pep* denatures in pH 7 or above is that the denaturation is triggered by the ionization of a group with pK_a near neutrality. *Pep* contains only two basic groups; at its N-terminus and His^{57}. Both of these groups are located at the surface of the 3-D structure and there is little reason why their ionization would effect structural stability. *Pep* also contains 23 carboxyl groups. The normal pK_a of these groups is near pH 3-4. However, partially buried and hydrogen bonded carboxyls have the potential for abnormal pK_a's. The best candidates from the electrostatic potential calculations (Lin *et al.*, 1993) are the interacting pair Asp^{11} and Asp^{159}. These carboxyl groups interact in a hydrogen bond in a similar manner as the active-site groups. Also, Asp^{11} is partially buried in the 3-D structure. It seems probable that Asp^{159} has a low pK_a while Asp^{11} has an abnormally high pK near pH 7. In a neutral solution, the repulsion from these two closely situated negatively charged groups would be unfavorable and could trigger the denaturation. Other groups with high probability are Glu^4, Glu^{13}, and Asp^{118} which all have high values in the calculated electrostatic potentials.

REARRANGING PEPSINOGEN

Since the domains in pepsinogen can fold independently, it is possible to consider the rearranging of the order of these domains. Native pepsinogen has the domains in the order of *pro-pep-sin*. Using the 3-D structure of pepsinogen, we have designed a rearranged molecule with the domain order of *sin-pro-pep*. The rearrangement is illustrated in Fig. 7. We have constructed *sin-pro-pep* gene, expressed, purified and refolded the rearranged pepsinogen. *Sin-pro-pep* is stable as a zymogen and proteolytically active upon acidification. The properties of this protein will be described elsewhere (X. Lin, G. Koelsch, J. Loy and J. Tang, manuscript in preparation).

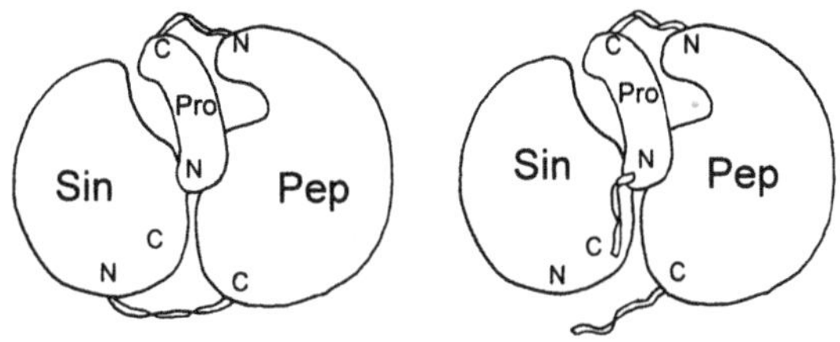

Figure 7. Schematic presentation of domain arranged pepsinogen (*sin-pro-pep*) as compared to that of the native pepsinogen (*pro-pep-sin*). For the convenience of description, the following abbreviations will be used to describe the pepsinogen fragments: *pro*, pro peptide; *pep*, the N-terminal lobe of pepsin; *sin*, the C-terminal lobe of pepsin; *propep**, a fraction of propep with proteolytic activity; *pep.pep*, homodimer of the N-terminal lobe of pepsin; *sin.sin*, homodimer of the C-terminal lobe of pepsin; *pep.sin*, two-chain pepsin, *sin-pro-pep*, single-chain internally rearranged pepsinogen.

CONCLUDING REMARKS

The observations that the domains of pepsin can independently fold and can form, albeit low yield, homodimer are consistent with the hypothesized evolutionary relationships of the single-chain eukaryotic aspartic protease and the homodimeric retroviral aspartic proteases. It is important to point out that protein and enzyme engineering can not prove the evolutionary process, however. Our studies seems to raise several interesting questions. First, whether one can improve the stability and yield of propep or pep homodimers by either protein engineering or by genetic screening or selection. Success in this will open detailed study of structure-function relationships in this interesting homodimer model. Second, can other domains in single-chain aspartic proteases form homodimers? Several eukaryotic enzymes have been cloned and expressed so the answering of this question should be technically very feasible. Third, how far can we go in the rearrangement of pepsinogen domains? Possibly to make hybrid zymogens and enzymes or even 'zebra' enzymes from different aspartic protease origins to produce interesting models for structure and function studies. These and other questions will be the challenge of future studies.

ACKNOWLEDGMENTS

This work is supported by NIH grant DK-01107. The authors wish to thank Gerald Koelsch for his contributions in rearranging pepsinogen molecule.

REFERENCES

1. Tang, J., James, M.N.G., Hsu, I.-N., Jenkins, J.A., and Blundell, T.L. (1978) "Structural Evidence for Gene Duplication in the Evolution of the Acid Protease" *Nature* **271**:618-621.
2. Rao, J.K.M., Erickson, J.W., and Wlodawer, A. (1991) Structural and Evolutionary Relationships between Retroviral and Eucaryotic Aspartic Proteinases" *Biochemistry* **30**:4663-4671.
3. Lin, X.L., Wong, R.N.S., and Tang, J. (1989) "Synthesis, Purification and Mutagenesis Studies of Recombinant Porcine Pepsinogen" *J. Biol. Chem.* **264**:4482-4489.
4. Lin, Y.-Z., Fusek, M., Lin, X.L., Hartsuck, J.A., Kezdy, F.J., and Tang, J. (1992a) "pH Dependence of Kinetic Parameters of Pepsin, Rhizopuspepsin, and Their Active-site Hydrogen Bond Mutants" *J. Biol. Chem.* **267**:18413-18418.

5. Lin, X.L., Lin, Y., Koelsch, G., Gustchina, A., Wlodawer, A., and Tang, J. (1992) "Enzymic Activities of Two-chain Pepsinogen, Two-chain Pepsin, and the Amino-terminal Lobe of Pepsinogen" *J. Biol. Chem.* **267**:17257-17263.
6. Marciniszyn, J., Jr., Huang, J.S., Hartsuck, J.A., and Tang, J. (1976) "Mechanism of Intramolecular Activation of Pepsinogen" *J. Biol. Chem.* **251**:7095-7102.
7. Lin, X.L., Loy, J.A., Sussman, F., and Tang, J. (1993) "Conformational Instability of the N- and C-terminal Lobes of Porcine Pepsin in Neutral and Alkaline Solutions" *Protein Sci.* **2**:1383-1390.
8. Al-Janabi, J., Hartsuck, J.A., and Tang, J. (1972) "Kinetics and Mechanism of Pepsinogen Activation" *J. Biol. Chem.* **247**:4628-4632.

COMPARATIVE INVESTIGATIONS ON PIG GASTRIC PROTEASES AND THEIR ZYMOGENS

Bent Foltmann,[1] Kenneth Harlow,[1] Gunnar Houen,[1] Peter K. Nielsen,[1] and Per Sangild[2]

[1]Dept. of Protein Chemistry
University of Copenhagen
Øster Farimagsgade 2A, DK-1353 Copenhagen K, Denmark
[2]Dept. of Clinical Sciences
Royal Veterinary and Agricultural University
13 Bülowsvej, DK-1870, Frederiksberg C, Denmark

INTRODUCTION

Pepsin (pepsin A EC 3.4.23.1) is the predominant proteolytic enzyme in the gastric juice of most vertebrates. It was named by Schwann (1836) long before the term enzyme was introduced, and in the early days of biochemistry pepsin A was used as a model enzyme in numerous investigations. The minor proteolytic components of the gastric juice have received much less attention. Pepsin B (EC 3.4.23.2) and gastricsin (EC 3.4.23.3) were isolated by Ryle and Porter (1959) and first designated parapepsin I and II. The corresponding zymogens were subsequently isolated (Ryle, 1960; Ryle, 1965) but no structural data were reported. Chymosin (EC 3.4.23.4) was for many years regarded as a milk-clotting enzyme that was characteristic for young ruminants (Foltmann, 1966), but immunochemical cross reactivity between calf chymosin and extracts of stomachs from newborn piglets showed that the gastric juice of piglets also contains chymosin (Foltmann et al., 1978).

The general literature on the gastric proteases is covered in reviews by Ryle (1970), Tang (1970), Fruton (1976) and Foltmann (1981). The present review summarizes a series of recent comparative investigations on the pig gastric proteases and their zymogens together with a detailed investigation on the activation of pepsinogen A.

ONTOGENY

A developmental change in the production of gastric zymogens and the corresponding enzymes has been observed in all mammals that have been investigated (Foltmann and Axelsen, 1980). The first detailed investigations on the development of the gastric proteases

Aspartic Proteinases: Structure, Function, Biology, and Biomedical Implications
Edited by Kenji Takahashi, Plenum Press, New York, 1995

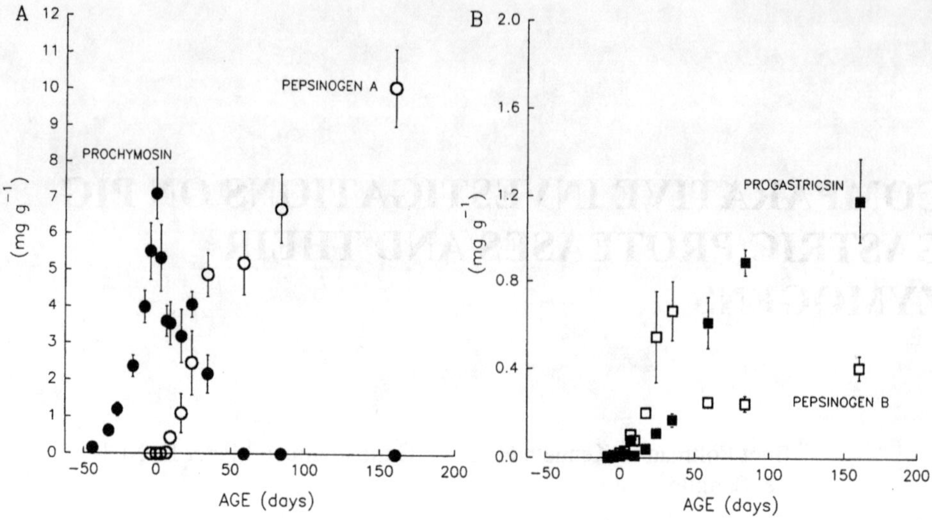

Figure 1. Ontogeny of zymogens for pig gastric proteases. Quantitation by rocket immunoelectrophoresis. The amounts are expressed as mg per g wet weight fundic tissue. Prochymosin, (•); pepsinogen A, (□); pepsinogen B, (□); progastricsin, (■) (Sangild, Turvey, Cranwell and Foltmann, in preparation).

in pigs comprised only chymosin and pepsin, and the fetal material was not very well defined. (Foltmann et al., 1981).

The present investigations include all four known gastric proteases from pig, and the fetal material is from fetuses of controlled age. All the gastric proteases have similar enzymic properties, but as in the previous investigations we have taken advantage of the immunochemical differences among the enzymes. Thus quantitative determinations were carried out by rocket immunoelectrophoresis with the use of monospecific antisera raised in rabbits (Axelsen, 1983).

The results are summarized in Fig.1. It is seen that the production of prochymosin reaches a maximum at birth, a rapid decline occurs during the first weeks of life and production comes to a stop after about 8 weeks. In some animals traces of pepsinogen A are observed at birth, but the major production of pepsinogen A starts after two weeks, approaching the adult level after about 8 weeks. The production of pepsinogen B starts at birth, a maximum is observed after 4 weeks of age and then levels off. The production of progastricsin starts slightly after the production of pepsinogen A and likewise it approaches adult level after about 8 weeks.

The investigations that are reported in Fig. 1 were mainly concentrated on developmental aspects. Adult animals also have considerable variations in the production of gastric zymogens. From our experience with preparation of the gastric proteases we estimate the following order of magnitude for zymogens in fundic gastric mucosa from adult animals: pepsinogen A 8-12 mg/g; pepsinogen B 0.15-0.3 mg/g and progastricsin 1-3 mg/g.

PURIFICATION

All zymogens and enzymes used in these investigations were purified at the Department of Protein Chemistry. The initial steps were essentially carried out according to Ryle (1970). From extracts of fundic mucosa the zymogens and enzymes were concentrated by batchwise adsorption to DEAE cellulose (Foltmann et al., 1992). For fractionation of pepsinogen A, pepsinogen B and progastricsin the DEAE cellulose was then packed in a chromatographic column, and eluted with a linear gradient from 0.02 M sodium phosphate, pH 6.9 to 0.02 M sodium phosphate, pH 6.9 containing 0.6 M NaCl. For preparation of chymosin, a linear gradient from 0.05 M sodium phosphate, pH 6.0 to to 0.4 M sodium phosphate, pH 5.4 was used. Thoughout the purifications the fractionation was monitored by SDS-PAGE, immunoelectrophoreses against monospecific antisera, and by agar gel electrophoresis at pH 6.0 followed by development as zymograms (Foltmann et al., 1985).

Pepsinogen A and progastricsin were further separated by chromatography on DEAE-Sepharose CL6B and final purification was obtained by MonoQ FPLC using a linear gradient from 0.05 M to 1.5 M ammonium acetate, pH 5.6 (Foltmann et al., 1992). It is noteworthy that progastricsin and pepsinogen B have almost the same mobilities by agar gel electrophoresis at pH 6.0, but by chromatography on DEAE-cellulose progastricsin elutes slightly after pepsinogen A, whereas pepsinogen B elutes with the front of pepsinogen A. A pool which contained equal amounts of pepsinogen A and B was subjected to gel filtration on a column of Sephacryl S-100 with 0.05 M sodium phosphate/0.15 M NaCl, pH 6.8 (Nielsen, Ryle and Foltmann, submitted). Final purification of pepsinogen B took place by MonoQ FPLC and a linear gradient from 0.02 M piperazine/HCl, pH 5.7 to 0.02 piperazine/HCl/1 M NaCl, pH 5.7. It should be observed that minor amounts of non-covalent complexes between propart peptides and enzymes could not be separated from progastricsin and pepsinogen B (see also section on activation). Chymosin was purified from selected peak fractions of the first DEAE-chromatography by MonoQ FPLC and a linear gradient from 0.05 M ammonium acetate pH 5.3 to 0.5 M ammonium acetate, pH 5.3 (Foltmann, 1992).

STRUCTURES

The homology among the four pig gastric proteases is exemplified in Fig. 2 which shows an alignment of the prosegments together with the N-terminal amino acid sequences of the active enzymes. The overall identity among the four prosegments is about 22% and the distribution of apolar and basic amino acid residues follows a common pattern. The inactive conformation of pepsinogen A is stabilized by salt bridges from Lys3p, Arg8p, Arg13p and Lys36p (James and Sielecki, 1986); at these positions basic residues are found in all of the prosegments. It is well known that the tertiary structures of homologous proteins are more conserved than the primary structures, thus it is likely that all the prosegments in principle have the same folding as that described for pepsinogen A (Sielecki et al., 1991). Among the differences it should be observed that pepsinogen B has Ile at position 5p, whereas all other known structures of gastric protease zymogens show Pro at this position (Foltmann, 1988).

Relative to the other zymogens, pepsinogen B has an insertion of a Lys-residue after position 30p. This is located in a bend between the second and third helix in the prosegment of pepsinogen A, and this will probably not distort the common folding. Lys36p has previously been regarded as an absolutely conserved residue which interacts with the aspartic acid residues 32 and 215 in the active site of the enzyme moiety of the molecule. In progastricsin Lys36p is substituted with an Arg, but it has been possible to model an arginine side chain at this position in the structure of pepsinogen A. Thus, this substitution does not contradict the assumption of a common folding for the prosegments (Foltmann et al., 1992).

Considerable differences are observed in the junction region between the prosegments and the mature enzymes, and the positions of the cleavage sites which lead to the mature enzymes are also different. This region probably represents a relatively loosely ordered part of the molecules, and only a weak electron density was found in the final electron density map by X-ray crystallography of pepsinogen A.

```
               1p       10p       20p       30p       40p
               .         .         .         .         .
Pepsinogen A   LVKVPLVRKKSLRQNLIKNGKLKDFLKTHK-HNPASKYFPEAAAL

Pepsinogen B   MERIILRKGKSIREAMEEQGVLEKFLKNRPKIDPAAKYHFNND

Progastricsin SVIKVPLKKLKSIRQAMKEKGLLEEFLKTHK-YDPAQRYRFGDF

Prochymosin  SGITRVPLRKGKSLRKELKERGLLEDFLQKQP-YALSSKYSSF


               1        10        20
               .         .         .
Pepsin A       IGDEPLENYLDTEYFGTIGIGTP

Pepsin B     AVAYEPFTNYLDSFYFGEISIGTP

Gastricsin   SVALEPM-AYLEAAYFGEISIGTP

Chymosin   GEVASEPLTNYLDTQYFGKIYIGTP
```

Figure 2. Alignment of the N-terminal amino acid sequences of pig gastric zymogens and enzymes. Numbering follows the sequence of pepsinogen A and pepsin A. Pepsinogen A and pepsin A (PIR protein database, acc. no. A 00982); pepsinogen B and pepsin B (Nielsen, Ryle and Foltmann, submitted); progastricsin and gastricsin (PIR protein database, acc. no. S 21754); prochymosin and chymosin (Houen, Madsen, Harlow, Lønblad and Foltmann, submitted).

The entire amino acid sequence of pig prochymosin was deduced from the nucleotide sequence of its cDNA. Fig. 3 shows an alignment between pig and calf chymosin with pig pepsin A as a reference. The two chymosins are about 82% identical in their primary structures; this is consistent with the pronounced immunochemical cross reactivity which lead to the discovery of pig chymosin. As pointed out below, however, there are considerable differences both between the milk-clotting activities and the general proteolytic actvities of the two chymosins. Thus pig chymosin has a very weak general proteolytic activity, which was probably the reason that this enzyme remained unobserved for many years.

Table 1 summarizes some of the differences between pig and calf chymosin. Possible subsite interactions are indicated on the basis of the structures of other aspartic proteases (Sielecki et al., 1990; Cooper et al., 1987). Some of the positions in which the two chymosins differ have been studied by site directed mutagenesis in calf chymosin. Val111 was substituted with Phe which lead to local changes in the conformation and increased K_M values, whereas k_{cat} values were only slightly altered (Strop et al., 1990). Lys220 has been changed to Leu (Suzuki et al., 1989). This resulted in a shift towards the acidic side for the pH optimum for hydrolysis of acid denatured hemoglobin, together with a marked increase of general proteolytic activity; at position 220 pig chymosin has a methionyl residue, and relative to calf chymosin the pH-optimum is somewhat broader towards the acidic side; but the general proteolytic activity of pig chymosin towards acid denatured hemoglobin is only 10% of that of calf chymosin. These properties may be considered relative to the substitution at position 303. Here all other known structures of gastric proteases have an Asp-residue whereas pig chymosin has Val. In the other proteases, the Asp-residue is hydrogen bonded to Thr218 near the active site, and is considered to be responsible in part for determining the state of ionization of the catalytic carboxylates. Renin, which also has a very weak general

```
               1       10        20        30        40        50
               .        .         .         .         .         .
PIG PEPSIN     IGDEPLENYLDTEYFGTIGIGTPAQDFTVIFDTGSSNLWVPSVYCSSLAC
PIG CHYMOS GEVASEPLTNYLDTQYFGKIYIGTPPQEFTVVFDTGSSELWVPSVYCKSDAC
BOV CHYMOS GEVASVPLTNYLDSQYFGKIYLGTPPQEFTVLFDTGSSDFWVPSIYCKSNAC

                        60        70        80        90       100
                         .         .         .         .         .
PIG PEPSIN     SDHNQFNPDDSSTFEATSQELSITYGTGSMTGILGYDTVQVGGISDTNQI
PIG CHYMOS     QNHHRFNPSKSSTFQNLDKPLSIQYGTGSIQGFLGYDTVMVAGIVDAHQT
BOV CHYMOS     KNHQRFDPRKSSTFQNLGKPLSIHYGTGSMQGILGYDTVTVSNIVDIQQT

                       110       120       130       140       150
                         .         .         .         .         .
PIG PEPSIN     FGLSETEPGSFLYYAPFDGILGLAYPSISASGATPVFDNLWDQGLVSQDL
PIG CHYMOS     VGLSTQEPSDIFTYSEFDGILGLGYPELASEYTVPVFDNMMHRHLVAQDL
BOV CHYMOS     VGLSTQEPGDVFTYAEFDGILGMAYPSLASEYSIPVFDNMMNRHLVAQDL

                       160       170       180       190       200
                         .         .         .         .         .
PIG PEPSIN     FSVYLSSNDDSGSVVLLGGIDSSYYTGSLNWVPVSVEGYWQITLDSITMD
PIG CHYMOS     FAVYMSRNDE-GSMLTLGAIDPSYYTGSLHWVPVTMQLYWQFTVDSVTIN
BOV CHYMOS     FSVYMDRNGQ-ESMLTLGAINPSYYTGSLHWVPVTVQQYWQFTVDSVTIS

                       210       220       230       240       250
                         .         .         .         .         .
PIG PEPSIN     GETIACSGGCQAIVDTGTSLLTGPTSAIANIQSDIGASENSDGEMVISCS
PIG CHYMOS     GVVVACNGGCQAILDTGTSMLAGPSSDILNIQMAIGATESQYGEFDIDCG
BOV CHYMOS     GVVVACEGGCQAILDTGTSKLVGPSSDILNIQQAIGATQNQYGEFDIDCD

                       250       270       280       290       300
                         .         .         .         .         .
PIG PEPSIN     SIDSLPDIVFTINGVQYPLSPSAYILQDDDSCTSGFEGMDVPTSSGELWI
PIG CHYMOS     SLSSMPTVVFEISGRMYPLPPSAYTNQDQGFCTSGFQG----DSKSQHWI
BOV CHYMOS     NLSYMPTVVFEINGKMYPLTPSAYTSQDQGFCTSGFQS----ENHSQKWI

                       310       320
                         .         .
PIG PEPSIN     LGDVFIRQYYTVFDRANNKVGLAPVA    326
PIG CHYMOS     LGVVFIQEYYSVFDRANNRVGLAKAI    323
BOV CHYMOS     LGDVFIREYYSVFDRANNLVGLAKAI    323
```

Figure 3. Alignment of the amino acid sequences of pig and calf chymosin. Numbering as pepsin A which is included as a reference. Porcine pepsin (PIR protein database, acc. no. A 00982); pig chymosin (Houen et al. submitted); bovine chymosin (PIR protein database, acc. no. A 00985).

Table 1. Comparison between Pig and Calf Chymosins

Residue no.	Pig	Calf	Sub-sites
12	Thr	Ser	S4
30	Val	Leu	S1/S3
74	Gln	His	S'2
111	Ile	Val	S1/S3
190	Leu	Gln	Near S'3
293	Asp	Glu	All near S'1
294	Ser	Asn	
295	Lys	His	
298	His	Lys	

The two enzymes are approximately 82% identical in their amino acid sequences: some differences may be important for subsite interaction.

proteolytic activity, has Ala at position 303. Thus, the substitution of Asp303 to Val303 may contribute to an explanation of the differences between the activities of calf and pig chymosin.

ACTIVATION

All gastric proteases are synthesized and secreted as zymogens which are converted into active enzymes in the acidic gastric juice. Investigations on the activation of calf prochymosin (Foltmann, 1966) lead to the assumption that the initial stage during the activation of gastric zymogens consists of a pH-dependent conformational change which uncovers the active site without cleavage of peptide bonds. The general validity of this assumption has been sustained in several subsequent investigations (McPhie, 1972; Sanny et al., 1975; Glick et al., 1986) and X-ray crystallographic analysis of pig pepsinogen A (Sielecki et al., 1991) has provided a structural elucidation of the conformational changes. The activation then proceeds by a limited proteolysis which finally removes from 42 to 47 amino acid residues from the N-terminus of the zymogen (the prosegment or propart).

Pepsinogen A

Activation at pH 2, 0°C at concentrations of pig pepsinogen A less than 0.2 mg/ml gives rise primarily to proteolytic cleavage between Leu16p and Ile17p (Christensen et al., 1977). The covalent intermediate is called pseudopepsin, and this is also formed by activation in presence of pepstatin (Dykes and Kay, 1976). At concentrations of pepsinogen A above 0.5 mg/ml the bond Leu44p - Ile1 is predominantly cleaved without production of intermediates (Kageyama and Takahashi, 1983).

Pepsinogen A is stable at pH 8.5 whereas pepsin rapidly loses activity when incubated at pH 8.5. Thus, for investigations on the activation kinetics of pepsinogen A, aliquots of the reaction mixtures were brought to pH 8.5 and the loss of capacity for activation at pH 2 was taken as a measure for the cleavage of pepsinogen into pepsin (Herriott, 1938; Al Janabi et al., 1972; Sanny et al., 1975; Marciniszyn et al., 1976; Twining et al., 1981; Glick et al., 1989).

During our general investigations on the zymogens for the gastric proteases we observed an alkali resistance of intermediates during the activation of pepsinogens. Furthermore, when the activation of pepsinogen A was monitored by zymograms at pH 6, we observed a component with electrophoretic mobility between that of pepsinogen and pepsin, and appreciable amounts of this component persisted for more than one hour after activation at pH 2 and room temperature. This prompted us to isolate the intermediates and analyse their structures and properties (Nielsen and Foltmann, 1993).

A solution of pepsinogen A (1 mg/ml) was activated at pH 2 and 0°C. After 20 min pH was raised to 7 and the reaction mixture was applied to a column of MonoQ equlibrated with 0.75 M ammonium acetate, pH 7; elution took place with linear gradient up to 1.5 M ammonium acetate, pH 7 in 30 min. This way we purified two homogeneous components as assesed by agar gel electrophoresis at pH 6.0. The first component showed a mobility like pepsinogen A, whereas the second was the intermediate with electrophoretic mobility between that of pepsinogen A and pepsin A. By SDS-PAGE both components yielded protein-bands with M_r equal to that of mature pepsin. The inherent enzymatic activities of the intermediates were determined by milk-clotting (Foltmann, 1970). Relative to mature pepsin, the milk-clotting activities of the first and the second components were 3% and 18%, respectively.

After amino acid analyses, N-terminal amino acid sequencing and mass spectrometry, it was concluded that the first intermediate consisted of a non-covalent complex between the entire prosegment, Leu1p - Leu44p and pepsin, whereas the second was a complex between Leu1p - Leu16p and pepsin. The first complex has previously been isolated by Kageyama and Takahashi (1983).

Treatment of activation mixtures at pH 8.5 and 28°C for 20 min has previously been used for investigations on the activation kinetics of pepsinogen A. Preliminary analyses on the stability of the complexes showed that 20 to 25% of the activity at pH 2 remained after the two complexes had been treated under such conditions. McPhie (1975) has reported that alkali unfolded pepsinogen may be refolded at pH 7. Consequently it was investigated how the pH 8.5 treated intermediates would behave if they were kept at a pH near neutrality before they were tested for proteolytic activity at pH 2. The purified intermediates were tested in parallel with an activation mixture which contained both intermediates but no uncleaved pepsinogen (Fig. 4). The results show that the complex Leu1p - Leu44p/pepsin and the activation mixture (which also contains this intermediate) regain most of the capacity for activation at pH 2, whereas the properties of the complex Leu1p - Leu16p/pepsin are unchanged.

Finally, it was examined whether stability at pH 8.5 could be obtained if a synthesized prosegment peptide, Leu1p-Leu16p, was added to mature pepsin and the mixture then treated as the activation mixtures. No increase in stability at pH 8.5 was observed.

These results may be evaluated relative to the tertiary structures of pepsinogen A (James and Sielecki, 1986; Sielecki et al. 1991) and pepsin A (Sielecki et al., 1990; Cooper et al., 1990; Abad-Zapatero et al., 1990). As indicated above, the prosegment of pepsinogen covers the active site cleft at neutral pH. Under acidic condition the prosegment moves out of the cleft and uncovers the active site, but the formation of the mature enzyme is accompanied by a more extensive conformational change. In pepsinogen, residues Val2p to Leu6p are located in a six-stranded β-sheet at the back of the molecule relative to the active site cleft. In mature pepsin residues Gly2 to Leu6 are topologically equivalent to residues Val2p to Leu6p in pepsinogen, but from the static X-ray crystallographic investigations it is not possible to determine at which stage the exchange of the N-terminal peptide segments occurs. The activation experiments may provide a clue to this. Since it was not possible to reconstruct a complex between Leu1p-Leu16p and mature pepsin, it seems likely that the complex which was isolated from the activation mixtures has the prosegment peptide bound

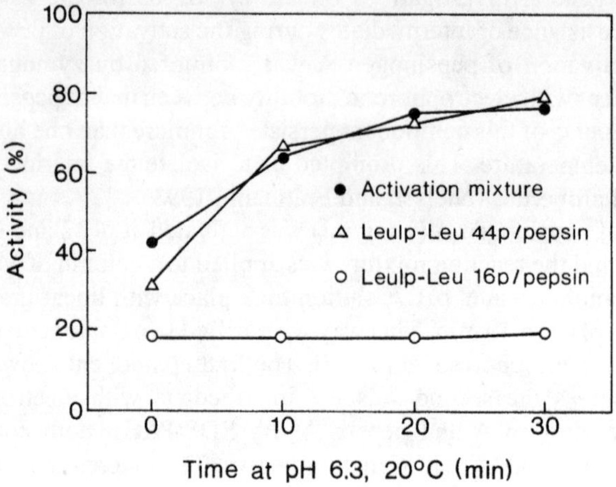

Figure 4. Stability at pH 8.5 of activation intermediates formed during activation of pepsinogen A. The samples were first brought to pH 8.5 for 20 min and 28°C, pH was then lowered to 6.3, and the samples were left at 20°C before determination of the proteolytic activity at pH 2. The results are expressed as percent of the activity of mature pepsin A. Activation mixture (1 mg of pepsinogen A in 1 ml activated for 10 min. at pH 2, 0°C), (Δ); complex Leu1p - Leu44p/pepsin, (•) complex Leu1p - Leu16p/pepsin, (□) (Nielsen and Foltmann, 1993).

in its original position in the six-stranded β-sheet. This assumption also contributes to an explanation of the observed stabilities at pH 8.5.

In mature pepsin, Glu4 is partly burried but still accessible to water. Ionization of this residue may destabilize the β-sheet; Lin et al. (1993) also found that Glu4 is one of the residues that may contribute to the instability of mature pepsin at neutral or alkaline conditions. In pepsinogen, the topologically equivalent residue is the apolar Val4p. The conservation of the location of Val4p in the activation intermediates may explain why these intermediates at pH 8.5 have stabilities between those of pepsinogen A and pepsin A.

Pepsinogen B

As indicated in the section on purification the preparations of pepsinogen B included some cleaved zymogen in complex with propart peptides. Activation of the preparation at pH 2 and room temperature was monitored by SDS-PAGE. The starting material showed bands of pepsinogen B together with minor amounts of pepsin B. A band of intermediate mobility appeared after 30 min. Whereas the band of pepsinogen B disappeared after 22 h the band of intermediate mobility persisted for 72 h. Electroblotting and sequencing showed that the intermediate had been formed by cleavage at the bond Met16p - Glu17p corresponding to the cleavage which gives rise to pseudopepsin A (Nielsen et al., submitted). Like calf pseudochymosin (Pedersen et al., 1979), pseudopepsin B was unstable at pH 5.5 and most of the intermediate was converted into pepsin B by incubation at pH 5.5 for 4 h at 37°C. The results show that in the absence of other proteases pepsinogen B is activated much slower than any other of the known gastric zymogens. This may be because of the very low general proteolytic activity of the enzyme (see below).

Progastricsin

Similarly to pepsinogen B, minor amounts of cleaved zymogen in the starting material prevented a quantitative evaluation of the activation. It occurs, however, at a rate which is of the same order of magnitude as that observed for pepsinogen A. SDS-PAGE followed by electroblotting and sequencing (Foltmann et al., 1992) showed, however, a transient intermediate which had been cleaved between Phe25p and Leu26p.

General Remarks

The activation of the gastric zymogens represents a series of complex and overlapping reactions which is a challenge to the protein chemist. With purified zymogens the reactions to a large extent depend on experimental conditions such as the concentration of zymogens, temperature, pH and ionic strength of the buffers. Thus, quantitative activation kinetics should be considered with caution. Furthermore, the sequence Phe25p - Leu26p is found in all the pig gastric zymogens but it is only cleaved during the activation of progastricsin. This illustrates the importance of subsite interactions for the specificities of the aspartic proteases. Finally it should be kept in mind, that under physiological conditions the activation occurs with a mixture of zymogens, which makes the process even more complicated.

ACTIVITY

For comparison of the proteolytic activities of the gastric proteases we have mainly used the clotting of milk and digestion of acid denatured hemoglobin as measures of these activities. These analyses represent a simulation of the natural activities of the enzymes, but in order to carry out the comparison, standardized conditions are required and these represent an approximation only to the physiological conditions of the digestion.

The clotting of milk is very pH-dependent. This reflects differences in the pH-optima of the enzymes as well as the properties of the casein micelles which are destabilized when the pH of the milk is lowered. Furthermore, as pointed out below milk from different species may show different properties in the clotting process. For practical reasons we have used clotting of milk reconstituted from bovine skim-milk powder as described previously (Foltmann, 1970). Under these conditions pepsin A, gastricsin and chymosin all showed a milk-clotting activity which was about 25% of that of calf chymosin, whereas about 10% was found for the milk-clotting activity of pepsin B. The cleavage site of bovine kappa-casein by pepsin A and gastricsin was found to be the same Phe-Met bond as that cleaved by calf chymosin (Drøhse and Foltmann, 1989). Porcine milk has not been available in amounts which allowed development of a standardized clotting analysis, but the results indicate that the clotting activity of pig pepsin A against porcine milk is about 2 to 3 times that of the clotting of bovine milk. The value for pig chymosin is 5 to 8 times greater with porcine than bovine milk. Conversely, the activity of calf chymosin against porcine milk is only about 60% of that against bovine milk. This shows that a species specific adaptation between the proteases and their natural substrates has occurred, (Foltmann et al., 1981).

Pig pepsin B and chymosin have very low general proteolytic activities, and since small amount only were available, the activities were determined by a modified Anson-test. The results and experimental details are presented in Fig. 5. It should be borne in mind that the relative activites indicate order of magnitudes. Determinations of general proteolytic activites depend on the experimental conditions (i.e. substrates, composition and ionic strength of buffers together with temperature and duration of the experiment). In a comparison between the activities of pepsin A and gastricsin Tang (1970) found that with hemoglobin as substrate the maximum activity of pepsin A was about 75% of that of gastricsin, whereas the relative activity of pepsin A was about 200% with ovalbumin as substrate. At the

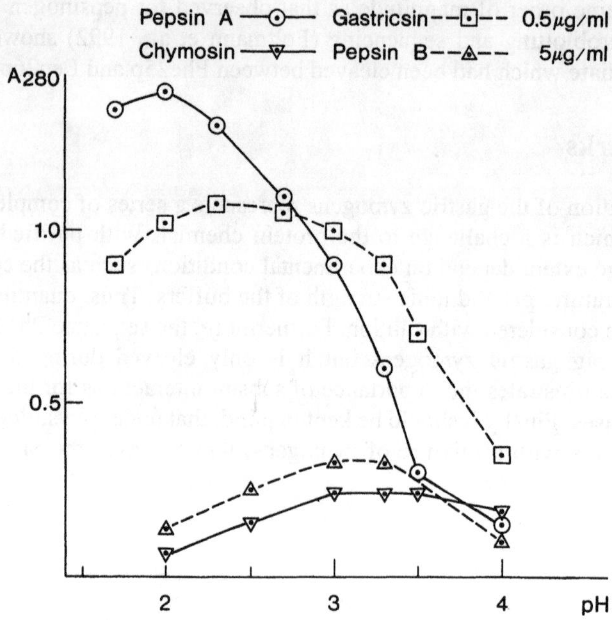

Figure 5. pH dependence of the relative proteolytic activities of pepsin A, pepsin B, gastricsin and chymosin. The assay took place at 37°C, 500 µl substrate (2% acid denatured hemoglobin in citrate buffer, ionic strength 0.1) was mixed with 50 µl of enzyme solution, after 2 h 400 µl of 0.5 M trichloroacetic acid was added, the precipitate was removed by centrifugation, and A_{280} of the supernatant was determined. Blank values were determined at each pH. The experiments were carried out with concentrations of enzymes that gave readings of 0.3 to 0.8 absorbancy unit. In order to compare the activities the results are calculated as activities corresponding to 0.5 and 5 µg of enzymes in the digestion mixtures.

Deptartment of Protein Chemistry we have found that at the same pH digestions in buffers of formate, citrate and acetate could give results which differed by up to 50%.

CONCLUDING REMARKS

Much information about the gastric proteases has been accumulated since the discovery of pepsin 157 years ago. Yet we have relatively little detailed knowledge about the physiological significance of the individual proteases. It is obvious that these proteases are important in the first steps of the digestion of proteins, and in their differences in pH-optima and specificities they may be supplementary to each other. After a meal, for instance, the pH of the gastric contents may rise to 5 or even higher. At pH 4 the general proteolytic activity of gastricsin is about 2 to 3 times that of pepsin A; thus, although the amount of gastricsin is 10% to 20% of pepsin A, gastricsin may still play a significant role in the digestion. We know almost nothing about a possible cause for the developmental change. So far it is a riddle why the production of a minor component like pepsin B passes through a maximum at 4 weeks of age, and we don't know if this observation has a general significance or if it is unique for pigs.

As regards chymosin, it has been pointed out (Foltmann and Axelsen, 1980) that chymosin-like proteases occur in all mammals with postnatal uptake of IgG. This provided indirect evidence for the hypothesis that the chymosins have evolved as neonatal proteases which have sufficient proteolytic activity to clot the milk, but with so weak a general proteolytic activity that the immunoglobulins are not damaged. The direct evidence is, however, still lacking.

REFERENCES

1. Abad-Zapatero, C., Rydel, T.J. and Erickson, J. (1990), Proteins: Struct. Funct. Genet. **8**, 62-81.
2. Al-Janabi, J., Hartsuck, J.A. and Tang, J. (1972), J. Biol. Chem. **247**, 4628-4632.
3. Axelsen, N.H. (ed.) (1983), Handbook of Immunoprecipitation in Gel Techniques, Scand. J. Immunol. **17**, suppl. 10.
4. Christensen, K.A., Pedersen, V.B. and Foltmann, B. (1977), FEBS Lett. **76**, 214-218.
5. Cooper, J., Foundling, S., Hemmings, A., Blundell, T.L., Jones, D.M., Hallett, A. and Szelke, M. (1987), Eur. J. Biochem. **169**, 215-221.
6. Cooper, J.B., Khan, G., Taylor, G., Tickle, I.J. and Blundell, T.L. (1990), J. Mol. Biol. **214**, 199-222.
7. Drøhse, H.B. and Foltmann, B. (1989), Biochim. Biophys. Acta **995**, 221-224.
8. Dykes, C.W. and Kay, J. (1976), Biochem. J. **153**, 141-144.
9. Foltmann, B. (1966), C.R.Trav. Lab. Carlsberg **35**, 143-231.
10. Foltmann, B. (1970), Methods Enzymol. **19**, 421-436.
11. Foltmann, B. (1981), Essays in Biochemistry **17**, 52-84.
12. Foltmann, B. (1988), Biol. Chem. Hoppe-Seyler **369**, Suppl., 311-314.
13. Foltmann, B. (1992), Scand. J. Clin. Lab. Invest. **52**, (Suppl. 210), 65-79.
14. Foltmann, B. and Axelsen, N.H. (1980), FEBS Proceedings, **60**, 271-280.
15. Foltmann, B., Lønblad, P. and Axelsen, N.H. (1978), Biochem. J. **169**, 425-427.
16. Foltmann, B., Jensen, A.L., Lønblad, P., Smidt, E. and Axelsen, N.H. (1981), Comp. Biochem. Physiol. **68B**, 9-13.
17. Foltmann, B. Szecsi, P.B. and Tarasova, N.I. (1985), Anal. Biochem. **146**, 353-360.
18. Foltmann, B., Drøhse, H.B., Nielsen, P.K. and James, M.N.G. (1992), Biochim. Biophys. Acta **1121**, 75-82.
19. Fruton, J.S. (1976), Adv. Enzymol. **44**, 1-36.
20. Glick, D.M., Auer, H.E., Rich, D.H., Kawai, M. and Kamath, A. (1986), Biochemistry **25**, 1858-1864.
21. Glick, D.M., Shalitin, Y. and Hilt, C.R. (1989), Biochemistry **28**, 2626-2630.
22. Herriott, R.M. (1938), J. Gen. Physiol. **22**, 65-78.
23. James, M.N.G. and Sielecki, A.R. (1986), Nature (London), **319**, 33-38.
24. Kageyama, T. and Takahashi, K. (1983), J. Biochem. (Tokyo), **93**, 743-54.
25. Lin, X., Loy, J.A., Sussman, F. and Tang, J. (1993), Prot. Sci. **2**, 1383-1390.
26. Marciniszyn, J., Huang, J.S., Hartsuck, J.A. and Tang, J. (1976), J. Biol. Chem. **251**, 7095-7102.
27. McPhie, P. (1972), J. Biol. Chem. **247**, 4277-4281.
28. McPhie, P. (1975), Biochemistry **14**, 5353-5356.
29. Nielsen, F.S. and Foltmann, B. (1993), Eur. J. Biochem. **217**, 137-142.
30. Pedersen, V.B., Christensen, K.A. and Foltmann, B. (1979), Eur. J. Biochem. **94**, 573-580.
31. Ryle, A.P. (1960), Biochem. J. **75**, 145-150.
32. Ryle, A.P. (1965), Biochem. J., **96**, 6-16.
33. Ryle, A.P. (1970), Methods Enzymol. **19**, 316-336.
34. Ryle, A.P. and Porter, R.R. (1959), Biochem. J. **73**, 75-86.
35. Sanny, C.G., Hartsuck, J.A. and Tang, J. (1975), J. Biol. Chem. **250**, 2635-2639.
36. Schwann, Th. (1836), Müllers Archiv. J. Anat. Physiol. 90-138.
37. Sielecki, A.R., Fedorov, A.A., Boodhoo, A., Andreeva, N.S. and James, M.N.G. (1990), J. Mol. Biol. **214**, 143-170.
38. Sielecki, A.R., Fujinaga, M., Read, R.J. and James, M.N.G. (1991), J. Mol. Biol. **219**, 671-692.
39. Strop, P., Sedlacek, J., Stys, J., Kaberabkova, Z., Blaha, I., Pavlickova, L., Pohl, J., Fabry, M., Kostka, V., Newmann, M., Frazao, C., Shearer, A., Tickle, I.J. and Blundell, T.L. (1990), Biochemistry **29**, 9863-9871.
40. Suzuki, J., Sasaki, K., Sasao, Y., Hamu, A., Kawasaki, H., Nishiyama, M., Horinouchi, S. and Beppu, T. (1989), Prot. Eng. **2**, 563-569.
41. Tang, J. (1970), Methods Enzymol. **19**, 406-421.
42. Twining, S.S., Sealy, R.C. and Glick, D.M. (1981), Biochemistry **20**, 1267-1272.

NON-MAMMALIAN VERTEBRATE PEPSINOGENS AND PEPSINS: ISOLATION AND CHARACTERIZATION

Kenji Takahashi,[1] Masao Tanji,[1] Etsuko Yakabe,[1] Akira Hirasawa,[1] Senerath B. P. Athauda,[1] and Takashi Kageyama[2]

[1]Department of Biophysics and Biochemistry
Faculty of Science
The University of Tokyo, Tokyo 113, Japan
[2]Department of Cellular and Molecular Biology
Primate Research Institute
Kyoto University, Inuyama, Aichi 484, Japan

INTRODUCTION

Pepsinogen, the precursor of pepsin, is secreted from the gastric mucosa into the gastric lumen, where it is autocatalytically activated to pepsin under acidic conditions. It is classified into three major types, that is, pepsinogen A, pepsinogen C (or progastricsin), and prochymosin (or neonatal pepsinogen). Further, pepsinogens, especially pepsinogens A, are often composed of several isozymogens. So far, a number of pepsinogens were purified from various sources, characterized and sequenced at the protein level and/or DNA level. Thus the complete amino acid sequences of mammalian pepsinogens are known for human (1), monkey (2), porcine (3-6), and rabbit (7) pepsinogens A, human (8), monkey (9), rat (10) and guinea pig (11) pepsinogens C, and bovine prochymosin (12).

As for non-mammalian vertebrates, however, only the complete sequences of chicken pepsinogen A (13) and embryonic pepsinogen (14) and partial sequences of tuna pepsinogens (15) had been reported previously and biochemical characterization had been performed with only chicken (16) and tuna (15) pepsinogens and pepsins. In order to study the structure-function relationships and the molecular evolution of pepsinogens and pepsins, it is desirable to isolate more pepsinogens and pepsins from various sources, especially from non-mammalian species, and to elucidate their properties, including the amino acid sequences. In a series of our recent studies, we have isolated pepsinogens and pepsins from turtle (A. Hirasawa *et al.*, to be published) and frog (17) as representatives of reptile and amphibian species, respectively, and also extended further the structural studies on tuna pepsinogens (M. Tanji *et al.*, to be published) as a representative of fish species. These pepsinogens and pepsins had unique molecular and enzymatic properties. This paper summarizes our recent studies on their isolation and properties. The actual animal species used in these studies were

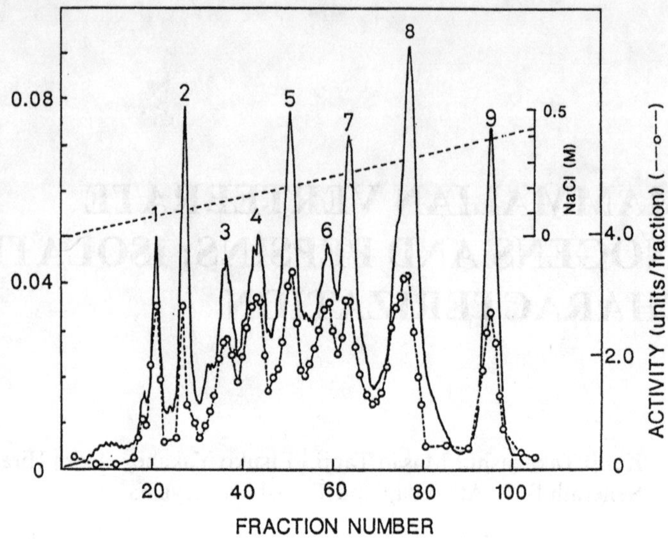

Figure 1. Fractionation of turtle pepsinogens by chromatography on Q-Sepharose. The pepsinogen fraction partially purified from the crude extract of gastric mucosa (wet weight, 84g) by chromatography on DE-52 cellulose and Sephadex G-100 was applied to a column (2.0 x 12 cm) of Q-Sepharose Fast Flow equilibrated with 50 mM piperazine-HCl buffer, pH 6.0. The protein was eluted with an NaCl gradient (0 - 0.45M) in the same buffer at a flow rate of 30 ml/h and 10-ml fractions were collected. Appropriate aliquots of the effluent were used for assay of potential pepsin activity with bovine hemoglobin as a substrate.

soft-shelled turtle (*Trionix sinensis*), bullfrog (*Rana catesbeiana*) and North Pacific bluefin tuna (*Thunnus Thynuus orientalis*).

TURTLE PEPSINOGENS AND PEPSINS

Isolation and Properties of Pepsinogens

The whole pepsinogen fraction was obtained by chromatography on DEAE-cellulose and Sephadex G-100 of the crude extract supernatant of the turtle gastric mucosa. Upon subsequent chromatography on Q-Sepharose, nine isozymogen fractions were separated as shown in Fig. 1.

They were further purified by rechromatography. Their molecular weights were estimated to be about 43,000 by SDS-PAGE. The amino acid compositions of these 9 isozymogens were roughly similar to one another and are not so much different from those of mammalian pepsinogens. They did not contain phosphate, but pepsinogens 4 through 9 were shown to contain carbohydrate (about 2% each). Multiple forms of pepsinogen isozymogens have so far been found for various animal species. Notably, seven isozymogens were found in human (18, 19), six in monkey (20), and eight in rabbit (7) stomachs. The nine isozymogens in turtle stomach thus exceeds these values. The same number of isozymogens was also found when three individual turtle stomachs were examined separately. Therefore, this multiplicity does not appear to be due to individual differences. Since these isozymogens were separated from one another fairly well by Q-Sepharose chromatography, there should be marked differences in net charge among the isozymogens. None of them contained phosphate, so these charge differences are suggested to be primarily due to some amino acid substitutions that would affect the net charge of the proteins. It is interesting to note that isozymogens 4 through 9 were found to be glycoproteins. Carbohydrate-containing pepsinogens are rather rare. We previously reported the occurrence of carbohydrate-containing pepsinogens in monkey stomach (21), but those pepsinogens were minor isozymogen components. In addition, only the major chicken pepsinogen A was reported to be a glycoprotein (13). Thus, to our knowledge, this is the first case that so many isozymogens from one species were separated and shown to be glycoproteins. The activation process of pepsinogens was investigated by SDS-PAGE, using the major isozymogens 7, 8 and 9 as examples. The activation took place very rapidly ; it took only a few minutes for complete conversion to the corresponding pepsins under the conditions used at pH 2.0 and 4 °C. The activation of turtle pepsinogens generally appeared to be very much faster than that of pepsinogens from other sources (17, 19, 22). In each case, apparently two intermediate forms were produced during the course of activation, indicating that the activation took place stepwisely.

Amino Acid Sequences of Pepsinogens and Pepsins

NH_2-terminal sequence analysis was performed with isozymogens 2 through 9. The results indicated that these isozymogens possess the same or very similar NH_2-terminal sequences although amino acid substitutions were found at a few positions. In addition, the NH_2-terminal sequences of some pepsin isozymes were also analyzed and the same or nearly the same amino acid sequences were obtained among the isozymes. In order to get further information about the amino acid sequences of turtle pepsinogens, we have cloned their cDNAs, determined the nucleotide sequence of one of the clones obtained and deduced the amino acid sequence. The cDNA clone sequenced was not a full-length cDNA and encoded part (19 residues) of the activation peptide and the complete amino acid sequence (315 residues) of the pepsin part of a pepsinogen. However, by combining this sequence with that obtained at the protein level for the NH_2-terminal sequences, we tentatively deduced the complete sequence of a turtle pepsinogen as shown in Fig. 2.

The deduced amino acid sequence is compared with the sequences of some other typical pepsinogens in Fig. 2. It is clear that the sequences around the two active-site aspartic acid residues and the locations of the three disulfide bonds have been conserved in turtle pepsinogen. The identity of the deduced amino acid sequence was approximately 71% with porcine pepsinogen A (3-6), 69 % with human pepsinogen A (1), 63 % with chicken pepsinogen A (13), 59 % with chicken embryonic pepsinogen (14), 56 % with bovine prochymosin (12) and 51 % with human pepsinogen C (8). Thus, the deduced sequence clearly belongs to that of the A-type pepsinogens. Since the NH_2-terminal sequences of the turtle pepsinogens and pepsins are very similar among the isozymogens (data not shown), all the isozymogens may be classified structurally as A-type rather than C-type pepsinogens. It is interesting to note that turtle pepsinogen A shows higher homology with porcine and human pepsinogens A than with chicken pepsinogen A. This might indicate that chicken pepsinogen A evolved at a faster rate than turtle pepsinogen A after their divergence. Further, it may be worth pointing out that in turtle pepsinogen, 5-residue deletions were observed at positions between 283 and 287 and between 330 and 331 (Fig. 2). A similar deletion in the first region was observed only with chicken pepsinogen A whereas that in the second region was observed only with chicken embryonic pepsinogen and bovine prochymosin. The turtle pepsinogen sequence contains four potential N-glycosylation sites. Interestingly, chicken embryonic pepsinogen contains five potential N-glycosylation site, and two sites (Asn^{102} and Asn^{115}) are common to turtle pepsinogen. On the other hand, chicken pepsinogen A and bovine prochymosin have two potential N-glycosylation sites, human pepsinogen C has one, and human and porcine pepsinogens A have none. Thus, in these properties, turtle pepsinogen A most resembles chicken pepsinogen A and/or chicken embryonic pepsinogen, possibly reflecting their evolutionary relationship.

```
                                 1           11         21             31         41           51         61
Turtle pepsinogen A          :  ·LVTKVPLKKG KSLRQNLKEH GLLE(D)F(N)KKH PYNPASRYFP SLGD··EF|ATEP LTNYMDAEYF GTISIGTPAQ
Human pepsinogen A           :  ·IMY----IRK ----RT-S-R ---K - - L --- NL---RK--- QWEAPTL|VDEQ- -E--L-M--- ---G------
Porcine pepsinogen A         :  ··LV----VRK -------IKD ---K - - L -T- KH----K--- ··EAAAL|IGD-- -E--L-T--- ---G------
Chicken pepsinogen A         :  ·SIHR------ ----KQ--D- ---- - - L --- ------K-H- V-|TA··TESY-- M------S-Y --------Q-
Chicken embryonic pepsinogen :  ··I-RL--ER- -K--EI-R-K ---H H - L QH- R-DIGTKFPH AFP-VLTVV--- -L-TL-M--Y --------P-
Bovine prochymosin           :  AEI-RI--Y-- ----KA---- ---- - - L Q-Q Q-GIS-K-SG ·F|····-V-SV- ----L-SQ-- -K-YL---P-
Human pepsinogen C           :  ·A-V------F --I-ETM--K ---G E - L RT- K-D--WK-R· ·F--L·|SVTY-- MA·----A-- -E------P-

71         81         91         101        111        121        131        141        151        161
DFTVVFDTGS SNLWVPSVTC SSAACTQHNR FNPSDSSTYR ATSQNLSIQY GTGSMTGILG YDNVQVGGLV DTNQIFGLSE TEPGSTFYYA PMDGILGLAY
---------- ---------- --L---N--- ---E-----Q S--ETV--T- ---------- --T-----IS ---------- -----FL--- -F--------
----I----- ---------- --L---D--Q ---D----FE ----E---T- ---------- --T-----IS ---------- -----FL--- -F--------
--S-I----- -------I-- K-S--SN-K- -D--K----V S-NETVY-A- -----S---- --T-A-SSID VQ-------- -----F---C NF-------F
---------- --------S- T-P--QS-QM ---Q-----K S-G-----H- ---D-E-TV- C-T-T-AS-M ----L----T S---QF-V-V KF------G-
E---L----- -DF----I-- K-N--KN-Q- -D-RK---FQ NLGKP---H- -----Q---- --T-T-SNI- -IQ-TV---T Q---DV-TT- EF-----M--
N-L-L----- ---------- Q-Q---S-S- ----E----S TNG-TF-L-- ----L--FF- --TLT-QSIQ VP--E----- N---TN-V-- QF---M----

171        181        191        201         211        221        231        241         251        261
PSIASSGATP VFDNMMSEGL VSQDLFSVYL SSDEQSGS·FV MFGGNDTSYY SGSLNWIPLS AETYWEITMD SITMNGQTIA· CSGGCQAIID TGTSLLAGPP
---S------ ----IWNQ-- ---------- -A-D----·V- I---I-S--- T-----V-VT V-G--Q--V- ------EA--· -AE-----V- ------T--T
---SA----- ----LWDQ-- ---------- --NDD---·V- LL--I-S--- T-----V-V- V-G--Q--L- ----D-E---· --------V- ------T--T
---S------ -------QH- -A-------- -K-GET--·-- L---I-PN-T TKGIY-V--- -----Q---- RV-VGNKYV-· -FFT----V- ------VM-Q
--L-AD-I-- -----VN-S- LE-N------ -R·-PM--·M- V---I-E--F T--I----V- YQG--Q-S-- --IV-K-E--· --S------- -----V---A
--L--EYSI- ------NRH- -A-------M DR-G-ESM·LT L·-AI-P--- T---H-V-VT VQQ--QFYV- -V-IS-VVV-· --E------L- ----K-V--S
-ALSVDE--T AMQG-VQ--A LTSPV----- -NQQG-SGGA- V---V-S-L- T-QIY-A-VT Q-L--Q-GIE EFLIG--ASGW --E-----V- ------TV-Q

271        281             291        301        311         321         331             341        351        361
SDVSNINSYI GAS···NGE··VSCS SMSSLPEIVF NINGIAFPVP ASAYIINDS·S SCSSS·FESMD ·····QGLWILGDVF IRLYYVVFDR ANNQVGLASLA
-PIA--Q-D- ---ENSD-DMV---- AI----D--- T---VQY--- P----LQSE·G --I-G··QG-N LPTESGE-------- --Q-FT---- --------PV-
-AIA--Q-D- ---ENSD--MVI--- -ID---D--- T-D-VQY-LS P----LQ-D·D --T-G·--G-- VPTSSGE-------- --Q--T---- ---K----PV-
GAYNR-IKDL -V-··SD--··I--D DI-K--DVT- H---H--TL- ----VL-ED·G --MLG·--N-G TPTELGEQ------- --E---I--- ---K---SP-S
--IND-Q-AV --NQNTY--YS-N-- HILAM-DV-- V-G--QY--- -L--TEQNGQG T-M--·-QNSS ·····AD-------- --V--SI--- ---R----KAI
--IL--QQA- --TQNQY--FDID-D NL-YM-TV-- E---KMY-LT P---TSQ-Q·G F-T-G·-Q-EN H····SQK------- --E--S---- ---L----KAI
QYM-ALLQAT ---EDEY-QFL-N-N -IQN--SLT- I---VE--L- P-S--LSNN·G Y-TVGVEPTYL SSQNG-P-------- L-S--S-Y-L G--R--F-TA-
```

Figure 2. Comparison of the amino acid sequence of turtle pepsinogen A with those of some other typical pepsinogens. Amino acids are shown in one-letter code. The sequences are aligned so as to maximize the homology. The numbering is that of turtle pepsinogen A. The NH_2-terminal sequence of pepsinogen 9 is used in the turtle pepsinogen sequence. Each vertical line indicates the boundary between the propart and the pepsin moiety. The boundary in chicken embryonic pepsinogen has not been determined. The positions of the active-site aspartic acids are shown with stars and the potential N-glycosylation sites in each zymogen with closed triangles. The dots indicate deletions, and the dashes indicate residues identical with those of turtle pepsinogen A.

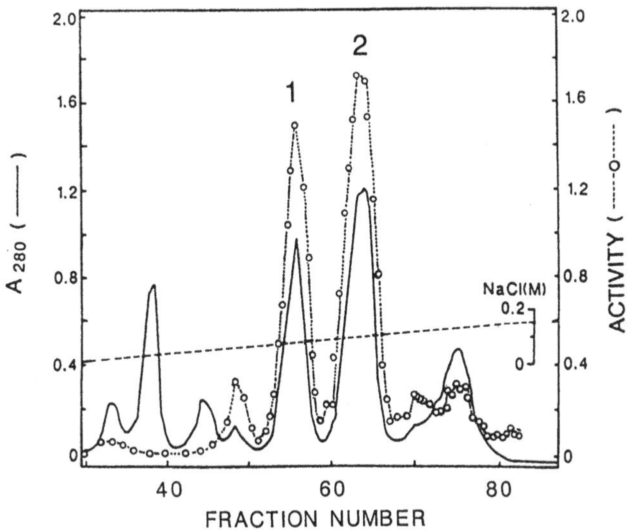

Figure 3. Fractionation of frog pepsinogens by chromatography on Q-Sepharose. The pepsinogen fraction partially purified from the crude extract of esophageal mucosa (wet weight, 130 g) by ammonium sulfate fractionation, followed by chromatography on DE-52 cellulose and Sephacryl S-200 was applied to a column (2.0 x 12cm) of Q-Sepharose Fast Flow equilibrated with 20 mM triethanolamine - HCl buffer, pH 7.5. The protein was eluted with an NaCl gradient (0 - 0.25 M) in the same buffer at a flow rate of 38 ml/h and 10-ml fractions were collected. Appropriate aliquots of the effluent were used for assay of potential pepsin activity with bovine hemoglobin as a substrate.

Enzymatic Properties of Pepsins

Each pepsin isozyme more or less hydrolyzed N-acetyl-L-phenylalanyl-L-diiodotyrosine (APDT), a good substrate for A-type pepsinogen. On the other hand, most isozymes were less sensitive to pepstatin than human pepsin A and more sensitive to pepstatin than human pepsin C. Thus, some of their properties were intermediate between those of mammalian pepsins A and C. As to the cleavage specificity toward oxidized insulin B chain, isozyme 1 resembled human pepsin A whereas isozyme 2 resembled human pepsin C rather than human pepsin A. On the other hand, isozymes 3-9 appeared to have similar specificity, resembling both human pepsins A and C. These results indicate that the discrimination between pepsin(ogen)s A and C is not always clear in non-mammalian vertebrates as based on enzymatic properties. Although mammalian pepsins A and C are known to possess somewhat different specificity from each other (23), it is interesting that further variation in specificity was found among turtle pepsin isozymes derived from a single species.

FROG PEPSINOGENS AND PEPSINS

Isolation and Properties of Pepsinogens

Two pepsinogens were found in both esophageal and gastric mucosae of bullfrog. However, the content was eight times higher in the esophageal mucosa than in the gastric mucosa. Therefore, the esophageal mucosa was used for isolation of pepsinogens. By ammonium sulfate fractionation followed by a series of chromatography on DEAE-cellulose,

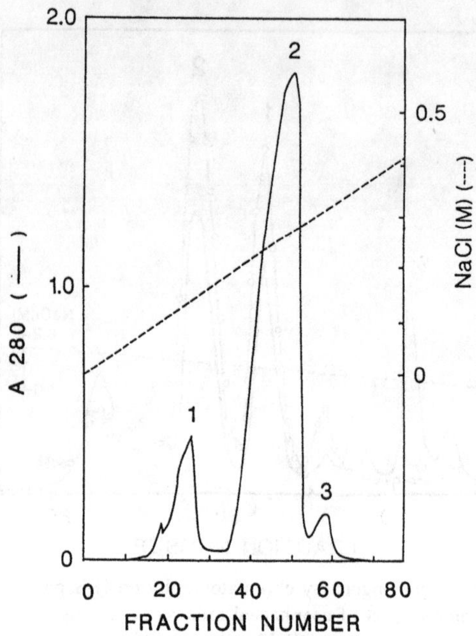

Figure 4. Fractionation of tuna pepsinogens by chromatography on Q-Sepharose. The pepsinogen fraction partially purified from the crude extract of gastric mucosa (wet weight, 92g) by ammonium sulfate fractionation, followed by chromatography on DE-52 cellulose, and Sephadex G-150 was applied to a column (2.0 x 12 cm) of Q-Sepharose Fast Flow equilibrated with 20 mM triethanolamine - HCl buffer, pH 7.5. The protein was eluted with as NaCl gradient (0 - 0.25 M) in the same buffer at a flow rate of 38 ml/h and 10-ml fractions were collected..

Sephacryl A-200 and Q-Sepharose, the two pepsinogens (pepsinogens 1 and 2) were purified as shown in Fig. 3. Their molecular weights were estimated to be both about 40,000 by SDS-PAGE and their amino acid compositions were very similar. They were both glycoproteins, and the carbohydrate contents were determined to be 1.5 and 1.3 mol per mol of pepsinogens 1 and 2, respectively. On the other hand, practically no phosphate was detected in these zymogens. The course of activation of each pepsinogen was analyzed by SDS-PAGE. The activation profiles of frog pepsinogens 1 and 2 were essentially the same. Under the conditions used at pH 2.0 and 4 °C, activation was rapid as in the case of turtle pepsinogen and complete within 5 min to produce the corresponding pepsins directly. No intermediate protein species was found. On the other hand, the activation of human pepsinogen A-5, used for comparison under the same conditions, proceeded through an intermediate form, and it took 120 min for complete conversion to pepsin. NH_2-terminal analysis of an activation mixture obtained in the early phase of activation revealed only the NH_2-terminal residues of the pepsinogens and pepsins. These results together with those obtained by SDS-PAGE indicated that activation took place essentially by a one-step mechanism.

Amino Acid Sequences of Pepsinogens and Pepsins

Analyses of the NH_2-terminal amino acid sequences of the pepsinogens and pepsins established the NH_2-terminal 70- and 78-residue sequences of pepsinogens 1 and 2, respectively. The NH_2-terminal 70-residue sequences of both pepsinogens are identical, and the entire activation peptides (*i.e.* propeptides) of pepsinogens 1 and 2 are both composed of 36 amino acid residues. Thus, the activation peptide is about 10 residues shorter than those of

```
                   1          10            20            30              40
                   .          .             .             .               .
Tuna    PG1   --LLQV  PL EKGQ S  A R EYLQEQ G L W EQYRLKYPYNPMA  K  FDPSF---A-
        PG2   --FHKL  PL IKG  K TA R EELQER G L W EDYRKQYPYHPMA  K  F-------YQ
        PG3   ---INV  PL TRH  KS M R ESLREK G I - E--L-PYQ-DPAI  KY RPEF---AT

Frog    C     --IIKV  PL KKF  KS M R EVMRDH G I - ----KAPVVDPAT  KY YNNF----A

Human   C     -AVVKV  PL KKF  KS I R ETMKEK G L L EDFLRTHKYDPAS  KY RFGD---LS
Monkey  C     -AVVKV  PL KKF  KS I R ETMKEK G L L GEFLRTHKYDPAW  KY HFGD---LS
Bovine  C     --LVKI  PL KKF  KS I R EIMKEK G L L ZBFLRTYKHNPAE  KY RFGD---FI
Rat           -SLLRV  PL RKM  KS I R QTMKEQ G V L KDFLKTHKYDPGQ  KY HFGNFGDYS

Turtle  A     -LVTKV  PL KKG  KS L R QNLKEH G L L EDFNKKHPYNPAS  RY FPSLGDEFA

Chicken A     -SIHRV  PL KKG  KS L R KQLKDH G L L EDFLKKHPYNPAS  KY HPVL-T-AT
Human   A     -IMYKV  PL IRK  KS L R RTLSER G L L KDFLKKHNLNPAR  KY FPQWEAPTL
Monkey  A     -IIYKV  PL VRK  KS L R RNLSEH G L L KDFLKKHNLNPAS  KY FPQAEAPTL
Rabbit  A     -VIHKV  PL VRK  KS L R KNLIEK G L L QDYLKTHTPNPAT  KY FPKE-TFAT
Bear    A     -FIIKV  PL VKK  KS L R KNLKEH G L L KDFLKKHSPNPAS  KY FPQE-A-AV
Porcine A     --LVKV  PL VRK  KS L R QNLIKD G K L KDFLKTHKHNPAS  KY FPEA-A-AL
Bovine  A     -SVVKI  PL VKK  KS L R QNLIEN G K L KEFMRTHKYNLGS  KY IREA-A-TL

Bov proch     AEITRI  PL YKG  KS L R KALKEH G L L EDFLQKQQYGISS  KY SGFG----E

Human   CD    --LVRI  PL HKFT S  I R RTMSEV G G S VEDL-IAK-GPVS  KY SQAVPAVTE
```

Figure 5. Comparison of the amino acid sequences of the activation segments. The sequences other than that of turtle pepsinogen are cited from references showing the sequences of human (25), monkey (26), rabbit (27), bear (28), porcine (4), bovine (29), chicken (13) pepsinogens A, human (30), monkey (31), bovine (32), rat (10) and frog (17) pepsinogens C, bovine prochymosin (12) , tuna pepsinogens 1, 2 and 3 (15), and human cathepsin D (33). The amino acids are shown in one-letter code. When two or more types of residues were identified at one position, the major one is shown. The numbering is based on the sequence of porcine pepsinogen. The hyphens indicate deletions. The vertical lines show the boundaries between the activation segments and the pepsin moieties. The residues in boxes are those common to pepsinogens of higher vertebrates.

mammalian pepsinogens A and C and turtle pepsinogen A. The complete amino acid sequence of one of the frog pepsinogens was deduced by nucleotide sequencing of its cDNA (see Fig. 6) (17). The cDNA thus isolated and analyzed codes for the full length of the pepsinogen molecule including a 16-residue signal sequence, a 36-residue activation segment, and a 332-residue pepsin sequence. The locations of and the sequences around the two active-site aspartic acid residues and the locations of the three disulfide bonds are well conserved and there is one potential *N*-glycosylation site at residue 94 of frog pepsinogen.

Previously, Shugerman and co-workers purified the major pepsinogen from the esophageal mucosa of bullfrog and reported that bullfrog pepsinogen is a unique "mini"pepsinogen with a molecular weight about 25 % lower than those of ordinary pepsinogens(24). Therefore, this "mini" pepsinogen was thought to be very interesting from the standpoint of molecular evolution. In the present study, however, we could not find a pepsinogen with such a low molecular weight. It seems that the molecular weight of the pepsinogen was underestimated in their study. Indeed, frog pepsinogen showed an anomalous behavior on gel filtration. Both pepsinogens were eluted at the position corresponding to a molecular weight of about 25,000. The reason for this is not certain at present, but may

```
                              1P        10P           20P           30P           40P     44P 1          10
                              :          :             :             :             :       : :          :
Tuna     Pg 2       --FHKLPLIKGKTAREELQERGLWEDYRKQYPYHPMAKF|YQDG---T-----EPMTNDA
Human    Pg A       -IMYKVPLIRKKSLRRTLSERGLLKDFLKKHNLNPARKYFPQWEAPTL-|VDEQPLENYL
Porcine  Pg A       --LVKVPLVRKKSLRQNLIKNGKLKDFLKTHKHNPASKYFPEA-A-AL-|IGDEPLENYL
Chicken  Pg A       -SIHRVPLKKGKSLRKQLKDHGLLEDFLKKHPYNPASKYHPVL|TA-T--ESYEPMTNYM
Bovine   Prochy     AEITRIPLYKGKSLRKALKEHGLLEDFLQKQQYGISSKYSGF|GEV-A---SV-PLTNYL
Human    Pg C       -AVVKVPLKKFKSIRETMKEKGLLGEFLRTHKYDPAWKYR-FG---DL|SVTYEPMA-YM
Frog     Pg C       --IIKVPLKKFKSMREVMRDHG-----IKAPVVDPATKYYNNF-|A-T---AFEPLANYM
Human    D          --LVRIPLHKFTSIRRTMSEVGGSVEDLIAK--GPVSKYSQAVPAVTE|GPIPEVLKNYM
                          PL  KS R     G L  FL         KY               P   Y

                 20        30            40          50          60             70        80
                 :         : *           :           :           :              :         :
         oo                o o oo o                                              o ooo
T2       DLSYYGVVSIGTPPQSFKVIFDTGSSNLWVPSVYCSSQ--ACQNHKKFNPQQSSTFKWGDQPLSIQYGTGSM
HA       DMEYFGTIGIGTPAQDFTVVFDTGSSNLWVPSVYCSSL--ACTNHNRFNPEDSSTYQSTSETVSITYGTGSM
PA       DTEYFGTIGIGTPAQDFTVIFDTGSSNLWVPSVYCSSL--ACSDHNQFNPDDSSTFEATSQELSITYGTGSM
CA       DASYYGTISIGTPQQDFSVIFDTGSSNLWVPSIYCKSS--ACSNHKRFDPSKSSTYVSTNETVYIAYGTGSM
BCH      DSQYFGKIYLGTPPQEFTVLFDTGSSDFWVPSIYCKSN--ACKNHQRFDPRKSSTFQNLGKPLSIHYGTGSM
HC       DAAYFGEISIGTPPQNFLVLFDTGSSNLWVPSVYCQSQ--ACTSHSRFNPSESSTYSTNGQTFSLQYGSGSL
FC       DMSYYGEISIGTPPQNFLVLFDTGSSNLWVPSTYCQSQ--ACTNHPQFNPSQSSSYSSNQQQFSLQYGTGSL
HD       DAQYYGEIGIGTPPQCFTVVFDTGSSNLWVPSIHCKLLDIACWIHHKYNSDKSSTYVKNGTSFDIHYGSGSL
         D  Y G I   GTP Q F V FDTGSS  WVPS YC S    AC  H  F P  SST                YG GS

                 90                    100        110        120        130       140
                 :                     :          :          :          :         :
                                                    oo        o  o        o o
T2       TGRLASDIVEV------------GGISVNNQVFGISQSEA-PFMAYMKAFDGILGLAFQSIASDNVVPVFDNM
HA       TGILGYDTVQV------------GGISDTNQIFGLSETEPGSFLYYAPFDGILGLAYPSISSSGATPVFDNI
PA       TGILGYDTVQV------------GGISDTNQIFGLSETEPGSFLYYAPFDGILGLAYPSISASGATPVFDNL
CA       SGILGYDTVAV------------SSIDVQNQIFGLSETEPGSFFYYCNFDGILGLAFPSISSSGATPVFDNM
BCH      QGILGYDTVTV------------SNIVDIQQTVGLSTQEPGDVFTTAEFDGILGMAYPSLASEYSIPVFDNM
HC       TGFFGYDTLTV------------QSIQVPNQEFGLSENEPGTNFVYAQFDGIMGLAYPALSVDEATTAMQGM
FC       TGILGYDTVQI------------QNIAISQQEFGLSVTEPGTNFVYAQFDGILGLAYPSIAEGGATTVMQGM
HD       SGYLSQDTVSVPCQSASSASALGGVKVERQVFGEATKQPGITFIAAKFDGILGMAYPRISVNNVLPVFDNL
          G   GYDT  V               I     Q  GLS  EPG         FDGI G A P

                 150       160             170       180       190       200
                 :         :               :         :         :         :
                                                                  oo
T2       VSQGLVSQPLFSVYLSSNSQQ-----GSEVVFGGIDSSHYTGQITWIPLTSATYWQIQMDSVTIN
HA       WNQGLVSQDLFSVYLSADDQS-----GSVVIFGGIDSSYYTGSLNWVPVTVEGYWQITVDSITMN
PA       WDQGLVSQDLFSVYLSSNDDS-----GSVVLLGGIDSSYYTGSLNWVPVSVEGYWQITLDSITMD
CA       MSQHLVAQDLFSVYLSKDGET-----GSFVLFGGIDPNYYTKGIYWVPLSAETYWQITMDRVTVG
BCH      MNRHLVAQDLFSVYMDRDGQ------ESMLTLGAIDPSYYTGSLHWVPVTVQQYWQFYVDSVTIS
HC       VQEGALTSPVFSVYLS-NQQ--GSSGGAVV-FGGVDSSLYTGQIYWAPVTQELYWQIGIEEFLIG
FC       IQQNLINQPLFAFYLS-GQQ-NSQNGGEVA-FGGVDQNYYSGQIYWTPVTSETYWQIGIQGFSVN
HD       MQQKLVDQNIFSFYLSRDP--DAQPGGELM-LGGTDSKYYKGSLSYLNVTRKAYWQVHLDQVEVA
                   FSVY                  G  D     T    W P      YWQ

                 210        220        230        240        250        260       270
                 :     *    :          :          :          :          :         :
                     o   oooo o
T2       GQTVA-CSGGCQAIIDTGTSLIVGPSRDIYNMNAWVGASTTQNGDATVNCQNIQSMPEVTFTLNGHAFTIP
HA       GEAIA-CAEGCQAIVDTGTSLLTGPTSPIANIQSDIGASENSDGDMVVSCSAISSLPDIVFTINGVQYPVP
PA       GETIA-CSGGCQAIVDTGTSLLTGPTSAIANIQSDIGASENSDGEMVISCSSIDSLPDIVFTINGVQYPLS
CA       NKYVA-CFFTCQAIVDTGTSLLVMPQGAYNRIIKDLGVS--SDGE--ISCDDISKLPDVTFHINGHAFTLP
BCH      GVVVA-CEGGCQAILDTGTSKLVGPSSDILNIQQAIGATQNQYGEFDIDCDNLSYMPTVVFEINGKMYPLT
HC       GQASGWCSEGCQAIVDTGTSLLTVPQQYMSALLQATGAQEDEYGQFLVNCNSIQNLPSLTFIINGVEFPLP
FC       GQATGWCSQGCQGIVDTGTSLLTAPQSVFSSLMQSIGAQQDQNGQYAVSCSNIQSLPTISFTLSGVSFPLP
HD       SGLTL-CKEGCEAIVDTGTSLMVGPVDEVRELQKAIGAVPLIQGEYMIPCEKVSTLPAITLKLGGKGYKLS
               C    CQAI DTGTS L  P          G      G     C       P   F ING

                 280          290          300       310       320
                 :            :            :         :         :
                          o o o o        oo
T2       ASAYVSQSYYG----CRTGFGGEG----NQQ-LWILGDVFIRQYYSIFDTQGQRIGLAQAV--
HA       PSAYILQSEGS----CISGFQGMNLPTESGE-LWILGDVFIRQYFTVFDRANNQVGLAPVA--
PA       PSAYILQDDDS----CTSGFEGMDVPTSSGE-LWILGDVFIRQYYTVFDRANNKVGLAPVA--
CA       ASAYVLNEDGS----CMLGFENMGTPTELGE-QWILGDVFIREYYVIFDRANNKVGLSPLS--
BCH      PSAYTSQDQGF----CTSGFQSEN---HS-Q-KWILGDVFIREYYSVFDRANNLVGLAKAI--
HC       PSSYILSNNGY----CTVGVEPTYLSSQNGQPLWILGDVFLRSYYSVYDLGNNRVGFATAA--
FC       PSAYVLQQNSG---YCTIGIMPTYLPSQNGQPLWILGDVFLRQYYSVYDLGNNQVGFAAAA--
HD       PEDYTLKVSQAGKTLCLSGFMGMDIPPPSG-PLWILGDVFIGRYYTVFDRDNNRVGFAEAARL
          S Y             C  G               WILGDVF R  Y     D   NN VG
```

be related to the glycoprotein nature of the pepsinogens. Taken together, we are led to conclude that their is no mini-pepsinogen in the bullfrog esophageal gland. When the amino acid sequence of frog pepsinogen was compared with those of some typical pepsinogens, the identities in frog pepsinogen were 65 % with human pepsinogen C, 51 % with human pepsinogen A, 48 % with chicken pepsinogen A, and 43 % with bovine prochymosin. Thus, frog pepsinogen is thought to belong to the pepsinogen C group. To our knowledge, this is the first finding of a carbohydrate-containing pepsinogen C.

Enzymatic Protperties of Pepsins

The two pepsins had essentially the same pH profile of activity and the optimal pH values were both around 3.0 with hemoglobin as a substrate. They had higher relative activity below pH 2 and between pH 3 and 5 as compared with porcine pepsin. Thus their pH profiles of activity more resemble that of pepsin C than pepsin A. Inhibition profiles of the two pepsins by pepstatin were essentially the same and their sensitivity to pepstatin was like that of mammalian pepsin C and significantly lower than that of porcine pepsin. On the other hand, neither pepsin was inhibited by diazoacetyl-DL-norleucine methyl ester (DAN) under the conditions where porcine pepsin was effectively inhibited. This low sensitivity to DAN is clearly different from that of mammalian pepsin C. However, they were inhibited by 1,2-epoxy-3-(*p* -nitrophenoxy)propane (EPNP) like porcine pepsin. These susceptibilities of frog pepsinogens toward the two pepsin inhibitors were rather similar to those observed with tuna pepsins, especially pepsin 3.

TUNA PEPSINOGENS AND PEPSINS

Isolation and Properties of Pepsinogens

The whole pepsinogen fraction was prepared from the crude extract supernatant of tuna gastric mucosa by ammonium sulfate fractionation followed by chromatography on DEAE-cellulose and Sephadex G-50, and the pepsinogens were separated into three fractions, i.e., pepsinogens 1, 2 and 3, by Q-Sepharose chromatography as shown in Fig. 4. Pepsinogen 2 was the major component and accounted for nearly 90 % of the total pepsinogen content. Their molecular weights were determined to be about 39,000 to 40,000 by SDS-PAGE. The activation profile examined at pH 2.0 and 4 °C differed significantly among the three pepsinogens. Pepsinogen 1 was converted to pepsin 1 in a stepwise manner, via four intermediate forms. The generation of the first intermediate form was rapid and completed within a few seconds. The generation of the successive intermediate forms and of pepsin 1 was slower. Over 1 h was needed for complete conversion of pepsinogen 1 to pepsin 1. This stepwise cleavage of the activation segment is unique since stepwise activation

---

Figure 6. Comparison of the amino acid sequences of tuna pepsinogen 2 with those of human (1), porcine (3-6), and chicken (13) pepsinogens A, bovine prochymosin (12), human (8) and frog (17) pepsinogens C, and human cathepsin D (33). The amino acid residues are expressed in one-letter code. The number is based on the sequence of porcine pepsinogen A (3). The residues at least common among higher vertebrate pepsinogens are shadowed, and also shown below the sequences. The hyphens indicate deletions and the asterisks indicate the active site aspartic acid residues. The vertical lines show the boundaries between the propeptides and the pepsin moieties. The pepsin residues in contact with residues of an ideal heptapeptide substrate in the case of porcine pepsin (34) are indicated with circles. Pg, pepsinogen; Prochy, prochymosin; D, cathepsin D.

has been known mainly as a two-step process in other pepsinogens. Pepsinogen 2 was converted to pepsin 2 by an apparently two-step process through an intermediate form. The generation of the intermediate form was as rapid as that of the first intermediate form from pepsinogen 1. The conversion of the intermediate to pepsin was slower, but was largely completed within 20 min. Pepsinogen 3 was converted directly to pepsin 3. The rate was rather slow and a few hours were needed for complete activation.

Amino Acid Sequences of Pepsinogens and Pepsins

The NH_2-terminal amino acid sequences of each pepsinogen, intermediate form and pepsin were determined. The sequences of these protein species enabled us to deduce the NH_2-terminal 58-72 residue sequences of the three pepsinogens. The activation segments were composed of 41, 37 and 35 residues in pepsinogens 1, 2 and 3, respectively. As shown in Fig. 5, deletions were notable in these tuna pepsinogens when compared with mammalian ones, especially in the COOH-terminal region of the activation segment. Therefore, the activation segments of tuna pepsinogens were 2 to 12 residues shorter than those of mammalian pepsinogens. The number of common residues in the activation segments, among the 3 tuna pepsinogens, was 11, showing that the homology amongst them is relatively low. The number of common residues in the activation segments, among the three tuna pepsinogens, is reduced to 4 when the sequences of the activation segment of 16 pepsinogens, including the three tuna pepsinogens, are compared (Fig. 5). These four common residues are Pro-5, Leu-6 Arg 12, and Gly-20, and these are presumably essential residues for the activation segment of pepsinogens. Basic amino acid residues in the activation segment have been shown to be important in the stabilization of the pepsin moiety at neutral pH. The number of basic residues in the segment was much smaller in pepsinogen 1 than in pepsinogens 2 and 3 and other animal pepsinogens. The number of basic residues equalled that of acidic residues in this segment of pepsinogen 1. Therefore, the most essential basic residues are likely to be conserved in pepsinogen 1. The less basic character of the activation segment of pepsinogen 1 also indicates that the overall basic character is not always necessary for the activation segment. Several residues are deleted in tuna pepsinogens as compared with those of other animals, resulting in the shorter activation segments. The 35-residue activation segment of pepsinogen 3 is the shortest one known. This length may be close to the minimum one essential for the function of the activation segment.

The complete amino acid sequences of pepsinogen 2 was determined by the conventional method of protein chemistry as shown in Fig. 6. The protein is composed of a 37-residue propeptide and a 323-residue pepsin moiety. The sequence was further confirmed by analysis of a cDNA clone for tuna pepsinogen 2. In this analysis, the presence of a 16-residue prepeptide was also demonstrated. The identity of the pepsinogen 2 sequence is 52 % with human pepsinogen A, 51 % with porcine pepsinogen A, 52 % with chicken pepsinogen A, 48 % with bovine prochymosin, 45 % with human pepsinogen C, and 46 % with frog pepsinogen C. Therefore, although the differences among these values are rather small, tuna pepsinogen 2 may be tentatively classified as an A-type pepsinogen. Figure 7 shows a phylogenetic tree based on the amino acid sequences of these pepsinogens and human cathepsin D, which appears to indicate that tuna pepsinogen 2 diverged from the common ancestor of pepsinogens at around (or even before) the time of separation of the pepsinogen A group and the prochymosin group. A phylogenetic tree based on the nucleotide sequences is also consistent with this tree (data not shown). Therefore, strictly speaking, it seems difficult to include tuna pepsinogen 2 in the pepsinogen A group. They might have to be treated as a distinct group of pepsinogen. Very recently, we could isolate and sequence cDNA clones for tuna pepsinogens 1 and 3, and could deduce the complete amino acid sequences of these pepsinogens (data not shown). Tuna pepsinogen 1 is most homologous

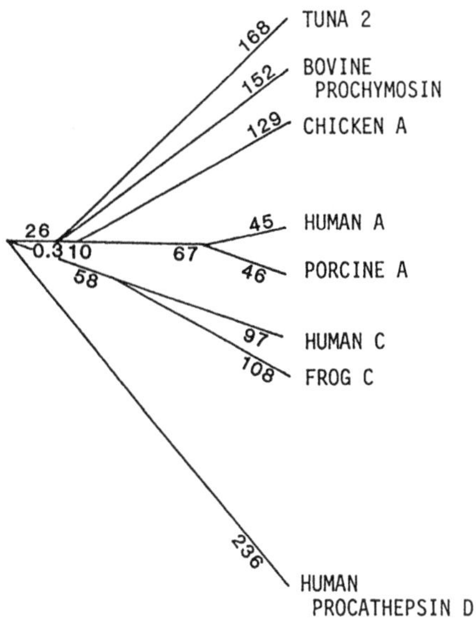

Figure 7. A phylogenetic tree of pepsinogens based on the amino acid sequences. The tree was constracted using the sequence data in Fig. 6 by the method of Fitch and Margoliash (*35*). Human cathepsin D is also included for comparison. The value on each branch indicates a reconstructed mutation distance.

with tuna pepsinogen 2 (66 % identity), whereas tuna pepsinogen 3 is less homologous with pepsinogen 2 (44 % identity) and is most homologous with the pepsinogen C group.

Enzymatic Properties of Pepsins

The three pepsinogens were converted to the corresponding pepsins and some enzymatic properties of these pepsins were investigated. The optimal pH values of pepsins 1, 2 and 3 for hemoglobin digestion were 2.9, 2.4 and 2·4, respectively, which were higher than that obtained for porcine pepsin A. The specific activity of pepsin 2 was two to five times higher than that of the other two, when measured under the standard conditions. The pH profiles of hemoglobin digestion activity of these pepsins were nearly the same and they had much higher relative activity in the region of pH about 2.5-5 than porcine pepsin A. Thus, their pH/activity profiles resemble those of mammalian pepsins C. It is interesting that they have significant activity at pH 5 where ordinary pepsins A and C are inactive. In this point, they somewhat resemble cathepsin D. As for the temperature dependence of hemoglobin digestion activity, the relative activities of pepsins 1 and 3 were much higher at lower temperatures than those of pepsin 2 and porcine pepsin A. These results with pepsins 1 and 3 are interesting considering the adaptation of fish to low temperatures.

Pepsins 1 and 2 were strongly inhibited by pepstatin, similar to porcine pepsin A, whereas pepsin 3 was less sensitive to pepstatin. In this respect, pepsin 3 resembles mammalian pepsin C. Each tuna pepsin was less sensitive to diazoacetyl-DL-norleucine methylester than was porcine pepsin A. The effects of EPNP varied among tuna pepsins although it was more or less inhibitory. Similar to mammalian pepsin C, each tuna pepsin scarcely hydrolyzed *N*-acetyl-L-phenylalanyl-L-diiodotyrosine, a good substrate for mammalian pepsin A. Thus, tuna pepsins 1 and 2 appear to have properties intermediate between those of pepsins A and C in several respects, and also bear some resemblance with cathepsin D.

REFERENCES

1. K. Sogawa, Y. Fujii-Kuriyama, Y. Mizukami, Y. Ichihara and K. Takahashi, Primary Structure of Human Pepsinogen Gene, *J. Biol. Chem.* 258 : 5306-5311 (1983).
2. T. Kageyama and K. Takahashi, The Complete Amino Acid Sequence of Monkey Pepsinogen A, *J. Biol. Chem.* 261 : 4395-4405 (1986).
3. J. Tang, P. Sepulveda, J. Marciniszyn, Jr., K. C. S. Chen, W-Y. Huang, N. Tao, D. Liu and J. P. Lanier, Amino-Acid Sequence of Porcine Pepsin, *Proc. Natl. Acad. Sci. U. S. A.* 70 : 3437-3439 (1973).
4. E. B. Ong and G. E. Perlmann, The Amino-Terminal Sequence of Porcine Pepsinogen, *J. Biol. Chem.* 243 : 6104-6109 (1968).
5. V. B. Pederson and B. Foltmann, The Amino Acid Sequence of A Hitherto Unobserved Segment from Porcine Pepsinogen Preceding the *N*-Terminus of Pepsin, *FEBS Lett.* 35 : 255-256 (1973).
6. V. M. Stepanov, L. A. Baratova, I. B. Pugacheva, L. P. Belyanova, L. P. Revina and E. A. Timokhina, N-Terminal Sequence of Swine Pepsinogen and Pepsin. The Site of Pepsinogn Activation, *Biochem. Biophys. Res. Commun.* 54 : 1164-1170 (1973).
7. T. Kageyama, K. Tanabe and O. Koiwai, Structure and Development of Rabbit Pepsinogens. Stage-Specific Zymogens, Nucleotide Sequence of cDNAs, Molecular Evolution, and Gene Expression during Development, *J. Biol. Chem.* 265 : 17031-17038 (1990).
8. T. Hayano, K. Sogawa, Y. Ichihara, Y. Fujii-Kuriyama and K. Takahashi, Primary Structure of Human Pepsinogen C Gene, *J. Biol. Chem.* 263 : 1382-1385 (1988).
9. T. Kageyama and K. Takahashi, The Complete Amino Acid Sequence of Monkey Progastricsin, *J . Biol. Chem.* 261 : 4406-4419 (1986).
10. T. Ishihara, Y. Ichihara, T. Hayano. I. Katsura, K. Sogwa, Y. Fujii-Kuriyama and K. Takahashi, Primary Strucutre and Transcriptional Regulation of Rat Pepsinogen C. Gene, *J. Biol. Chem.* 264 : 10193-10199 (1989).
11. T. Kageyama. M. Ichinose, S. Tsukada, K. Miki, K.Kurokawa, O. Koiwai, M. Tanji, E. Yakabe, S. B. P. Athauda and K. Takahashi, Gastric Prochathepsin E and Progastricsin from Guinea Pig. Purification, Molecular Cloning of cDNAs, and Characterization of Enzymatic Properties, with Special Reference to Procathepsin E, *J. Biol. Chem.* 267 : 16450-16459 (1992).
12. B. Foltmann, V. B. Pedersen. H. Jacobsen. D. Kauffman and G. Wybrandt, The Complete Amino Acid Sequence of Prochymosin, *Proc. Natl. Acad. Sci. U. S. A.* 74 : 2321-2324 (1977).
13. M. Baudyš and V. Kostka, Covalent Structure of Chicken Pepsinogen, *Eur. J. Biochem.* 136 : 89-99 (1983).
14. K. Hayashi, K. Agata, M. Mochii, S. Yasugi, G. Eguchi and T. Mizuno, Molecular Cloning and the Nucleotide Sequence of cDNA for Embryonic Chicken Pepsinogen : Phylogenetic Relationship with Prochymosin, *J. Biochem.* 103 : 290-296 (1988).
15. M. Tanji, T. Kageyama and K. Takahashi, Tuna Pepsinogens and Pepsins. Purification, Characterization and Amino-Terminal Sequences, *Eur. J. Biochem.* 177 : 251-259 (1988).
16. Z. Bohak, Purification and Characterization of Chicken Pepsinogen and Chicken Pepsin, *J. Biol. Chem.* 244 : 4638-4648 (1969).
17. E. Yakabe, M. Tanji, M. Ichinose, S. Goto, K. Miki, K. Kurokawa, H. Ito, T. Kageyama and K. Takahashi, Purification, Characterization, and Amino Acid Sequences of Pepsinogens and Pepsins from the Esophageal Mucosa of Bullfrog (*Rana catesbeiana*), *J. Biol. Chem.* 266 : 22436-22443 (1991).
18. I. M. Samloff, Slow Moving Protease and the Seven Pepsinogens. Electrophoretic Demonstration of the Existence of Eight Proteolytic Fractions in Human Gastric Mucosa,*Gastroenterology* 57 : 659-669 (1969).
19. S. B. P. Athauda, M. Tanji, T. Kageyama and K. Takahashi, A Comparative Study on the NH_2-Terminal Amino Acid Sequences and Some Other Properties of Six Isozymic Forms of Human Pepsinogens and Pepsins, *J. Biol. Chem.* 106 : 920-927 (1989).
20. T. Kageyama and K. Takahashi, Pepsinogen C and Pepsin C from Gastric Mucosa of Japanese Monkey, *J. Biochem.* 80 : 983-992 (1976).
21. T. Kageyama and K. Takahashi, Pepsinogens and Pepsins from Gastric Mucosa of Japanese Monkey. Punfication and Characterization, *J. Biochem.* 79 : 455-468 (1976).
22. K. Takahashi and T. Kageyama, Multiplicity and Intermediates of the Activation Mechanism of Zymogens of Gastric Aspartic Proteinases, *in* : "Aspartic Proteinases and Their Inhibitors", V. Kostka, ed., pp.265-282, Walter de Gruyter, Berlin and New York (1985).

23. O. Matsuzaki and K. Takahashi, Improved Purification of Slow Moving Protease from Human Gastric Mucosa and Its Action on the B Chain of Oxidized Bovine Insulin, *Biomed Res.* 9 : 515-523 (1988).
24. R. P. Shugerman, B. I. Hirschowitz, A. S. Brown, R. E. Schrohenloher and J. G. Spenney, A Unique "Mini" Pepsinogen Isolated from Bullfrog Esophageal Glands, *J. Biol. Chem.* 257 : 795-798 (1982).
25. T. Kageyama and K. Takahashi, Isolation of an Activation Intermediate and Determination of the Amino Acid Sequence of the Activation Segment of Human Pepsinogen A, *J. Biochem.* 88 : 571-582 (1980).
26. T. Kageyama and K. Takahashi, Monkey Pepsinogens and Pepsins. IV. The Amino Acid Sequence of the Activation Peptide Segment of Japanese Monkey Pepsinogen, *J. Biochem.* 88 : 9-16 (1980).
27. T. Kageyama and K. Takahashi, Rabbit Pepsinogens. Purification, Characterization, Analysis of the Conversion Process to Pepsin, and Determination of the NH_2-Terminal Amino Acid Sequences, *Eur. J. Biochem.* 141 : 261-269 (1984).
28. T. Kageyama, A. Moriyama and K. Takahashi, Purification and Characterization of Pepsinogens and Pepsins from Asiatic Black Bear, and Amino Acid Sequence Determination of the NH_2-Terminal 60 Residues of the Major Pepsinogen, *J. Biochem.* 94 : 1557-1567 (1983).
29. M. Harboe, P. M. Anderson and B. Foltmann, The Activation of Bovine Pepsinogen - Sequence of the Peptides Released, Identification of a Pepsin Inhibitor, *J. Biol. Chem.* 249 : 4487-4494 (1974).
30. B. Foltmann and A. L. Jansen, Human Progastricsin - Analysis of Intermediates during Activation into Gastricsin and Determination of the Amino Acid Sequence of the Propart, *Eur. J. Biochem.* 128 : 63-70 (1982).
31. T. Kageyama and K. Takahashi, Monkey Pepsinogens and Pepsins. VII. Analysis of the Activation Process and Determiantion of the NH_2-Terminal 60-Residue Sequence of Japanese Monkey Progastricsin, and Molecular Evolution of Pepsinogens, *J. Biochem.* 97 : 1235-1246 (1985).
32. B. Foltmann and V. B. Pedersen, Comparison of the Primary Structures of Acidic Proteinases and of Their Zymogens, *in* : "Acid Proteinases" J. Tang, ed., pp. 3-22, Plenum Press, New York (1977).
33. P. L. Faust, S. Kornfeld and J. M. Chirgwin, Cloning and Sequence Analysis of cDNA for Human Cathepsin D, *Proc. Natl. Acad. Sci. U. S. A.* 82 : 4910-4914 (1985).
34. A. R. Sielecki, A. A. Fedorov, A. Boodhoo, N. S. Andreeva and M. N. G. James, Molecular and Crystal Structures of Monoclinic Porcine Pepsin Refined at 1·8 Å Reslolution, *J. Mol Biol.* 214 : 143-170 (1990)
35. W. M. Fitch and E. Margoliash, Construction of Phylogenetic Trees.- A Method Based on Mutation Distances as Estimated from Cytochrome *C* Sequences is of General Applicability, *Science* 155 : 279-284 (1967).

TRANSCRIPTION REGULATION OF HUMAN AND PORCINE PEPSINOGEN A

G. Pals,[1] P. H. S. Meijerink,[1] J. Defize,[1] J. P. Bebelman,[1] M. Strunk,[1]
F. Arwert,[1] A. Timmerman,[3] and W. H. Mager[2]

[1]Institute of Human Genetics
[2]Department of Biochemistry and Molecular Biology
Vrije Universiteit
v.d. Boechorststraat 7, 1081 BT Amsterdam, The Netherlands
[3]Royal Institute of Public Health and Environmental Protection
P.O. Box 1 3720 ME Bilthoven, The Netherlands

INTRODUCTION

The gene regulation of pepsinogens is scientifically interesting, because of the high specificity of expression in gastric chief cells. Also from a clinical point of view pepsinogen gene regulation is interesting. Pepsinogen, activated by acid to pepsin, is one of the aggressive factors in the stomach that play a role in peptic ulcer disease (1). A dominantly inherited high pepsinogen A (PGA) output is associated with duodenal ulcer (2). On the other hand, a low production of pepsinogen A appears to be associated with atrophic gastritis and gastric cancer (3). Changes in gene expression of different isozymogens have been demonstrated in Barrett's oesophagus (4) and gastric cancer (5). Defize et al. (6) showed, that these changes in gene expression may be associated with methylation of DNA. They demonstrated that N-methyl-N'nitro-N-nitrosoguanidine (MNNG), a methylating agent, caused changes in pepsinogen C (PGC) gene expression in rats and in PGA gene expression in cultured human gastric chief cells. In rats the changes in PGC expression were associated with the development of gastric cancer. Other studies have shown hypomethylation of the pepsinogen genes in pepsinogen producing tissue, i.e. fundic mucosa (7,8). A decrease of PG synthesis in gastric tumours was correlated with a different methylation pattern (9).

Molecular Genetics

Pigs have only a single PGA gene (10). The human pepsinogen A (hPGA) group shows extensive genetic heterogeneity and is encoded by a gene cluster located on chromosome 11q13, containing a variable number of PGA genes (11,12).

The human pepsinogen A genes are named after their *Eco*RI-generated DNA fragments and an asterisk indicates a gene with a high expression. The 15.0 gene codes for Pg5, the 16.6 and 20.0 genes for Pg4 and the 15.0* gene for Pg3. The 15.0* gene shows a high

Aspartic Proteinases: Structure, Function, Biology, and Biomedical Implications
Edited by Kenji Takahashi, Plenum Press, New York, 1995

level of expression. Furthermore, two 12.0 genes exist, both with a low expression, and it is therefore still uncertain for which isozymogen they code. The two 12.0 genes can be distinguished by an *Ava*II polymorphism. Involvement of *cis*-acting elements on the hPGA promoter and *trans*-acting nuclear protein factors is indicated by the fact that these genes are tissue-specifically expressed in the chief and neck cells of gastric fundic mucosa.

To identify *cis*-acting elements we have employed a transient expression assay in primary cultures of gastric chief cells from pig and from monkey (*Macaca fascicularis*).

MATERIALS AND METHODS

Plasmid Construction

Transient expression analyses of the human PGA genes were performed with the promoter region of a human pepsinogen A gene encoding Pg5. This gene is characterized by an *Eco*RI-generated DNA fragment of 15.0 kb (12). The chloramphenicol acetyl transferase gene was used as a reporter(13). All recombinant plasmids were constructed according to established procedures (14). Constructs containing the chloramphenicol acetyltransferase (CAT) reporter gene were made in SuperCAT, a derivative of pSV2CAT (15) containing the bacterial CAT gene and the simian virus 40 (SV40) splice and polyadenylation signals on a *Hind*III-*Bam*HI fragment (positions 3370 to 5003 in PSV2CAT) cloned into the *Hind*III and *Nde*I sites of pUC12. SV40CAT used as a control in transfection studies, contains the SV40 early promoter region on a *Pvu*II-*Hind*III fragment of pSV2CAT (15) cloned into SuperCAT digested with *Sma*I and *Hind*III.

Human PGA genes were isolated from cosmids cgHGP-4,-15,-23 and -25 (12,16). The upstream regions, starting at the *Xho*I site at +30 relative to the cap site, were subcloned into pIC20R (17) and subsequently cloned upstream of a promoterless CAT gene in SuperCAT.

The porcine PGA gene was subcloned from cosmid POC4, as described before (10). The upstream regions, starting at +21 relative to the cap site, were cloned upstream of a promoterless CAT gene in SuperCAT.

Small upstream fragments were constructed using restriction digestion and partial *Bal*31 digestion, followed by blunt end formation with T4 DNA polymerase and religation of the vector with T4 DNA ligase. The exact endpoints were determined by nucleotide sequencing (18). Plasmid DNA to be used for transfections was purified through centrifugation to equilibrium in cesium chloride-ethidium bromide density gradients (14).

Transient Expression

Primary cultures of gastric chief cells, derived from crab eating monkey (*Macaca fascicularis*) or pig, were used for transfection experiments. Chief cells were isolated from gastric mucosa and grown as monolayers, as described previously (19,20). Peptic cells were identified by immuno-histological staining, using antihuman pepsinogen A antibodies (21). After reaching 50 to 80% confluence, cells were transfected with the constructs described above (5 μg DNA), using lipofectin (Life Technologies Inc) as the transfecting agent (22) in a serum free medium (Optimem; Life Technologies Inc) overnight. Transfection was stopped by adding serum containing medium. Post-transfection culture time was 24-48 hrs, during which period the cells were incubated with sodium butyrate ($5 \cdot 10^{-3}$M). Dibutyryl-cyclic-AMP (10^{-5}M) was added to stimulate PGA synthesis (23). Cells were harvested by scraping, lysed and assayed for CAT by ELISA (Boehringer Mannheim).

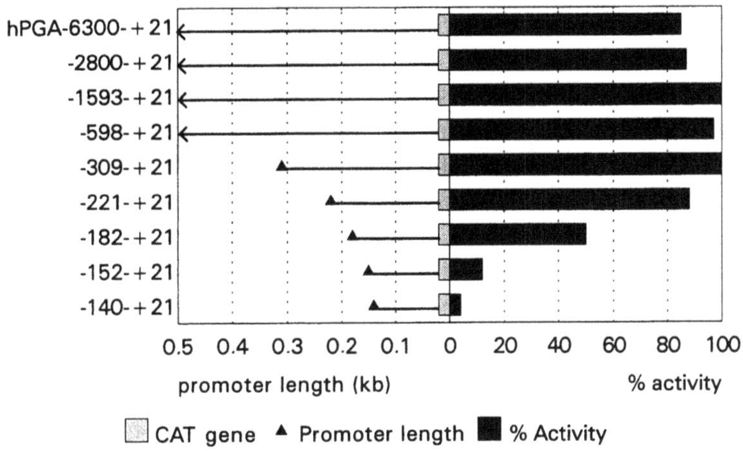

Figure 1. Transient expression analysis of hPGA promoter fragments in monkey gastric chief cells. Bars indicate relative CAT activity in chief cells transfected with *Bal*31 generated hPGA promoter fragments, ranging from -6300-+21 (relative to cap site) to -140-+21. CAT activity was determined by ELISA and set at 100% for the hPGA 15.0 1593bp/CAT construct.

Cell specific expression of hPGA genes was assessed by transfection of human cells (HeLa AND hEPg2) and monkey hepatocytes with the 1593 and 598 bp human promoter fragments of the 15.0 (Pg5) hPGA gene essentially as described for the chief cells.

Methylation of DNA

Recombinant plasmid DNA was methylated, using four methylases; M.*Sss*I, M.*Hpa*II, M.*Msp*I and M.*Hha*I, having methylation sites: CmG ; CCmGG, CmCGG and GCmGC respectively, according to the manufacturers protocol (New England Biolabs Inc). Completeness of DNA methylation was checked by determining protection against cleavage by the corresponding methylation sensitive restriction enzyme; *Hpa*II for M.*Sss*I and M.*Hpa*II methylated DNA, *Hha*I for M.*Hha*I and *Msp*I for M.*Msp*I methylated DNA.

RESULTS AND DISCUSSION

Transient Expression; Human PGA

Regulation of pepsinogen A gene expression is as yet poorly understood. We previously cloned a 1.7 kb fragment of the 5' upstream region of a human PGA gene, encoding Pg5 (24). This gene is part of a haplotype, containing several PGA genes, and is characterized on Southern blot by an *Eco*RI fragment of 15 kb (17). In the present study, transient expression analysis was performed with this fragment and other 5' upstream fragments from the same PGA gene. Expression of the CAT reporter gene was demonstrated in primary cultures of monkey gastric chief cells, but not in the human cell lines HepG2 and HeLa, nor in primary cultures of monkey liver cells (results not shown). The control SV40-CAT plasmid was transcriptionally active in all cell types tested. The results indicate that analysis of the human PGA promoter in this test system is not hampered by the species difference with the *Macaca fascicularis*, since the extended human PGA promoter constructs showed almost the same activity as the control SV40 promoter. Despite the high sequence

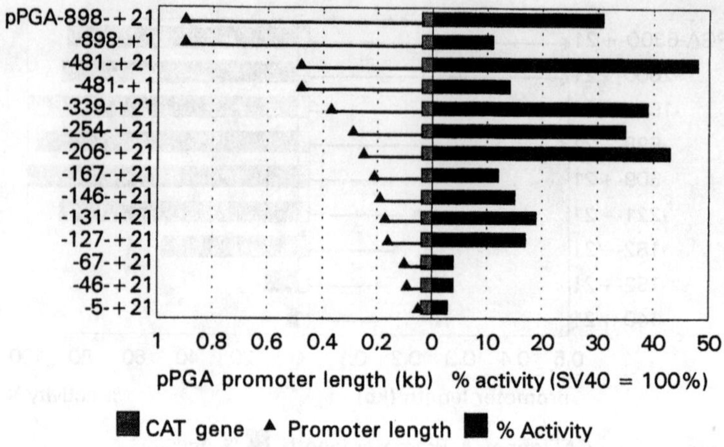

Figure 2. Transient expression analysis of pPGA promoter fragments in gastric chief cells from pig. Bars indicate relative CAT activity in chief cells transfected with Bal31 generated pPGA promoter fragments, ranging from 920 to 26 bp. CAT activity was determined by ELISA and set at 100% for the control SV40/CAT construct.

homology between human and porcine PGA promoters (10), however, the human PGA promoter was found to be inactive in porcine gastric chief cells and only a low level of transcription activation was found in some experiments (results not shown).

Regulatory Elements; Human PGA

A 5'-deletion series of the human PGA promoter region, ranging from -6300 to -140 bp, provided evidence that the proximal part of the promoter up to -221 is necessary and sufficient for chief cell specific transcription activation (figure 1). Deletion of the 5'-region to -309 bp had no effect on transcription activation and deletion down to -221 resulted only in a slight decrease. Deletion down to -182, however, resulted in a marked decrease, whereas deletion down to -152 almost completely abolished transcription activation. These results suggest that this proximal part of the promoter, between -152 and -221 contains *cis*-acting elements that represent binding sites for *trans*-acting factors, that are essential in tissue specific transcription activation of this gene. However, previous studies using DNAse I footprinting, show that several putative protein binding sites are also present in the region between the transcription start site and position -145 of the human PGA promoter region (24), suggesting that also more proximal elements of the promoter are involved in transcription activation or regulation. Identification of all elements, composing the complete PGA promoter has to await further studies.

Regulatory Elements; Porcine PGA

Results of transient expression of porcine PGA-CAT constructs in porcine gastric chief cells (10) also indicated an important role for the proximal promoter. As shown in figure 2, progressive 5' deletion of the promoter down to -206 had only a small effect. However, a large reduction in transient promoter activity was found after deletion of the region from -206 to -167. Further deletion to -67 resulted in almost complete loss of promoter activity. Deletion of part of the noncoding sequence of the first exon, from +2 to +21, also resulted in a reduction of promoter activity. From these data it was concluded that in the porcine PGA promoter at least three regulatory domains are located in the regions from -206

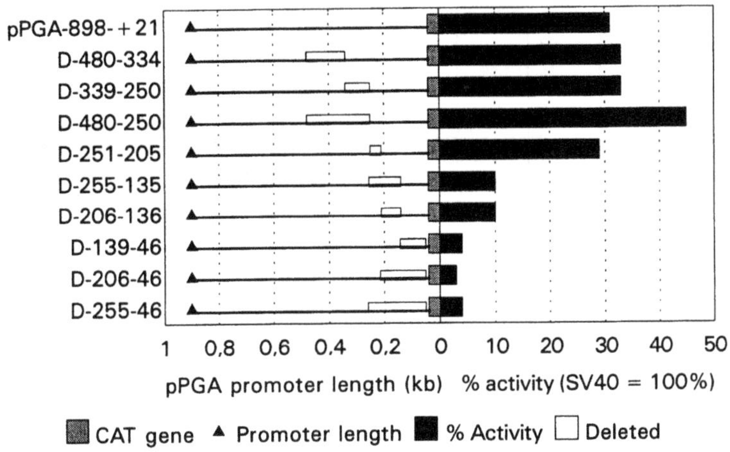

Figure 3. Transient expression analysis of pPGA promoter fragments in gastric chief cells from pig. Bars indicate relative CAT activity in chief cells transfected with pPGA promoter fragments with internal deletions. CAT activity was determined by ELISA and set at 100% for the control SV40/CAT construct.

to -167, from -127 to -67 and from +2 to +21. This was confirmed by experiments with internal deletions, showing that deletion of the regions -206 to -46 or -139 to -46 results in complete loss of activity, whereas deletion of -206 to -137 reduces transcription about 3-fold (figure 3). These results indicate an important role for the region from -136 to -46, whereas the entire region from -206 to +21 is necessary for full cell-specific promoter activity. DNAseI footprint analysis of the porcine PGA promoter showed a gastric chief cell specific protection of the sequence from -134 to -127 (10).

Different Human PGA Genes

We studied the transient expression of the most common human PGA promoters (figure 4). Most of these genes have comparable activities, but we found an approximately twofold higher transcriptional activity for the 15.0* promoter as compared with the 1593 bp 15.0 promoter. This higher expression level of the 15.0* gene is in accordance with previous family and population studies.

The results with the 15.0* gene, coding for the human Pg3 isozymogen, indicate that this gene either contains extra enhancer elements or that a repressor element is deactivated. Deletion of the 15.0* promoter to 0.6 kb reduced the activity to the same level as the 15.0 gene. This indicates that the putative enhancer element is present between -600 and -1600. Further experiments with deletion constructs of this promoter are in progress to delineate the position of putative enhancer elements.

CpG methylation

The human PGA promoter contains many CpG dinucleotides, being putative sites for methylation. As discussed above, the human PGA promoter region is probably unmethylated in gastric tissue and methylated in other tissues (11). Moreover, earlier studies have shown that *in vivo* methylation by MNNG resulted in loss of expression a specific PGA isozymogen (13). These observations may imply CpG methylation renders the PGA gene transcriptionally silent. Therefore, *in vitro* methylation of several PGA promoter-CAT fusion plasmids was performed, to study the effect of methylation on the human PGA promoter.

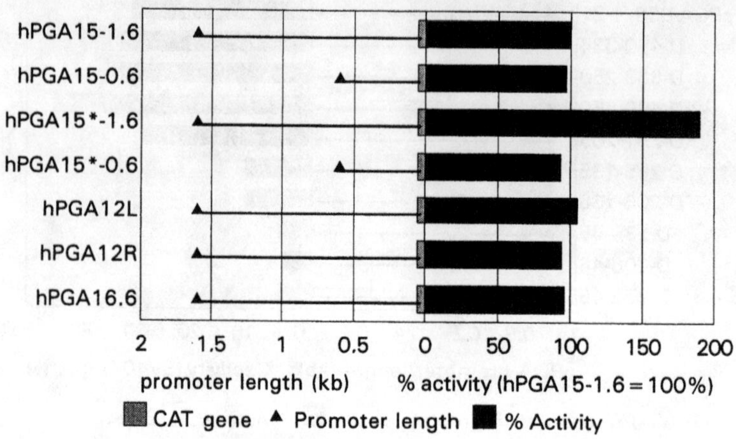

Figure 4. Transient expression analysis of hPGA promoter fragments from different human PGA genes. CAT activity was determined by ELISA and set at 100% for the hPGA 15.0 1593bp/CAT construct.

Methylation of the CAT gene itself had no effect on transcription or transfection efficiency, demonstrated by identical CAT activity in cells transfected with a methylated or non-methylated SV40-CAT construct (figure 5). The SV40 promoter was insensitive to methylation. Neither did methylation of the whole plasmids interfere with their integrity as checked by electrophoresis. Digestion by the corresponding methylation sensitive restriction enzymes showed complete protection of methylated DNA against digestion (results not shown).

Our data demonstrate that critical 5'cytosine methylation may play an important role in pepsinogen gene expression. As shown in figure 5, methylation of all CpG sequences by M.*Sss*I incapacitated the human PGA promoter. *Msp*I methylation of the 5'C in CCGG sites had no effect on activity, whereas *Hpa*II methylation of the 3'C in the same recognition sequences resulted in significant loss (50%) of transcription activity, indicating that the CpG methylation is critical.

Most remarkably, methylation of the single *Hha*I site at position -451 in the 598 bp fragment from the 15.0 and 12.0R genes caused an even greater reduction in activity (80%) (figure 6), although loss of this site did not affect activity: deletion of the 598 bp fragment to -450, removing the *Hha*I site, did not affect promoter activity. The resulting 450 bp fragment was found to be insensitive to *Hha*I methylation, indicating that the methylation effect is not due to other *Hha*I sites in the plasmid. These results indicate, that the loss of transcriptional activity is not caused by hampered binding of an activator protein due to methylation. Therefore the *Hha*I site may be positioned in a negative regulatory element, which after methylation, becomes accessible for binding with a repressor protein. This supports the indirect model of transcription inhibition by methylation, in which methylated DNA is bound by nuclear proteins, which secondarily prevents transcription factors from interacting with the gene or as in our case, act as repressors. A similar mechanism has been described by Boyd and Bird (25), although in their study they found that at least 15 methylated CpG's were needed to observe inhibition.

The 15.0* gene also contains a *Hha*I site at the same position and methylation of this site reduced expression about twofold, which brought the expression level down to approximately the normal level for the 15.0 gene (figure 6). The 12.0L gene has a 130 bp insertion on the 3' side of the *Hha*I site. Methylation of this site had no effect on promoter activity. This could be caused either by disruption of the putative repressor binding site or by the larger distance to the transcription initiation site.

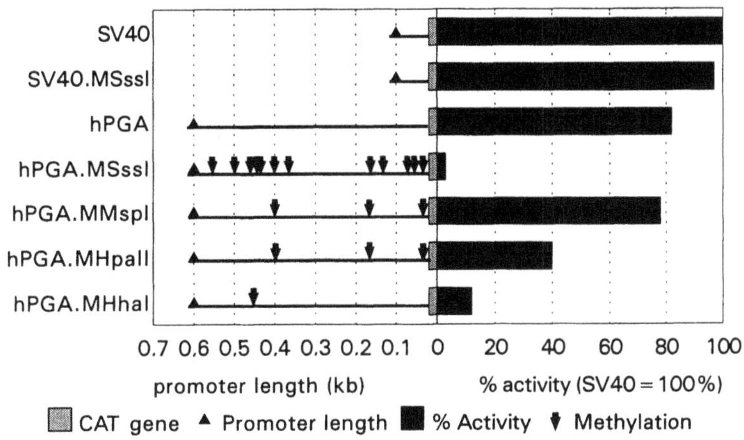

Figure 5. The effect of methylation on transient expression of human PGA promoter fragments. CAT activity was determined by ELISA and set at 100% for the control SV40/CAT construct.

To study protein binding to the putative repressor binding site, a fragment of 40 bp around the *Hha*I site was amplified by PCR and part of it was methylated with M.*Hha*I. In preliminary experiments we found that only the methylated fragment formed protein complexes as visualized on gel shift assays after incubation of methylated and non-methylated fragments with HeLa cell nuclear extracts (results not shown).

The observation, that *in vitro* methylation of cytosine in CpG sequences of the human PGA promoter blocks transcription completely, may indicate that this mechanism serves to silence the expression of the PGA genes in non-gastric tissue. It is very unlikely that normal regulation of the PGA gene expression in pepsinogen producing cells is controlled by methylation.

CONCLUSIONS

This study shows that pepsinogen synthesis is regulated by tissue-specific regulatory elements. The proximal part (about 200 bp) of the promoter is necessary and sufficient for gastric chief cell specific transcription. In one human gene, a putative enhancer element is present between 0.6 and 1.6 kb upstream of the cap site.

Our data strongly support a regulatory role for methylation in pepsinogen A synthesis in non-gastric tissue by means of a silencer specifically binding at methylated DNA.

Abbreviations

hPGA: human pepsinogen A; pPGA: pig pepsinogen A; PGC: pepsinogen C; CAT: chloramphenicol acetyltransferase; dbcAMP: di-butyryl cyclic AMP; bp: base pairs; kb: kilobases; MNNG: N-methyl-N'nitro-N-nitrosoguanidine.

ACKNOWLEDGMENTS

This work was financially supported by Glaxo Group Research Ltd, England.

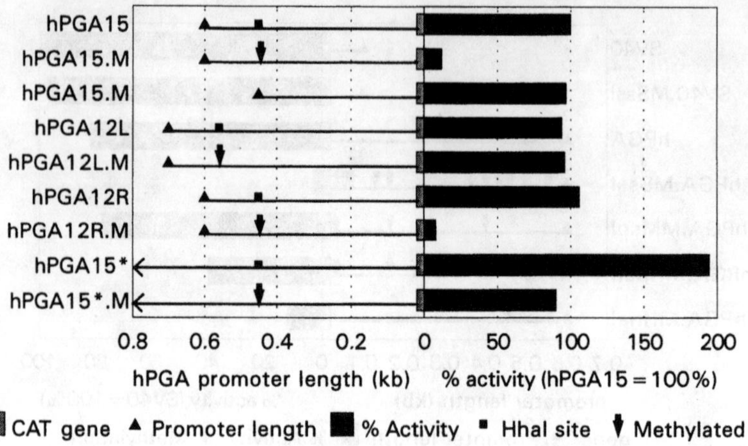

Figure 6. The effect of *HhaI* methylation on transient expression of human PGA promoter fragments. CAT activity was determined by ELISA and set at 100% for the hPGA 15.0 598bp/CAT construct. M indicates methylation with M.*HhaI*.

REFERENCES

1. I.M. Samloff, Peptic ulcer: The many proteinases of aggression, Gastroenterology 96:586-95(1989).
2. J.I. Rotter, I.M. Samloff, G.M. Petersen, Pepsinogen genetics and duodenal ulcer disease, Gut 27:224-226(1986).
3. B.D. Westerveld, G. Pals, C.B.H.W. Lamers, J. Defize, J.C. Pronk, R.R. Frants, E.C.M. Ooms, J. Kreuning, P. Kostense, A.W. Eriksson, S.G.M. Meuwissen, Clinical significance of pepsinogen A isozymogens, serum pepsinogen A and C levels and serum gastrin levels, Cancer 59:952-958(1987).
4. G. Pals, A.W. Eriksson, J.C. Pronk et al. Differential expression of pepsinogen isozymogens in a patient with Barrett's oesophagus, Clin Genet 34:90-7(1988).
5. G. Pals, B.D. Westerveld, J.C. Pronk, A. Bosma, A.W. Eriksson, S.G.M. Meuwissen, Pepsinogen A isozymogens in normal mucosa and tumor tissue of gastric cancer patients, Neth. J. Med. 28:461-467(1985).
6. J. Defize, J. Derodra, R.H. Hunt, N-methyl-N'nitro-N-nitroso-guanidine induces changes in rat and human pepsinogen phenotypes, Cancer 62:1958-61(1988).
7. M. Ichinose, K. Miki, M. Tatematsu, et al. Cell-specific hypomethylation of the pepsinogen gene in pepsinogen-producing cells, Biochem. Biophys. Res. Commun. 155(2):670-7(1988).
8. M. Ichinose, K. Miki, M. Tatematsu, et al. Hypomethylation and expression of pepsinogen A genes in the fundic mucosa of human stomach, Biochem. Biophys. Res. Commun. 151:275-82(1988).
9. M. Ichinose, K. Miki, R.N.S. Wong, et al. Methylation and expression of human genes in normal tissues and their alteration in stomach cancer, Jpn. J. Cancer. 82:686-92(1991).
10. P.H.S. Meijerink, J.P. Bebelman, A.M. Oldenburg, J. Defize, R.J. Planta, A.W. Eriksson, G. Pals, W.H. Mager, Gastric chief cell-specific transcription of the pepsinogen A gene, Eur. J. Biochem. 213:1283-9(1993).
11. R.T. Taggart, T.K. Mohandas, T.B. Shows, G.I. Bell, Variable numbers of pepsinogen genes are located in the centromeric region of human chromosome 11 and determine the high frequency electrophoretic polymorphism, Proc. Natl. Acad. Sci. 82:6240-4(1985).
12. B. Zelle, M.P. Evers, P.C Groot, et al. Genomic structure and evolution of the human pepsinogen A multigene family. Hum Genet 1988;78(1):79-82.
13. C.M. Gorman, LF. Moffat, B.H. Howard, Recombinant genomes which express chloramphenicol acetyltransferase in mammalian cells, Mol. Cell. Biol. 2:1044-51(1982).
14. T. Maniatis, F. Fritsch, J. Sambrook, Molecular cloning: a laboratory manual. Cold Spring Harbor Laboratory 1982; Cold Spring Harbor, N.Y.
15. CM. Gorman, High efficiency gene transfer into mammalian cells. In: DM Glover (ed), DNA cloning, vol 11. IRL press Oxford 1985; pp143-190.

16. J.P. Bebelman, M.P.J. Evers, B. Zelle, R. Bank, J.C. Pronk, S.G.M. Meuwissen, W.H. Mager, R.J. Planta, A.W. Eriksson, R.R. Frants, Family and population studies on the human pepsinogen A multi gene family, Hum, Genet, 82:142-6(1989).
17. J. Marsh, M. Ertie, E. Weykes, The pIC plasmid and phage vectors with versatile cloning sites for recombinant selection by insertional inactivation, Gene 32:481-5(1984).
18. F. Sanger, S. Nicklen, A.R. Coulson, DNA sequencing with chain terminating inhibitors, Proc. Natl. Acad. Sci. USA. 74:5463-7(1977).
19. J. Defize, R.H. Hunt, Effect of acid and pepsin on pepsinogen synthesis and secretion in chief cell monolayer cultures, Gut 30:774-81(1989).
20. J. Defize, F. Arwert, H. Kortbeek, R.R. Frants, S.G.M. Meuwissen, A.W. Eriksson, Pepsinogen synthesis and secretion in monolayer culture of human and rabbit gastric mucosal cells, Virchows Arch. B 49:225-30(1985).
21. S.G.M. Meuwissen, H. Mullink, A. Bosma, et al. Immunocytochemical localization of pepsinogen A and C in the human stomach, Progr. Clin. Biol. Res. 173:185-198(1985).
22. P.L. Felgner, T.R. Gadek, M. Holm, et al. Lipofectin: a highly efficient, lipid-mediated DNA-transfection procedure, Proc. Natl. Acad. Sci. USA, 84:7413-7(1987).
23. J. Defize, RH. Hunt, Control of pepsinogen synthesis and secretion in chief cell monolayers, Dig. Dis. Sci. 33:-1588-91(1988).
24. P.H.S. Meijerink, J.P. Bebelman, G. Pals, F. Arwert, R.J. Planta, A.W. Eriksson, WH. Mager, Analysis of the promoter of a human pepsinogen A gene, Adv. Exp. Med. Biol. 306:87-90(1992).
25. J. Boyd, A. Bird, DNA methylation inhibits transcription indirectly via a methyl-CpG binding protein Cell -64:1123-34(1991).

A COMPARATIVE STUDY ON AMINO ACID SEQUENCES OF THREE MAJOR ISOFORMS OF HUMAN PEPSIN A

Nadezhda Tarasova,[1] Nancy D. Denslow,[2] Benny F. Parten,[2] Nha Tran,[2] Hung P. Nhuyen,[2] Arwyn Jones,[3] Norman B. Roberts,[3] and Ben M. Dunn[2]

[1]ABL-Basic Research Program, NCI, FCRDC
P.O. Box B, Frederick, MD, 21702-1201
[2]Department of Biochemistry and Molecular Biology
University of Florida
Gainesville, FL, 32610-0245
[3]Department of Clinical Chemistry
University of Liverpool
Liverpool L7 8XP (UK)

INTRODUCTION

Human pepsin A was shown to consist of up to four isoforms with differing enzymatic properties [1, 2, 3]. Unfortunately, the absence of complete sequences has hindered progress to any clear explanation for these differences. The sequence data available has been based mainly on N-terminal amino-acid and cDNA analysis of human pepsinogens [2, 4, 5]. We have therefore undertaken further more detailed amino-acid sequence studies of isolated human pepsin isoforms in order to clarify the structural basis for the differences observed in enzymatic activities.

MATERIALS and METHODS

Purification of Human Pepsins

Stomachs were obtained during dissection no later than 3 h after death certification and kept at -70° C before use. To start the purification the stomachs were thawed at room temperature. Mucosa was separated with scissors, homogenized with 4 vol. of 0.02 M Tris-HCl buffer pH 7.2 and centrifuged at 50 000 x g. Pepsinogens were batch-absorbed on DEAE-cellulose DE-23 (Whatman) and eluted in a column with 0.5 M NaCl. The zymogen peak was passed through Cibacron Blue F3GA-agarose to remove the main contaminant serum albumin, and dialyzed against 0.02 M Tris-HCl

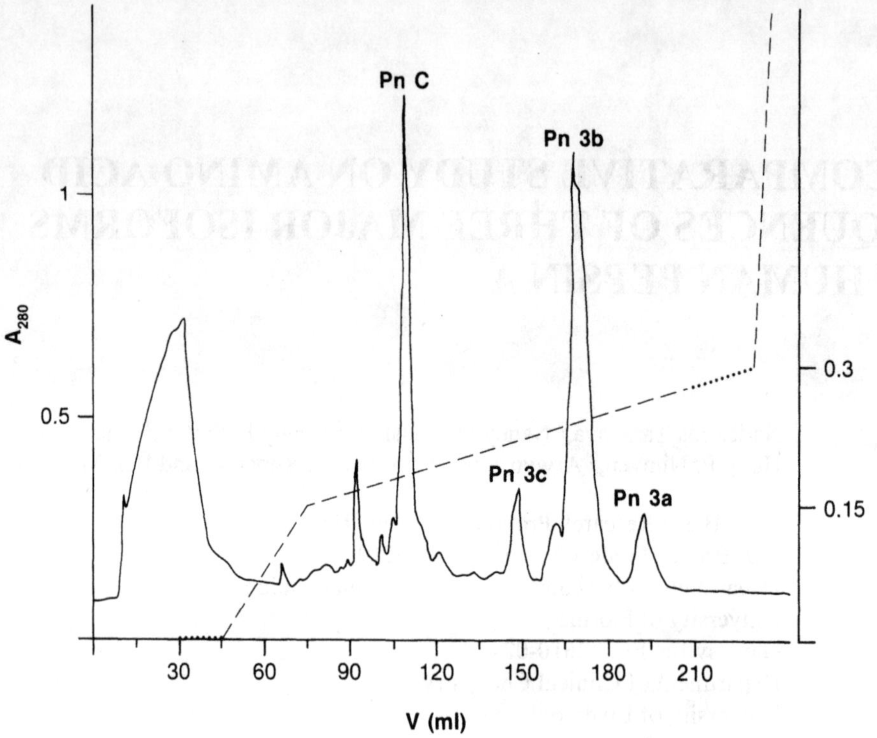

Figure 1. FPLC of a mixture of human pepsins. 25 mg of pepsinogens activation mixture was applied on 1 x 20 cm MonoQ column and eluted in 150 ml gradient of 0.15-0.3 M NaCl in 0.02 M piperazine - HCl buffer pH 5.0.

pH 7.2. The mixture of pepsinogens and progastricsin was obtained after chromatography on DEAE-cellulose DE-52 with elution in a gradient of 0-0.5 M NaCl. The zymogens were dialyzed against water and freeze-dried. For activation pepsinogens were dissolved in distilled water to give a concentration of 3 mg/ml and pH was adjusted to 3.0 with 1 M HCl. After incubation for 15 min at 37°C, pH was brought to 5.0 with 1 M Na acetate and the protein was dialyzed against 0.02 M piperazine-HCl pH 5.0. Final purification of isoforms was achieved by FPLC on 1 x 20 cm MonoQ column in a gradient of 0.15-0.3 M NaCl in 0.02 M piperazine-HCl buffer pH 5.0 (Fig.1).

Sequencing of Human Pepsin Isoforms

All three isoforms isolated were separately hydrolyzed with trypsin, chymotrypsin, V8 protease and cyanogen bromide. The resulting fragments were fractionated either by SDS-PAGE with subsequent blotting onto PVDF membrane, or by FPLC. Edman degradation was performed on a MilliGen/Biosearch 6600 ProSequencer (Millipore-Waters, Wasford, UK) or ABI-473 Protein Sequencer (Applied Biosystems, Foster City, CA). The phenylthiohydantoin - amino acids were identified by reverse phase HPLC on a Waters 600 system (Waters SequeTag C18 column) or on-line using an Applied Biosystems Model 130 Analyzer.

```
            1                          10                            20
Pepsin 3a   V D E Q P L E N Y L D M E Y F G T I G I G T P A Q D F
Pepsin 3b   V D E Q P L E N Y L D M E Y F G T I G I G T P A Q D F
Pepsin 3c   V D E Q P L E N Y L D M E Y F G T I G I G T P A Q D F

                30                   40                   50
            T V V F D T G S S N L W V P S V Y C S S L A C T N H N
            T V V F D T G S S N L W V P S V Y C S S L A C T N H N
            T V V F D T G S S N L W V P S V Y C S S L A C T N H N

                      60                 PO4 70                   80
            R F N P E D S S T Y Q S T S E T V S I T Y G T G S M T
            R F N P E D S S T Y Q S T S E T V S I T Y G T G S M T
            R F N P E D S S T Y Q S T S E T V S I T Y G T G S M T

                            90                 100
            G I L G Y D T V Q V G G I S D T N Q I F G L S E T E P
            G I L G Y D T V Q V G G I S D T N Q I F G L S E T E P
            G I L G Y D T V Q V G G I S D T N Q I F G L S E T E P

             110                 120                 130
            G S F L Y Y A P F D G I L G L A T P S I S S S G A T P
            G S F L Y Y A P F D G I L G L A T P S I S S S G A T P
            G S F L Y Y A P F D G I L G L A T P S I S S S G A T P

                    140                 150                 160
            V F D N I W N Q G L V S Q D L F S V Y L S A D D Q S G
            V F D N I W N Q G L V S Q D L F S V Y L S A D D Q S G
            V F D N I W N Q G L V S Q D L F S V Y L S A D D K S G

                            170                 180
            S V V I F G G I D S S Y Y T G S L N W V P V T V E G Y
            S V V I F G G I D S S Y Y T G S L N W V P V T V E G Y
            S V V I F G G I D S S Y Y T G S L N W V P V T V E G Y

           190                  200                 210
            W Q I T V D S I T M N G E A I A C A E G C Q A I V D T
            W Q I T V D S I T M N G E A I A C A E G C Q A I V D T
            W Q I T V D S I T M N G E A I A C A E G C Q A I V D T

                  220                 230                 240
            G T S L L T G P T S P I A N   I Q S D I G A S E N S D
            G T S L L T G P T S P I A N   I Q S D I G A S E N S D
            G T S L L T G P T S P I A K/R I Q S D I G A S E N S D

                          250                 260                  270
            G D M V V S C S A I S S L P D I V F T I N G V Q Y P V
            G D M V V S C S A I S S L P D I V F T I N G V Q Y P V
            G D M V V S C S A I S S L P D I V F T I N G V Q Y P V

                                280                 290
            P P S A Y I L Q S E G S C I S G F Q G M N L P T E S G
            P P S A Y I L Q S E G S C I S G F Q G M N L P T E S G
            P P S A Y I L Q S E G S C I S G F Q G M N L P T E S G

                300                 310                 320
            E L W I L G D V F I R Q Y F T V F D R A N N Q V G L A
            E L W I L G D V F I R Q Y F T V F D R A N N Q V G L A
            E L W I L G D V F I R Q Y F T V F D R A N N Q V G L A

            P V A
            P V A
            P V A
```

Figure 2. Human pepsin isoforms sequences. Sequenced parts of primary structures are shadowed. The residues that differ in different isoforms are underlined.

RESULTS AND CONCLUSIONS

Three major human pepsin isoforms were isolated from human gastric juice and stomach mucosa. The pepsins were designated Pn3a, Pn3b and Pn3c (Fig. 1) in accordance with nomenclature used in [1]. The identity of pepsin isoforms from gastric juice and from stomach mucosa was confirmed by running them on ion-exchange HPLC column [1] in

Table 1. Differences in Primary Structures between Human Pepsin Isoforms

Isoform	Total number of residues sequenced	Residue No.				
		30	68	160	230	291
3a	192	Val	PO4 Ser	Gln	Asn	
3b	278	Val	Ser	Gln	Asn	Leu
3c	187	Val	Ser	Lys	Lys/Arg	
cDNA [4,5]		Val/Leu	Ser	Gln/Lys	Asn	Leu/Val

identical condition. The minor isoform Pn1 isolated from gastric juice [1] was not found during purification of pepsins from mucosa, probably due to coelution of corresponding pepsinogen with pepsins on the step of DEAE-cellulose chromatography.

Up to 278 amino acid residues were determined (Fig. 2). Partial sequences for each isoform were compared to each other and to published cDNA sequences. Table 1 summarizes the differences that were detected. Leu/Val substitution in position 30 revealed by cDNA sequencing was not found in the isoforms that were studied. Val was discovered in all isoforms. In contrast to the previous finding, that human pepsinogens contain no phosphate groups [2], Ser 68, appeared to be phosphorylated in Pepsin 3a, like same residue in porcine

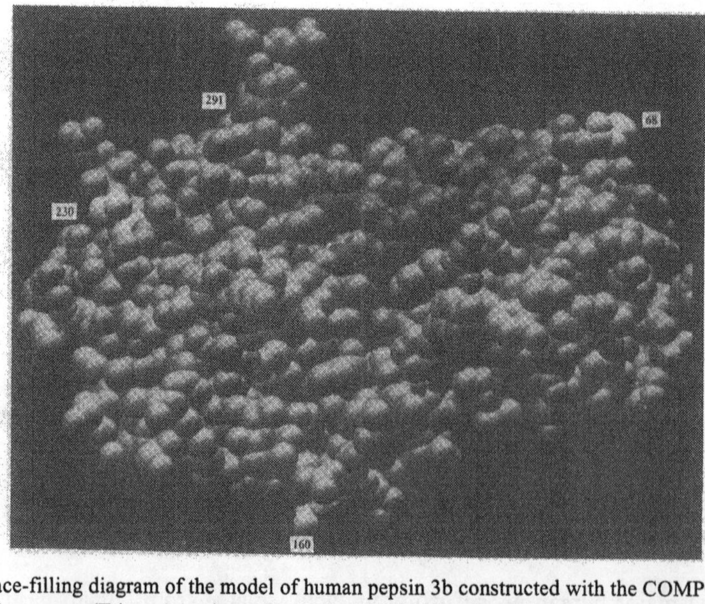

Figure 3. Space-filling diagram of the model of human pepsin 3b constructed with the COMPOSER module of the SYBYL program (Tripos Associates, St. Louis, MO, USA). The molecule is oriented so that the active site cleft appears on the upper half of the molecule. The location of the catalytic site is indicated by the use of dark grey to highlight the two Asp residues and the prominent "flap" (residues Val 71 to Thr 81). Residues shown in this study to vary between a, b and c isoforms of pepsin are marked with numbers.

pepsin. The differences in position 160 (Lys in Pn3c and Gln in Pn3a and Pn3b) were in good correlation with reported cDNA structures; however the presence of a cationic residue in position 230 in Pn3c was not revealed by cDNA structure determination. The disagreements between the results of protein and cDNA sequencing could be attributed to individual variations. However this is unlikely, as mucosa and gastric juices, from different subjects, were pooled prior to purification.

The differences in primary structures of the isoform so far discovered do not explain the variations in activities. The positions of all variable residues are located far from the active site cleft in the three-dimensional model of human pepsin molecule (Fig. 3). Further studies of human pepsin primary and tertiary structures that are now in progress can provide the rationale for the differences in enzymatic properties of the isoforms that have been revealed.

ACKNOWLEDGMENTS

Research sponsored in part by the National Cancer Institute, DHHS, under contract No. NO1-CO-74101 with ABL and by NIH Grant DK 18865.

REFERENCES

1. K. Peek, N.B. Roberts, W.H. Taylor, J. Chromatography **476**:291-297 (1989).
2. S.B.P. Athauda, M. Tanji, T. Kageyama, K. Takahashi, J. Biochem. **10**:920-927 (1989).
3. B.M.Dunn et al. unpublished observation.
4. K. Sogawa, Y. Fuji-Kuriyama, Y. Mizukami, Y. Ichinara, K. Takahashi, J. Biol. Chem. **258**: 5306-5311 (1983).
5. M.P.J.Evers, B. Zelle, J.P.Bebelman, V. van Beusechem, L. Kraaman, M.J.V. Hoffer, J.C. Pronk, W.H. Mager, R.J. Planta, A.W. Eriksson, R.R. Frants, Genomics **4**:232-239 (1989).

MOLECULAR MASS DETERMINATION BY ELECTROSPRAY MASS SPECTROMETRY OF HUMAN PEPSINS, GASTRICSIN, AND PORCINE PEPSIN A VARIANTS

A.T. Jones,[1] B. N. Green,[2] S. P. Wood,[1] and N. B. Roberts[3]

[1]Department of Crystallography
Birkbeck College
London, UK
[2]Fisons Instruments
VG BioTech
Altrincham, UK
[3]Department of Clinical Chemistry
University of Liverpool
Liverpool, UK

INTRODUCTION

Proteins are normally analysed by Electrospray Mass Spectrometry (ESMS) as their multiply charged positive ions, when their molecular masses are typically measured to within 0.01% of the values calculated from the primary sequences. The aspartic proteinases obtained from gastric secretions, however, have an insufficient number of basic amino acids to be analysed in this way, but should be well suited to negative ion analysis because they contain a large number of acidic residues. The amount of available data on the analysis of acidic proteins by ESMS is sparse and in order to assess the efficacy of negative ion ESMS, a commercial sample of porcine pepsin A and several proteinases isolated from human gastric juice were analysed and their measured molecular masses compared with values calculated from cDNA sequences (1-3). Attempts were made from mass data, to characterise purified human pepsin A variants with the published cDNA sequences(2); and particular emphasis was placed on obtaining a true molecular mass for human pepsin 1 (fast moving pepsin) for which there is evidence of carbohydrate attachment(4).

Aspartic Proteinases: Structure, Function, Biology, and Biomedical Implications
Edited by Kenji Takahashi, Plenum Press, New York, 1995

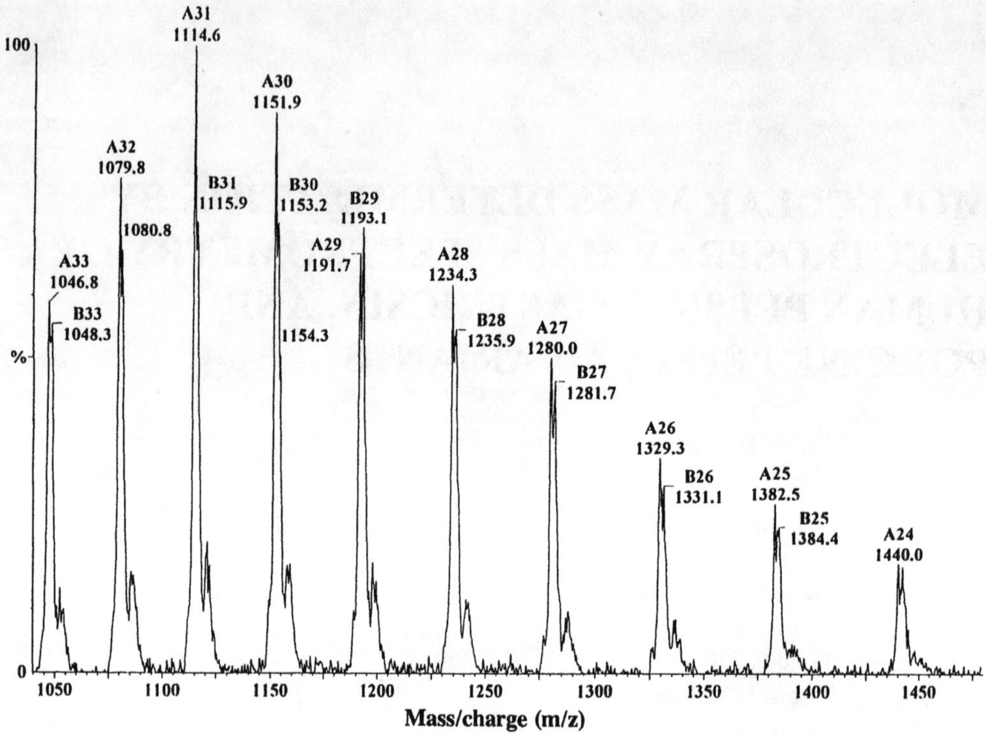

Figure 1. Original mass/charge data for commercial porcine pepsin A. The numbers over the peaks signify the number of negative charges on the molecule.

EXPERIMENTAL

Pepsin and Gastricsin Purification by High Performance Ion Exchange Chromatography (HPIEC)

Human gastric juice was filtered and dialysed at 4°C for 24 hours against 50mM sodium acetate buffer pH 4.1. The dialysate was then centrifuged at 4000rpm for 10 minutes prior to ultrafiltration in a stirred cell containing a 10,000 molecular weight cut-off membrane. 1 ml aliquots of the retained fraction were injected on to a 7.5 x 0.75 cm TSK DEAE-5PW anion exchange column (Anachem, Luton, UK). Pepsin and gastricsin variants were separated using a linear gradient of up to 0.8M NaCl. Commercial porcine pepsin A (Sigma Co. Ltd., Poole, UK) was found to contain two variants A1 and A2 (approximate contents 80% and 20% respectively), and these were separated and collected as above. For ESMS analysis, the collected fractions were desalted by extensive dialysis against 100M HCl solution, pH 4.0, at 4°C.

Mass Spectrometry

Negative ion electrospray data were acquired on a VG Quattro-BQ (Fisons Instruments, VG BioTech, Altrincham, UK), a triple quadrupole instrument with a mass range for

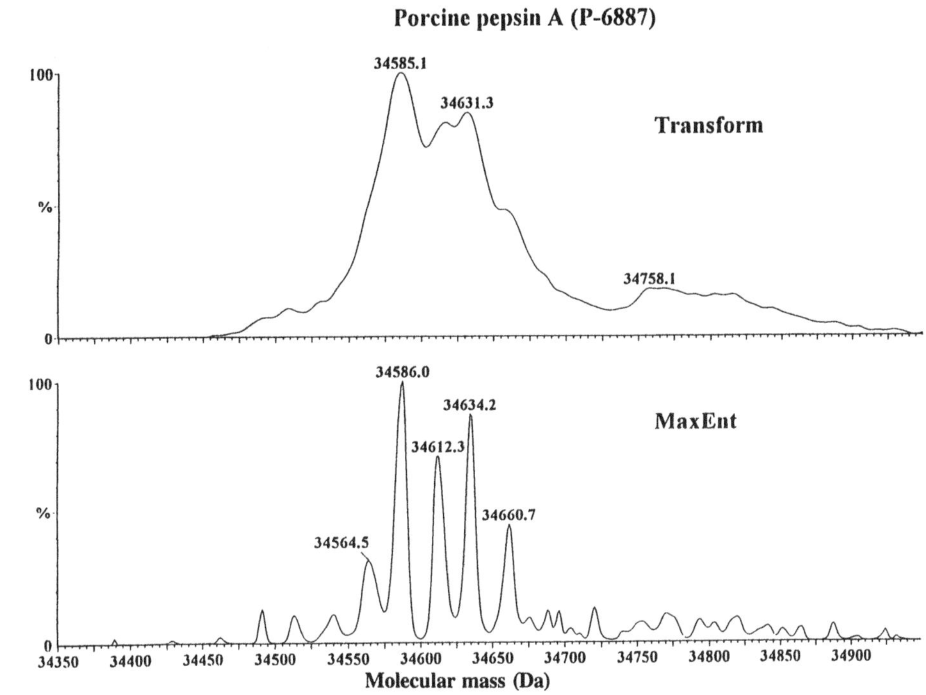

Figure 2. Transform (upper) and maximum entropy (lower) processed data for porcine pepsin A from the original mass/charge data shown.

singly charged ions of 4000. Samples as purified above were initially dissolved in water, then diluted to give final concentrations of 5-10 pM of sample per μl of 50/50 2-propanol/water containing 0.3% ammonia. 10 μl aliquots of these solutions were introduced into the electrospray source at 5 μl/min., whilst scanning over the mass to charge (m/z) ratio range 1040-1480 at 10 seconds per scan. Typically, data were summed from several 10 μl introductions of sample over 5-10 minutes to obtain the final spectra. These spectra, on a m/z scale were transformed onto a true molecular mass scale using the software supplied with the instrument. Mass scale calibration employed the multiply charged negative ion series from separate introductions of horse heart myoglobin (average molecular mass 16,951.5).

RESULTS AND DISCUSSION

Porcine Pepsin

Figure 1 shows typical data from commercial porcine pepsin A, as originally produced by the spectrometer on a m/z ratio scale. The two main components in the sample are labelled A and B. Figure 2 compares the result of processing the original data by two methods in order to present them on a true molecular mass scale. The transform method

Table 1. Summary of Molecular Mass Measurements on Human and Porcine Pepsins

Enzyme	Measured mass (Da) (No. of measurements)	Tentative assignments with cDNA sequences(1,2) cDNA variant	Predicted mass (Da)	Mass difference (Da)
Human				
Gastricsin	35458.5 (2)	Gastricsin	35455.3	3.2
Pepsin 3c	34641.8 (2)	Pepsin A4 Pepsin A5	34652.3 34635.0	-10.5 6.8
Pepsin 3b minor	34723.7 (1)	Pepsin A4 + phosphate Pepsin A5 + phosphate	34732.2 34715.0	-8.6 8.7
Pepsin 3b	34629.8 (3)	Pepsin A3 Pepsin A5	34622.2 34635.0	7.6 -5.2
Pepsin 3a	34705.5 (2)	Pepsin A3 + phosphate	34702.2	3.3
Pepsin 1	34635.0 (2)	Pepsin A3 Pepsin A5	34622.2 34635.0	12.8 0.0
Porcine				
Pepsin A	34586.3 (5)	Pepsin A	34583.8	2.5
Pepsin A1	34585.3 (1)	Pepsin A	34583.8	1.5
Pepsin A2	34613.5 (1)	None	None	-.-

reproduces the resolution in the original m/z data. However, the maximum entropy (MaxEnt) based method enhances the resolution and allows otherwise unresolved components to be measured. Hence all the results presented in this paper are from MaxEnt processed data(5). The mass data from porcine pepsin A (Fig. 2) indicate the presence of several components. Five repeat analyses of this sample gave a mean molecular mass for the major component of 34586.3, only 2.5 Da (0.007%) higher than the value (34583.8) calculated from the revised sequence of Lin et al.(1) The major porcine pepsin A variant (A1 Table 1) had a molecular mass close to the major component from the unseparated sample. However the minor variant porcine pepsin A2 had a significantly higher mass of 34,613.5 and was equivalent to peak 34,612.3 of the parent mixture.

The components observed with higher masses (i.e. > 34,612.3) would appear by analogy with positive ion data, to be due to adduct formation.

Human Pepsins

The chromatographic separation of 4 human pepsin A variants and gastricsin is shown in Figure 3. The corresponding mass spectra for gastricsin and 3 pepsin A variants are shown in Figure 4. Pepsin 3b minor (Table 1) was rarely observed in gastric juice samples and when present appeared as a shoulder peak on pepsin 3b.

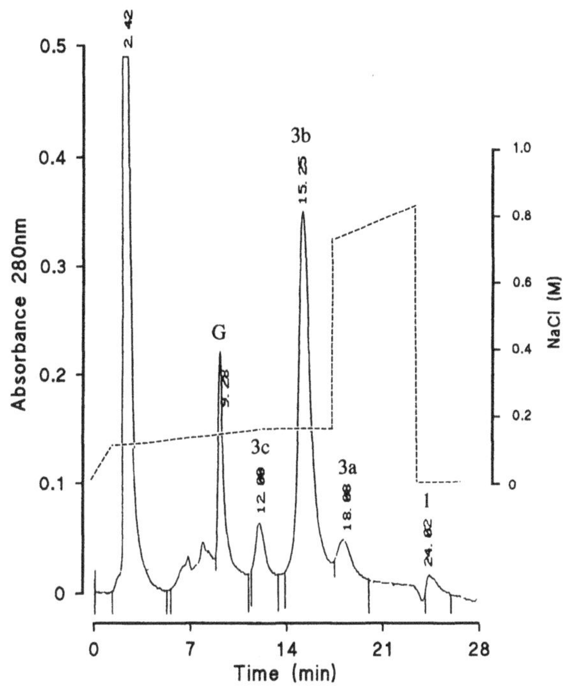

Figure 3. Separation of four human pepsins (3c, 3b, 3a, 1) and gastricsin (G) from gastric juice by high performance ion exchange chromatography.

Tentative assignments of calculated masses of human pepsin A variants with predicted cDNA masses (Table 1) suggests pepsin 3a to be phosphorylated pepsin A3, and that pepsin 3b minor is phosphorylated pepsin A4 or A5. Pepsins 3b and 3c cannot confidently be assigned to any of the three sequenced genes (A3, A4 or A5) as the predicted and measured masses consistently disagree.

Human pepsin 1 had a similar mass to the other human pepsin variants (Table 1) infering no mass spectrometric evidence of covalently attached carbohydrate. Recent data(6) suggest that human pepsin 1 is a mixture of pepsin A variants non-covalently associated with a glycoprotein of molecular mass > 1 x 10^6 Da which would be unsuitable for analyses by ESMS.

CONCLUSIONS

ESMS has been successfully applied in negative ion mode for the mass analysis of negatively charged proteins. Porcine pepsin A consists of at least two components with significantly different masses. Two of the human pepsin A variants appear to be phosphorylated but there is no evidence of covalently attached carbohydrate to human pepsin 1.

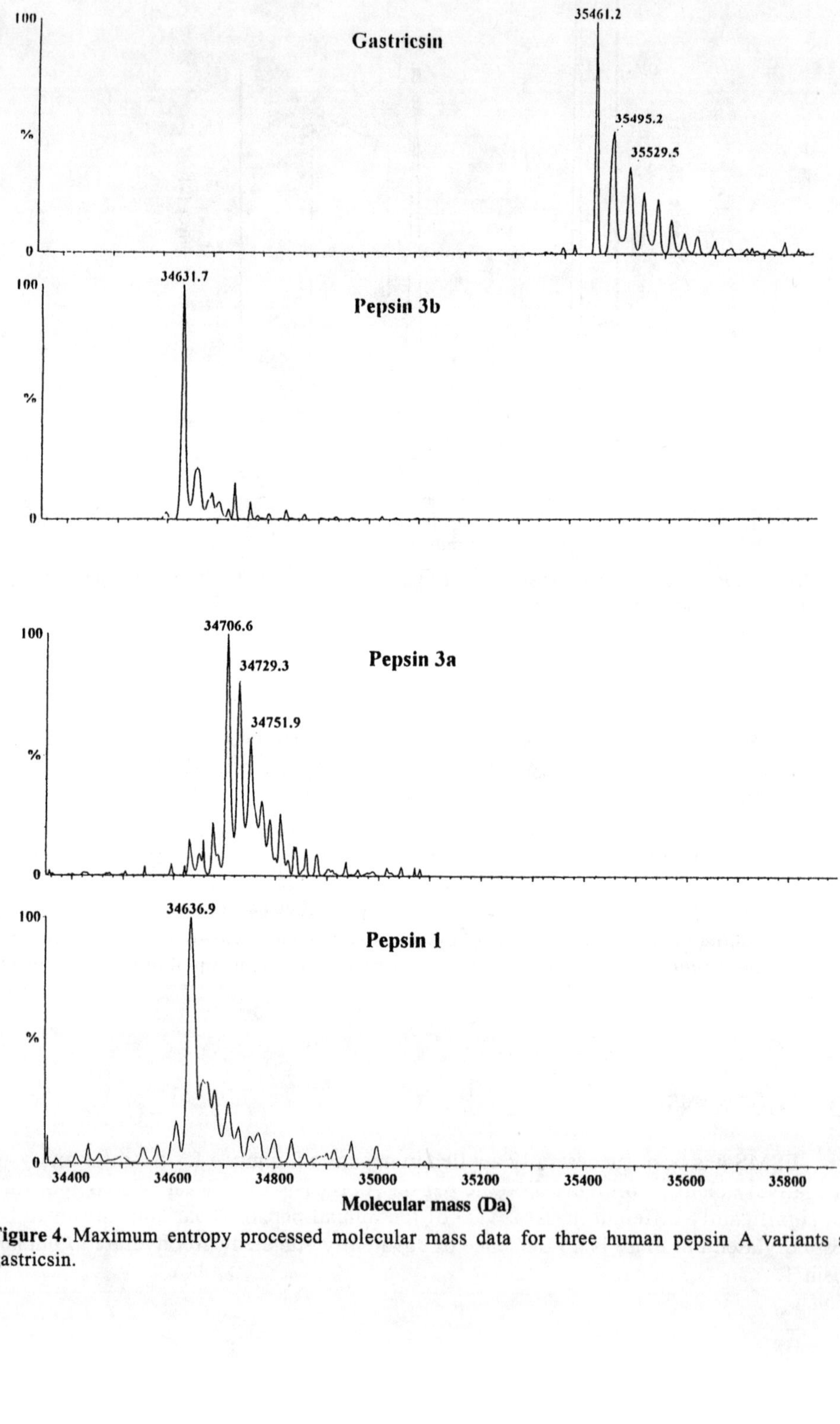

Figure 4. Maximum entropy processed molecular mass data for three human pepsin A variants and gastricsin.

REFERENCES

1. Lin, X-L., Wong, R.N.S., Tang, J., Synthesis, purification and active site mutagenesis of recombinant porcine pepsinogen, J. Biol. Chem. 264 4482:(1989).
2. Evers, M.P.J., Zelle, B., Bebelman, J.P., van Beusechem, V., Kraakman, L., Hoffer, M.J.V., Pronk, J.C., Mager, W.H., Planta, R.J., Eriksson, A.W., Frants, R.R., Nucleotide sequence comparison of five human pepsinogen A (PGA) genes:evolution of the PGA multigene family, Genomics 4 232:(1989).
3. Hayano, T., Sogawa, K., Ichihara, Y., Fujii-Kuriyama, Y., Takahashi, K., Primary structure of human pepsinogen C gene, J. Biol. Chem. 263 1382:(1988).
4. Pearson J.P., Allen, A., Roberts, N.B., Taylor, W.H., The carbohydrate component of human pepsin 1, Clinical Sciences 72 Suppl. 16. 33:(1987)
5. Ferrige, A.G., Seddon, M.J., Green, B.N., Jarvis, S.A, Skilling, J., Disentangling electrospray data with maximum entropy, Rapid Commun. Mass Spectrom., 6 707:(1992)
6. Jones A.T. Unpublished data

EVIDENCE FOR ELECTROSTATIC INTERACTIONS IN THE S_2 SUBSITE OF PORCINE PEPSIN

Chetana Rao and Ben M. Dunn

Department of Biochemistry and Molecular Biology
University of Florida, Gainesville, FL 32610-0245

A convenient system for studying protein-ligand interactions and structure-activity relationships has been the Aspartic proteinase family of enzymes. Pepsin, a member of this family involved in the digestion of proteins in the stomach, has been a valuable model system for investigation into enzymatic specificity. The active site cleft of porcine pepsin is made of distinct subsites S_5-S_3' (I. Schecter and A. Berger, 1967, B. B. R. C. **27**: 157) all of which may contribute to the specificity of the enzyme. Herein, we report the cation binding preference of the S_2 subsite of porcine pepsin, utilizing chromogenic octapeptide substrates at pH values ranging from 3 to 6. Residue 287 is within the S_2 region of the active site cleft, and varies widely among the aspartic proteinase family, with Glu, Gln, Met, and Gly occurring in various enzymes.

These studies are indicative of electrostatic interactions in the S_2 subsite of porcine pepsin involving the Glu287 residue. The changes in the kinetic parameter k_{cat}/K_m as a function of pH can be attributed to the deprotonation of the Glu287 side chain in the S_2 subsite of the enzyme promoting electrostatic attractions between a positively charged residue such as lysine, arginine or histidine in the P_2 position of the substrate and the negatively charged side chain of the glutamic acid at a higher pH. This kinetic data supports previous studies carried out by Pohl and Dunn (1988, Biochemistry, **27**: 4827).

MATERIALS AND METHODS

1. Substrate cleavage: All the chromogenic substrates under study were synthesized by the solid phase method using an Applied Biosystems Model 430A (Protein Core Facility at the University of Florida). Stock peptide solutions were made in distilled water and quantified by amino acid analysis. The purity of the peptides (≥ 90%) was verified by HPLC analysis.

The hydrolysis of the substrates was monitored at 37°C, pH 3.5 in 0.1 M sodium formate buffer by averaging the decrease in absorbance over a range of 284-324 nm using a Hewlett Packard Diode Array Spectrophotometer (Dunn et al, 1994). The cleavage of the peptides was also monitored on an HPLC column followed by the sequence determination of the respective peaks to ensure that the site of cleavage by the enzyme (Phe * Nph dipeptide

Table 1.

P_5	P_4	P_3	P_2	P_1	P_1'	P_2'	P_3'	k_{cat} (sec^{-1})	K_m (μM)	k_{cat}/K_m ($\mu M^{-1}s^{-1}$)
K	P	I	L	F	Nph	R	L	20 ± 6	14 ± 5	1.40 ± 0.60
K	P	I	A	F	Nph	R	L	30 ± 9	25 ± 8	1.20 ± 0.50
K	P	I	Nle	F	Nph	R	L	15 ± 4	13 ± 3	1.14 ± 0.40
K	P	I	E	F	Nph	R	L	72 ± 9	70 ± 11	1.02 ± 0.20
K	P	I	S	F	Nph	R	L	57 ± 17	63 ± 18	0.91 ± 0.40
K	P	I	D	F	Nph	R	L	46 ± 12	52 ± 9	0.88 ± 0.30
K	P	I	R	F	Nph	R	L	20 ± 5	24 ± 4	0.82 ± 0.30
K	P	I	I	F	Nph	R	L	14 ± 4	17 ± 6	0.81 ± 0.50
K	P	I	V	F	Nph	R	L	17 ± 5	31 ± 8	0.56 ± 0.20
K	P	I	N	F	Nph	R	L	22 ± 6	43 ± 4	0.51 ± 0.10
K	P	I	Q	F	Nph	R	L	24 ± 8	72 ± 28	0.33 ± 0.10
K	P	I	H	F	Nph	R	L	8 ± 2	33 ± 5	0.24 ± 0.07
K	P	I	T	F	Nph	R	L	17 ± 5	80 ± 12	0.22 ± 0.06
K	P	I	K	F	Nph	R	L	5 ± 1	28 ± 7	0.18 ± 0.06
K	P	I	P	F	Nph	R	L	NC	NC	NC

Nph = *p*-Nitrophenylalanine.

NC = Not cleaved under standard assay conditions (0.1M sodium formate, pH 3.5).

bond) was correct. The K_m and V_{max} values were determined from the initial rates at six different substrate concentrations. The pH dependence studies were carried out using the same procedure with formate/NaCl, acetate/NaCl, or MES/NaCl buffer (I = 0.1M) at a pH values ranging from 3 to 6.

2. pK_a determination: The pK_a values were calculated using the plots of k_{cat}, k_{cat}/K_m or K_m versus pH and fitting these plots in the Enzfitter program using the following equation

$$X = \frac{X(\min) + X(\max) \times 10^{(pH-pK_a)}}{1 + 10^{(pH-pKa)}}$$

where X = k_{cat}, k_{cat}/K_m or K_m , X(min) and X(max) are estimates of lower and upper limits of X.

Table 2.

	P_5	P_4	P_3	P_2	P_1	P_1'	P_2'	P_3'	k_{cat} vs. pH	k_{cat}/K_m vs. pH
1	K	P	I	K	F	X	R	L	-----	4.9 ± 0.1
2	K	P	I	R	F	X	R	L	-----	5.4 ± 0.3
3	K	P	I	A	F	X	R	L	-----	-----

RESULTS/DISCUSSION

In an attempt to understand the detailed interactions in the S_3 subsite of selective aspartic proteinases, of animal and microbial origin, our laboratory carried out studies with the peptide K(P_5)-P(P_4)-X(P_3)-E(P_2)-F(P_1)*Nph(P_1')-R(P_2')-L(P_3') with substitutions in the P_3 position (Dunn et al., 1986). The most effective substrates for porcine pepsin were the alanine and isoleucine replacements in the P_3 position. In order to understand the S_2 subsite preferences we have now designed chromogenic octapeptide substrates of the form K(P_5)-P(P_4)-I(P_3)-E(P_2)-F(P_1)*Nph(P_1')-R(P_2')-L(P_3') with systematic substitutions in the P_2 position shown in Table 1.

As evident from table 1, the S_2 subsite seems to accommodate hydrophobic substitutions readily which include residues such as leucine, alanine, norleucine, as well as β-branched isoleucine and valine. The negatively charged aspartate also fits about as well with k_{cat}/K_m values approaching the control peptide. These observations can be correlated to the crystal structure solved by Abad-Zapatero *et al.* Based on the three-dimensional structure one can see a predominantly hydrophobic pocket in the S_2 subsite consisting of Thr77, Thr218, Met289 and Ile300. Another residue which borders the region between the S_2 and the S_4 subsite is Glu287, which potentially could be involved in electrostatic interaction.

The peptide containing histidine and lysine substitutions exhibited relatively low k_{cat}/K_m numbers. One point was the acceptance of arginine to a favorable extent as opposed to the poor acceptance of histidine and lysine. pH dependence studies were carried out with both the peptides with arginine and lysine in the P_2 position to see whether any of these residues had any observable control over the speculated deprotonation of the Glu287 residue positioned between the S_2 and S_4 subsites (see Table 2).

Proline was totally unacceptable in the S_2 subsite which may be explained due to the unique conformation of the proline side chain leading to unproductive binding. This is in agreement with the studies on pepsin carried out by Powers *et al.*

pH DEPENDENCE STUDIES

To understand the specific cation binding preferences in the S_2 subsite pH dependence studies were carried out with the peptides outlined in Table 2.

Table 2 presents the pK_a values obtained from graphs of k_{cat} or k_{cat}/K_m versus pH for the three of the peptides under study for porcine pepsin.

The k_{cat}/K_m parameter for peptides 1 and 2 exhibits an apparent pK_a of around 5.2. As discussed earlier, the S_2 subsite is predominantly composed of hydrophobic amino acids with the exception of the negatively charged Glu287. At this juncture, this apparent pK_a can be attributed to the Glu287 residue which has the ability to provide electrostatic attraction for positively charged residues in the P_2 position of the substrate at a higher pH. Generation of Glu287 mutants is now underway to test this hypothesis. No pH dependence was observed with the alanine containing control peptide (peptide 3).

ACKNOWLEDGMENTS

We would like to thank Ruth J. Davenport and Joe Geber for the peptide synthesis, Benne Parten and Robin Proveaux for amino acid analysis and sequencing. This work was supported by NIH grant AM18865.

REFERENCES

1. I. Schecter and A. Berger, Biochem. Biophy. Res. Comm. **27**:157-162 (1967).
2. J. Pohl and B. M. Dunn, Biochem., **27**:4827-4834 (1988).
3. S. L. Meyer, *Data Analysis for Scientists and Engineers*, John Wiley and Sons, Inc., New York (1975).
4. C. Abad-Zapatero, T. J. Rydel, D. J. Neidhart, J. Luly and J. W. Erickson, in Aspartic Proteinases: Genetics, Structure and Mechanisms, B. M. Dunn, ed., Plenum Press New York (1991).
5. B. M. Dunn, P. E. Scarborough, R. Davenport, and W. Swietnicki, "Analysis Of Proteinase Specificity By Studies Of Peptide Substrates: The Use Of Ultraviolet And Fluorescence Spectroscopy To Quantitate Rates Of Enzymatic Cleavage", in Peptide Protocols: Synthesis Purification, and Characterization, Ben M. Dunn and Michael Pennington, eds., (a volume of the METHODS IN MOLECULAR BIOLOGY series, John M. Walker, series editor), Humana Press, Clifton (1994).
6. Powers, J.C., A.D. Harley, and D.V. Myers, "Adv.Exp.Med.Biol. **95**: p. 141-157 (1977).
7. B. M Dunn, M. Jiminez, B. F. Parten, M. J. Valler, C. E. Rolph and J. Kay, Biochem. J. **237**: 899-906 (1987).

PROTEIN ENGINEERING OF SURFACE LOOPS: PRELIMINARY X-RAY ANALYSIS OF THE CHY155-165RHI MUTANT

Raj R.V. Dhanaraj,[1] Jim E. Pitts,[1*] Phil Nugent,[1] Poonsook Orprayoon,[1] Jon B. Cooper,[1] Tom L. Blundell,[1] Janna Uusitalo,[2] and Merja Penttilä[2]

[1]Laboratory of Molecular Biology
Department of Crystallography
Birkbeck College
University of London
Malet Street, London WC1E 7HX
United Kingdom
[2]VTT Biotechnical Laboratory
P.O. Box 202, SF-02151, Espoo, Finland

INTRODUCTION

The protein engineering programme at Birkbeck seeks to develop generic methods based on a design cycle involving biochemical preparation and characterization, determination and comparative analysis of three-dimensional structures, rule-based design, site-directed mutagenesis and expression of the mutants. Knowledge of the tertiary structure of proteins and the use of molecular modelling techniques provide a powerful approach for the design of novel biomolecules with specifically engineered properties. The aspartic proteinase family have been studied in detail by X-ray analysis and provide a suitable data base for the development of protein engineering design principles[1-6]. This family of endopeptidases belonging to a wide range of biological species with varying substrate specificities. The members of the family are of considerable commercial importance since enzymes such as chymosin and *Mucor pusillus* pepsin have been exploited by the food industry in cheese and soya processing, while renins, cathepsins and the retroviral proteinases are prime targets of the pharmaceutical industries in drug design.

We have chosen chymosin as a model enzyme to undertake specific alterations to a number of properties including size, zymogen activation, specificity, thermal stability and pH dependence. In the case of alteration of size, we are interested in maintaining activity while reducing surface loops to allow more efficient penetration of complex substrates used in food processing. We have identified a surface loop region from residues 155-165 predicted to be suitable for replacement with both sequences from other family members and also with novel loops of variable chain length designed by knowledge-based modelling. We report

Table 1. Comparison of the Amino Acid Sequence of the Chymosin 155-165 Loop with a Selection of Aspartic Proteinases Analyzed by X-Ray Analysis

CHYMOSIN	M	D	R	N	-	-	-	-	G	Q	E	S	M	L
RHIZOPUSPEPSIN	I	G	K	A	K	N	-	-	G	G	G	G	E	L
ENDOTHIAPEPSIN	M	G	Y	H	-	-	-	-	-	A	P	G	T	L
MUCORPEPSIN	M	N	T	N	-	-	-	-	D	G	G	G	Q	L

here our preliminary results of the structural analysis of the 155-165 rhizopuspepsin loop substitution into chymosin.

METHODS

Comparison of the three-dimensional structures of chymosin with those of the other members of the aspartic proteinase family indicated that a surface loop (q^Nr^N) formed by residues 155-165 (pepsin numbering) may be a suitable target for modification[7,8]. Possible replacements with homologous sequences from the other members of the family are indicated in Table 1. Below we describe the expression, crystallization and structural analysis of such a loop exchange mutant where the target loop was substituted by the corresponding loop from *Rhizopus chinensis* pepsin.

SITE-DIRECTED MUTAGENESIS OF CHYMOSIN

The introduction of the loop sequence from rhizopuspepsin for residues 155-165 into the chymosin cDNA was performed using a phosphorothioate based site-directed mutagenesis method[9] (Table 2). The mutagenic 61-mer oligonucleotide (Table 3) was synthesised using phosphoramidite chemistry and purified by reverse phase chromatography on a FPLC system. The *Eco* RI-*Msc* I restriction fragment containing the mutation was cloned into the vector pAMH104E-Δ*Bam* HI in the correct orientation[10], the DNA was then cleaved with

Table 2. DNA and Amino Acid Sequence Comparison of Chymosin and Rhizopuspepsin 155-165 Loops

```
CHYMOSIN
   155
   MET ASP ARG ASN GLY GLN --- --- GLU SER MET LEU

5' ATG GAC AGG AAT GGC CAG         GAG AGC ATG CTC 3'

5' ATC GGC AAG GCT AAG AAC GGA GGT GGC GGA GAG CTC 3'

   ILE GLY LYS ALA LYS ASN GLY GLY GLY GLY GLU LEU
   155
RHIZOPUSPEPSIN
```

The base changes introduced by the mutagenic 61-mer oligonucleotide are indicated in bold

Table 3. Mutagenic 61-base Oligonucleotide Used to Introduce the Loop Replacement Chy155-165Rhi

```
3' GACAAGAGCCAAATGTA
                    GCCGTTCCGATTCTTGCCTCCACCGCCTCT
                                                  CGAGTGCACCCCC 5'
```

The central region in bold contains the mutagenic sequence which relates by base pairing and opposite polarity to that shown in Table 2

Eco RI, dephosphorylated and combined with the large *Eco* RI fragment of PAMH104 to produce the mutant expression vector pCHY-RHI115. The correct insert was confirmed by DNA sequencing using the M13 dideoxy chain termination method of Sanger[11].

EXPRESSION OF CHYMOSIN MUTANTS IN *Trichoderma reesei*

The filamentous fungus *Trichoderma reesei* has been used successfully for heterologous expression of chymosin and antibody fragments[12,13]. The fungal strain Rut C30 was transformed with the expression plasmid pCHY-RHI155 containing the mutant DNA and plated onto selective minimal agar containing as a sole nitrogen source. The correctly folded zymogen undergoes secretion into the culture media under the direction of the strong cellobiohydrolase I (*cbhI*) promoter. The low pH of the growth media leads to autocatalytic activation of the enzyme. The best producer/transformant was grown at the 10 l scale in an Applikon bioreactor using cellulose as the sole carbon source. Chymosin activity was detectable in the liquid culture after three days of growth. The milk clotting activity was maximal at day five, while the amount of total secreted proteins continued to increase. The yield of the mutant Chy155-165Rhi was 27 mg/l. Enzyme kinetics using a synthetic chromogenic peptide indicated that the activity of the mutant is similar to that of wild-type chymosin B.

CRYSTALLIZATION AND DATA COLLECTION

The purified mutant was concentrated to 10 mg/ml in sodium phosphate buffer at pH 5.6 and crystallised by the hanging drop vapour diffusion method using sodium chloride as the precipitant. Plate-like crystals of dimensions 0.20 X 0.15 X 0.05 mm, grown over a period of 3 to 4 weeks, were found to be isomorphous with the native crystals. X-ray intensity data were collected by the oscillation technique with a MAR RESEARCH image plate detector using the Daresbury Laboratory Synchrotron Radiation Source. Although the X-ray diffraction data extended to 2.2 Å resolution, the images showed splitting due to the existence of multiple crystals that were impossible to separate. The intensities were weak due to the small size of the crystals at resolutions beyond 2.5 Å. Oscillation images from two differently-oriented crystals were processed using the CCP4 software suite and merged to a unique set of reflections (R_{merge} 12.4%) extending to a resolution of 2.5 Å.

REFINEMENT AND RESULTS

The atomic coordinates of native chymosin B[5], after the removal of the mutated loop, were used to obtain the initial difference Fourier maps. Interactive model building on an Evans and Sutherland PS300 with the graphics software FRODO[14], followed by least squares refinement cycles using RESTRAIN[15], gradually revealed the atomic positions of the

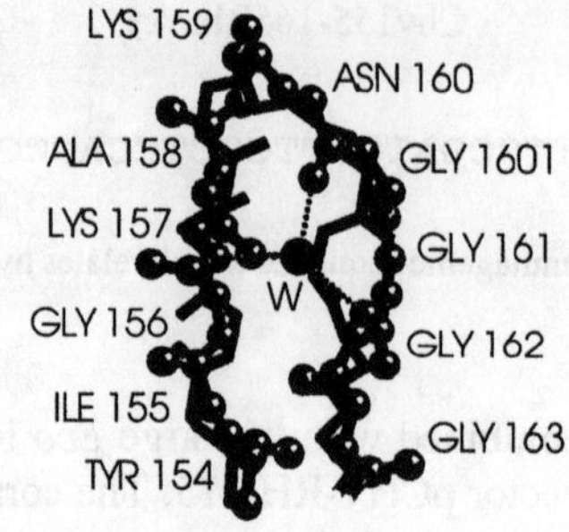

Figure 1. The main-chain atoms corresponding to the mutated loop (in ball and stick representation) and the water molecule (W) which stabilises the β-hairpin conformation of the mutant; the equivalent loop in rhizopuspepsin structure (smooth cylindrical joints) has been superposed for comparison. The pepsin numbering scheme has been used for labelling.

residues in the altered loop. The current crystallographic agreement factor and correlation coefficient for the refined model are 0.206 and 0.917 respectively, for reflections in the resolution range 8.0-2.5 Å. The superposition of the equivalent loop from rhizopuspepsin on that of the mutant is shown for comparison (Figure 1). Although the overall conformation of this mutant enzyme is similar to that of chymosin, clear conformational differences are found for the swapped loop when compared with the same sequence in the rhizopuspepsin structure. As the loop sequence is identical in the rhizopuspepsin and Chy155-165Rhi mutant, this indicates a strong influence of the local side-chain environment on the conformation, most significantly at residues Gly 161 and Gly 162. Interestingly the mutant loop exhibits the presence of a water molecule in the vicinity of the hairpin (Figure 1). This water molecule stabilises the β-hairpin conformation by making hydrogen bonds with the main-chain nitrogen and carbonyl oxygen atoms of the residues in the loop. This is in contrast to the casein rhizopuspepsin which does not have a water molecule in the equivalent position.

CONCLUSION

The loop mutant Chy155-165Rhi was produced in the filamentous fungus *trichoderma reesei* and exported into the growth medium to yield a correctly folded and highly active product. Structural and kinetic analyses confirm that the mutant adopts a native fold. Therefore surface loops may be successfully swapped between enzymes of different species without loss of catalytic activity, although they might subtly be altered by their local environment. Such studies will provide further insights into the evolution of loop and their impact on structure and function.

REFERENCES

1. K. Suguna, R.R. Bott, E.A. Padlan, E. Subramanian, S. Sheriff, G.H. Cohen and D.R. Davies, Structure and refinement at 1.8Å of the aspartic proteinase from *Rhizopus chinensis*, *J. Mol. Biol.* **196**:877 (1987).
2. J.B. Cooper, G. Khan, G. Taylor, I.J. Tickle and T.L. Blundell, X-ray analyses of aspartic proteinase. II. Three-dimensional structure of the hexagonal crystal form of porcine pepsin at 2.3Å resolution, *J. Mol. Biol.* **214**:199 (1990).

3. M. Newman, F. Watson, P. Roychowdhury, H. Jones, M. Badasso, A. Cleasby, S.P. Wood, I.J. Tickle and T.L. Blundell, X-ray analyses of aspartic proteinases. V. Structure and refinement at 2.0Å of the Aspartic Proteinase from *Mucor pusillus*, *J. Mol. Biol.* **230**:260 (1993).
4. T.L. Blundell, J.A. Jenkins, B.T. Sewell, L.H. Pearl, J.B. Cooper, I.J. Tickle, B. Veerapandian and S.P. Wood, X-ray analyses of aspartic proteinase. The three-dimensional structure at 2.1Å resolution of endothiapepsin, *J. Mol. Biol.* **211**:919 (1990).
5. M. Newman, M. Safro, C. Frazeo, G. Khan, A. Zdanov, I.J. Tickle, T.L. Blundell and N. Andreeva, X-ray analyses of aspartic proteinases. VI. Structure and refinement at 2.2Å of the Aspartic Proteinase of bovine chymosin, *J. Mol. Biol.* **221**:1295 (1991).
6. V. Dhanaraj, C.G. DeAlwis, C. Frazeo, M. Badasso, B.L. Sibanda, I.J. Tickle J.B. Cooper, H.P.C. Driessen, M. Newman, C. Aguilar, S.P. Wood, T.L. Blundell, P.M. Hobart, K.F. Geoghegan, M.J. Ammirati, D.E. Danley, B.A. O'Connor and D.J. Hoover, X-ray analyses of peptide-inhibitor complexes define the structural basis of specificity for human and mouse renins, *Nature* **357**:466 (1992).
7. V. Barkholt, Amino acid sequence of endothiapepsin, *Eur. J. Biochem.*, **167**:327 (1987).
8. J.E. Pitts, V. Dhanaraj, C.G. DeAlwis, D. Mantafounis, P.G. Nugent, P. Orprayoon, J.B. Cooper, M. Newman and T.L. Blundell, Multidisciplinary cycles for protein engineering: site-directed mutagenesis and X-ray structural studies of aspartic proteinases, *Scan. J. of Clinical and Laboratory Investigation* **52**:39 (1992).
9. J.W. Taylor, J. Ott and F. Eckstein, The rapid generation of oligonucleotide-directed mutations at high frequency using phosphorothioate-modified DNA, *Nucleic Acids Res.* **13**:8764 (1985).
10. J.E. Pitts, D. Mantafounis, P.G. Nugent, D. Quinn, P. Orprayoon, J.M. Uusitalo and M. Penttilä, Expression and characterisation of chymosin pH optima mutants produced in *Trichoderma reesei*, *J. Biotechnology* **28**:69 (1993).
11. F. Sanger, S. Nicklen and A.R. Coulson, DNA sequencing with chain-terminating inhibitors, *Proc. Natl. Acad. Sci. USA* **74**:5463 (1977).
12. A. Harkki, J. Uusitalo, M. Bailey, M.E. Penttilä and J.K.C. Knowles, A novel fungal expression system:secretion of active calf chymosin from the filamentous fungus *Trichoderma reesei*, *Bio/Technology* **7**:596 (1989).
13. E. Nyyssönen, M.E. Penttilä, A. Harkki, A. Saloheimo, J.K.C. Knowles and S. Keränen, Efficient production of antibody fragments by filamentous fungus *Trichoderma reesei*, *Bio/Technology* **11**:591 (1993).
14. T.A. Jones, A graphics model building and refinement system for macromolecules, *J. Appl. Crystallogr.* **11**:268 (1978).
15. I. Haneef, D.S. Moss, M.J. Stanford and N. Borkakoti, Restrained structure-factor least-squares refinement of protein structures using a vector processing computer, *Acta Crystallogr. A* **41**:426 (1985).

SEMINAL PROGASTRICSIN

Pal Bela Szecsi and Hans Lilja

Department of Clinical Chemistry
Malmö General Hospital, Lunds University
Malmö, Sweden

INTRODUCTION

Lundquist and Seedorff [1] demonstrated the presence of a zymogen in human seminal fluid, that could be activated into an active enzyme by acidification to pH 2. As the enzyme was purified, it became evident that it was in fact equivalent to the minor gastric enzyme originally named as progastricsin (EC 3.4.23.3) [2-6].This report provides a short review on the similarities and differences between human gastric and seminal progastricsin, and the possible function of the seminal progastricsin.

MATERIAL AND METHODS

Enzymes and N-terminal Sequencing

Human seminal progastricsin from pooled semen samples and gastric progastricsin was purified by anion-exchange, gelfiltration and fast performance liquid chromatography [7]. Automatic Edman degradation was carried out using an Applied Biosystems 477A Sequencer equipped with an on–line PTH analyser model 120A.

RNA and Northern Blot

Human seminal vesicles and prostate gland were obtained from post mortem organ donors, and mRNA was isolated with guanidine thiocyanate, fractionated on oligo (dT)–cellulose, denatured and subjected to electrophoresis on 1.5% agarose gel containing glyoxal. Hybridization was performed overnight at 65°C with a 1.35kb human gastric progastricsin probe, previously obtained from a human gastric mucosa cDNA library [8].

Library Construction, Screening and DNA Sequencing

A human prostate cDNA library was constructed in λgt10, containing 7.2×10^5 independent recombinant phages and screeened at high stringency with the 1.35 kb probe of

Aspartic Proteinases: Structure, Function, Biology, and Biomedical Implications
Edited by Kenji Takahashi, Plenum Press, New York, 1995

progastricsin. An additional human prostate λgt11 library was obtained from Clontech (CAT#HL1051b) and screened as above.

Immunohistochemistry and *In Situ* Hybridization

Human tissue from prostatic glands, seminal vesicles, epidydimis, testis and gastric mucosa (positive control) were fixed in Bouin's solution and embedded in paraffin. Immunohistochemistry was performed with the ABC-technique and a polyclonal antibody toward gastric progastricsin. *In situ* hybridization was performed using a 3′-labelled digoxigenin labelled synthetic 38 mer oligonucleotide complementory to human progastricsin. Control staining of the gastric mucosa both by immunohistochemistry and *in situ* hybridization showed positive staining in the chief cells and mucous neck cells similar to that previously described [9].

Degradation of Seminal Proteins

Series of seminal fluid were incubated at 37°C, at pH corresponding to the conditions found in the vagina before and shortly after intercourse [10]. At different time intervals, an aliquot was removed, solubilized and SDS-PAGE was performed[11]. The following digestion series were carried out: *A*, seminal fluid at authentic pH 8.0; *B*, seminal fluid at pH 3.4 (adjusted with hydrochloric acid); *C*, seminal fluid at pH 3.4 in presence of 20 μg/ml pepstatin; *D*, seminal fluid at pH 3.4 in presence 1 mM PMSF, 1mM benzamidine, 1 mM antipain, 1 mM leupeptin, 10 mM EDTA and 10 mM EGTA; *E*, seminal fluid acidified to pH 2.2 for 2 min and alkalized at pH 9.2 for 5 min followed by incubation at pH 3.4; *F*, progastricsin depleted seminal fluid at pH 8.0; *G*, progastricsin depleted seminal fluid at pH 3.4; *H*, progastricsin depleted seminal fluid at pH 3.4 with addition of purified human progastricsin (100 μg/ml final concentration); *I*, progastricsin depleted seminal fluid at pH 3.4 with addition of purified human progastricsin (100 μg/ml final concentration) in presence of 20 μg/ml pepstatin.

RESULTS AND DISCUSSION

Structural Difference between Seminal and Gastric Progastricsin

Human progastricsin appears as three components differing by approximately one charge unit and all are converted to gastricsin with identical electrophoretic mobility [12]. This indicates that a charge difference is located to the propart. Monoclonal antibodies toward progastricsin may react only with a noncarbohydrate moiety of the seminal zymogen, being able to distinguish between gastric and seminal progastricsin [7]. Purified seminal and gastric progastricsin showed no differences in the amino acid composition, and N-terminal protein sequencing of all 43 residues of the propart and 34 residues of the mature enzyme, both sequences in agreement of that deduced from the nucleotide sequences [13-15]. A 704 bp clone derived from the 3′-end was found in the λgt10 library, the 5'-end was recovered by PCR of reverse transcriptized mRNA. The combined nucleotide sequences showed only few differences from the published gastric cDNA and genomic sequences [13-15]. Two substitutions resulted in a change in amino acids, neither could account for the charge difference between extra-gastric and gastric progastricsin. The λgt11 library revealed a nearly full length clone with 3 conservative nucleotide substitutions of which one was identical with the λgt10 clones. It is most likely that seminal and gastric progastricsin arise from a single gene, with posttranslational modifications yet to be determined[15].

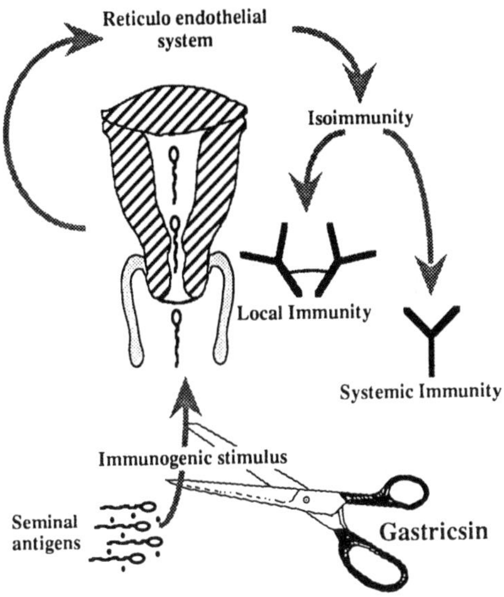

Figure 1. Degradation of seminal proteins. A pool of seminal fluid was incubated at 37°C, and the pH was lowered from 8.12 to 3.50 with HCl. Aliquots were removed at intervals and subjected to SDS-PAGE.

Origin of Seminal Progastricsin

Northern blot showed a transcript of approximately 1.8 kb strongly hybridizing signal both in mRNA isolated from the seminal vesicles and the prostatic gland. By immunohistochemistry and *in situ* hybridization progastricsin positive staining was observed in both the epithelium of seminal vesicles and prostatic glands. In the prostate progastricsin-positive cells were confined to the epithelium of few acini with uneven distribution among the individual cells. In contrary to the seminal vesicles, progastricsin staining in the prostatic gland often showed an apical enhancement. Some cells stained very intensely some times alternating with positive and negative cells. No positive cells were found in the stroma of the prostate or seminal vesicles, testis or any part of the epididymis. Our results favour the view that seminal progastricsin arise from both the seminal vesicles and the prostate [16,17].

Function of Seminal Progastricsin

The acidic enviroment of the human vagina (pH 3.93, range 3.68-4.22) activates seminal progastricsin in average 3 h (range 2-7 h) after deposition of semen [10]. The gastricsin activity could be detected after 5 hours (range 4-7 hours) in the vagina and persisted for at least 24 h after coitus.[10]. In contrast to previous reports [18,19], the concentration of progastricsin (mean 42.2 μg/ml) in semen did not correlate with the volume of the ejaculate, the appearance of semen, or concentration, motility, viability, morphology, or penetration ability of the spermatozoa[10]. Two individuals had considerably lower progastricsin concentration than the others, one of these had only trace amount (0.09 μg/ml). These findings suggested that the progastricsin amount in seminal fluid did not correlate with male infertility and that the majority of males had fairly high concentrations of the zymogen in semen.

In presence of high concentrations of benzamidine were seminal fluid proteins not degraded during 2 hours incubation at pH 8.0 at 37°C, but degraded within 30 min into

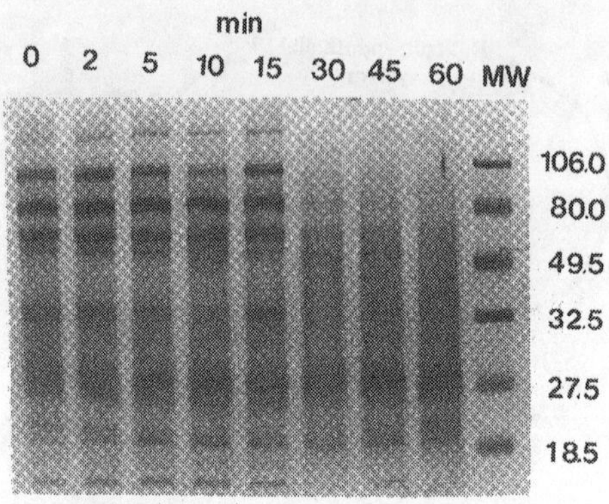

Figure 2. Hypothesis on the function of seminal progastricsin. The seminal fluid containing the inactive progastricsin neutralize the acidic vagina after ejaculation. During the following 4-7 hours vaginal pH returns to acidic levels activating progastricsin to gastricsin. The formed active enzyme subsequently degrades the seminal proteins left in the vagina, preventing exposure of these potential antigens to the woman.

smaller sized molecular fragments upon acidification of semen to pH 3.4-3.5 [11] (Fig. 1). This degradation was not inhibited by inhibitors of serine-, thiol- or metallo-proteases, but pepstatin prevented any degradation. Removing (by immunoaffinity chromatography) or denaturation (alkaline denaturation) of the progastricsin in the seminal fluid also strongly prevented the proteolysis of seminal proteins. Readdition of purified progastricsin to progastricsin-depleted seminal fluid restored the extensive degradation. Again, pepstatin could prevent this proteolysis. The degradation rate of the individual seminal fluid proteins was investigated by Western blot employing antibodies to several of the major proteins. The β-microseminoprotein and the low molecular fragments of semenogelin I and II seem relatively resistant to gastricsin-mediated degradation at acidic pH. Prostate specific antigen, prostate acid phosphatase, and Zn-α2-glycoprotein are rapidly degraded, whereas albumin, transferrin and lactoferrin are degraded more slowly, but are clearly degraded within 2 hours [11].

The genital tract in the human female, in particular the uterus, is well endowed with immunologically competent cells in order to prevent genital infections. Seminal antigens are an immunogenic stimulus for the female, and are presented for recognition by the immune defence system. However, since not all females become infertile due to deposition of seminal plasma antigens during coitus, a mechanism blocking the formation of antibodies must exist.

The site of function of the seminal progastricsin is most likely the vagina, without direct relation to the fertilization process. Once activated, seminal progastricsin is responsible for degradation of seminal fluid proteins at acidic conditions equivalent to that observed in the vagina (Fig. 2). Furthermore, by degrading the large amount of proteins found in seminal fluid, progastricsin might have a role in preventing immunoinfertility.

REFERENCES

1. F. Lundquist F and H.H. Seedorff, Pepsinogen in human seminal fluid *Nature* **170**:1115 (1952).
2. I.M. Samloff and W.M. Liebman, Purification and immunochemical characterization of group II pepsinogens in seminal fluid. *Clin Exp Immunol* **11**:405 (1972).
3. H. Hirsch-Marie and M. Conte, Étude de la protéase acide du liquide séminal humain. *Bull Soc Chim Biol* **49**,147 (1967).

4. I. Samloff and W. Liebman, Purification and immunochemical characterization of group II pepsinogens in seminal fluid. *Clin Exp Immunol* **11**,405 (1972).
5. L. Chiang, L. Contreras, J. Chiang and P. Ward , Human prostatic gastricsinogen: The precursor of seminal fluid acid proteinase. *Arch Biochem Biophys* **210**:14 (1981).
6. V. Richmond, J. Tang, S. Wolf, R.E. Trucco and R. Caputto, Chromatographic isolation of gastricsin, the proteolytic enzyme from gastric juice with pH optimum 3.2. *Biochim Biophys Acta* **29**,453 (1958).
7. P.B. Szecsi, C. Kock and B. Foltmann,Seminal pepsinogen C is not identical with though very similar to gastric pepsinogen C. *FEBS Lett* **238**:101 (1988).
8. J. Tang and R. Wong, Evolution in the structure and function of aspartic proteases. *J Cell Biochem* **33**:53 (1987).
9. K.S. Varis, Peptic cells. *in*: "Pepsinogens in man. Clinical and genetic advances", J. Kreuning, I.M. Samloff, J.I. Rotter. and A.W. Eriksson, eds., Alan R. Liss, New York, (1983).
10. P.B. Szecsi, D. Dalgaard, G. Stakemann, G. Wagner and B. Foltmann, The concentration of pepsinogen C in human semen and the physiological activation of zymogen in vagina. *Biol Reprod* **10**:653 (1989).
11. P.B. Szecsi and H. Lilja, Gastricsin-mediated proteolytic degradation of human seminal fluid proteins at pH levels found in the human vagina.*J Andrology* **14**:351 (1993).
12. B. Foltmann, A. Jensen, Human progastricsin Analysis of intermediates during activation into gastricsin and determination of amino acid sequence of the propart. *Eur J Biochem* **128**:63 (1982).
13. T. Hayano, K. Sogawa, Y. Ichihara, Y. Fujii-Kuriyama and K. Takahashi, Primary structure of human pepsinogen C gene. *J Biol Chem* **263**:1382 (1988).
14. R.T. Taggart, L.G. Cass, T.K. Mohandas , Human pepsinogen C (progastricsin). Isolation of cDNA clones, localization to chromosome 6, and sequence homology with pepsinogen A. J *Biol Chem* **264**:375 (1989).
15. G. Pals, T. Azuma, T.K. Mohandas, Human pepsinogen C (progastricsin) polymorphism: Evidence for a single locus located at 6p21.1-pter. *Genomics* **4**:137 (1989).
16. W.A. Reid, L. Vongsorak, J. Svasti, M.J. Valler and J. Kay, Identification of the acid proteinase in human seminal fluid as a gastricsin originating in the prostate. *Cell Tissue Res* **236**:597 (1984).
17. J.H. Reese, J.E. McNeal, E.A. Redwine, I.M. Samloff and T.A. Stamey, Differential distribution of pepsinogen II between the zones of the human prostate and the seminal vesicle. *J Urol* **136**:1148 (1986).
18. M. Hirsch-Marie and D. Delafontaine, Seminal pepsinogen: Quantitation and correlation with the spermiogram. *Int J Androl* **1**:397 (1978).
19. M. Seiffers, H. Segal and L. Miller, Preliminary appraisal of the role of seminal pepsinogen I in human sterility. *Fertil Steril* **16**,202 (1965).

EFFECTS OF HYDROCORTISONE ON THE PEPSINOGEN-PRODUCING CELLS IN RAT STOMACH MUCOSA

Masao Ichinose,[1] Shinko Tsukada,[1] Kazumasa Miki,[1] Nobuyuki Kakei,[1] Masashi Matsushima,[1] Naohisa Yahagi,[1] Satoshi Ishihama,[1] Yasuhito Shimizu,[1] Kiyoshi Kurokawa,[1] Hiroshi Fukamachi,[2] Satoshi Yonezawa,[3] Senarath B. P. Athauda,[5] Takashi Kageyama,[4] and Kenji Takahashi[5]

[1]First Department of Internal Medicine
Faculty of Medicine
University of Tokyo
Tokyo 113
[2]Laboratory of Molecular Embryology
Zoological Institute
Faculty of Science
University of Tokyo
Tokyo 113
[3]Department of Embryology
Institute for Developmental Research
Aichi Prefectural Colony, Kasugai 480-03
[4]Department of Molecular and Cellular Biology
Primate Research Institute
Kyoto University
Inuyama 484, Japan
[5]Department of Biophysics and Biochemistry
Faculty of Science
University of Tokyo
Tokyo 113

INTRODUCTION

Pepsinogen is a marker of the terminal differentiation of stomach mucosa. At present, controlling mechanisms of differentiation in stomach mucosa is not fully understood. Previous studies demonstrated that administration of hydrocortisone to developing rats induces a precocious increase in the mucosal pepsinogen level in the stomach (1-3), indicating that glucocorticoids are somehow involved in the differentiation of the stomach

Aspartic Proteinases: Structure, Function, Biology, and Biomedical Implications
Edited by Kenji Takahashi, Plenum Press, New York, 1995

mucosa. However, the physiological significance of glucocorticoids in the regulation of pepsinogen gene expression is not well understood. In addition, the effects of glucocorticoids on pepsinogen-producing cells in fully-differentiated stomach mucosa are less clear (4,5). In this study, we examined the effects of hydrocortisone on infant and adult rat stomach mucosa, especially on pepsinogen gene expression and the morphology of pepsinogen-producing cells.

MATERIALS AND METHODS

Animals and Hydrocortisone Treatment

Wistar strain rats were used in all experiments. Young rats were nursed by their mothers in separate cages at constant temperature. Infant rats were given subcutaneous injections of hydrocortisone 21-sodium succinate (6.25-100 mg/kg body weight; Upjohn Co.) every 6 hours from day 7 to 9 after birth. Control animals received injections of physiological saline. For studies, all infant rats were anesthetized with ethyl ether and the stomach was quickly removed. Small pieces of each stomach were used for histological analysis. The mucosa of the remaining stomach was scraped off the muscle layer with a thin piece of glass and stored in liquid nitrogen until use.

Adrenalectomy and Hydrocortisone Replacement

Young adult Wistar-strain male rats (8 weeks old) were anesthetized with ethyl ether and were either adrenalectomized or sham-operated by the dorsal approach. The adrenalectomized rats received subcutaneous injection of either physiological saline or various doses of hydrocortisone 21-sodium succinate every 8 hours (12.5-100 mg/kg body weight) after the operation for the analysis of hydrocortisone replacement. Rats were sacrificed periodically after the operation and their stomachs were analyzed biochemically and histologically.

Isolation and Analysis of RNA from the Stomach Mucosa

RNA was isolated from the stomachs by the guanidium/cesium chloride method (6). A sample of 10 μg of RNA was denatured and subjected to electrophoresis on agarose formalin gel by the method of Goldberg (7). Then, the RNA was transferred to a nitrocellulose filter, baked and hybridized as described (8). The *Pvu*II-*Xba*I fragment of pRPC1, which contains nearly full length cDNA coding for rat pepsinogen (9) was labeled with [α-^{32}P] dCTP by nick-translation (10) and used as a hybridization probe.

Pepsinogen Level

Rat stomach mucosa was minced, homogenized and was centrifuged at 105,000 *g* as described (1-3) The supernatant was used for assays of the potential peptic activity of pepsinogens by the method of Anson with a slight modification (1-3). Protein was measured by the method of Lowry et al. (11).

Histological Examination

For histological examinations, tissues were fixed in 10% phosphate-buffered formalin and embedded in paraffin. The avidin-biotin-peroxidase complex (ABC) method (12)

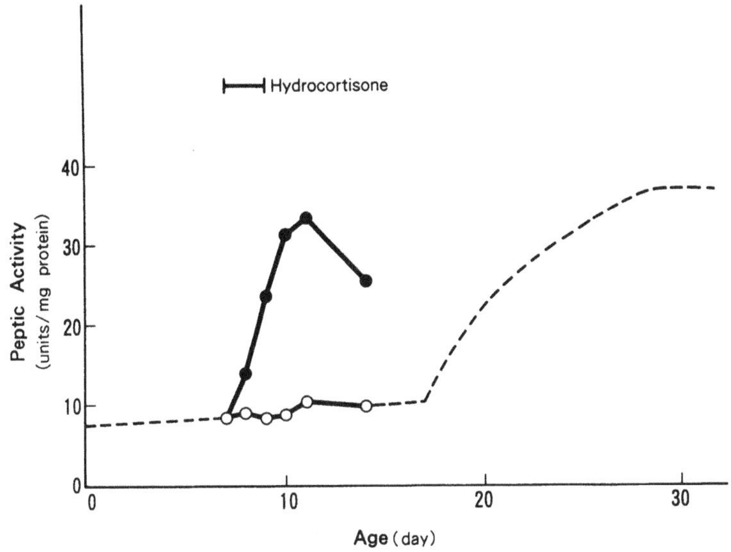

Figure 1. Premature induction of pepsinogen in infant rat stomach mucosa by hydrocortisone treatment. Hydrocortisone (50 mg/kg) was injected subcutaneously into the back of infant rats every 6 hours from day 7 to 9 after birth. Values for enzyme activity are averages for 5 rats. The dashed line indicates the activities during normal development. •, hydrocortisone treatment; O, control.

with the rabbit anti-rat pepsinogen antibody described previously (13) was used to detect pepsinogen-producing cells in the rat stomach mucosa. The mucosal cells containing class III mucin were detected by paradoxical concanavalin A (Con-A) staining (14). The replicating cells were labelled by injecting BrdU intraperitoneally and incorporated BrdU was immunohistochemically detected.

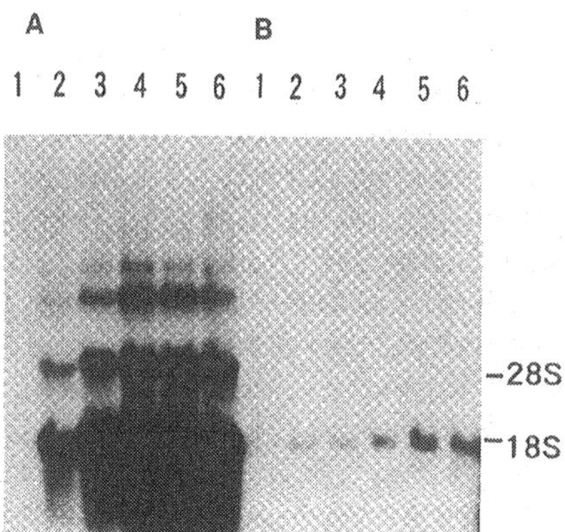

Figure 2. Pepsinogen mRNA levels in stomach mucosa of hydrocortisone-treated infant rats. Total RNA (10 μg/lane) from hydrocortisone-treated (A) and control (B) stomach mucosa of infant rats was analyzed by Northern blot hybridization with rat pepsinogen cDNA probe. Lanes 1-6 show results on days 7, 8, 9, 10, 11 and 14, respectively, after birth.

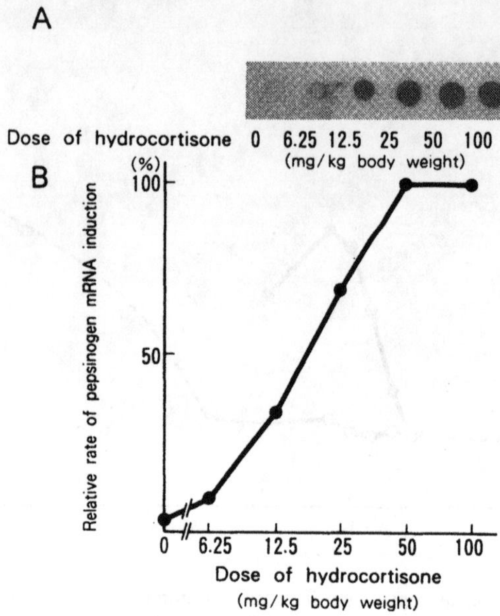

Figure 3. Effect of the dose of hydrocortisone on pepsinogen mRNA expression in infant rat stomach mucosa. Infant rats (7 days old) were given various doses of hydrocortisone subcutaneously every 6 hours. The rats were sacrificed 36 hours later and total RNA from the stomach mucosa was isolated and analyzed by dot blot hybridization with ^{32}P-labeled pepsinogen cDNA probe (A). Dots were photographed and scanned with a densitometer. Results were normalized to the optical density of RNA from rats treated with 100 mg/kg body weight of hydrocortisone every 6 hours (B).

RESULTS AND DISCUSSION

In the newborn rat stomach, mucosal pepsinogen level is less than 20% of that in adult rat stomach. The mucosal pepsinogen level is relatively constant for the first 17 days after birth and then increases rapidly, reaching the adult level about 28 days after birth (1-3). As shown in Figure 1, administration of hydrocortisone to rats 2 to 9 days after birth caused a precocious increase in the mucosal pepsinogen level. This increase was associated with change from the immature to the mature pattern of pepsinogen isozymogens and morphological maturation of the stomach mucosa as described previously (1-3). To determine if the mucosal pepsinogen mRNA level is increased by the hydrocortisone treatment, we analyzed RNAs from control and hydrocortisone-treated rat stomach mucosa by Northern blot hybridization with rat pepsinogen cDNA as a probe. The stomach mucosa from 7-day-old control rats gave a weak signal of pepsinogen mRNA of about 1.8 kb and the mRNA level showed a slight but steady increase between day 7 and 14 (Fig. 2B). In contrast, when hydrocortisone was injected every 6 hours from day 7 to 9 after birth, increase in pepsinogen mRNA expression was noted on day 8 and reached a maximal level on day 11 (Fig. 2A). The increase in the mRNA level after hydrocortisone injection showed a lag of about 6 hours. Northern blot analysis revealed that hydrocortisone stimulated the expressions of various sized RNA transcripts together with pepsinogen mRNA, suggesting the induction of pepsinogen mRNA precursors and the transcriptional regulation of pepsinogen genes by hydrocortisone. This effect of hydrocortisone on pepsinogen mRNA induction was dose-dependent and was maximal on injection of about 50 mg/kg body weight every 6 hours (Fig. 3). With this treatment, the induced levels of mucosal pepsinogen and its mRNA on day 11 were almost the same as those of adult rat stomach mucosa.

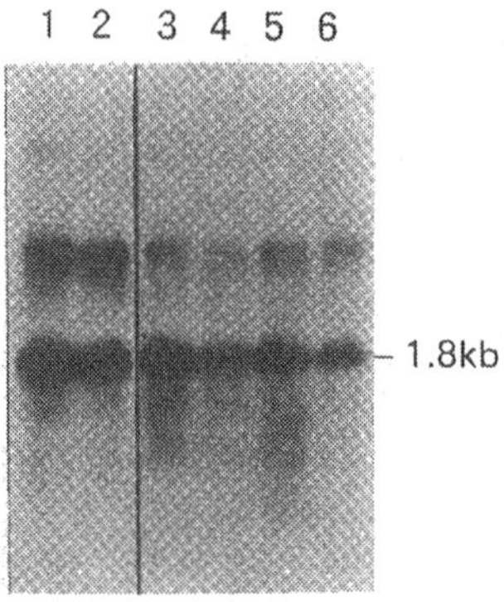

Figure 4. Effect of adrenalectomy on pepsinogen mRNA expression in adult rat stomach mucosa. The rats were sacrificed one week after the operation. A sample of 10 μg of RNA was sujected to electrophoresis, transferred to a nylon membrane and was analyzed by Northern blot hybridization with ^{32}P-labeled pepsinogen cDNA probe. Lanes 1-3 correspond to the sham-operated control and lanes 4-6 to adrenalectomized rats (1 week after the operation).

To investigate the effects of adrenal glucocorticoids on the function of pepsinogen-producing cells in adult rat stomach mucosa, rats were adrenalectomized to reduce endogenous glucocorticoids levels. In the operated animals, serum corticosteron level was significantly low as compared with the sham-operated control, consistent with successful operation. In adrenalectomized rats, there was a slight, but significant decrease in the mucosal pepsinogen content [control rats, 0.57 U/mg protein; adrenalectomized rats, 0.38 U/mg protein, 1 week after the operation ($P < 0.05$)]. In good accordance with the reduction in the mucosal pepsinogen content, mucosal pepsinogen mRNA level was also reduced (Fig. 4).

The adrenalectomy-induced reduction in the mRNA level was relatively slow and a maximal reduction was observed after 48 hours. Adrenalectomy did not have any significant influence on the numbers of pepsinogen-producing cells one week after the operation.

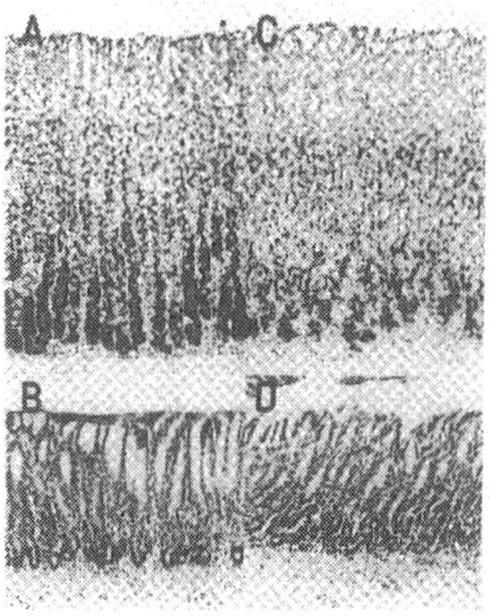

Figure 5. Effect of adrenalectomy on pepsinogen-producing cells in adult rat stomach mucosa. Stomach mucosa from the sham-operated control (A and B) and adrenalectomized rat (C and D) was immunochemically stained with anti-rat pepsinogen antibodies by the avidin-biotin peroxidase complex (ABC) method. Cells staining with anti-rat pepsinogen antibodies are seen. A and C: fundic gland mucosa, ×200. B and D: pyloric gland mucosa, ×200.

However, immunohistochemical analysis using anti-pepsinogen antibodies revealed that pepsinogen content in the producing cells was decreased throughout the stomach mucosa and the reduction was especially remarkable in the immature cells located in the glandular neck of the fundic mucosa (Fig. 5). The adrenalectomy did not have any significant influence on the overall morphology of the stomach mucosa one week after the operation. However, the number of the mitotic cells labelled by BrdU was increased. In addition, expression of both cathepsin E in the surface mucous cells and class III mucin in the mucous cells located in the glandular neck was also reduced, indicating that these cells are not fully differentiated. The basic character of the adrenalectomy-induced changes was the same even after one month; however, the number of the pepsinogen-producing cells tended to be slightly decreased. These changes were reversed by hydrocortisone replacement within 48 hours and in adrenalectomized rats, hydrocortisone dose-dependently increased pepsinogen gene expression with the maximal effect attained at a dose of 50 mg/kg.

The results of the present study clearly indicate that glucocorticoids are deeply involved in the regulation of the function of pepsinogen-producing cells in both developing and fully-differentiated stomach mucosa. The mechanisms whereby glucocorticoids exert their effects on pepsinogen gene is unknown. The observed effects of glucocorticoids are not selective for the pepsinogen-producing cells. Instead, they have global effects on the proliferation and differentiation of the mucosal cells in the stomach. Thus, it is probable that glucocorticoids modulate the expression of pepsinogen through the regulation of the differentiation program of the pepsinogen-producing cells.

REFERENCES

1. C.Furihata, T.Kawachi, and T.Sugimura, Premature induction of pepsinogen in developing rat gastric mucosa by hormones, Biochem. Biophys. Res. Commun. 47: 705 (1972).
2. C.Furihata, Y.Iwasaki, T.Sugimura, M.Tatematsu, and M.Takahashi, Differentiation of pepsinogen-producing cells in the fundic and pyloric mucosa of developing rats,Cell Differentiation 2:179 (1973).
3. M.Tatematsu. M.Takahashi, H.Tsuda, M.Hirose, C.Furihata, and T.Sugimura, Precocious differentiation of immature chief cells in fundic mucosa of infant rats induced by hydrocortisone, Cell Differentiation 4:285 (1975).
4. A.R.Cooke, Role of adrenocortical steroids in the regulation of gastric secretion, Gastroenterol. 52:272 (1976).
5. I.M.Samloff. Pepsinogens, pepsin and pepsin inhibitors. Gastroenterol. 60:586 (1971).
6. A.Ullrich, J.Shine, J.Chirgwin, R.Pictet, E.Tischer, W.J.Rutter, and H.M.Goodman, Rat insulin genes: construction of plasmids containing the coding sequence, Science 196:1313 (1977).
7. D.A.Goldberg, Isolation and partial characterization of the Drosophila alcohol dehydrogenase gene, Proc. Natl. Acad. Sci. U.S.A. 77:5794 (1980).
8. M.Ichinose, K.Miki, C.Furihata, M.Tatematsu, Y.Ichihara, T.Ishihara, I.Katsura, K.Sogawa, Y.Fujii-Kuriyama. M.Tanji, H.Oka, T.Matsushima, and K.Takahashi, DNA methylation and expression of the rat pepsinogen gene in embryonic, adult and neoplastic tissues, Cancer Res. 48:1603-1609 (1988).
9. Y.Ichihara, K.Sogawa, K.Morohashi, Y.Fujii-Kuriyama and K.Takahashi, Nucleotide sequence of a nearly full-length cDNA coding for pepsinogen of rat gastric mucosa,Eur. J. Biochem. 161:7 (1986).
10. P.W.Rigby, M.Dieckmann, C.Rhodes, and D.Berg, Labeling deoxyribonucleic acid to high specific activity in vitro by nick translation with DNA polymerase,J.Mol. Biol. 113:237 (1977).
11. O.H.Lowry, N.J.Rosebrough, A.L.Farr, and R.J.Randall, Protein measurement with the Folin phenol reagent, J. Biol. Chem. 193:265 (1951).
12. S.M.Hsu, L.Raine, and H.Fanger, Use of avidin-biotin-peroxidase complex (ABC) in immunoperoxidase techniques: a comparison between ABC and unlabeled antibody (PAP) procedures. J. Histochem. Cytochem. 29:577 (1981).
13. C.Furihata, D.Saito, H.Fujiki, Y.Kanai, T.Matsushima, and T.Sugimura, Purification and characterization of pepsinogens and a unique pepsn from rat stomach, Eur. J. Biochem. 105:43 (1980).

14. T.Katsuyama, and S.S.Spicer, Histochemical differentiation of complex carbohydrates with variants of the concanavalin A-horseradish peroxidase method, J. Histochem. Cytochem. 26:233 (1978).

EFFECTS OF OMEPRAZOLE, A PROTON PUMP INHIBITOR, ON PEPSINOGEN-PRODUCING CELLS, WITH SPECIAL REFERENCE TO NEONATAL DEVELOPMENT

Nobuyuki Kakei,[1] Masao Ichinose,[1] Shinko Tsukada,[1] Masae Tatematsu,[2] Noriaki Tezuka,[2] Naohisa Yahagi,[1] Masashi Matsushima,[1] Satoshi Ishihama,[1] Yasuhito Shimizu,[1] Masahiro Kido,[1] Kazumasa Miki,[1] Kiyoshi Kurokawa,[1] Kenji Takahashi,[3] and Hiroshi Fukamachi[4]

[1]First Department of Internal Medicine
Faculty of Medicine
University of Tokyo, Tokyo 113
Japan
[2]Laboratory of Pathology
Aichi Cancer Center Research Institute
Nagoya 464, Japan
[3]Department of Biophysics and Biochemistry
Faculty of Science
University of Tokyo, Tokyo 113
Japan
[4]Zoological Institute
Faculty of Science
University of Tokyo, Tokyo 113
Japan

INTRODUCTION

Omeprazole, a proton pump inhibitor, exerts its effects by binding covalently to the H^+/K^+-ATPase that is located in the apical membrane of gastric parietal cells. However, its effects on pepsinogen-producing cells have not been investigated in detail and results of previous relevant studies on pepsinogen secretion are conflicting. Gastric mucosal cells proliferate and differentiate under an acidic environment in the gastric lumen. Thus, it is probable that omeprazole-induced reduced acidity in the gastric lumen has some influences on the proliferation and differentiation of the mucosal cells. As the first step to elucidate the effects of omeprazole on pepsinogen-producing cells, we analyzed the processes from the

Aspartic Proteinases: Structure, Function, Biology, and Biomedical Implications
Edited by Kenji Takahashi, Plenum Press, New York, 1995

gene expression to the secretion of pepsinogen *in vivo* in adult rat glandular stomach. Next, we investigated the effects of omeprazole on the development of rat gastric mucosa biochemically and histologically, with special reference to pepsinogen expression as a marker of terminal differentiation.

MATERIALS AND METHODS

Animals and Treatment with Omeprazole

Male Wistar 8-week-old rats weighing 250-280 g were given subcutaneous injections of homogenized omeprazole (courtesy of Fujisawa-Astra, Osaka, Japan) in a 0.5% solution of sodium carboxymethyl cellulose (pH 7.6) at a dose of 0.1, 2 or 12.5 mg/kg body weight every 24 hours for five days. Control animals received vehicle alone. At the end of the treatment, some rats were fasted for 18 hours and were subjected to pyloric ligation for gastric analysis. Other rats were killed after anesthesia with ethyl ether and their stomachs were removed for biochemical analysis. To investigate the sequential changes induced by omeprazole, rats were given subcutaneous injections of omeprazole at a dose of 12.5 mg/kg body weight and were then killed 6, 12, 18, 24, 48, 72 and 96 hours after the first injection of omeprazole for gastric analysis and biochemical analysis.

Next, Wistar-strain female rats were mated and the birthday of pups was set on day 0. Pups were divided into four groups. The first group received subcutaneous injections of homogenized omeprazole in 0.5 % sodium carboxymethyl cellulose solution (pH 7.6) (25 mg/kg body weight once a day) from days 1 to 28 (1). The second group received ranitidine (30 mg/kg body weight twice a day) administrated subcutaneously. The third group were treated with subcutaneous injections of recombinant rat gastrin (0.8 mg/kg body weight twice a day, Sigma, St. Louis, USA). The control group were injected with vehicle alone. On days 5, 10, 13, 15, 17, 19, 21, 24 and 28, rats were killed for biochemical and morphological analyses.

Analysis of Gastric Secretion

After 18 hours of fasting, adult rats were subjected to pylorus ligation for gastric analysis. Two hours later, the stomach was removed, and the gastric contents were collected and centrifuged. The volume of each sample was measured, the pH was recorded, and the acid content was titrated to pH 7.0 with 0.1 N NaOH. Potential activity of pepsinogen in gastric juice was measured with hemoglobin as substrate.

Northern Blot Hybridization

RNA was extracted from the stomachs by the guanidium isothiocyanate method and purified by CsCl density gradient ultracentrifugation and was transferred to a nylon membrane and subjected to hybridization analysis as described elsewhere (2). The *Pvu*II-*Xba*I fragment of pRPC1, which contains nearly full-length cDNA coding for rat pepsinogen was used as the probe for hybridization. Pepsinogen mRNA levels were determined by densitometric scanning of autoradiograms.

Assay of Pepsinogen

Rat gastric mucosa was minced, homogenized and centrifuged as described previously (2). The supernatant was used for assays of the potential activity of pepsinogen, with hemoglobin as substrate (2).

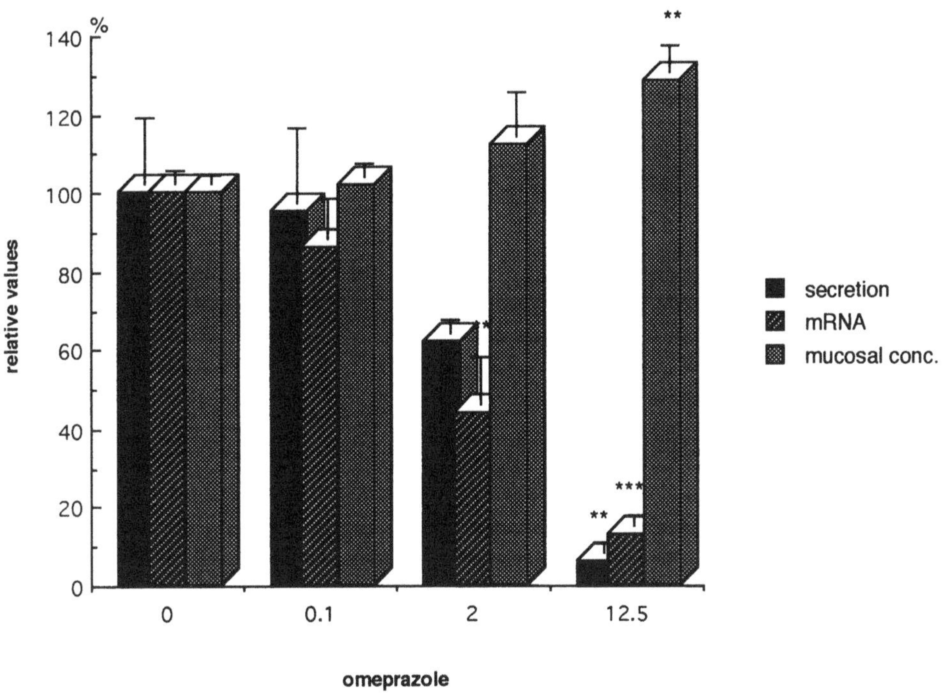

Figure 1. Effects of omeprazole at a dose of 0.1, 2 or 12.5 mg/kg body weight on secretion, mRNA levels and mucosal concentrations of pepsinogen. Values are means ±SEM. **p<0.01 and ***p<0.001 compared with the control levels.

Histological Studies

For histological examinations, the removed stomachs were fixed in pure acetone and embedded in paraffin. The avidin-biotin-peroxidase complex (ABC) method with anti-pepsinogen antibodies and anti-cathepsin E antibodies was used to detect pepsinogen-producing cells and surface mucous cells, respectively. Paradoxical concanavalin A (Con-A) staining was also used to detect mucosal cells containing class III mucin.

Studies on Cell Proliferation

For studies on mucosal cell proliferation, three infant rats on each day were intraperitoneally injected with 5-bromo-2'-deoxyuridine (BrdU, Sigma, St. Louis, USA, 100 mg/kg body weight) 1 hour before sacrifice. Nuclei in the cells that incorporated BrdU were immunohistochemically detected.

RESULTS AND DISCUSSION

Omeprazole inhibited secretion and gene expression of pepsinogen (Fig. 1) as well as acid secretion in a dose-dependent manner. The reduction in the secretion was rapid and was followed by a decrease in pepsinogen mRNA levels (Fig. 2). Concurrently,

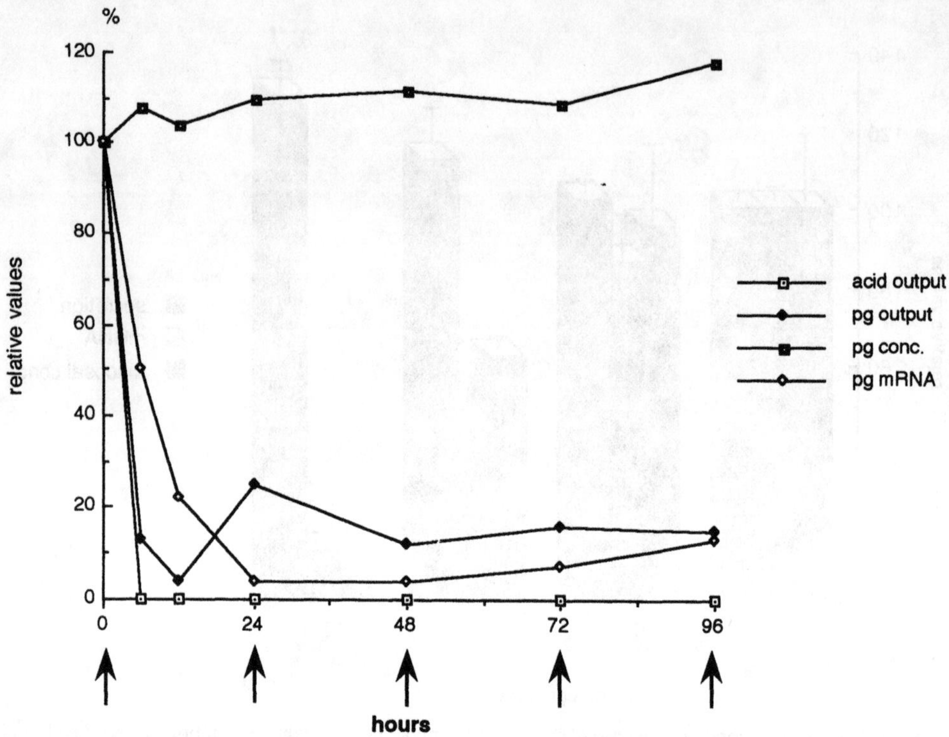

Figure 2. Sequential effects of omeprazole (12.5 mg/body weight) on acid secretion, pepsinogen secretion, mucosal pepsinogen concentrations and pepsinogen mRNA levels. Time zero represents the first injection of omeprazole. Arrows represent injections of omeprazole. The results are shown relative to values at time zero.

the mucosal pepsinogen concentrations increased dose-dependently with time (Fig. 1 and 2). Immunohiustochemical study revealed the relative proportions of cells that could be stained with the pepsinogen-specific antibodies were almost unchanged after treatment with omeprazole. However, these cells in both fundic and pyloric mucosa were more intensely stained with the antibodies in omeprazole-treated rats than in the controls (Fig. 3), suggesting that pepsinogen secretion was more severely inhibited by omeprazole than was its synthesis. Consistent with this, ultrastructural study revealed the pepsinogen granules were irregular in shape and appeared to become more fragile upon omeprazole treatment, suggestive of the long term storage of pepsinogen in these granules within the cells (Fig. 4). It has been proposed that pepsinogen secretion exerts a regulatory influence on its synthesis through a feedback mechanism that involves changes in the size of intracellular stores of pepsinogen (3). However, it is unclear whether the secretion and synthesis of pepsinogen are coupled. In addition, the effect of altered rates of secretion on expression of the gene for pepsinogen has not been studied. In our experiment, inhibition of pepsinogen secretion was followed by a reduction in the mRNA levels. Moreover, the relationship between the dose of omeprazole and its effect on pepsinogen secretion corresponded closely to that for the gene expression. These data support the model for control of the synthesis of pepsinogen and suggest that pepsinogen secretion and the gene expression are regulated

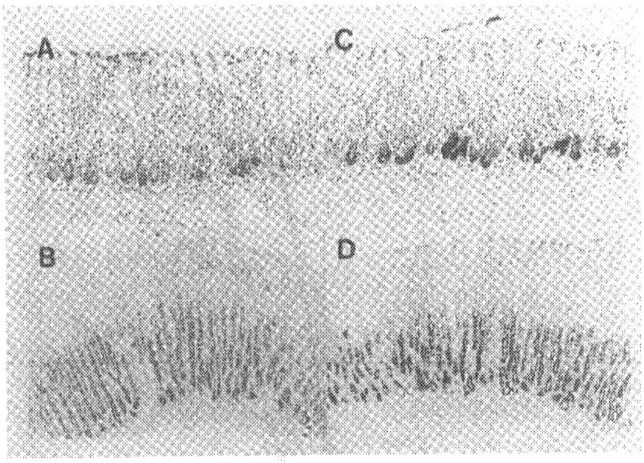

Figure 3. Immunohistochemical staining of control (A and B) and omeprazole-treated (C and D) rat pyloric glands (A and C) and fundic glands (B and D) with pepsinogen-specific antibodies (avidin-biotin peroxidase complex method).

coordinately. The mechanism by which omeprazole inhibits pepsinogen secretion and the gene expression is unclear and requires further study. Although potent, the inhibitory effect of this drug on the secretion, synthesis and gene expression of pepsinogen was incomplete. In addition, immunohistochemical analysis revealed that some pepsinogen-producing cells did not respond to omeprazole treatment, while most cells responded quickly (Fig. 3). Thus, the observed effects of omeprazole might not be the results of direct interaction of the drug with pepsinogen-producing cells. Luminal application of hydrocholic acid has been repeatedly reported to stimulate pepsinogen secretion *in vivo* in stomach (4). Thus, it is possible that the inhibited pepsinogen secretion by omeprazole is the results of the reduced acidity in the gastric lumen.

Next, we have demonstrated that administration of omeprazole to infant rats caused an elevation of mucosal pH and suppressed an increase in mucosal pepsinogen and its mRNA levels during stomach development (Fig. 5). The suppression was especially remarkable from days 15 to 21. Reflecting the reduced acidity in the gastric lumen, a marked increase in the serum gastrin levels was observed in omeprazole-treated rats; they

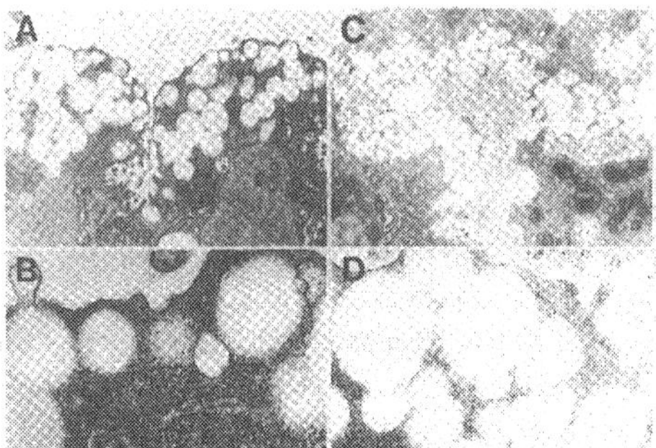

Figure 4. Electron micrographs of chief cells of control (A and B) and omeprazole-treated (C and D) rat. B and D are higher magnifications of a part of A and C, respectively.

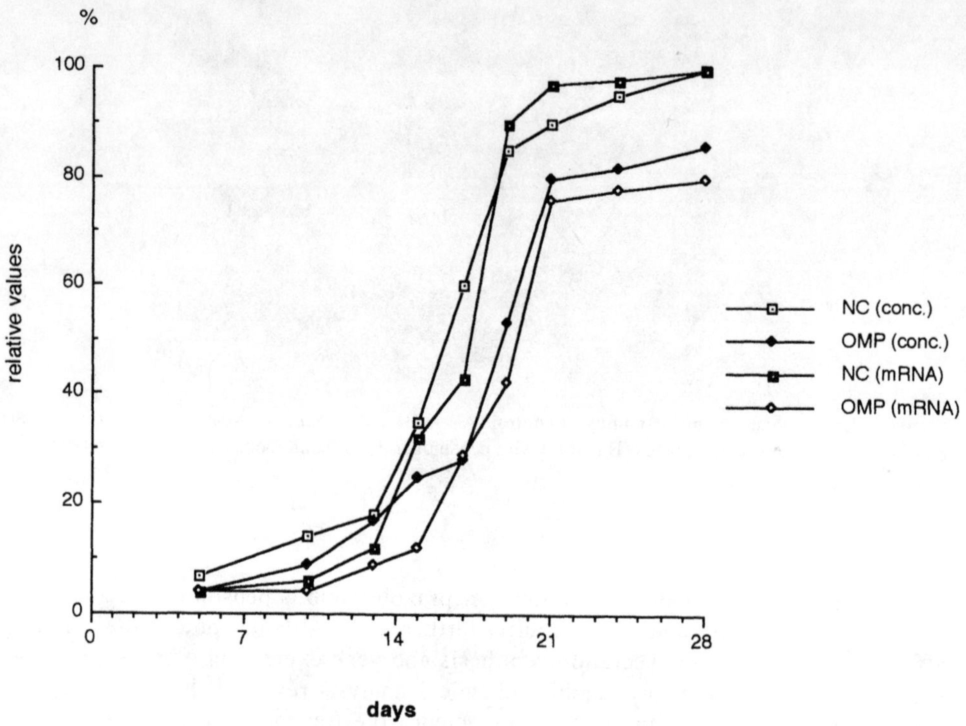

Figure 5. Mucosal pepsinogen and its mRNA levels during development in control and omeprazole-treated rats. The values are shown relative to the values of 28-day-old control rats.

gradually increased after birth and were 2-3 fold higher compared with those in control rats (Fig. 6).

Omeprazole treatment increased the number of BrdU-labeled cells per a pit in both fundic and pyloric glands, indicating an increase in the mucosal cell proliferation (data not shown). This led to a gradual increase in the mucosal thickness of the treated animals throughout the experimental period, resulting in 10-20 % increase as compared with the controls on day 28. Immunohistochemistry with the specific antibodies for rat pepsinogen revealed various pepsinogen-producing cells throughout gastric mucosa. In fundic glands, treatment with omeprazole markedly reduced the number of mature chief cells strongly immunoreactive for pepsinogen (Fig. 7). Instead, the number of cells weakly immunoreactive for pepsinogen was increased. Paradoxical Con-A staining revealed that these cells were positive for class III mucin that mature chief cells do not contain in their cytoplasm (Fig. 7). Thus, the cells could be identified as immature chief cells that synthesize both class III mucin and small amount of pepsinogen. In control rats, pyloric gland cells contained both class III mucin and pepsinogen, and the amount of them appeared to increase with the development judging from the intensities of staining. Though omeprazole treatment did not alter the proportion of these cells, it reduced the intensities of staining for both, indicating a retardation in maturation. Immunohistochemistry with anti-rat cathepsin E antibodies revealed that omeprazole also reduced the expression of cathepsin E in surface mucous cells, indicating a delayed maturation of the cells (Fig.7). The mechanisms of action of ome-

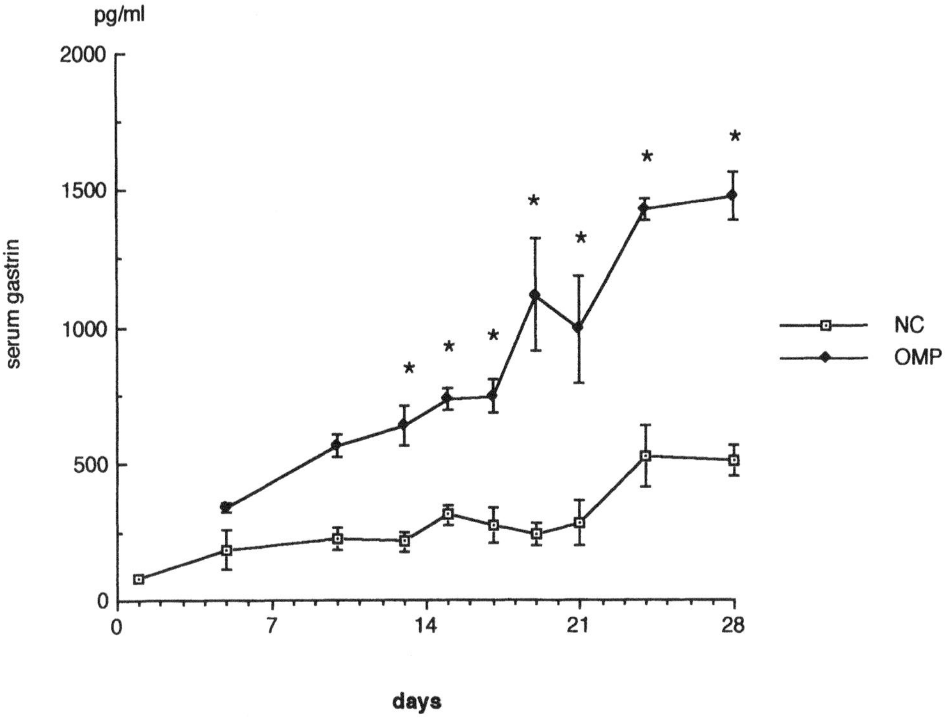

Figure 6. Effect of omeprazole on serum gastrin levels during development. Values are means±SEM. *$p<0.05$ compared with the control levels.

prazole remain to be elucidated. Previous studies have demonstrated that H^+/K^+-ATPase activities can be detected in 18-day-old fetal rat stomach and acid secretion begins on day 20 of gestation (5). Acid secretion increases slowly after birth and significant development of acid secretion is observed from days 15 to 21 after birth when weaning takes place (6). In our experiment the effects of omeprazole are most evident during this period. In addition, essentially the same effects were obtained by administration of ranitidine (data not shown). Thus, it appears that the observed effects are not caused by a direct action of omeprazole. It is probable that omeprazole-induced achlorhydria plays an important role in delaying maturation and differentiation of gastric mucosa. Achlorhydria induces an elevation of the serum gastrin levels, leading to an observed increase in the mucosal thickness. Thus, hypergastrinemia could be involved in the observed effects of omeprazole. In the present study, however, exogenous administration of rat gastrin did not induce notable changes on gastric mucosal cells (data not shown). In addition, gastric mucosa has been reported to have no gastrin receptor detected before postnatal days 20 (7). Thus, luminal acid or intraglandular acid itself might be one of the factors deeply involved in normal differentiation of gastric mucosa.

In conclusion, omeprazole reduced the activity of pepsinogen-producing cells *in vivo* in adult rat stomach and it altered the developmental program of infant rat gastric mucosa probably by reducing the gastric luminal acidity. Elucidation of the roles of luminal acid in maintaining the normal function of pepsinogen-producing cells and in the development of gastric mucosa is a problem for future study.

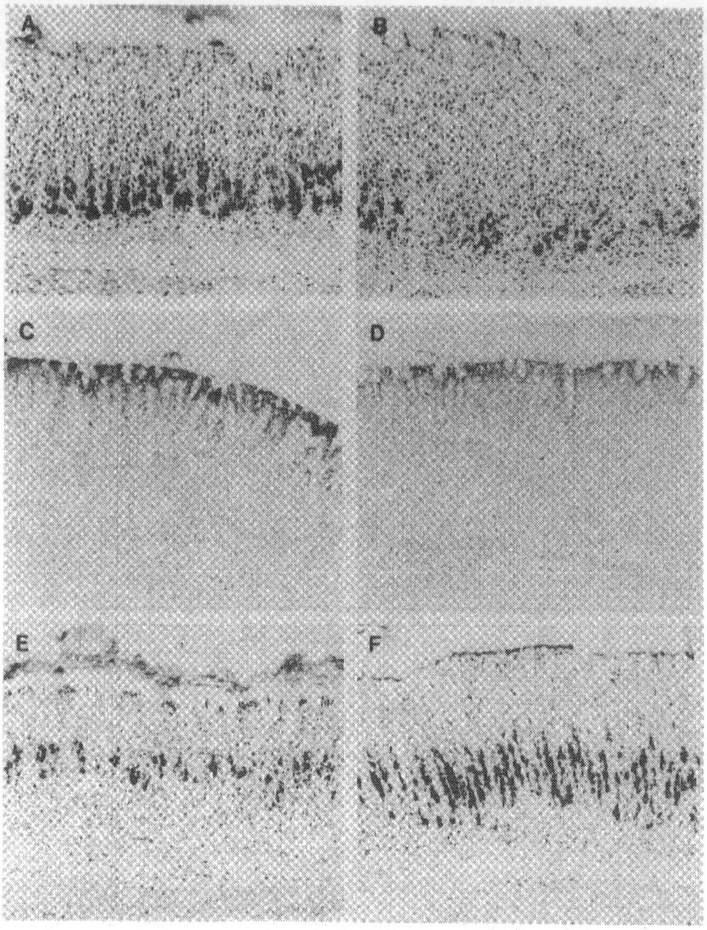

Figure 7. Gastric mucosal morphology in control (A, C and E) and omeprazole-treated (B, D and F) 19-day-old rats. A and B; immunohistochemical staining with pepsinogen-specific antibodies. C and D: immunohistochemical staining with cathepsin E-specific antibodies. E and F: paradoxical Con-A staining.

REFERENCES

1. N. Kakei, M. Ichinose, S. Tsukada, M. Tatematsu, N. Tezuka, C. Furihata, N. Yahagi, M. Matsushima, K. Miki, K. Kurokawa, K. Takahashi, S. Yonezawa and H. Fukamachi, Effects of Omeprazole on Gastric Mucosal Growth and Differentiation in Developing Rat, *Biochem. Biophys. Res. Commun.* 197:310 (1993).
2. N. Kakei, M. Ichinose, S. Tsukada, M. Tatematsu, N. Tezuka, N. Yahagi, M. Matsushima, K. Miki, K. Kurokawa, K. Takahashi and H. Fukamachi, Omeprazole, a Proton Pump Inhibitor, Reduces the Secretion, Synthesis and Gene Expression of Pepsinogen in the Rat Stomach, *Biochem. Biophys. Res. Commun.* 195:997 (1993).
3. B.J. Hirschowitz, Secretion of Pepsinogen. Handbook of Physiology, Alimentary Canal. Washington DC., American Physiology Society (C.F. Code ed.) 889 (1967).
4. L.R. Johnson, Effect of Gastric Mucosal Acidification on the Action of Pepsinogogues, *Am. J. Physiol.* 225:1411 (1973).
5. F. Hervatin, F. Benkouka, J.C. Robert, G. Pranzi and A. Soumarmon, The Ontogeny of Rat Gastric H^+/K^+-ATPase, *Biochim. Biophys. Acta* 985:320 (1989).

6. B. Garzon, R. Ducroc, J.P. Onolfo, J.F. Desjeux and J.P. Geloso, Biphasic Development of Peptagastrin Sensitivity in Rat Stomach, *Am. J. Physiol.* 242:G111 (1982).
7. K. Takeuchi, W. Peitsch and L.R. Johnson, Mucosal Gastrin Receptor. V. Development in Newborn Rats, *Am. J. Physiol.* 240:G163 (1981).

TRANSCRIPTION OF EMBRYONIC CHICK PEPSINOGEN GENE IS AFFECTED BY MESENCHYMAL SIGNALS THROUGH ITS 5'-FLANKING REGION

Kimiko Fukuda,[1,2] Hidetoshi Saiga,[2] and Sadao Yasugi[2]

[1]Zoological Institute
Faculty of Science
University of Tokyo
Hongo, Tokyo 113, Japan
[2]Department of Biology
Faculty of Science
Tokyo Metropolitan University
Minamiohsawa, Hachiohji, Tokyo 192-03, Japan

INTRODUCTION

It is now well established that epithelial-mesenchymal interactions are essential in the spatio-temporally regulated differentiation processes in organogenesis in higher vertebrates [1, 2]. The morphological and the functional differentiation of an organ necessarily involves the local expression or suppression of sets of genes in the cells constituting the organ.

Expression of embryo-specific chick pepsinogen (ECPg) gene which is specifically transcribed in proventricular (PV, glandular stomach) gland epithelial cells in embryonic period [3] has been most extensively studied as an example of gene expression regulated by the epithelial-mesenchymal interactions [4]. Its expression is induced under the influence of the PV mesenchyme in PV and gizzard (GZ, muscular stomach) epithelial cells and is inhibited by the GZ mesenchyme [5, 6, 7]. The ECPg gene expression was shown to be regulated by the mesenchyme at the transcriptional level [8, 9]. We have no information however, as to the transcriptional regulation of ECPg gene by mesenchymal signals.

In the present experiment, we demonstrated that mesenchymal signals affect transcription of ECPg gene through its 5'-regulatory region.

MATERIAL AND METHODS

Endodermal epithelial fragments and mesenchymal fragments were isolated from the PV and the GZ of 6 day chick embryos (stage 28-29 of Hamburger and Hamilton, [10]) with

Aspartic Proteinases: Structure, Function, Biology, and Biomedical Implications
Edited by Kenji Takahashi, Plenum Press, New York, 1995

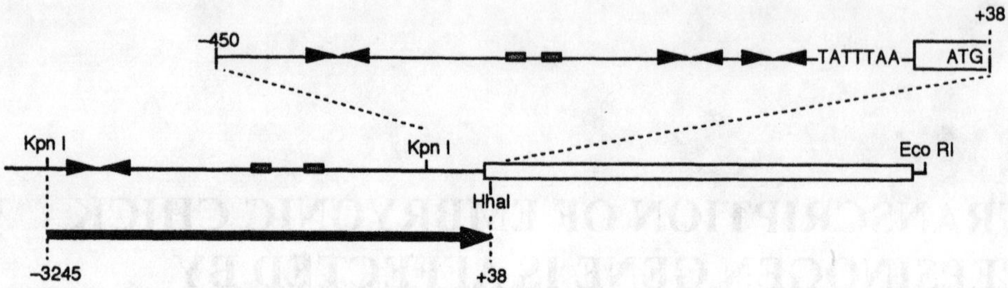

Figure 1. The restriction map of the ECPg gene and its 5'-flanking region. Horizontal bar in the middle indicates the isolated DNA [8]. The cleavage sites of *Eco*R I, *Kpn* I, *Pst* I and *Hha* I are shown. Upper horizontal bar indicates 5'-flanking region of the ECPg gene. Open box represents the region of ECPg gene. Hatched arrow (bases -3245 to +38) shows the subcloned fragment linked to luciferase reporter gene (see Materials and Methods). Face-to face arrowheads indicate inverted repeats, and shadowed boxes indicate the consensus viral enhancer core sequences in positive or negative orientation (see Discussion).

using 0.03% collagenase in Tyrode's solution for 1 hour at 37°C. The isolated epithelial and mesenchymal fragments were then treated with trypsin (0.5%) in PBS for 5 minutes at 25°C and were dissociated by gentle passing through micropipet chip. By this procedure, mesenchymal fragments were completely dissociated into single cells and epithelial fragments became single or very small clusters of cells.

Plasmid DNA was introduced into epithelial cells of PV and GZ by lipofection for 6 hours. Transfected epithelial cells were mixed with mesenchymal cells, centrifuged, and the resultant cell aggregates were cultured on porous membranes (8μm pore size) , rested on a stainless-steel grid placed within a small culture dish. The culture dishes were incubated for 6 days at 37°C in 95% air and 5% CO_2 (cell aggregate culture). The culture medium was as described previously [7].

Vector DNA used for construction of luciferase vectors was PGV-C or PGV-B that contains the luciferase gene with or without simian virus-40 (SV40) promoter-enhancer (Toyo Inki), respectively. The plasmid PGV-3.3 was constructed by inserting 3245 bp 5'-flanking region followed by the transcription start site and 38 bp of first exon of the ECPg gene into *Sac*I-*Xho*I sites of PGV-B (Fig. 1).

RESULTS

During culture of cell aggregates epithelial cells were sorted out from mesenchymal cells, so that both epithelial cells and mesenchymal cells were reorganized. When PV and GZ epithelial cells were mixed with PV mesenchymal cells, epithelial cysts and gland-like structures was formed and ECPg was expressed in many epithelial cells (Fig. 2 A, B). When epithelial cells were mixed with GZ mesenchymal cells, epithelial cells formed only epithelial cysts and never formed gland-like structures, and ECPg expression in the epithelial cells was not observed (Fig. 2 C, D). Thus, epithelial cells retained reactive potency to the inductive influence of PV and GZ mesenchymes and mesenchymal cells also retained the inductive ability according to their origin as in the *in vitro* organotypic culture [6].

PGV-3.3, which contains 3.2 kb 5'-flanking region of the ECPg gene and luciferase gene was introduced into dissociated PV or GZ epithelial cells and cultivated with the PV or GZ mesenchymal cells in the cell aggregate culture and transient expression of luciferase activity was measured. There was no difference in the transfection efficiency between PV and GZ epithelial cells (data not shown). Plasmid PGV-C which contains SV40 enhancer-

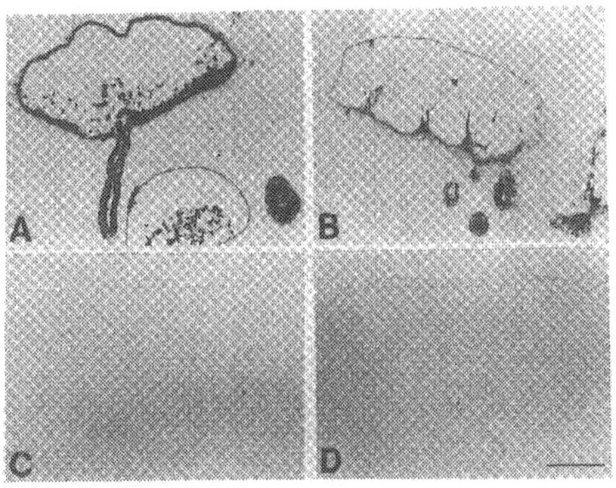

Figure 2. The structures formed by epithelial cells and ECPg expression in the cell aggregates. The cell aggregates were sectioned and subjected to the detection of the ECPg protein by a specific antibody. Scale bar equals 100 μm.

promoter and the luciferase reporter gene was transfected as a positive control and result obtained were used to normalize the expression of luciferase in epithelial cells.

Results in Fig. 3 show that the expression of luciferase activity in epithelial cells transfected with PGV-3.3 was regulated by mesenchymal cells. Thus, in the aggregates consisting of PV epithelial cells and PV mesenchymal cells, high luciferase activity was observed. In contrast, only basal level of luciferase activity was detected in the aggregates of GZ epithelial cells and GZ mesenchymal cells. Thus the 3.2 kb 5'-flanking sequence appears to be transcriptionally competent to direct expression of the luciferase gene only in ECPg expressing epithelial cells. When PGV-3.3-introduced PV epithelial cells were mixed with GZ mesenchymal cells luciferase activity was low. On the other hand, when PGV-3.3-introduced GZ epithelial cells were mixed with PV mesenchymal cells the activity was almost as high as that in the aggregates of PV epithelial cells and PV mesenchymal cells. These results unequivocally demonstrate that the 3.2 kb 5'-flanking sequence can mediate

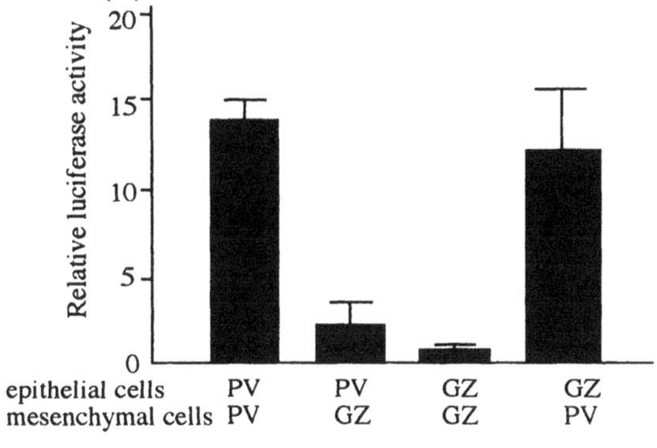

Figure 3. Activation and repression of transcription of the luciferase reporter gene under the influence of PV or GZ mesenchymal cells. Plasmid containing the luciferase gene under the control of 3.3 kb of 5'-flanking region of the ECPg gene was transfected into PV or GZ epithelial cells. After cultivation of epithelial cells with PV or GZ mesenchymal cells, luciferase activity was measured in cell extracts. All values are expressed as percents to luciferase activity in the cells transfected with the PGV-C plasmid which contains enhancer and promoter sequences from SV40. Values are represented as mean±S. D. of 3 determinations.

inductive and suppressive influences of PV and GZ mesenchymes, respectively, in the heterologous cell aggregate cultures.

The plasmid pmiwZ, containing β-actin promoter and Lac-Z, was introduced into epithelial cells and, after cultivation, we counted the Lac-Z-positive cells in each aggregate. The numbers of positive cells were not significantly different among four types of aggregates (data not shown). Therefore the difference in the luciferase activity among aggregates is not due to the different proliferation rates of epithelial cells.

DISCUSSION

In this study we established a new culture method, cell aggregate culture, in which both epithelium and mesenchyme were dissociated into small clusters of cells, mixed and cultivated. With this method we could detect the expression of the luciferase reporter gene under the control of 5'-flanking sequence of the ECPg gene. We also showed that under the conditions of this method, PV and GZ epithelial cells retain the ability of morphogenesis and cytodifferentiation. Their differentiation was properly induced by PV mesenchymal cells and suppressed by GZ mesenchymal cells even if the normal organization of these epithelia and mesenchymes was once destroyed.

The present study clearly demonstrated that the 5'-flanking region of the ECPg gene contains organ-specific regulatory region within 3245 bases from the start site of transcription and this region can mediate inductive or suppressive influences of PV or GZ mesenchymal cells. The sequence of this region (data not shown) reveals that there are two sequences at 1853 bp and 1403 bp upstream from the start site, respectively, which are identical with the inverted consensus viral enhancer core sequence, C(A,T)(A,T)(A,T)CCA (Fig. 1). It is noted that there are two other sequences identical with this sequence or with the inverted sequence of this at 272 bp and 361 bp upstream from the start site [9]. In the region within 3245 bases from the start site of transcription, there are four inverted repeats consisting of 8 or 9 bases with one mismatch (Fig. 1). Whether these sequences function as promoters or enhancers is open to question at present. Nuclear factors extracted from PV but not from liver were found to interact with 5'-flanking region (-398 to -195) of the ECPg gene (Fukuda and Yasugi, unpublished data). This suggests that the putative trans-acting factors are produced in the PV epithelium and these factors bind to cis-elements of the ECPg gene to enhance transcriptional activity of the gene. To examine whether the expression of these trans-acting factors is modulated by the influences of the mesenchyme is very important in understanding the molecular mechanisms of tissue interactions.

In mammals, prochymosin production in gastric mucosa is replaced by pepsinogen production during several weeks of the neonatal period [11]. The expression of the prochymosin gene is thus similar to the ECPg gene expression which occurs transiently in embryonic period and is converted to expression of adult chick pepsinogen. So one can expect that there is a homology of the 5'-flanking sequences between the prochymosin gene and ECPg gene. However, when the 5'-flanking sequence of ECPg gene is compared with that of calf prochymosin gene [12], no significant similarities are observed. Some similar sequences are present in the 5'-flanking regions of calf prochymosin gene and human pepsinogen A gene [13].

Previously we investigated the methylation state of DNA of the 5'-flanking region of ECPg gene isolated from ECPg-expressing and non-expressing organs and found no significant difference in the degree of methylation [15].We suppose, therefore, that the methylation state of 5'-flanking region itself is not involved in the regulation of organ-specific expression of ECPg gene, and that some factors which bind to this region are important to modulate the activity of this gene.

Recently some mesenchymal factors involved in the epithelial-mesenchymal interactions are identified, such as epimorphin [15] and hepatocyte growth factor [16], but the precise molecular mechanism of reactivity of the epithelium to mesenchymal factors has been less analyzed. The present paper shows that such mesenchymal signals affect the transcriptional activity of the ECPg gene through its 5'-flanking region, shedding some light on the molecular mechanism of the reactivity in the epithelium.

REFERENCES

1. N.K. Wessells, Tissue Interactions and Development, Menlo Park: W. A. Benjamin (1977).
2. R.H. Sawyer, and J.F. Fallon, Epithelial Mesenchymal Interactions in Development: Praeger Publishers, New York (1983).
3. S. Yasugi, and T. Mizuno, Developmental changes in acid proteases of the avian proventriculus, J. Exp. Zool. 216:331 (1981).
4. S. Yasugi, Role of epithelial-mesenchymal interactions in differentiation of epithelium of vertebrate digestive organs, Dev. Growth Differ. 35:1 (1993).
5. K. Takiguchi, S. Yasugi, and T. Mizuno, Gizzard epithelium of chick embryos can express embryonic pepsinogen antigen, a marker protein of proventriculus, Wilhelm Roux's Arch. Dev. Biol. 195:475 (1986).
6. K. Takiguchi, S. Yasugi, and T. Mizuno, Developmental changes in the ability to express embryonic pepsinogen in the stomach epithelia of chick embryos, Wilhelm Rouxs Arch. Dev. Biol. 197:56 (1988).
7. K. Takiguchi, S. Yasugi, and T. Mizuno, Pepsinogen induction in chick stomach epithelia by reaggregated proventricular mesenchymal cells *in vitro*, Dev. Growth Differ. 30:241 (1988).
8. K. Hayashi, K. Agata, M. Mochii, S. Yasugi, G. Eguchi, and T. Mizuno, Molecular cloning and the nucleotide sequence of cDNA for embryonic chicken pepsinogen; Phylogenetic relationship with prochymosin, J. Biochem. 103:290 (1988).
9. K. Hayashi, S. Yasugi, and T. Mizuno, Pepsinogen gene transcription induced in heterologous epithelial-mesenchymal recombinations of chicken endoderms and glandular stomach mesenchyme, Development 103:725 (1988).
10. V. Hamburger, and H.L. Hamilton, A series of normal stages in the development of chick embryo, J. Morph. 88:49 (1951).
11. B. Foltmann, A.L. Jensen, P. Lonblad, E. Smidt, and N.H. Axelsen, A developmental analysis of the production of chymosin and pepsin in pigs, Comp. Biochem. Physiol. 68B:9 (1981).
12. M. Hidaka, K. Sasaki, T. Uozumi, and T. Beppu, Cloning and structural analysis of the calf prochymosin gene, Gene 43:197 (1986).
13. K. Sogawa, Y. Fujii-Kuriyama, Y. Mizukami, Y. Ichihara, and K. Takahashi, Primary structure of human pepsinogen gene, J. Biol. Chem. 258:5306 (1983).
14. K. Fukuda, M. Ichinose, H. Saiga, K. Shiokawa, and S. Yasugi, Developmental changes of DNA methylation pattern of embryonic chick pepsinogen gene, J. Biochem. 155:502 (1994).
15. Y. Hirai, K. Takabe, M. Takashina, S. Kobayashi, and M. Takeichi, Epimorphin: A mesenchymal protein essential for epithelial morphogenesis, Cell 69:471 (1992).
16. R. Montesano, K. Matsumoto, T. Nakamura, and L. Orci, Identification of a fibroblast-derived epithelial morphogen as hepatocyte growth factor, Cell 67:901 (1992).

SERUM PEPSINOGEN VALUES AS POSSIBLE MARKERS FOR EVALUATING THE POSSIBILITY OF PEPTIC ULCER RECURRENCE UNDER H_2-BLOCKER HALF-DOSE MAINTENANCE THERAPY

Masashi Matsushima, Kazumasa Miki, Masao Ichinose, Nobuyuki Kakei, Naohisa Yahagi, Masahiro Kido, Yasuhito Shimizu, Satoshi Ishihama, Shinko Tsukada, Kiyoshi Kurokawa and Kenji Takahashi[1]

The First Department of Internal Medicine
Faculty of Medicine
University of Tokyo
Bunkyo-ku, Tokyo, 113, Japan
[1]Department of Biophysics and Biochemistry
Faculty of Science
University of Tokyo
Bunkyo-ku, Tokyo, 113, Japan

INTRODUCTION

The introduction of H_2-blockers and proton pump inhibitors has made peptic ulcer a curative disease without surgery. However, prevention of the ulcer recurrence is still a difficult problem. Various maintenance therapies have been set up and tried. But there has been almost no indicator to select a proper therapy for each case. Thus, a marker for predicting ulcer recurrence has been long desired. Serum pepsinogen (PG) values, PG I originating from fundic gland (1) and PG II from throughout stomach mucosa and Brunner gland (2), have been reported to reflect morphology and function of the gastroduodenal mucosa (3,4). Some of the previous studies suggested their possible roles as markers for peptic ulcer diseases and recurrence. For the ulcer incidence, a high PG II value and a low PG I/II ratio in gastric ulcer patients and high PG I value in duodenal ulcer patients were reported to be observed against normal control subjects (5). For the ulcer recurrence, high PG I values in recurrent peptic ulcer patients were reported to be observed against no recurrent ones (6). For evaluating the clinical usefulness as a marker of ulcer recurrence,

Aspartic Proteinases: Structure, Function, Biology, and Biomedical Implications
Edited by Kenji Takahashi, Plenum Press, New York, 1995

Table 1. Serum Pepsinogen Values in Peptic Ulcer Patients with or without Recurrence

Case No.	Sex	Age	Organ with ulceration	Recurrence	PG I	PG II	PG I + II	PG I/II
		yr			ng/ml	ng/ml	ng/ml	
1	M	56	S	NR	51.3	19.0	70.3	2.7
2	M	68	S	NR	33.8	16.0	49.8	2.1
3	M	52	S	NR	28.9	10.7	39.6	2.7
4	M	39	S	NR	84.5	18.8	103.3	4.5
5	M	48	S	NR	46.6	11.6	58.2	4.0
6	M	52	S	NR	39.0	21.9	60.9	1.8
7	M	42	S	NR	51.9	22.4	74.3	2.3
8	M	59	S	NR	42.8	10.7	53.5	4.0
9	M	40	S	NR	51.6	18.1	69.7	2.9
10	M	54	S	NR	57.6	18.3	75.9	3.1
11	M	42	S	NR	52.7	14.1	68.8	3.7
12	M	42	S	NR	40.4	10.5	50.9	3.9
13	M	56	S	NR	42.8	15.2	58.0	2.8
14	M	38	S	R	98.6	25.2	123.8	3.9
15	M	55	S	R	83.0	32.5	115.5	2.6
16	M	45	S	R	95.0	16.5	111.5	5.8
17	M	38	S	R	59.3	14.3	73.6	4.1
18	M	54	S	R	84.2	30.0	114.2	2.8
19	M	62	S	R	111.0	38.2	149.2	2.9
20	M	57	S	R	55.7	27.7	83.4	2.0
21	M	51	D	NR	51.2	13.9	65.1	3.7
22	M	46	D	NR	47.5	8.0	55.5	5.9
23	M	63	D	NR	48.5	15.0	63.5	3.2
24	M	46	D	NR	51.9	13.5	65.4	3.8
25	M	58	D	NR	43.1	9.6	52.7	4.5
26	M	39	D	NR	61.9	17.4	79.3	3.6
27	M	64	D	NR	95.6	19.2	114.8	5.0
28	M	56	D	NR	39.8	23.1	62.9	1.7
29	M	49	D	R	64.6	17.0	81.6	3.8
30	M	54	D	R	87.9	17.7	105.6	5.0
31	M	63	D	R	121.4	31.7	153.1	3.8
32	M	45	D	R	200.0	60.1	260.1	3.3
33	M	62	D	R	93.1	21.6	114.7	4.3

S, stomach; D, duodenum; NR, no recurrence; R, recurrence.

there must be done a follow-up study of peptic ulcer patients under a fixed, widely used maintenance therapy according to a rigid follow-up protocol. In the past, there were a few follow-up studies with serum pepsinogens under H_2-blocker half-dose therapy, a representative maintenance therapy, only of duodenal ulcer patients, which suggested the possibility of PG I or total pepsinogen being a recurrence marker (7,8). Here, we report that PG I, II and I+II are useful recurrence markers not only in duodenal ulcer, but also in gastric ulcer patients under H_2-blocker half-dose maintenance therapy.

Table 2. Mean Levels of Serum Pepsinogens in Gastric, Duodenal and Peptic Ulcer Patients with or without Recurrence

	Recurrence	Case	Age	PG I	PG II	PG I + II	PG I/II
			Mean, SE	Mean, SE	Mean, SE	Mean, SE	Mean, SE
			yr	ng/ml	ng/ml	ng/ml	
Gastric ulcer	no recurrent	13	50.0, 2.3	48.0, 3.6	15.9, 1.1	64.1, 4.2	3.12, 0.22
	recurrent	7	49.9, 3.3	83.8, 7.1	26.3, 3.0	110.1, 8.8	3.44, 0.44
	significance of difference		ND[a]	p[b] <0.01	p<0.01	p<0.01	ND
Duodenal ulcer	no recurrent	8	52.9, 2.9	54.9, 5.9	15.0, 1.6	69.9, 6.5	3.93, 0.42
	recurrent	5	54.6, 3.2	113.4, 21.0	29.6, 7.2	143.0, 28.1	4.04, 0.25
	significance of difference		ND	p<0.05	p<0.05	p<0.05	ND
Peptic ulcer	no recurrent	21	51.1, 1.9	50.6, 3.3	15.6, 0.9	66.3, 3.7	3.42, 0.22
	recurrent	12	51.8, 2.4	96.2, 10.6	27.7, 3.5	123.9, 13.6	3.69, 0.29
	significance of difference		ND	p<0.01	p<0.01	p<0.01	ND

[a]ND, no different, determined as p>0.05 in Student's t-test.
[b]Calculated in Student's t-test between no recurrent and recurrent cases.

MATERIALS AND METHODS

Patients

The 33 patients studied, all male, ranged in age from 39 to 68 years. The subjects were extracted as follows. Employees in a workplace received annual gastrointestinal health check by an indirect x-ray examination as screening. The employees with some abnormalities in the screening examination were further examined by gastroduodenal endoscopy. The endoscopically diagnozed peptic ulcer patients were treated with an H_2-blocker of roxatidine acetate hydrochloride (Teikoku-Zouki, Co. Ltd.) by daily dose of 150 mg, 75 mg each after breakfast and before sleep for 2 months. When starting the therapy, serum pepsinogen values were measured in the ulcer patients. After 2 months, they were re-examined by endoscopy and only the patients with ulcer scarring in the examination were registered as the subjects for this study. No hepatorenal disease was pointed out in the subject patients through annual health check performed at the same time of the screening indirect x-ray examination.

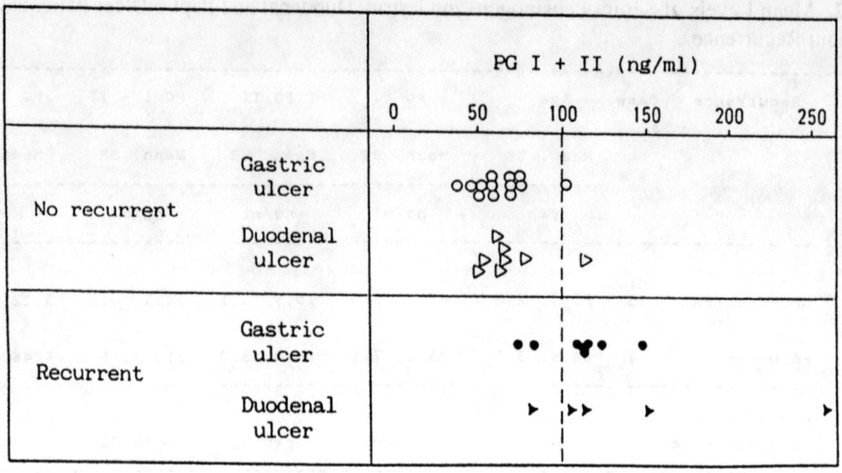

Figure 1. Serum total pepsinogen (PG I + PG II) values in patients with recurrent or no recurrent ulcer under H_2-blocker half-dose maintenance therapy.

Protocol

The 33 patients received a maintenance therapy with 75 mg of the same drug before sleep daily for 6 months. At 3 and 6 months of the maintenance therapy, they received follow-up endoscopic examinations.

Procedure

Serum PG I and II were measured at the time of endoscopic diagnosis of the peptic ulcer using Pepsinogen I/II RIA bead kit (Dainabot, Co. Ltd.). All the endoscopic examinations were performed by the same doctor with experience of the examination of more than

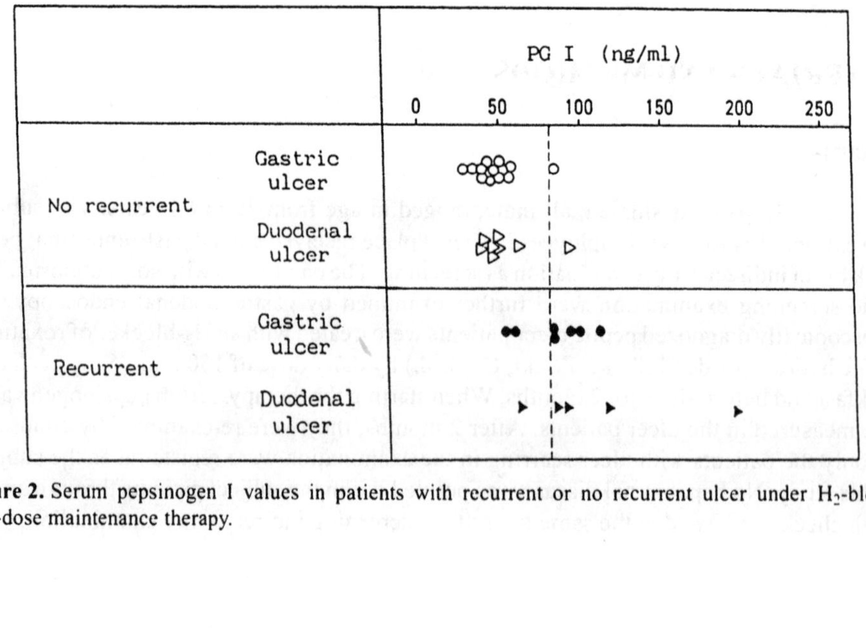

Figure 2. Serum pepsinogen I values in patients with recurrent or no recurrent ulcer under H_2-blocker half-dose maintenance therapy.

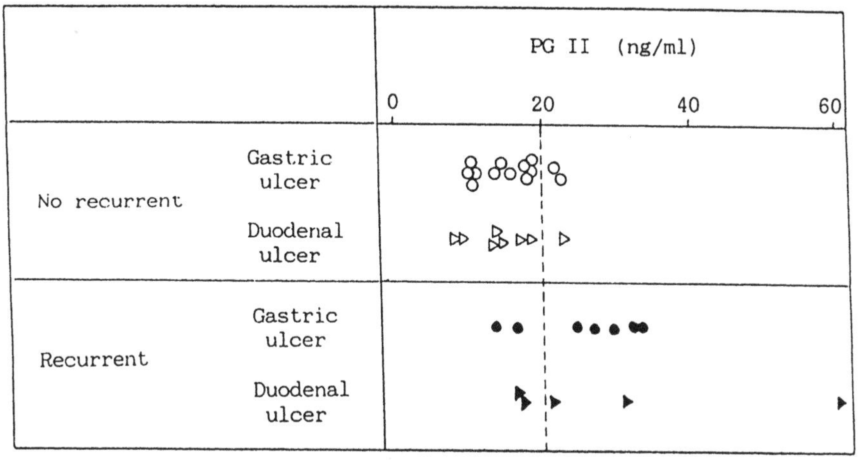

Figure 3. Serum pepsinogen II values in patients with recurrent or no recurrent ulcer under H_2-blocker half-dose maintenance therapy.

20 years using GIF-P20 (Olympus, Co. Ltd.). The statistical analyses were performed using Student's t-test for unpaired samples.

RESULTS

Thirteen no recurrent and 7 recurrent gastric ulcer and 8 no recurent and 5 recurrent duodenal ulcer patients were examined (see Table 1). There was no statistically significant difference of age between recurrent and no recurrent group in duodenal, gastric or peptic ulcer (see Table 2). While, in each of gastric, duodenal and peptic ulcer cases, PG I, PG II and PG I + PG II values in recurrent group were higher than those in no recurrent one with statistic significance ($p<0.01$ in gastric and peptic ulcer; $p<0.05$ in duodenal ulcer, shown also in Table 2). These data suggest a possibility that these pepsinogen values may be a clinically useful marker in recurrence of duodenal, gastric and peptic ulcer. Distribution of PG I + II, PG I and PG II in no recurrent and recurrent groups is shown in figure 1, 2 and 3, respectively. Adopting the border lines to make false-positive and false-negative at least, as 100, 80 and 20 ng/ml in PG I + II, PG I and PG II, respctively, the efficiency values, true / total, were 84.8% in PG I + II and PG I (shown in Table 3) and 78.8% in PG II. Therefore, PG I + II and PG I were better than PG II as a peptic ulcer recurrence marker under H_2-blocker half-dose maintenance therapy judged from this study.

DISCUSSION

Since Samloff *et al.* (5) and Ichinose *et al.* (9), several papers pointed out the relation between serum PG I and peptic ulcer incidence and recurrence. Appearance of H_2-blocker and proton pomp inhibitor has made almost all peptic ulcers curative without surgery and now, prevention of the recurrence, that is, selection and period of the maintenance therapy, is an important problem. Thus, a marker to evaluate peptic ulcer in susceptibility to relapsing has been hoped and studies on the markers under representative maintenance therapies, such as H_2-blocker half-dose maintenance therapy, have been required. Battaglia *et al.* reported

Table 3. Peptic Ulcer Recurrence and Serum Pepsinogen Values with the Cut-Off Value in PG I + II as 100 ng/ml[a]

	PG I + II	
	<100 ng/ml	>100 ng/ml
recurrent	3	9
no recurrent	19	2

Efficiency, 84.8 %

[a]The same result was given with the cut-off value in PG I as 80 ng/ml.

that duodenal ulcer patients with elevated total serum pepsinogen and maximum acid output (MAO), blood group A, early onset of the disease, male sex, alcohol intake and cigarette consumption are more prone to relapse, under H_2-blocker half-dose maintenance therapy for 1 year (8). Inbe *et al.* also reported PG I is a useful marker in duodenal ulcer recurrence under H_2-blocker half-dose maintenance therapy for 1 year (7). MAO and serum pepsinogen I or I + II were reported to be highly correlated, and are recognized to reflect morphology and function of the gastric mucosa (4). Thus, these three are of almost the same value as the marker. But serum pepsinogen values are more easily measurable than MAO, so that we think serum pepsinogens are preferable among them. As for gastric ulcer, Jorde *et al.* reported that recurrent ulcer patients had significantly higher MAO than no recurrent ulcer patients, under a maintenance therapy with ranitidine 150 mg/day, one of H_2-blocker of half-dose, for 1 year (10). And in this study, we reported serum PG I and PG I + II were good recurrence markers not only in duodenal, but also in gastric ulcer patients.

In clinical application, we recommend measurement of serum pepsinogen values at diagnosis of peptic ulcer. The patients with PG I more than 80 or PG I + II more than 100 ng/ml will receive more effective maintenance therapy than H_2-blocker half-dose therapy, such as H_2-blocker full-dose, and H_2-blocker half-dose plus other drugs, *etc.* What is the best maintenance therapy for the peptic ulcer patients with high risk of recurrence? Next, we, gastroenterologists, will have to solve this problem.

CONCLUSION

Serum PG I, PG II and PG I + II are useful markers for susceptibility of peptic ulcer recurrence under H_2-blocker half-dose maintenance therapy and are appliable to clinical use.

REFERENCES

1. I.M. Samloff, Cellular Localization of Group I Pepsinogens in Human Gastric Mucosa by Immunofluorescence, *Gastroenterology* 61:185 (1971)
2. I.M. Samloff and W.M. Liebman, Cellular Localization of the Group II Pepsinogens in Human Stomach and Duodenum by Immunofluorescence, *Gastroenterology* 65:36 (1973)
3. I.M. Samloff, K. Varis, T. Ihamaki, M. Siurala and J.I. Rotter, Relationships among Serum Pepsinogen I, Serum Pepsinogen II, and Gastric Mucosal Histology, *Gastroenterology* 83:204 (1982)
4. K. Miki, M. Ichinose, A. Shimizu, S.C. Huang, H. Oka, C. Furihata, T. Matsushima and K. Takahashi, Serum Pepsinogens as a Screening Test of Extensive Chronic Gastritis, *Gastroenterol. Jpn.* 22:133 (1987)

5. I.M. Samloff, G.N. Stemmermann, L.K. Heilbrun and A. Nomura, Elevated Serum Pepsinogen I and II Levels Differ as Risk Factors for Duodenal Ulcer and Gastric Ulcer, *Gastroenterology* 90:570 (1986)
6. I.M. Samloff, D.M. Secrist and E. Passaro Jr., Serum Group I Pepsinogen Levels and Their Relation to Gastric Acid Secretion in Patients with and without Recurrent Ulcer, *Gastroenterology* 70:309 (1976)
7. A. Inbe, K. Sumii, N. Uemura, M. Kimura, K. Haruma, K. Tokumo, H. Ogoshi, H. Teshima, I. Fukuhara, S. Murakami and G. Kajiyama, Recurrence of Duodenal Ulcer under Maintenance Treatment with Special Reference to Serum Pepsinogen I, *Gastroenterol. Endosc.* 29:2200 (1987) [in Japanese]
8. G. Battaglia, R. Farini, F. Di Mario, A. Piccoli, M. Plebani, F. Vianello, A. Burlina and R. Naccarato, Recurrence of Duodenal Ulcer under Continuous Antisecretory Treatment: an Approach to the Detection of Predictive Markers, *Am. J. Gastroenterol.* 79:831 (1984)
9. M. Ichinose, K. Miki, C. Furihata, T. Kageyama, R. Hayashi, H. Niwa, H. Oka, T. Matsushima and K. Takahashi, Radioimmunoassay of Serum Group I and Group II Pepsinogens in Normal Controls and Patients with Various Disorders, *Clin. Chim. Acta* 126:183 (1982)
10. R. Jorde, P.G. Burhol and T. Hansen, Ranitidine 150 mg at Night in the Prevention of Gastric Ulcer Relapse, *Gut* 28:460 (1987)

THE CLINICAL APPLICATION OF THE SERUM PEPSINOGEN I AND II LEVELS AS A MASS SCREENING METHOD FOR GASTRIC CANCER

Kazumasa Miki,[1] Masao Ichinose,[1] Nobuyuki Kakei,[1] Naohisa Yahagi,[1] Masashi Matsushima,[1] Shinko Tsukada,[1] Satoshi Ishihama,[1] Yasuhito Shimizu,[1] Takehisa Suzuki,[1] Kiyoshi Kurokawa,[1] and Kenji Takahashi[2]

[1]First Department of Internal Medicine
Faculty of Medicine
Faculty of Science
University of Tokyo
Tokyo, Japan 113
[2]Department of Biophysics and Biochemistry
Faculty of Science
University of Tokyo
Tokyo, Japan 113

INTRODUCTION

Gastric cancer, despite a recent decline in the incidence, is still a leading cause of death in Japan. For this reason, much effort has been directed to the early detection of the cancer through mass screening programs throughout the country. In most workplaces in Japan, an indirect X-ray examination, using 10 cm square film, is the conventional first screening step, after which those suspected of having some abnormalities in the gastric mucosa are further investigated either with a higher quality X-ray examination or by endoscopy. However, the sensitivity of the conventional X-ray screening step is by no means high. To improve the effectiveness of gastric cancer screening, we have devised a new screening method that utilizes measurement of serum pepsinogen (pepsinogen I and II) levels. This new screening system is based on the facts that a considerable part of gastric cancers develop in gastric mucosa affected by severe and extensive atrophic gastritis (1) and that serum pepsinogen levels serve as a sensitive marker of chronic atrophic gastritis (2). We report the first application of serum pepsinogen measurement for mass screening of gastric cancer at a certain workplace and compare the results with those of the conventional X-ray screening method.

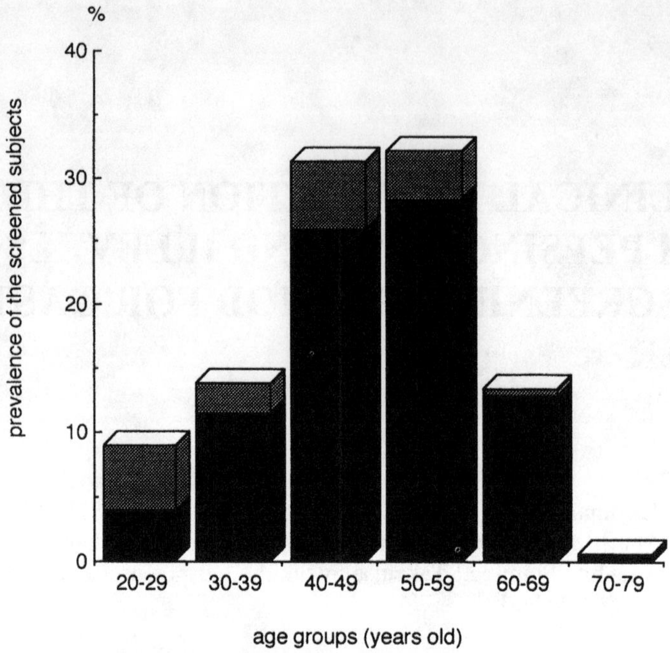

Figure 1. Distribution of the serum pepsinogen-screened subjects, according to age and sex in 1993. Black boxes represent the percentage of the male and gray boxes denotes that of the female.

METHODS

The numbers of the company employees who received an annual periodic health check-up were 4647, 4758 and 5457 in 1991, 1992 and 1993, respectively. The blood samples for routine laboratory tests were taken after fasting and aliquots of the separated sera were stored at -20°C until the serum pepsinogen levels were measured by using pepsinogen I and II Riabead Kits (Dainabot Co. Ltd.), a modified radioimmunoassay method that we have previously established (4). The cutoff level should be such that the positive rate in the screened subjects be less than 20 % and that the sensitivity for detecting cancers be as high as possible, since nearly 20 % of the screened subjects by the conventional X-ray examinations require more thorough follow-up examinations with endoscopy. Thus, in 1991, we used a pepsinogen I level of less than 50 µg/litter, together with a I/II ratio of less than 3.0, as the cutoff level (4). In 1992 and 1993, however, we set the cutoff level at 70 µg/litter of pepsinogen I and 3.0 of a I/II ratio to improve the sensitivity of this method. According to the cutoff levels, subjects showing low serum pepsinogen levels were considered to have extensive chronic atrophic gastritis and were further investigated by endoscopy (Olympus GIF-P20) as the second step screening. In 1992, those who had undergone endoscopy within a year were excluded from the second step screening by endoscopy unless their pepsinogen I levels were less than 30 µg/litter and I/II ratios were less than 2.0. In 1993, all those who had undergone endoscopy within a year were excluded from the second step screening regardless of their pepsinogen levels.

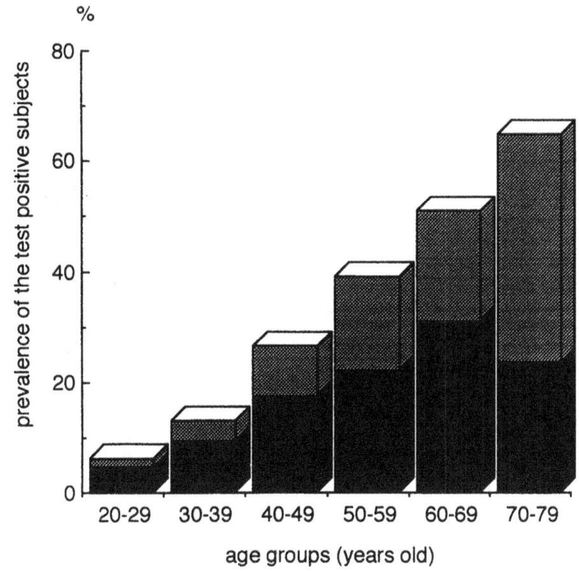

Figure 2. Age distribution of serum pepsinogen test-positive subjects in 1993. Black boxes represent the incidence of positive subjects who need second step screening. Gray boxes represent the incidence of positive subjects who are excluded from second step screening due to the experience of endoscopy within a year.

RESULTS

Age distribution of the serum pepsinogen-screened subjects in 1993 is shown in Fig. 1. The proportions of the subjects positive for serum pepsinogen tests increased with age (Fig.2), which is in agreement with previous studies. According to our criteria, the overall subjects who need second screening by endoscopy were 875 (18.8 %), 1106 (23.2 %) and 1005 (18.4 %), and among them, 676 (14.5 %), 918 (19.3 %) and 780 (14.3 %) underwent endoscopic examination in 1991, 1992 and 1993, respectively. As a result, 4 (0.086 %), 9 (0.189 %) and 12(0.220%) cases of gastric cancer and 4 (0.086 %), 3 (0.063 %) and 3 (0.055 %) cases of gastric adenoma were detected in 1991, 1992 and 1993, respectively. Fig. 3 shows the results of gastric mass screening using the conventional indirect X-ray examinations from 1986 to 1990 and those of new serum pepsinogen tests from 1991. Using the conventional method, the mean of the cancer detection rates was 0.066 %. On the other hand, the mean of the cancer detection rates of the new serum pepsinogen screening method was 0.168 %, suggesting that the cancer detection rate of the new method was superior to the conventional one.

All the subjects who were found to have a cancer were free from symptoms. Of total 25 cases of cancer screened by the serum pepsinogen tests from 1991 to 1993, 21 cases (84 %) were in the early stage, indicating that the proportion of early-stage cancers detected by the new serum pepsinogen screening method was higher than that detected by the conventional screening method (70 %). It is interesting that a considerable part of the detected cancers were so small that it was quite difficult to visualize these lesions in further high-quality X-ray examinations. Histologically, 17 cancers (68 %) were well-differentiated adenocarcinomas and 8 cancers (32 %) were poorly differentiated adenocarcinomas. It is noteworthy that in 1993, the third year since the introduction of the new screening method, 6 cancers (50 %) out of 12 cancers were endoscopically resected, because they were

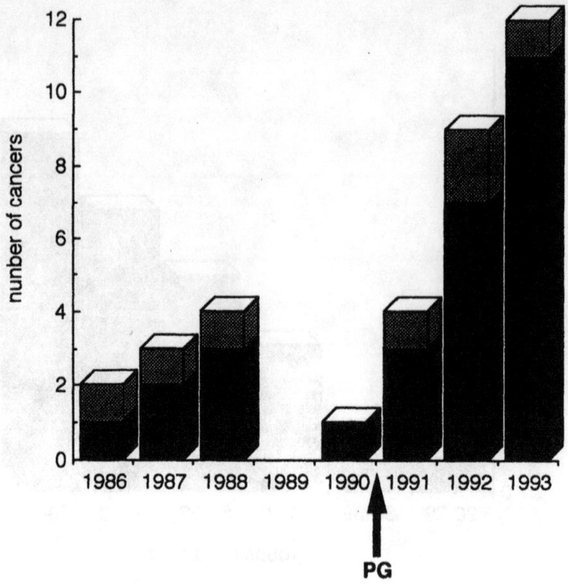

Figure 3. Number of cancers detected by conventional X-ray screening and serum pepsinogen method. Conventional X-ray examination had been used for screening until 1990. Arrow represents the introduction of the new serum pepsinogen screening method in 1991. Black boxes represent early gastric cancer and gray boxes represent advanced gastric cancer.

adenocarcinomas limited to the mucosa with less than 1 cm in diameter without ulcer scar in the lesions. The other cancers were surgically resected and the postoperative course of each patient was uneventful.

As for gastric adenomas that were considered as precancerous lesions, 10 adenomas (0.067 %) were detected by the serum pepsinogen screening method from 1991 to 1993. With the conventional screening method, no adenomas were pointed out and 3 adenomas (0.020 %) were accidently found by the second step screeining with endoscopy.

DISCUSSION

In Japan, mass screenings for gastric cancers have been conducted for over a quarter of a century, and now more than six million people were screened annually, mainly by indirect X-ray examinations. However, the efficiency of the conventional screening method is not high. According to the Annual Report on Gastric Mass Screenings in Japan from the Japanese Society of Gastroenterological Mass Survey in 1989, the mean value for the incidence of cancers detected in workplaces in all over Japan is 0.03 % (5). To improve the efficiency of the mass screening for gastric cancer, it should be focused on subjects at high risk of developing gastric cancer. There is a hypothesis that chronic atrophic gastritis is a possible precursor condition of gastric cancer, especially of intestinal type. Indeed, there are evidences that indicate the risk of gastric cancer significantly increases among populations where chronic atrophic gastritis is widespread. On the other hand, serum pepsinogen levels provide a precise measure of the extent of chronic atrophic gastritis as we have previously demonstrated (2) (3). Thus, the high reliability of this serum test to predict chronic atrophic gastritis led us to consider its application to mass screening for gastric cancer, since the pepsinogen I level and the I/II ratio of gastric cancer patients are significantly lower than those of cancer-free subjects. Recently, two independent reports have been published that prevalence rates of atrophic gastritis diagnosed with the criteria using a serum pepsinogen I level of less than 70 μg/litter and a I/II ratio of less than 3.0 show a high correlation with standardized mortality ratios (6) or age-adjusted mortality rates for gastric cancer among prefectures (7). These reports seem to support our hypothesis that serum pepsinogen level could be an objective marker that selects the high risk population of gastric cancer.

This paper reports the results of the application of the serum pepsinogen screening method to mass screening of total 14,862 employees of a Japanese company in the first three years. In these screening series, 25 gastric cancers (0.168 %) were detected including 21 at early stage by the new screening method during this period. Workers at the same company had previously been screened by the conventional X-ray method, and the mean rate of cancer detection was 0.066 % in the last five years. Superficially, these data indicate the serum pepsinogen screening method is superior to the conventional one, though a possibility that the change in the screening method from conventional morphological screening to biochemical one affects the cancer detection rate cannot be denied. However, this seems unlikely, because since the introduction of the new screening method in 1991, the cancer detection rate has not been decreasing in these three years. Thus, the high cancer detection rate achieved by the serum pepsinogen screening method is more likely to result from its high sensitivity.

In summary, we have devised a mass screening method that utilizes the measurement of the serum pepsinogen levels to initially identify subjects with a high risk of gastric cancer, and established the gastric cancer screening system, named "stomach dry dock". The results of this study suggest the new screening method is superior to the conventional X-ray mass screening method. Moreover, it has many advantages: (1) the procedure is easy and subjects experience no discomfort; (2) no X-ray hazards are incurred and no side effects are experienced due to barium ingestion; (3) it is less expensive (about a half the cost of X-ray screening); (4) the procedure is quick and many serum samples can be measured simultaneously.

REFERENCES

1. P. Correa, The Gastric Precancerous Process, *Cancer Surv.* 2:437 (1983)
2. K. Miki, M. Ichinose, A. Shimizu, S. C. Huang, H. Oka, C. Furihata, T. Matsushima and K. Takahashi, Serum Pepsinogens as a Screening Test of Extensive Chronic Gastritis, *Gastroenterol. Jpn.* 22:133 (1987)
3. K. Miki, M. Ichinose, N. Kawamura, M. Matsushima, H. B. Ahmad, M, Kimura, J. Sano, T. Tashiro, N. Kakei, H. Oka, C, Furihata and K. Takahashi, The Significance of Low Serum Pepsinogen Levels to Detect Stomach Cancer Associated with Extensive Chronic Gastritis in Japanese Subjects, *Jpn. J. Cancer Res.* 80, 114 (1989)
4. M. Ichinose, K. Miki, C. Furihata, T. Kageyama, R. Hayashi, H. Niwa, H. Oka, T. Matsushima and K. Takahashi, Radioimmunoassay of Serum Group I and Group II Pepsinogens in Normal Controls and Patients with Various Disorders, *Clin. Chim. Acta* 126:183 (1982)
5. K. Miki, M. Ichinose, K. B. Ishikawa, N. Yahagi, M. Matsushima, N. Kakei, S. Tsukada, M. Kido, S. Ishihama, Y. Shimizu, T. Suzuki and K. Kurokawa, Clinical Application of Serum Pepsinogen I and II Levels for Mass Screening to Detect Gastric Cancer, *Jpn. J. Cancer Res.* 84:1086 (1993)
6. A. Fukao, S. Hisamichi, N. Ohsato, N. Fujino, N. Endo and M. Iha, Correlation between the Prevalence of Gastritis and Gastric Cancer in Japan, *Cancer Causes and Control* 4:17 (1992)
7. M. Kabuto, H. Imai, S. Tsugane and S. Watanabe, Correlation between Atrophic Gastritis Prevalance and Gastric Cancer Mortality among Middle-aged Men in 5 Areas in Japan, *J. Epidemiol.* 3:35 (1993)

A MINUTE GASTRIC CANCER DETECTED BY A NEW SCREENING METHOD USING SERUM PEPSINOGEN I AND II

Naohisa Yahagi, Kazumasa Miki, Masao Ichinose, Nobuyuki Kakei, Masashi Matsushima, Masahiro Kido, Yasuhito Shimizu, Satoshi Ishihama, Shinko Tsukada, Kiyoshi Kurokawa and Kenji Takahashi[1]

The First Department of Internal Medicine
Faculty of Medicine
University of Tokyo
Bunkyo-ku,Tokyo, 113, Japan
[1]Department of Biophysics and Biochemistry
Faculty of Science
University of Tokyo
Bunkyo-ku,Tokyo, 113, Japan

INTRODUCTION

A gastric cancer, despite a recent decline in the mortality rate, is still a leading cause of cancer death in Japan, so that considerable effort continues to be focused on the early detection of the cancer through mass screening programs throughout the country. In most workplaces, an indirect barium x-ray examination combined with endoscopic inspection has been the standard screening method used in Japan. However, this method has certain drawbacks in detecting early gastric cancers, for exsample due to the low sensitivity of the barium x-ray procedure, which is the first step of the screening process. Therefore, to improve the effectiveness of gastric cancer screening, we have established a new screening system that uses serum pepsinogen tests instead of the indirect barium x-ray method, since it has been found that the serum tests are better able to detect advanced stages of chronic atrophic gastritis [1][2], which is thought to be a precancerous condition that lead to the development of gastric cancer. In 1991, the first gastric mass screening using this system was carried out in the workplace and here we report on the first case of minute gastric cancer detected by this new screening method.

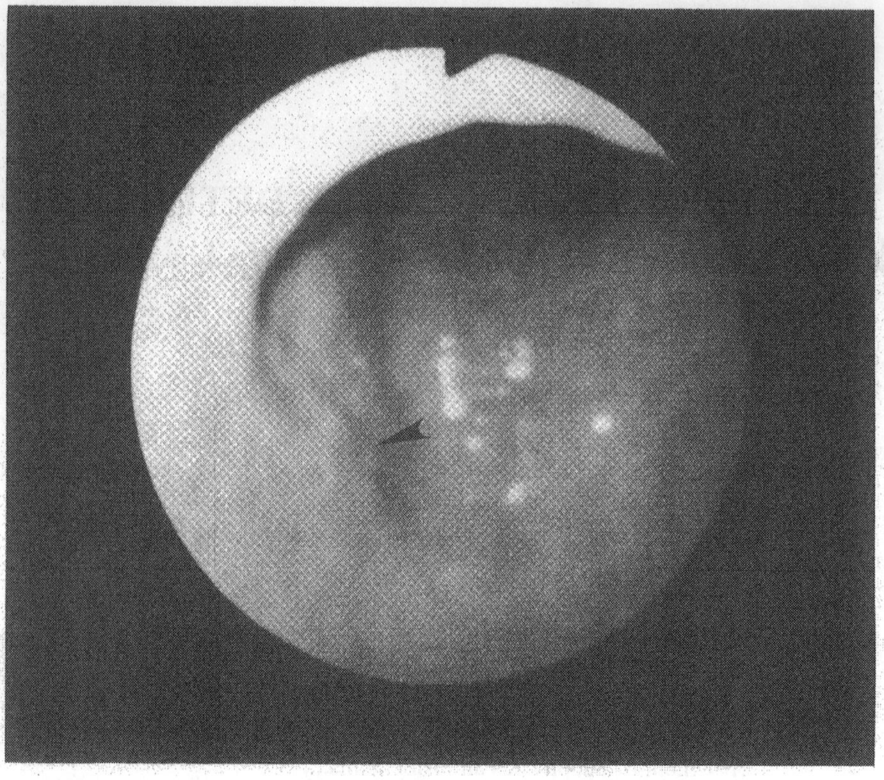

Figure 1. Endoscopic findings in the stomach. Microerosion can be seen (arrow) on the top of a thin fold converging into an ulcerous scar on the anterior wall of the proximal antrum.

A CASE REPORT

In 1991, a 61-year-old male, underwent an annual health check-up at the clinic of a company. The results of his physical and laboratory data, including ECG and chest x-ray were all normal, except for serum pepsinogen (PG) levels; that is PG I 31.3 μg/l, PG II 22.0 μg/l and the PG I/II ratio 1.4 respectively. Thus, according to our cut-off values[3], subjects with a PG I level of less than 50 μg/l and a PG I/II ratio of less than 3.0 are considered to have reacted positively to the presence of advanced stage of chronic atrophic gastritis. The subject, however, had no physical complaints and had no history of previous illnesses. His immediate family also had no unusual medical history. Based on the results of his pepsinogen levels, he underwent a second screening a few weeks later, at which time endoscopy was used. Endoscopy revealed that his gastric mucosa showed severe and extensive chronic atrophic gastritis, with a thin, discolored mucosal surface, and mucosal folds that were less prominent and very scarce. Further, a microerosion on the top of a thin fold converged into an ulcerous scar was found on the anterior wall of the proximal antrum(see Fig. 1). On air-inflation, this fold promptly disappeared, suggesting that this microerosion was limited to the mucosal layer. Suspecting that this lesion might be a type IIc gastric carcinoma, biopsy specimens were taken from the microerosion. The results of histological anlysis of the sample revealed it to be a well-differentiated adenocarcinoma and surgery was indicated. The subject underwent a preoperative barium x-ray examination, at which time thorough visualization of the lesion proved difficult. Further, no particular abnormalities were detected from other

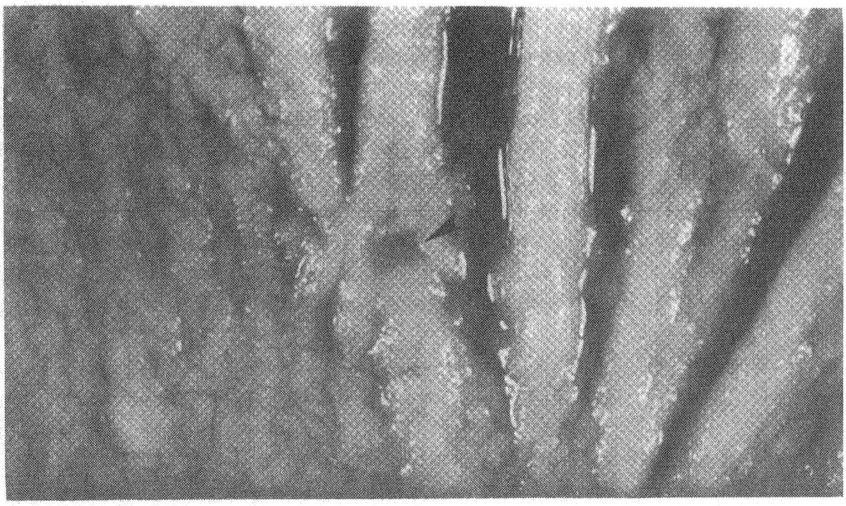

Figure 2. Resected specimen, showing a microeroded area, 5 mm in diameter, (arrow) on the top of a fold.

preoperative check-ups, including an abdominal ultrasonography and computed tomography. The subject then was given a subtotal gastrectomy (Billroth-I). A freshly resected stomach showed marked atrophy and intestinal metaplasia and there was a microerosion on a mucosal fold at the anterior wall of the proximal antrum. The size of the lesion was 5 mm in diameter and no difference in color was seen between the lesion and the adjacent mucosa (see Fig. 2). No abnormality was seen on the serosal surface. A cross section of the specimen showed a slight cancerous invasion into the submucosal layer, but not into the muscle proper. In addition, no histological evidence of metastasis was detected in the seven lymph nodes that were resected. Therefore, given these findings and the minute size of the lesion, which was limited only to the slightly depressed surface of the mucosa, the pathological diagnosis was an early gastric carcinoma, type IIc.

DISCUSSION

Pepsinogens, the precursors of pepsins, consist of two immunochemically distinct groups [4]: pepsinogen I (PG I) and pepsinogen II (PG II) [5], the former deriving from the fundic gland mucosa, and the latter from throughout the stomach mucosa and Brunner's gland [5]. They are mostly secreted into the gastric lumen, but trace amounts of these enzymes are also seen in the blood stream [6]. Previous studies have shown that the serum pepsinogen levels reflect the morphological and functional status of the gastric mucosa [7,8]. Indeed, the serum PG I and the PG I/II ratio provide better information with respect to the extent of chronic atrophic gastritis than gastric intubation [1]. Since chronic atrophic gastritis is considered to be a precancerous lesion of the stomach [9-11], we have established radioimmunoassay (RIA) for serum pepsinogens [12,13] and designed a new mass screening to identify possible gastric cancer cases. This system measures the levels of the serum pepsinogens, PG I and PG II, of each subject by means of RIA bead kit (DAINABOT Co, Ltd, Japan), and those that show low serum pepsinogen levels are then examined by endoscopy. This initial screening method has proven effective and requires less manpower than the conventional screening by barium x-ray, and based on trials that have been conducted,those who show a serum PG I level of less than 50 μg/l and a PG I/II ratio of less than 3.0 are considered positive [3]. Using this new method, we screened 4,647 employees of a company in 1991, and 875 subjects (18.8%) had positive serum pepsinogen tests. Among the positive subjects of the company's employee, 676 (14.5%) underwent an endoscopic examination and 4 gastric cancer cases (incidence: 0.086%) and 4 precancerous lesion cases were detected. The patient we report herein is the first gastric cancer case detected by the new screening system using the serum pepsinogen tests. What is most significant is that the detected gastric cancer was only 5 × 5 mm in size, since it is difficult to visualize such a minute lesion by the conventional indirect barium x-ray examination that employs 10 × 10 cm-sized film. Although the number of the screened subjects was relatively small, the results, especially with respect to cancer detectability, were comparable to the usual mass screening systems.

In conclusion, we would like to emphasize the merits of this new system . They are: (1) its low cost, (2) no x-ray hazards, (3) ease of testing, (4) quantative measurement are achieved, and (5) the results are quickly known. It thus is hoped that this new gastric mass screening system will help to reduce the gastric cancer mortality rate in Japan.

REFERENCES

1. K. Miki, M. Ichinose, A. Shimizu, et al., Serum pepsinogens as a screening test of extensive chronic gastritis, *Gsatroenterol. Jpn.* 22:133 (1987)
2. K. Miki, M. Ichimose, N. Kawamura, et al., The significance of low serum pepsinogen levels to detect stomach cancer associated with extensive chronic gastritis in Japanese subjects, *Jpn. J. Cancer Res.* 80:111 (1989)
3. K. Miki, M. Ichimose, K.B. Ishikawa, et al., Clinical application of serum pepsinogen I and II levels for mass screening to detect gastric cancer, *Jpn. J. Cancer Res.* 84:1086 (1993)
4. I. Kushner, W. Rapp, P. Brutin, *et al.* Electrophoretic and immunochemical demonstration of the existence of four human pepsinogens, *J. Clin. Invest.* 43:1983 (1964)
5. I.M. Samloff, Pepsinogens, pepsins and pepsin inhibitors, *Gastroenterology* 60:586 (1971)
6. B.I. Hirschowitz, Pepsinogen; its origin, secretion and excretion, *Physiol. Rev.* 37:475 (1957)
7. O.A.A. Bock, G. Arapakis, L.J. Witts and C.D. Richards, The serum pepsinogen level with special reference to the histology of the stomach mucosa, *Gut* 4:106 (1963)
8. I.M. Samloff, K. Varis, T. Ihamaki, M. Siurala and J.I. Rotter, Relationship among serum pepsiongen I, serum pepsinogen II and gastric mucosal histology, *Gastroenterology* 83:204 (1982)
9. P. Correa, The gastric precancerous process, *Cancer Surveys* 2:437 (1983)
10. T. Imai, T. Kubo and H. Watanabe, Chronic gastritis in Japanese with reference to high incidence of gastric carcinoma, *J. Natl. Cancer Inst.* 47:179 (1971)
11. R. Cheli,L. Simon, H. Aste, et al., Atrophic gastritis and intestinal metaplasia in asymptomatic Hungarian and Italian populations, *Endoscopy* 12:105 (1980)
12. M. Ichinose, K. Miki, C. Furihata, et al., Radioimmunoassay of group II pepsinogens in human serum, *Clin. Chim. Acta* 122:61 (1982)
13. M. Ichinose, K. Miki, C. Furihata, et al., Radioimmunoassay of group I and group II pepsinogens in normal controls and patients with various disorders, *Clin. Chim. Acta* 126:183 (1982)

TWO CASES OF EARLY COLORECTAL CANCER ASSOCIATED WITH GASTRIC ADENOMA DETECTED BY SERUM PEPSINOGEN SCREENING METHOD

Yasuhito Shimizu,[1] Nobuyuki Kakei,[1] Tomonori Wada,[1] Naohisa Yahagi,[1] Masahiro Kido,[1] Satoshi Ishihama,[1] Shinko Tsukada,[1] Masashi Matsushima,[1] Masao Ichinose,[1] Kazumasa Miki,[1] Kiyoshi Kurokawa,[1] and Kenji Takahashi[2]

[1]First Department of Internal Medicine
Faculty of Medicine
University of Tokyo, 7-3-1 Hongo, Bunkyo-ku, Tokyo, Japan 113
[2]Department of Biophysics and Biochemistry
Faculty of Science
University of Tokyo, 7-3-1 Hongo, Bunkyo-ku, Tokyo, Japan 113

INTRODUCTION

It is widely accepted that the serum pepsinogen levels reflect the morphological and functional status of gastric mucosa and the diagnostic value in atrophic gastritis is now established (1). Since a considerable number of gastric cancer develop in gastric mucosa affected by severe and extensive atrophic gastritis, it is considered as a precancerous condition that leads to the development of gastric cancer (2). Indeed, the prevalence rates of atrophic gastritis diagnosed by the serum pepsinogen test show a high correlation with standardized mortality ratios (3) or age-adjusted mortality rates for gastric cancer among prefectures (4). Thus, the serum pepsinogen levels could be an objective marker that selects the high risk population of gastric cancer (5). On the basis of the above mentioned facts, mass screening that utilizes the measurement of the serum pepsinogen levels has been performed in a certain workplace in Tokyo since 1991. Subjects showing low serum pepsinogen levels (pepsinogen I of less than 70 ng/ml and a I/II ratio of 3.0) were considered to have advanced stage of atrophic gastritis and further investigated by endoscopy as the second step of screening. As a result, 13 gastric adenocarcinomas and 7 gastric adenomas were found during the first two years. It is interesting that of 7 patients with gastric adenomas, two colorectal adenocarcinomas were accidentally found. Here we report the two cases of colorectal adenocarcinomas associated with gastric adenoma screened by the serum pepsinogen test.

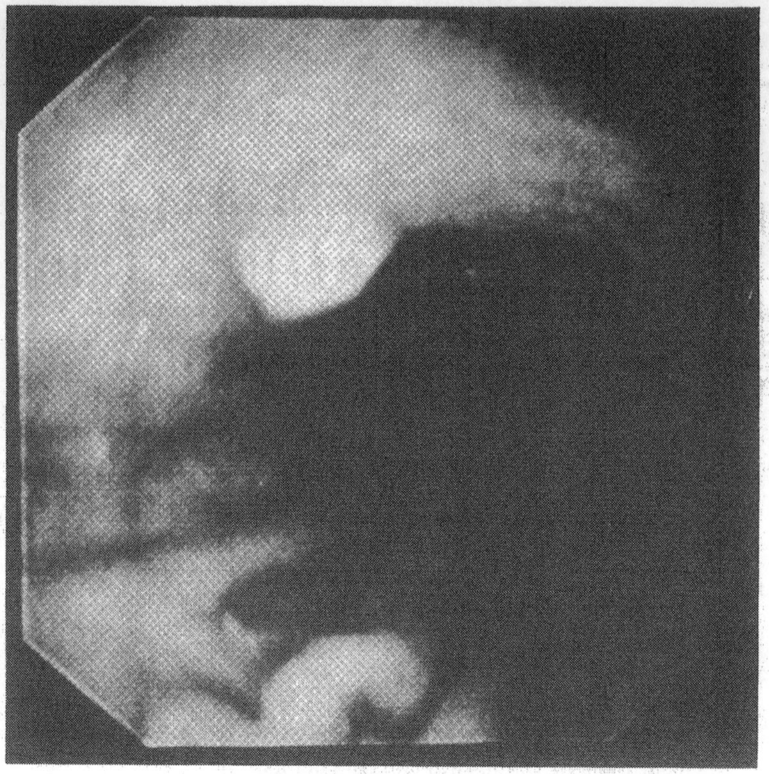

Figure 1. Endoscopic appearance of gastric adenoma in the lower body of the stomach in CASE 1.

CASE 1

In 1990, a 55-year-old female underwent an annual health check-up at the clinic of the company, and anemia was first picked up. She also noticed occasional blood attachment on stool. Next year, anemia and low pepsinogen levels (Pepsinogen I 25.0 ng/ml, Pepsinogen II 22.4 ng/ml and a I/II ratio 1.1) were pointed out at an annual health check-up, and a further investigation by endoscopy was advised. Esophagogastroduodenoscopy revealed a small whitish elevated lesion with 5 mm in diameter at the anterior wall of lower part of the body of stomach (Fig. 1). As expected, her gastric mucosa was affected by severe and extensive atrophic gastritis. Biopsied specimens taken from the elevated lesion revealed the histologic picture of gastric adenoma. Stool was positive for occult blood test. In July 1991, she was admitted to our hospital for endoscopic mucosal resection of the gastric adenoma and investigation of anemia. On admission, the patient appeared to be a well-nourished-female. On rectal examination, a hard polypoid lesion was palpable with the finger tip on the posterior wall and blood clots were noted on the examining finger. The other physical examinations were normal. The results of laboratory tests were all normal except for hemoglobin level of 8.8 g/dl. The gastric adenoma was successfully resected with endoscopy. Colonoscopy revealed a sessile polypoid lesion with 1 cm in diameter in the rectum (Fig. 2). Pathologically, it was papillo-tubular adenocarcinoma. The patient was taken to surgery and local mucosal resection per anus was performed. The post-operative course was uneventful.

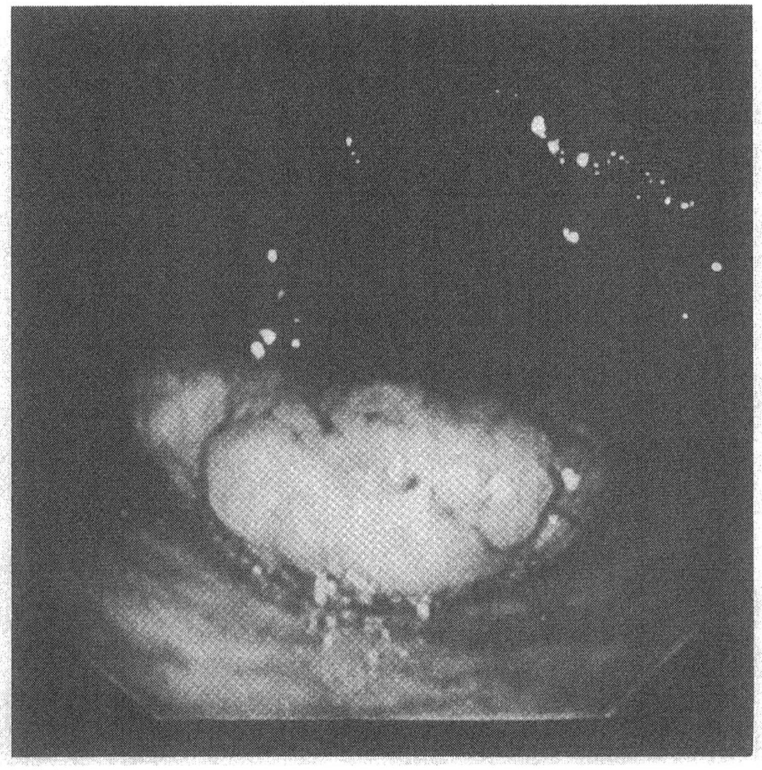

Figure 2. Colonoscopic appearance of rectal cancer in CASE 1.

CASE 2

In 1992, a 56-year-old male underwent an annual health check-up at the same clinic described in CASE 1. The results of the physical and laboratory data were all normal except for serum pepsinogen levels (Pepsinogen I 5.5 ng/ml, Pepsinogen II 5.3 ng/ml and a I/II ratio 1.0). On the basis of the result of his pepsinogen levels, he was further investigated by endoscopy. It revealed a flat elevated lesion with 5 mm in diameter at the anterior wall of the antrum (Fig. 3). His gastric mucosa was affected by severe and extensive atrophic gastritis. Histologically, the elevated lesion was adenoma. The patient was admitted to our hospital to resect the lesion endoscopically. He had lost 7.5 kg in weight in a few months. The physical examination on admission showed no abnormalities. Laboratory tests were normal except for hemoglobin level of 13.4 g/dl and CEA (carcinoembrionic antigen) of 11ng/ml (normal< 5 ng/ml). Stool was positive for occult blood test. The gastric adenoma was successfully resected with endoscopy. Then he underwent colonoscopy because of the high value of CEA and positive occult blood test in stool. It revealed three colonic polyps all of which were treated by endoscopic removal. Of three polyps, a focal cancer in an adenoma was observed in the pedunculated polyp present in the sigmoid colon with 12 mm in diameter (Fig. 4).

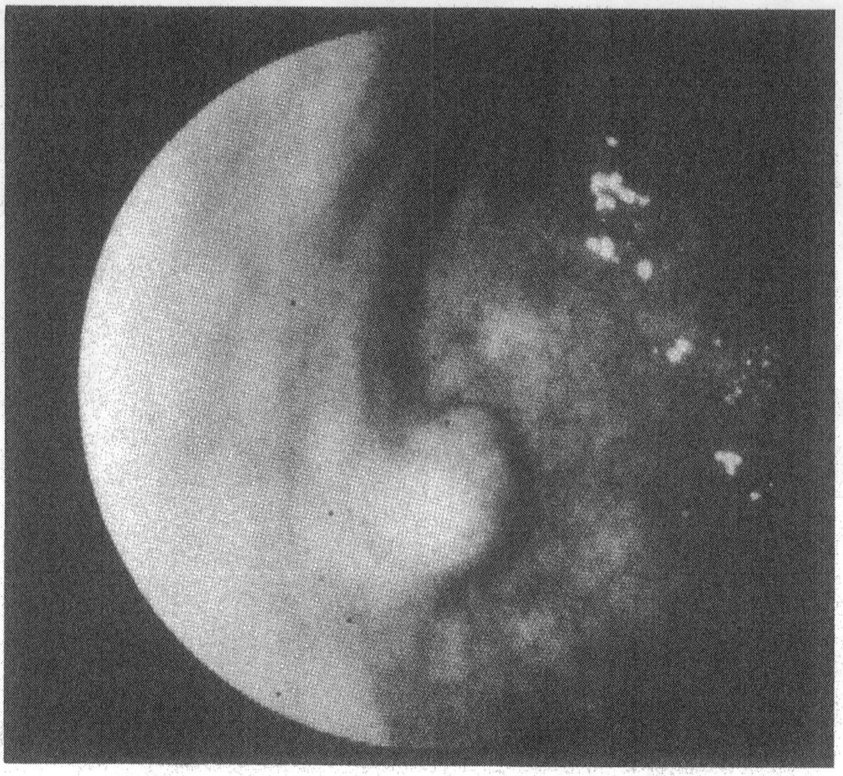

Figure 3. Endoscopic appearance of gastric adenoma in the antrum of the stomach in CASE 2.

DISCUSSION

Out of 7 patients with gastric adenoma screened by the serum pepsinogen method, at least two patients have been found to have colorectal carcinomas. It is unknown whether the other five subjects have colorectal cancer because they have not undergone colonoscopy. Is the association of colorectal cancer in the two patients merely fortuitous? Fabry *et al.* reported that of 6 patients with multiple gastric adenomas two had synchronous colonic cancer (6). Shemesh and his colleagues demonstrated that patients with multiple colonic adenomas or with colorectal cancers are at high risk of developing additional gastric neoplasms, particularly gastric adenomas (7). In addition, Ishimori and Kawamura reported the modified combination assay using pepsinogen I and/or pepsinogen II in addition to tumor markers such as CEA and AFP (alphafeto-protein) had a higher sensitivity for colorectal cancers than tumor markers alone (8). This result suggests the possibility that the patients with colorectal cancer show lower pepsinogen levels than the age-adjusted normal control subjects and leads to the hypothesis that high-risk subjects of gastric cancer also belong to the high-risk group of colorectal cancer. In this regard, it is interesting that Newbold have evoked a hypothesis that intestinal-type gastric carcinoma and colonic carcinoma have a common pathogenesis, since the former is histologically indistinguishable from the latter and undifferentiated cells capable of division are found on mucosal surface in precancerous conditions of stomach and colon; atrophic gastritis and adenomatous colonic polyp (9). In this line, we need to investigate whether the patients with colorectal cancer really show lower

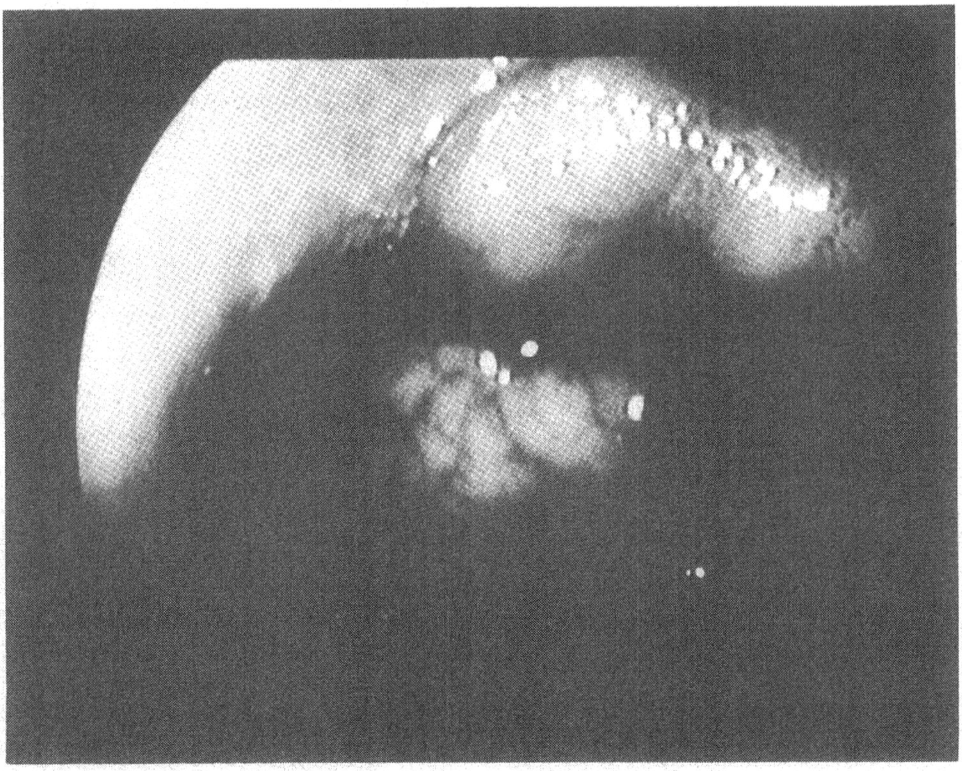

Figure 4. Colonoscopic appearance of colonic polyp in the transverse colon in CASE 2. Focal carcinoma was demonstrated in the removed polyp.

pepsinogen levels and compare the prevalence rates of colorectal cancer in subjects of lower pepsinogen levels with those in subjects of normal pepsinogen levels. Whether the serum pepsinogen levels could be a marker selecting the high risk group of colorectal cancer is a problem for future study.

REFERENCES

1. K. Miki, M. Ichinose, A. Shimizu, S. C. Huang, H. Oka, C. Furihata, T. Matsushima and K. Takahashi, Serum Pepsinogens as a Screening Test of Extensive Chronic Gastritis, *Gastroenterol. Jpn.* 22:133 (1987)
2. K. Miki, M. Ichinose, N. Kawamura, M. Matsushima, H. B. Ahmad, M, Kimura, J. Sano, T. Tashiro, N. Kakei, H. Oka, C, Furihata and K. Takahashi, The Significance of Low Serum Pepsinogen Levels to Detect Stomach Cancer Associated with Extensive Chronic Gastritis in Japanese Subjects, *Jpn. J. Cancer Res.* 80, 114 (1989)
3. A. Fukao, S. Hisamichi, N. Ohsato, N. Fujino, N. Endo and M. Iha, Correlation between the Prevalence of Gastritis and Gastric Cancer in Japan, *Cancer Causes and Control* 4:17 (1992)
4. M. Kabuto, H. Imai, S. Tsugane and S. Watanabe, Correlation between Atrophic Gastritis Prevalance and Gastric Cancer Mortality among Middle-aged Men in 5 Areas in Japan, *J. Epidemiol.* 3:35 (1993)
5. K. Miki, M. Ichinose, K. B. Ishikawa, N. Yahagi, M. Matsushima, N. Kakei, S. Tsukada, M. Kido, S. Ishihama, Y. Shimizu, T. Suzuki and K. Kurokawa, Clinical Application of Serum Pepsinogen I and II Levels for Mass Screening to Detect Gastric Cancer, *Jpn. J. Cancer Res.* 84:1086 (1993)
6. T. L. Fabry, A. Frankel and J. D. Waye, Gastric Polyps, *J. Clin. Gastroenterol.* 15:144 (1983)

7. E. Shemesh, A. Czerniak, A. Pines and L. Bat, Is There an Association between Gastric Polyps and Colonic Neoplasms?, *Digestion* 42: 212 (1989)
8. A. Ishimori and T. Kawamura, The New Examinations for Gastrointestinal Disorders, *J. Jpn. Soc. Int. Med.* 82:512 (1993)
9. K. M. Newbold, Intestinal-type Gastric Carcinoma and Colonic Carcinoma: A Common Pathogenesis?, *Lancet* i:131 (1989)

COMPARISONS OF THE THREE-DIMENSIONAL STRUCTURES, SPECIFICITIES AND GLYCOSYLATION OF RENINS, YEAST PROTEINASE A AND CATHEPSIN D

C.F. Aguilar, V. Dhanaraj, K. Guruprasad, C. Dealwis, M. Badasso, J.B. Cooper, S.P. Wood, and T.L. Blundell

Laboratory of Molecular Biology
Department of Crystallography
Birkbeck College
London WC1 7HX, UK

ABSTRACT

The crystal structures of complexes of the aspartic proteinases, human and mouse renins, yeast proteinase A and cathepsin D, with peptide analogue inhibitors are compared. Differences occur in the relative positions of the domain comprising residues 190-302 (pepsin numbering) compared to the remaining structure and in the nature and position of the irregular regions joining the ß-strands and α-helices. The first three of the five residues of the oligosaccharide structures attached to Asn 67 of yeast proteinase and cathepsin D cover the same region of the protein surface. All enzymes have an unusual, proline-rich region (292-297) which acts as a second flap (in addition to that involving residues 72-81). This covers the active site cleft, but can be very close to the substrate/inhibitor at P_3' and P_4' only in the renins.

INTRODUCTION

The first aspartic proteinases to be studied by X-ray analysis were digestive enzymes of broad specificity such as pepsin (1,2,3) and chymosin (4,5) from mammalian sources, and endothiapepsin (6), penicillopepsin (7), rhizopuspepsin (8) and mucorpepsin (9) from fungi. These bilobal enzymes have rather open active site clefts, consistent with binding a wide range of amino acid sequences.

Recently the three-dimensional structures of several other enzymes, some of which have more specific roles, have been reported. These include renin, the most specific enzyme

Aspartic Proteinases: Structure, Function, Biology, and Biomedical Implications
Edited by Kenji Takahashi, Plenum Press, New York, 1995

```
                 4         14            24          34          44
catx   G p̆ i p E v L k Ñ ÿ m d̄ a q Ÿ Y g ē I G I G ĩ p̆ p q ç F ĩ V v F d̃ T̃ G s̄ S ñ L W̃ V P S̃ i h̄
lypa   g G h d V p L t ñ y l ñ a q Ÿ Ÿ T d I t L G ĩ p̆ p q n F k V i L d̃ T̃ G s̄ S N L W̃ V P S̃ n ē
1bbs   t t s s V i L t Ñ y m d̄ t q Ÿ y G e I g I G ĩ p̆ p q t F k V v F d̃ T̃ g s̄ S N V W̃ V P S̃ s̄ k
mren   a d l i s p v v L t Ñ y l ñ s q Ÿ y g e I g I G ĩ p̆ p q t F k V i F d̃ T̃ g s̄ A N L W̃ V P S̃ ĩ k̄
       ββββββββφφ ββββββββφ ω ββββββββ βββ
       •  •  •   • •  •   •  •  •  •  ••••••

        52        62          72          82          91a
catx   Ç k l l D̃ i A Ç w i h h k Ÿ ñ s d̄ k S̃ s ĩ y v k ñ g ĩ s F d̄ i h̄ ÿ g s g s̄ L s̄ g ÿ L S̃ q D̃ T V S V p̆
lypa   Ç g - - s̄ l A Ç f l H s̄ k Ÿ d h̄ e a S̃ s̄ s̄ ÿ k a n g t e F a i q ÿ g t g s L e G ÿ I S̃ q D̃ t L s I
1bbs   Ç s r l y t A Ç v y h k l F d̄ a s̄ d̄ S̃ s s̄ y k h ñ g ĩ e l ĩ L a ÿ s t g t V s̄ g f L S̃ q D̃ i I t V
mren   Ç s r l y l A Ç g i H s l Ÿ ē s s̄ d S̃ s s̄ y m̄ e ñ g d̄ d̄ F ĩ i h ÿ g s g r̄ V k G f L S̃ q̄ D̃ s V ĩ V
                    3 3 3    βββ ββββββ     βββ ββββββββββ
       •    ••    •    •     •     •  •  •• •  •  •

        92       98          108            118          128          138
catx   ç a  g G v k V e r Q̃ v F G Ē A t k Q̃ p g i t F i a a k̄ f D̃ G I L G M A Ÿ p r i S̃ v n n v l P V F D̃
lypa   - -  g d l t I p k Q d F A Ē A t s Ē p g l t f a f g k̄ f D̃ G i L G L G Ÿ d t i s v d k V v P P F ÿ
1bbs   - -  g g i t V t - Q̃ m F G Ē V t e M p a l p̆ f m l a e f d G v V G M G F i e q A i g a v t P I F d̄
mren   - -  g g I ĩ V t - Q̃ t F G Ē V t q̄ L p l i p̆ F m l A q f D G v L G M G F p a q A v g g v t P V F d̄
          ββββ ββββββββ  φ  α     ββββ   3 3 3        ααα
          •  • ••••  •     •• •••••   •     •••

        148          158          164          174          184
catx   ñ L m̄ q q k l V d q ñ i F S F y L s r̄ d p - d   a q p̆ G G ē L m L G G ĩ d̄ s k̄ y ÿ k̄ g s l s y l n V
lypa   ñ a i q q̄ d l L d ē k r̄ F A F y L g d̄ t s k d   t e n g G ē A T̃ F G G i d̄ e s̄ k̄ f k g d i t w L p V
1bbs   ñ I i s q g v L k e d V F S̃ F Ÿ Y ñ r d s - -   - a l g G q̄ I V L G g s̄ d p q h̄ y e g n f h̄ ÿ i n L
mren   h̄ I l s q g v L k e k V F S̃ V Ÿ Y ñ r g p - -   a l l G G ē V V L G g s̄ d̄ p q h y q g d f h̄ ÿ v s L
       αααα φ    βββββ        ββββφ       φ ββββ
       •   •   ••• •         • • ••           •

        194          204          214          224          234
catx   ĩ r k A y W Q̃ V h̄ L d̄ q̄ V ē V a s g l t L ç k e G ç ē A I V d̃ T̃ g ĩ s l M V G p v d e V r e L Q̃ k a
lypa   r r̄ k̄ a y W Ē V k F e G I G L g d ē y - a e l e s h̄ G A A I d̃ T̃ g ĩ s l I T L p s̄ g l A ē m̄ I Ñ a ē
1bbs   i k ĩ G v W q̄ I q M k G V s̄ V g s s t l L ç e d g ç l A l V d̃ T̃ G a s y I s G s ĩ s s l e k L M e a
mren   s̄ k t d s̄ W q̄ I ĩ M k g V s V g s s t l l Ç e ē g ç ē V v V d̃ T̃ G s s f I S A p ĩ s s L k l I M q a
          ββββ ββββφ   β φ φφββββββ    βββ αααααααααα
          • • •   •        • • •   •••  •  ••
```

	244 254 264 274 280
catx	I *g* A v p l i *q g* ē y m I p ç ē k v s t̲ L p a I t L k̃ L *g g* k g y k L s̄ p e d̃ Y̲ T̲ L k v s̲ q a *g k t̃*
1ypa	I *g* A k k̃ g w t g q̃ Ỹ t̲ L d̃ Ç n t̲̃ r̲̃ d̲ n l p d L i F ñ̲ F *n g* y ñ F t I G p y d̃ Y̲ T̲ l ē v s - - - - g
1bbs	L *g* A a a a - l f d̃ y v V a Ç a e G p t̲ L p d I s̄ F h̲̃ L *g g* k̃ ē Y t̃ L t s a d Y̲ V f q̲ a *s̃* y a a a k
mren	L *g* A a e k r̃ l *h* ē y V V s̲̄ ç s̄ q V p t̲ l p d I s̄ F n L *g g* r a ỹ t̃ L s̄ S̲ t̃ d̃ Y̲ V l *a* a a - - - - *k*
	α ϕ β β β β β 3 3 3 β β β β β ϕ ϕ β β β β β α α α α β β
	• • • • • • • • •

	290 300 310 320
catx	l Ç l S̃ G F m̃ G m d i p p p̆ s G p̆ L W̃ i L G D̲ V F I G r̲̃ y̲ **Y̲** T̲ V F **D̃** r̃ d̃ ñ *ñ* r̃ V G F a ē a a
1ypa	s̄ Ç i S̲̄ A I t̃ p m d̃ f p e p̆ v g p̆ L A i V G D̲ A F L R̲̃ k̃ **Y̲ Y̲** S̃ I Ỹ d̲ i g n *ñ̲* A V G L A k A i
1bbs	l Ç t̃ L A I h̲̃ a m d̃ i p p p̆ t G p̆ T̲̃ W a L G A T̲ F I r̲̃ k̲ f **Y** T̲ ē F d̃ r̃ r n *ñ* r̃ I G F a l A r
mren	l ç t V A L h a m d̃ i p p p̆ t G p̆ V W̃ v L G A T̲ F I r̲̃ k̲̃ f **Ỹ** T̲ ē F **D̃** r̃ h ñ *n̲* r̃ I G F A l A r
	β β β β β β β β β β α α α α β β β β β β ϕ β β β β β β
	• •

Figure 1. The alignment of the sequences of cathepsin D (catx), yeast proteinase A (1ypa), human renin (1bbs) and mouse renin (mren). Pepsin numbering is used. Local structural features are also illustrated: solvent inaccessible residues, uppercase; solvent accessible, lower case; positive phi, italic; cis-peptide, breve; hydrogen bond to other sidechain, tilde; hydrogen bond to mainchain amide, bold; hydrogen bond to mainchain carbonyl, underline; disulphide bond, cedilla.

of the group, which is involved in the formation of angiotensin II, a potent pressor and aldosterogenic substance controlling vascular tone, fluid volume and sodium excretion. Because of the unique specificity of renin, its inhibition is widely expected to provide selective therapy for hypertension, congestive heart failure and associated degenerative disorders linked to angiotensin II. Several crystal structures of human renin have been reported at medium resolutions both for the uncomplexed enzyme (10,11) and various complexes with inhibitors (11,12). A higher resolution structure of a mouse renin complex with a decapeptide inhibitor is also reported (12,13). Crystal structures have been reported for cathepsin D (14,15), a lysosomal proteinase implicated in the invasiveness of breast cancer which is also a target for drug design. The crystal structure of a related yeast aspartic proteinase has also been described (16). Yeast proteinase A is found in the vacuoles of *Saccharomyces cerevisiae* and is involved in proteolysis during nitrogen starvation and proteolytic maturation of other vacuolar enzymes (17).

These structures have a similar general topology, but in addition a further flap covers the active site in a way that might contribute to specific interactions with the substrate. In this paper we review the crystal structures of the renins (12,13) and yeast proteinase A (16), defined by X-ray analysis in our laboratory, and compare them with the crystal structure of human cathepsin D, the coordinates of which were kindly made available by Dr John Erickson (15).

Comparisons of Sequences and 3-D Structures

Figure 1 gives an alignment of the sequences of renins, cathepsin D and proteinase A and also illustrates local structural features derived from the crystal structures. The phyletic tree based on the sequences alone (Figure 2) shows that cathepsin D clusters slightly more closely to pepsin and chymosin than to the renins. Yeast proteinase is closely related to all

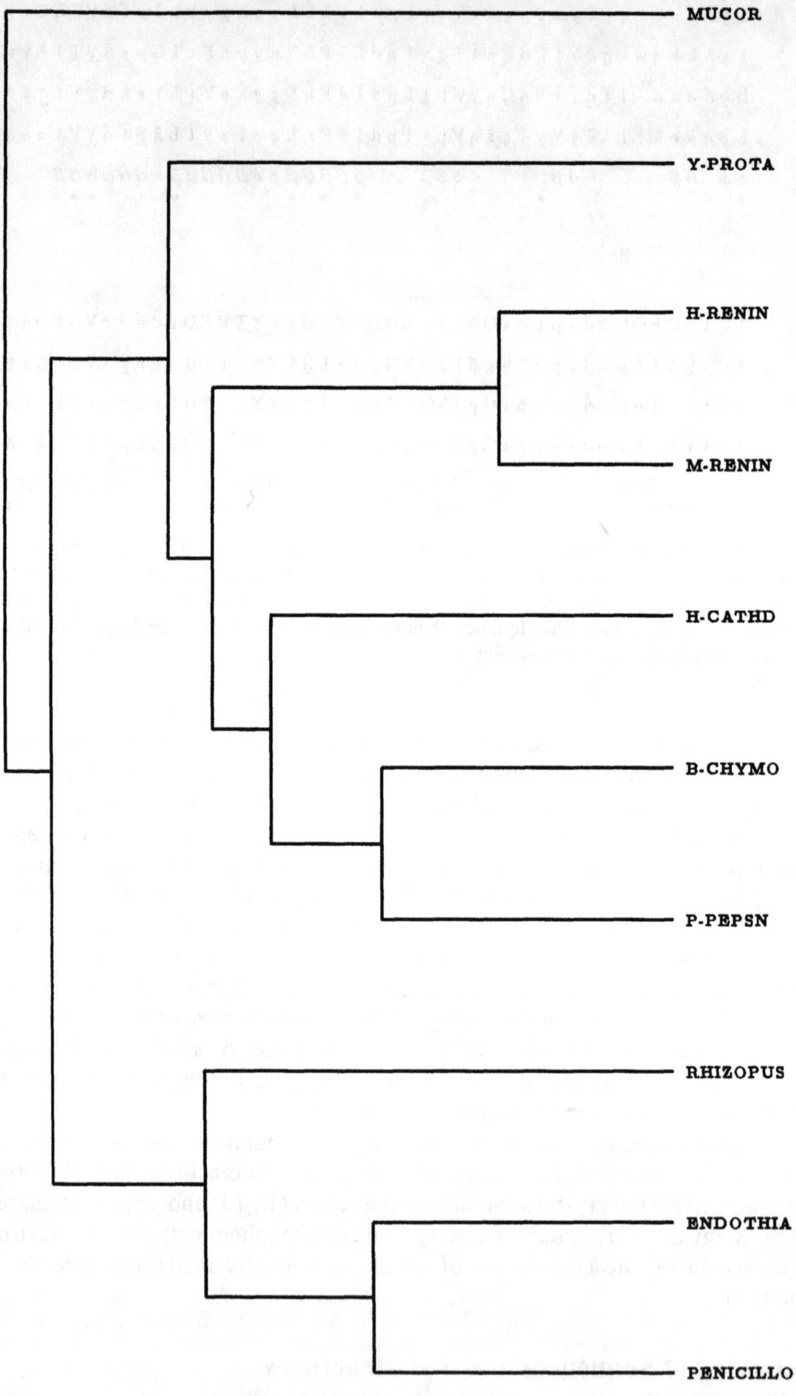

Figure 2. Phyletic tree based on sequences of the family of aspartic proteinases of known three-dimensional structure: MUCOR, Mucor pusillus pepsin; Y-PROTA, Yeast proteinase A; H-RENIN, human renin; M-RENIN, mouse renin; H-CATHD, human cathepsin D; B-CHYMO, bovine chymosin; P-PEPSN, porcine pepsin; RHIZOPUS, Rhizopuspepsin; ENDOTHIA, Endothiapepsin; PENICILLO, Penicillopepsin.

mammalian aspartic proteinases (approximately 40% sequence identity) and more distantly related (about 25%) with the fungal enzymes.

The four enzymes have two topologically similar lobes (Figure 3) with a local two-fold axis passing through the plane containing the active site aspartates. Šali *et al.*(18) found that endothiapepsin exists in two forms that differ in orientation of part of the carboxyl-terminal lobe with respect to the rest of the molecule. The same type of structural variability also accounts for some of the differences in other aspartic proteinases (2,3,19,20) and those compared here as shown in Figure 4. The first of the two rigid bodies comprises residues 2 to 189 and 303 to 326 (pepsin numbering after ref 2). The second comprises residues 190 to 302. Rigid body comparison indicates that the largest rotation shift occurs between proteinase A and endothiapepsin, (mean=18.0°). The rotation shifts for mouse renin and pepsin were 6.6° and 4.7° with respect to yeast proteinase. The translation shifts are negligible.

Unique to the renins is a *cis*-proline, Pro 111, which caps the helix h_{N2} and contributes to the subsites S_3 and S_5. On an equivalent loop in the C-lobe (related by the intra-molecular pseudo 2-fold symmetry), there is a sequence of prolines: Pro 292, Pro 293, Pro 294 and Pro297. Pro 294 and Pro 297 are in a *cis*-conformation. This structure is also found in cathepsin D and a similar structure, but with a glutamic residue replacing Pro 293, is found in yeast proteinase. Although these loops have similar conformations, they have different positions relative to the active site clefts, partly due to the rigid body movements and partly due to different residues at positions 291 and 295. The equivalent loops are very different in other aspartic proteinases such as pepsin, chymosin and the fungal proteinases (see Figure 5).

In renins, cathepsin D and proteinase A, the proline-rich loop and the loop comprised of residues 241-250, lie on either side of the 'flap' formed by residues 72-81. Thus, the cleft is covered by 'flaps' from both lobes rather than from the N-lobe alone as in other pepsin-like aspartic proteinases (Figure 6); this gives a superficial similarity to the dimeric, retroviral proteinases where each subunit provides an equivalent flap that closes down on top of the inhibitor.

Human renin, cathepsin D and yeast proteinase A are all glycoproteins. In yeast proteinase A, five residues, NAG , NAG, MAN, MAN and NAG of the oligosaccharide structure attached to Asn 67 are clearly visible in the electron-density maps. This five-residue sequence constitutes one of the branches of the postulated trimannoside structure. The carbohydrate chain is stabilized through two hydrogen bonds between the protein and the carbohydrate moiety: Asp131 O-N2 NAG2 (3.0 Å), Asp 100 OD1-O6 NAG1 (2.8 Å). There are also two hydrogen bonds between Asp67 and Thr69: 67 OD1-OG1 69 (2.6 Å) and 67 OD1-N 69 (2.7 Å). The first three saccharide residues of the oligosaccharide moiety are attached to the equivalent residue (Asn67) in cathepsin D and cover the same region of the protein surface (see Figures 3a and 3b). This degree of similarity in saccharide binding between the two enzymes is quite unexpected. Although the oligosaccharide in cathepsin D is involved in targetting to the lysosome, proteinase A is targetted to the vacuole via a different mechanism and so targetting cannot account for the similarity of oligosaccharide structures. The second carbohydrate attachment site, 266 in proteinase A and 183 in cathepsin D, is disordered in both enzymes.

Human renin is N-glycosylated at two sites: Asn -2 is fully exposed to the solvent and the disorder is expected, and although the GlcNAc residue attached to Asn 67 is well-defined, little of the other saccharide groups are visible. The mouse submaxillary renin has no N-glycosylation site and it appears that glycosylation is not essential for activity.

Active Site Cleft and Inhibitor Binding

In the crystal structures, cathepsin D, yeast proteinase A and mouse and human renins are complexed with peptide analogues that mimic the transition state. The inhibitors can be

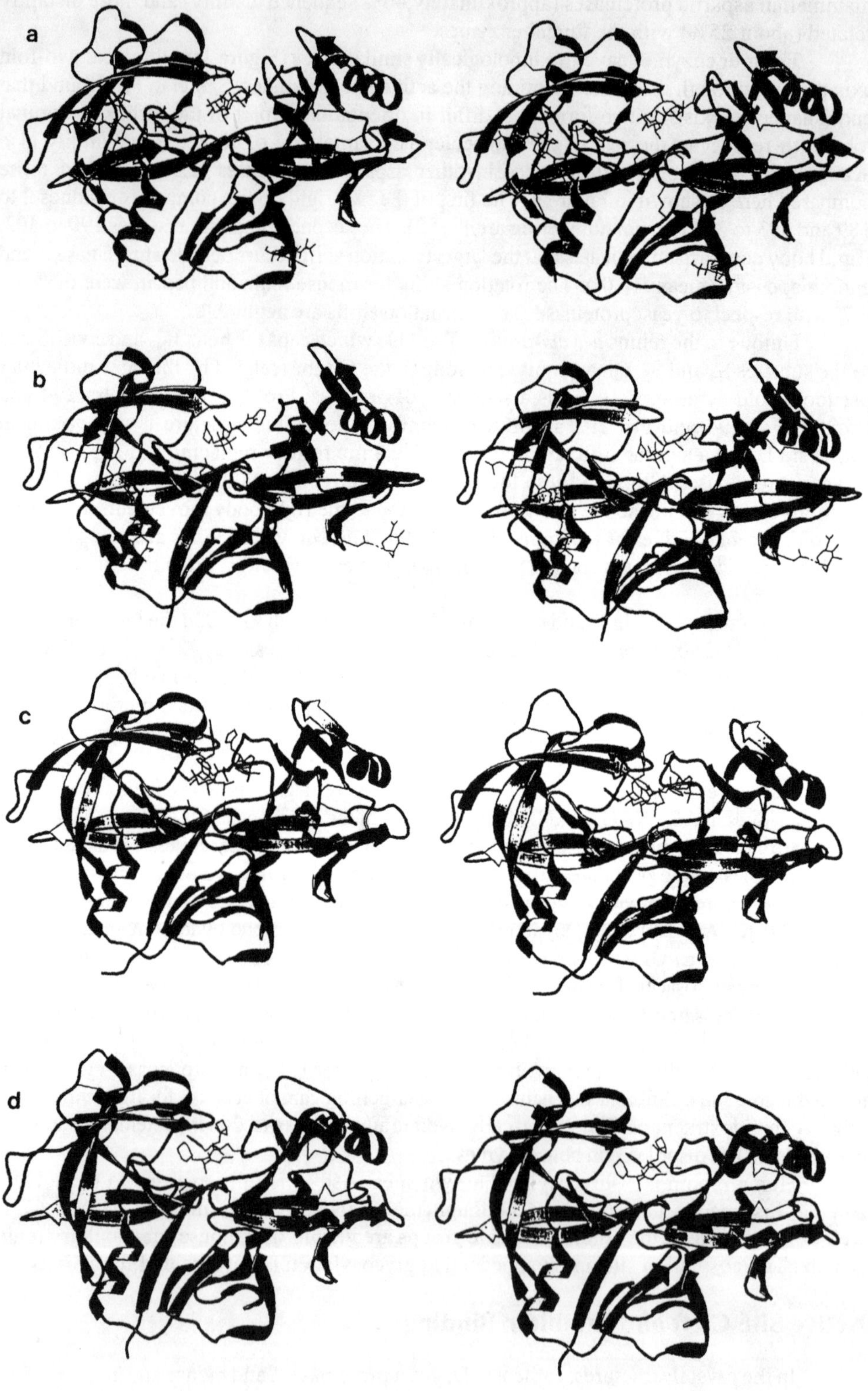
a
b
c
d

compared to those bound to other aspartic proteinases such as rhizopuspepsin (21), penicillopepsin (22), endothiapepsin (23, 24, 25, 26) and pepsin (19).

The difluorostatone-containing tripeptide inhibitor CP-81,282 (4(S)-[[[N-(morpholino-1-carbonyl)-L-phenylalanyl]-L-norleucyl]-5-cyclohexyl-2,2-difluoro-3-oxopenta noyl]-methylamine) of proteinase A can be directly compared to the structure of a complex of the same inhibitor with endothiapepsin (27). Superposition of the proteinase A-CP-81,282 and endothiapepsin-CP-81,282 complexes indicate that they are equivalent in position and hydrogen bonds. This demonstrates how inhibitor/substrate binding is conserved over very great evolutionary distances even if the enzyme evolves to a very different sequence.

Human renin is complexed with a P_4 to P_1' norstatine-containing inhibitor CP-85,339 (Pro-Phe-SMeCys-norCStaOiPr hydrochloride) and mouse renin is complexed with a decapeptide hydroxyethylene isostere inhibitor CH-66 (Piv-His-Pro-Phe-His-Leu-ψ[CH(OH)CH_2]-Leu-Tyr-Tyr-Ser-NH_2). Although the human renin inhibitor extends only from P_4 to P_1, the cyclohexyl norstatine residue at P_1 mimics a dipeptide analogue with its isopropyloxy group occupying the subsite for the side-chain of P_1'. Fortunately the mouse renin inhibitor extends further at both ends, from P_6 to P_4', with all residues clearly defined.

All four inhibitors are bound in the extended conformation that characterises other aspartic proteinase-inhibitor complexes and make extensive hydrogen bonds with the enzymes as shown in Figure 7. The conformations and interactions are most similar from P3 to P1'. However, the hydrogen bond observed between P_3 N and 219 O_γ in most aspartic proteinases is not made in the human renin complex, apparently as a consequence of a repositioning of the protonated Pro P_4 at the N-terminus of this short inhibitor.

Let us first consider renin-inhibitor interactions. In modelling exercises (28-32), it was assumed that specificities derive from differences in the sizes of the residues in the specificity pockets (S_n) which complement the corresponding side chains at positions P_n in the substrate/inhibitor (33). However, this simple assumption only partly accounts for the steric basis of specificity. For example, in the specificity subsite S_3 the phenyl rings of Phe P_3 occupy almost identical positions in both human and mouse renin inhibitor complexes. Modelling studies have predicted the specificity subsite S_3 to be larger in renins than in other aspartic proteinases due to substitution of smaller residues, Pro 111, Leu 114 and Ala 115, in place of larger ones in mammalian and fungal proteinases. However, a compensatory movement (~5Å) of the helix h_{N2} makes the pocket quite compact and complementary to the aromatic ring. Thus, the positions of secondary structures differ between renin and other aspartic proteinases with a consequent important difference in the specificity pocket.

Differences in the positions of secondary structural elements may be particularly important to the specificity differences at P_2'. Mouse submaxillary and other non-primate renins do not appreciably cleave human angiotensinogen or its analogues (34) which have an isoleucine at P_2', although they do cleave substrates with a valine at this position. In contrast, human renin not only cleaves the human and non-primate substrates but also the rat angiotensinogen with tyrosine at P_2', albeit rather slowly (35). This can be explained in terms of the three-dimensional structures. In the mouse renin complex, the P_2' tyrosyl ring is packed parallel to the adjoining helix h_3 in a narrow pocket and there is only limited space

---

Figure 3. Schematic stereo-representation of the polypeptide-chains of the four aspartic proteinases. (a) human cathepsin D complexed with pepstatin inhibitor, (b) yeast proteinase A complexed with a difluorostatone-containing inhibitor (CP-81,282), (c) mouse renin complexed with a decapeptide hydroxyethylene isostere inhibitor (CH-66) and (d) human renin complexed with a norstatine-containing inhibitor (CP-85,339). Helices are shown as cylinders and sheets as flat ribbons. The catalytic aspartates are shown in all the enzymes. The glycosylation of cathepsin D and yeast proteinase is also indicated. Lys187 implicated in lysosomal targetting in the case of cathepsin D is also shown.

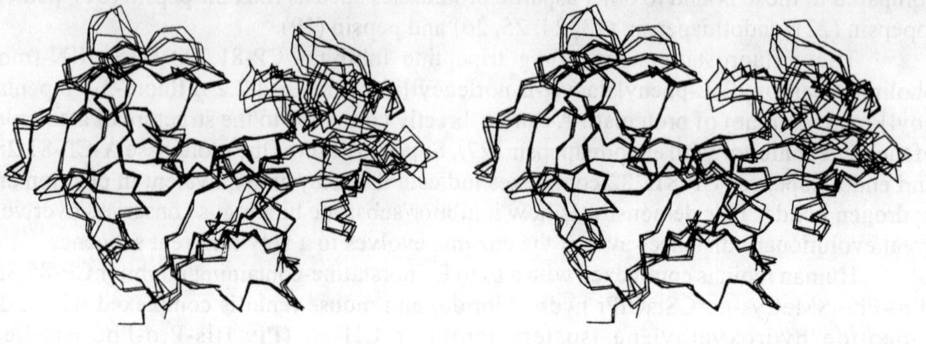

Figure 4. Least-squares superposition of the Cα backbones of the four aspartic proteinases showing the relative rigid body shifts in the C-domain.

available beyond the C_β methylene group. This is able to accommodate a valine, but not the larger isoleucine at P_2'. On the other hand in the human renin structure, differences in the orientation and position of helix h_3 bring it closer (by ~0.5Å) to the substrate-binding site than in mouse renin. It is orientated in such a fashion in human renin that, although it can accommodate the isobutyl side chain of isoleucine at P_2', aromatic rings of phenylalanine and tyrosine will have severe short contacts with the side chain of Ile 130 (valine in mouse renin). Thus, the reorientation of a helix, coupled with subtle differences in the shapes of the side chains, makes significant changes in the substrate specificity at this subsite.

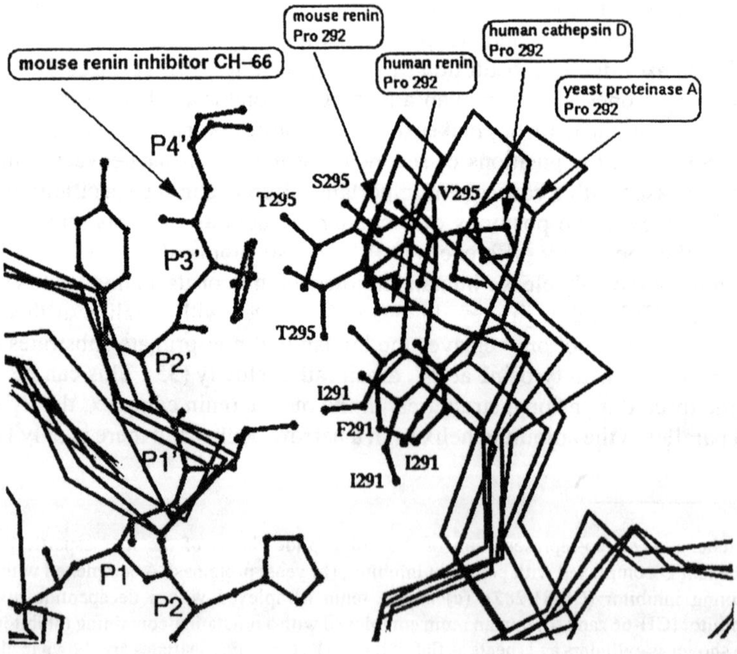

Figure 5. Least-squares superposition of the enzymes in the proline rich loop region showing interactions with the inhibitor near the prime site.

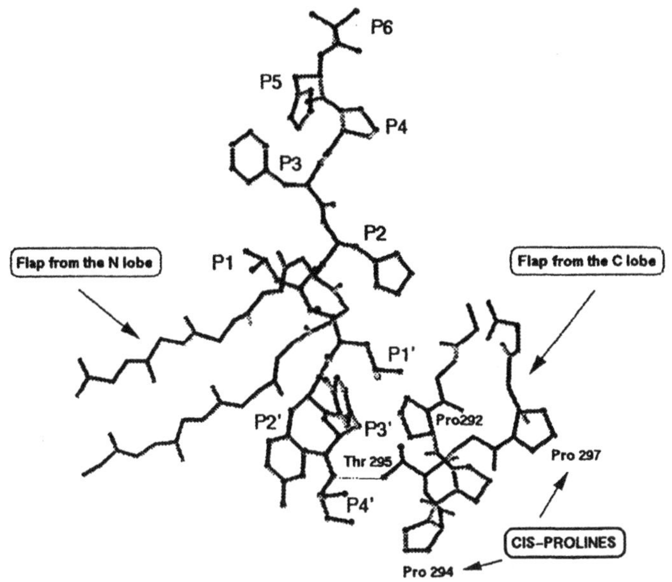

Figure 6. Flaps in the inhibitor-bound mouse renin showing interactions at the prime site.

There are significant differences between the renins and the other enzymes, cathepsin D and yeast proteinase at P2'. The small residues at 189 (valine and serine in human and mouse renins respectively) are replaced by the larger aromatic sidechain of tyrosine in both cathepsin D and yeast proteinase A and leads to differences in the relative positions of the inhibitors and the enzymes.

The three-dimensional structures of human and mouse renins show that the situation at S_1' is more complex. Differences in the S_1'subsites for the human and mouse renins are complementary to the valine and leucine at P_1' in human and mouse angiotensinogens (28,29); residue 213 is leucine in human renin and valine in mouse renin. The S_1' pockets of chymosin, pepsin and endothiapepsin have an aromatic side-chain at residue 189 whilst the renins have amino acids with smaller side-chains (valine in human and serine in mouse renins). This would be expected to make the pocket larger in renins. However, the structure of the mouse renin complex shows that the substrate moves closer to the enzyme in renins as a result of the smaller residue at 189 and the pocket is made even more compact due to changes in the position and composition of the poly-proline loop (residues 290-297). Thus, the specificity difference at this site arises not only from the movement of a secondary structure, in this case a loop region, but also from the substitution of an enzyme residue that allows the substrate to come closer to the body of the enzyme.

The specificity pockets are also influenced by the elaboration of loops on the periphery of the binding cleft in renins. This is most marked at P_3' and P_4'. For instance, in endothiapepsin, which has been the subject of the greatest number of studies, different conformations are adopted at P_3' and the residue at P_4' is generally disordered. In contrast these residues are clearly defined in mouse renin. This is mainly a consequence of the poly-proline loop, illustrated in Figure 5. The X-ray analysis of the mouse renin complex shows that the S_3' and S_4' subsites are formed by the poly-proline loop together with residues of the 'flap' and a similar situation is likely to occur in human renin. The well-defined interactions of P_3' described in the mouse renin complex explains the significant affinity when inhibitors have phenylalanine or tyrosine at P_3' as well as the importance of a P_3' residue

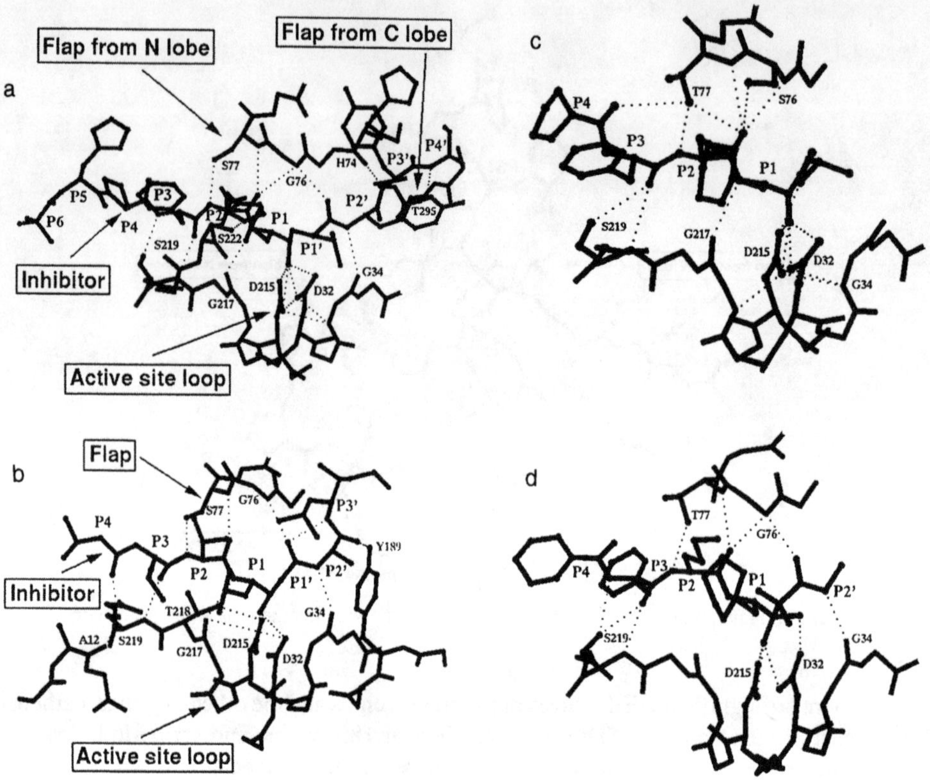

Figure 7. Interactions of the inhibitors and enzymes. (a) mouse renin, (b) cathepsin D, (c) human renin and (d) yeast proteinase A.

for catalytic cleavage of a substrate by renin. In cathepsin D and yeast proteinase, this proline-rich loop has a similar conformation (see above) but occupies a position further away from the inhibitors. This probably accounts for their decreased specificity for substrates at S_3' and S_4'.

The locations of the five oxygen atoms of the catalytic aspartyl side chains and the inhibitor hydroxyl group are essentially superimposable in all complexes. The C-OH bonds lie at identical positions when the structures of inhibitor complexes of several aspartic proteinases are superposed, in spite of the differences in the sequence and secondary structure. The complex array of hydrogen bonds is common to the enzymes with the exception that the hydrogen bond to threonine or serine at 218 is replaced by alanine in human renin. These observations emphasise that the enzymes must have the same *modus operandi* although their pH optima are quite different.

The detailed analyses of inhibitor complexes reported here indicate the general structural features that may contribute to specificity, especially of renin but demonstrate the need for careful, high resolution X-ray analyses for more confidence in drug design. In particular, they show that even minor alterations in the positions of secondary structural elements can lead to major changes in the disposition of the subsites. Since these define the species-specificity and determine the catalytic efficiency of the enzymes, a thorough understanding of their composition is indispensable for the synthesis of suitable inhibitors. The specificity pockets are further modified by elaboration of surface loops, which can be disordered in the uncomplexed enzymes and difficult to model with precision from homologous structures. These data establish a new foundation for the rational design of renin and cathepsin D inhibitors in the search for clinically successful candidates.

REFERENCES

1. N.S. Andreeva, A.S. Zdanov, A.E. Gustchina and A.A. Fedorov. *J. Biol. Chem.*, 259:11353 (1984).
2. A.R. Sielecki, A.A. Fedorov, A. Boodhoo, N.S. Andreeva and M.N.G. James. *J. Mol. Biol.*, 214:143 (1990).
3. J.B. Cooper, G. Khan, G. Taylor, I.J. Tickle and T.L. Blundell. *J. Mol. Biol.*, 214:199 (1990).
4. G.L. Gilliland, E.L. Winbourne, J. Nachman and A. Wlodawer. *Proteins*, 8:82 101 (1990).
5. M.P. Newman, M. Safro, C. Frazao, G. Khan, A. Zdanov, I.J. Tickle, T.L. Blundell and N. Andreeva *J. Mol. Biol.*, 221:1295 (1991).
6. T.L. Blundell, J.A. Jenkins, B.T. Sewell, L.H. Pearl, J.B. Cooper, I.J. Tickle, B. Veerapandian and S.P. Wood. *J. Mol. Biol.*, 214:199 (1990).
7. M.N.G. James and A.R. Sielecki. *J. Mol. Biol.*, 163:299 (1983).
8. K. Suguna, R.R. Bott, E.A. Padlan, E. Subramanian, S. Sheriff, G.H. Cohen and D.R. Davies. *J. Mol. Biol.*, 196:877 (1987a).
9. M. Newman, F. Watson, P. Roychowdhury, H. Jones, M. Badasso, A. Cleasby, S.P. Wood, I.J. Tickle and T.L. Blundell. *J. Mol. Biol.*, 230:260 (1993).
10. A.R. Sielecki, K. Hayakawa, M. Fujinaga, M.E.P. Murphy, M. Fraser, A.K. Muir, C.T. Carilli, J.A. Lewicki, J.D. Baxter and M.N.G. James. *Science*, 243:1346 (1989).
11. J. Rahuel, J.P. Priestle, M.G. Grütter. *J. Struct. Biol.*, 107:227 (1991).
12. V. Dhanaraj, C.G. Dealwis, C. Frazao, M. Badasso, B.L. Sibanda, I.J. Tickle, J.B. Cooper, H.P.C. Driessen, M. Newman, C. Aguilar, S.P. Wood, T.L. Blundell, P.M. Hobart, K.F. Geoghegan, M.J. Ammirati, D.E. Danley, B.A. O'Connor and D.J. Hoover. *Nature*, 357:466 (1992).
13. C. Dealwis, C. Frazao, M. Badasso, J.B. Cooper, K. Murakami, M. Szelke, S.P. Wood and T.L. Blundell. *J.Mol.Biol.*, in press (1993).
14. P. Metcalf and M. Fusek. *EMBO J.*, 12:1293 (1993).
15. E.T. Baldwin, T.N. Bhat, S. Gulnik, M.V. Hosur, R.C. Sowder, R.E. Cachau, J. Collins, A.M. Silva and J.W. Erickson. *Proc. Natl. Acad. Sci. USA.*, 90:6796 (1993).
16. C. Aguilar, M. Badasso, J. Cooper, T. Dreyer, D. Hoover, S.P. Wood and T.L. Blundell. Manuscript in preparation (1994).
17. T. Dreyer. *Carlsberg Res. Commun.*, 54:85 (1989).
18. A. Šali, B. Veerapandian, J.B. Cooper, S.I. Foundling, D.J. Hoover and T.L. Blundell. *EMBO J.*, 8:2179 (1989).
19. C. Abad-Zapatero, T.J. Rydel and J. Erickson. *Proteins*, 8:62 (1990).
20. A. Šali, B. Veerapandian, J.B. Cooper, D.S. Moss, T. Hofman and T.L. Blundell. *Proteins, Structure, Function and Genetics*, 12:158 (1992).
21. K. Suguna, E.A. Padlan, R.B. Bott, J. Boger, K.D. Parris and D.R. Davies. *Proteins*, 13:195 (1992).
22. M.N.G. James, A.R. Sielecki, K. Hayakawa, M.H. Gelb. *Biochemistry*, 31:3872 (1992).
23. D. Bailey, J.B. Cooper, B. Veerapandian, T.L. Blundell, B. Atrash, D.M. Jones and M. Szelke. *Biochem. J.*, 289:363 (1993).
24. S.I. Foundling, J.B. Cooper, J. Watson, F.E. Cleasby, L.H. Pearl, B.L. Sibanda, A. Hemmings, S.P. Wood, T.L. Blundell, T.L. Valler, J. Kay, J. Boger, B.M. Dunn, B.J. Leckie, D.M. Jones, B. Atrash, A. Hallet and M. Szelke. *Nature*, 327:349 (1987).
25. J.B. Cooper, S.I. Foundling, A. Hemmings, T.L. Blundell, D.M. Jones, A. Hallet and M. Szelke. *Eur. J. Biochem.*, 169:215 (1987).
26. J.B. Cooper, S.I. Foundling, T.L. Blundell, J. Boger, R.A. Jupp and J. Kay. *Biochemistry*, 28:8596 (1989).
27. B. Veerapandian, J.B. Cooper, A. Sali, T.L. Blundell, R.L. Rosati, B.W. Dominy, D.B. Damon and D.J. Hoover. *Protein Science*, 1:322 (1992).
28. T.L. Blundell, B.L. Sibanda and L.H. Pearl. *Nature*, 304:273 (1983).
29. B.L. Sibanda, T.L. Blundell, P.M. Hobart, M. Fogliano, J.S. Bindra, B.W. Dominy and J.M. Chirgwin. *FEBS Lett.*, 174:102 (1984).
30. K. Akahane, H. Umeyama, S. Nakagawa, I. Moriguchi, S. Hirose, K. Iizuka and K. Murakami. *Hypertension*, 7:3 (1985).

31. W. Carlson, M. Karplus and E. Haber. *Hypertension*, 7:13 (1985).
32. C. Hutchins and J. Greer. *Critical Reviews in Biochemistry and Molecular Biology*, 26:77 (1991).
33. I. Schechter and A. Berger. *Biochem. Biophys. Res. Commun.*, 27:157 (1967).
34. M. Poe, J.K. Wu, Tsau-Yen Lin, K. Hoogsteen, H.G. Bull and E.E. Slater. *Analyt. Biochem.*, 140:459 (1984).
35. F. Cumin, D. Le-Nguen, B. Castro, J. Menard and P. Corvol. *Biochim. Biophys. Acta.*, 913:10 (1987).

DISCOVERY OF INHIBITORS OF HUMAN RENIN WITH HIGH ORAL BIOAVAILABILITY

Dennis J. Hoover,[1*] Bruce A. Lefker,[1*] Robert L. Rosati,[1]
Ronald T. Wester,[1] Edward F. Kleinman,[1] Jasjit S. Bindra,[1]
William F. Holt,[2] William R. Murphy,[2] Michael L. Mangiapane,[2]
Gregory M. Hockel,[2] Ian H. Williams,[2] Ward H. Smith,[3]
Michael J. Gumkowski,[3] Richard M. Shepard,[4] Mark J. Gardner,[4] and
Mark R. Nocerini[4]

[1]Department of Medicinal Chemistry
[2]Department of Cardiovascular and Metabolic Diseases
[3]Department of Pharmaceutical Research and Development
[4]Department of Drug Metabolism
Central Research Division
Pfizer Inc, Groton, CT 06340

INTRODUCTION

Oral bioavailability has been a formidable barrier to the development of drugs which mimic the structures of biologically active peptides. This report describes the discovery of potent renin inhibitors having high oral bioavailability and good plasma pharmacokinetics. These are substances derived by modification of the P4-P1' sequence of angiotensinogen, the renin substrate. These results demonstrate a successful peptide mimetic approach to the design of orally bioavailable pharmacologically active agents.

Hypertension is a huge medical problem, by some estimates affecting up to 15% of the population in the Western World. An increasingly preferred strategy for treating hypertension is to reduce levels of the pressor hormone angiotensin II (AII), by blockade of the renin-angiotensin system (RAS, Figure 1). In the RAS, the aspartic protease renin cleaves the plasma glycoprotein angiotensinogen at the Leu^{10}-Val^{11} bond, generating angiotensin I (AI), a decapeptide. AI is converted by angiotensin converting enzyme (ACE), to AII. AII mediates a myriad of receptor-mediated events including blood vessel constriction, fluid retention, and proliferation of vascular tissue.

Three approaches to blocking the RAS have been extensively explored. These are inhibition of renin, inhibition of ACE, and antagonism of AII receptors. Presently, the only approved drugs are ACE inhibitors (ACEI). Many AII antagonists are in clinical trials. ACEI

Aspartic Proteinases: Structure, Function, Biology, and Biomedical Implications
Edited by Kenji Takahashi, Plenum Press, New York, 1995

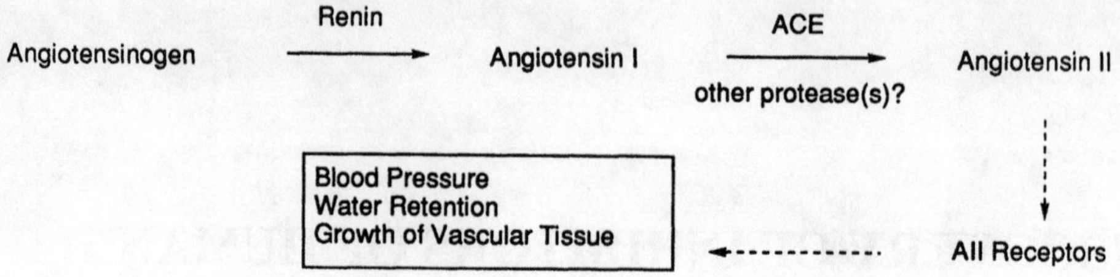

Figure 1. The renin-angiotensin system.

have proved widely successful in treating hypertension and are more recently showing promise in reducing risks of myocardial infarction and diabetic nephropathy.

Renin remains an important target for blocking the RAS. Renin inhibitors (RI) could more effectively lower AII levels, by preventing formation of AI. ACE-independent pathways for conversion of AI to AII (e.g., chymase) remain uninhibited during ACEI therapy. RI's could therefore emerge superior to ACEI's in reducing the incidence of complications related to unchecked AII levels. Moreover, RI's may show an improved side effect profile over ACEI's. Renin has only one known natural substrate, whereas ACE degrades other peptide hormones such as bradykinin, which is potentiated during ACEI therapy.

Oral administration is an absolute requirement for a successful RI drug, and low oral bioavailability has been a main obstacle to this approach. Renin has been effectively inhibited to date only by substances which mimic a portion of the protein substrate which binds in the active site. In general, peptidomimetics have hitherto shown little promise as oral drugs.

In this report we describe the discovery of orally bioavailable renin inhibitors. Two orally bioavailable renin inhibitors have also been recently reported by the Abbott [1] and Fujisawa [2] groups.

RESULTS AND DISCUSSION

Potent Pentapeptide Renin Inhibitors

Pepstatin, a natural product [3] was the smallest potent renin inhibitor known to us at the inception of this effort. Structure-activity studies around pepstatin indicated it to be a transition-state inhibitor of aspartic proteases with the central statine imitating the putative tetrahedral intermediate in substrate hydrolysis [4].

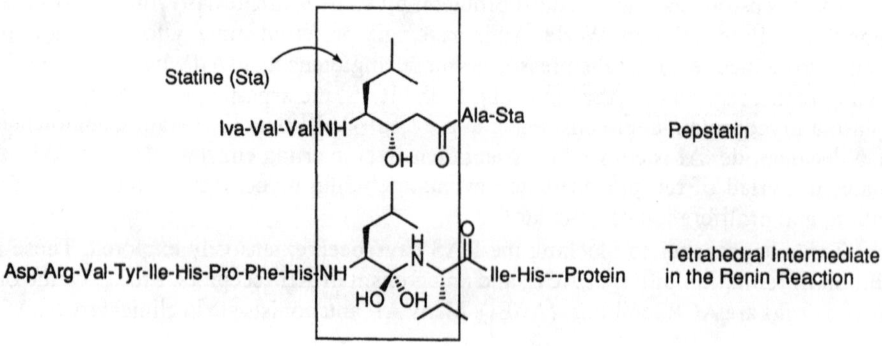

Table 1. Statine-Containing Renin Inhibitors

Compound	*Structure*	*Human Plasma Renin IC_{50}, nM*[a]
1	Pro-Phe-His---Sta---Val-Phe	7,000
2	Phe-His---Sta---Val-Phe	71,000
3	His---Sta---Val-Phe	570,000
4	Boc-Phe-His---Sta---Val-Phe	2,100
5	Boc-Phe-His---Sta---Ile-Phe	160
6	Boc-Phe-His---Sta---Ile(R)Phe[b]	50
7	Boc-Phe-His--CSta-Ile(R)Phe[c]	5

a, pH 7.4, 37°C, RIA (ref. 23); b, reduced bond (-CH_2-NH-) isostere; c, CSta - analog of statine bearing cyclohexylmethyl sidechain.

A potent inhibitor of pepsin and cathepsin D, pepstatin is a relatively weak inhibitor of renin [5].One reason for this might be that the sidechains of pepstatin correspond poorly to those of the renin substrate (hog sequence -Pro-Phe-His-Leu^{10}-Leu^{11}-Val-Tyr-). Thus, statine-containing renin inhibitor **1** was prepared (Table 1), and found to be a potent inhibitor of human plasma renin. Deletion of N-terminal residues dramatically reduced potency (**2** and **3**). The t-Boc derivative **4**, a synthetic precursor to **2**, was, surprisingly, a better inhibitor than either **1** or **2**. Substitution of Ile for Val in **4**, following the report of ^{12}Ile in the N-terminal sequence of human angiotensinogen [6], gave **5**, of submicromolar potency. To further reduce peptide character, amine **6** was prepared, and a further increase in potency was realized. In an X-ray study of a pepstatin fragment bound to penicillopepsin, the leucyl sidechain of the statyl residue made many nonbonded (hydrophobic) contacts with the enzyme [7]. This suggested a further way to increase potency, by increasing hydrophobic contacts at the inhibitor P_1 [8] residue.

Initially, the analog of **5**, having the isobutyl sidechain of the statine residue replaced by benzyl, was prepared and showed poorer binding (IC_{50} 460 nM). The corresponding cyclohexylmethyl inhibitor (**7**) was, however tenfold more potent than statine **6**. This activity difference translated similarly in other structures, and the P_1 cyclohexylmethyl sidechain has since become a characteristic of potent renin inhibitors [9]. Independent discoveries of other potent renin inhibitors based on the angiotensinogen sequence were reported during completion of this work [10,11].

The ability of compounds to inhibit renin in living animals was initially assessed in rats infused with exogenous hog renin (Table 2).

Pentobarbital-anesthetized rats, treated i.v. with renin inhibitor or control vehicle, were also treated i.v. 5, 15, and 30 minutes later with hog renin. In the absence of inhibitor, a short-lived pressor rise resulted from formation of angiotensin II, initiated by hog renin cleavage of rat angiotensinogen. In rats pretreated with **6** and **7** (10 mg/kg, i.v.), the hog renin-induced pressor rise was significantly attenuated for only 15 minutes. A much more complete and longer-lasting response was found in P2' lysine-containing inhibitor **8**. This increase was apparently unrelated to intrinsic inhibitory potency, as the $P_{2'}$·Ile analog (**5**, hog renin IC_{50} 1000 nM) showed no inhibition in this model (10 mg/kg i.v.). This result shifted our focus to lysine containing inhibitors.

Inhibitor **8** was unstable, *in vitro*, in rat liver homogenate, the primary degradation being hydrolysis at the Lys-Phe bond. In contrast, inhibitor **9**, having C-terminal Lys-Sta, proved stable. Though Phe-His cleavage in **8** or **9** was not seen in this *in vitro* system, high

Table 2. Inhibition of the Porcine Renin-Induced Pressor Rise in Anesthetized Rats by Pentapeptide Renin Inhibitors[a]

		Plasma renin IC50, nM		*% inhibition (A-B/A x 100)*			*I.V. Dose*
Cpd	*Structure*	*Hu*[b]	*Hg*[c]	*t = 5*	*15*	*30min*	*mg/kg*
6	Boc-Phe-His--Sta---Ile(R)Phe	50	3700	16[d]	41[d]	10	10
7	Boc-Phe-His--CSta--Ile(R)Phe	5	80	49[d]	48[d]	31	10
8	Boc-Phe-His---Sta---Lys-Phe	2900	400	100[d]	100[d]	60[d]	10
9	Boc-Phe-His--CSta--Lys-Sta	10	10	96[d]	58[d]	32[d]	1
10	Boc-Phe-MeHis-CSta-Lys-Sta	3000	120	75[d]	53[d]	50[d]	10

a, mean arterial pressure (MAP) at carotid, A - average MAP rise (20-30 mm Hg) after 40-80 mU/kg i.v. of hog renin (Sigma), no inhibitor, B - MAP rise, post-inhibitor, from equivalent hog renin injection, t - time after administration of inhibitor; b, see Table 1; c, hog renin and rat plasma, pH 7.4; [d]$P< 0.05$.

levels of the metabolite Boc-Phe were detected by HPLC in the plasma of rats dosed intravenously with both inhibitors. Consequently, inhibitor **10**, N-methylated at the Phe-His amide bond, proved stable to rat liver homogenate, and generated no detectable plasma levels of Boc-Phe on i.v. administration to rats. The increased proteolytic stablility of **9** and **10**, however, was not accompanied by increased duration of action of either substance in *in vivo*, relative to **8**. To understand the short duration of action of **8-10**, further metabolism and disposition studies were conducted. Radiolabelled **8** was prepared from uniformly ^{14}C-labelled phenylalanine. Inhibitors **9** and **10** were directly tritiated to high specific activity (3.2 and 2.7 Ci/mmol, respectively), by exchange of the C2 imidazole proton with tritium [12] catalyzed by Pd/Al_2O_3 in methanol (exposure time and catalyst ratio optimized for each compound using deuterium and ^{1}H NMR assessment). Key results obtained after i.v. administration of these inhibitors to rats are shown in Table 3.

The half-life of each inhibitor was similar, correlating with the similar durations of action observed *in vivo*. The cause of the short half-life, however, varied with inhibitor. Compound **8** was almost completely metabolized. Compound **9**, stabilized at the P2'-P3' bond (liver homogenate) was partially cleaved at Phe-His, but cleared primarily by excretion into the bile. Compound **10** was almost completely excreted as intact drug. The oral bioavailabilities of **8**,**9**, and **10** were expectedly found to be low in the rat(<2%).

Table 3. Plasma Half-life and Disposition of Radiolabelled Pentapeptide Renin Inhibitors[a]

				% of Dose Composition[g]	
Cpd	*Structure*[f]	*Plasma $T_{1/2}$, min*	*Excreted Radioactivity*[b]	*Metabolites*	*Drug*
8	Boc-Phe*His---Sta---Lys*Phe	8	95	91[c]	4 (urine)
9	Boc-Phe*His--CSta--Lys-Sta	12	75	13[d]	58 (bile) 4 (urine)
10	Boc-Phe-MeHis-CSta-Lys-Sta	13	82	5[e]	76 (bile) 1 (urine)

a, after i.v. administration to anesthetized male SD rats at 10 mg/kg; b, 0-12 or 0-24h; c, 84% in urine as Boc-Phe, 7% in feces and bile as Boc-Phe-His-Sta-Lys; d, 7% in bile as His-CSta-Lys-Sta, 6% in urine (unidentified); e, total in urine and bile; f, * indicates labile bond; g, identity as indicated by HPLC comparison with authentic standard.

Table 4. Hydroxyethylene Isostere Inhibitors

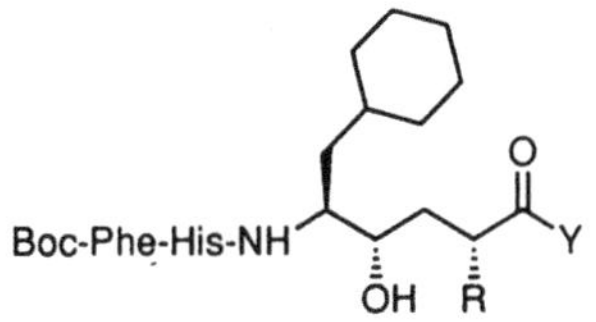

Cpd	R	Y	Human Plasma Renin IC50, nM[a]
11(CP-71,362)	$CH_2(CH_3)_2$	Lys-Phe	23
12	$CH_2(CH_3)_2$	$NHCH_3$	5
13	CH_2Ph	$NHCH_3$	4
14	CH_2Cy^b	$NHCH_3$	530
15	$CH_2(CH_3)_2$	$NH(CH_2)_3CH_3$	1

a, see Table 1; b, Cy = cyclohexyl

These data suggested that pentapeptide-like inhibitors offered little promise to be the sought after long-acting oral agents. Similar independent reports augmented this bias [13].

Potent Tripeptide Renin Inhibitors

Effort was then focused on finding smaller inhibitors. Incorporation of a hydroxyethylene isostere [14], made available by a convenient lactone alkylation [15], into the lysine series produced **11** (CP-71,362, Table 4), having high affinity to human renin, and unique potency against rat and dog plasma renins (1 nM and 12 pM, respectively) [16]. Immediate structure-activity studies around CP-71,362 revealed *increased* renin affinity in the *smaller* monomethylamide **12** [17]. $P_{1'}$ benzyl analog **13** was also potent, but cyclohexylmethyl compound **14** was not. Increased affinity was found in GABA analog **15**. The ability of **12** and **15** to inhibit endogenous renin was explored in the sodium-depleted, anesthetized marmoset monkey. After i.v. administration of either **12** or **15** at 3 or 1 mg/kg, respectively, blood pressure fell approximately 30-35 mm (equivalent to the maximum reduction achieved by captopril) and returned to control over approximately 3 hours. No blood pressure lowering was observed, however, on intraduodenal administration of **12** to renin-dependent guinea pigs at 100 mg/kg. Inhibitor **12** showed 5% oral bioavailability in the rat and a plasma elimination half-life of 95 minutes (drug labelled by catalytic exchange tritiation). Clearance of drug-related radioactivity was incomplete (34% in 48 h), composed mostly (33%) of metabolites formed by pathways other than Phe-His cleavage (no Boc-Phe detected).

Serendipity: Norstatine, and Oral Activity

The success in eliminating C-terminal residues (e.g., **12** vs. **11**) suggested this as a strategy towards even simpler inhibitors, oral activity and improved bioavailability. Carbonyl hydrates such as **16** (Y an electron withdrawing group) could have high renin affinity because of a close resemblance to the putative tetrahedral intermediate in the renin reaction. This hypothesis was stimulated by the NMR observation that pepsin bound a ketone-containing peptide in the higher-energy hydrate form [18], and further supported by the derivation of peptide aldehydes (e.g., CBZ-Phe-His-leucinal, IC_{50} 750 nM [19]), which were doubtless bound to renin as the hydrates.

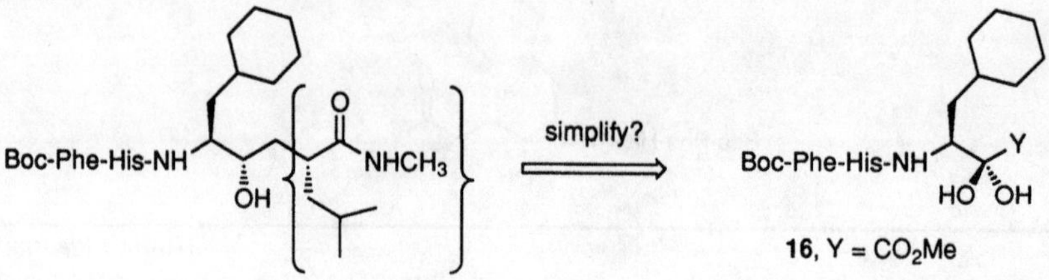

16, Y = CO_2Me

Keto ester **16** was synthesized and showed 1 μM (IC_{50}) inhibition of human plasma renin. Unexpectedly, alcohol **17** (Table 5), the synthetic precursor to **16**, was ten-fold more potent. This alcohol, unlike keto ester **16**, was also configurationally stable at the P_1 stereocenter. Stability and enhanced potency made this series even more attractive for further study. Introduction of a morpholino group [20] increased potency three-fold in **18**. Substitution of the P_2 histidine by norleucine, the latter explored earlier in pentapeptide renin inhibitors [21], gave the inhibitor **19**, unexpectedly of nanomolar potency. The corresponding methylamide (**20**) was less potent. Further modification at P2, inspired by the wide variation observed in the bound conformation of this sidechain in different endothiapepsin-inhibitor complexes [22] resulted in the norvaline and S-methylcysteine inhibitors **21** and **22**, respectively.

"Oral" activity was now observed in these substances. Methylamide **20** reduced mean arterial pressure for 5 hours on intraduodenal (i.d.) administration to conscious, sodium-depleted marmosets at 50 mg/kg. Norvaline analog 21 produce blood pressure lowering for 4 hours after a 10 mg/kg i.d. dose. S-Methylcysteine **22** proved superior, lowering blood

Table 5. Norstatine Renin Inhibitors

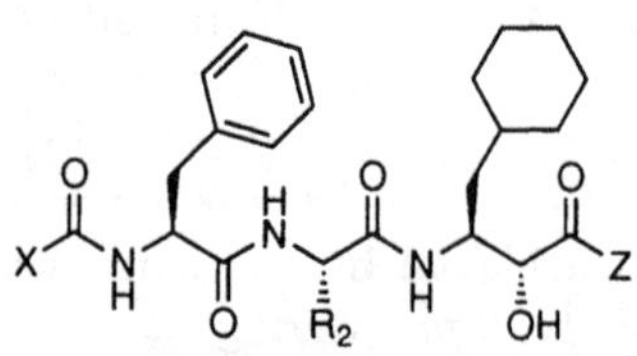

Cpd	X	R2	Z	Human Plasma Renin IC50 (nM)[a]
17	$(CH_3)_3C$-O	CH_2-(4-imidazolyl)	OCH_3	100
18	4-morpholino	CH_2-(4-imidazolyl)	OCH_3	35
19	4-morpholino	$(CH_2)_3CH_3$	OCH_3	2
20	4-morpholino	$(CH_2)_3CH_3$	$NHCH_3$	58
21	4-morpholino	$(CH_2)_2CH_3$	$OCH(CH_3)_2$	8
22[b]	4-morpholino	CH_2SCH_3	$OCH(CH_3)_2$	1
23	4-oxo-1-piperidino	CH_2SCH_3	$OCH(CH_3)_2$	1

a, see Table 1; b, CP-80,794, Terlakiren

Table 6. Oral Bioavailability of Inhibitor **22** in Suspension and Solution[a]

Condition	*Vehicle*	*Route*	*Dose*[b]	*Oral Bioavailability*
suspension	0.5% methylcellulose	p.o.	25 mg/kg	2.9 ±3.4 %
solution	glycerol formal[c]	p.o.	50 mg/kg	10 ± 3.6 %

[a], Species: beagle dogs; [b], orally, i.v. leg dosed at 1 mg/kg; [c]16% glycerol formal, 19% propylene glycol, 20% polysorbate 80, 19% Cremophor EL, 26% water

pressure for 4 hours after a 3 mg/kg i.d. dose, and for >6 hours following oral (p.o.) dosing at 10 mg/kg as an aqueous suspension. These and other hemodynamic effects of **22** (CP-80,794, Terlakiren) in various animals are reported [23]. Other orally active inhibitors of human renin containing norstatine have been independently reported [24].

Renin Inhibitors with High Oral Bioavailability

While showing good oral activity, **22** showed low oral bioavailability in animals. In the dog, the oral bioavailability of **22** was about 3% with considerable variability, when dosed as an oral suspension (Table 6). Two contributing factors were, possibly, low aqueous solubility (0.08 mg/mL in a pH 6.5 isotonic buffer at 37 °C), and susceptibility to chymotryptic cleavage at the Phe-SMeCys bond.

Two key observations, both related to inhibitor solubility, influenced subsequent efforts. First, when inhibitor **22** was dosed *in solution*, oral bioavailability increased significantly and became less variable (Table 6). Second, an analog of **22**, ketone **23** (Table 5), when dosed *in solution*, produced a sustained and pronounced oral hypotensive response in renin-dependent guinea pigs [25]. In contrast, **23** did not show reproducible blood pressure lowering in this model after an equivalent dose *as an aqueous suspension*. Thus, different oral properties of two poorly soluble inhibitors, oral bioavailability of **22**, and oral efficacy of **23**, were enhanced when each inhibitor was dosed in solution.

From these observations this hypothesis was formed: renin inhibitors with higher water solubility, still related in structure to **22** and **23**, could have much higher oral bioavailability. Water solubility, alone, would be an insufficient criterion, as soluble renin inhibitors of more distant structure were known to have low oral bioavailability [26].

We chose to explore amine-containing analogs of **22** and **23**, and to position the amine functionality at the N-terminus, where structural variation and the presence of polar functionality should be best tolerated by renin. These inhibitors were expected to bind with the N-terminus situated in the S4 binding pocket of renin, where substantial access to water should begin. This understanding of the renin active site was supported by early renin modelling [27] and aspartic protease crystal structures [7,28,29]. From much more recent X-ray data [30,31], Figure 2 predicts the binding of a morpholine-containing renin inhibitor to human renin, with a solvent-exposed N-terminus.

Results of evaluation of some amine-containing inhibitors are shown in Table 7 [32]. Inhibitors **24** and **25**, very close analogs of **22** and **23**, respectively, were weaker inhibitors, eliciting very short hypotensive responses after i.v. dosing in the guinea pig. Prolyl inhibitor **26** (CP-85,339 [31]), also showed very short-lived activity on i.v. administration. Aniline **27** was a poor inhibitor, *in vitro*, with extremely low aqueous solubility. Pyridine **28** showed long-lived blood pressure lowering activity after i.v., but was not p.o., dosing, perhaps

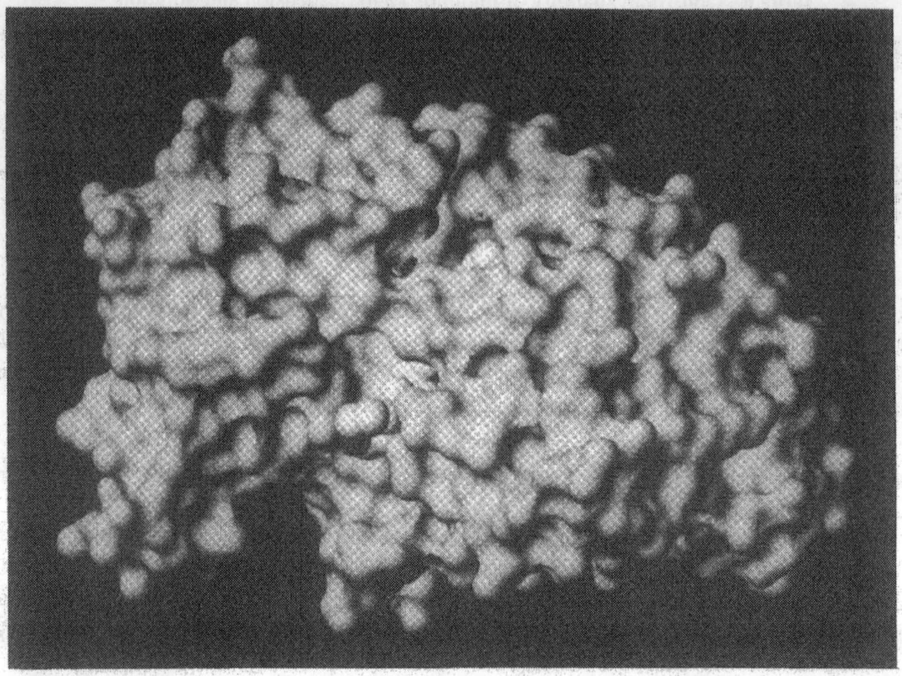

Figure 2. Predicted binding of a morpholine-containing inhibitor to human renin. Inhibitor CP-81,282 (Mor-CO-Phe-Nle-DifluoroC-StatoneNHMe, endothiapepsin-bound X-ray coordinates [30]), modelled into the X-ray coordinates of recombinant human renin (from the renin/CP-85,339 crystal structure [31]). View shows the solvent-accessible (Connoly) surface of renin (green) and the solvent-exposed N-terminus of the inhibitor (yellow, with morpholine oxygen in red). The N-terminus of the inhibitor is solvent-exposed. (See color plate preceding page 1.)

because of low solubility. The more soluble methylpiperazine **29** was orally active. In contrast, dimethylaminopiperidine **30**, subnanomolar *in vitro*, produced exceptional blood pressure lowering activity after i.v. dosing [33,34]. As the hydrochloride or methanesulfonate salt, **30** readily dissolved (>10 mg/mL) in water, and was active at 10 mg/kg, p.o., a dose at which **22** and **23** were not.

Inhibitor **30** showed 9% oral bioavailability in the dog. This finding validated the working hypothesis that inhibitors related in structure to **22** and **23**, but with higher solubility, could have higher oral bioavailability. Inhibitor **30** was synthesized by reductive amination of ketone **23**, a reaction which readily allowed the synthesis of other analogs. For example, reductive amination with the appropriate amine hydrochloride (5 equiv), sodium acetate (10 equiv), and sodium cyanoborohydride (1.3 equiv) in methanol at 25 °C produced the expected amino, methylamino, pyrrolidino, piperidino, and N-methylpiperazino reductive amination products in 32-82% yield. These amines are 0.4-3 nM inhibitors of human and guinea pig renins. Like **30**, each has high intrinsic aqueous solubility and is freely water-soluble as the hydrochloride, exhibiting oral activity at a dose (either 10 or 20 mg/kg) where inhibitor **22** proved inactive. Dimer **31** was isolated as a byproduct in the reductive amination of **23** with ammonium chloride, **31** is also a potent renin inhibitor (2 nM). These *in vitro* observations are consistent with the working hypothesis that the N-terminal portion of these inhibitors remains solvent exposed on binding to renin.

Table 7. *In Vitro* and *In Vivo* Activity of Amine-Containing Renin Inhibitors

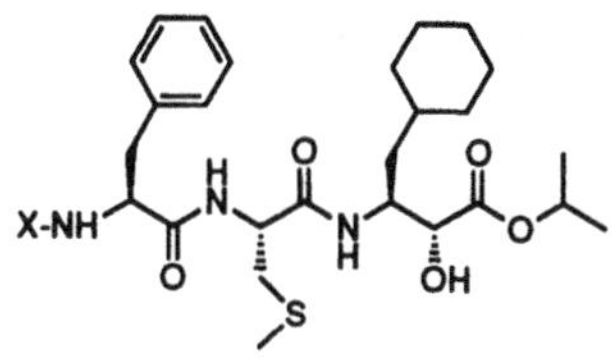

Cpd	X	Plasma Renin IC50, nM[a] Hu	GPig	Blood Pressure Lowering in the Sodium-depleted Guinea Pig[b] Route	Dose[d] mg/kg	Duration[e] (min)	Equilibrium Solubility[c] pH 7 buffer (mG/mL)
24	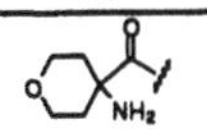	15	6	i.v.	1	10	0.48
25	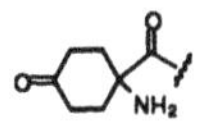	26		i.v.	3	15	0.34
26[f]	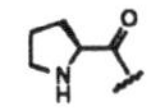	37	8	i.v.	2	10	
27	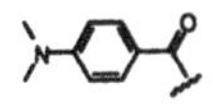	55	21				<0.001
28	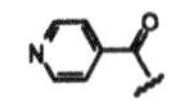	0.3	2	i.v. p.o.	1 30	80 0	<0.03
29	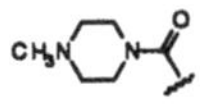	5	4	p.o.	30	45	0.73
30	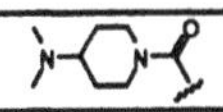	0.3	0.3	i.v. p.o.	1 10	>150 60	3.3

a, see Table 1; b, see ref. 23; c, by filtration and HPLC after equilibration for 7-12 days at 25 °C; d, inhibitor dosed in aqueous methylcellulose as the hydrochloride; e, time between dosing and return of blood pressure to within 5 mmHg of control; f, CP-85,339.

Inhibitors **22**, **23** and **30** are susceptible to chymotrypsin hydrolysis at the P_3-P_2 bond (Phe-SMeCys). Digestion of inhibitor prior to absorption could also limit oral bioavailability. Two strategies to achieve chymotrypsin-stable renin inhibitors are to increase steric bulk of the P_3 sidechain, precluding binding in the S_1 pocket of chymotrypsin, and to eliminate the P_3 NH, precluding hydrogen bonding between this donor in the renin inhibitor and the Ser 214 carbonyl oxygen of chymotrypsin [35]. Table 8 shows the effect of these modifications on important properties of aminopiperidine renin inhibitors, including oral bioavailability. Incorporation of O-methyltyrosine [35] gave chymotrypsin stable inhibitor **32**, which, showed good *in vitro* and *in vivo* activity. Inhibitor **33**, having a more bulky cyclohexylmethyl sidechain was more, but not completely stable to chymotrypsin, but produced a high level of oral activity. Oral bioavailability increased from 9% in **30** to 17% in **33**. Finally, incorporation of the (*R*)-benzylsuccinic acid moiety [35] produced chymotrypsin-stable inhibitor **34**, having excellent oral efficacy and record oral bioavailability (34%).

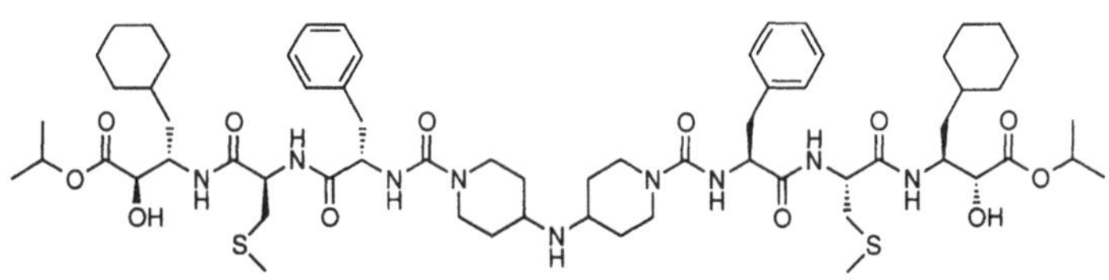

31, Hu IC_{50} = 2 nM

Table 8. Effect of P_3 Modification on Renin Affinity, Chymotrypsin Stability, Oral Activity, and Oral Bioavailability

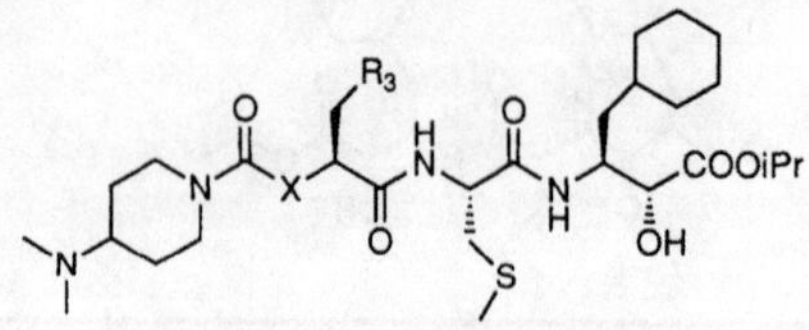

Cpd	X	R3	Renin Plasma IC50, nm Human[b]	Chymotrypsin Stability % Intact Drug[c]	Guinea Pig Efficacy[a] Oral Route Dose (mg/kg)	Duration (min)	Oral Bioav. (Dog)
30	NH	Phenyl	0.3	28	10	60	9 %
32	NH	p-$CH_3OC_6H_4$	3	99	20	150	
33	NH	Cyclohexyl	2	54	10	330	17%[d]
34	CH_2	Phenyl	7	90	10	>300	34%[e]

a, see Table 7; b, see Table 1; c, after 40 min with 2.5×10^{-7} M Chymotrypsin at 37 °C; d, n = 2; e, n=4

After dosing of inhibitor **34** to dogs, a major metabolite of similar HPLC retention time was seen in plasma. The concentration of this metabolite achieved the micromolar level within 30 minutes following a 20 mg/kg oral dose of **34**, equalled that of **34** at 2 hours, and declined thereafter with an essentially identical elimination profile to **34**. This metabolite was identified by mass spectroscopy as the corresponding monomethylamine (**35**, CP-108,671). Independent synthesis and HPLC comparison confirmed its identity.

Table 9. Selected Properties of Inhibitor **35** (CP-108,671)

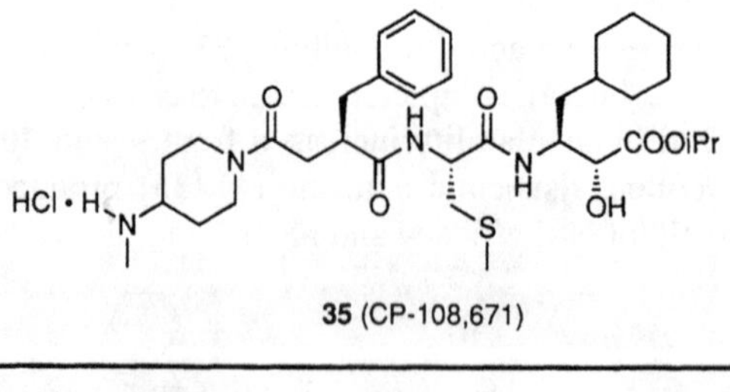

35 (CP-108,671)

Human Plasma Renin IC50	Dog Pharmacokinetics[a] Bioavailability	I.V. T1/2	AUC(0-inf) μg/h.mL	Bioavailability Marmoset[b]	Cyno[c]	Equilibrium Solubility pH 7
4 nm	60 ± 10%	2.4 h	10.4 ± 3.2	61%	>27%	>20 mg/mL

a, 10 mg/kg p.o. vs. 5 mg/kg i.v.; b, 2.8 mg/kg p.o. vs. 2.8 mg/kg i.v.; c, 9.4 mg/kg p.o. vs. 4.7 mg/kg i.v.

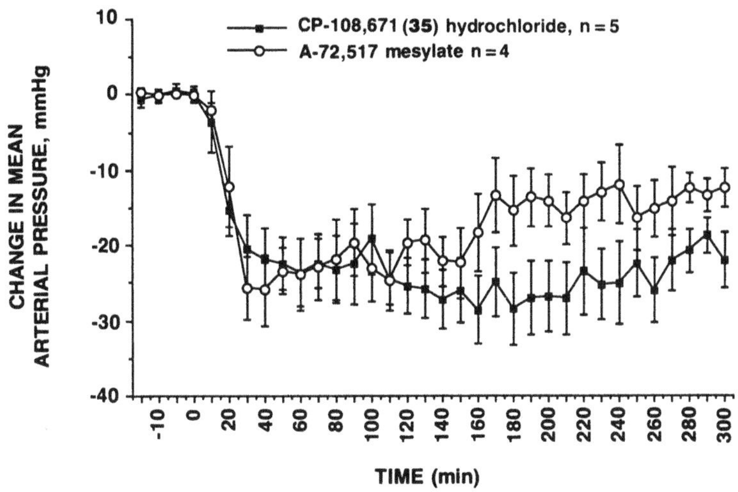

Figure 3. Blood pressure lowering activity of **35** (CP-108,671, hydrochloride salt) and A-72,517 (mesylate salt, synthesized as described [1]) on dosing at 3 mg/kg, p.o., in the conscious, sodium-depleted marmoset

Inhibitor **35** was found to have 60% oral bioavailability in the dog, determined from the plasma exposure profiles following oral and i.v. administration. Other pharmacokinetic properties of **35** are summarized in Table 9. This inhibitor also has high oral bioavailability in two primate species (59%, marmoset and >27%, cynomolgus monkey). High primate bioavailability thus distinguishes **35** from two other bioavailable renin inhibitors [1,2]. In the sodium-depleted guinea pig, **35** reduced blood pressure for greater than 5 hours following an oral dose of 10 mg/kg, with a nearly superimposable profile to that of **34**. In sodium-depleted marmosets, **35** lowered blood pressure for greater than 5 hours following a 3 mg/kg oral dose, comparing very favorably with the recently reported [1] bioavailable inhibitor A-72,517 (Figure 3).

Inhibitor **35** is also a potent inhibitor of other species plasma renins (IC_{50}'s 3.4 nM (cyno), 3.4 nM (marmoset), 2.2 nM (guinea pig), and 3.5 nM (dog)). It is a competitive inhibitor of porcine renin (Ki = 4.9 nM), and is highly selective for renin over other aspartic proteases (human cathepsin D, 550 nM, porcine pepsin 2100 nM). In the guinea pig, a high (10 mg/kg) i.v. dose does not affect cardiac output or heart rate, and doses as low as 0.1 mg/kg, i.v., reduce total peripheral resistance [32].

SUMMARY

Knowledge of the sequence of a bioactive protein (angiotensinogen) and the availability of a natural product inhibitor lead (pepstatin) were the starting point for discovery of potent penta- and hexapeptide renin inhibitors. Study of the metabolism and disposition of these substances forced the discovery of simpler inhibitors leading to the discovery of oral activity in Terlakiren (**22**). Modification of physical properties led to the synthesis of aminopiperidine **30**, which was identified by oral efficacy profiling. Structural modification to give enzymatic stability produced the bioavailable benzylsuccinate inhibitor **34**. Its bioactive monomethylamine metabolite (**35**, CP-108,671) was subsequently found to have uniformly high oral bioavailability and activity in various species including primates.

This report illustrates a successful peptide-based approach to orally active and bioavailable series of pharmacologic agents.

ACKNOWLEDGMENTS

The following individuals performed much of this work: P. A. Benoit, M. P. Carta, R. J. Chambers, K. Chausse', D. B. Damon, M. L. De La Mata, C. F. Ebbinghaus, E. S. Eberhardt, S. S. Ellery, M. J. Foley, M. L. Gillaspy, W. A. Hada, P. J. Hiza, J. A. Jacobsen, R. L. Kaye, J. T. MacAndrew, C. A. Mularski, I. M. Purcell, H. Sherman, K. A. Simpson, J. Wentland, T. A. Appleton, and J. Glidden. We are grateful to C. A. Harbert, N. J. Hutson, H.-J. Hess, J. L. LaMattina, L. S. Melvin, W. Curatolo, M. Morville and F. C. Falkner for their support. It is our pleasure to acknowledge a stimulating collaboration with Prof. T. L. Blundell and colleagues, including S. F. Foundling, J. B. Cooper, V. Dhanaraj, and B. Veerapandian, of the X-ray Crystallography group, Birkbeck College, University of London. We thank S. F. Sneddon for assistance in preparing Figure 2.

REFERENCES

1. S. H. Rosenberg, K. P. Spina, S. L. Condon, J. Polakowski, Z. Yao, P. Kovar, H. H. Stein, J. Cohen, J. L. Barlow, V. Klinghofer, D. A. Egan, K. A. Tricarico, T. J. Perun, W. R. Baker, and H. D. Kleinert, Studies directed toward the design of orally active renin inhibitors. 2. Development of the efficacious, bioavailable renin inhibitor (2S)-2-benzyl-3-[[(1-methylpiperazin-4-yl)sulfonyl]-proprionyl]-3-thiazol-4-yl-L-alanine amide of (2S,3R,4S)-2-amino-1-cyclohexyl-3,4-dihydroxy-6-methylheptane (A-72,517), *J. Med. Chem.* **36**: 460-467 (1993).
2. T. Ogihara, J. Higaki, M. Nagano, K. Higashimori, K. Masuo, and H. Mikami, A novel human renin inhibitor with a clinically applicable bioavailability, FK906, *Hypertension* **20**:S423 (1992).
3. H. Umezawa, T. Aoyagi, H. Morishima, M. Matsuzaki, M. Hamada, and T. Takeuchi, Pepstatin, a new pepsin inhibitor produced by actinomycetes, *J. Antibiot.* **23**:259 (1970).
4. D. H. Rich and E. T. O. Sun, Mechanism of the inhibition of pepsin by pepstatin, *Biochem. Pharmacol.* **29**:2205-2212 (1980), and references therein.
5. T. Aoyagi, H. Morishima, R. Nishizawa, S. Kunimoto, T. Takeuchi, H. Umezawa, and K. Ikezawa, Biological activity of pepstatins, pepstatone A and partial peptides on pepsin, cathepsin D, and renin, *J. Antibiot.* **25**: 689-694 (1972).
6. D. A. Tewksbury, R. A. Dart, and J. Travis, The amino terminal amino acid sequence of human angiotensinogen, *Biochem. Biophys Res. Commun.*, **99**: 1311-1315 (1981).
7. M. N. G. James, A. Sielecki, F. Salituro, D. H. Rich, and T. Hofmann, Conformational flexibility in the active sites of aspartyl proteinases revealed by a pepstatin fragment binding to penicillopepsin, *Proc. Natl. Acad. Sci. USA* **79**:6137-6141 (1982).
8. Nomenclature: J. Schechter and A. Berger, On the size of the active site in proteases. I. Papain, *Biochem. Biophys. Res. Commun.* **27**: 157-162 (1967).
9. J. Boger, L. S. Payne, D. S. Perlow, N. S. Lohr, M. Poe, E. H. Blaine, E. H. Ulm, T. W. Schorn, B. I. LaMont, T-Y. Lin, M. Kawai, D. H. Rich, and D. F. Veber, Renin Inhibitors. Synthesis of subnanomolar, competitive, transition-state analogue inhibitors containing a novel analogue of statine, *J. Med. Chem.* **28**: 1779-1790 (1985).
10. M. Szelke, B. Leckie, A. Hallett, D. M. Jones, J. Sueiras, B. Atrash, and A. F. Lever, Potent new inhibitors of human renin, *Nature* **299**:555-557 (1982).
11. J. Boger, N. S. Lohr, E. H. Ulm, M. Poe, E. H. Blaine, G. M. Fanelli, T-Y. Lin, L. S. Payne, T. W. Schorn, B. I. LaMont, T. C. Vassil, I. I. Stabilito, D. F. Veber, D. H. Rich, and A. S. Boparai, Novel renin inhibitors containing the amino acid statine, *Nature* **303**: 81-84 (1983).
12. E. Klauschenz, M. Bienert, H. Egler, U. Pleiss, H. Niedrich, and K. Nikolics, Tritium labelling of gonadotropin releasing hormone in its proline and histidine residues, *Peptides* **2**:445-452 (1981).
13. J. Boger, C. D. Bennett, L. S. Payne, E. H. Ulm, E. H. Blaine, C. F. Homnick, T. W. Schorn, B. I. Lamont and D. F. Veber, Design of proteolytically-stable, peptidal renin inhibitors and determination of their fate *in vivo*, *Regul. Pept.* **S4**: 8-13 (1985).

14. M. Szelke, D. M. Jones, B. Atrash, A. Hallett, and B. Leckie, Novel transition-state analogue inhibitors of renin, *Peptides: Structure and Function. Proceedings of the 8th American Peptide Symposium;* V. J. Hruby and D. H. Rich, Eds., Pierce Chemical Co.: Rockford, IL, 579-582 (1983).
15. A. H. Fray, R. L. Kaye, and E. F. Kleinman, A short, stereoselective synthesis of the lactone precursor to 2R,4S,5S hydroxyethylene dipeptide isosteres, *J. Org. Chem.* **51**:4828-4833 (1986).
16. E. F. Kleinman, A. H. Fray, W. F. Holt, M. A. R. Kiron, W. R. Murphy, I. M. Purcell, and R. L. Rosati, CP-71,362, an unusually potent inhibitor of rat and dog renin, *Bioorg. Med. Chem. Lett.*, **4**, 589 (1994).
17. Other hydroxyethylene inhibitors of this class are reported: P. Buhlmayer, A. Caselli, W. Fuhrer, R. Goschke, V. Rasetti, H. Rueger, J. L. Stanton, L. Criscione, and J. M. Wood, Synthesis and biological activity of some transition-state inhibitors of human renin, *J. Med. Chem.* **31**:1839-1846 (1988).
18. D. H. Rich, M. S. Bernatowicz, and P. G. Schmidt, Direct 13C NMR evidence for a tetrahedral intermediate in the binding of a pepstatin analogue to porcine pepsin, *J. Am. Chem. Soc.* **104**:3535-3536 (1982).
19. T. Kokubu, K. Hiwada, Y. Sato, T. Iwata, Y. Imamura, R. Matsueda, Y. Yabe, H. Kogen, M. Yamazaki, Y. Iijima, and Y. Baba, Highly potent and specific inhibitors of human renin, *Biochem. Biophys. Res. Commun.* **118**:929-933 (1984).
20. M. Miyazaki, N. Toda, Y. Etoh, T. Kubota, and K. Iizuka, Newly synthesized, potent human renin inhibitor, Presented at the 59th General Meeting of the Japanese Pharmacological Society, April 1-4, 1986, Niigata, Japan, *Jpn. J. Pharmacol.* 40S: 70p (1986).
21. R. Guegan, J. Diaz, C. Cazaubon, M. Beaumont, C. Carlet, J. Clement, H. Demarne, M. Mellet, J-P. Richaud, D. Segondy, M. Vedel, J-P. Gagnol, R. Roncucci, B. Castro, P. Corvol, G. Evin, and B. P. Roques, Pepstatin analogues as novel renin inhibitors, *J. Med. Chem.* **29**:1152-1159 (1986).
22. T. L. Blundell, personal communication.
23. W. R. Murphy, R. T. Wester, R. L. Rosati, D. J. Hoover, I. M. Purcell, J. T. MacAndrew, T. M. Schelhorn, D. E. Wilder, A. H. Smith, and W. F. Holt, Hemodynamic effects of the renin inhibitor CP-80,794 in several species, *Amino Acids: Chemistry, Biology, and Medicine*, Lubek, G.; Rosenthal, G.A., eds., ESCOM, pp. 676-688 (1990).
24. K. Iizuka, T. Kamijo, T. Kubota, K. Akahane, H. Umeyama, and Y. Kiso, New human renin inhibitors containing an unnatural amino acid, norstatine, *J. Med. Chem.* **31**:701 (1988).
25. Dosed as a solution in 0.1M sodium dodecyl sulfate (SDS) 20 mg/kg p.o., effect lasting >5h (maximum drop 25 mm Hg), . Solubility of **23** in 0.1M SDS at 27 °C, 8.7 mg/mL, in pH 6.5 isotonic buffer at 37 °C, 0.02 mg/mL.
26. For example, see H. D. Kleinert, J. R. Luly, B. A. Bopp, K. M. Verburg, P. A Hoyos, M. D. Karol, J. J. Plattner, R. R. Luther, H. H. Stein, Profile of the renin inhibitor, enalkiren, *Cardiovascular Drug Rev.* **8**: 203-219 (1990)
27. B.L. Sibanda, T. Blundell, P. M. Hobart, M. Fogilano, J. S. Bindra, B. W. Dominy, and J. M. Chirgwin, Computer graphics modelling of human renin: specificity, catalytic activity, and intron-exon junctions, *FEBS Lett.* **174**:102-110 (1984).
28. R. Bott, E. Subramanian, and D. R. Davies, Three-dimensional structure of the complex of the rhizopus chinensis carboxyl protease and pepstatin at 2.5 resolution, *Biochemistry* **21**:6956-6962 (1982).
29. L. Pearl and T. L. Blundell, The active site of aspartic proteinases, *FEBS Lett.* **174**: 96-101 (1984).
30. B. Veerapandian, J. B. Cooper, A. Sali, T. L. Blundell, B. W. Dominy, R. L. Rosati, D. B. Damon, Hoover, D.J., Direct Observation by X-ray Analysis of the Tetrahedral 'Intermediate' of Aspartic Proteinases, *Protein Science* **1**, 322-328 (1992).
31. V. Dhanaraj, C. DeAlwis, C. Frazao, M. Badasso, B. L. Sibanda, I. J. Tickle, J. B. Cooper, H. P. C. Driessen, M. Newman, C. Aguilar, S. P. Wood, T. L. Blundell, P. M. Hobart, K. F. Geoghegan, M. J. Ammirati, D. E. Danley, B. A. O'Connor, and D. J. Hoover, X-ray analyses of peptide-inhibitor complexes define the structural basis of specificity for human and mouse renins, *Nature* **357**: 466-472 (1992).
32. Complete details of the synthesis and evaluation of compounds in this series will be submitted for publication elsewhere.
33. Other aminopiperidine renin inhibitors are independently reported. See: P. Raddatz, A. Jonczyk, K-O. Minck, C. J. Schmitges, and J. Sombroek, Substrate analogue renin inhibitors containing replacements of histidine in P2 or isosteres of the amide bond between P3 and P2 sites, *J. Med. Chem.* **34**:3267-3280 (1991), and reference 34.

34. P. Raddatz, A. Jonczyk, K-O. Minck, F. Rippmann, C. Schittenhelm, and C. J. Schmitges, Renin inhibitors containing new P1-P1' dipeptide mimetics with heterocycles in P1', *J. Med. Chem.* **35**:3525-3536 (1992).
35. J. J. Plattner, P. A. Marcotte, Hollis D. Kleinert, H. H. Stein, J. Greer, G. Bolis, A. K. L. Fung, B. A. Bopp, J. R. Luly, H. L. Sham, D. J. Kempf, S. H. Rosenberg, J. F. Dellaria, B. De, I. Merits, and T. J. Perun, Renin inhibitors. Dipeptide analogs of angiotensinogen utilizing a structurally modified phenylalanine residue to impart proteolytic stability, *J. Med. Chem.* **31**: 2277-2288 (1988).

STRUCTURE OF HUMAN CATHEPSIN D: COMPARISON OF INHIBITOR BINDING AND SUBDOMAIN DISPLACEMENT WITH OTHER ASPARTIC PROTEASES

John W. Erickson, Eric T. Baldwin, T. Narayana Bhat, and Sergei Gulnik

Structural Biochemistry Program
PRI/DynCorp
NCI-Frederick Cancer Research and Development Center
Frederick, Maryland 21702

INTRODUCTION

Cathepsin D (EC 3.4.23.5) (CatD) is an intracellular aspartic protease (AP) that is normally found in the lysosomes of higher eukaryotes[1]. CatD shares two structural features with lysosomal enzymes that distinguish it from most secreted, extracellular APs. First, the mature enzyme from humans is found as a two-chain enzyme as a result of post-translational cleavage and removal of an insertion loop in the N-domain[2,3]. Second, CatD contains phosphorylated, N-linked oligosaccharides that target the enzyme to lysosomes via mannose-6-phosphate receptors (MPR)[4,5]. Interest in CatD as a target for drug design results from its association with several biological processes of therapeutic significance, particularly breast cancer[6] and Alzheimer's disease[7] (Table 1). Recent studies have implicated CatD in the processing of β-amyloid precursor protein to promote amyloid plaque formation in Alzheimer's brain[7]. Numerous studies of primary breast cancers have demonstrated that elevated levels of CatD were correlated with an increased risk of metastasis and shorter relapse-free survival[6]. High levels of CatD produced in the vicinity of the growing tumor may degrade the extracellular matrix, either directly or indirectly by potentiating the activity of other tissue proteases, such as collagenase, and thereby promote tumor growth and metastasis[8,9]. While a role for CatD or other APs in metastasis has been questioned by recent results that show a negative effect of pepstatin A in an *in vitro* invasion assay using breast cancer-derived MCF7 cells[10], new studies have revealed that elevated levels of CatD secretion in breast cancers correlate with tumor aggressiveness[11].

The availability of potent and specific inhibitors of Cat D is important for the further elucidation of the roles of this enzyme in human disease. As a step towards the structure-based design of CatD inhibitors, we reported the structures of human CatD in its native and pepstatin-inhibited forms[12]. The structures of native CatD, from human spleen, and pepstatin-inhibited CatD, from bovine liver, were elucidated simultaneously by Metcalf and

Aspartic Proteinases: Structure, Function, Biology, and Biomedical Implications
Edited by Kenji Takahashi, Plenum Press, New York, 1995

Table 1. Possible Biological Roles of Cathepsin D

Protein catabolism[1]
Connective tissue disease[21]
Muscular dystrophy[22]
Neurodegenerative disease[23,24]
Antigen processing of MHC peptides[25-27]
Scrapie[28]
Alzheimer's disease[7]
Breast cancer and metastasis[6]

co-workers[13]. Results from both studies were in good agreement overall with most aspects of the structures, including the interpretation of carbohydrate structure for the human enzymes. In this paper, we summarize our structural results obtained with human liver CatD and present a comparative analysis of the subdomain structures of CatD and several other APs.

RESULTS AND DISCUSSION

The purification and crystallization of human liver CatD, and its co-crystallization with pepstatin were reported earlier[2,12]. Both forms crystallized in the same space group, $P6_5$, with similar unit cell dimensions (a = b = 125.9 Å, c = 104.1 Å) despite their different crystal morphologies (Fig. 1). Both structures are well-refined at 2.5 Å resolution, and their coordinates have been deposited in the Protein Data Bank with reference names 1LYA and 1LYB for the native and inhibited structures, respectively. The dimeric nature of the crystallographic asymmetric unit probably results from crystal packing, and there is no evidence for specific association of CatD molecules in dilute solution.

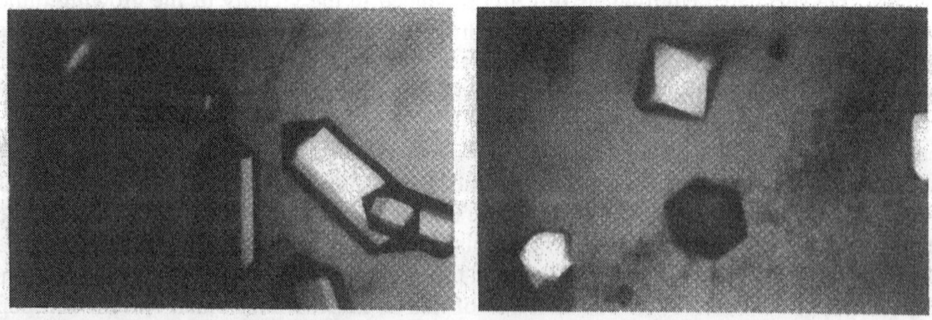

Figure 1. Crystals of human liver CatD showing the different morphologies of the native (left) and pepstatin-inhibited (right) forms of the enzyme. The hexagonal rods are about 0.5 mm in length.

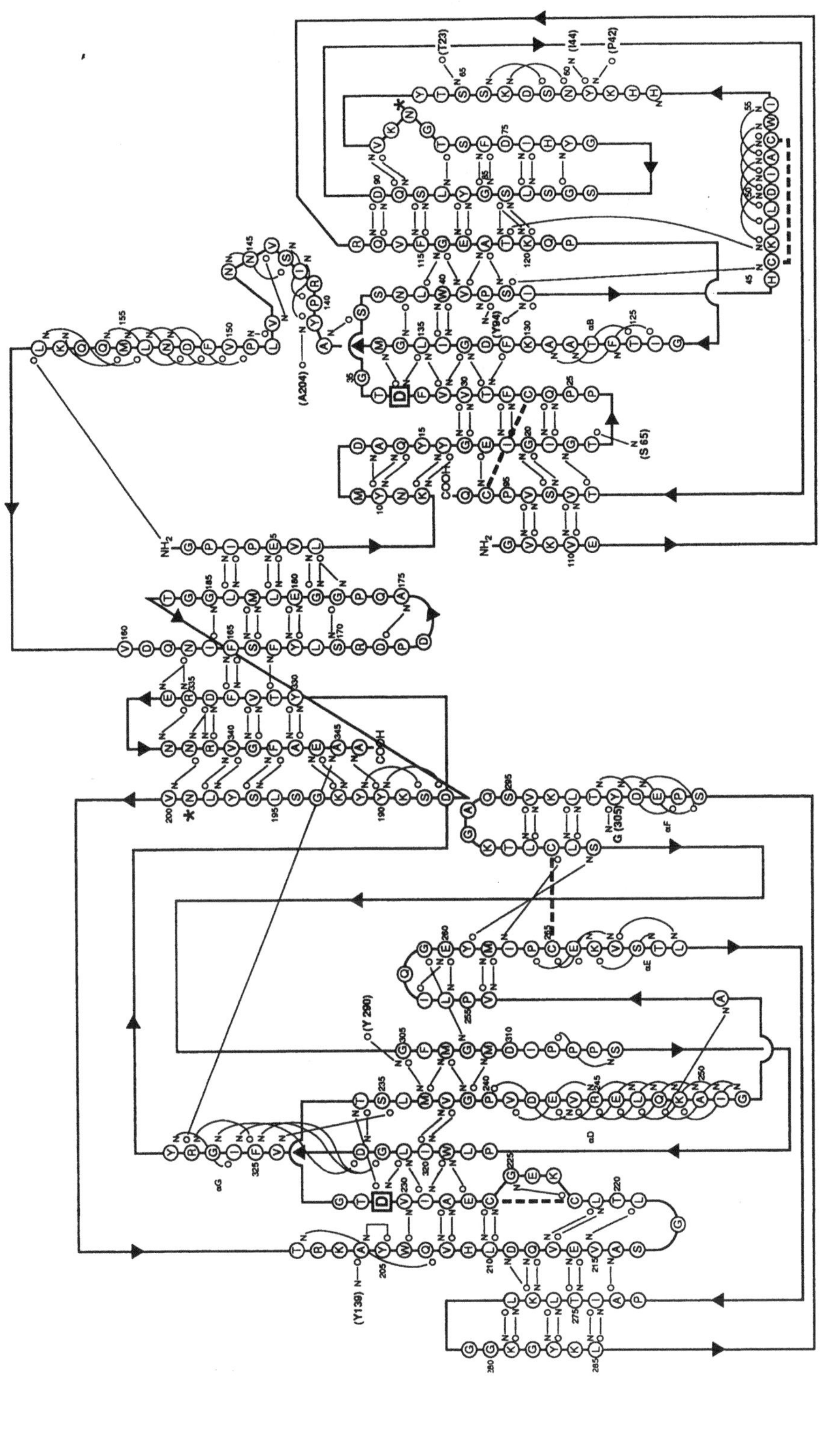

Figure 2. Hydrogen bonding and secondary structure diagram for the main chain of human CatD drawn to emphasize the pseudo two-fold symmetry of the enzyme. Each domain contains one active site aspartic acid (boxed), two disulfide bridges (dashed lines), and one N-linked glycosylation site (*).

The secondary structure and overall fold of CatD is very similar to that of other APs (Fig. 2). Superposition of all equivalent Ca atoms leads to rms deviations of around 0.9 Å with the mammalian enzymes for renin, pepsin and chymosin[12]. We prepared primary sequence alignments for human and porcine CatD, human renin and porcine pepsin using their three-dimensional structural homologies (Fig. 3). The positions of CBH attachment are shown, along with the N-domain insertions and cleavages based on peptide sequencing data. Continuous electron density extended from Gly1 through Gln97, and from Gly106 to Ala346. The C-terminal Ser107 of the light chain was disordered in both crystal structures. The heavy chain C-terminal Arg347-Leu348 residues were either proteolytically processed or disordered since they were not visible in the electron density maps. Other notable features of the CatD structure include the existence of three cis-peptide bonds involving prolines at positions 24, 314 and 317, and a disulfide bond connecting Cys96, near the C-terminus of the light chain, with Cys27. The latter was predicted from modeling studies[14]. The N-termini of the two chains are within 5 Å of each other in the mature enzyme; thus, internal cross-linking studies to analyze light and heavy chain re-folding and association might be feasible.

Binding to pepstatin A had little effect on the overall structure of the enzyme; the two structures agree to within 0.40 Å rms deviation for all backbone atoms (Fig. 4). However, the tip of the flap, residues 78-81, moves in towards pepstatin by about 1.7 Å (Fig. 4), and other, more subtle alterations in subsite atom positions were also observed. The refined crystal structure of pepstatin A bound to rhizopuspepsin[15] afforded us an opportunity to compare the structures of the two complexes in detail. Using the fluorogenic substrate Ac-Glu-Glu(Edans)-Lys-Pro-Ile-Cys-Phe-Phe-Arg-Leu-Gly-Lys(Dabcyl)-Glu-NH2, which was kindly supplied by Dr. Grant Krafft, Abbott Laboratories, we determined the Ki's for pepstatin with CatD and rhizopuspepsin to be 3.8 and 150 pM, respectively[12]. The bound conformations of the highly flexible inhibitor molecule were nearly identical despite the structural and affinity differences between the two enzymes (Fig. 5A).

CatD forms two more hydrogen bonds to the pepstatin A backbone than rhizopuspepsin - Tyr205 OH with P2' Ala CO, and Ser235 Oγ with P4 Iva CO (Fig. 5B). The P4 and P2' carbonyl groups in the rhizopuspepsin complex are neither hydrated nor bound to the enzyme and are, thus, unfavorably desolvated relative to the CatD complex. We also analyzed the van der Waals contacts for both complexes; CatD forms about 50% more contacts than rhizopuspepsin with pepstatin, 93 vs 67, with the greatest differences being in the S2, S2' and S3' subsites, 38 vs 20 contacts (Table 2). The looser packing of the rhizopuspepsin/pepstatin complex is a consequence of a larger gap between the large and small subdomains that results from the different subdomain displacements in the two enzymes (see below). The combination of looser packing and unsatisfied hydrogen bonding could account for the weaker binding affinity of pepstatin for the fungal enzyme. Despite the tight binding of pepstatin to CatD in solution, our analysis of the crystal structure indicates that the complementarity between inhibitor residues and enzyme subsites is suboptimal; we remain puzzled by the very high affinity of this naturally-occurring inhibitor for CatD. It will be of interest to compare other APs complexed with the same inhibitor in order to evaluate the structural correlates of affinity and specificity for this important class of enzymes.

Based on three-dimensional superposition analysis, we and others have noted that the size and shape of the active site clefts of AP structures may be modulated by the disposition of two subdomains that can act as independent rigid bodies[16,17] - a large subdomain, consisting of the N-domain and mainly the active site "psi-loop" portion of the C-domain; and a small subdomain comprising the rest of the C-domain, mainly residues 220-298 (pepsin numbering). Displacements have been observed upon complexation for pepsin and endothiapepsin, as well as between pepsin and its zymogen structure. Based on these observations, we and others have proposed that the subdomains are flexible to some

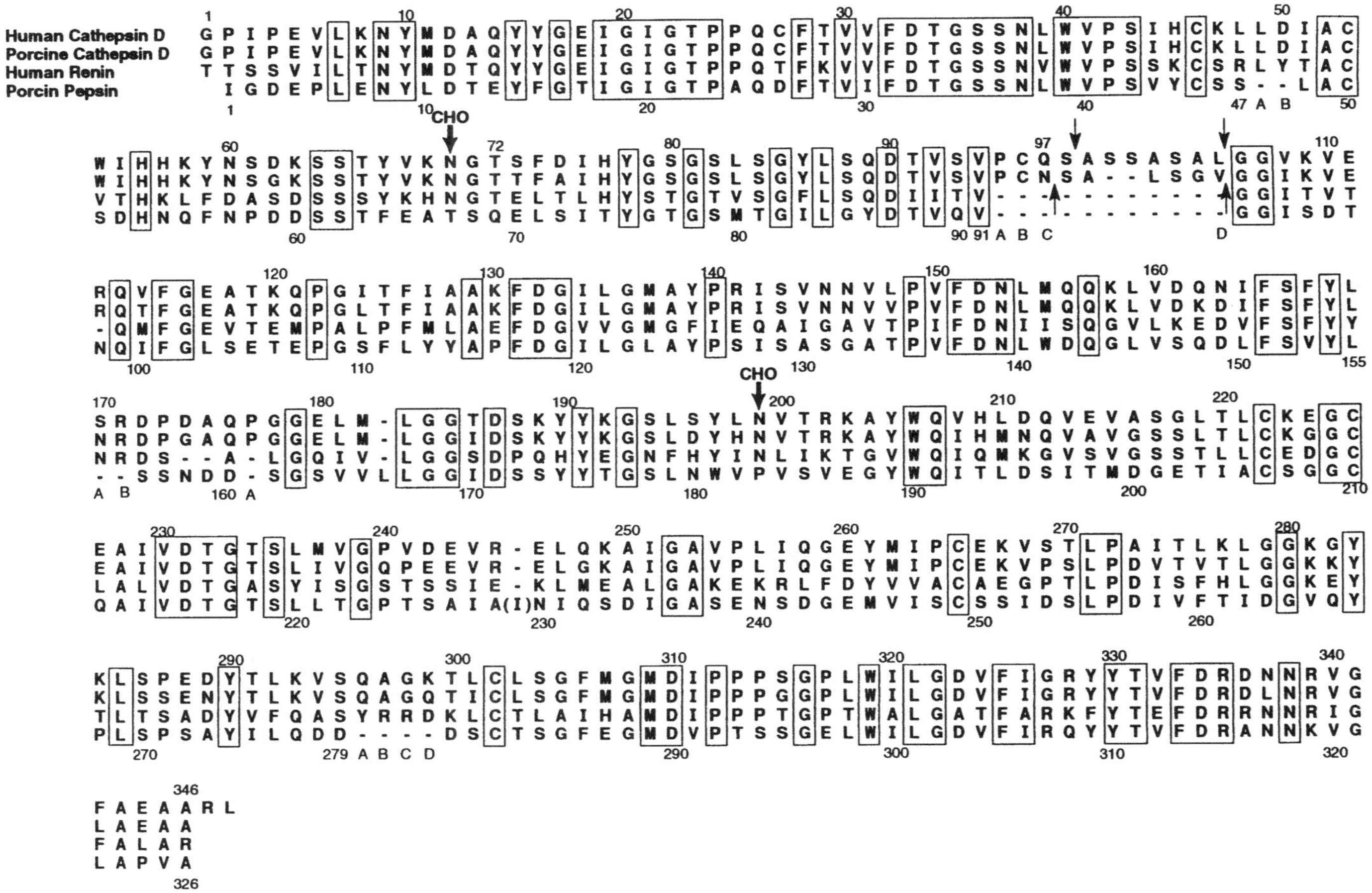

Figure 3. Sequence alignments based on 3D structural superpositions for human CatD, human renin and porcine pepsin. The alignment of porcine CatD was based on primary sequence homology to the human enzyme. Residue numbering for CatD is shown above, and that for porcine pepsin is shown below. Arrows indicate the main processing sites in the insertion loop of CatD; residues modified by carbohydrate addition are indicated by CHO. Conserved residues are boxed.

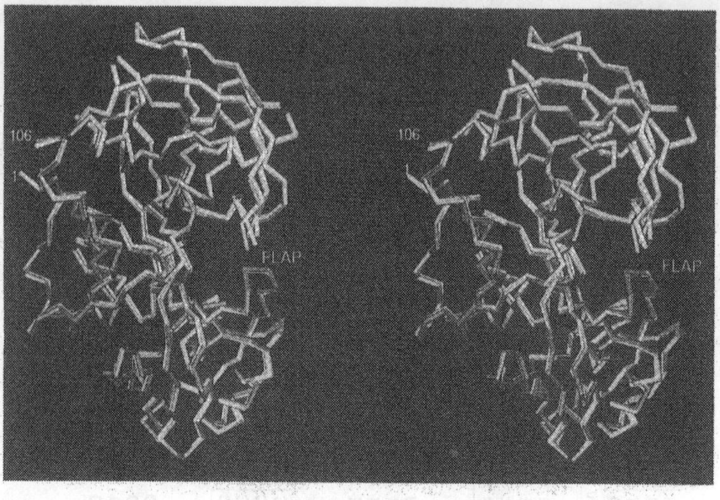

Figure 4. Stereo superposition of Ca backbones of native (white) and inhibited (gray) CatD. N-termini of light and heavy chains are numbered. The diameter of the tubes corresponds to 0.4 Å.

degree. We further hypothesized that the disposition of the subdomains may determine the specificity and affinity of a particular AP for substrates and inhibitors[17]. Earlier we reported no observable subdomain displacement differences between the native and inhibited forms of CatD[12]. However, we have recently performed a systematic comparison of the refined subdomain structures for the two crystallographically-independent molecules for both native and inhibited CatD using three-dimensional superposition analysis (Table 3). There is a slight but consistent trend towards an improved fit of the native and inhibited structures using the large subdomains alone over the total structure with a concomitant increase in the rms deviation for the small, C-terminal subdomains. This result indicates that there is a minor, 0.1-0.2 Å subdomain displacement of the inhibited relative to the native structure even though the two crystal forms are practically isomorphous. The displacement upon inhibitor binding for CatD is much less than that observed for either pepsin[18]or endothiapepsin[16]. As a control, the two crystallographically independent molecules from both crystal structures were compared similarly and showed no evidence of subdomain displacements (Table 3). Thus, inhibitor binding, but not crystal environment, seems to influence the subdomain positions for CatD.

We also compared the subdomain positions in CatD with those for other APs (Table 4). Comparison of CatD with human renin[19] reveals that the 0.9 Å overall superposition is unevenly distributed between the two domains. Superposition of CatD with rhizopuspepsin revealed even more striking differences. The large subdomain of CatD could be superposed with that of renin and rhizopuspepsin to give rms fits of 0.76 and 0.80 Å, respectively. When aligned on the large subdomains, 1.64 and 2.62 Å deviations, respectively, were observed for the small subdomains. The latter fits could be improved to 1.07 and 2.42 Å by using only the β-sheet portions (Table 4). However, the small subdomain β-sheet of CatD could be fitted independently to within 0.80 and 1.21 Å for renin and rhizopuspepsin, respectively. The lack of simultaneous correspondence for the two subdomains results from different relative subdomain displacements for the three structures. These displacements are probably best conceptualized in terms of quaternary structure interactions; the local tertiary and secondary structures of individual segments of the small subdomains, especially those near the active

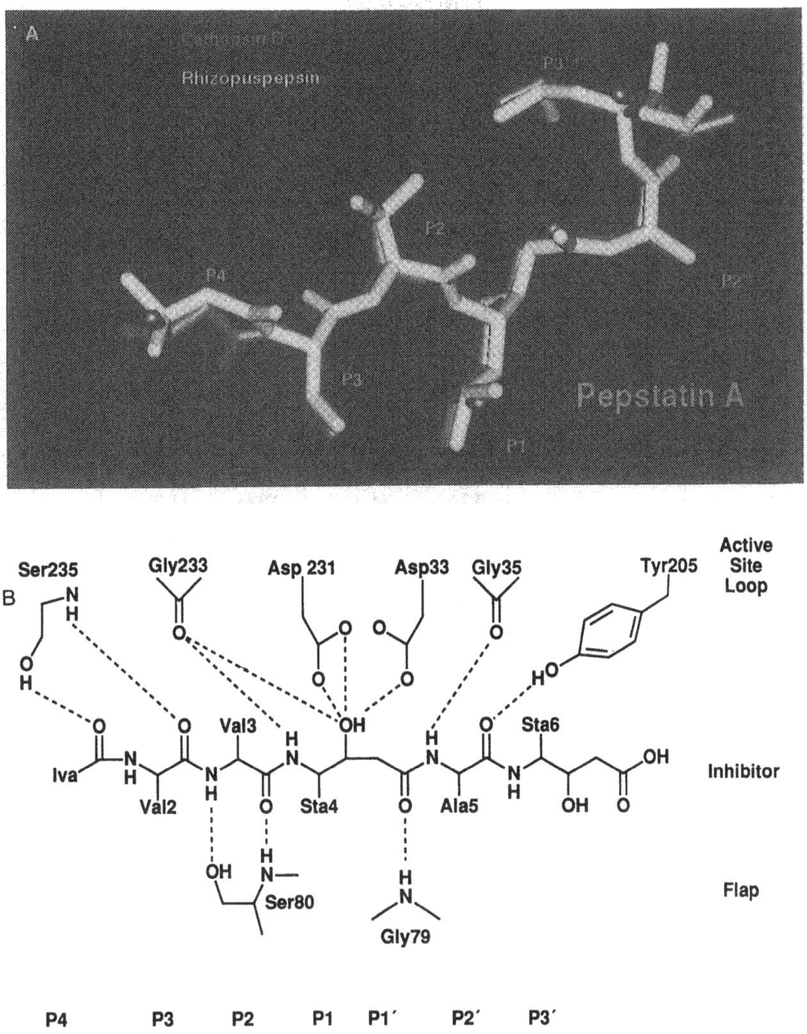

Figure 5. (A) Superposition of the pepstatin A molecules from the crystal structures of human CatD and rhizopuspespin[15]. (B) Hydrogen bonding interactions between pepstatin and CatD. Iva and Sta represent isovaline and statine residues, respectively. Reproduced with permission.[12]

site, may be highly conserved despite large alterations in the positions that these elements occupy relative to the large subdomain. This is perhaps best illustrated by the structures of the proline-rich loops in renin and CatD. These regions exhibit highly similar sequences and three-dimensional structures in the two enzymes (Fig. 6). In renin, this segment interacts with the P' region of inhibitors, whereas in CatD, as a consequence of a wider subdomain displacement, it is several Å distant from inhibitor, and does not play a directly observable role in its binding (Fig. 7A).

Comparison of the subdomain displacements of CatD, renin and rhizopuspepsin offer insights into the structural reason for the looser inhibitor packing with rhizopuspepsin than with CatD, and also suggests a structural explanation for the different range of substrate selectivity afforded by these enzymes. Rhizopuspepsin, with the largest subdomain displace-

Table 2. Van der Waals Contacts with Pepstatin A

Pepstatin A Subsite	Pepstatin A Residue	Cathepsin D Residue	Cathepsin D (Contacts, Ave Dist)	Rhizopuspepsin Residue	Rhizopuspepsin (Contacts, Ave Dist)
S4	IVA-1I	S235	(2,3.5)	T222	(1,4.1)
		L236	(1,3.5)	L223	(1,3.9)
				F278	(4,3.9)
S3	VAL-2I	S80	(1,3.1)	D79	(2,3.5)
		G233	(3,3.6)	G220	(1,3.6)
		T234	(4,3.7)	T221	(1,3.3)
		S235	(4,3.9)	T222	(3,3.8)
S2	VAL-3I	Y78	(2,3.7)	Y77	(1,3.7)
		G79	(4,3.6)	G78	(2,3.8)
		S80	(8,3.7)	D79	(5,3.7)
		G233	(1,4.1)	G220	(1,4.0)
		T234	(3,3.8)	T221	(3,3.7)
S1	STA-4I	V31	(2,3.7)	D33	(1,3.2)
		D33	(7,3.8)	D35	(6,3.2)
		G35	(4,3.8)	G37	(3,3.8)
		Y78	(9,3.8)	Y77	(11,3.6)
		G79	(2,3.9)	G78	(1,3.7)
		S80	(1,4.0)	D79	(1,3.9)
		I134	(1,3.9)	L122	(2,3.9)
				F114	(1,4.1)
		D231	(4,3.2)	D218	(3,3.3)
		G233	(6,3.6)	G220	(2,3.5)
	STA-6I	I311	(2,3.9)	W294	(1,3.9)
		I320	(1,3.6)	I298	(1,3.4)
		G79	(1,3.9)		
S2′	ALA-5I	G35	(3,3.7)		
		S36	(2,3.7)	S38	(1,3.8)
		H77	(1,4.1)	I75	(1,4.1)
		Y78	(3,4.0)	Y77	(2,3.9)
		Y205	(3,3.7)	W194	(1,3.7)
S3′	STA-6I	H77	(5,3.6)	S76	(1,3.1)
		Y205	(3,3.7)	W194	(2,3.8)

Van der Waals contacts were measured by the number of heavy atom contacts within 4.2 Å on a residue-by-residue basis.

Table 3. Subdomain Superposition Analysis for CatD

	CN mol2			CP mol1			CP mol2		
CN mol1	all	322/338	0.30	all	326/338	0.28	all	324/338	0.36
	dom1	230/236	0.27	dom1	232/236	0.23	dom1	230/236	0.31
	dom2	95/111	0.31	dom2	98/111	0.47	dom2	92/111	0.46
CN mol2				all	328/338	0.42	all	324/338	0.34
				dom1	230/236	0.33	dom1	225/236	0.25
				dom2	83/111	0.56	dom2	91/111	0.48
CP mol1							all	326/338	0.32
							dom1	231/236	0.28
							dom2	96/111	0.33

CN and CP indicate native and pepstatin-inhibited CatD, respectively. Mol1 and Mol2 indicate the two crystallographically independent molecules in the asymmetric unit. Dom1 indicates the large subdomain comprising residues 1-207, 229-238, 320-346 (1-220 and 298-326, pepsin numbering); dom2 indicates the small subdomain comprising residues 208-228, 239-319 (221-297, pepsin numbering) as defined by [17]. Each box contains the following information: aligned regions; residues used in the alignment/total number of residues for that region; rms deviation in Å. The rmsd for all Ca atoms (all) or for domain 1 Ca atoms only (dom1) were calculated using ALIGN20 with outliers excluded as indicated in the second column of each entry. The rmsd between domain 2 Ca atoms only (dom2) was calculated using RMS (T.N.B., unpublished) after domain 1 Ca atoms had been first superimposed with ALIGN. The displacement appears to be vectoral and results in a slight narrowing of the active site cleft upon inhibitor binding (Figure 4).

ment and the widest active site cleft (Fig. 7B), exhibits weaker binding to pepstatin than does CatD[12]. Renin, with the narrowest cleft, actually displays the lowest affinity for pepstatin among the three enzymes. While a structure for a renin-pepstatin complex has not been solved, examination of the renin structure indicates that pepstatin would make several steric contacts with the proline-rich loop and flap regions; thus, inhibitor or enzyme or both would have to rearrange a portion of its structure to allow binding to occur.

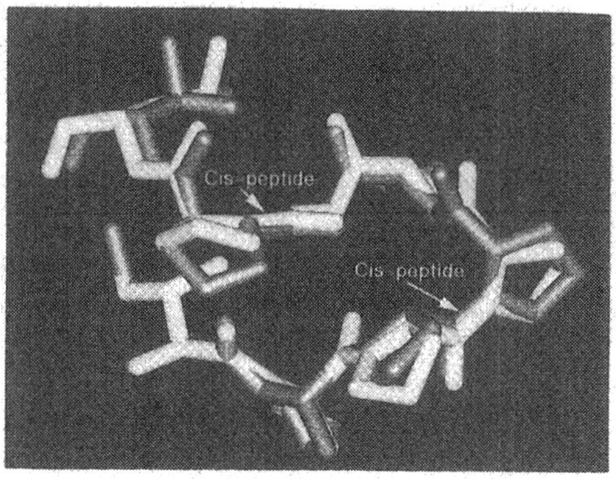

Figure 6. Superposition of the proline-rich loops, residues 312-318, for CatD (light) and renin (gray).

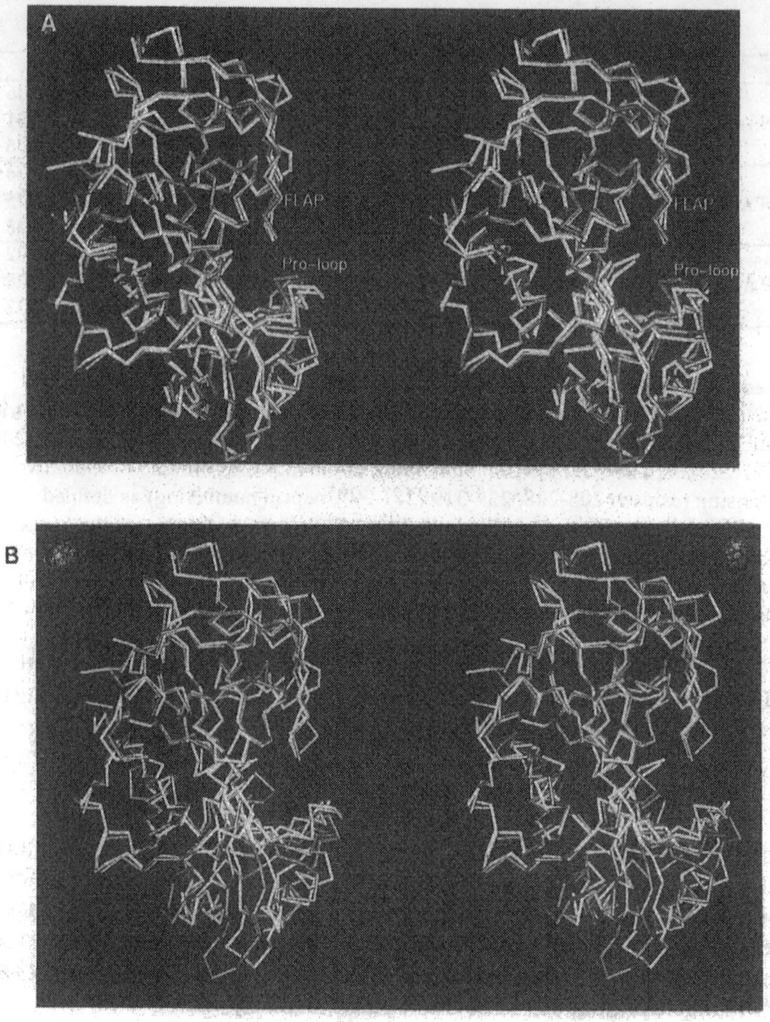

Figure 7. (A) Stereo superposition of CatD (light) and renin[19] (gray) Cα backbones. The different subdomain displacements result in a narrower active site cleft for renin in which the proline-rich loop is several Å closer to the N-domain flap. (B) Stereo superposition of CatD (light) and rhizopuspepsin (gray) Cα backbones. The different subdomain displacements result in a wider active site cleft for rhizopuspepsin. Both sets of structures were superposed by equivalencing only the large subdomain residues to emphasize the different subdomain displacements.

The ordering of the three enzymes with respect to their degree of substrate selectivity is roughly correlated with their relative subdomain displacements. These displacements along with the nature of the subsite side chains define the size and shape of the active site cleft: the narrow active site cleft for renin is consistent with the stringent substrate specificity of this enzyme; the wide cleft in rhizopuspepsin suggests weak specificity; and, the intermediate-sized gap for CatD correlates with an intermediate degree of substrate specificity. Thus, attempts to explain different substrate preferences for the APs purely on the basis of subsite residue differences may be misleading in those instances where the subdomain displacement is uncertain, such as in the case of modeled structures of APs. Knowledge

Table 4. Subdomain Superposition Analysis for CatD, Renin, and Rhizopuspepsin

		Renin		Rhizopus	
CP	all	301/332	0.87	296/325	1.34
	dom1	221/232	0.76	208/234	0.80
	dom2	93/97	1.64	85/90	2.62
	βsheet	57/57	1.07	49/55	2.42
	align	57/57	0.80	53/55	1.21

The pepstatin complex of CatD was compared to crystal structures of an inhibited form of human renin[19] and the pepstatin/rhizopuspepsin complex[15]. Each box contains the following information: aligned regions; residues used in the alignment/total number of residues for that region; rms deviation in Å. See Table 3 for definitions of CP, all, dom1 and dom2, and for alignment details. The rmsd between equivalent atoms of domain 2 is larger than might be expected due to large positional differences in equivalenced loop regions which cannot be easily accounted for by a single rigid-body motion. However, the core β-sheet of domain 2 (residues 208-214, 221-226, 239-253, 273-290, 304-310, 318-319; CatD numbering) could be aligned with an rmsd on the order of that obtained for all Cα atoms of domain 1. Visual inspection of the superimposed structures reveal that subdomain 2 of renin and rhizopuspepsin are both rotated relative to CatD resulting in narrower and wider active site clefts, respectively (Figure 7).

of the subdomain structure is necessary for a complete description of subsite structure, particularly the S' subsites. A fuller understanding of the significance of these subdomain displacements and the ability to accurately model them are important research challenges.

ACKNOWLEDGMENTS

The authors wish to thank Drs. Suguna and David Davies, NIH, for the coordinates of the refined pepstatin/rhizopuspepsin complex and Dr. Tom Blundell, Birkbeck College, for the inhibitor/renin coordinates. The content of this publication does not necessarily reflect the views or policies of the Department of Health and Human Services (DHHS), nor does mention of trade names, commercial products, or organizations imply endorsement by the U.S. Government.

REFERENCES

1. A. J. Barrett, *in*: "Proteinases in Mammalian Cells and Tissues", A. J. Barrett, ed., pp. 209-248, Elsevier/North-Holland Biomedical Press (1977)
2. S. Gulnik, E. T. Baldwin, N. Tarasova and J. Erickson, *J. Mol. Biol.* **227**:265-270 (1992)
3. T. Kobayashi, K. Honke, S. Gasa, T. Fujii, S. Maguchi, T. Miyazaki and A. Makita, *Int. J. Biochem.* **24**:1487-1491 (1992)
4. S. Kornfeld and I. Mellman, *Ann. Rev. Cell Biol.* **5**:483-525 (1986)
5. K. von Figura and A. Hasilik, *Annu. Rev. Biochem.* **55**:167-193 (1986)
6. H. Rochefort, *Semin. Cancer Biol.* **1**:153-160 (1990)
7. R. Siman, S. Mistretta, J. T. Durkin, M. J. Savage, T. Loh, S. Trusko and R. W. Scott, *J. Biol. Chem.* **268**:16602-16609 (1993)
8. L. A. Liotta and W. G. Stetler-Stevenson, *Semin. Cancer Biol.* **1**:99-106 (1990)
9. L. A. Liotta, *Sci. Am.* 54-63 (1992)
10. M. D. Johnson, J. A. Torri, M. E. Lippman and R. B. Dickson, *Cancer Res.* **53**:873-877 (1993)

11. L. M. Sanchez, A. A. Ferrando, I. Diez-Itza, F. Vizoso, A. Ruibal and C. Lopez-Otin, *Br. J. Cancer* **67**:1076-1081 (1993)
12. E. T. Baldwin, T. N. Bhat, S. Gulnik, M. V. Hosur, R. C. Sowder, R. E. Cachau, J. Collins, A. M. Silva and J. W. Erickson, *Proc. Natl. Acad. Sci. USA* **90**:6796-6800 (1993)
13. P. Metcalf and M. Fusek, *EMBO J.* **12**:1293-1302 (1993)
14. S. Yonezawa, T. Takahashi, X.-j. Wang, R. N. S. Wong, J. A. Hartsuck and J. Tang, *J. Biol. Chem.* **263**:16504-16511 (1988)
15. K. Suguna, R. R. Bott, E. A. Padlan, E. Subramanian, S. Sheriff, G. H. Cohen and D. R. Davies, *J. Mol. Biol.* **196**:877-900 (1987)
16. A. Sali, B. Veerapandian, J. B. Cooper, S. I. Foundling, D. J. Hoover and T. L. Blundell, *EMBO J.* **8**:2179-2188 (1989)
17. C. Abad-Zapatero, T. J. Rydek and J. Erickson, *Proteins* **8**:62-81 (1990)
18. C. Abad-Zapatero, T. J. Rydel, D. J. Neidhart, J. Luly and J. W. Erickson, *in*: "Structure and Function of Aspartic Proteinases", B. M. Dunn, ed., pp. 9-21, Plenum Press, New York (1991)
19. V. Dhanaraj, C. G. Dealwis, C. Frazao, M. Badasso, B. L. Sibanda, I. J. Tickle, J. B. Cooper, H. P. Driessen, M. Newman, C. Aguilar, S. P. Wood, T. L. Blundell, P. M. Hobart, K. F. Geoghegan, M. J. Ammirati, D. E. Danley, B. A. O'Conner and D. J. Hoover, *Nature* **357**:466-472 (1992)
20. Y. Satow, G. H. Cohen, E. A. Padlan and D. R. Davies, *J. Mol. Biol.* **190**:593-604 (1986)
21. J. F. Woessner Jr., *in*: "Tissue Proteinases", A. J. Barrett and J. T. Dingle, ed., pp. 291-311, North-Holland Publishing Co. (1971)
22. P. Gopalan, M. J. Dufrense and A. H. Warner, *Canadian Journal of Physiology and Pharmacology* **65**:124-129 (1987)
23. A. Matus and G. D. Green, *Biochemistry* **26**:8083-8086 (1987)
24. M. Banay-Schwartz, T. DeGuzman, A. Kenessey, M. Palkovits and A. Lajtha, *J. Neurochem.* **58**:2207-2211 (1992)
25. J. M. van Noort and A. C. M. van der Drift, *J. Biol. Chem.* **264**:14159-14164 (1989)
26. L. E. Guagliardi, B. Koppelman, J. S. Blum, M. S. Marks, P. Cresswell and F. M. Brodsky, *Nature* **343**:133-139 (1990)
27. P. J. Peters, J. J. Neefjes, V. Oorschot, H. L. Ploegh and H. J. Geuze, *Nature* **349**:669-676 (1991)
28. J. F. Diedrich, H. Minnigan, R. I. Carp, J. N. Whitaker, R. Race, W. Frey and A. T. Haase, *J. Virol.* **65**:4759-4768 (1991)

CATHEPSIN D CRYSTAL STRUCTURES AND LYSOSOMAL SORTING

Peter Metcalf[1] and Martin Fusek[2*]

[1]EMBL, postfach 10.2209, D69012 Heidelberg, Germany
[2]Institute of Organic Chemistry and Biochemistry
Flemingovo n. 2, Prague, Czech Republic

INTRODUCTION

The D is not for dull! In this chapter we discuss the features of cathepsin D which are interesting because of their relevance to studies on lysosomal targeting. We will not describe the active site cleft, the centre of attention for aspartic proteinases with greater specificity or with more obvious medical significance. Mechanisms of protein targeting are of fundamental interest, and increasingly implicated in medicine. Unlike the aspartic proteases with previously known structures, cathepsin D is intracellular, and we solved crystal structures for human spleen cathepsin D and pepstatin inhibited bovine liver cathepsin D partly to learn more about what special features of this molecule cause it to be targeted to lysosomes[1-3]. There is one other solved structure of a lysosomal enzyme; the cysteine proteinase cathepsin B[4]. We are currently comparing its surface with that of cathepsin D to search for expected common features, as discussed in the following paragraphs.

The analysis of missorting or secretion mutants has been used to identify many molecules involved in intracellular sorting pathways. In the case of lysosomal targeting, a rare human genetic disease, mucolipidosis II, in which cells secrete all lysosomal enzymes, led to the discovery of the mannose 6-phosphate mediated lysosomal pathway (reviewed[5]). Mammalian aspartic proteases are transported from their site of synthesis in the endoplasmic reticulum to the Golgi and it is only after the *trans* -Golgi that cathepsin D is clearly located differently from secreted aspartic proteases (figure 1). However by this stage, cathepsin D, like other lysosomal molecules in the exocytic pathway, has already been tagged for lysosomal transport. In the region of the *cis* -Golgi, carbohydrates of only lysosomal molecules are selectively phosphorylated by a still poorly characterized phosphotransferase, which was shown to be non-functional in the mucolipidosis II patients. This addition of phosphate to the 6 position of terminal mannose residues of carbohydrates of soluble

[*]Current address: OMRF, 825 NE 13th St., Oklahoma City, 73104.

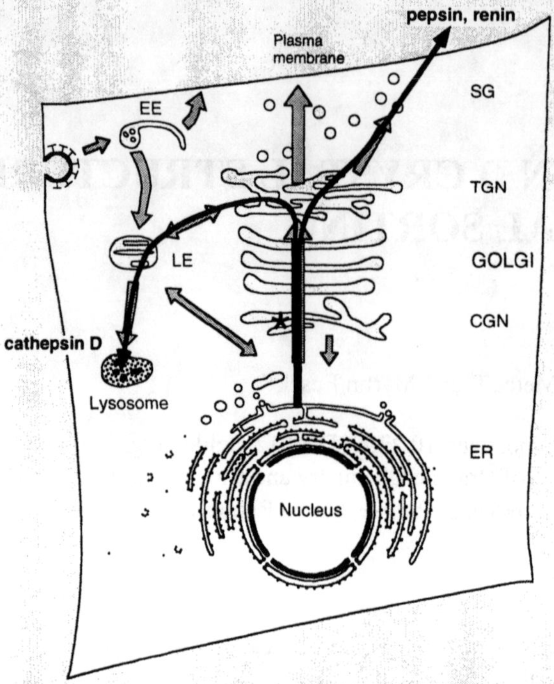

Figure 1. Intracellular sorting of mammalian aspartic proteases. Schematic diagram of a cell including the main organelles and transport pathways. The dark arrows show intracellular routes taken by aspartic proteases. Cathepsin D, pepsinogen and renin are synthesized in the endoplasmic reticulum (ER) and enter the exocytic pathway. Procathepsin D is phosphorylated in the *cis*-Golgi (CGN) and transported via late endosomes(LE) to lysosomes by the mannose 6-phosphate dependent pathway. Pepsinogen and prorenin are targeted to secretory granules (SG) in the *trans*-Golgi (TGN) and are secreted from cells via the regulated exocytic pathway. The asterisk shows the *cis*-Golgi location of the phosphotransferase which initially tags soluble molecules like cathepsin D destined for the lysosome.

lysosomal molecules is the earliest known targeting event for lysosomal proteins in the exocytic pathway. Two mannose 6-phosphate receptor molecules are involved in the *trans* -Golgi to late-endosome part of the pathway, and cell lines lacking one or both of the receptor types hypersecrete lysosomal enzymes, confirming the importance of the mannose 6-phosphate mediated pathway for lysosomal enzyme transport[6-8].

Experiments with the partially purified *cis* -Golgi phosphotransferase (it is still not cloned) indicated that it recognizes soluble lysosomal enzymes by binding to a three-dimensional structural motif common to only those molecules destined to lysosomes[9] (figure 2). How the phosphotransferase manages to phosphorylate carbohydrates of the many and varied lysosomal enzymes to which it binds is obviously a major conceptual problem. Lysosomal soluble enzymes share no evident sequence motif analogous to known targeting motifs. Consequently a structural and mutational analysis of lysosomal proteins will be necessary to identify the three-dimensional lysosomal targeting signal recognized by the phosphotranferase and responsible for lysosomal targeting.

A strategy for locating targeting signals in protein sequences involves expression studies with chimeric proteins made from homologous, but differently targeted proteins.

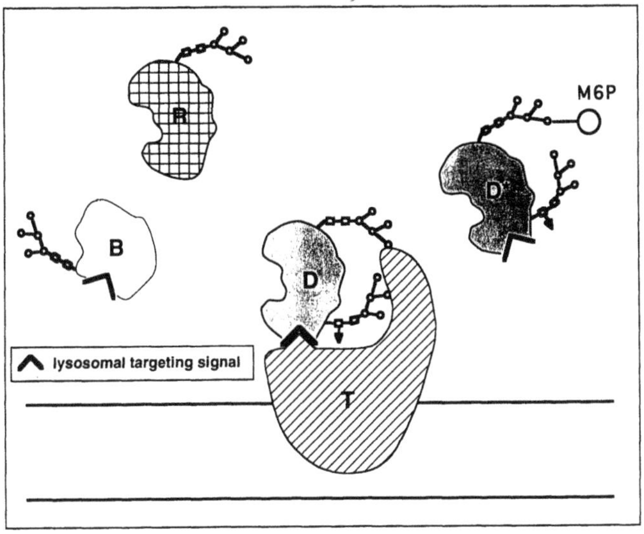

Figure 2. The initial step of lysosomal soluble enzyme targeting: specific phosphotransferase recognition in the *cis*-Golgi. The figure illustrates current understanding of the first known point at which soluble lysosomal enzymes are distinguished from other molecules in the exocytic pathway. The *cis*-Golgi phosphotransferase (T), a ~250KD membrane bound protein complex, recognizes only molecules containing the hypothetical lysosomal targeting signal, such as cathepsin D (**D**) shown here attached to the phosphotransferase via the targeting signal. The result of the interaction is that terminal mannose residues on the carbohydrates of only lysosomal molecules are phosphorylated. These bind to mannose 6-phosphate receptors in later steps of the lysosomal sorting pathway. **D*** indicates a phosphorylated cathepsin D molecule (M6P = mannose 6-phosphate). **B** represents a cathepsin B molecule, also containing the targeting signal and phosphorylated by the transferase. **R** shows a non-lysosomal renin molecule, otherwise similar to cathepsin D, but lacking the targeting signal necessary for phosphorylation.

Because of the homology, one can expect that the chimeras fold correctly, so that their expression can then be used to focus in on the parts of the sequence sufficient to alter targeting. In the case of lysosomal targeting, cathepsin-D/pepsinogen chimeras have been used to identify a minimal region of the cathepsin D sequence sufficient for phosphorylation and consequently for lysosomal targeting[10]. Similar work is underway to locate a targeting signal in the cathepsin B sequence. The 29 residue minimal targeting region for human cathepsin D includes K203 and the segment C265-L292.

Surprisingly, the situation in yeast appears to be very different. Vacuolar and lysosomal transport appear to have little in common. While there are interesting homologies between some molecules involved in intracellular transport in yeast and in mammalian cells[11] there is no evidence for mannose 6-phosphate or mannose 6-phosphate receptor genes in yeast. Mutational studies have clearly shown that the vacuolar targeting signal in the yeast cathepsin-D analog proteinase-A is in the prosegment, nowhere near the lysosomal cathepsin D 28-residue C-terminal targeting region[12].

METHOD

Recombinant correctly glycosylated procathepsin D, as seen by the *cis* -Golgi phosphotransferase would have been ideal for this project. Because of the difficulty of producing crystallizable amounts of procathepsin D we purified and crystallized mature

Table 1. Cathepsin D Diffraction Data and Model Statistics

	Human spleen	Bovine liver / pepstatin A
Diffraction Data		
Number of unique reflections to 3.0Å	15,735	18,332
Diffraction agreement		
R(free)	25.7%	24.4%
R	16.5%	16.7%
Stereochemical Ideality		
Bonds	0.005Å	0.006Å
Angles	1.1°	1.14°

cathepsin D from a number of sources and obtained useful crystals from human spleen and pepstatin inhibited bovine liver protein. We carried out many purification experiments to try and improve crystal quality, and were able to purify single chain bovine cathepsin D but not in sufficient yields for crystals. The details of the purification and crystallization for the two structures have already been described[1,2]. Briefly, both proteins were crystallized using PEG at about pH4 - 4.5 and both crystallized in space group $P2_12_12_1$ with two molecules in the asymmetric unit, but with different packing. The human protein crystals contain two-chain protein (missing a ~4 residue loop near residue 100) and the bovine crystals are about 50% single chain. The first bovine protein crystals were obtained in November 1989 and the first human protein crystals were obtained in April 1990, using protein purified at the Institute of Organic Chemistry and Biochemistry in Prague. The structures were solved by the molecular replacement method with pepsin as a search model using the program X-PLOR[13]. The molecular replacement solutions were verified by the observation of carbohydrate density in both maps and of pepstatin density in the bovine/pepstatin map. Data quality and model statistics are given in Table 1.

The free R factor criteria[14] for crystallographic models was essential for this work to produce a believable model from comparatively low resolution data. Since the concept is new we will describe it briefly here. The improvement of crystallographic refinement programs has meant that it has become relatively easy to refine an acceptable protein model with good geometry and a low conventional R factor, which is however wrong. If the resolution is poor, or if too many water molecules have been added or obvious non-crystallographic symmetry omitted, the ratio of model parameters to diffraction data is high and modern refinement programs can then fit the data with a realistic but incorrect structure. In the case of cathepsin D there was a worry that one would produce pepsin-like loops (from

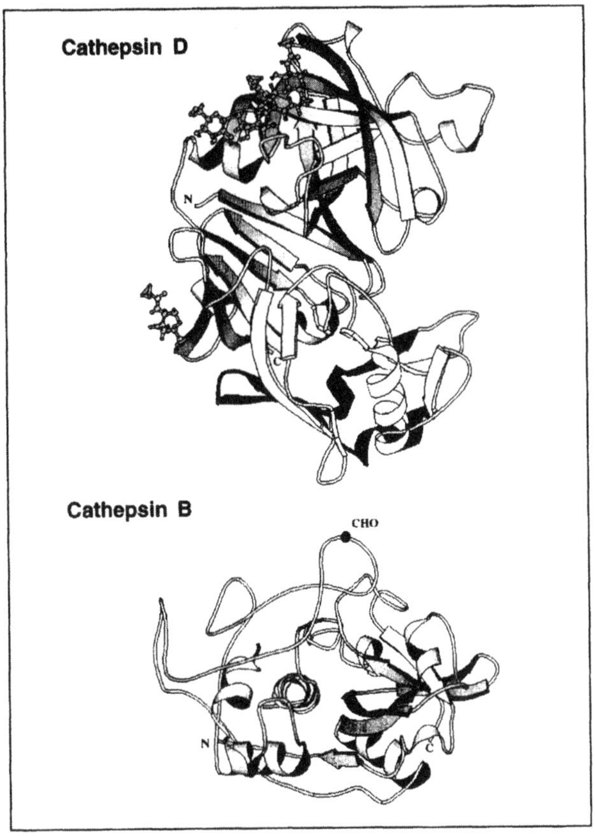

Figure 3. The first two lysosomal enzyme structures: ribbon diagrams of cathepsin B and cathepsin D. The cathepsin D figure shows carbohydrate visible in the bovine map attached to the N-terminal structural domain, shown shaded grey. The smaller C-terminal structural domain shown in white at the lower/front part of the figure contains the 28 residue minimal sequence region (black) required for mannose 6-phosphorylation and lysosomal targeting. The figure was made using the program MOLSCRIPT19. Cathepsin B coordinates were provided by W.Bode[4].

the search model) that would still refine with the cathepsin D data. To guard against this over-fitting, one removes a control fraction (typically 10%) of the data from the modeling process and then checks if the model also fits the omitted fraction of the data. The more stringent match to the control fraction of the data is the 'free R value'. For the cathepsin D structures values of 25% were obtained in the 6-3Å range, which appear to be reasonably good, when compared to values given in recent papers where the statistic is given. The free R criteria has also been used to demonstrate that the current best high resolution crystallographic model of an aspartic protease (as judged by this criteria) looks like an NMR structure, with multiple conformations and large variations in the loop regions.

RESULTS AND DISCUSION

A ribbon diagram showing one of the bovine cathepsin D molecules is shown in figure 3, along with cathepsin B, the first lysosomal protein structure, which should share

```
            |         |         |         |         |         |         |         |         |           Ncd
---GPIPEVLKNYMDAQYYGEIGIGTPPQCFTVVFDTGSSNLWVPSIHCKLLDIACWIHHKYNSDKSSTYVKNGTSFDIHYGSGSLSGYLSQDTVSVPCQ  CDh
---GPIPELLKNYMDAQYYGEIGIGTPPQCFTVVFDTGSANLWVPSIHCKLLDIACWTHRKYNSDKSSTYVKNGTTFDIHYGSGSLSGYLSQDTVSVPCN  CDb
             |         |         |         |           |         |         |         |         |      Npep
---IG-DEPLENYLDTEYFGTIGIGTPAQDFTVIFDTGSSNLWVPSVYCS--SLACSDHNQFNPDDSSTFEATSQELSITYGTGSMTGILGYDTVQV---  PEP
                                       + active D 1
                                                                            + sugar 1

                                                  ----------                            ----------                                  ----  loops
                                                      1                                  2 'flap'

 100
 |         |         |         |         |         |         |         |         |         |           Ncd
SASSASALGGVKVERQVFGEATKQPGITFIAAKFDGILGMAYPRISVNNVLPVFDNLMQQKLVDQNIFSFYLSRDPDAQP--GGELMLGGTDSKYYKGSL  CDh
PSSSSP--GGVTVQRQTFGEAIKQPGVVFIAAKFDGILGMAYPRISVNNVLPVFDNLMQQKLVDKNVFSFFLNRDPKAQP--GGELMLGGTDSKYYRGSL  CDb
             |         |         |         |         |         |           |           |            Npep
-------GGISDTNQIFGLSETEPGSFLYYAPFDGILGLAYPSISASGATPVFDNLWDQGLVSQDLFSVYLSSN--DDS--GSVVLLGGIDSSYYTGSL  PEP

----------          ----------                                                                      loops
3 'processed'           4

    200
    |         |         |         |         |         |         |         |         |         |       Ncd
SYLNVTRKAYWQVHLDQVEVASGLTLCKEGCEAIVDTGTSLMVGPVDEVRELQKAIGAVPLIQGEYMIPCEKVSTLPAITLKLGGKGYKLSPEDYTLKVS  CDh
MFHNVTRQAYWQIHMDQLDVGSSLTVCKGGCEAIVDTGTSLIVGPVEEVRELQKAIGAVPLIQGEYMIPCEKVSSLPEVTVKLGGKDYALSPEDYALKVS  CDb
|         |         |         |         |         |         |         |         |         |         Npep
NWVPVSVEGYWQITLDSITMDGETIACSGGCQAIVDTGTSLLTGPTSAIANIQSDIGASENSDGEMVISCSSIDSLPDIVFTIDGVQYPLSPSAYILQ--  PEP
       *                                                                ****************************       target
  + sugar 2
                                  + active D 2
                    --------                                  ---------                            ----  loops
                       5                                         6                                   7

    300
    |         |         |         |         |                                                         Ncd
QAGKTLCLSGFMGMDIPPPSGPLWILGDVFIGRYYTVFDRDNNRVGFAEAARL                                               CDh
QAETTVCLSGFMGMDIPPPGGPLWILGDVFIGRYYTVFDRDQNRVGLAEAARL                                               CDb
    |         |         |         |         |                                                         Npep
--DDDSCTSGFEGMDVPTSSGELWILGDVFIRQYYTVFDRANNKVGLAPVA--                                               PEP

------     ----------                                                                               loops
          8 'poly-proline'
```

Figure 4. Sequence alignment of human and bovine cathepsin D and pepsin. Key: Ncd - human cathepsin D numbering; CDh and CDb - human and bovine cathepsin D (bovine sequence from J. Chirguin); Npep - porcine pepsin numbering; PEP - porcine pepsin; loops - regions which differ most between aligned bovine and human cathepsin D, and between other aspartic proteases; target - the minimal lysosomal targeting region identified by chimera expression studies[10]. The active site aspartates and carbohydrate binding sites are marked.

the three-dimensional lysosomal targeting signal. Some features of the cathepsin D structures are described below:

Dimers

The first results of molecular replacement revealed that cathepsin D is packed in both the bovine and human crystals as 'mouth-to-mouth' dimers with the active site cleft of molecule A facing the active site cleft of the non-crystal symmetry related molecule B. The bovine A and B molecules are translated from each other significantly compared to the human A and B molecules. Because of the possibility that the dimerization seen in the crystals may reflect a role for oligomerization *in-vivo* we analyzed the cathepsin D samples by gel-filtration and analytical ultracentrifugation. We found no evidence for dimerization of cathepsin D in solution.

Loops

When mammalian aspartic proteinase crystal structures are compared the most obvious differences are in the loops. A sequence alignment of cathepsin D and pepsin (figure 4) shows the different numbering for the amino acids and the location of the variable loops. One might expect the loops of the two cathepsin D structures to be more like each other than like loops of other aspartic proteases, and if they were not one would question how well they were defined in the crystal structure. We compared the bovine and human loops after aligning the structures taking into account the C-terminal structural subdomain[15-17] and found that, in general, the corresponding alpha-carbon distances differ as much as between loops of other aspartic proteases.

The loop regions which varied most between the two cathepsin D structures are the parts of the molecules with high B factors (i.e. with low contrast in the electron density map). One exception is the 'flap' in the bovine/pepstatin structure which is closed down over the pepstatin with low B factors but which is much more flexible in the human structure. Another loop with a strikingly different conformation in the two structures is the small disulphide bonded loop C46-C53, which is rotated 30deg about an axis passing though the disulphide bond.

The analysis of the loops suggests that most of them are mobile, and that the different conformations seen in the crystal structures and the 'multi-conformer' models represent snap shots of the dynamic behaviour of the protein in solution. Since sensitivity to proteolysis is presumably correlated with flexibility it seems noteworthy that cathepsin D is as stable as it is in the protease rich lysosomal environment and that only the loop near residue 100 is partially removed *in-vivo*.

Carbohydrate

The composition of the N and C-terminal carbohydrates of bovine spleen cathepsin D has been determined (Easton, R. L., Dell, A. and Morris, H. R. in preparation). The expected population of Asn-$(GlcNac)_2(Man)_{3-6}$ was observed at the N-terminal site but a significant proportion of the C-terminal carbohydrate has a fucose side chain attached to the proximal GlcNac. The carbohydrate electron density is clearest for the bovine cathepsin D structure, where density corresponding to $(GlcNac)_2(Man)_3$ can be seen at the N terminal site. Only a single carbohydrate can be seen at the C-terminal site and there is no visible density corresponding to a fucose side chain.

The Hypothetical Lysosomal Targeting Region

The minimal region of cathepsin D implicated in lysosomal targeting by the chimera expression studies[10] consists of the 28 residue disulphide bonded loop C265-L292 and K203 (Q203 in bovine cathepsin D). The region is well-defined in the electron density maps and comprises about half of the C-terminal structural domain. The end of the loop is adjacent to the C-terminus of cathepsin D (figure 3), which is intriguing because cathepsin D has a two residue (-RL) extension not present on other aspartic proteases. The significance of this C-terminal extension for targeting requires further mutant expression studies. It is curiously similar to the targeting signal for import of cytoplasmic proteins into peroxisomes (-SKL)[18]. Cathepsin D has many more positively charged surface residues than other aspartic proteases, although only K203 has so far been implicated in targeting. We are currently comparing the calculated surface charge patterns of cathepsins B and D to see if there is any common feature that may correspond to the expected shared phosphotransferase binding site.

REFERENCES

1. F. Bieber, V. Brachvogel, R. Arni, M. Fusek and P. Metcalf, Crystallization and initial crystallographic results for pepstatin A inhibited bovine cathepsin D, *J.Mol.Biol.* **227**, 1265-1268 (1992).
2. M. Fusek, M. Baudys and P. Metcalf, Purification and crystallization of human cathepsin D, *J.Mol.Biol.* **226**, 555-557 (1992).
3. P. Metcalf and M. Fusek, Two crystal structures for cathepsin D: the lysosomal targeting signal and active site, *EMBO J.* **12**, 1293-302 (1993).

4. D. Musil, D. Zucic, D. Turk, I. Mayr, R. A. Engh, R. Huber, T. Popovic, V. Turk, T. Towatari, N. Katunuma and W. Bode, The refined 2.15 Å X-ray crytal structure of human liver cathepsin B: the structural basis for its specificity., *EMBO J.* **10**, 2321-2330 (1991).
5. Kornfeld, Structure and function of the mannose 6-phosphate/insulin like growth factor II receptors, *Annu. Rev. Biochem.* **61**, 307-330 (1992).
6. T. Ludwig, C. E. Ovitt, J. Remmler, P. Lobel, U. Ruther and B. Hoflack, Targeted disruption of the mouse cation-dependent mannose 6-phosphate receptor results in partial missorting of multiple lysosomal enzymes, *EMBO J.* **12**, 5225-5235 (1993).
7. T. Ludwig, H. Munier-Lehmann, M. Hollinshead, C. Ovitt, P. Lobel and B. Hoflack, Differential sorting of lysosomal enzymes in mannose 6-phosphate receptor-deficient fibroplasts (submitted), (1994).
8. A. Köster, P. Saftig, U. Matzner, K. von Figura, C. Peters and R. Pohlmann, Targeted disruption of the Mr 46 000 mannose 6-phsophate receptor gene in mice results in misrouting of lysosomal proteins, *EMBL J.* **12**, 5219-5223 (1993).
9. S. Kornfeld, Trafficking of lysosomal enzymes in normal and disease states, *J. Clin. Invest.* **77**, 1-6 (1986).
10. T. J. Baranski, P. L. Faust and S. Kornfeld, Generation of lysosomal enzyme targeting signal in the secretory protein pepsinogen, *Cell* **63**, 281-291 (1990).
11. P. Novick and P. Brennwald, Friends and Family: The role of Rab GTPases in vesicular traffic, *Cell* **75**, 597-601 (1993).
12. D. J. Kilonsky, L. M. Banta and S. D. Emr, Intracellular sorting and processing of a yeast vacuolar hydrolase: proteinase A propeptide contains vacuolar targeting information., *Mol. Cell. Biol.* **8**, 2105-2116 (1988).
13. A. T. Brünger, Extension of molecular replacement: A new search strategy based on patterson correlation refinement, *Acta Cryst.* **A46**, 46-57 (1990).
14. A. T. Brünger, Free R Value: a novel statistical quantity for assessing the accuracy of crystal structures, *Nature* **355**, 472-475 (1992).
15. C. Abad-Zapatero, T. H. Rydel and J. Erickson, Revised 2.3Å structure of porcine pepsin: Evidence for a Flexible Subdomain, *Proteins* **8**, 62-81 (1990).
16. A. Sali, B. Veerapandian, J. Coper, D. Modd, T. Hofmann and T. Blundell, Domain flexibility in aspartic proteases, *Proteins* **12**, 158-170 (1992).
17. A. R. Sielecki, A. A. Federov, A. Boodhoo, N. S. Andreeva and M. N. G. James, Molecular and crystal structures of monoclinic porcine pepsin refined at 1.8Å resolution, *J. Mol. Biol.* **214**, 143-170 (1990).
18. S. J. Gould, G.-A. Keller, M. Schneider, S. H. Howell, L. J. Garrard, J. M. Goodman, B. Distel, H. Tabak and S. Subramani, Peroxisomal protein import is conserved between yeast, plants, insects and mammals., *EMBO J.* **9**, 85-90 (1990).
19. P. J. Kraulis, MOLSCRIPT: a program to produce both detailed and schematic plots of protein structures, *J. Appl. Crystallogr.* **24**, 946-950 (1991).

ISOLATION AND CHARACTERIZATION OF HUMAN GASTRIC PROCATHEPSIN E AND CATHEPSIN E

Senarath B. P. Athauda,[1,2] Takashi Kageyama,[3] Takayuki Takahashi,[1] Hideshi Inoue,[1] Masao Ichinose,[1] Masanori Ukai,[4] and Kenji Takahashi[1]

[1]Department of Biophysics and Biochemistry
Faculty of Science
The University of Tokyo
Hongo, Bunkyo-ku, Tokyo 113, Japan
[2]Department of Biochemistry
Faculty of Medicine
University of Peradeniya
Peradeniya, Sri Lanka
[3]Department of Cellular and Molecular Biology
Primate Research Institute
Inuyama, Aichi 484, Japan
[4]Ukai Hospital
Nagoya 453, Japan

INTRODUCTION

Human gastric cathepsin E is an aspartic proteinase present in human gastric mucosa and different from pepsinogens A and C and cathepsin D. It has been localized not only to stomach mucosa, but also to erythrocyte membranes and several lymphoid associated tissues and cells including thymus, spleen, macrophages and polymorphonuclear lymphocytes (1,2). However its physiological function and the process of activation remain obscure. Due to its intracellular localization in lymphoid associated tissues and cells it has been suggested to have a role in immune function (3,4). Studies on cathepsin E has not progressed like those on other aspartic proteinases (pepsinogens and cathepsin D), because of difficulties in obtaining a sufficient amount of the purified native enzyme. Therefore, to understand the structure-function relationship of cathepsin E and to clarify its physiological function, it is important to establish a purification procedure of procathepsin E and cathepsin E for further characterization.

In this study procathepsin E was purified from human gastric mucosa and activation of procathepsin E to cathepsin E was studied at different pHs. Cathepsin E was isolated from the activation mixture and its NH_2-terminal sequence was analyzed with that of procathepsin

E. Procathepsin E was activated at pH 4.0 and the resulting mature cathepsin E was purified from the activation mixture.

Since the optimum pH of cathepsin E toward hemoglobin is 3.0-3.5 and its proteolytic activity and substrate specificity were investigated previously only at acidic pHs (5-8). As far as cathepsin E can be active only at acidic pHs, its physiological roles *in vivo* would be highly restricted. Therefore, in order to shed light on the physiological roles of cathepsin E, its potential proteolytic activity toward the B chain of oxidized insulin and cleavage specificity were investigated at different pHs. In addition, the stability of cathepsin E was investigated at different pHs.

MATERIALS AND METHODS

Specimens of human stomachs were generously supplied by Dr. M. Ukai. DEAE cellulose (DE-52) was a product of Whatman. Sephacryl S-200 and a Mono Q HR5/5 column was purchased from Pharmacia. Polyvinylidene difluoride (PVDF) membrane was purchased from Millipore. Pyroglutamate aminopeptidase was from Boehringer Mannheim Biochemica, pepstatin A was from Peptide Institute (Osaka), the B chain of oxidized bovine insulin was from Sigma and reagents for automated amino acid sequencing were obtained from Applied Biosystems. Other reagents used were of the highest grade available.

Purification of Procathepsin E

Gastric procathepsin E was purified from human gastric mucosa as described previously (9) with additional chromatography on Concanavalin(Con) A-Sepharose and FPLC on Mono Q as follows. The potential activity toward haemoglobin was assayed as described previously (9). The procathepsin E fraction from the second DE-52 chromatography was adjusted to 0.2 M NaCl and applied to a column (1.0 x 4.0 cm) of Con A-Sepharose equilibrated with 0.02 M phosphate buffer-0.2 M NaCl at pH 7.5 (buffer A). The column was washed with 600 ml of the above buffer and eluted with buffer A containing 0.4 M methyl-α-D-mannopyranoside at pH 7.5. Fractions with potential proteolytic activity were pooled and concentrated by using a small DE-52 column (1.0 x 1.0 cm). The concentrated sample was gel-filtered on a Sephacryl S-200 column (3.8 x 130 cm) equilibrated with buffer A. An aliquot (2 mg) of the pooled fraction from the Sephacryl S-200 gel filtration was dialyzed and injected into a Mono Q (HR 5/5) column (0.5 x 5 cm) equilibrated with 0.02M Tris-HCl buffer, pH 7.8, and eluted with a linear gradient of 0-1 M NaCl in a total volume of 30 ml in the FPLC system. The fractions with potential proteolytic activity were further purified by rechromatography on a Mono Q column.

Activation Profile of Procathepsin E

Procathepsin E solution (0.16 mg/ml) in 0.1 M sodium formate buffer, pH 3.5, was incubated at 14 °C. Aliquots were withdrawn at appropriate time intervals and mixed with a 1/3 volume of the sample buffer for sodium dodecyl sulfate-polyacrylamide gel electrophoresis(SDS-PAGE). Aliquots of about 20 μl were subjected to SDS-PAGE according to Laemmli (10). Further, procathepsin E solution (0.25 mg/ml) was incubated at different pHs and at 37 °C to investigate the pH dependence of activation. Aliquots were withdrawn at different time intervals and analyzed by SDS-PAGE as above.

Purification of Cathepsin E

Procathepsin E was activated to cathepsin E by dialyzing its solution against 0.05 M sodium acetate buffer, pH 4.0, at 4 °C for 3 h and the activation mixture was applied to a pepstatin-Sepharose column (1.0 x 1.0 cm) equilibrated with the same buffer. The protein was eluted with 0.05 M Tris/HCl buffer, pH 8.0, containing 1 M NaCl and further purified by FPLC on a Mono Q column.

Amino Acid Sequence Analysis

The NH_2-terminal amino acid sequences of procathepsin E and cathepsin E were determined by using an Applied Biosystems pulse-liquid protein sequencer model 477A/120A. The NH_2-terminal sequence of procathepsin E was analyzed before and after treatment with pyroglutamate aminopeptidase. In the later case, procathepsin E (200 μg) was dissolved in 0.25 ml of 20 mM N-ethylmorpholine acetate buffer, pH 8.0, containing 100 mM EDTA and 10 mM DTT and mixed with 10 μg of pyroglutamate aminopeptidase. The mixture was incubated at 4 °C for 18 h, followed by incubation at 25 °C for 4 h. The reaction mixture was mixed with the SDS-PAGE sample buffer and subjected to electrophoresis according to Laemmli (10). Proteins in the gel were electroblotted onto PVDF membrane with 10 mM 3-(cyclohexylamino)-1-propanesulfonic acid buffer, pH 11, and the membrane was stained with Coomassie Brilliant Blue R-250. Protein bands were cut out, washed with distilled water, and subjected to the automated sequencing.

Digestion of Peptides and Analysis of the Digest

The B chain of oxidized insulin (100 nmol) was digested at 37 °C with cathepsin E (0.2 nmol, 8 μg) in 300 μl of each of the following buffers: 0.5 M sodium formate buffer, pH 3.0, 0.5 M sodium acetate buffer, pH 5.5, and 0.5 M Tris-HCl buffer, pH 7.4. Aliquots of 100 μl were removed at appropriate time intervals and submitted to high performance liquid chromatography (HPLC) using a Hitachi 655A-11 system on a column (0.46 x 25 cm) of TSKgel ODS-120T (Tosoh Co.). The peptides were eluted with a gradient of acetonitrile (0-50% in 60 min) in 0.1% trifluoroacetic acid at a flow rate of 0.8 ml/min. The effluent was monitored by measuring the absorbance at 215 nm and the peptide peak fractions were pooled and lyophilized. An aliquot of each peptide fraction dissolved in water was submitted to amino acid analysis using an Applied Biosystems automated derivatizer analyzer 420A/130A.

Isolation of Glycopeptide

The glycopeptide fraction was prepared by digestion of procathepsin E with thermolysin and *B. subtilis* aminopeptidase followed by chromatography on a Sephadex G-25 column (1.6 x 4 cm) as described previously(9).

RESULTS AND DISCUSSION

Purification of Procathepsin E

In this study a significant amount of procathepsin E was purified from human gastric mucosa by a modification of the previous method (9) in a yield of 19% (Table I). The procedure was improved mainly by introducing Con A-Sepharose chromatography based on

Table 1. Purification of Procathepsine E

Step	Total protein (mg)	Total activity (units)	Relative activity (units/mg)	Yield (%)
Crude extract	21000	36464	1.73	100
DEAE-cellulose(1st)	3928	31374	7.96	86
DEAE-cellulose (2nd)				
pepsinogen	3108	30512	9.81	83
procathepsin E	528	525	0.94	1.4(100)
Con A- Sepharose	69	428	6.20	1.2(82)
Sephacryl S-200	15.6	179	11.50	0.5(34)
FPLC(Mono Q)	2.4	98	40.83	0.3(19)

the finding of mannose and glucosamine in cathepsin E (9) and by including FPLC on a Mono Q column as the last step of purification. After removing contaminating proteins in the first DEAE-cellulose chromatography, pepsinogens were separated from procathepsin E in the second DEAE-cellulose chromatography (Table I).

Procathepsin E fractions were further purified by Con A -Sepharose chromatography, Sephacryl-200 gel filtration and FPLC (Fig. 1a).

The resulting procathepsin E fractions in peak 3 was further purified by rechromatography on a Mono Q column (Fig. 1b). The enzyme preparation was maintained at pH above 7.5 and 4 °C throughout the purification procedure to obtain procathepsin E with a high recovery. The purified procathepsin E showed a single band on SDS-PAGE. The molecular weight of the enzyme was determined to be approximately 80,000(dimer, under non-reducing conditions) and 40,000 (monomer, under reducing conditions) (Fig. 2).

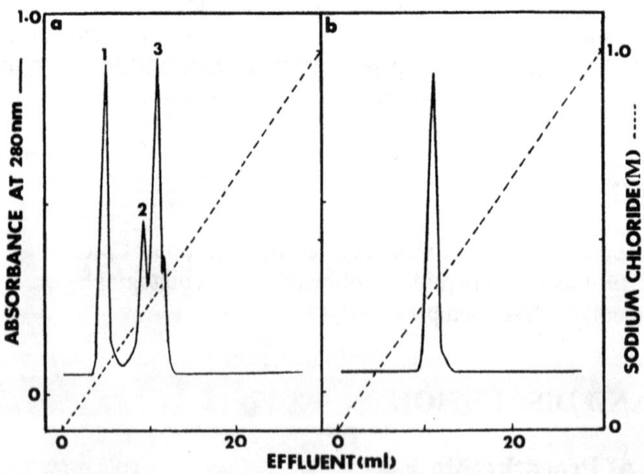

Figure 1. Mono Q chromatography (FPLC) of procathepsin E. a. An aliquot(2 mg) of the pooled and dialyzed fraction from the Sephacryl S-200 chromatography was injected into a Mono Q column(0.5 x 5.0 cm) and eluted with a linear gradient of NaCl in 0.02 M Tris-HCl buffer, pH 7.8. Potential proteolytic activity was detected in peak 3. b. Rechromatography of the peak 3 fraction on the Mono Q column.

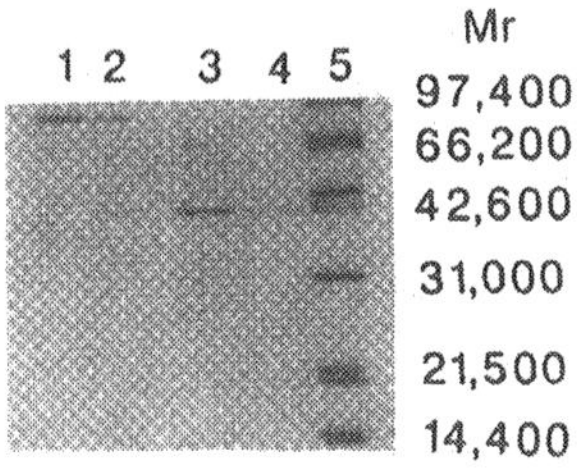

Figure 2. SDS-PAGE of procathepsin E. Lanes 1 and 2, procathepsin E without 2-mercaptoethanol; lanes 3 and 4, procathepsin E with 2-mercaptoethanol; lane 5, molecular weight marker proteins(Pharmacia).

Activation of Procathepsin E to Cathepsin E

The time course of actiation of procathepsin E at pH 3.5 and 14 °C as analyzed by SDS-PAGE is shown in Fig. 3.

Apparently direct or one-step conversion of procathepsin E into cathepsin E was observed, whereas porcine cathepsin D(11) and human pepsinogens A and C (12) were shown to be activated through intermediate forms. The conversion was rapid and complete within 10 min and this conversion was inhibited completely by 0.1 mM pepstatin A (data not shown). These results indicate that procathepsin E is converted autocatalytically to its mature form, cathepsin E, at pH 3.5. Further,the processing of procathepsin E to cathepsin E was investigated at different pH values (2.0-7.4) at 37 °C (data not shown). The results demonstrated that autocatalytic fast conversion of procathepsin E into cathepsin E occurs at the acidic pH (2.0-4.0) within few minutes and rather slow conversion at pH 5.0 and very slow conversion at neutral pHs (6.0-7.4). Complete conversion was observed after 6-h incubation at pH 5.0 and after 96-h incubation at pH 6.5 and 7.4. This fast activation rate of procathepsin E at the acidic pH rather resembles that of human pepsinogen C (12). The molecular weight of the dimeric form of cathepsin E was estimated to be 76,000 under non-reducing conditions and that of the monomeric form produced under mild reducing conditions to be 38,000. The differences in molecular weight between procathepsin E and cathepsin E were consistent with the removal of the activation peptide segment from the zymogen upon conversion to its mature form.

Purification of Cathepsin E

Procathepsin E was converted into mature cathepsin E by lowering the pH of the solution to 4.0. Cathepsin E was purified from the resulting activation mixture by pepstatin-Sepharose chromatography and FPLC on a Mono Q column (Fig. 4).

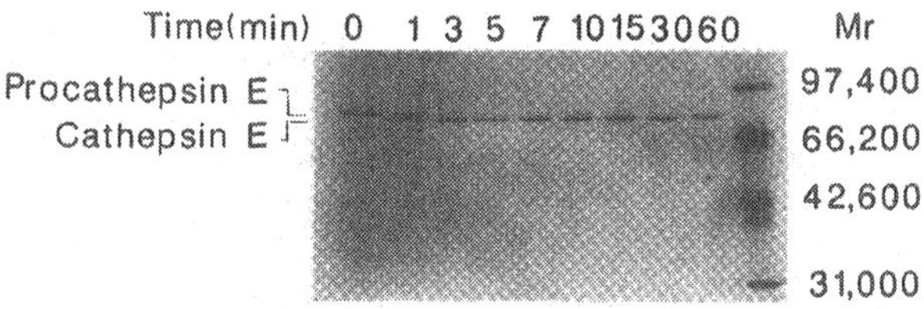

Figure 3. Time course of autocatalytic activation of procathepsin E analyzed by SDS-PAGE. Procathepsin E solution (0.16 mg/ml) was incubated at pH 3.5 and 14 °C. Aliquots were removed at appropriate time intervals and subjected to SDS-PAGE under non-reducing conditions. PCE, procathepsin E; CE, cathepsin E.

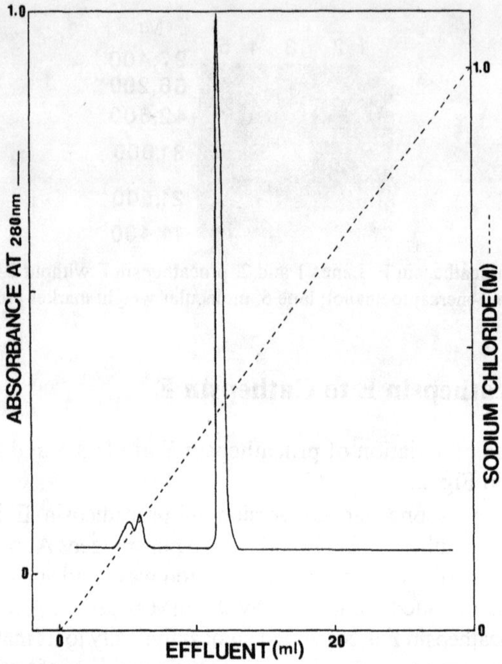

Figure 4. Mono Q chromatography (FPLC) of cathepsin E. An aliquot(1 mg) of pooled and dialyzed fraction from the pepstatin-Sepharose chromatography was injected into a Mono Q column(0.5 x 5.0 cm) and eluted with a linear gradient of NaCl in 0.02 M Tris-HCl buffer, pH 7.8. Proteolytic activity was detected in the major peak.

Procathepsin E and cathepsin E were eluted from the Mono Q column at different ionic strengths, suggesting that the FPLC on the Mono Q column could be used to separate procathepsin E from cathepsin E.(Figs. 1 and 4).

Stability of cathepsin E at different pHs was investigated by incubating the purified cathepsin E preparaion (0.25 mg/ml) at different pHs at 37 °C and analyzing the aliquots collected at different time intervals by SDS-PAGE. The cathepsin E band disappeared completely after 6-h incubation at pH 3.0-5.5, whereas rather slower reduction of band intensity was observed at neutral pHs (6.0-7.4). This result clearly suggests moderate stability of cathepsin E at neutral pH values.

NH_2-Terminal Amino Acid Sequences of Procathepsin E and Cathepsin E

When the procathepsin E preparation was analyzed directly by the protein sequencer, no phenylthiohydantion(PTH)-amino acid could be detected in a significant amount, indicating that the NH_2-terminus of the protein may be blocked. This is consistent with the putative NH_2-terminal glutamine residue of procathepsin E(13), which may be easily converted into pyroglutamic acid residue, resistant to Edman degradation. Therefore, procathepsin E was first treated with pyroglutamate aminopeptidase, then purified by SDS-PAGE, electroblotted and sequenced. Thus, we could obtain the NH_2-terminal 20-residue amino acid sequence of procathepsin E (Fig. 5). This result confirms the occurrence of glutamine residue at the NH_2-terminus of procathepsin E, as suggested by Azuma *et al.*(13).

```
                   1          10                  20
Procathepsin E    :<E x S L H R V P L R R H P S L K K K L R A R--
                   ↴   ¬ ¬ ¬ ¬ ¬ ¬ ¬ ¬ ¬ ¬ ¬ ¬ ¬ ¬ ¬ ¬ ¬ ¬ ¬ ¬
                   1          10                  20                  30        36
cDNA              : Q G S L H R V P L R R H P S L K K K L R A R S Q L S E F W K S H N L D M

                   1          10                  20                  30        36
Cathepsin E(Major): I Q F T E S ? S M D Q S A K E P L I N Y L D M E Y F G T I S I G S P P Q---
                    ¬ ¬ ¬ ¬ ¬ ¬ ¬ ¬ ¬ ¬ ¬ ¬ ¬ ¬ ¬ ¬ ¬ ¬ ¬ ¬ ¬ ¬ ¬ ¬ ¬ ¬ ¬ ¬ ¬ ¬ ¬ ¬ ¬ ¬ ¬ ¬

                         1          10        16
Cathepsin E(Minor):      T E S ? S M D Q S A K E P L I N---
                         ¬ ¬ ¬ ¬ ¬ ¬ ¬ ¬ ¬ ¬ ¬ ¬ ¬ ¬ ¬ ¬
                   37   40                  50                  60                  70
cDNA              : I Q F T E S C S M D Q S A K E P L I N Y L D M E Y F G T I S I G S P P Q N F T V I F---

      <E : Pyroglutamic acid residue,
       x : deletion,
       ¬ : Each cycle of Edman degradation,
       ↴ : Released by the action of pyroglutamate aminopeptidase
```

Figure 5. Comparison of the NH_2-terminal amino acid sequences of human gastric procathepsin E and cathepsin E determined at the protein level and those predicted from its cDNA sequence (13).

However, the observed amino acid sequence was found to start from the third amino acid residue, Ser^3, of the sequence predicted from the cDNA sequence (13). This suggests the deletion of the second amino acid residue, Gly^2, in the protein sequence. This might be due to the occurrence of isozymogens, like pepsinogens(12). However, the posibility cannot be excluded that the second amino acid residue was also removed from the protein by concomitant action of some contaminating proteinase(s) that might be present in the pyroglutamate aminopeptidase preparation used.

Sequencing of the purified cathepsin E revealed the presence of one major and one minor sequences in a molar ratio of approximately 3:1. The NH_2-terminal 36-residue sequence of the major one and 15-residue sequence of the minor one could be sequenced simultaneously (Fig. 5) and the results appeared to be consistent with the sequence predicted from the cDNA sequence(13). The difference in the two sequences was eventually found to be due to the production of two protein species through autocatalytic cleavages at two different sites, Met^{36}-Ile^{37} and Phe^{39}-Thr^{40}, during acid treatment.

Further, when the cathepsin E sample similarly prepared as above was stored at pH 8.0 and 0 °C about three months and then sequenced, the NH_2-terminal sequence was found to start mostly from Ser^{42} of procathepsin E (13). This suggested the occurrence of autodigestion of cathepsin E during storage. To minimize such a possibility of autodigestion, if any, cathepsin E was obtained by incubating procathepsin E for a shorter period (3 min) at pH 3.5 and 14 °C, then purified by SDS-PAGE, electroblotted and sequenced. Again one major and one minor sequences were obtained in an approximate molar ratio of 3:2. This result was essentially the same as that obtained previously, thus confirming the cleavages at two different sites. The NH_2-terminal of cathepsin E is thus positioned six or three residues before Cys^{43} which is thought to be involved in disulfide bond formation between the two monomers. Therefore, the dimeric form of the enzyme is thought to be maintained throughout the activation.

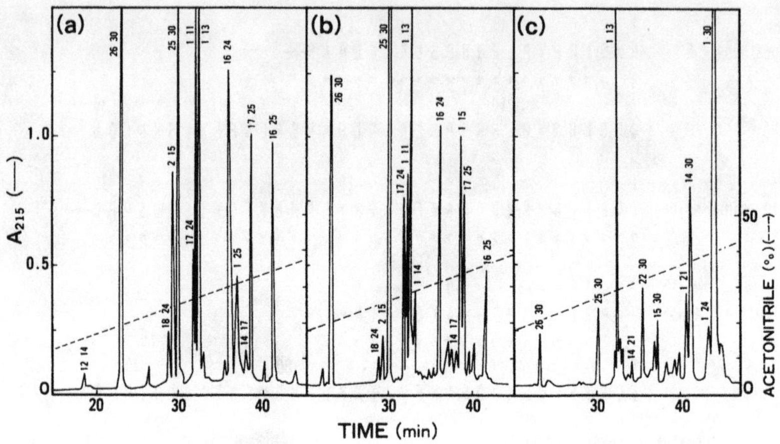

Figure 6. HPLC patterns of 3-h cathepsin E digests of the B chain of oxidized insulin at pH 3.0 (a), 5.5 (b) and 7.4 (c) on a TSK gel ODS-120T column. The identity of each peak is shown with the residue numbers in the B chain of oxidized insulin.

The Site of Carbohydrate Attachment

A single glycopeptide fraction was isolated in a yield of over 80% from a thermolysin-aminopeptidase digest of cathepsin E by chromatography on a Sephadex G-25 column as described previously (9). The amino acid sequence analysis of this peptide fraction revealed that the peptide fraction was a mixture of two major peptides, I and II, in an approximate molar ratio of 3:2. Their sequences were analyzed as a mixture and could be deduced as shown below.

I : Ile-Gly-Ser-Pro-Pro-Gln-Asx

II: Gly-Ser-Pro-Pro-Gln-Asx

Upon Edman degradation, each residue could be identified as a PTH-amino acid except for the COOH-terminal residue. No PTH-amino acid was obtained from this residue. However, it could be identified as dansyl-aspartic acid by dansylation followed by acid hydrolysis. Therefore, this residue is thought to be originally asparagine to which the carbohydrate chain is bound. The sequences of the glycopeptides coincide with the residue 67(or 68) to 73 in the predicted procathepsin E sequence (13). Therefore, the results establish the glycosylation site to be the aspargine residue at position 73 in the procathepsin E sequence. This explain why the aspargine at position 73 could not be identified as a PTH-amino acid in the Edman degradation of the mature protein. Interestingly the same asparagine residue at the corresponding position is also the only site of carbohydrate attachment in Japanese monkey pepsinogen I (14).

Proteolytic Activity and Cleavage Specificity of Cathepsin E Toward the B Chain of Oxidized Insulin at Different pHs

The HPLC patterns of the 3-h digests of the B chain of oxidized insulin at pH 3.5, 5.5 and 7.4 are shown in Fig. 6.

The B chain was cleaved rapidly and extensively at pH 3.0 (Fig. 6a), but was also hydrolyzed to a considerable extent at pH 5.5 (Fig. 6b) and 7.4 (Fig. 6c). As can seen from the figures, the HPLC patterns were significantly different among the three pH values.

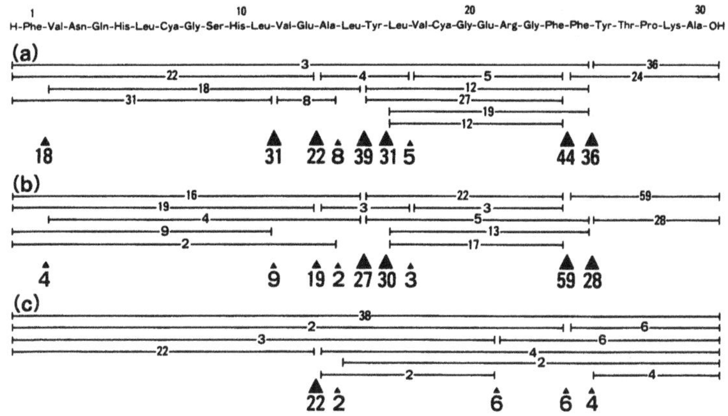

Figure 7. Yields of peptides produced from the B chain of oxidized insulin by cathepsin E and the sites of cleavage at pH 3.0 (a), 5.5 (b) and 7.4 (c). Bars indicate peptide fragments produced and the number under each arrowhead shows the estimated extent of cleavage in percent of the peptide bond. Large, medium and small arrowheads indicate major, secondary and minor cleavage sites, respectively.

Particularly the HPLC patterns of the digest at pH 7.4 appeared to be much simpler than those of the digests at pH 3.0 and 5.5. Upon longer incubation, hydrolysis proceeded further, but the cleavage pattern at each pH value(data not shown) appeared to be roughly similar to that obtained with the 3-h digest.

The cleavage sites and the extents of hydrolysis estimated from amino acid analysis are shown in Fig. 7 for the digests at pH 3.0, 5.5 and 7.4. At pH 3.0 and 5.5 several peptide bonds, especially Phe-X, Tyr-X and Leu-X bonds, were simultaneously cleaved to marked extents, whereas, at pH 7.4, the Glu^{13}-Ala^{14} bond was cleaved fairly selectively. This rather selective cleavage at pH 7.4 suggests that cathepsin E may be involved in limited proteolysis of certain proteins and/or peptides under neutral(physiological) conditions. The reason for the change in cleavage specificity with pH is not very clear, but the difference in ionization state in the enzyme active site region and/or the substrate molecule would be primarily responsible for this change. The enzyme appears to prefer the ionized form of glutamic acid residue in the cleavage of Glu-X bonds. The extent of cleavage of this bond was estimated to be 22% at pH 3.0, 19% at pH 5.5, 22% at pH 7.4. By summing up the estimated extents of hydrolysis of individual peptide bonds, the total extent of hydrolysis could be calculated. These values relative to that at pH 3.0 (taken as 100%) were 77% at pH 5.5, 17% at pH 7.4.

The proteolytic activity of cathepsin E toward the B chain of oxidized insulin was completely inhibited by 0.1 mM pepstatin at pH 3.0 and 5.5, whereas 0.1 mM pepstatin had no inhibitory effect when used at pH 7.4. Further, when cathepsin E, which had been inhibited by incubation with 0.1 mM pepstatin at pH 3.0 or 5.5, was brought to pH 7.4, the enzyme restored full activity and essentially the same HPLC pattern as that obtained without pepstatin was obtained. Further, the following proteinase inhibitors, 0.1 mM diisopropylfluorophosphate, 1 mM phenylmethanesulfonyl fluoride, 10 mM *o*-phenanthroline, 0.05 mM diprotin A, 0.1 mM leupeptin, 0.1 mM antipain, 0.1 mM E-64, and 0.1 mM amastatin, did not affect the hydrolysis of the B chain of oxidized insulin at pH 7.4. Therefore the activity at pH 7.4 toward the B chain of oxidized insulin was not inhibited by any of the serine, cysteine and metallo- proteinase inhibitors tested, indicating that this activity is due to cathepsin E itself and not to a contaminating proteinase(s).

Gastric cathepsin E was optimally active at pH around 3.0 and has been thought to be devoid of proteolytic activity at neutral pH values. Indeed it was reported to have rather low activity at pH 5.0 and practically no activity at pH above 6.0 toward hemoglobin as substrate (9,15). In this study, however, human gastric cathepsin E was found to possess rather high activity at pH 5.5 and still significant activity at pH 7.4 toward the B chain of oxidized insulin. Reduced and carboxymethylated ribonuclease A and several other peptides were also shown to be hydrolyzed to significant extents by cathepsin E at pH 7.4 (data not shown). These results suggest that cathepsin E may function *in vivo* in a wider physiological pH range than was hitherto considered and it may be involved in a limited proteolysis of certain proteins and /or peptides under neutral(physiological) conditions.

ACKNOWLEDGMENTS

We thank Drs. Masao Tanji and Masaru Tanokura for their useful suggestions. This study was supported in part by Grants-in-Aid for Scientific Research from the Ministry of Education, Science and Culture of Japan.

REFERENCES

1. D.J. Etherington and W.J. Talyor, *Clin.Sci.* 42:79-90 (1972).
2. S. Yonezawa, K. Fuji, Y. Maejima, K.Tamato, Y. Mori and N. Muto, *Arch. Biochem. Biophys.* 267:176-183 (1988).
3. I.M. Samloff, R.T. Taggart, T. Shiraishi, T. Branch, W.A. Reid, R. Heath, R.W. Lews, M.J. Valler and J. Kay, *Gastroenterology* 93:77-84 (1987).
4. N.Muto, M. Yamamoto, S. Tani and S. Yonezawa, *J. Biochem.* (Tokyo) 103:629-632(1988).
5. O. Matsuzaki and K. Takahashi, *Biomed. Res.* 9:515-523(1988).
6. H. Rangel and C. Lapresle, *Biochim. Biophys. Acta* 128:372-379(1966).
7. S. Yonezawa, T. Tanaka and T. Miyauchi, *Arch. Biochem. Biophys.* 256:499-508(1987).
8. R.A. Jupp, A.D. Richards, J. Kay, B.M. Dunn, J.B. Wyckoff, I.M. Samloff and K. Yamamoto, *Biochem. J.* 254:895-898(1988).
9. T. Kageyama and K. Takahashi, *J. Biochem.* (Tokyo)87:725-735(1980).
10. U.K. Laemmli, *Nature* 227:680-685(1970).
11. G.E. Conner, *Biochem. J.* 263:601-604(1989).
12. S.B.P. Athauda, M. Tanji, T. Kageyama and K. Takahashi, *J. Biochem. (Tokyo)* 106:920-927(1989).
13. T. Azuma, G. Pals, T. K. Mohandas, J.M. Couvreur and R.T. Taggart, *J. Biol. Chem.* 264:16748-16753(1989).
14. T. Kageyama and K. Takahashi, *Biochem. Biophys. Res. Commun.* 74:789-795(1977).
15. N. Muto, K.M. Arai and S. Tani, *Biochim. Biophys. Acta* 745:61-69(1983).

ISOLATION, CHARACTERIZATION, AND STRUCTURE OF PROCATHEPSIN E AND CATHEPSIN E FROM THE GASTRIC MUCOSA OF GUINEA PIG

Takashi Kageyama,[1] Masao Ichinose,[2] Kazumasa Miki,[2] Akihiko Moriyama,[3] Satoshi Yonezawa,[4] Masao Tanji,[5] Senarath B. P. Athauda,[5] and Kenji Takahashi[5]

[1]Department of Cellular and Molecular Biology
Primate Research Institute
Kyoto University
Inuyama, Aichi 484, Japan
[2]First Department of Internal Medicine
Faculty of Medicine
The University of Tokyo
Bunkyo-ku, Tokyo 113, Japan
[3]Department of Chemistry
College of General Education
Nagoya City University
Mizuho-ku, Nagoya 467, Japan
[4]Department of Embryology
Institute for Developmental Research
Aichi Prefectural Colony, Kasugai, Aichi 480-03, Japan
[5]Department of Biophysics and Biochemistry
Faculty of Science
The University of Tokyo
Bunkyo-ku, Tokyo 113, Japan

INTRODUCTION

Cathepsin E [EC 3.4.23.34] is a non-secretory intracellular aspartic proteinase, and its precursor is known as procathepsin E (1,2). (Pro)cathepsin E is a dimeric enzyme different from other aspartic proteinases, consisting of two identical subunits of about 40 kDa (3,4). The enzymatic properties of cathepsin E have been shown to resemble those of pepsins [EC 3.4.23.1]; for example, it has hydrolytic activity at acidic pH, with an optimum at pH 2-3 against protein substrates, and is sensitive to various pepsin inhibitors (5-10).

Aspartic Proteinases: Structure, Function, Biology, and Biomedical Implications
Edited by Kenji Takahashi, Plenum Press, New York, 1995

Table 1. Purification of Procathepsin E and Progastricsin from the Gastric Mucosa of Guinea Pigs (14)

Step		Protein	Activity	Specific activity	Yield
		mg	units	units/mg protein	%
1. Supernatant of crude homogenate		176	161	0.91	100
2. DEAE-Sephacel					
Procathepsin E		6.5	31	4.8	19
Progastricsin		11	85	7.7	53
3. Purification of procathepsin E					
Sephadex G-100		2.0	20	10	12
FPLC		0.67	15	22	9.3
4. Purification of progastricsin					
Sephadex G-100		4.5	49	11	30
FPLC	1	1.4	27	19	17
	2	0.92	16	17	9.9

Proteolytic activity was determined by the hemoglobin-digestion method of Anson and Mirsky (29).

Structural studies of (pro)cathepsin E have not yet progressed to a level comparable with those of pepsins mainly because of the difficulty in obtaining a sufficient amount of the native enzyme. Azuma et al. reported the primary structure of human cathepsin E which was deduced from the molecular cloning and analysis of its cDNA (11). The structural analysis suggested that the dimeric form is produced by covalent association of two monomers through a disulfide bridge. To understand the structural features of (pro)cathepsin E, it seems appropriate to elucidate the structures of (pro)cathepsin E from different sources and compare them with those of other aspartic proteinases.

Although the physiological roles of (pro)cathepsin E is still unclear, it has been suggested to play an important role in intracellular processing of proteins and/or peptides (4,12). Since cathepsin E releases endothelin from its precursor by a specific cleavage (13), it seems likely that the enzyme preferentially processes or degrade certain types of biologically active peptides or their precursors.

In this paper, we describe a series of studies including purification, molecular cloning, and elucidation of enzymatic properties of guinea-pig (pro)cathepsin E.

```
                                        1
                                        GGAAGACGGGCCTCTGCAGGAGAAAAGCTGCTGCTCGGACC
ACG ATG AAA ACG TTT CTG CTG TTA CTG CTG GTG CTC CTG GAG CTG GGA CAG GCC CCA GGA
    Met Lys Thr Phe Leu Leu Leu Leu Leu Val Leu Leu Glu Leu Gly Gln Ala Pro Gly
    -19                                                                      -1
GCC CTT CAC AGG GTG CCC CTC AGC CGG CGT GAG TCC CTC CGG AAG AAG CTG CGA GCC CAG
Ala Leu His Arg Val Pro Leu Ser Arg Arg Glu Ser Leu Arg Lys Lys Leu Arg Ala Gln
 →   →   →   →   →   →   →   →   →   →   →   →   →   →   →   →   →       →   →
 1                                  10                                       20
GGA CAG CTC ACT GAG CTC TGG AAA TCC CAG AAC TTG AAT ATG GAC CAG TGC TCC ACC ATC
Gly Gln Leu Thr Glu Leu Trp Lys Ser Gln Asn Leu Asn Met Asp Gln Cys Ser Thr Ile
 →   →   →   →   →   →       →   →               ▲
                                    30                                       40
CAG AGT GCC AAT GAG CCC CTC ATC AAC TAC CTG GAT ATG GAG TAC TTC GGC ACT ATC TCC
Gln Ser Ala Asn Glu Pro Leu Ile Asn Tyr Leu Asp Met Glu Tyr Phe Gly Thr Ile Ser
                                    50                                       60
ATT GGC TCT CCA CCA CAA AAC TTC ACT GTC ATC TTC GAC ACC GGC TCC TCC AAT CTC TGG
Ile Gly Ser Pro Pro Gln Asn Phe Thr Val Ile Phe Asp Thr Gly Ser Ser Asn Leu Trp
                                    70                                       80
GTT CCC TCT GTC TAC TGC ACC AGC CCA GCC TGC CAA ACG CAC CCC GTG TTC CAC CCT TCC
Val Pro Ser Val Tyr Cys Thr Ser Pro Ala Cys Gln Thr His Pro Val Phe His Pro Ser
                                    90                                      100
CTG TCC AGC ACC TAC AGG GAG GTG GGG AAT TCT TTC TCC ATT CAG TAC GGG ACC GGG AGC
Leu Ser Ser Thr Tyr Arg Glu Val Gly Asn Ser Phe Ser Ile Gln Tyr Gly Thr Gly Ser
                                    110                                     120
TTG ACT GGG ATC ATC GGA GCC GAC CAA GTC TCT GTG GAA GGA CTC ACT GTG GTT GGC CAG
Leu Thr Gly Ile Ile Gly Ala Asp Gln Val Ser Val Glu Gly Leu Thr Val Val Gly Gln
                                    130                                     140
CAA TTT GGA GAA AGT GTC CAA GAA CCG GGC AAG ACC TTT GTG CAT GCA GAG TTT GAT GGG
Gln Phe Gly Glu Ser Val Gln Glu Pro Gly Lys Thr Phe Val His Ala Glu Phe Asp Gly
                                    150                                     160
ATT CTG GGC CTA GGC TAT CCC TCC TTG GCT GCT GGA GGG GTG ACC CCT GTG TTT GAC AAC
Ile Leu Gly Leu Gly Tyr Pro Ser Leu Ala Ala Gly Gly Val Thr Pro Val Phe Asp Asn
                                    170                                     180
ATG ATG GCC CAG AAC CTG GTG GCT CTG CCA ATG TTT TCT GTC TAC ATG AGC AGT AAC CCA
Met Met Ala Gln Asn Leu Val Ala Leu Pro Met Phe Ser Val Tyr Met Ser Ser Asn Pro
                                    190                                     200
GGT GGC TCA GGC AGT GAA CTG ACC TTC GGA GGC TAT GAC CCC TCC CAT TTC TCG GGG AGC
Gly Gly Ser Gly Ser Glu Leu Thr Phe Gly Gly Tyr Asp Pro Ser His Phe Ser Gly Ser
                                    210                                     220
CTG AAT TGG GTC CCA GTC ACC AAG CAG GCC TAC TGG CAG ATT GCC CTG GAT GGA ATC CAG
Leu Asn Trp Val Pro Val Thr Lys Gln Ala Tyr Trp Gln Ile Ala Leu Asp Gly Ile Gln
                                    230                                     240
GTG GGG GAC AGT GTG ATG TTC TGC TCC GAG GGC TGC CAG GCC ATT GTG GAC ACA GGA ACT
Val Gly Asp Ser Val Met Phe Cys Ser Glu Gly Cys Gln Ala Ile Val Asp Thr Gly Thr
                                    250                                     260
TCC CTC ATC ACT GGC CCT CCT GGA AAG ATC AAG CAG CTG CAA GAA GCT CTC GGG GCC ACA
Ser Leu Ile Thr Gly Pro Pro Gly Lys Ile Lys Gln Leu Gln Glu Ala Leu Gly Ala Thr
                                    270                                     280
TAC GTG GAC GAA GGA TAT TCT GTG CAG TGT GCC AAC CTC AAC ATG ATG CTG GAT GTC ACT
Tyr Val Asp Glu Gly Tyr Ser Val Gln Cys Ala Asn Leu Asn Met Met Leu Asp Val Thr
                                    290                                     300
TTC ATC ATC AAC GGG GTC CCC TAC ACG CTC AAC CCA ACC GCC TAC ACC CTG CTG GAC TTC
Phe Ile Ile Asn Gly Val Pro Tyr Thr Leu Asn Pro Thr Ala Tyr Thr Leu Leu Asp Phe
                                    310                                     320
GTG GAC GGC ATG CAG GTC TGC AGC ACT GGC TTT GAA GGA CTT GAA ATC CAG CCT CCA GCT
Val Asp Gly Met Gln Val Cys Ser Thr Gly Phe Glu Gly Leu Glu Ile Gln Pro Pro Ala
                                    330                                     340
GGA CCC CTC TGG ATC CTC GGG GAT GTC TTC ATC CGG CAG TTC TAC GCC GTC TTT GAT CGT
Gly Pro Leu Trp Ile Leu Gly Asp Val Phe Ile Arg Gln Phe Tyr Ala Val Phe Asp Arg
                                    350                                     360
GGG AAT AAC CGC GTG GGG TTG GCC CCA GCA GTC CCC TGA GTAGAGTTTTATGTCCCTGCCACCTGA
Gly Asn Asn Arg Val Gly Leu Ala Pro Ala Val Pro
                                    370
CACTTGTGGGCTTGTTAGGTTGGGGCATTGTTTACACTAAAAAAAAAAAAAATTCATTTTCAGTAAAAAAAAAAAAAAA
                                                                            1326
```

Figure 1. The nucleotide sequences and the deduced amino acid sequence of cDNA for guinea-pig procathepsin E. The two active-site aspartic acids are shown by shading. The horizontal arrows show the residues determined by Edman degradation of native procathepsin E. The vertical arrowhead indicates the site of cleavage for the activation that generates cathepsin E. Residues in italics, i. e., Asn67 and Asn311, are potential N-glycosylation sites.

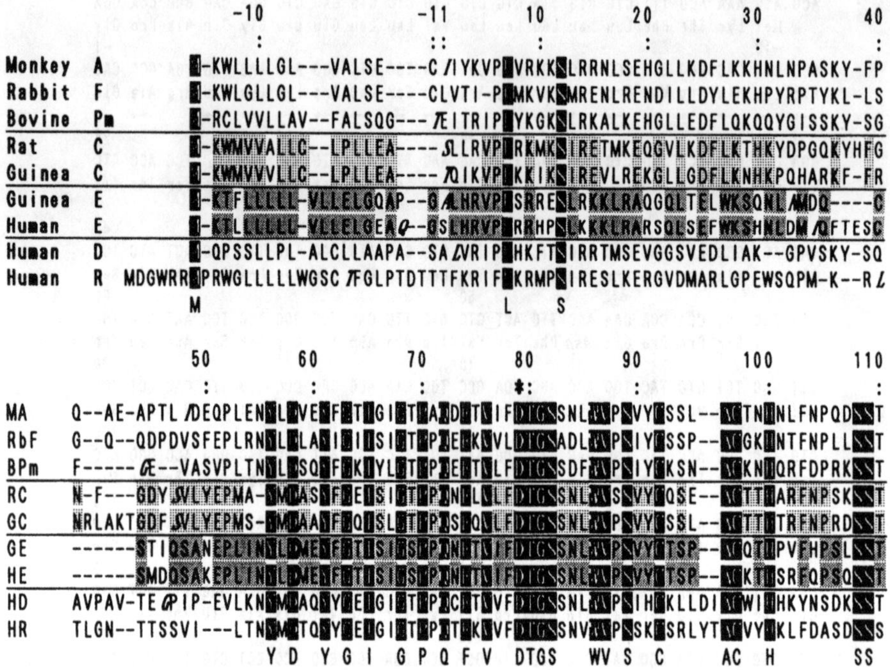

Figure 2. Comparison of the NH_2-terminal amino-acid sequences of guinea-pig preprocathepsin E with those of typical mammalian aspartic proteinases. Amino acid sequences were deduced from the cDNA or genomic DNA sequences of monkey pepsinogen A (31), rabbit pepsinogen F (32), bovine prochymosin (33), rat progastricsin (34), human procathepsin E (11), human procathepsin D (35), and human prorenin (36). Bars represent deleted residues. The italic letters around positions 1 and 50 show the NH_2-terminal amino acids of the proenzymes and the active forms, respectively. Common residues between guinea-pig and human procathepsins E are shown by deep shading. Asterisks show the aspartic acids at the active sites. Residues in filled boxes are conserved in mammalian aspartic proteinases.

PURIFICATION OF PROCATHEPSIN E

Procathepsin E was purified from guinea-pig gastric mucosa by a four-step procedure including DEAE-Sephacel chromatography, Sephadex G-100 gel filtration, and FPLC on a Mono Q column (Table 1) (14). The purified proenzyme gave a single protein band upon nondenaturing and denaturing PAGE. The molecular mass determined by SDS-PAGE under reducing conditions was about 43 kDa. By contrast, the native procathepsin E was eluted at the position corresponding to a molecular mass of about 80 kDa on gel filtration and gave a band of protein with a similar molecular mass on SDS-PAGE under non-reducing conditions (Figure 3). Therefore, procathepsin E was deduced to be a dimer. Procathepsin E was a glycoprotein, and the content of carbohydrate was estimated to be about 4% by weight. The NH_2-terminal sequences of about 30 residues of the proenzyme were determined by Edman degradation (Figure 1). Only a single residue was identified at each step. Thus, procathepsin E appears to be composed of identical subunits.

The level of procathepsin E in the gastric mucosa of guinea pig was 4- to 10-times higher than that in human gastric mucosa (6,9,10) and was the highest among those reported to date for various animal tissues. The reason for this high level is not clear, but it seems that the gastric mucosa of the guinea pig may serve as a good source of procathepsin E for future studies at the protein level.

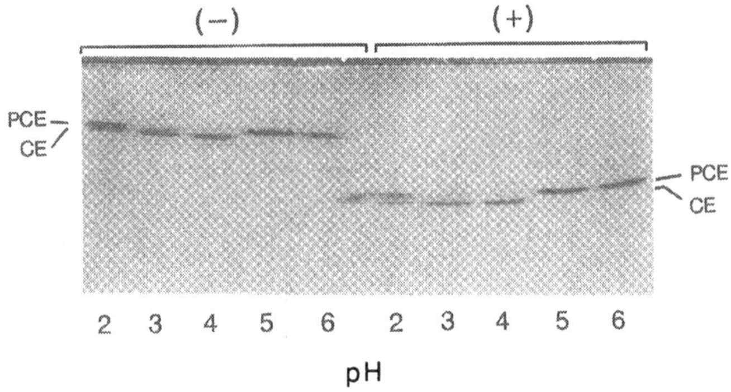

Figure 3. Activation of procathepsin E at various pH values analyzed by SDS-PAGE under non-reducing (-) and reducing (+) conditions. Activation was carried out at an initial concentration of procathepsin E of 0.05 mg/ml at 14°C. The time of activation reaction was 15 min. The right half of the band of dimeric procathepsin E in lane 5 (from the left) migrated to the position of monomeric procathepsin E as a result of the effect of 2-mercaptoethanol in lane 6. PCE, CE, and Int stand for procathepsin E, cathepsin E, and the intermediate form(s), respectively.

STRUCTURE OF PROCATHEPSIN E

The amino acid sequence of procathepsin E was deduced from its cDNA (Figure 1). The proenzyme consisted of three regions, i.e., the pre-peptide (signal peptide), the pro-peptide (activation segment), and the active enzyme of 19, 33, and 339 residues, respectively. The amino acid residues that are conserved in other mammalian aspartic proteinases were also well conserved. Asn67 and Asn311 were found to be the potential N-glycosylation sites (15). Some deletions and insertions were noted in the pro-peptide and the NH_2-terminal region of the cathepsin E moiety as compared with the sequences of other mammalian aspartic proteinases (Figure 2). The molecular mass of procathepsin E was calculated to be 40,086 Da.

Procathepsin E has some unique structural features among the aspartic proteinases (Figure 2). The Lys at position 37 (numbering for pepsinogen A from man and monkey), which is common to various aspartic proteinases and is thought to be important for the function of the activation segment, was absent. There was a Cys residue at position 37. A disulfide bridge involving Cys37 between the two monomers was found to be responsible for generating the dimer as described in a later section.

The structure of guinea-pig procathepsin E was most similar to that of human procathepsin E with 86% and 84% identity at the nucleotide and the amino acid levels, respectively. The identity with other aspartic proteinases was less than 60%. We constructed a phylogenic tree to examine the relationships among various aspartic proteinases including procathepsin E, pepsinogens, procathepsin D, and prorenin (14). It was proven that procathepsin E belong to an independent group within the family of aspartic proteinases.

ACTIVATION OF PROCATHEPSIN E

The profile of activation of the proenzyme was analyzed by SDS-PAGE (Figure 3). Activation of procathepsin E proceeded autocatalytically under acidic conditions (16); the rate of activation was maximal at pH 4.0 and decreased gradually as the pH was lowered to

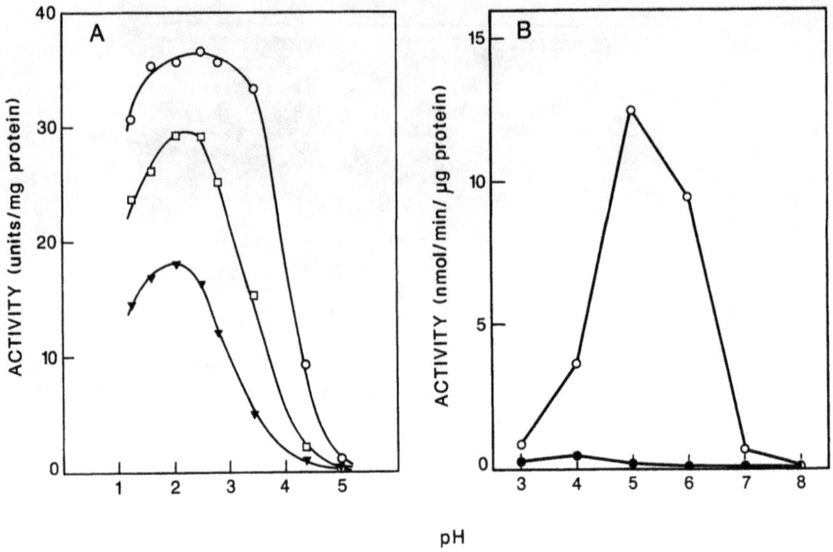

Figure 4. Dependence on pH of the hydrolysis of hemoglobin (A) and substance P (B). The concentrations of hemoglobin and substance P of the assay mixtures were 2% and 50 μM, respectively. (o), Guinea-pig cathepsin E; (□), guinea-pig gastricsin; (▼), porcine pepsin A; (•), bovine cathepsin D.

2.0. In addition, appreciable activation occurred both at pH 5.0 and 6.0 upon prolonged incubation. The rate of activation did not change when monomeric procathepsin E was activated under the same conditions. Procathepsin E appeared to be converted to cathepsin E directly, since the intermediate form(s) was generated at a very low level. The bands of procathepsin E and cathepsin E were detected at positions of 82 kDa and 76 kDa, respectively, after SDS-PAGE under non-reducing conditions, whereas the pro- and active forms gave a band of 43 kDa and a band of 39 kDa, respectively, after SDS-PAGE under reducing conditions. Therefore, the dimeric form was maintained throughout the activation. Isolation and structural analysis of the active form revealed that the site of cleavage upon activation was the Leu32-Asn33 bond. Thus, the NH_2-terminus of cathepsin E is located four residues before Cys37. The cleavage site was the same when monomeric procathepsin E was activated under the same conditions.

As shown in the preceding section, Lys37 of the propeptide (activation segment), which is present in other mammalian aspartic proteinase zymogens, was not found in guinea pig nor in human procathepsin E (Figure 2). The positive charge of the Lys residue has been shown to provide electrostatic stabilization via hydrogen bonding to one of the net negative charges of the two aspartic acids at the active site. Therefore, the Lys residue has been suggested to be essential for maintaining the proenzyme in an inactive form, thereby playing an important role in the activation of these aspartic proteinases (17). The activation of procathepsin E proceeded most rapidly at pH 4.0, and appreciable activation occurred at even higher pH. This phenomenon was markedly different from the case of pepsinogens which are activated most rapidly at pH 2.0 and below (18). The maximal activation at weakly acidic pH may be correlated with the absence of Lys37 in procathepsin E, since the electrostatic stabilization is thought to be weak in procathepsin E. Since procathepsin E is a non-secretory intracellular proteinase and since its activation would occur at physiological pH, the maximum rate of activation at weak acidic pH seems to be well adapted to the physiology of the proenzyme.

Table 2. Relative Hydrolytic Activities of Cathepsin E and Cathepsin D against Biologically Active Peptides (30)

Peptide	Activity	
	Guinea-pig cathepsin E	Bovine cathepsin D
	$nmol \cdot min^{-1} \cdot \mu g\ protein^{-1}$	
(Brain-gastrointestinal peptides)		
Substance P	12.5	0.097
Neurokinin A	11.8	0.019
Eledoisin	5.3	0.16
Kassinin	3.3	0.010
Cholecystokinin-octapeptide	0.009	0.003
Neuromedin C	UC[a]	0.001
Bombesin	UC	0.007
Neurotensin	UC	0.018
(Growth factor fragments)		
Acidic FGF 102-111	1.0	0.012
Basic FGF 106-120	1.0	0.18
(Opioid peptides)		
Beta-Endorphin	0.47	18.6
Dynorphin A	0.046	0.005
(Vasoconstrictor peptides)		
Porcine renin substrate	13.6	4.2
Human renin substrate	0.55	0.066
Big endothelin-1	0.14	0.057

[a]UC, uncleaved.

ENZYMATIC PROPERTIES OF CATHEPSIN E

Cathepsin E was optimally active at around pH 2.5 toward hemoglobin as a substrate (Figure 4A) (5-10,19). Cathepsin E had higher specific activity than those of pepsin A and gastricsin. Cathespin E was inhibited by pepstatin, a specific inhibitor of aspartic proteinases (6,14). Susceptibility of cathepsin E to pepstatin was the same as that of porcine pepsin A, the inhibition profile indicating the strong equimolar binding of pepstatin to the active site.

Cathepsin E showed a unique hydrolytic specificity for biologically active peptides (Figure 5). The best peptide substrates for cathepsin E were tachykinins, such as substance P and neurokinin A (Table 2). Tachykinins contain an essential sequence of Phe-Xaa-Gly-Leu-Met as their five COOH-terminal residues. This sequence is known to be necessary for the biological activity of tachykinins. The cleavage of the Phe-Xaa bond was highly specific, and it resulted in the effective inactivation of these peptides. The optimal pH for hydrolysis was 5.0 (Figure 4B). This pH value was much higher than that for hydrolysis of hemoglobin or other protein substrates. In contrast to the rapid hydrolysis of tachykinins by cathepsin E, cathepsin D hydrolyzed tachykinins very slowly, although the cleavage sites were the same as those found in the case of cathepsin E. The substance P-hydrolyzing activity of cathepsin D was 500-fold lower than that of rabbit cathepsin E. Cathepsin D did, however, hydrolyze several peptides other than tachykinins, showing broad substrate-specificity. These suggest that the role in intracellular proteolysis is different between cathepsin E and cathepsin D.

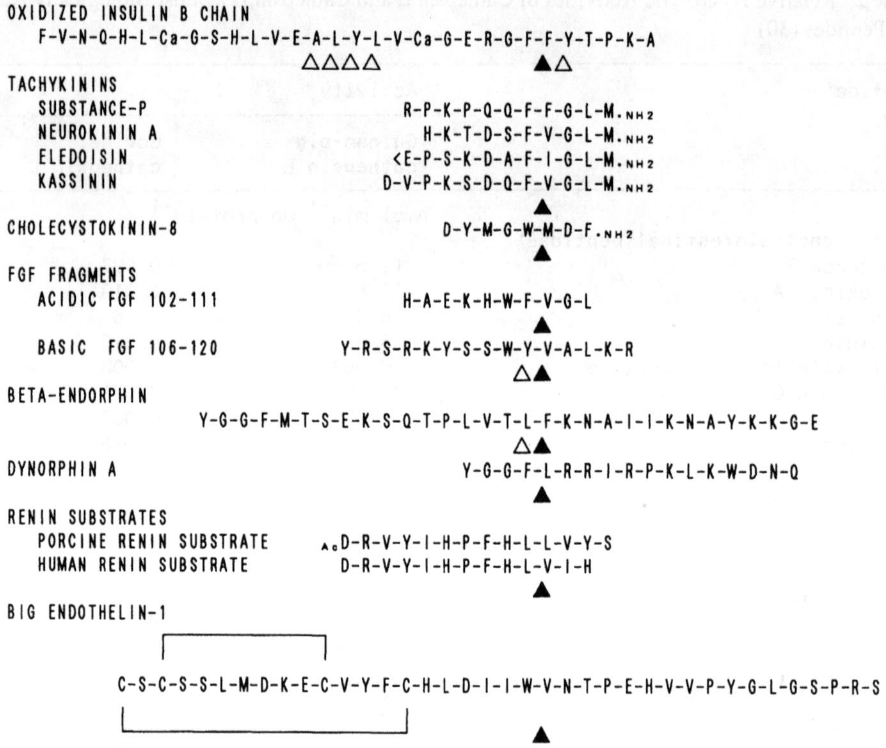

Figure 5. The sites of cleavage of various biologically active peptides by cathepsin E and cathepsin D. (▲), Major cleavage sites. (Δ), Intermediate or minor cleavage sites. NH_2 shown by reduced letters indicates an amide on the carboxyl terminus of the peptides

Basic and acidic fibroblast growth factors (FGFs) have functional domains, and one of these domains has been shown to be similar to a domain in tachykinins. Peptide fragments containing such homologous sequences were hydrolyzed by cathepsin E at pH 5.0. The cleavage sites corresponded to those of tachykinins (Figure 5).

Other biologically active peptides were also hydrolyzed by cathepsin E, although the rate of hydrolysis was lower than the rates of hydrolysis of tachykinins (Table 2). These peptides includes beta-endorphin, dynorphin A, human renin substrate, and big endothelin-1. Exceptionally, porcine angiotensin I was generated effectively from its precursor by cathepsin E. The pH optimum of the reaction was around 4.

The P1 and P'1 positions of peptide substrates for cathepsin E were occupied exclusively by hydrophobic and aromatic amino acids (Figure 5) (8,10,20), as is commonly observed for substrates of aspartic proteinases (2). Since rates of hydrolysis and sites of cleavage of various biologically active peptides differed significantly for different aspartic proteinases, the specificity must be caused by variation in the interactions between features of substrates, other than those at P1 and P'1 positions, and the unique structures of the subsites that form the active site clefts of the respective proteinases (21,22).

The optimal pH for hydrolysis of biologically active peptides was much higher than that for hydrolysis of protein substrates. The differences between pH optima for different substrates may due to several factors, such as changes in the conformation and ionization of substrates at different pH values. Furthermore, in the case of a protein substrate, it is reasonable to suppose that destruction of the tertiary structure of a protein at extremely low pH may be partly responsible for the maximal rate of hydrolysis at such pH values. From the perspective of intracellular proteolysis, the relatively high pH optimum for the hydrolysis of biologically active peptides by cathepsin E is quite important since a weakly acidic pH is easily generated within the microenvironment of a cell.

Various biologically active peptides exhibit a high degree of functional diversity at the levels of both peptide production and peptide reception. Peptides bind to cell-surface receptors and activate a chain of events for signal transduction within the cell. Degradation or processing of biologically active peptides by cathepsin E is thought to modulate signal transduction, with resultant regulation of the behavior of target cells. Distribution of procathepsin E in adult contractile tissues, such as urinary bladder, stomach and colon (23,24), is well correlated with the distribution of neurokinin-2 tachykinin receptor (25). Since substance P and related tachykinins are clarified to be the best substrates for cathepsin E, this shows the high possibility of the involvement of cathepsin E in the regulation of the biological effect of substance P on these tissues.

INTERCONVERSION BETWEEN THE DIMER AND THE MONOMER

The conversion of the dimeric (pro)cathepsin E to the monomeric form occurred in the presence of a low concentration of a reducing reagent (3,4,14). The dimer was converted to the monomer to the extent of 30-50% by incubation of the former with 1 mM 2-mercaptoethanol, L-cysteine, or reduced glutathione at 37°C for 20 min. The conversion was complete with any one of these reagents at 10 mM under the same conditions. The conversion was reversible, since the dimeric form was regenerated after removing the reducing reagent by ammonium sulfate precipitation or dialysis. When the monomer was carboxymethylated, carboxymethyl-Cys was determined to be 0.88 mol/mol of monomeric (pro)cathepsin E. This partially modified monomeric (pro)cathepsin E retained the complete proteolytic activity. Carboxymethyl-Cys was identified at position 4 from the NH_2-terminus of cathepsin E (position 37 of procathepsin E) by Edman degradation. The result thus provided the direct evidence that the dimeric form was generated by formation of a disulfide bridge at Cys37 between the two monomers.

Such high susceptibility of (pro)cathepsin E to a reducing agent is thought to be due to the tertiary structure of procathepsin E, in which the region around Cys37 is presumed to be on the surface of the protein molecule as expected from the tertiary structures of other aspartic proteinase (26). Therefore, it may be reasonable to consider that a reducing agent, such as glutathione, may regulate the interconversion between the two forms *in vivo*. Indeed, the occurrence of the monomeric form of procathepsin E was evidenced in the human gastric mucosa (Athauda, S. B. P. and Takahashi, K., unpublished results). The difference in enzymatic properties between the dimeric and the monomeric forms of cathepsin E was investigated. Although the hydrolytic activity against hemoglobin at pH 2.0 was the same for both forms, a slight increase in activity was observed with the monomer at pH 5.0 as compared to the dimer. Such an increase was not observed, however, when the enzyme was assayed with other protein substrates.

The interconversion between the two forms may have little significance in the case of procathepsin E, since no difference in properties was found between the two forms. On the other hand, the conversion of the dimer to the monomer seems critical in the case of cathepsin E. The monomeric cathepsin E is more unstable than the dimeric cathepsin E at weakly alkaline pH (Figure 6). This characteristic is very similar to that of pepsin, although

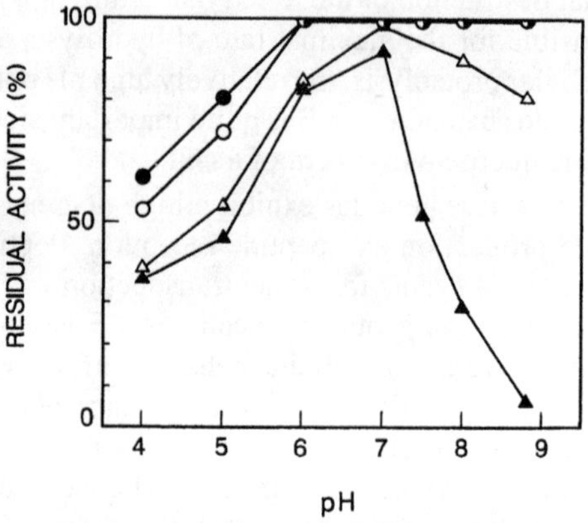

Figure 6. Stability of procathepsin E and cathepsin E at various pH values. The native dimeric forms of procathepsin E (o) and cathepsin E (Δ) were incubated at 28°C for 2 h in 0.1 M buffer at various pH values, and the residual activity was determined by the hemoglobin-digestion method. The monomeric forms of procathepsin E (•) and cathepsin E (▲), generated by treatment with 2-mercaptoethanol, were subjected to the same analysis. The assay mixture of the monomeric form contained 0.2 mM 2-mercaptoethanol to prevent regeneration of the dimer.

pepsin is inactivated more rapidly at weakly alkaline pH (27). Considering that a drastic conformational change is involved in the process of alkali denaturation of pepsin (28), monomeric cathepsin E might be more susceptible to conformational change at weakly alkaline pH than is dimeric cathepsin E. Therefore, the dimeric form is thought to be essential for stabilizing cathepsin E. Thus, the generation of the monomeric form may be important in the degradation of cathepsin E *in vivo*. On the other hand, cathepsin E, as well as procathepsin E, was rather unstable at weakly acidic pH in both dimeric and monomeric forms, presumably due to autodigestion since the enzyme has appreciable proteolytic activity under weakly acidic conditions.

REFERENCES

1. C. Lapresle, in "Tissue Proteinases" (A. J. Barrett and J. T. Dingle, eds.) p. 135, North-Holland Publishing Co., Amsterdam (1971).
2. A. J. Barrett, in "Proteinases in mammalian cells and tissues" (A. J. Barrett, ed.) p. 209, Elsevier/North-Holland Biomedical Press, Amsterdam (1977).
3. S. Yonezawa, K. Fujii, Y. Maejima, K. Tamoto, Y. Mori, and N. Muto, Arch. Biochem. Biophys. 267:176-183 (1988).
4. K. Yamamoto, H. Yamamoto, M. Takeda, and Y. Kato, Biol. Chem. Hoppe-Seyler 369(Suppl.):315 (1988).
5. K. Yamamoto, N. Katsuda, and K. Kato, Eur. J. Biochem. 92:499 (1978).
6. T. Kageyama and K. Takahashi, J. Biochem. 87:725 (1980).
7. N. Muto, K. Murayama-Arai, and S. Tani, Biochim. Biophys. Acta 745:61 (1983).
8. S. Yonezawa, T. Tanaka, and T. Miyauchi, Arch. Biochem. Biophys. 256:499 (1987).
9. I. M. Samloff, R. T. Taggart, T. Shiraishi, T. Branch, W. A. Reid, R. Heath, R. W. Lewis, M. J. Valler, and J. Kay, Gastroenterology 93:77 (1987).
10. O. Matsuzaki and K. Takahashi, Biomed. Res. 9:515 (1988).
11. T. Azuma, G. Pals, T. K. Mohandas, J. M. Couvreur, and R. T. Taggart, J. Biol. Chem. 264:16748 (1989).
12. K. Bennett, T. Levine, J. S. Ellis, R. J. Peanasky, I. M. Samloff, J. Kay, and B. M. Chain, Eur. J. Immunol. 22:1519 (1992).

13. W. E. Lees, S. Kalinka, J. Meech, S. J. Capper, N. D. Cook, and J. Kay, FEBS Lett. 273:99 (1990).
14. T. Kageyama, M. Ichinose, S. Tsukada, K. Miki, K. Kurokawa, O. Koiwai, M. Tanji, E. Yakabe, S. B. P. Athauda, and K. Takahashi, J. Biol. Chem. 267:16450 (1992).
15. S. B. P. Athauda, O. Matsuzaki, T. Kageyama, and K. Takahashi, Biochem. Biophys. Res. Commun. 168:878 (1990).
16. S. B. P. Athauda, T. Takahashi, T. Kageyama, and K. Takahashi, Biochem. Biophys. Res. Commun. 175:152 (1991).
17. M. N. G. James and A. R. Sielecki, Nature 319:33 (1986).
18. J. Al-Janabi, J. A. Hartsuck, and J. Tang, J. Biol. Chem. 247:4628 (1972).
19. S. Yonezawa, T. Tanaka, N. Muto, and S. Tani, Biochem. Biophys. Res. Commun. 144:1251 (1987).
20. S. B. P. Athauda, T. Takahashi, H. Inoue, M. Ichinose, and K. Takahashi, FEBS Lett. 292:53 (1991).
21. A. R. Sielecki, A. A. Fedorov, A. Boodhoo, N. S. Andreeva, and M. N. G. James, J. Mol. Biol. 214:143 (1990).
22. J. Kay, and B. M. Dunn, Scand. J. Clin. Lab. Invest. 52:23 (1992).
23. N. Muto, M. Yamamoto, S. Tani, and S. Yonezawa, J. Biochem. 103:629 (1988).
24. H. Sakai, T. Saku, Y. Kato, and K. Yamamoto, Biochim. Biophys. Acta 991:367 (1989).
25. K. Tsuchida, R. Shigemoto, Y. Yokota, and S. Nakanishi, Eur. J. Biochem. 193:751 (1990).
26. A. R. Sielecki, M. Fujinaga, R. J. Read, and M. N. G. James, J. Mol. Biol. 219:671 (1991).
27. R. M. Herriott, J. Gen. Physiol. 45:57 (1962).
28. P. McPhie, Biochem. Biophys. Res. Commun. 158:115 (1989).
29. M. L. Anson, and A. E. Mirsky, J. Gen. Physiol. 16:59 (1932).
30. T. Kageyama, Eur. J. Biochem. 216:717 (1993).
31. T. Kageyama, K. Tanabe, and O. Koiwai, Eur. J. Biochem. 202:205 (1991).
32. T. Kageyama, K. Tanabe, and O. Koiwai, J. Biol. Chem. 265:17031 (1990).
33. M. Hidaka, K. Sasaki, T. Uozumi, and T. Beppu, Gene 43:197 (1986).
34. Y. Ichihara, K. Sogawa, K. Morohashi, Y. Fujii-Kuriyama, and K. Takahashi, Eur. J. Biochem. 161:7 (1986).
35. P. L. Faust, S. Kornfeld, and J. M. Chirgwin, Proc. Natl. Acad. Sci. U. S. A. 82:4910 (1985).
36. T. Imai, H. Miyazaki, S. Hirose, H. Hori, T. Hayashi, R. Kageyama, H. Ohkubo, S. Nakanishi, and K. Murakami, Proc. Natl. Acad. Sci. U. S. A. 80:7405 (1983).

CATHEPSIN E AND CATHEPSIN D: BIOSYNTHESIS, PROCESSING AND SUBCELLULAR LOCATION

Kenji Yamamoto

Department of Pharmacology
Kyushu University Faculty of Dentistry
Fukuoka 812, Japan

INTRODUCTION

Intracellular protein turnover occurs via two major pathways, lysosomal and non-lysosomal proteolytic pathways. Cathepsins D (CD) and E (CE) are two major intracellular aspartic proteinases in mammalian cells. Besides their structural and immunological distinction (1-5), they are different in both tissue distribution and intracellular localization (4, 6), suggesting that they possess different physiological roles in various cell types. CD is known as a representative lysosomal enzyme and to be widely distributed in almost all the mammalian cells. On the other hand, it has been shown that CE has a limited distribution in some cell types (4, 6) and is not localized in the lysosome (7-11). Therefore, it is suggested that these two analogous enzymes CD and CE are involved in different intracellular protein catabolism; lysosomal and extralysosomal proteolysis, respectively.

Most of the aspartic proteinases are thought to be synthesized as the high-molecular-weight zymogens which are activated by an acid-triggered, autocatalytic proteolysis. Recent structural studies of CD (3) and CE (5, 12) revealed that the proenzymes consist of the following three parts: the signal peptides, the propeptides (activation segments), and the active enzymes. In addition, CD and CE are known as glycoproteins (13-15). These findings suggest that both enzymes are initially synthesized as high molecular mass precursors on the membrane-bound ribosomes and then transported to the Golgi complex. However, it is unknown how CE is sorted from CD as well as other secretory proteins and targeted to its final destinations and where the proteolytic processing of CE occurs in cells. In particular, the biosynthetic pathway of CE, as compared with CD, remains unclear. Also, little is known about the post-translational processing of CE, including the modification of its carbohydrate chains during this intracellular sorting. The major problems for the CE study are thought to be the difficulty in purification of the enzyme due to its low cellular concentration and the lack of a specific substrate that is discriminated only by CE. To solve these problem, we have recently succeeded in preparing heterologous cells expressing human CE at a high rate and isolating the precursor and mature forms of the recombinant CE.

Aspartic Proteinases: Structure, Function, Biology, and Biomedical Implications
Edited by Kenji Takahashi, Plenum Press, New York, 1995

In this report, we describe the intracellular localization, biosynthesis and processing of CE as compared with CD, in various cell types.

INTRACELLULAR LOCALIZATION OF CATHEPSIN E AS COMPARED WITH CATHEPSIN D

We have previously demonstrated that the distribution of CE in various rat tissues and blood cells is markedly different from that of CD by using the discriminative antibodies specific for each enzyme. CD was found in all of the tissues and cells except for erythrocytes, whereas CE had a limited distribution in certain cell types such as lymphoid tissues, gastrointestinal tracts, and urinary organs (4, 6). We also reported that the immunohistochemical localization of the two enzymes at the light microscopy level was also different (4). Furthermore, we have analyzed the exact intracellular localization of CE in various cell types of human gastric mucosa and of rat tissues at the electron microscopy level and compared with that of CD (10). The results indicated that the intracellular localization of CD and CE was noticeably distinct in all of the cell types examined. CD was exclusively confined to lysosomes, although the concentration of this enzyme varied in each cell type. In contrast, CE was not detected in typical lysosomal structures and its subcellular localization varied with different cell types. Microvillous localization of CE was found in the intracellular canaliculi of human and rat gastric parietal cells, rat renal proximal tubule cells, the bile canaliculi of rat hepatic cells, and the ruffled border of rat osteoclasts. The localization of the cytosolic compartment was shown in various cell types including human and rat gastric cells and rat neutrophils. CE was also found in the cisternae of the endoplasmic reticulum and the dilated nuclear envelope of human and rat gastric cells. Interestingly, in rat kidney, CE was detected only in the proximal tubule cells, while CD was found mainly in the lysosomes of the distal tubule cells but not in those of the proximal tubule cells. These findings indicate that CE plays a role in extralysosomal proteolysis related to specialized functions of each cell type, such as absorption and secretion.

At this time, it is unclear whether CE can exert its catalytic activity in such neutral cytoplasmic compartments. However, since it has been shown that CE is stabilized by physiological concentrations of ATP or RNAs and exhibits its catalytic activity even at around neutral pH (16, 17), it seems likely that the enzyme is regulated by such cellular factors to play a part in intracellular extralysosomal proteolysis. On the other hand, it has been more recently reported that CE can hydrolyze limitedly such polypeptides as insulin B-chain (18) and spectrin (Yamamoto K. personal communication) at neutral pH without such intracellular factors. Thus, it is more likely that protein degradation by CE *in vivo* is influenced by the type and nature of protein substrates.

PROTEOLYTIC AND GLYCOLYTIC PROCESSING OF CATHEPSIN E

To study the molecular basis of the biosynthesis and processing of CE, we isolated both the natural and recombinant proenzymes and the respective mature enzymes, and compared their structural and catalytic properties. In addition, we studied the biosynthesis and intracellular kinetics of CE, as well as CD, by pulse-chase analyses in murine Friend erythroleukemia (FEL) cells and rat macrophages.

Pro-CE was firstly isolated from human erythrocyte membranes by a simple method involving immunoaffinity chromatography on anti-human cathepsin E IgG-Sepharose and

Table 1. Comparison of the N-terminal Amino Acid Sequences of the Natural and Recombinant Human Pro- and Natural CE with the Corresponding Sequences of Human Gastric CE predicted from Its cDNA Sequence

	5 10 15
Erythrocyte pro-CE	S-L-H-R-V-P-L-R-R-H-P-S-L-K-K
Recombinant pro-CE	S-L-H-R-V-P-L-R-R-H-P-S-L-K-K
cDNA	Q-G-S-L-H-R-V-P-L-R-R-H-P-S-L-K-K
	30 35 40
Recombinant Intermediate CE	K-S-H-N-L-D-M-I-Q-F-T
cDNA	K-S-H-N-L-D-M-I-Q-F-T
	37 40 45 50
Erythrocyte mature CE(I)	I-Q-F-T-E-S-C-S-M-D-Q-S-A-K-E-P-L-I
(II)	T-E-S-C-S-M-D-Q-S-A-K-E-P-L-I
Recombinant mature CE	I-Q-F-T-E-S-C-S-M-D-Q-S-A-K-E-P-L-I
cDNA	I-Q-F-T-E-S-C-S-M-D-Q-S-A-K-E-P-L-I

The numberings refer to the sequence of human gastric CE predicted from its cDNA sequence (5).

chromatography on concanavalin A-Sepharose, Mono Q and Superdex 200 (19). The N-terminal amino acid sequence analysis revealed that the proenzyme started with the third amino acid residue, Ser^3, of the sequence of human gastric CE predicted from its cDNA sequence (5) and was lacking the two amino acid residues Gln^1-Gly^2. The following amino acid sequence was identical with the corresponding sequence of human gastric CE. When the N-terminal sequence of the mature CE simultaneously purified from human erythrocyte membranes was determined, two amino acid residues were found in the first several steps of sequencing; one species started with Ile^{37} of the sequence predicted from the cDNA sequence and the other started with Thr^{40}. Therefore, the conversion from the proenzyme to the mature form was found to occur by the cleavage of Met^{36}-Ile^{37} and Phe^{39}-Thr^{40} bonds. More recently, we have succeeded in the expression of human CE cDNA in Chinese hamster ovary (CHO) cells and the purification of the recombinant enzymes (20). In this study, pro-CE (90-kDa), the intermediate form (84-kDa) and the mature CE (82-kDa) were identified and their N-terminal sequences were shown to start with Ser^3, Lys^{30} and Ile^{37}, respectively, of the sequence of human gastric CE predicted from its cDNA sequence. These results were consistent with the data obtained with the natural enzymes as described above (Table 1).

Both natural and recombinant pro-CEs were rapidly converted into the mature enzymes by brief acid treatment at 4.0 and 37 °C for 5 min. Under the same conditions, the intermediate form of the recombinant enzyme was also completely converted into the mature form. The rate of conversion was a little slower at 4 °C than at 37 °C. This conversion was inhibited by 10 μM pepstatin. The N-terminal sequences of the mature enzymes generated from both natural and recombinant proenzymes by acid treatment were identical with those of the respective mature enzymes. It is generally accepted that two different types of reactions, intramolecular and intermolecular reactions, are involved in the activation process of pepsinogens (21, 22). The intramolecular reaction is thought to be important in the initial phase of activation to produce the active enzymes, whereas the intermolecular reaction is

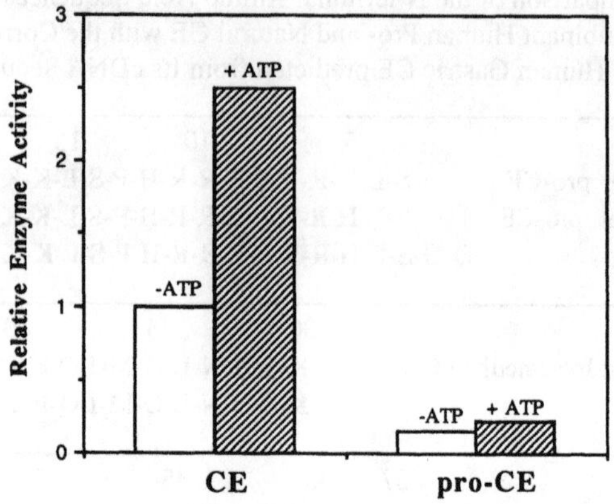

Figure 1. Effect of ATP on the hydrolysis of hemoglobin by pro-CE and the mature enzyme from human erythrocyte membranes at pH 5.5. Aliquots of each enzyme sample in 0.1 M sodium acetate buffer (pH 5.5) (3 μg/ml for each) were preincubated at 37 °C for 5 min with or without ATP (6.25 mM) and then added to the hemoglobin solution (final 0.1%, pH 5.5). The reaction was stopped by addition of 5% trichloroacetic acid and acid-soluble products were determined by fluorescamine method. The values are expressed as the relative ratio of the activities to that of the mature enzyme in the absence of ATP.

likely to be essential for acceleration and completion of the activation reaction. Therefore, it would be of interest to know which reaction is involved in the activation process of pro-CE. Although no precise studies have been made to analyze the activation process of pro-CE, our preliminary experiments have demonstrated that the conversion from pro-CE to the mature enzyme is dependent on the protein concentration, suggesting the predominance of intermolecular reaction.

Since pro-CE is rapidly activated at pH values below 4.0 to convert the mature enzyme, it is difficult to determine under acidic assay conditions whether the proenzyme is catalytically active or not. Recently, it has been demonstrated that, although CE from human erythrocyte membranes is essentially inactive at pH values above 5.5 by itself, it is stabilized by ATP (16) and RNA (17) in such a way as to maintain its active conformation and exhibits its activity on hemoglobin and synthetic peptide substrates even at pH 5.8. Thus we examined the effect of ATP at pH 5.5 on the proenzyme, as well as the mature enzyme. As expected, the mature enzyme was activated by ATP to apparently exhibit its activity on hemoglobin, which corresponded to 250% of the enzyme activity measured in the absence of ATP (Fig. 1). In contrast, regardless of the presence of ATP, the proenzyme showed very little activity. The results indicate that the proenzyme is catalytically inactive.

In order to determine the type of carbohydrate moieties, deglycosylation of the proenzyme and the mature enzyme was performed by using endoglycosidase H, which is specific for high mannose-type oligosaccharide chains and digests between the two proximal *N*-acetylglucosamine residues of asparagine-linked carbohydrate chains. As shown in Fig. 2, both the proenzyme and the mature enzyme from human erythrocyte membranes were found to be resistant to the endo H treatment. Similarly, the mature enzyme from rat erythrocyte membranes showed resistance to the endo H treatment. In contrast, all of the forms of the recombinant human CE were found to be sensitive to endo H. The mature enzyme from rat spleen was also sensitive to the endo H treatment. These results indicate that CE exists in either the high mannose-type or the complex type and strongly suggest that the type of oligosaccharide chain is cell specific or varies with the cellular localization.

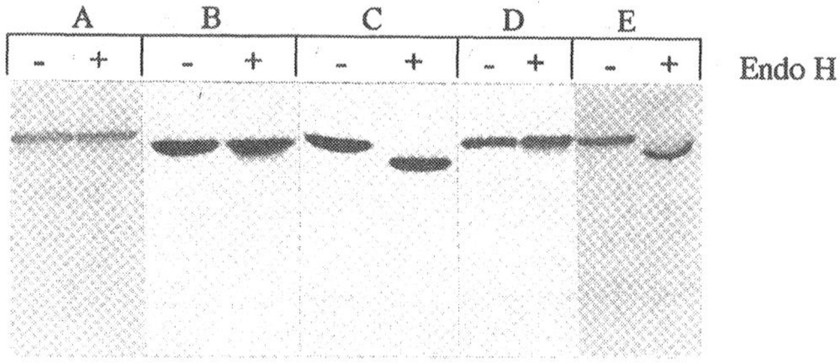

Figure 2. Sensitivity to endo H of pro-CE from human erythrocyte membranes and CE from various sources. The samples of purified pro-CE (A) and CE (B) from human erythrocyte membranes, the recombinant human CE (C) and the mature enzymes from rat erythrocyte membranes (D) and spleen (E) were incubated at 37 °C for 18 h with or without endo H followed by SDS-polyacrylamide gel electrophoresis and immunoblotting.

PULSE-CHASE EXPERIMENTS

To study the biosynthesis and processing of CE as well as CD, the pulse-chase experiments were carried out in mouse Friend erythroleukemia (FEL) cells and rat macrophages which both were labeled with [^{35}S]methionine for 30 min and chased for different periods with an excess of unlabelled methionine. At each time point, extracts of the respective cells were immunoprecipitated by the antibodies specific for each enzyme followed by sodium dodecyl sulfate/polyacrylamide gel electrophoresis under reduced conditions and fluorography. In FEL cells, CD was initially synthesized as a precursor with a molecular mass of 46-kDa after 30 min of labeling and subsequently processed into the mature form of 42-kDa, whereas CE was first synthesized as a 46-kDa proform and did not undergo significant change of its molecular mass throughout the chase period until 24 h (Fig. 3). Whether the 46-kDa form of CE was the proenzyme was further confirmed by autocatalytic conversion to the 42-kDa mature enzyme. To clarify whether CE in FEL cells undergoes the glycolytic processing during the biosynthesis, the nature of carbohydrate chains of this protein was determined by endo H and endo F. Within 2 h chase, CE was clearly sensitive to both Endo H and F, but it was converted to the Endo H- and F-resistant form after 4 h chase. The results indicate that CE undergoes the modification of its carbohydrate moiety from the high-mannose-type to the complex-type during its biosynthetic pathway. On the other hand, CD was found to be sensitive to both Endo H and F during the entire chase period under the same conditions.

In rat macrophages, CE was initially synthesized as a precursor with a molecular mass of 46-kDa after 30 min of labeling and subsequently proteolytically processed via a 42-kDa intermediate to a 38-kDa mature form in early stages of the biosynthesis. On the other hand, all the forms of CE in these cells were found to be completely sensitive to Endo H even after 24 h chase, indicating that this enzyme persisted as the form bearing the high-mannose-type carbohydrate moieties. On the other hand, CD in these cells underwent the time-dependent proteolytic processing in the course of intracellular transport. The results were consistent with the data obtained with FEL cells. These findings obtained with the macrophages indicate that the proteolytic processing machinery and the mode of glysylation of CE vary with different cell type.

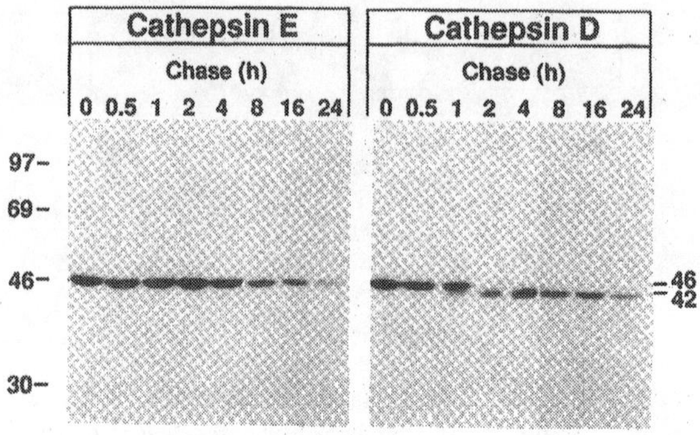

Figure 3. Pulse-chase analysis of CE and CD in FEL cells. Cultured FEL cells were pulsed for 30 min with [^{35}S]methionine and chased with cold methionine for the indicated times (0-24 h). Labeled CE and CD prepared from the cell lysates during the pulse-chase experiments were immunoprecipitated and analyzed by SDS-polyacrylamide gel electrophoresis followed by fluorography.

REFERENCES

1. K. Yamamoto, O. Kamata, N. Katsuda and K. Kato, Immunochemical Difference Between Cathepsin D and Cathepsin E-Like Enzyme from Rat Spleen, *J. Biochem.* **87:** 511 (1980)
2. V.C. Puizdar, C. Lapresle and V. Turk, Evidence for Presence of Large Amounts of Cathepsin E in Rat Spleen, *FEBS Lett.* **185:** 236 (1985)
3. P.L. Faust, S. Kornfeld and J.M. Chirgwin, Cloning and Sequence Analysis of cDNA for Human Cathepsin D, *Proc. Natl. Acad. Sci. USA* **82:** 4910 (1985)
4. H. Sakai, T. Saku, Y. Kato and K. Yamamoto, Quantitation and Immunohistochemical Localization of Cathepsins E and D in Rat Tissues and Blood Cells, *Biochim. Biophys. Acta* **991:** 367, 1989
5. T. Azuma, G. Pals, T.K. Mohandas, J.M. Couvreur and R.T. Taggart, Human Gastric Cathepsin E: Predicted Sequence, Localization to Chromosome 1, and Sequence Homology with Other Aspartic Proteinases, *J. Biol. Chem.* **246:** 16748 (1989)
6. N. Muto, M. Yamamoto, S. Tani and S. Yonezawa, Characteristic Distribution of Cathepsin E Which Immunologically Cross-Reacts with the 86 kDa Acid Proteinase from Rat Gastric Mucosa, *J. Biochem.* **103:** 629 (1988)
7. K. Yamamoto, M. Takeda, H. Yamamoto, M. Tatsumi and Y. Kato, Human Erythrocyte Membrane Acid Proteinase (EMAP): Sidedness and Relation to Cathepsin D, *J. Biochem.* **97:** 821 (1985)
8. S. Yonezawa, K. Fujii, Y. Maejima, K. Tamoto, Y. Mori and N. Muto, Further Studies on Rat Cathepsin E: Subcellular Localization and Existence of the Active Subunit Form, *Arch. Biochem. Biophys.* **267:** 176 (1988)
9. E. Ichimaru, H. Sakai, T. Saku, K. Kunimatsu, Y. Kato, I. Kato and K. Yamamoto, Characterization of Hemoglobin-Hydrolyzing Acidic Proteinases in Human and Rat Neutrophils, *J. Biochem.* 108, 1009 (1990)
10. T. Saku, H. Sakai, Y. Shibata, Y. Kato and K. Yamamoto, An Immunocytochemical Study on Distinct Intracellular Localization of Cathepsin E and Cathepsin D in Human Gastric Cells and Various Rat Cells, *J. Biochem.* **110:** 956 (1991)
11. H. Sakai, Y. Kato and K. Yamamoto, Synthesis and Intracellular Distribution of Cathepsins E and D in Differentiating Murine Friend Erythroleukemia Cells, *Arch. Biochem. Biophys.* **294:** 412 (1992)
12. T. Kageyama, K. M. Ichinose, S. Tsukada, K. Miki, K. Kurokawa, O. Koiwai, M. Tanji, E. Yakabe, S.B.P. Athauda and K. Takahashi, Gastric Procathepsin E and Progastricsin from Guinea pig: Purification, Molecular Cloning of cDNAs, and Characterization of Enzymatic Properties, with Special Reference to Procathepsin E. *J. Biol. Chem.* **267:** 16450 (1992)
13. K. Yamamoto, N. Katsuda and K. Kato, Affinity Purification and Properties of Cathepsin E-Like Acid Proteinase from Rat Spleen, *Eur. J. Biochem.* **92:** 499 (1978)
14. M. Takeda, E. Ueno, Y. Kato and K. Yamamoto, Isolation, and Catalytic and Immunochemical Properties of Cathepsin D-Like Acid Proteinase from Rat Erythrocytes. *J. Biochem.* **100:** 1269 (1986)

15. T. Kageyama and K. Takahashi, A cathepsin D-Like Acid Proteinase from Human Gastric Mucosa, *J. Biochem.* **87:** 725 (1980)
16. D.J. Thomas, A.D. Richards, R.A. Jupp, E. Ueno, K. Yamamoto, I.M. Samloff, B.M. Dunn and J. Kay, Stabilisation of Cathepsin E by ATP, *FEBS Lett.* **243:** 145 (1989)
17. K. Yamamoto, H. Sakai, E. Ueno and Y. Kato, in: *Structure and Function of The Aspartic Proteinases: Genetics, Structures, and Mechanisms* (B.M. Dunn, ed) pp. 297-306, Plenum, New York
18. S.B.P. Athauda, T. Takahashi, H. Inoue, M. Ichinose and K. Takahashi, Proteolytic Activity and Cleavage Specificity of Cathepsin E at the Physiological pH as Examined towards the B Chain of Oxidized Insulin, *FEBS Lett.* **292:** 53 (1991)
19. M. Takeda-Ezaki and K. Yamamoto, Isolation and Biochemical Characterization of Procathepsin E from Human Erythrocyte Membranes, *Arch. Biochem. Biophys.* **304:** 352 (1992)
20. T. Tsukuba, H. Hori, T. Azuma, T. Takahashi, R.T. Taggart, A. Akamine, M. Ezaki, H. Nakanishi, H. Sakai and K. Yamamoto, Isolation and Characterization of Recombinant Human Cathepsin E Expressed in Chinese Hamster Ovary Cells, *J. Biol. Chem.* **268:** 7276 (1993)
21. T. Kageyama and K. Takahashi, Activation Mechanism of Monkey and Porcine Pepsinogen A: One-Step and Stepwise Activation Pathways and Their Relation to Intramolecular and Intermolecular Reactions, Eur. *J. Biochem.* **165:** 483 (1987)
22. T. Kageyama, Analysis of the Activation of Pepsinogen in the Presence of Protein Substrates and Estimation of the Intrinsic Proteolytic Activity of Pepsinogen, *Eur. J. Biochem.* **176:** 543 (1988)

GLYCOPROTEINS OF THE ASPARTYL PROTEINASE GENE FAMILY SECRETED BY THE DEVELOPING PLACENTA

R. Michael Roberts,[1,2] Sancai Xie,[1] Robert J. Nagel,[1] Bonn Low,[1] Jonathan Green,[2] and Jean-François Beckers[3]

[1]Department of Animal Sciences
University of Missouri
Columbia MO 65211
[2]Department of Biochemistry
University of Missouri
Columbia MO 65211
[3]Faculty of Veterinary Medicine
University of Liege, Belgium

SUMMARY

Pregnancy in cattle and sheep can be diagnosed by the presence of placentally-derived antigens (pregnancy-associated glycoproteins or PAG-1) in maternal serum soon after implantation begins at about Day 20 following conception. Molecular cloning of their cDNA has revealed that PAG-1 belong to the aspartic proteinase gene family and have about 50% amino acid sequence identity to pepsin. However, critical amino acid substitutions at the active site regions suggest that both bovine and ovine PAG-1 are enzymatically inactive. PAG-1 expression has been shown by *in situ* hybridization and immunocytochemistry to be localized to the trophoblast binucleate cells, which invade maternal uterine endometrium during implantation. The glycoproteins are concentrated in dense cytoplasmic granules that are discharged after the binucleate cells have migrated to the maternal side of the placental barrier. We suggest, therefore, that the PAG-1 might have an endocrine function either as carriers of other bioactive peptides or by acting as hormones themselves. Recently screening of placental libraries with nucleic acid probes has identified additional cDNA that are very abundant and code for polypeptides (PAG-2 and PAG-3) related to, but antigenically and structurally distinct from PAG-1 described above. These molecules have sequences of amino acids at their catalytic centers that are consistent with their being potentially functional proteinases but their role during pregnancy, like that of PAG-1, is unclear.

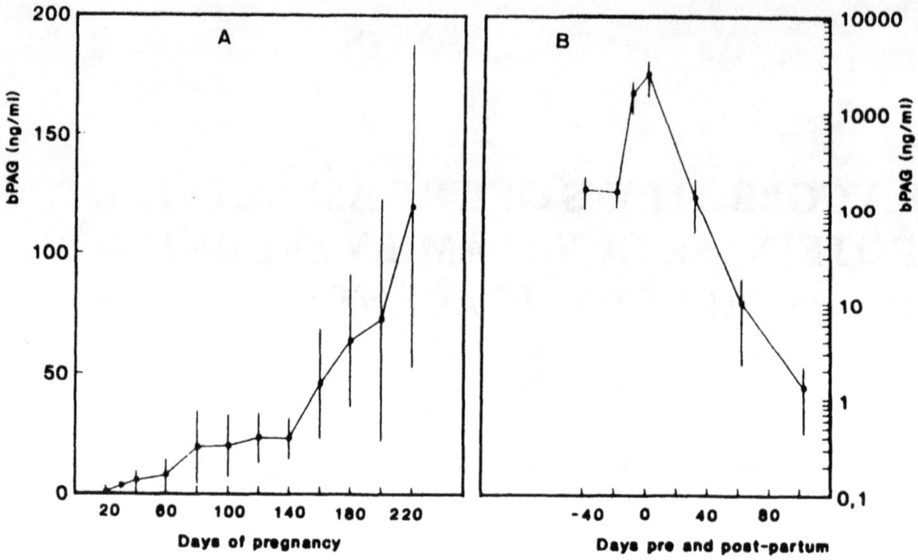

Figure 1. Profile of the concentrations (Mean±SD) of bPAG in serum of 20 cows bled from Day 20 of pregnancy until 100 days after parturition. Note the linear (A) and the logarithmic (B) scales of the Y axis respectively from 20 to 220 days of gestation and from Day 40 prepartum to Day 80 postpartum. From Zoli *et al.* (10) with permission.

PAG: A BRIEF HISTORY

Pregnancy in humans can be detected soon after the conceptus begins to implant by the presence of a placentally-produced hormone-human chorionic gonadotrophin (hCG)-first in the maternal bloodstream and a little later in urine (1). However, CG expression in gestation is a unique feature of primates and equids (2,3) and, as a consequence, a simple, unequivocal serum test for pregnancy based on CG is not available for other mammals, including the economically valuable farm species. In 1982 Butler *et al.* (4) reported the partial purification and characterization of a pregnancy-specific protein (PSP-B) in cattle. They first raised an antiserum against antigens in homogenates of placental membranes and used tissue and blood from nonpregnant cattle to adsorb out immunoglobulins not specific to pregnancy. The resulting antiserum was employed to monitor the partial purification of PSP-B, which was initially reported to have a molecular weight of 47,000 to 53,000 and a pI of ~4.0. Over the next decade PSP-B was never thoroughly characterized, although a range of molecular weights was published (4-6).

In 1991 Zoli *et al.* (7) reported the purification of a pregnancy-associated glycoprotein (PAG, but here called PAG-1) from bovine placental tissue. The methodology was similar to that of Butler *et al.* (4) in that an antiserum was first generated against placental extracts from which antibodies against common maternal antigens had been removed by immunoadsorption. This reagent was then used to track PAG-1 purification through a series of chromatographic steps that led to the isolation of an acidic glycoprotein product of M_r ~67,000. It is now clear that PSP-B and PAG-1 are identical products (8).

The presence of PAG-1 (or PSP-B) in blood serum has provided the basis of a potentially useful pregnancy test in cattle. The antigen generally becomes detectable by about Day 20 post-breeding and all pregnant cows appear to have measurable levels by Day 24 (Fig. 1) (5,9,10). This period coincides with the time that the trophoblast attaches

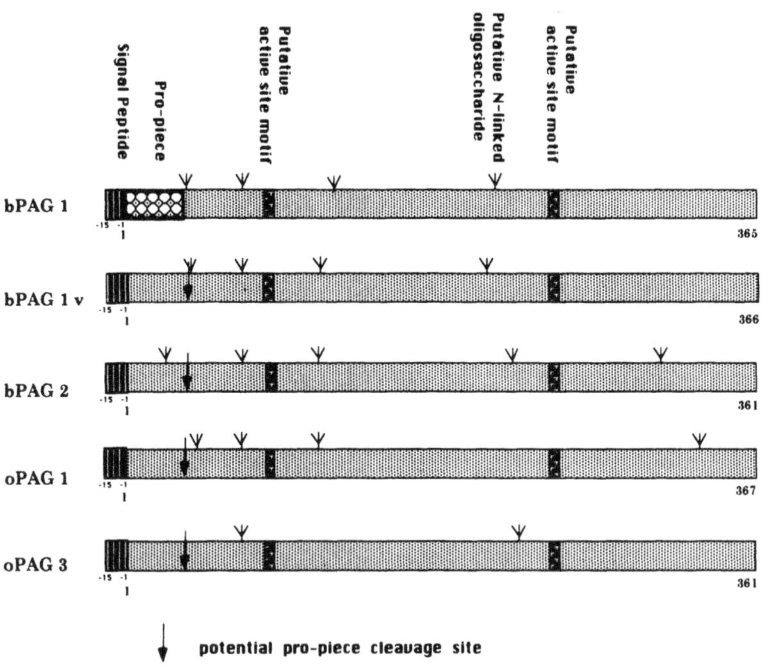

Figure 2. Diagram showing the predicted structures of the five PAG polypeptides whose cDNA have so far been cloned. Of these, only bPAG-1 has been purified and its NH_2-terminal amino acid sequence confirmed. Data indicate that a pro-peptide has been removed at position +38. For each PAG the putative signal sequence (bars); potential sites of attachment of carbohydrate chains to Asn residues ( ) and the polypeptide regions associated with the catalytic center are shown. The lengths of the signal sequences in each case were 15 residues. The lengths of the pre-proteins ranged from 367 (oPAG-1) to 361 (bPAG-2 and oPAG-3).

itself firmly to the uterine wall and when placentation begins (11). Concentrations of the antigen rise gradually during gestation and peak just prior to parturition. The antiserum raised against PSP-B (PAG-1) has since been used to monitor pregnancy in sheep (12), goats (13), deer (14) and wood bison (15).

The reason for the relatively late appearance of PAG-1 in cattle, relative to hCG in women, probably relates to the fact that implantation is initiated correspondingly later in cows, and PAG-1 is produced exclusively by the invading trophoblast binucleate cells (16,17). Although a very early pregnancy test is desirable in the cattle industry, the blood test for PAG-1 can be performed earlier than the presently used alternative (rectal palpation). One major shortcoming of the blood test, however, lies in the apparent long half-life (~9 days) of PAG-1 in maternal blood. Because concentrations at term may be well above 1 μg/ml, it requires at least 3 months for levels to drop back to threshold values (Fig. 1), and cows are customarily bred within 2 to 3 months after calving.

THE cDNA CLONING OF PAG

The antiserum generated by Zoli *et al.* (7) was used to screen cDNA libraries prepared from mid-pregnant placental tissues of cattle and sheep (16). The cDNA so isolated were approximately 1.7 kbp in length and coded for polypeptides of 380 and 382 amino acids in cattle and sheep respectively (Fig. 2). Each polypeptide had a predicted signal sequence of

	bPAG 1	bPAG 1v	bPAG 2	oPAG 1	oPAG 3	hPepA	hCatD	hCatE
bPAG 1	100							
bPAG 1v	86.5	100						
bPAG 2	57.9	58.5	100					
oPAG 1	72.0	73.8	58.7	100				
oPAG 3	59.7	58.9	64.7	59.8	100			
hPepA	49.5	50.4	49.1	51.2	49.9	100		
hCatD	37.5	38.8	37.4	41.4	41.7	43.3	100	
hCatE	42.1	41.8	43.7	41.8	44.4	53.5	51.6	100

Figure 3. Primary amino acid sequence similarities among the various PAG and selected aspartyl proteinases. The sequences for human pepsinogen A (hPepA), human cathepsin D (hCatD) and human cathepsin E (hCatE) were derived from Sogawa *et al.* (22), Faust *et al.* (23) and Azuma *et al.* (24).

15 residues and, in the case of the bovine PAG where an amino terminal sequence of the purified glycoprotein had already been obtained (7), it was clear that a pro-piece of 38 residues had also been removed.

Ovine PAG-1 has yet to be isolated and chemically sequenced. Most surprisingly the PAG clearly belonged to the aspartic proteinase gene family (Fig. 3) and especially resembled pepsin, where the extent of sequence identity was close to 50%. Although there are considerable stretches of sequence where PAG and pepsin sequences differ markedly (Fig. 4), examination of the corresponding nucleotide sequence of these genes indicates that many of these differences arose from sudden frameshifts. Despite the resemblance to pepsin, the purified bPAG-1 product had no detectable proteolytic activity towards denatured hemoglobin, yet it bound tightly to immobilized pepstatin and so presumably had an intact substrate-binding cleft (16).

Examination of the normally conserved regions in the amino terminal and carboxyl terminal lobes, which together constitute the catalytic centers of aspartic proteinases revealed yet further surprises. Ovine PAG-1 (oPAG-1) lacked one of the invariant aspartyl residues considered essential for catalysis and could not, therefore, be active. In contrast, bPAG-1 had an alanine substituted for a normally invariant glycine residue (Fig. 5). A molecular modeling study performed in Dr. Jordan Tang's laboratory has indicated that the presence of the relatively bulky methyl side chain of alanine would probably cause considerable displacement of the water molecule that resides symmetrically between the two adjacent aspartic acid residues (Fig. 6). Because this water molecule is involved in peptide bond cleavage, there is good reason to suppose that neither ovine nor bovine PAG can be catalytically active. Consequently, the removal of the pro-piece from bPAG-1 is probably not achieved autocatalytically. More likely some other proteinase is responsible.

If the PAG-1 molecules have any function, therefore, it seems unlikely that it is in proteolysis. Two alternative roles, both of which capitalize on the apparently intact substrate-binding cleft, immediately suggest themselves. One is that the PAG-1 bind and transport bioactive peptides, possibly in order to neutralize their effects or to extend the half-lives of such molecules in much the same way as specific-binding proteins interact with the insulin-like growth factors (18). Alternatively, the PAG-1 may themselves be hormones which bind to specific amino acid sequences on receptor molecules.

```
        -15                           -1 1
bPAG 1  M K W L V L L G L V A F S E C I V   K I P L R R L K T M R N V V S G K N M L N N   24
hPEP A  - - - - L - - - - - - - L - - - - M Y - V - - I - K - S L - R T L - E R G L - K D

    25  F L K E H A Y S L S Q I S F R G S N L T T H         P L R N I K D L V Y M G N I   60
        - - - K - N L N P A R K Y - P Q W E A P - L V D E Q - - E - Y L - M E - F - T -

    61  T I G T P P Q E F Q V[V F D T A S S]D L W V P S D F C T S P A C S T H V R F R H  100
        G - - - - A - D - T -[- - - - G - -]N - - - - - - V Y - S - L - - T N - N - - N P

   101  L Q S S T F R L T N K T F R I T Y G S G R M K G V V V H D T V R I G N L V S T D  140
        E D - - - Y Q S - S E - V S - - - - T - S - T - I L G Y - - - Q V - G I S D - N

   141  Q P F G L S I E E Y G   F E G R I Y   D G V L G L N Y P N I S F S G A I P I F D  178
        - I - - - - E T - P - S - L Y Y A P F - - I - - - A - - S - - S - - - T - V - -

   179  K L K N Q R A I S E P V F A F Y L S K D E R E G S V V M F G G V D H R Y Y E G E  218
        N I W - - G L V - Q D L - S V - - - A - D Q S - - - - I - - - I - S S - - T - S

   219  L N W V P L I Q A G D W S V H M D R I S I E R K I I A C S D G C K A[L V D T G T  258
        - - - - - V T V E - Y - Q I T V - S - T M N G E A - - - A E - - Q -[I - - - - -

   259  S]D I V G P R R L V N N I H R L I G A I P R G S E H Y V P   C S E V N T L P S I  297
        -]L L T - - T S P I A - - Q S D - - - S E N S D G D M - V S - - A I S S - - D -

   298  V F T I N G I N Y P V P G R A Y I L K D D R G R C Y T T F Q E   N R V S S S T E  336
        - - - - - - V Q - - - - P S - - - - Q S E   - S - I S G - - G M - L P T E - G -

   337  T W Y L G D V F L R L Y F S V F D R G N D R I G L A R A V  365
        L - I - - - - - I - Q - - T - - - - A - N Q V - - - P V A
```

Figure 4. Comparison of the amino acid sequences of bPAG-1 and human pepsinogen A. Dashes in the hPepA sequence indicate identity. The sequences in the two catalytic site regions are shaded and boxed. Gaps have been placed to maximize alignments.

EXPRESSION OF PAG-1 DURING PREGNANCY

Immunocytochemical studies have shown that PAG-1 in both cattle and sheep is localized to the binucleate cells in the outer epithelial layer (trophectoderm) of the placenta (Fig. 7).

These cells, which probably arise by endomitosis from precursor mononucleate cells in the same epithelium, start to appear just prior to the time that the pre-placenta (trophoblast) attaches to the uterine wall (around Day 17 in cattle). They constitute the invasive components of the placenta of ruminant species, and, once formed, migrate from the trophoblast and fuse with maternal uterine epithelial cells. At this stage the dense secretory granules discharge their contents towards the capillaries that feed the basolateral surface of the uterine epithelium. The PAG-1 co-localize to these dense granules with placental lactogen (19-20), which is an established hormone, thus reinforcing the notion that they may have an endocrine function.

The PAG-1 mRNA is expressed abundantly from the time binucleate cells first form, just prior to implantation, until term (~145 days in sheep, ~280 days in cattle). In this regard their expression differs markedly from that of another trophoblast-specific product of ruminant species, the interferon-tau (IFN-τ). The latter is localized to the mononucleate cells of the trophectoderm and is expressed for only a few days prior to the beginning of implantation (21).

IDENTIFICATION OF cDNA FOR ADDITIONAL PAG

Recently nucleic acid screening of the bovine cDNA libraries under low conditions of stringency has revealed cDNA for additional PAG molecules. The first to be characterized

PROTEIN	N-TERMINUS	C-TERMINUS
Bovine PAG-1	F D T A S S	V D T G T S
Ovine PAG-1	F D T G S S	V G T G T S
Pepsin (rhesus)	F D T G S S	V D T G T S
Chymosin (bovine)	F D T G S S	V D T G T S
Renin (mouse)	F D T G S A	V D T G S S
Cathepsin D (human)	F D T G S S	V D T G T S
Cathepsin E (human)	F D T G S S	V D T G T S
Consensus	F D T G S S	V D T G T S

Figure 5. Alignment at the active site residues found in a variety of aspartyl proteinases. The open boxes denote substitutions that do not alter enzymatic activity. Amino acid substitutions that are likely to render the enzyme inactive are underlined. Data for aspartyl proteinase sequences are derived from Evers *et al.* (25), Tang and Wong (26), Hidaka *et al.* (27), Miyazaki *et al.* (28) and the references for Figure 3.

encoded a protein with 87% amino acid sequence identity to bPAG-1. It has been named bPAG-1_{var} and is not discussed further here. A second highly abundant cDNA, which represents a 376 amino acid polypeptide (PAG-2) with only 58% amino acid sequence identity to bPAG-1, was isolated by a similar screening strategy (Fig. 2). Curiously, the sequences comprising the catalytic center are consistent with PAG-2 being a potentially active aspartyl proteinase (Fig. 5), although such proteolytic activity remains to be demonstrated. This laboratory is currently producing bPAG-2 in *E. coli* by recombinant techniques. Provided the molecule can be successfully renatured and activated, we should soon be able to test the ability of PAG-2 to cleave a range of substrates.

Nucleic acid screening of the ovine cDNA library has revealed the transcripts for yet another PAG produced by the ovine placenta that is sufficiently distinct in sequence from oPAG-1 and bPAG-2 to constitute yet another subtype (oPAG-3). This molecule also possesses an apparently intact catalytic center (Fig. 5).

Southern genomic blot analysis of bovine DNA with exon-specific cDNA probes strongly suggests that there are many (probably at least ten) PAG-related genes (data not shown). Although it is unclear whether these genes are expressed in placenta or elsewhere, it seems possible that the PAG family of proteins may be extensive.

POST-TRANSLATIONAL PROCESSING OF PAG-1

After pro-piece removal, the polypeptide portion of PAG-1 has a molecular weight of about 37,000, yet when purified PAG-1 is analyzed on SDS-polyacrylamide gels it migrates with an apparent M_r close to 70,000. Part of this mass is undoubtedly derived from carbohydrate as PAG-1 is known to be a glycoprotein with at least 10% by weight neutral sugar (7). Treatment of PAG-1 with N-glycanase to remove oligosaccharides linked to asparagine residues reduces the apparent M_r to about 60,000 (data not shown), and biosynthetic studies carried out on cultured explants of placenta in presence of tunicamycin, an inhibitor of N-glycosylation, have shown a similar modest reduction in PAG-1 size. Whether this "extra" mass of PAG-1 is due to O-linked carbohydrate or to some other form of post-translational modification is unclear. It does not appear to result from any unusual physical properties of the polypeptide itself, because, when synthesized as a glycosylated recombinant product in CHO or Cos-1 cells, its mobility in SDS-gels is consistent with an

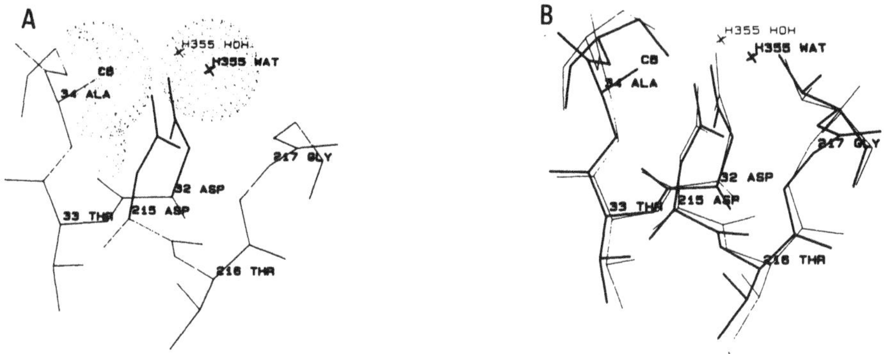

Figure 6. Computer modeling of the highly conserved regions of PAG, homologous to the active sites of aspartic proteinases. (A) Overlay of the modeled bPAG structure with the active site of porcine pepsin (29). Bold lines depict the structure of the bPAG residues homologous to the active site of porcine pepsin, with α-carbons labeled according to pepsin numbering. Thin lines are the active site of the crystal structure of porcine pepsin. The water molecule from the PAG model is labeled as H355 WAT, and the solvent from the crystal structure of porcine pepsin is labeled as H355 HOH. (B) Computer model of the elements of bPAG homologous to the active site of pepsin. The "active-site" Asp-32 and Asp-215 (pepsin numbering) are plotted with bold lines. The van der Waals surface of water molecule H355 WAT and Ala-34 from the PAG model are depicted as stippled areas. The native position of H355 HOH from the crystal structure of porcine pepsin is drawn and labeled with light lines. The native position of the solvent is altered by the presence of the methyl group of the Ala-34β-carbon, labeled CB. From Xie *et al.* (16) with permission.

M_r of 43,000. It seems likely, therefore, that binucleate cells of the placenta process PAG in some unusual manner.

Even though PAG is first synthesized as a ~70K product in explant cultures and that form tends to predominate in the tissues, smaller sized molecules accumulate with time, particularly in the medium (Fig. 8). Recent pulse-chase experiments with labeled amino acids have clearly established that the 70K form precedes the formation of the 47K and 53K forms. To date the amino terminal sequences of these secondary products have not been obtained and so the nature of the processing events remains unclear. Nevertheless their presence probably explains the range of sizes that has been reported for PAG-1 (or PSP-B) in the years following its first description in 1982.

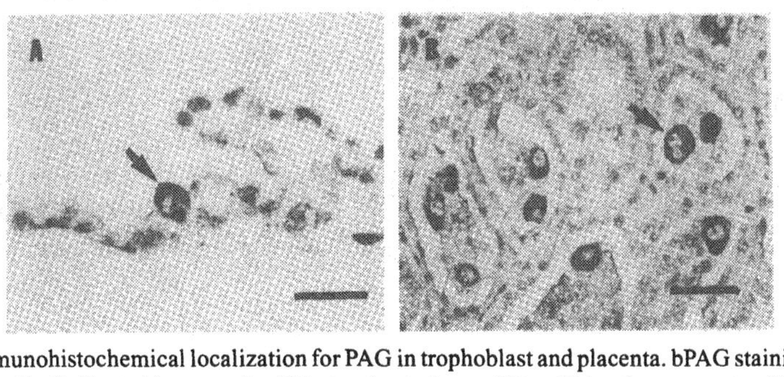

Figure 7. Immunohistochemical localization for PAG in trophoblast and placenta. bPAG staining was detected in binucleate cells (indicated by arrow) within trophectoderm of Day 18 bovine conceptus (**A**) and in bovine placenta of midgestation fetal cotyledonary villi (**B**) that interdigitate with maternal uterine tissue. (Bars = 50 μm) From Xie *et al.* with permission.

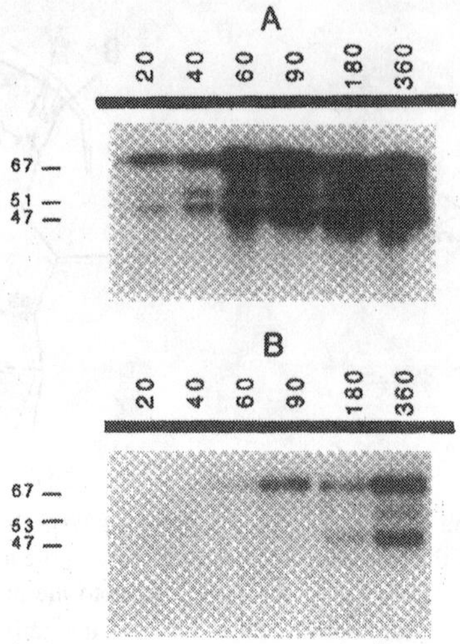

Figure 8. Time course of oPAG biosynthesis in ovine placental explants *in vitro*. Trophoblast tissue from a Day 25 conceptus was incubated for the time periods (in min) shown above the lanes. Immunocomplexes from tissue extract (A) and medium (B) were analyzed in 10% polyacrylamide gels containing SDS and radioactive polypeptides were detected by fluorography. Molecular weights on the left ($x10^{-3}$) were calculated from standards electrophoresed on each side of the gel. From Xie *et al.* (16) with permission.

DISCUSSION

The realization that the placenta-specific antigens PAG-1 from cattle and sheep were related in sequence to aspartyl proteinases, and especially to pepsin, but lacked functional catalytic centers was surprising, yet consistent with the hypothesis that members of proteinase gene families need not necessarily display their function through the hydrolysis of peptide bonds. It is possible, of course, that PAG-1 is expressed promiscuously in placenta as a functionless product, but we prefer the alternative hypothesis that it uses its apparently intact substrate-binding cleft to associate specifically with some other protein or polypeptide. The presence of other abundant transcripts representing additional aspartyl proteinases resembling PAG, yet possessing apparently intact catalytic centers, constitutes a further puzzle. It remains to be determined whether these molecules are catalytically active and, even if they are, whether their physiological function is that of proteinases.

ACKNOWLEDGMENTS

The authors thank Mrs. Gail Foristal for preparing the manuscript and Mr. Jim Bixby for technical assistance. The work was supported, in part, by NIH Grant HD29483 and is contribution 12,046 from the Missouri Agricultural Experiment Station Journal Series.

REFERENCES

1. G.D. Hodgen and J. Itskovitz, Recognition and maintenance of pregnancy. *in*: "*The Physiology of Reproduction*," E. Knobil and J. Neill, eds., Raven Press, New York, pp 1995-2021 (1988).

2. F. Talamantes and L. Ogren, The placenta as an endocrine organ: Polypeptides. *in*: "*The Physiology of Reproduction*," E. Knobil and J. Neill, eds., Raven Press, New York, pp 2093-2144 (1988).
3. J.H. Nilson, J.A. Bokar, C.M. Clay, T.A. Farmerie, R.A. Fenstermaker, D.L. Hamernik, and R.A. Keri, Different combinations of regulatory elements may explain why placenta-specific expression of the glycoprotein hormone -subunit gene occurs only in primates and horses. *Biol. Reprod.* 44:231 (1991).
4. J.E. Butler, W.C. Hamilton, R.G. Sasser, C.A. Ruder, G.M. Hass, and R.J. Williams, Detection and partial characterization of two bovine pregnancy-specific proteins. *Biol. Reprod.* 26:925 (1982).
5. R.G. Sasser, C.A. Ruder, K.A. Ivani, J.E. Butler, and W.C. Hamilton, Detection of pregnancy by radioimmunoassay of a novel pregnancy-specific protein in serum of cows and a profile of serum concentrations during gestation. *Biol. Reprod.* 35:936 (1986).
6. R.G. Sasser, J. Crock, and C.A. Ruder-Montgomery, Characteristics of pregnancy-specific protein B in cattle. *J. Reprod. Fert.* Suppl. 37:109 (1989).
7. A.P. Zoli, J.F. Beckers, P. Woutters-Ballman, J. Closset, P. Falmagne, and F. Ectors, Purification and characterization of a bovine pregnancy-associated glycoprotein. *Biol. Reprod.* 45:1 (1991).
8. R.A. Lynch, B.M. Alexander, and R.G. Sasser, The cloning and expression of the bovine pregnancy specific protein B (bPSPB) gene. *Biol. Reprod.* 46(Suppl 1), Abstract 89 (1992).
9. P. Humblot, S. Camous, J. Martal, J. Charlery, N. Jeanguyot, M. Thibier, and R.G. Sasser, Diagnosis of pregnancy by radioimmunoassay of a pregnancy-specific protein in the plasma of dairy cows. *Theriogenology* 30:257 (1988).
10. A.P. Zoli, L.A. Guilbault, P. Delabaut, W.B. Ortiz, and J.F. Beckers, Radioimmunoassay of a bovine pregnancy-associated glycoprotein in serum.: its application for pregnancy diagnosis. *Biol. Reprod.* 46:83 (1992).
11. F.B.P. Wooding, Current topic: the synepitheliochorial placenta of ruminants: binucleate cell fusions and hormone production. *Placenta* 13:101 (1992).
12. C.A. Ruder, J.N. Stellflug, J.J. Dahmen, and R.G. Sasser, Detection of pregnancy in sheep by radioimmunoassay of sera for pregnancy-specific protein B. *Theriogenology* 29:905 (1988).
13. P. Humbolt, G. DeMontigry, N. Jeanguyot, F. Tetedoie, B. Payen, M. Thibier, and R.G. Sasser, Pregnancy-specific protein B and progesterone concentrations in French Alpine goats throughout gestation. *J. Reprod. Fert.* 89:205 (1990).
14. A.K. Wood, R.E. Short, A-E. Darling, G.L. Dusek, R.G. Sasser, and C.A. Ruder, Serum assays for detecting pregnancy in mule and white-tailed deer. *J. Wildl. Manage.* 50:684 (1986).
15. J.C. Haigh, C. Gates, A. Ruder, and R. Sasser, Diagnosis of pregnancy in wood bison using a bovine assay for pregnancy-specific protein B. *Theriogenology* 36:749 (1991).
16. S. Xie, B.G. Low, K.K. Kramer, R.J. Nagel, R.V. Anthony, A.P. Zoli, J-F. Beckers, and R.M. Roberts, Identification of the major pregnancy-specific antigens of cattle and sheep as inactive members of the aspartic proteinase family. *Proc. Natl. Acad. Sci. U.S.A.* 88:10247 (1991).
17. A.P. Zoli, P. Demez, J.F. Beckers, M. Reznik, and A. Beckers, Light and electron microscopic immunolocalization of bovine pregnancy-associated glycoprotein in the bovine placentome. *Biol. Reprod.* 46:623 (1992).
18. R.H. McCusker and D.R. Clemmons, The insulin-like growth factor binding proteins: structure and biological functions. *in*: "*The Insulin-Like Growth Factors,*" P.N. Schofield, ed., Oxford University Press, Oxford, pp 110-150 (1992).
19. F.B.P. Wooding, Localization of ovine placental lactogen in sheep placentomes by electron microscope immunocytochemistry. *J. Reprod. Fert.* 62:15 (1981).
20. S.M. Kappes, W.C. Warren, S.L. Pratt, R. Liang, and R.V. Anthony, Quantification and cellular localization of ovine placental lactogen messenger ribonucleic acid expression during mid- and late gestation. *Endocrinology* 131:2829 (1992).
21. R.M. Roberts, J.C. Cross, and D.W. Leaman, Interferons as hormones of pregnancy. *Endocrine Rev.* 13:432 (1992).
22. K. Sogawa, Y. Fujii-Kuriyama, Y. Mizukami, Y. Ichihara, and K. Takahashi, Primary structure of human pepsinogen gene. *J. Biol. Chem.* 258:5306 (1983).
23. P.L. Faust, S. Kornfeld, and J.M. Chirgwin, Cloning and sequence analysis of cDNA for human cathepsin D. *Proc. Natl. Acad. Sci. U.S.A.* 82:4910 (1985).
24. T. Azuma, G. Pals, T.K. Mohandas, J.M. Couvreur, and R.T. Taggart, Human gastric cathepsin E: Predicted sequence, localization to chromosome 1, and sequence homology with other aspartic proteinases. *J. Biol. Chem.* 264:16748 (1989).

25. M.P.J. Evers, B. Zelle, J.P. Bebelman, V. Van Beusechem, L. Kraakman, M.J.V. Hoffer, J.C. Pronk, W.H. Mager, R.J. Planta, A.W. Ericksson, and R.R. Frants, Nucleotide sequence comparison of five human pepsinogen A (PGA) genes: evolution of the PGA multigene family. *Genomics* 4:232 (1989).
26. J. Tang and R.N.S. Wong, Evolution in the structure and function of aspartic proteases. *J. Cell. Biochem.* 33:53 (1987).
27. M. Hidaka, K. Sasaki, T. Uozumi, and T. Beppu, Cloning and structural analysis of the calf prochymosin gene. *Gene* 43:197 (1986).
28. H. Miyazaki, A. Fukamizu, S. Hirose, T. Hayashi, H. Hori, H. Ohkubo, S. Nakanishi, and K. Murakami, Structure of the human renin gene. *Proc. Natl. Acad. Sci. U.S.A.* 81:5999 (1984).
29. N.S. Andreeva, A.S. Zdanov, A.E. Gustchina, A.A. Fedorov, Structure of ethanol-inhibited porcine pepsin at 2-Å resolution and binding of the methyl ester of phenylalanyl diiodotyrosine to the enzyme. *J. Biol. Chem.* 259:11353 (1984).

STRUCTURE AND POSSIBLE FUNCTION OF ASPARTIC PROTEINASES IN BARLEY AND OTHER PLANTS

Jukka Kervinen,[1] Kirsi Törmäkangas,[1] Pia Runeberg-Roos,[1]
Kunchur Guruprasad,[2] Tom Blundell,[2] and Teemu H. Teeri[1]

[1]Institute of Biotechnology
P.O. Box 45, FIN-00014 University of Helsinki, Finland
[2]Department of Crystallography
Birkbeck College, University of London
Malet Street, London WCIE 7HX, UK

INTRODUCTION

The aim of this short article is to review the literature on plant aspartic proteinases in general as well as to present more detailed information on our studies on the barley aspartic proteinase. Aspartic proteinases (EC 3.4.23) are widely dispersed in the plant kingdom including monocotyledonous and dicotyledonous species as well as gymnosperms. These enzymes have been detected in seeds, leaves and flowers in different plants as well as in the digestive fluid of some insectivorous species. Like their animal and microbial counterparts (1-3), plant aspartic proteinases are inhibited by pepstatin and preferentially cleave peptide bonds between amino acid residues with large hydrophobic side chains. The biological functions of plant aspartic proteinases are not as well characterized as those of animal aspartic proteinases. In the insectivorous plant *Nepenthes*, aspartic proteinase may participate in the digestion of insects trapped in the digestive fluid of the pitcher. In the leaves of tomato and tobacco, aspartic proteinases may contribute to the hydrolysis of extracellular pathogenesis-related proteins. Aspartic proteinases have also been reported to be associated with the intracellular storage protein bodies of hemp and buckwheat seeds and to take part in the hydrolysis of storage proteins in wheat and cocoa *in vitro*. In addition, aspartic proteinases from seeds have been shown to process the *Arabidopsis* 2S albumin storage proteins *in vitro*. The cDNA-derived primary structures of barley, rice, and cardoon aspartic proteinases are known. The open reading frame of the barley enzyme consists of a 66-amino acid preprosequence and a 442 residues long mature protein, about the same size as recently also revealed for rice and cardoon enzymes. The mature plant enzymes are most homologous (~50 %) to mammalian lysosomal cathepsin D. The homology is dispersed over N- and C-terminal regions leaving an internal nonhomologous plant specific region of about 100 amino acid residues. The assumed processing pattern of the barley and cardoon enzymes differs from that of animal and microbial aspartic

Table 1. Aspartic Proteinases Purified from Different Plants

Plant, *Latin name*, part	**Mol wt** (kDa); **subunits**	**pH-optima** (Substrate)	**Other names**	**Ref.**
Monocots				
Wheat, *Triticum aestivum*, resting seeds	50[b], 58[c]; 1	3.0 (Albumin), 4.5 (Gliadin)		6
Wheat, *Triticum aestivum*, green leaf	89[b]; n.d.	4.5 (Hemoglobin)		9-11
xHaynaldoticum sardoum, dry seeds	66[b]; n.d.	3.1 (Chromophoric peptide[e])		8,12
Rice, *Oryza sativa*, resting seeds	60-65[b], 48[d]; n.d.	2.5-4.2 (Hemoglobin, casein/urea)	Carboxyl proteinase	13,14
Sorghum, *Sorghum vulgare*, germinated seeds	80[b]; n.d.	3.6 (Albumin)	Acid protease	15,16
Barley, *Hordeum vulgare*, resting seeds	(48 and 40)[c,d]; 2	3.5-3.9 (Hemoglobin) 3.7-4.1(Chromophoric peptide[e])		53,57,58
Dicots				
Hemp, *Cannabis sativa*, resting seeds	20[a]; 1-4 (?)	3.4 (Hemoglobin), 4.3 (Edestin)	Edestinase	19,21,22
Buckwheat, *Fagopyrum esculentum*, seeds	28[b]; n.d.	3.5 (Albumin)		23
Lotus, *Nelumbo nucifera*, resting seeds cotyledons	36-37[a,b]; n.d.	2.4-3.8 (Casein, Hemoglobin)		25,26
Nepenthes macferlanei, pitcher secretion	59[b]; n.d.	2.2 (Albumin, Fibrin)	Nepenthesin	28-31
Drosera peltata, leaves, stem and trapping organs	n.d.; n.d.	2-3 (Casein)		28
Cucumber, *Cucumis sativus*, seeds	40-42[b]; n.d.	3.2 (Hemoglobin)		32
Squash, *Cucurbita maxima*, seeds	40-42[b]; n.d.	3.6 (Hemoglobin)		32
Tomato, *Lycopersicon esculentum*, leaf	37[c]; 1	2.5-3.5 (FTC-hemoglobin)	Aspartyl endoproteinase	34
Tobacco, *Nicotiana tabacum*, leaf	36-40[b]; 1 (?)	3.0 (FTC-hemoglobin)		35
Cardoon, *Cynara cardunculus*, flower	46-49[c,d]; 2	5.1 (FTC-casein)	Cynarase, Cardosin	36-41
Rape, *Brassica napus*, seeds[f]	30[c] (?); n.d.	3.5 (2S albumin precursor)		42
Cocoa, *Theobroma cacao*, seeds	45[c]; 2	3.4 (Albumin)		44-45,(g)
Gymnosperms				
Scots pine, *Pinus sylvestris*, seeds[f]	n.d.; n.d.	3.7 (Hemoglobin)		49,50
Jack pine, *Pinus banksiana*, seeds	n.d.; n.d.	3.5 (Hemoglobin)		52

Determined by (a) ultracentrifugation, (b) gel filtration, (c) SDS-PAGE, (d) cDNA. Substrate: (e) Pro-Thr-Glu-Phe-Phe(4-NO_2)-Arg-Leu. (f) Only partially purified. (g) Personal communication, see text. n.d = not determined.

proteinases. Immunocytochemical and *in situ* hybridization studies show that the barley aspartic proteinase (HvAP, *Hordeum vulgare* aspartic proteinase) is mainly present in the embryo, aleurone layer and scutellum but not in the starchy endosperm. Intracellularly, HvAP is located in the vacuoles corresponding to the lysosomal localization of cathepsin D. Vacuolar HvAP may participate in the processing of the C-terminus of barley prolectin. A molecular model of HvAP has been constructed using the rule-based comparative approach encoded in the COMPOSER suite of computer programs.

PROPERTIES OF PLANT ASPARTIC PROTEINASES

The main properties of the best known plant aspartic proteinases are compiled in Table I and discussed below.

Monocotyledonous Plants

Wheat. Resting wheat grains contain pepstatin-sensitive proteinase activity (4-8). The corresponding enzyme has been purified to homogeneity by affinity chromatography on bacilliquin-silochrome followed by ion-exchange chromatography, chromatofocusing, and preparative PAGE. The enzyme is monomeric with an isoelectric point value at pH 5.0, it partially hydrolyzes storage protein gliadin and the enzyme has apparently only endopeptidase activity (6). In wheat flour extracts, pepstatin- and DFP-sensitive proteases partially degraded wheat gluten proteins in acetic acid solution (4). In germinating wheat seeds, the degradation of gliadin, one of the gluten proteins, is mainly caused by *de novo*-synthesized cysteine proteinases. However, a preliminary modification of wheat gliadin by an aspartic proteinase and a carboxypeptidase enhanced the hydrolysis of gliadin 3 to 4-fold by cysteine proteinases *in vitro* (7). The data suggest that the aspartic proteinase together with other proteases enables the rapid and complete degradation of storage proteins during the seed germination (6,7).

An aspartic proteinase has also been purified from green wheat leaves using DEAE-cellulose ion exchange chromatography followed by hemoglobin-Sepharose affinity column. The proteinase is sensitive to aspartic proteinase inhibitors diazoacetylnorleucine methyl ester (DAN) and 1,2-epoxy-3-(*p*-nitrophenoxy)propane (EPNP) but over 50 % inhibition of the enzyme by serine proteinase inhibitor phenylmethylsulfonyl fluoride (PMSF) was also noticed. Sensitivity to pepstatin was not tested. The purified enzyme degraded the most abundant leaf protein ribulose-1,5-bisphosphate carboxylase/oxygenase (Rubisco) *in vitro* (9-11).

A partially purified aspartic proteinase has been isolated from the resting grain bran (roughly corresponding to the aleurone layer) of *xHaynaldoticum sardoum*, a hybrid between *Triticum durum* and *Haynaldia villosa*, by pepstatin A-agarose affinity chromatography. The estimated molecular weight of the isolated enzyme was between 60 and 100 kDa, apparently 66 kDa (8,12).

Rice. An aspartic proteinase has partially been purified from resting rice grains on DEAE-ion exchange column followed by gel chromatography. The enzyme was distributed mainly in the living parts of the grain: the embryo and the aleurone layer. Since storage globulins of rice have a similar distribution and they are rapidly hydrolyzed at the onset of germination, it was suggested that the rice aspartic proteinase could have a role in the primary hydrolysis of storage proteins providing amino acids for the rapid synthesis of new hydrolytic enzymes which are secreted into the starchy endosperm (13). Recently, a cDNA derived primary structure of an aspartic proteinase from rice suspension culture has become available (14). From the sequence, the calculated size of the putative active enzyme is about 48 kDa and the homology to the barley aspartic proteinases is 74 %. Interestingly, the rice aspartic proteinase also contains the internal sequence of 99 amino acid residues similarly to the barley enzyme but lacking from animal or microbial aspartic proteinases. In addition, a C-terminal active site Thr in a sequence Asp-Thr-Gly has been replaced by Ser.

Sorghum. An aspartic proteinase has been purified from the germinated seeds of sorghum using DEAE-cellulose chromatography as a main purification step. When synthetic peptides and ribonuclease A were used as substrates, the enzyme cleaved peptide bonds only on the carboxyl side of Asp and Glu residues (15,16).

Maize and Finger Millet. Aspartic proteinase activity has been detected from the seed extracts of maize (*Zea mays*) and finger millet (*Eleusine coracana*). In resting maize seeds, the aspartic proteinase activity is dominant but germination is accompanied by the appearance of cysteine proteinases which eventually become the prevailing endopeptidase group (17). In germinating millet seeds, the aspartic proteinase activity was highest on the 3rd day of germination and decreased almost to zero before the 6th day of germination. The millet enzyme hydrolyzed endogenous trypsin/α-amylase inhibitor *in vitro* (18).

Barley. See more detailed description later in the text.

Dicotyledonous Plants

Hemp. Dormant seeds of hemp contain aspartic proteinase activity intracellularly associated with protein bodies (19). Protein bodies, also known as aleurone grains, are vacuole/lysosome -like organelles, filled with storage proteins and several hydrolytic enzymes (20). As analyzed by gel electrophoresis, the purified enzyme preparation contained at least three protein components but the individual proteins were not further characterized. The purified enzyme was named edestinase since it hydrolyzed edestin, one of the main storage proteins of hemp seeds, located also in protein bodies (21,22).

Buckwheat. Two aspartic proteinases, therein named carboxyl proteinases I and II, have been purified from dormant buckwheat seeds (23). Both enzymes have the same optimal pH for the activity and apparently the same size but in isoelectric focusing the enzymes I and II separated into two active fractions with isoelectric points 4.2 and 6.3, respectively. Also the amino acid compositions of the two enzymes differed. The enzyme preparation consisting of both enzymes hydrolyzed buckwheat seed proteins albumin and 13S globulins *in vitro*. Globulins are the main storage protein form in buckwheat seeds. Elpidina et al. (24) have also shown that aspartic proteinase activity, with several other enzyme activities typical for lysosomal/vacuolar -like organelles, was accociated in isolated protein body fraction with 13S globulins. Thus, aspartic proteinases may take part in the degradation of 13S globulins in germinating buckwheat seeds as suggested by Belozerskii et al. (23).

Lotus. The purified aspartic proteinase from the cotyledons of dormant lotus seeds hydrolyzed casein and hemoglobin readily but several endogenous storage proteins of dicots such as edestin, zein, lotus seed globulin, and soybean casein were poor substrates. Many structural and catalytic properties of the lotus enzyme were found to be fairly similar to pepsin and cathepsin D such as its pH optimum, molecular weight, inhibition by potassium permanganate, and the behavior on acetylation (25,26).

***Nepenthes* and *Drosera*.** In many carnivorous plants, the organs suited for trapping and digestion of insects and other small invertebrates secrete aspartic proteinases among other hydrolyzing enzymes (27). In the genus *Nepenthes*, a group of tropical plants, the digestive fluid of the pitchers contains aspartic proteinase activity (28-30). The purified enzyme preparations have been obtained from the combined pitcher fluid collected from several species belonging to the genus *Nepenthes* (28,31) and from the secretion of *Nepenthes macferlanei* (30). The secretion contains one to two active aspartic proteinases with the pH optimum in the range 2 to 3. This pH corresponds to the pH of the digestive fluid in open pitchers and thus forms a suitable environment for the digestive activity (30). An aspartic proteinase closely resembling the *Nepenthes* enzyme has also been purified from *Drosera peltata* of the sundew family (28). The hydrolytic specificity tested for *Nepenthes macferlanei* (30) seems to be strikingly similar to that of animal stomach pepsin although more narrow specificities for *Nepenthes* (31) and *Drosera* (28) enzymes have also been suggested.

Cucumber and Squash. Aspartic proteinases have been purified and partially characterized from cucumber and squash seeds using affinity chromatography on a pepstatin-Sepharose column as the main purification step. The enzymes hydrolyzed serum and egg albumin, casein, edestin from hemp seeds and the endogenous cucumber and squash seed globulins. In studies following the hydrolysis of the oxidized B-chain of insulin, the squash enzyme preferentially cleaved peptide bonds between two bulky hydrophobic amino acid residues as Phe_1-Val_2 and Phe_{24}-Phe_{25} and Tyr containing bonds (32).

Tomato and Tobacco. Plants respond to different stress situations by expressing and accumulating intracellular and extracellular pathogenesis-related (PR) proteins which have been proposed to be part of the defence system against microbial infection. Many PR proteins are acid soluble and often very resistant to degradation by proteases (33). An extracellular aspartic proteinase, purified from leaves of tomato and tobacco, degrades secreted PR proteins *in vitro*, but has no activity towards intracellular proteins. The sensitivity of PR proteins to the aspartic proteinase activity and the codistribution of the enzyme and PR proteins indicate that the biological action of PR proteins might be regulated by aspartic proteinases. The enzyme probably prevents the overaccumulation of PR proteins in infected leaves. Tobacco and tomato aspartic proteinases are also present in healthy leaves at similar levels but the function in healthy tissues is unknown (34,35).

Cardoon. Calf chymosin (rennet) is universally used as one of the main milk-clotting enzymes in cheese manufacturing industry. However, in traditional cheese production, several other coagulating agents have also been used for a long time. In Portugal, an enzyme preparation obtained by aqueous extraction of cardoon flowers is widely exploited. Three closely related heterodimeric aspartic proteinases have been isolated from the senescent violet part (style) of a cardoon flower. The enzymes were named cynarases 1, 2, and 3 and according to SDS-PAGE the subunit sizes are 33 + 17 kDa, 34 + 17 kDa, and 36 + 14 kDa, respectively. The enzymes contain N-linked high mannose-type glycans (36). In addition, two heterodimeric aspartic proteinases have been isolated from the fresh flowers of a different variety of cardoon and named cardosins A (subunits 31 + 15 kDa) and B (34 + 14 kDa). The structural and kinetic differences between cardosin A and cardosin B, e.g. peptide substitutions, suggest that the enzymes are produced from two genes (37). Purified aspartic proteinases from both cardoon varieties specifically cleaved bovine κ-casein between Phe_{105}-Met_{106} resulting in milk clotting. The enzymes also showed preference to some other peptide bonds (38-40). When insulin A and B-chains were hydrolyzed with cardosins, the enzymes typically cleaved bonds next to the amino acid residues with bulky hydrophobic side chains (37,39). The expression pattern, studied for cynarases, is very tissue-specific. Cynarases are expressed in styles of the flower but not in leaves or seeds. Based on immunocytochemistry, cynarases seem to be present only in the epidermis of the style (41). Recently, the cDNA-derived primary structure of the cynarase 3 has partly been solved and the alignment shows 78 % and 63 % homology to barley and rice aspartic proteinases, respectively (41).

Rape. The seeds of *Arabidopsis* and rape (*Brassica napus*) contain large amounts of 2S storage albumins which undergo intensive postranslational processing during the maturation in developing seeds. After removal of the signal peptide, the propeptide is cleaved at four other bonds which are conserved in *Arabidopsis* and rape. D'Hondt et al. (42) have shown that an aspartic proteinase, which has been partially purified from rape seeds, processes translated 2S albumins *in vitro*. Based on the hydrolysis products, the amino terminal propeptide as well the internal propeptide linking the two subunits of the mature 2S albumin, are cleaved by an aspartic proteinase. However, 2S albumin precursor was incompletely processed and thus other kind of proteases are most probably needed for complete maturation. Secondary structure predictions from the 2S albumin internal processed fragment with changes of the amino acid residues in and around the cleavage sites suggest that both sequence and structural determinants are important in cleavage site recognition.

Cocoa. Specific precursors of aroma and taste ingredients of roasted cocoa seeds are known to be produced by limited proteolysis of seed storage proteins during industrial fermentation of cocoa. This post-mortem heterogeneous proteolysis, dependent on the pH-value (4.0 to 5.5), in seed extracts has been shown to be a result of the cooperative hydrolytic action of an aspartic proteinase and a serine carboxypeptidase. During fermentation the cocoa aspartic proteinase gives rise to peptides with hydrophobic amino acid residues in terminal positions which are subsequently cleaved by a serine carboxypeptidase (43-45). However, it should be noted that although aspartic proteinases seem to have an essential role during fermentation in the industrial explotation of cocoa seeds, the biological role of the enzyme in the hydrolysis of storage proteins in naturally germinating grains is not known. Recently, a heterodimeric aspartic proteinase of 45 kDa has been purified from resting cocoa seeds and according to SDS-PAGE the subunit sizes are 34 and 16 kDa (personal communication, Prof. Böle Biehl, Technische Universität, Braunschweig, Germany).

Kidney Bean and Soybean. Aspartic proteinase activity has been detected but not purified from resting seed extracts of two leguminous plants, kidney bean (*Phaseolus vulgaris*) and soybean (*Glycine max*). In kidney bean seeds, aspartic proteinase is present in resting and germinating seed cotyledons (46). In soybean, the aspartic proteinase, therein called carboxylendopeptidase, is present both in cotyledons and embryonic axis and in both tissues the highest activity was obtained at pH 4.0 using [^{125}I]iodoinsulin B-chain as an exogenous substrate. In studies on the autolytic hydrolysis of seed proteins, extracts prepared from resting seed cotyledons and embryonal axes were incubated for various times at pH 4.0. Most of the seed proteins remained intact and only the limited pepstatin-sensitive hydrolysis of unidentified 30 kDa polypeptide was noticed. Therefore, other kinds of proteases are obviously needed for the effective hydrolysis of soybean storage proteins (47).

Orange. Aspartic proteinase activity has been measured from the developing and senescing leaf extracts of orange tree (*Citrus sinensis*). In *in vitro* experiments, leaf extracts containing protease activity were incubated with purified Rubisco in the presence and absence of specific proteinase inhibitors and the reaction products were analyzed by

SDS-PAGE. Four types of Rubisco hydrolyzing activity were detected and the aspartic proteinase partially hydrolyzed both the large and small subunits of Rubisco (48).

Gymnosperms

Pine. Resting and germinating seeds of Scots pine (*Pinus sylvestris*) (49,50) and jack pine (*Pinus banksiana*) (51,52) contain a prominent aspartic proteinase activity. In the resting seeds of Scots pine, the activity is high in the embryo and it decreases about 50 % in the first days of germination and remains at a constant level thereafter. Low aspartic proteinase activity is also present in resting and germinating seed endosperm (50). A large oligomeric aspartic proteinase complex of at least 300 kDa has been purified to near electrophoretic homogeneity from resting jack pine seeds. The complex consisted of at least seven peptide subunits although some of them may be impurities coeluted with the enzyme (52).

TISSUE-SPECIFIC EXPRESSION OF THE BARLEY ENZYME AND STUDIES ON ITS FUNCTION AND THREE-DIMENSIONAL STRUCTURE

Grains (5,53,54) as well as green and senescing leaves (55) and roots (56) of barley contain aspartic proteinase activity. The corresponding enzyme has been purified from resting grains by affinity chromatography on a pepstatin-Sepharose column. During purification, two isoform enzymes of 48 and 40 kDa were found, the larger containing peptides of 32 and 16 kDa, and the smaller peptides of 29 and 11 kDa (53). Subsequent cDNA cloning (57) and comparison with the sequences of peptides derived from the various proteins purified, suggest that the two heterodimeric proteins are derived from the same gene and represent sequential processing products of the protein precursor. Sequence comparison shows that major parts of the barley aspartic proteinase, HvAP, are remarkably similar to other aspartic proteinases, especially to mammalian lysosomal cathepsin D and yeast aspartic proteinase A, but the C-terminal lobe contains a unique sequence of 104 amino acid residues carrying no similarity to animal or microbial aspartic proteinases. Part of this insert area is removed during processing of the enzyme. In addition, the processing pattern of the enzyme differs from that of mammalian and microbial aspartic proteinases, and the amino acid sequence next to the active site Asp_{215} (pepsin numbering) is Asp-Ser-Gly instead of the more common Asp-Thr-Gly (57). Like most animal and microbial aspartic proteinases, HvAP preferentially cleaves peptide bonds between amino acid residues with large hydrophobic side chains (58). We present here recent results on the expression of HvAP in different barley tissues, studies on its intracellular localization and possible function as well as data on the molecular modelling.

HvAP is Present in Developing and Germinating Grains

The proteases of the barley grain have been extensively studied because of their importance in the storage protein hydrolysis during grain germination (59) and, thus, *e.g.* in the malting process (60). A major part of the endoproteolytic activity is due to cysteine proteinases which are synthesized *de novo* during germination. However, barley grains also contain a prominent aspartic proteinase activity and the enzyme is present already in resting grains. It has been suggested that the aspartic proteinase may have a role at the onset of germination in cereals (13,59).

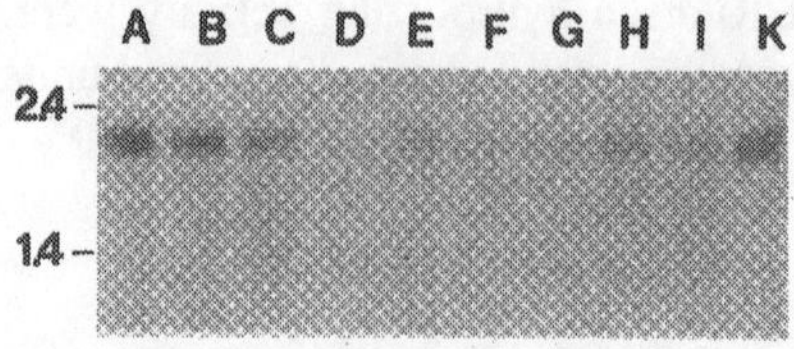

Figure 1. Northern blot analysis of barley aspartic proteinase. Total RNA (10 μg per lane) from various barley tissues was separated on a 1.5% denaturing agarose gel and blotted and hybridized with the ^{32}P-labelled 481 bp *Bam*HI-*Sca*I -fragment of HvAP-cDNA. Lanes: (A) developing flower (=carpel, stamen, lodicules) 1 to 2 days before anthesis, (B) developing palea, lemma (awn removed) and rachilla 1 to 2 days before anthesis, (C) developing testa and inner pericarp, 9 DAA, (D) developing starchy endosperm, 9 DAA, (E) developing embryo, 19 DAA, (F) embryo from 28 h germinated seeds, (G) scutellum from 28 h germinated seeds, (H) root tips from 3 d germinated seeds, (I) basal part of leaves from 6 to 7 d germinated seeds, (K) stem internodes from plants 3 to 7 days before anthesis.

We have recently studied (61) the expression of HvAP in developing and germinating grains as well as in other tissues (Fig. 1). Based on pepstatin-sensitive activities in different barley tissues, aspartic proteinase is present in an active form in developing, resting, and germinating grains. Immunocytochemical studies on developing and germinating seeds show that during seed development, HvAP is produced in the embryo, scutellum, aleurone layer, testa and pericarp but in the starchy endosperm HvAP is present only in the crushed and depleted area adjacent to the scutellum. Fig. 2 shows the immunocytochemical localization of HvAP in the aleurone layer of developing grain. During the maturation phase of development, HvAP disappears from the aleurone layer. However, the HvAP-encoding mRNA remains in the embryo and aleurone layer (Fig. 3). HvAP is present in the embryo and scutellum already at the onset of germination. Marttila et al. (62) have also shown that HvAP is already present in the scutellum and the proximal aleurone layer of ungerminated grain. During the first three days of germination the enzyme reappears in the aleurone layer cells but HvAP is probably not secreted into the starchy endosperm where the bulk hydrolysis of storage proteins occurs (61,62). The function of HvAP in the scutellum and aleurone layer is not known. However, since HvAP has a narrow hydrolytic specificity (58) and it is present in the aleurone layer and scutellum which produce and secrete several hydrolytic enzymes into starchy endosperm during germination, it is tempting to speculate that HvAP would process and activate enzyme precursors before their secretion into endosperm.

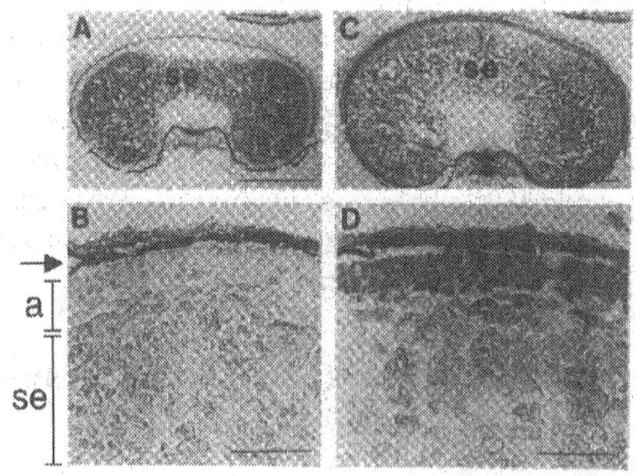

Figure 2. Immunocytochemical localization of HvAP in developing barley grains (15 days after anthesis, cv. Kustaa). Grains were embedded in paraffin and transverse sections were probed with (A,B) preimmuneserum or (C,D) antibody to HvAP, followed by secondary antibody linked to alkaline phosphatase and stained with New Fuchin Substrate System Kit (Daco Corp., Ca, USA). The boxed sections in A and C are magnified in B and D, respectively. se = starchy endosperm, a = aleurone layer. Arrow → testa. Scale bar = 1 mm (A and C) and 0.1 mm (B and D).

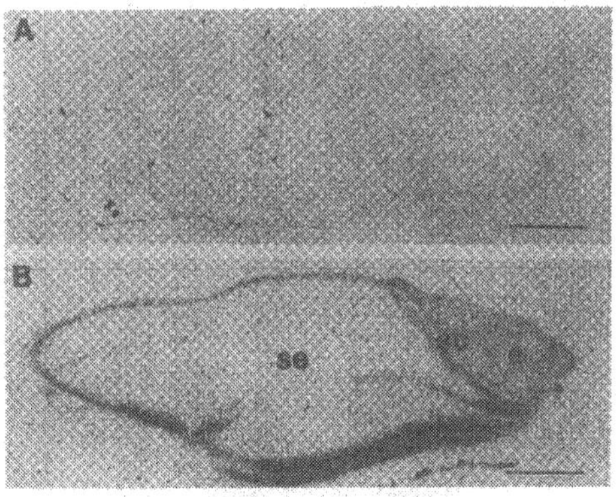

Figure 3. Localization of HvAP in developing barley grains (24 days after anthesis, cv. Kustaa) by *in situ* hybridization analysis. The 353 bp *Eco*RI-*Bam*HI fragment of HvAP-encoding cDNA, cloned in sense and antisense orientations under T7 promoter, was used as a probe. Grains were embedded in paraffin and longitudinal sections were treated with (A) sense or (B) antisense probes. The signal was detected by X-ray film autoradiography. e = embryo, sc = scutellum, se = starchy endosperm. Scale bar = 1 mm.

HvAP Is a Vacuolar Enzyme that Processes Barley Prolectin *in Vitro*

We have shown both by biochemical means and by immunocytochemistry that HvAP is localized to the vacuoles of barley leaf and root cells (56). First, we demonstrated that HvAP can be translocated into the lumen of microsomes with the concomitant loss of 25 amino acid residues at the N-terminus, the ER-signal sequence. Second, we isolated vacuoles from barley leaf protoplasts and showed that the vacuoles contained all of the aspartic proteinase activity found in the protoplasts. Third, we detected the HvAP protein exclusively in vacuoles by immunocytochemical electron microscopy of barley leaf and root cells (56). Based on the vacuolar localization and the previously reported hydrolytic specificity of HvAP (58), we hypothesized that the hydrophobic C-terminal vacuolar targeting signal of barley prolectin might be cleaved off by HvAP. Barley lectin is a 36 kDa homodimeric protein, which is localized in the vacuoles of root cap cells and embryonic root cells. It harbours a C-terminal proregion which has been shown to be necessary and sufficient for vacuolar targeting in tobacco. This targeting signal is missing from the mature vacuolar lectin. We tested our hypothesis in collaboration with Prof. Natasha Raikhel's group (Michigan State University, USA). Prolectin, expressed in bacteria, was used for *in vitro* hydrolysis with purified HvAP. Matrix Assisted Laser Desorption Mass Spectrometry, Western blotting and N-terminal sequencing show that 13 amino acid residues present at the C-terminal of prolectin were cleaved off whereas the N-terminus stayed intact during hydrolysis (Fig. 4). However, prolectin also seemed to be processed by carboxypeptidase Y *in vitro*. As immunocytochemical electronmicroscopy showed that HvAP and barley lectin *in vivo* are colocalized in the root cells of developing embryos and germinating seedlings, we propose that the vacuolar HvAP participates in the processing of the C-terminal vacuolar targeting signal of barley prolectin (56).

Computer Based Modelling of HvAP

A three-dimensional model of HvAP has been constructed using the rule based comparative approach encoded in the COMPOSER suite of computer programs (63). The model was based on the three-dimensional structures of the homologous pepsin, chymosin, renin, yeast proteinase A, human cathepsin D and other aspartic proteinases previously solved by high resolution X-ray

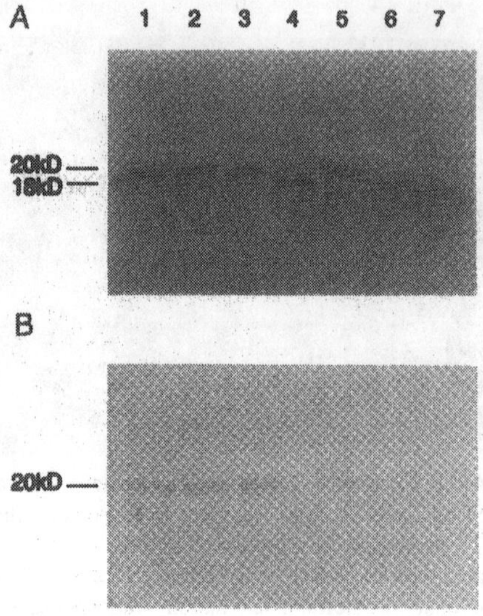

Figure 4. (A) Coomassie blue staining of prolectin digested with HvAP at 37°C. Lane 1, prolectin. Lane 2, prolectin incubated overnight in buffer. Lanes 3 and 5, prolectin incubated 5 and 15 min, respectively, with HvAP and pepstatin. Lanes 4 and 6, prolectin incubated 5 and 15 min, respectively, with HvAP. Lane 7, prolectin incubated overnight with HvAP. (B) Western blotting of prolectin digested with HvAP. The monoclonal antibody used, specifically recognises the C-terminal peptide in prolectin. Lanes as in Figure 4A.

analysis. A schematic representation of the model showing the overall topology and the arrangement of beta-strands and alpha-helixes is shown in Fig. 5. The site of barley specific insert is also indicated. In order to define the residues involved in the sub-site specificity, the inhibitor CH-66 (trimethylacetyl-His-Pro-Phe-His-Leu-ψ(CH(OH)CH_2)-Leu-Tyr-Tyr-Ser-NH_2) was modelled into the active site of HvAP using the coordinates of the inhibitor complex of mouse renin as a guide. The substrate Lys-Pro-Ile-Glu-Phe-Phe(4-NO_2)-Arg-Leu was also modelled into the active site of the crystal structures of porcine pepsin, yeast proteinase A, human cathepsin D and in the models of human cathepsin E and the barley grain aspartic proteinase, using the coordinates of the pepsin inhibitor (H256) complexed with the endothiapepsin as a guide, in order to provide a structural rationale for the observed specificity in these enzymes. The modelling work suggests that S3 and S5 pockets of HvAP are relatively large compared to those in mouse renin. The presence of Arg at position 295 is unique to the barley grain aspartic proteinase and the presence of a bulky charged group at this position may be involved in hydrogen bond interactions with the substrate stabilizing the complex of HvAP with the substrate. The 104 residues of the plant specific insert of HvAP were not included in the modelling.

CONCLUDING REMARKS

Many structural and catalytic properties of plant aspartic proteinases, such as acidic pH optimum, hydrolytic specificity and inhibition by pepstatin, are similar to the corresponding enzymes in animals and microbes. Sequence based three-dimensional modelling of the barley aspartic proteinase also shows that in spite of enzyme's different processing pattern, its chain folding closely resembles that of the other aspartic proteinases. All these features support the hypothesis that eukaryotic aspartic proteinases have evolved from the same origin. However, all three solved plant primary sequences (from barley, rice and cardoon)

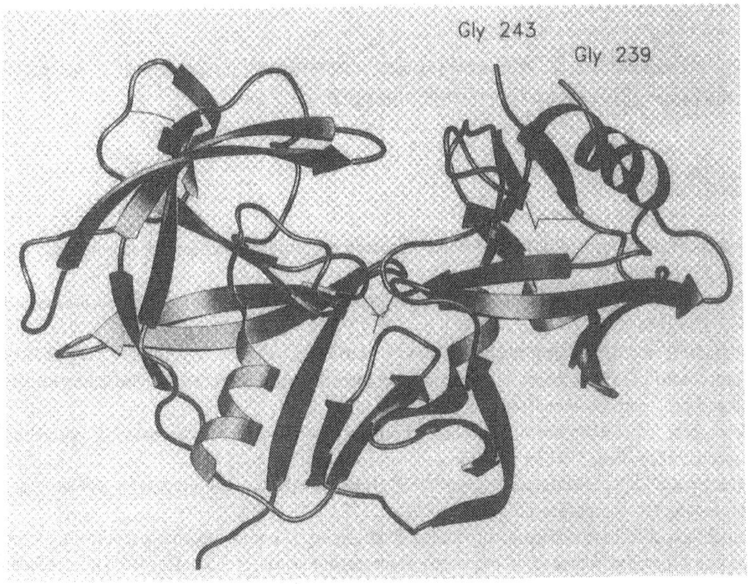

Figure 5. Schematic representation showing overall 'fold' of HvAP. Helices are shown as cylinders and sheets as flat ribbons. The remaining polypeptide is shown as a rope that represents the turn/coil regions in the molecule. The side chains of Thr_{33} and of the plant specific Ser_{216} in the second active site are shown. The three disulphide bridges are also indicated. The region of the plant-specific 'large insertion' is between Gly_{239} and Gly_{243} (not illustrated).

contain an internal unique region of about 100 amino acid residues carrying no similarity to other known aspartic proteinases. The C-terminal lobe of these three plant aspartic proteinases also contains Asp-Ser-Gly instead of the more common Asp-Thr-Gly in the active site. The evolutionary or biological importance of these specific features in plant enzymes is not known. Aspartic proteinases are also produced from plant pararetroviruses such as CaMV (64). Putative aspartic proteinase is encoded as well by plant retrotransposons such as the tobacco *Tnt*1 (65) and barley *BARE*-1 (66). Since aspartic proteinase is necessary for retrotransposon and retroviral activity (67,68), it will be interesting to see if the plant retrotransposon-encode aspartic proteinases are functional.

Resting seeds of many plants contain aspartic proteinase activity and several studies have been undertaken to elucidate the role of these enzymes in the initial hydrolysis of storage proteins at the onset of seed germination. However, so far no convincing evidence has been presented that aspartic proteinases are involved in the degradation of storage proteins during germination *in vivo*. In addition, aspartic proteinases have been found from other plant tissues. It is obvious that plant aspartic proteinases have a wide localization in diverse tissues in the plant kingdom and they seemingly have several functions appropriate to the needs of different tissues during the life cycle of plants. So far, it is not known if plants harbour only one type of aspartic proteinase which is differentially expressed or whether several aspartic proteinases are present in a similar manner as in animals. If the endogenous aspartic proteinase inhibitors, so far detected only in potato (69-71) and tomato (72), have a role in the regulation is not known either. Much remains to be learned about the structure and biological functions of the plant aspartic proteinases as well as the regulation of the activity of the individual enzymes in different tissues. What the natural substrates are and how these proteins are selectively recognized and hydrolyzed in plant cells are challenging questions waiting for their answers.

ACKNOWLEDGMENTS

We are grateful to all our colleagues in different laboratories who have provided (p)reprints and have allowed us to cite their unpublished results.

REFERENCES

1. J. Tang and R.N.S. Wong, Evolution in the structure and function of aspartic proteases, *J. Cell. Biochem.* 33:53 (1987).
2. D.R. Davies, The structure and function of the aspartic proteinases, *Annu. Rev. Biophys. Biophys. Chem.* 19:189 (1990).
3. P.B. Szecsi, ed., The aspartic proteases. *Scand. J. Clin. Lab. Invest.* 52:(Suppl. 210) (1992).
4. Y. Kawamura and D. Yonezawa, Wheat flour proteases and their action on gluten proteins in dilute acetic acid, *Agric. Biol. Chem.* 46:767 (1982)
5. P.C. Morris, R.C. Miller, and D.J. Bowles, Endopeptidase activity in dry harvest-ripe wheat and barley grains, *Plant Sci.* 39:121 (1985).
6. M.A. Belozersky, Sh.T. Sarbakanova, and Ya.E. Dunaevsky, Aspartic proteinase from wheat seeds: isolation, properties and action on gliadin, *Planta* 177:321 (1989).
7. Y.E. Dunaevsky, S.T. Sarbakanova, and M.A. Belozersky, Wheat seed carboxypeptidase and joint action on gliadin of proteases from dry and germinating seeds, *J. Exp. Bot.* 40:1323 (1989).
8. L. Galleschi and U. Andreoni, A rapid and sensitive method of assaying the purified aspartic proteinases in cereals, *Plant Physiol. Biochem.* 28:793 (1990).
9. G.J.T. Frith, K.H.J. Gordon, and M.J. Dalling, Proteolytic enzymes in green wheat leaves I. Isolation on DEAE-cellulose of several proteinases with acid pH optima, *Plant and Cell Physiol.* 19:491 (1978).
10. G.J.T. Frith, M.B. Peoples, and M.J. Dalling, Proteolytic enzymes in green wheat leaves III. Inactivation of acid proteinase II by diazoacetyl-DL-norleucine methyl ester and 1,2-epoxy-3-(*p*-nitrophenoxy)-propane, *Plant and Cell Physiol.* 19:819 (1978).
11. G.J.T. Frith, L.B. Swinden, and M.J. Dalling, Proteolytic enzymes in green wheat leaves II. Purification by affinity chromatography, and some properties of proteinases with acid pH optima, *Plant and Cell Physiol.* 19:1029 (1978).
12. L. Galleschi, A. Capocchi, P. Giannoni, and C. Floris, Proteinase activities in quiescent and germinating seeds of *xHaynaldoticum sardoum*, *Physiol. Plant.* 75:1 (1989).
13. E. Doi, D. Shibata, T. Matoba, and D. Yonezawa, Characterization of pepstatin-sensitive acid protease in resting rice seeds, *Agric. Biol. Chem.* 44:741 (1980).
14. H. Hashimoto, R. Nishi, H. Uchimiya, and A. Kato, Nucleotide sequence of a cDNA encoding aspartic proteinase in rice, Accession number D12777, submitted to the DNA database of Japan (August 1, 1992).
15. G.K. Garg and T.K. Virupaksha, Acid protease from germinated sorghum. 1. Purification and characterization of the enzyme, *Eur. J. Biochem.* 17:4 (1970).
16. G.K. Garg and T.K. Virupaksha, Acid protease from germinated sorghum. 2. Substrate specificity with synthetic peptides and ribonuclease A, *Eur. J. Biochem.* 17:13 (1970).
17. B. San Segundo, J.M. Casacuberta, and P. Puigdomènech, Sequential expression and differential hormonal regulation of proteolytic activities during germination in *Zea mays* L., *Planta* 181:467 (1990).
18. U. Vidyavathi, B. Shivaraj, and T.N. Pattabiraman, Proteases in germinating finger millet (*Eleusine coracana*) seeds, *J. Biosci.* 5:219 (1983).
19. A.J. St. Angelo, R.L. Ory, and H.J. Hansen, Localization of an acid proteinase in hempseed, *Phytochemistry* 8:1135 (1969).
20. J.D. Bewley and M. Black, eds., "Seeds - Physiology of development and germination," p. 253-303, Plenum Press, New York (1986).
21. A.J. St. Angelo, R.L. Ory, and H.J. Hansen, Purification of acid proteinase from *Cannabis sativa* L., *Phytochemistry* 8:1873 (1969).
22. A.J. St. Angelo and R.L. Ory, Properties of a purified proteinase from hempseed, *Phytochemistry* 9:1933 (1970).
23. M.A. Belozerskii, Ya.E. Dunaevskii, G.N. Rudenskaya, and V.M. Stepanov, Carboxyl proteinases from buckwheat seeds, *Biochemistry (USSR)* 49:401 (1984).

24. E.N. Elpidina, Y.E. Dunaevsky, and M.A. Belozersky, Protein bodies from buckwheat seed cotyledons: isolation and characteristics, *J. Exp. Bot.* 41:969 (1990).
25. S. Shinano and K. Fukushima, Studies on lotus protease. Part II. Purification and some properties, *Agric. Biol. Chem.* 33:1236 (1969).
26. S. Shinano and K. Fukushima, Studies on lotus protease. Part III. Some physicochemical and enzymic properties, *Agric. Biol. Chem.* 35:1488 (1971).
27. B.E. Juniper, R.J. Robins, and D.M. Joel, eds., "The carnivorous plants," p. 189-207, Academic Press, London (1989).
28. S. Amagase, Digestive enzymes in insectivorous plants III. Acid proteases in the genus *Nepenthes* and *Drosera peltata*, *J. Biochem.* 72:73 (1972).
29. K. Takahashi, W.-J. Chang, and J.-S. Ko, Specific inhibition of acid proteases from brain, kidney, skeletal muscle, and insectivorous plants by diazoacetyl-DL-norleucine methyl ester and by pepstatin, *J. Biochem.* 76:897 (1974).
30. Z.A. Tökés, W.C. Woon, and S.M. Chambers, Digestive enzymes secreted by the carnivorous plant *Nepenthes macferlanei* L., *Planta* 119:39 (1974).
31. S. Amagase, S. Nakayama, and A. Tsugita, Acid protease in *Nepenthes*. II. Study on the specificity of nepenthesin, *J. Biochem.* 66:431 (1969).
32. A. Polanowski, T. Wilusz, M.K. Kolaczkowska, M. Wieczorek, and A. Wilimowska-Pelc, Purification and characterization of aspartic proteinases from *Cucumis sativus* and *Cucurbita maxima* seeds, in: "Aspartic Proteinases and their Inhibitors," V. Kostka, ed., p. 49, Walter de Gruyter, Berlin, New York (1985).
33. D.J. Bowles, Defense-related proteins in higher plants, *Annu. Rev. Biochem.* 59:873 (1990).
34. I. Rodrigo, P. Vera, and V. Conejero, Degradation of tomato pathogenesis-related proteins by an endogenous 37-kDa aspartyl endoproteinase, *Eur. J. Biochem.* 184:663 (1989).
35. I. Rodrigo, P. Vera, L.C. Van Loon, and V. Conejero, Degradation of tobacco pathogenesis-related proteins. Plant Physiol., *Plant Physiol.* 95:616 (1991).
36. U. Heimgartner, M. Pietrzak, R. Geertsen, P. Brodelius, A.C. da Silva Figueiredo, and M.S.S. Pais, Purification and partial characterization of milk clotting proteases from flowers of *Cynara cardunculus*, *Phytochemistry* 29:1405 (1990).
37. C. Faro, P. Veríssimo, Y. Lin, J. Tang, and E. Pires, Cardosin A and B, aspartic proteinases from *Cynara cardunculus* L., The 5th International Conference on Aspartic Proteinases, Gifu, Japan, September 19-24, 1993, abstract book, P31.
38. M. Cordeiro, E. Jakob, Z. Puhan, M.S. Pais, and P.E. Brodelius, Milk clotting and proteolytic activities of purified cynarases from Cynara cardunculus - a comparison to chymosin, *Milchwissenschaft* 47:683 (1992).
39. C.J. Faro, A.J.G. Moir, and E.V. Pires, Specificity of a milk clotting enzyme extracted from the thistle *Cynara cardunculus* L.: action on oxidised insulin and κ-casein, *Biotech. Lett.* 14:841 (1992).
40. I.Q. Macedo, C.J. Faro, and E.M. Pires, Specificity and kinetics of the milk-clotting enzyme from cardoon (*Cynara cardunculus* L.) toward bovine κ-casein, *J. Agric. Food Chem.*, 41:1537 (1993).
41. M. Cordeiro, Z.-T. Xue, M. Pietrzak, M.S. Pais, and P. Brodelius, Plant aspartic proteinases from *Cynara cardunculus* (cardoon). Nucleotide sequence of a cDNA encoding cynarase and it's tissue specific expression, The 5th International Conference on Aspartic Proteinases, Gifu, Japan, September 19-24, 1993, abstract book, P30.
42. K. D'Hondt, D. Bosch, J. Van Damme, M. Goethals, J.Vandekerckhove, and E. Krebbers, An aspartic proteinase present in seeds cleaves *Arabidopsis* 2 S albumin precursors *in vitro*, *J. Biol. Chem.* 268:20884 (1993).
43. P.-M. Kirchhoff, B. Biehl, and G. Crone, Peculiarity of the accumulation of free amino acids during cocoa fermentation, *Food Chem.* 31:295 (1989).
44. B. Biehl, H. Heinrichs, H. Ziegeler-Berghausen, S. Srivastava, Q. Xiong, D. Passern, V.I. Senyuk, and M. Hammoor, The proteases of ungerminated cocoa seeds and their role in the fermentation process, *Angew. Bot.* 67:59 (1993).
45. J. Voigt, B. Biehl, H. Heinrichs, S. Kamaruddin, G. Gaim Marsoner, and A. Hugi, In-vitro formation of cocoa-specific aroma precursors: aroma-related peptides generated from cocoa-seed protein by co-operation of an aspartic endoprotease and a carboxypeptidase. *Food Chem.* 49:173 (1994).
46. A. Mikkonen, Activities of some peptidases and proteinases in germinating kidney bean, *Phaseolus vulgaris*, *Physiol. Plant.* 68:282 (1986).
47. H.M. Bond and D.J. Bowles, Characterization of soybean endopeptidase activity using exogenous and endogenous substrates, *Plant Physiol.* 72:345 (1983).

48. J.L. García-Martínez and J. Moreno, Proteolysis of ribulose-1,5-bisphosphate carboxylase/oxygenase in *Citrus* leaf extracts, *Physiol. Plant.* 66:377 (1986).
49. M.A. Salmia, S.A. Nyman, and J.J. Mikola, Characterization of the proteinases present in germinating seeds of Scots pine, *Pinus sylvestris*, *Physiol. Plant.* 42:252 (1978).
50. M.A. Salmia, Proteinase activities in resting and germinating seeds of Scots pine, *Pinus sylvestris*, *Physiol. Plant.* 53:39 (1981).
51. J. Bourgeois and L. Malek, Metabolic changes related to the acceleration of jack pine germination by osmotic priming, *Tree Physiol.* 8:407 (1991).
52. J. Bourgeois and L. Malek, Purification and characterization of an aspartyl proteinase from dry jack pine seeds, *Seed Sci. Res.* 1:139 (1991).
53. P. Sarkkinen, N. Kalkkinen, C. Tilgmann, J. Siuro, J. Kervinen, and L. Mikola, Aspartic proteinase from barley grains is related to mammalian lysosomal cathepsin D, *Planta* 186:317 (1992).
54. R. Wrobel and B.L. Jones, Appearance of endoproteolytic enzymes during the germination of barley, *Plant Physiol.* 100:1508 (1992).
55. J. Kervinen, M. Kontturi, and J. Mikola, Changes in the proteinase composition of barley leaves during senescence in field conditions, *Cereal Res. Commun.* 18:191 (1990).
56. P. Runeberg-Roos, J. Kervinen, V. Kovaleva, S. Gal, and N.V. Raikhel, The aspartic proteinase of barley is a vacuolar enzyme that processes probarley lectin in vitro, *Plant Physiol.* 105:321 (1994).
57. P. Runeberg-Roos, K. Törmäkangas, and A. Östman, Primary structure of a barley-grain aspartic proteinase - a plant aspartic proteinase resembling mammalian cathepsin D, *Eur. J. Biochem.* 202:1021 (1991).
58. J. Kervinen, P. Sarkkinen, N. Kalkkinen, L. Mikola, and M. Saarma, Hydrolytic specificity of the barley grain aspartic proteinase, *Phytochemistry* 32:799 (1993).
59. J. Mikola, Proteinases and peptidases in germinating cereal grains, in: "Fourth international symposium on pre-harvest sprouting in cereals," D.J. Mares, ed., p. 463, Westview Press/Boulder, Colorado (1987).
60. T.-M. Enari and T. Sopanen, Mobilisation of endospermal reserves during the germination of barley, *J. Inst. Brew.* 92:25 (1986).
61. K. Törmäkangas, J. Kervinen, A. Östman, and T. Teeri, Tissue-specific localization of aspartic proteinase in developing and germinating barley grains, *Planta* 195:116 (1994).
62. S. Marttila, B.L. Jones, and A. Mikkonen, Differential localization of two acid proteinases in germinating barley (*Hordeum vulgare*) seed, *Physiol. Plant.* 93:317 (1995).
63. K. Guruprasad, K. Törmäkangas, J. Kervinen, and T.L. Blundell, Comparative modelling of barley-grain aspartic proteinase: a structural rationale for observed hydrolytic specificity, *FEBS Lett.* 352:131 (1994).
64. M. Torruella, K. Gordon, and T. Hohn, Cauliflower mosaic virus produces an aspartic proteinase to cleave its polyproteins, *EMBO J.* 8:2819 (1989).
65. M.-A. Grandbastien, A. Spielmann, and M. Caboche, Tnt1, a mobile retroviral-like transposable element of tobacco isolated by plant cell genetics, *Nature* 337:376 (1989).
66. I. Manninen and A.H. Schulman, *BARE-1*, a *copia*-like retroelement in barley (*Hordeum vulgare* L.), *Plant Mol. Biol.* 22:829 (1993).
67. J. Wellink and A. van Kammen, Proteases involved in the processing of viral polyproteins, *Arch. Virol.* 98:1 (1988).
68. S.D. Youngren, J.D. Boeke, N.J. Sanders, and D.J. Garfinkel, Functional organization of the retrotransposon Ty from *Saccharomyces cerevisiae*: Ty protease is required for transposition, *Mol. Cell. Biol.* 8:1421 (1988).
69. M. Mareš, B. Meloun, M. Pavlík, V. Kostka, and M. Baudyš, Primary structure of cathepsin D inhibitor from potatoes and its structure relationship to soybean trypsin inhibitor family, *FEBS Lett.* 251:94 (1989).
70. D.B. Maganja, B. Štrukelj, J. Pungerŭuar, F. Gubenšek, V. Turk, and I. Kregar, Isolation and sequence analysis of the genomic DNA fragment encoding an aspartic proteinase inhibitor homologue from potato (*Solanum tuberosum* L.), *Plant Mol. Biol.* 20:311 (1992).
71. T. Hildmann, M. Ebneth, H. Peña-Cortés, J.J. Sánchez-Serrano, L. Willmitzer, and S. Prat, General roles of abscisic and jasmonic acids in gene activation as a result of mechanical wounding, *Plant Cell* 4:1157 (1992).
72. J.D. Hansen and D.J. Hannapel, A wound-inducible potato proteinase inhibitor gene expressed in non-tuber-bearing species is not sucrose inducible, *Plant Physiol.* 100:164 (1992).

ASPARTIC PROTEINASES (CYPROSINS) FROM *CYNARA CARDUNCULUS* SPP. *FLAVESCENS* CV. CARDOON; PURIFICATION, CHARACTERISATION, AND TISSUE-SPECIFIC EXPRESSION

P.E. Brodelius,[1] M. C. Cordeiro,[2] and M.S. Pais[3]

[1]Department of Plant Biochemistry
Lund University
P.O.Box 7007, S-22007 Lund, Sweden
[2]Departamento de Biologia
Universidade de Evora
Apartado 94, P-7001 Evora Codex, Portugal
[3]Centro de Biotecnologia Vegetal
Faculdade de Ciências de Lisboa
Bloco C2, Campo Grande, P-1700 Lisboa, Portugal

1. INTRODUCTION

Flowers of *C. cardunculus* (cardoon) have been used in Portugal since many centuries to make cheese with raw ewe or/and goat milk. Several types of cheeses are well known and highly appreciated for their taste and quality. Examples are Serra, Serpa and Azeitão. The skills of making these artisanal cheeses have been inherited over generations within the same families and are still used today. Farmers collect the mature flowers or upper part of styles during June and July and store them in dry places to be used for clotting of milk during the next seasons, autumn and winter. Traditionally the cheeses are made mainly with fresh ewe's milk. The proteinases are extracted from flowers with lukewarm water. This extract is filtered through a cloth which contains a spoon of salt. The milk (2-4 l per cheese) is also passed through the cloth and then left for coagulation during 30-45 minutes. The whey is removed by collecting the curd in another piece of cloth, and afterwards it is brought into shape with a mold. The ripening period is around 6 weeks at 12°C (room temperature in some regions during winter) in a humid and aerated environment.

The cardoon plant belongs to the Compositae family and grows wild in stony places and clay soils in the south of Portugal and Madeira and Canary Islands. It can also be found in the southern and western parts of the Mediterranean region and in northern Africa [1]. A new revision of the genus *Cynara* was recently presented and includes cardoon and the artichoke as different cultivars in the same subspecies, i.e. *C. cardunculus* spp. *flavescens* [1]. In Portugal the cardoon cultivar is commonly named "cardo do coalho" which basically means curdling thistle. Many other species in over 13 genera are also commonly referred to as "cardo" (=thistle) in Portugal [2,3]. This name has been given to plants having hard needles on their leaves and receptacle bracts.

In total eight different species are described for this genus. In the revision [1], only *C. cardunculus* is referred to be used in cheese making. However, *C. humilis* and *C. scolymus* (now *C. cardunculus* spp. *flavescens* cv. artichoke) have been found to possess clotting activity. In Spain flowers of *C. humilis* are used to make various cheeses. Examples are Serena, Torta del Casar, Pedroches and Grazalema. Outside the genus *Cynara* some other Compositae species have been used to clot milk. However, these plants generally show a much lower clotting activity. That is the case for *Centaurea calcitrapa* [Domingos *et al.*, unpublished], *Sylibum marianum* [4] and *Onopordum turcicum* [5].

The morphology of the flowers within the genus *Cynara* is very similar and therefore often a mixture of flowers from different species is used for cheese making. This together with differences in developmental stage at which flowers are collected, contributes to the variations in taste and quality often found in cheeses made with this clotting agent. To avoid problems of this kind, it is necessary to standardise the enzyme mixture used and in this way improve the quality of these artisanal cheeses.

During recent years we have been involved in studies of the proteolytic enzymes (cyprosins; previously called cynarases) present in flowers of *C. cardunculus* spp. *flavescense* cv. cardoon. A review of our results so far obtained will be presented here.

2. EXTRACTION AND PURIFICATION OF CYPROSINS

Proteins were extracted from dried or fresh flowers of *C. cardunculus* using 50 mM Tris-HCl buffer, pH 8.3. No significant difference has been observed for enzymes prepared from dried, fresh or senescent flowers. However, slight variations in the different forms of cyprosin (see below) have been observed between flowers collected at different locations and even between plants at a specific location. Furthermore, we have observed, as reported for clotting activity already in 1935 by Christen and Virasoro [6], that the enzyme is present in the violet part of the flower in high amounts while essentially no enzyme could be found in the white parts of flowers [7,8].

Three heterodimeric aspartic proteinases (cyprosin 1, 2 and 3) with milk-clotting activity have been purified using ammonium sulphate fractionation (30-80% saturation) and ion-exchange chromatography (DEAE-Sepharose and MonoQ) [7]. SDS-PAGE of the three purified cyprosins is shown in Figure 1.

3. CHARACTERISATION OF PURIFIED CYPROSINS

3.1. Molecular Weights

The subunit sizes as estimated by SDS-PAGE are 32.5 and 16.5 kD, 33.5 and 16.5 kD, and 35.5 and 13.5 kD for cyprosin 1, 2 and 3, respectively (Figure 1). A native M_r of around 49 kD was determined for the three cyprosins by native PAGE. Gel filtration chromatography indicated somewhat lower native M_rs, i.e., 41, 42 and 45 kD for cyprosin 1, 2 and 3, respectively.

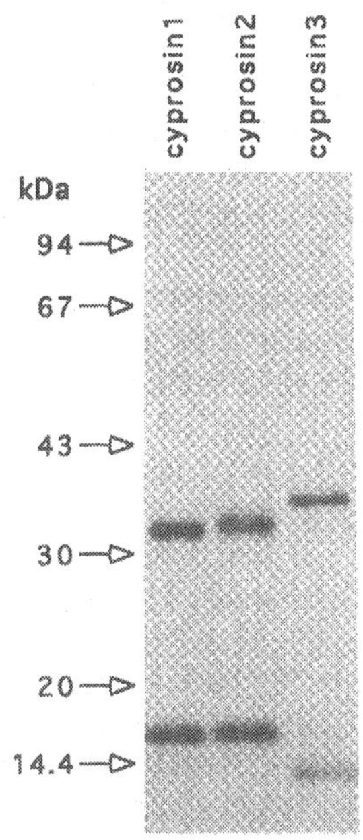

Figure 1. SDS-PAGE of purified cyprosins from *C. cardunculus.*

3.2. Immunological Properties

Polyclonal antibodies (rabbit), raised against the large subunit of cyprosin 3, react with the large subunits of the other two cyprosins indicating some common features of the three forms of the enzyme. Consequently, all three enzymes may be visualised on western blots using these antibodies.

3.3. Isoelectric Points

By two dimensional electrophoresis (IEF and SDS-PAGE) the apparently pure cyprosins 2 and 3 could each be separated into three isozymes [8]. The isoelectric points (3.85, 4.00 and 4.15) were the same for cyprosin 2 and cyprosin 3. Immunostained western blots of 2-D gels have shown that the three isozymes hybridise with the anti-cyprosin 3 antibodies. Reversed-phase high performance liquid chromatography of electrophoretically purified subunits of cyprosins 2 and 3 has also shown a microheterogeneity [unpublished].

3.4. Peptide Mapping and Protein Microsequencing

From the results summarised above it may be concluded that the natural population of cyprosins in flowers of *C. cardunculus* is very complex. They vary in subunit size and they exist in different isoforms. This complexity complicates detailed studies of the protein. In order to overcome this difficulty preparative SDS-PAGE has been used to prepare the various subunits for detailed analysis. First peptide mapping of cyprosin 2 and 3 was carried out by subjecting the electrophoretically purified subunits to BrCN and enzymatic cleavage [8].

The results obtained indicate that the various forms of cyprosins present in flowers of cardoon have common structural features and that they may be derived from common procyprosins. The occurrence of heterodimeric forms of the enzyme containing subunits of different sizes may be due to proteolytic cleavage of a procyprosin at different sites originating a group of closely related enzymes with slight differences in their catalytic properties. The occurrence of isozymes may be due to the expression of a gene family as discussed below (section 6).

An internal partial N-terminal sequence was obtained from a BrCN-peptide obtained by cleavage of cyprosin 2 [8]. This internal sequence (met-leu-asn-gln-gly-leu-val-gln-glu) has been of great importance for the identification of a cDNA clone coding for the cyprosin [9].

3.5. Glycosylation

Cyprosins may be adsorbed to a ConA-Sepharose column indicating that they are glycoproteins containing high mannose type glucans. In fact, endoglucosidase H treatment of cyprosins and subsequent SDS-PAGE has shown that all three forms of the enzyme contain high mannose type glucans [7]. We also have indications that a second glucan not removed by endoglucosidase H is bound to the cyprosins. This assumption is also supported by the fact that the deduced amino acid sequence of cyprosin contains two putative glycosylation sites [9].

3.6. Catalytic Properties

Initially, the proteolytic activity of various enzyme preparations was determined with a standard fluorometric proteolytic assay based on the release of TCA soluble FITC-labelled peptides from casein [7]. Using this assay we could establish that cyprosin 3 showed the highest and cyprosin 1 the lowest specific activity. Furthermore, we could establish that the cyprosins belong to the aspartic proteinase family. Addition of pepstatin to the assay mixture inhibited the activity of all three cyprosins almost completely while other protease inhibitors such as leupeptin, aprotenin, cystatin and iodoacetamide did not affect the activity to any great extent.

Our studies have recently been concentrated on the most active cyprosin, i.e. cyprosin 3. Preparative isoelectric focusing has been used to isolate relative large amounts of the three isozymes of this cyprosin for further characterisation. Some preliminary studies using four randomly chosen synthetic peptides (KPLEVF*RL, KPLELF*RL, PHLSFF*RIPPK and KYHPF*LTA; F*=p-NO_2-F) have shown that the three isozymes show different catalytic properties. KPLEVF*RL, KPLELF*RL and PHLSFF*RIPPK were all hydrolysed at different rates by the three isozymes while KYHPF*LTA, as expected, was not hydrolysed by any of the isozymes as shown in Figure 2. At present, more extensive studies on substrate specificity and catalytic properties of cyprosin isozymes are in progress.

3.7. pH Optimum of Cyprosins

Initially, the pH-optimum for the three cyprosins was determined using the FITC-casein assay. As casein is insoluble at acidic pH-values these experiments gave a relatively high pH-optimum (5.1) for the three cyprosins [7]. However, recently a synthetic peptide (KPLELF*RL) was used to estimate the pH-optimum of the three cyprosin 3 isozymes. A considerably lower pH-optimum (4.1) was obtained for the three isozymes in these experiments.

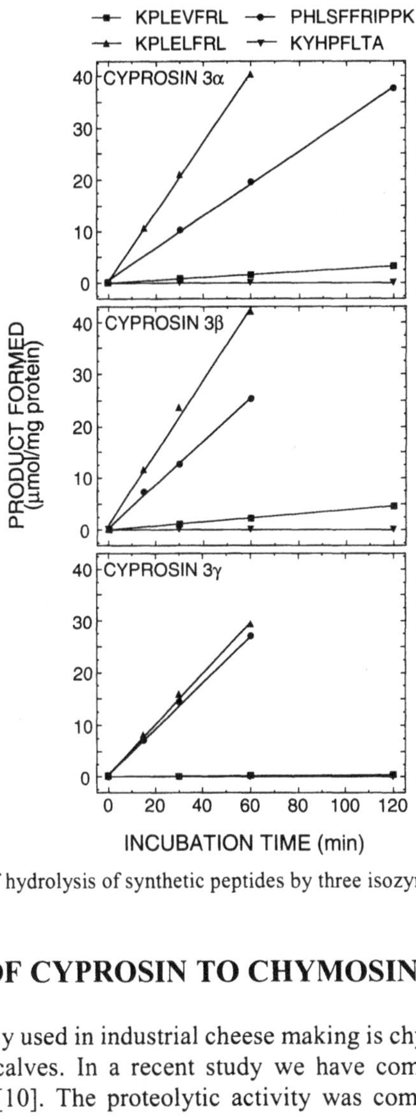

Figure 2. Rate of hydrolysis of synthetic peptides by three isozymes of cyprosin 3.

4. COMPARISON OF CYPROSIN TO CHYMOSIN

The enzyme mostly used in industrial cheese making is chymosin which is extracted from the abomasum of calves. In a recent study we have compared various aspects of cyprosin and chymosin [10]. The proteolytic activity was compared using the standard fluorometric proteolytic assay based on the hydrolysis of FITC-labelled casein or by investigation of the release of Non-Protein-Nitrogen (NPN) from milk. The clotting activity was studied using cow's and ewe's milk. The pattern of breakdown products after hydrolysis of milk proteins was determined by isoelectric focusing.

4.1. Proteolytic Activity

High proteolytic activities were observed for crude extracts of *C. cardunculus* flowers and for purified cyprosins 2 and 3, while chymosin and cyprosin 1 showed relatively low activities using FITC-casein as substrate. The low activity of chymosin is most likely due to the fact that this enzyme exhibits a high specificity towards only one peptide bond (Phe_{105}-Met_{106}) of κ-casein. The higher proteolytic activity measured for cyprosins 2 and 3 indicate a less specific action of these enzymes and the release of a number of peptides. The low activity of cyprosin 1 may simply be due to a less active form of the plant enzyme.

The NPN release has been determined as function of incubation time with milk from cow and ewe. Chymosin, crude extract and cyprosin 3 showed, based on the shape of NPN curves, very similar patterns during the initial phase of the hydrolysis using both types of milk. Afterwards, however, no increase in NPN was observed for chymosin when the incubation time was extended to 18 h while for crude extract and cyprosin 3 a continuous slow increase in NPN was observed. Cyprosins 1 and 2 showed lower NPN values but presented also a slow continuous increase upon extended incubation times. From these measurements it can be concluded that the plant enzymes show some unspecific hydrolysis of milk proteins.

4.2. Milk Clotting Activity

Milk clotting measurements was obtained with the formagraph. In these experiments the clotting times for chymosin and cyprosin 3 were adjusted to approximately the same value. Using 60 µg of cyprosin 3 and 94 µg of chymosin in the clotting assay, clotting times of 8.5 and 7.5 min was observed, respectively. The curd firmness is characterised by the A_{10}-value. This value is slightly lower for cyprosin 3 than for chymosin indicating a softer curd. The clotting times for cyprosins 1 and 2 were 4 and 1 h, respectively, at considerably higher protein concentrations.

Measurements of the clotting activity using milk from ewe showed that cyprosin 3 exhibited a considerably higher activity than chymosin. With addition of 71 µg of chymosin or 45 µg of cyprosin 3 to a clotting assay the clotting times were 22 and 10 min, respectively.

4.3. Isoelectric Focusing of Milk Proteins

For isoelectric focusing of milk proteins the samples used were taken five minutes after the milk was clotted and at a time equal to five times the clotting time for all enzyme preparations. These samples were compared to a sample of untreated milk in order to assess which proteins have been hydrolysed.

The presence of the para κ-casein peptide obtained from κ-casein was detected in all samples (Figure 3) showing that all enzymes cleave κ-casein at the Phe_{105}-Met_{106} bond. Samples taken immediately after clotting and at a later time showed no detectable difference. κ-Casein B appeared to be completely cleaved by the cyprosins. The presence of a weak band after hydrolysis at the position of κ-casein B is due to another protein which is not cleaved by the enzymes. One of the γ-caseins, which are products of hydrolysis of β-casein by the original milk protease (plasmin), is further cleaved by the cyprosins but not, however, by chymosin.

Due to the less specific activity of the cyprosins at least 5 new peptides could be identified in samples treated with these enzymes. These peptides were not present in untreated milk or milk treated with chymosin (Figure 3). The origin of two of these new peptides may be suggested, i.e. new peptides 1 and 5 are products from the hydrolysis of β-casein A^2 and α_{s1}-casein, respectively. The origin of the other three peptides cannot at present be suggested.

Isoelectric focusing of proteins after hydrolysis of ewe's milk in the same way as described above for cow's milk revealed no significant differences in the IEF patterns of the various enzyme preparations indicating a higher hydrolytic specificity of the cyprosins against proteins in ewe's milk.

4.4. Conclusions

In this comparison, purified cyprosin 3 and chymosin revealed similar clotting activities, in particular in the initial phase of the hydrolysis when curd is formed. Slight

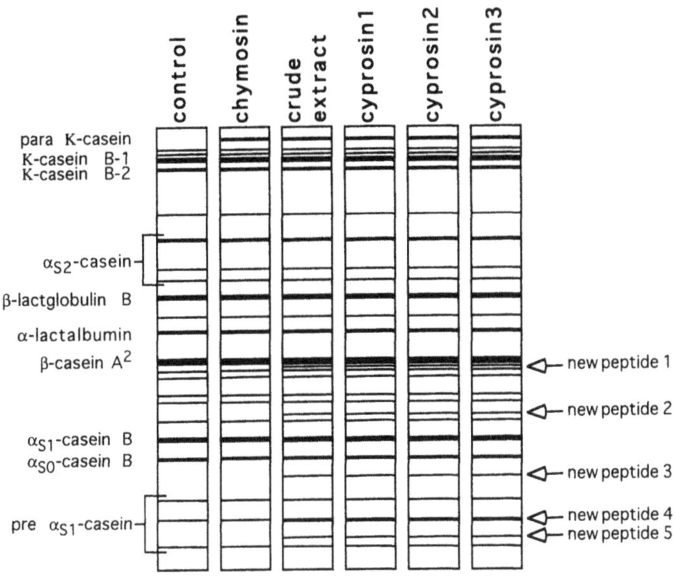

Figure 3. Schematic representation of isoelectric focusing of milk proteins after treatment with various enzyme preparations.

differences in curd firmness were detected which after the ripening period give rise to a softer cheese made with the plant enzyme. NPN and IEF studies with cow milk indicated that the cyprosins cleaved the Phe_{105}-Met_{106} bond of κ-casein which is the only bond hydrolysed by chymosin. However, in addition to this protein, cyprosins also hydrolysed β-casein A^2, α_{s1}-casein and some of the γ-caseins. Cyprosin 3 has shown a considerably higher clotting activity in ewe milk than chymosin, and also a more specific hydrolysis of proteins in this type of milk.

5. PRIMARY STRUCTURE OF CYPROSIN

5.1. Comparison of Cyprosin to Other Aspartic Proteinases

A cDNA clone (*cypro1s*) encoding a cyprosin has been isolated and characterised [9]. The nucleotide sequence contains a 1419 bp open reading frame coding for 473 aa including a putative full length mature protein (439 aa) and a partial prosequence (34 aa). The deduced amino acid sequence of cyprosin shows a relatively high homology to other plant aspartic proteinases; barley (78%) and rice (63%). Compared to other aspartic proteinases the plant enzymes contain an insert of around 100 amino acid (plant specific insert) in the C-terminal part of the protein molecule. Omitting this insert, the homology to mammalian aspartic proteinases is relatively high; human cathepsin D (55%) and bovine chymosin b (43%). A more detailed comparison of cyprosin to other aspartic proteinases may be found in this volume [11].

5.2. Phylogenetic Analysis of Aspartic Proteinases

The amino acid sequence of 34 aspartic proteinases have been used to establish a phylogenetic tree as shown in Figure 4. It is evident from this figure that the aspartic proteinases are found in groups containing related enzymes. For instance, the three plant enzymes are grouped together. The plant enzymes are mostly related to the cathepsin D enzymes. Furthermore, it is interesting to observe that the plant enzymes appear to be more related to vertebrate enzymes than to fungal enzymes. Finally, it is somewhat surprising that the barley (monocotyledon) enzyme appears to be more related to the cyprosin (dicotyledon) than to the rice (monocotyledon) enzyme.

5.3. Prediction of Three Dimensional Structure of Cyprosin

The crystal structure of a number of aspartic proteinases have been solved at high resolution [12,13]. Some examples are porcine pepsin, calf chymosin, rhizopuspepsin, penicillopepsin, endothiapepsin and HIV-1 aspartic proteinase. The crystal structures of these aspartic proteinases are closely related in overall shapes and the tracing of chain foldings even though the amino acid homology is low in some cases. Because of the similarities in primary and tertiary structures, it has been possible to model the tertiary structures of aspartic proteinases by fitting the amino acid sequences onto existing presumably homologous crystal structures. Using this approach it should thus be possible to model at least in part the three dimensional structure of plant aspartic proteinases. It will be interesting to study the influence of the extra loop (plant specific region) on the three dimensional structure of plant aspartic proteinases. If the homologous sequence of cyprosin (i.e. the N-and C-terminal regions) is superimposed on the crystal structures of other aspartic proteinases it is possible with a high probability to determine where the plant specific loop will appear in the structure. Using the primary structure of cyprosin and the crystal structure of rhizopuspepsin or bovine chymosin b, we can predict that the insert will appear on the outer surface of the protein molecule at a position corresponding to asp245/gly246 and tyr243/gly244 in rhizopuspepsin and chymosin, respectively. The function of this insert is not at present known.

6. GENOMIC ORGANISATION OF CYPROSIN GENES

Genomic DNA from young flower tissue of *C. cardunculus* was isolated, digested with restriction endonucleases (*Eco*RI and *Hind*III) and analysed by Southern hybridisation [9]. The *Eco*RI and *Hind*III digests each showed 4-5 strong hybridising bands and several minor bands when the entire 1.7 kb insert of *cypro1s* was used as probe. From these hybridisation patterns it may be suggested that the cyprosin genes are organised as a multi-gene family.

The simultaneous expression of a number of cyprosin genes can explain the heterogeneity observed in the gene product described above. Further studies are in progress to isolate cyprosin genes and to characterise this multi-gene family.

7. TISSUE-SPECIFIC ACCUMULATION OF CYPROSIN

Studies on the expression of cyprosins in plants of *C. cardunculus* were carried out by investigating the proteolytic and milk-clotting activities, as well as, the occurrence of the proteins in various extracts [8], the occurrence of cyprosin mRNA [9] and by immunogold-labelling of appropriate tissues [8].

7.1. Accumulation of Cyprosins in Flowers

Flowers in several stages of development (2 to 50 mm in length), styles and corollas from mature flowers, as well as, seeds and leaves have been used to study the accumulation

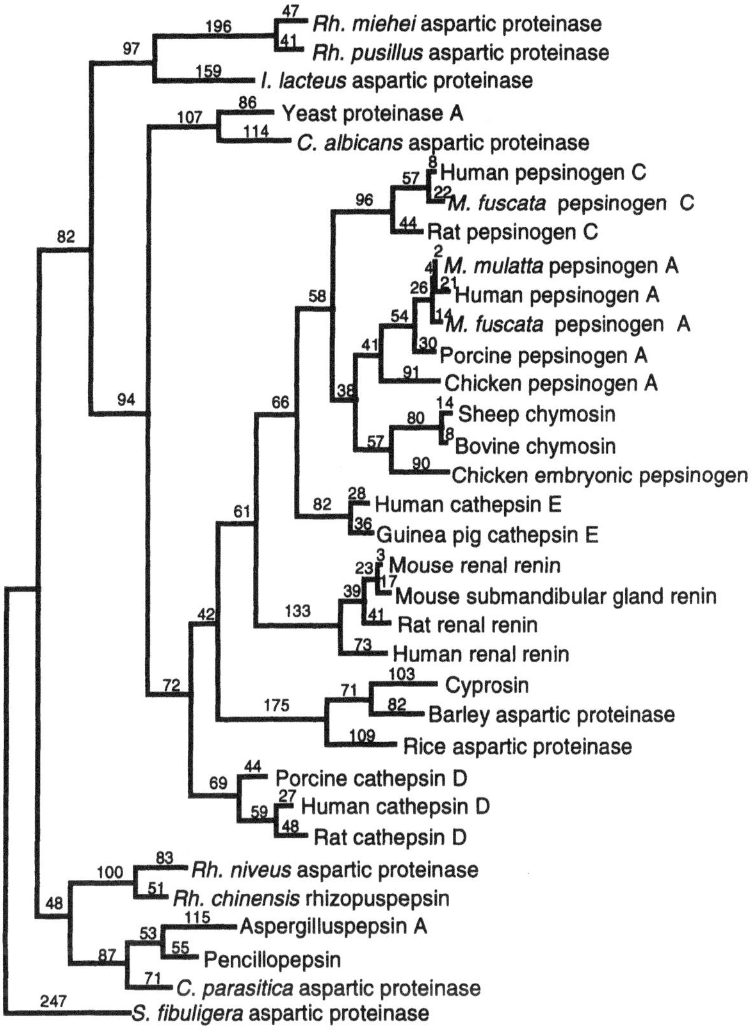

Figure 4. Phylogenetic tree for 34 different aspartic proteinases. The tree was obtained by using the computer programs PILEUP and PAUP.

of cyprosins. Early stages of flower development (2-7 mm) showed relatively low specific proteolytic activity (20 to 30% of mature flowers). During flower development a continuous increase of the proteolytic activity was observed. The highest activities were obtained in mature flowers and in isolated styles and corollas from such flowers. Seeds and leaves showed no significant proteolytic activity.

The earliest stage to show undoubtable clotting activity was flowers of 11 mm length. Extracts from younger flowers (2 to 7 mm) did not clot milk within the assay time (480 min). Along flower development clotting times were reduced due to increased cyprosin activities in extracts of later stages of flower development. The highest clotting activities were found in extracts of styles and corollas isolated from mature flowers. No clotting was observed with extracts made from seeds or leaves.

These results are confirmed by SDS-PAGE and immunoblots. The presence of cyprosin 3 can be detected in very early stages (flowers of 2 mm length) showing that the enzyme is formed already in the beginning of flower development. Cyprosins 1 and 2 are present at a later stage (flowers of 7 mm length).

The intensity of the bands corresponding to the cyprosins, in SDS-PAGE and immunostained western blots, shows an increase along with flower development, revealing that the enzymes are being accumulated in the flowers. The strongest band intensity is observed in samples from styles and corollas. No bands corresponding to the cyprosins can be found in seeds and leaves.

The plant clotting enzymes are specifically expressed in the flowers of cardoon, they have been detected only in the flower organs, and mainly in styles and corollas. Cyprosins are not present in leaves, midrib or seeds. The enzymes are present in an active form in flower tissues at very early stages of flower development (2 mm). Clotting activity is not detected at this stages probably due to the low concentration of cyprosins present in these flowers.

7.2. Isolation of RNA from Plant Tissue for Northern Blot Analysis

Total RNA, isolated from different organs, including several stages of flower development, styles from mature flowers, bract and leaf, was separated by agarose gel electrophoresis, blotted onto a GeneScreen membrane and hybridised with the ^{32}P-labelled *Eco*RI 1.7 kb insert of *cypro1s*. The cDNA clone hybridised to a 1.8 kb mRNA from flower and bract tissues but there was no detectable hybridisation to mRNA from leaves.

The intensity of hybridising transcripts increased from early stages of floral development (flowers 6-10 mm in length) to later stages of floral development (flowers up to 40 mm in length), while in the later stages of floral development (flowers 50 mm in length and styles from open flowers) no hybridisation signal was visualised indicating that the synthesis of mRNA encoding the cyprosin starts in early stages of floral development and switches off at maturation of the flower. These findings are in good agreement with enzyme activity (proteolytic and milk clotting) measured in extracts made from flowers at different stages of development and with the organ-specific expression of cyprosins.

The genes coding for the cyprosins appear to be expressed during flower development. They seem to be specific for the organs present in the flower and also in bracts.

7.3. Immunogold Labelling of Cyprosins

The localisation of cyprosins in flower tissues was performed by immunogold labelling using antibodies raised against the large subunit of cyprosin 3. Studies were carried out with flowers fixed before anthesis at several stages of development.

Silver enhancement of immunogold labelled flower cuts (1-2 μm) allowed the detection of specific labelling of the cyprosins by light microscopy. Figure 5 shows the localisation of cyprosins in transversal cuts made at the intermediate region of flowers (35-40 mm in length). The cardoon proteases appear to be specifically located in the epidermal cell layer of styles. No other tissues from styles and corollas have shown any labelling. The cortex, the transmitting tissue and vascular bundles of styles, as well as, the epidermal cell layer and cortex of corollas did not show any labelling.

The cuts obtained from the basal region of flowers show a less intense labelling when compared to cuts from the intermediate region of the same flowers. These data indicate a differential distribution of cyprosins along the flower, and a higher accumulation of the enzymes in the upper part of the flower.

Also in electron microscopy specific labelling of cyprosin antibodies was detected in the epidermal cell layer of styles. Gold particles (10 nm) are visualised in the hyaloplasm

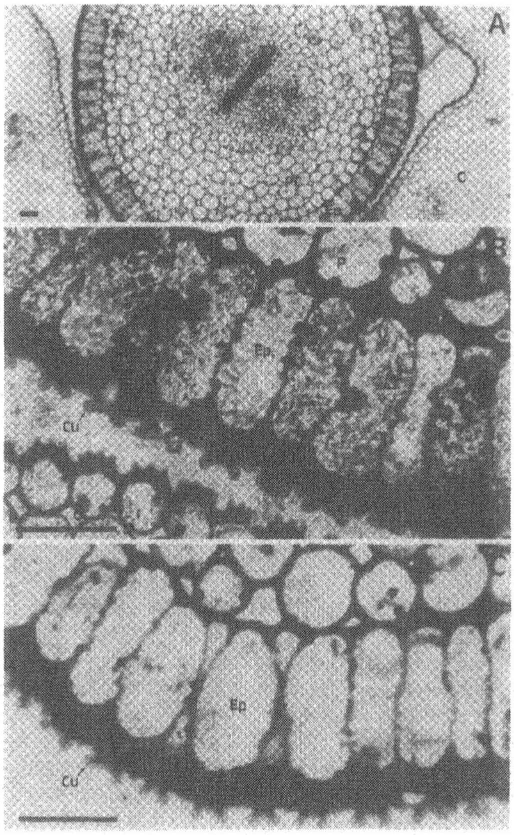

Figure 5. Localisation of cyprosins in styles of *C. cardunculus*. A. Silver enhancement showing the presence of cyprosins in the epidermal cell layer of styles; B. Amplified view of A; C. Amplified view of control. C = corolla; Cu = cuticule; Ep = epidermal layer; PC = parenchymatous cortex; TT = transmitting tissue; VB = vascular bundle. The bar correspondes to 50 μm.

concentrated in electron dense agglomerates disperse all over the cell. They are not concentrated in plastids, mitochondria or the nucleus. The structural preservation used in immunogold labelling techniques does not allow to clearly distinguish between the endoplasmic reticulum, Golgi apparatus, coated vesicles and cytosol, because osmium tetraoxide is not used during this fixation and for this reason cellular membranes are not stained.

Since cyprosins are glycoproteins and appear to contain a signal peptide it is unlikely that a cytosolic form exists. So, their detection in the cytosol can be explained by cyprosins being transiently present either in the endoplasmic reticulum, Golgi apparatus or coated vesicles during synthesis and processing at this stage of flower development (before anthesis). Further studies are in progress to confirm this hypothesis.

8. CONCLUDING REMARKS

A relatively complex mixture of cyprosins accumulate in flowers of *C. cardunculus*. This is most likely due to the expression of a gene family resulting in the existence of different isozymes. These isozymes are further processed to give rise to three different heterodimeric forms of the enzyme by hydrolysis of the peptide chain at different points.

The primary structure of cyprosin shows a relatively high homology to other plant aspartic proteinases. Cyprosin contains a plant specific insert (102 aa) in the C-terminal part of the enzyme.

The enzyme appears to be specifically expressed in the epidermal cell layer of styles. At present, we can only speculate about the function of these enzymes. They may be involved in defence of flowers against pathogens, involved in the senescence process and/or involved in the processing of specific proteins in the flowers.

9. REFERENCES

1. A.F.L.S. Wiklund, *The genus Cynara L. (Asteraceae-Carduae). Bot. J. Linn. Soc.* 109:75 (1992).
2. A.X. Pereira Coutinho, *Flora de Portugal (Plantas Vasculares)*, 2nd Ed. Bertrand (Irmaos) Lda, Lisboa (1939).
3. J.A. Franco, *Nova Flora de Portugal (Continente e Acores)* Vol II, Sociedade Astoria Lda, Lisboa (1984).
4. P. Fevereiro, J.M. Cabral, M.M.R. Fonseca, J.M. Novais, and M.S. Pais, Callus and suspension culture of *Silybum marianum* biosynthesis of proteins with clotting activity. *Biotechnol. Lett.* 8:19 (1986).
5. I.M. Tamer, Identification and partial purification of a novel milk clotting enzyme from Onopordum turcicum. *Biotechnol. Lett.* 15:427 (1993).
6. C. Christen and E. Virasoro. Présure vegetal. Extraction et proprietés. *Le Lait* 15:354 (1935).
7. U. Heimgartner, M. Pietrzak, R. Geertsen, P. Brodelius, A.C. da Silva Figueiredo, and M.S.S. Pais, Purification and partial characterization of milk clotting proteases from flowers of *Cynara cardunculus. Phytochemistry* 29:1405 (1990).
8. M. Cordeiro, M.S. Pais, and P.E. Brodelius, Tissue-specific expression of multiple forms of an aspartic proteinase in flowers of *Cynara cardunculus. Physiol Plant*, 92:645 (1994).
9. M. Cordeiro, Z.-T. Xue, M. Pietrzak, M.S. Pais, and P.E. Brodelius, Isolation and characterization of a cDNA from flowers of *Cynara cardunculus* encoding cyprosin (an aspartic proteinase) and its use to study the organ-specific expression of cyprosin. *Plant Mol. Biol.* 24:733 (1994).
10. M. Cordeiro, E. Jacob, Z. Puhan, M.S. Pais, and P. Brodelius, Milk clotting and proteolytic activities of purified cynarases from *Cynara cardunculus*: A comparison to chymosin. *Milchwissenschaft* 47:683 (1992).
11. M. Cordeiro, Z.-T. Xue, M. Pietrzak, M.S. Pais, and P.E. Brodelius, Plant aspartic proteinases from *Cynara cardunculus* spp. *flavesens* cv. cardoon; Nucleotide sequence of a cDNA encoding cyprosin and it's organ-specific expression. In: *Aspartic Proteinases: Structure, Function, Biology, and Biomedical Implications.* (K. Takahashi, ed.) p. 367, Plenum Press, New York, (1994).
12. J. Tang and R.N.S. Wong, Evolution in the structure and function of aspartic proteases. *J. Cell. Biochem.* 33:53 (1987).
13. A. Sali, B. Veerapandian, J.B. Cooper, D.S. Moss, T. Hofmann, and T.L. Blundell Domain flexibility in aspartic proteinases. *Proteins* 12:158 (1992).

ACID-ACTIVATION OF RAT PRORENIN FOLLOWING NON-PROTEOLYTIC ALTERATION

F. Suzuki, N. Tanaka, K. Takeuchi, Y. Muramoto, T. Inagami,[1] K. Murakami,[2] and Y. Nakamura

Department of Biotechnology
Gifu University
Gifu 501-11, Japan
[1]Department of Biochemistry
School of Medicine
Vanderbilt University
Nashville, TN 37232
[2]Institute of Applied Biochemistry
University of Tsukuba
Ibaraki 305, Japan

INTRODUCTION

Prorenin is an inactive precursor of renin (EC 3.4.23.15) which plays an important role in control of blood pressure and electrolyte balance. The inactive prorenin is irreversibly activated by trypsinization. The prorenin level in the plasma has been observed to be 5-10 times higher than that of renin (1-4). This fact has attracted the attention of many investigators. However, the mechanism of its activation or processing has not been clarified except for some reports on activation of human prorenin (2,5-7). It has been reported that human prorenin was activated by acidification as well as trypsinization (1,2,4-7). Recently we have observed that recombinant rat prorenin secreted from the Chinese hamster ovary cells transfected with a vector containing rat pre-prorenin cDNA (8) was also activated by dialysis at an acidic pH of 3.3 at 4°C (9). The amino acid sequence of rat prorenin prosegment, deduced from the cDNA (8), had a candidate scissile peptide bond for an autocatalytic activation by renin, a bond between L33 and L34. (Every amino acid residue for renin molecule was, in this paper, numbered from the N-terminal amino acid residue of pre-prorenin, M1.) Therefore, there is the possibility that rat prorenin might be altered to an intermediate form by the autocatalysis during the acid treatment. In this study, we purified the recombinant rat prorenin to elucidate the biochemical properties of the acid-activated prorenin and to examine whether rat prorenin was auto-activated at acidic pH.

Aspartic Proteinases: Structure, Function, Biology, and Biomedical Implications
Edited by Kenji Takahashi, Plenum Press, New York, 1995

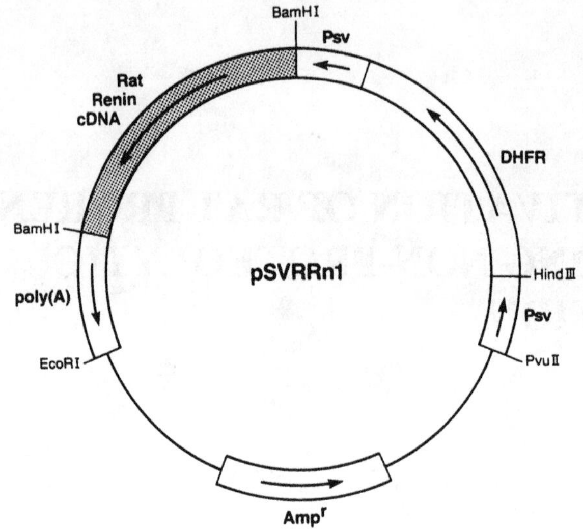

Figure 1. Structure of the expression plasmid pSVDRRn1. Psv: early promoter and replication origin of SV40, DHFR: mouse dihydrofolate reductase gene.

MATERIALS AND METHODS

Chinese hamster ovary cells, which were transfected with an expression vector (pSVRRn1, Fig. 1) containing rat pre-prorenin cDNA (8), secreted prorenin into a serum-free conditioned medium (10). The prorenin secretion level was increase by successive steps of screening and methotrexate treatment. This expression system, accordingly, produced about 2 mg/l of prorenin in a serum-free medium, S-Clone (Sanko Pure Chemical, Japan), for 5 days. Prorenin was acidified by dialysis against 0.1M sodium acetate at an acidic pH, containing 5 mM EDTA and 0.02% sodium azide, or by adjusting pH in the prorenin solution to an acidic pH with the concentrated buffer. The renin activity of acid-activated prorenin was determined by an angiotensin I ELISA system (11) after incubation of prorenins from rat and human (12) with crude preparations of recombinant rat angiotensinogen (13) and native sheep angiotensinogen (14), respectively, under the standard assay conditions (15). The N-terminal amino acid sequence of prorenin was determined by the method of Matsudaira (16) using a protein sequencer (471A with a gradient pump 140B, Applied Biosystems). Immunoblot analysis of prorenin before and after acidification was carried out under the same conditions as described previously (10), using anti-rat prorenin prosegment (G48-V-D-M-T-R-I-S-A-E-Y-G-E-F-I62) antibodies (17) and anti-rat renin antibodies (18).

RESULTS AND DISCUSSION

From 750 ml of culture media 0.58 mg of inactive prorenins (Fr. I and Fr. II) were purified as electrophoretically homogeneous proteins as examined by SDS-PAGE by CM-TOYOPEARL column chromatography and gel filtration on an Ultrogel ACA44 column (Table 1). Their molecular weights were estimated as 47,000 and 45,000, respectively by SDS-PAGE (Fig. 2). Fr. I was used as standard prorenin preparation in this study. The N-terminal amino acid sequence of the prorenin was analyzed to be L-P-T-D-T-A-S-F-G-R-I-L-L-K-K-M-P-S-V-R-. This sequence was identical to that of the residues from L22 to R41

Table 1. Purification of Recombinant Rat Prorenin Secreted from CHO Cells

Step	Total Activity# (mg AngI/h)	Protein (mg)	Specific Activity (mg AngI/mg/h)	Fold	Recovery (%)
Serum-free medium (S-clone)	184	64.5	2.85	1	100
Ultrafiltration	168	21.1	7.96	2.8	91
CM-Toyopearl column					
Fraction I	45.6	1.29	35.3	12.4	25
Fraction II	73.4	2.27	32.3	11.3	40
Ultrogel ACA44 column					
Fraction I	20.3	0.23	88.3	31.0	11
Fraction II	30.7	0.35	87.7	30.8	17

#Prorenin amount in each step was indicated as the renin activity which was measured after prorenin was activated by trypsinization under the same conditions as described previously(9).

deduced from the coding region of rat renin cDNA (9). These results suggested that the signal peptide sequence of rat renin was M1-G-G-R-R-M-P-L-W-A-L-L-L-L-W-T-S-C-S-F-S21. On the other hand, the N-terminal amino acid of the mature rat renal renin has recently been reported to be S72 (19,20). Accordingly, rat prorenin prosegment should localize in the amino acid sequence from L22 to T71 in the preprorenin molecule.

More than 90% of the original amount 0.3 μg of rat prorenin was activated in 24 days by dialysis at pH 3.3 and 4°C as shown in Fig. 3. Immunoblot analysis of rat prorenin showed that the migration distance of prorenin was identical on the SDS-PAGE before and after the acidification (Fig. 4). The N-terminal amino acid sequence of prorenin also did not change after the acidification, although the amino acid sequence of rat prorenin prosegment was reported to have a candidate scissile peptide bond for autocatalytic activation by renin (8). These results show that renin has a remarkably high degree of substrate-specificity, and indicate that a specific processing enzyme is essential for the activation of prorenin into mature renin. This processing mechanism must be different from that for pepsinogen which is completely auto-catalytic, and that for procathepsin D which was recently reported to include a formation of intermediate procathepsin D by autocatalysis (21).

Figure 2. SDS-PAGE of purified rat prorenin. Proteins were stained by Coomassie Blue R250.

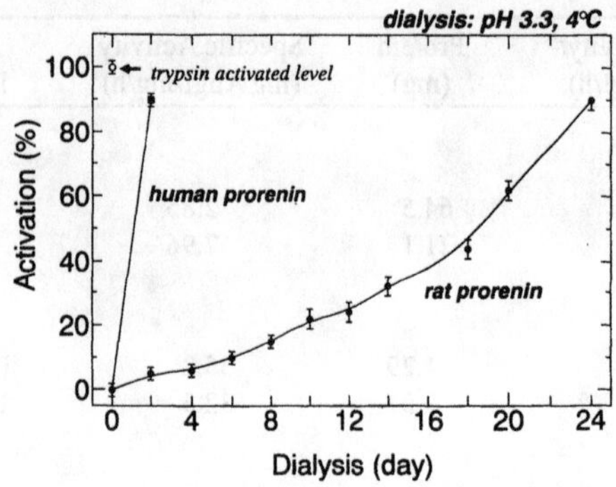

Figure 3. Acid-activation of purified rat prorenin. The purified prorenin solution containing 0.3 μg/ml rat or human prorenin, 0.1% bovine serum albumin, 5mM EDTA and 0.02% sodium azide, was dialysed at pH 3.3 and 4°C. The assay conditions for renin activity were described in the MATERIALS AND METHODS.

The speed of acid-activation of rat prorenin was about 10 times slower than that of human prorenin (Fig. 3). The activated prorenin was re-inactivated by neutralization to pH 6.8. The activation was accelerated by increase of the temperature from 4 to 25°C and of NaCl concentration from 0 to 0.2 M. The 90% of 0.3 μg of rat prorenin was activated in 24 h by the acidification at 25°C. The first-order rate constant for the acid-activation was dependent on the protonation of polar group(s) with pK 3.3. The Km value of the acid-activated prorenin for rat angiotensinogen was 2.4 μM which was slightly lower than that of trypsinized mature renin (22). These results indicate that the prosegment portion in the acid-activated prorenin had a minimum of obstructive effect on the interaction between angiotensinogen and renin molecules.

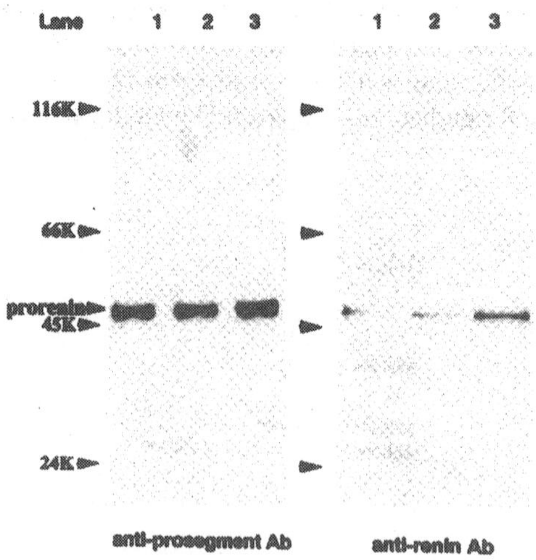

Figure 4. Immunoblot analysis of acidified and neutralised rat prorenin. The purified rat prorenin solution including 0.1% bovine serum albumin and proteinase inhibitors (Lane 1) was acidified at pH 3.3 and 25°C for 24 h (Lane 2), re-neutralised at pH 6.5 and 25°C for 48 h (Lane 3), and then their aliquats were analyzed according to the procedure as described previously (10) using anti-rat prorenin prosegment antibodies (17) and anti-rat renin antibodies (18).

Thus, it was concluded that the prorenin was activated at acidic pH, presumably due to a non-proteolytic alteration in its structure, and that salt-bridges between the portions of the prosegment and the mature renin played an important role in the inactivation of rat prorenin.

ACKNOWLEDGMENTS

This work was supported by Grants-in-Aid for Scientific Research from Ministry of Education, Science and Culture of Japan, Chichibu Cement Co. and Mitsui Norin Co., Ltd.

REFERENCES

1. J.E. Sealey, S.A. Atlas and J.H. Laragh, Prorenin and other large molecular weight forms of renin, Endocr. Rev. 1: 365-391 (1980).
2. W.A. Hsueh, Potential effects of renin activation on the regulation of renin production, Am. J. Physiol. 247: F205-F212 (1984).
3. S. Kim, M. Hosoi, K. Nakajima and K. Yamamoto, Immunological evidence that kidney is primary source of circulating inactive prorenin in rats, Am. J. Physiol. 260: E526-E536 (1991).
4. M. Hosoi, S. Kim, T. Tabata, H. Nishitani, Y. Nishizawa, H. Morii, K. Murakami and K. Yamamoto, Evidence for the presence of differently glycosylated forms of prorenin in the plasma of anephric man, J. Clin. Endocrinol. Metal. 74: 680-684 (1992).
5. F. Derek, M.P.A. Schalekamp and M.A.D.H. Schalekamp, Two-step prorenin-renin conversion. Isolation of an intermediary form of activated prorenin, J. Biol. Chem. 262: 2472-2477 (1987).
6. R.L. Heinrikson, J. Hui, H. Zürcher-Neely and R.A. Poorman, A structure model to explain the partial catalytic activity of human prorenin, Am. J. Hypertens. 2: 367-380 (1989).
7. T. Yamauchi, M. Nagahama, M. Watanabe, Y. Ishizuka, H. Hori and K. Murakami, Site-directed mutagenesis of human prorenin. Substitution of three arginine residues in the propeptide with glutamine residues yields active prorenin, J. Biochem. 107: 27-31 (1990).
8. M. Tada, A. Fukamizu, M.S. Seo, S. Takahashi and K. Murakami, Nucleotide sequence of rat renin cDNA, Nucl. Acid. Res. 16: 3576 (1988).
9. F. Suzuki, A. Takahashi, K. Murakami and Y. Nakamura, Rat renin is activable at acidic pH,Genetic Hypertens. 218: 499-501 (1992).
10. M. Hosoi, S. Kim, T. Yamauchi, T. Watanabe, K. Murakami, F. Suzuki, A. Takahashi, Y. Nakamura and K. Yamamoto, Similarity between physicochemical properties of recombinant rat prorenin and native inactive renin, Biochem. J. 275: 727-731 (1991).
11. F. Suzuki, S. Yamashita, A. Takahashi, M. Ito, S. Miyazaki, Y. Nagata and Y. Nakamura, Highly sensitive microplate-ELISA for angiotensin I using 3,3',5,5'-tetramethylbenzidine, Clin. Exper. Hyper. A12: 83-95 (1990).
12. R.A. Poorman, D.P. Palermo, L.E. Post, K. Murakami, J.H. Kinner, C.W. Smith, I. Reardon and R. Heinrikson, Isolation and characterization of native human renin derived from Chinese hamster ovary cells, Proteins: Struct. Func. Gene. 1: 139-145 (1986).
13. K. Kimoto, N. Iwai, K. Higuchi, T. Hatae, H. Hori, A. Fukamizu, K. Murakami and T. Inagami, Expression and characterization of recombinant rat angiotensinogen derived from Chinese hamster ovary cells, Biomedical Res. 13: 41-46 (1992).
14. F. Suzuki, A. Takahashi, A. Hyodoh, S. Miyazaki, Y. Ishizuka, K. Murakami and Y. Nakamura,A micloplate enzyme-linked immunosorbent assay system for human renin using 3,3',5,5'-tetramethylbenzidine: an application to human renin and prorenin study, Biomed. Res. 13: 321-326 (1992).
15. K. Murakami, F. Suzuki, N. Morita, H. Ito, K. Okamoto, S. Hirose and T. Inagami, High molecular weight renin in stroke-prone spontaneously hypertensive rats, Biochim. Biophys. Acta 622: 115-12 (1980).
16. P. Matsudaira, Sequence from picomole quantities of proteins electroblotted onto polyvinylidene difluoride membranes, J. Biol. Chem. 262: 10035-10038 (1987).

17. F. Suzuki, S. Miyazaki, M. Uozumi, N. Okamoto, S. Yamashita, K. Murakami and Y. Nakamura, Proteinic prorenin-releasing-stimulator (PRS) in the rat submandibular gland, FEBS Lett. 256: 47-50 (1989).
18. Y. Takii, A.F.S. Figueiredo and T. Inagami, Application of immunological methods to the identification and characterization of rat kidney inactive renin, Hypertension 7: 236-243 (1985).
19. S. Kim, M. Hosoi, N. Kikuchi and K. Yamamoto, Amino-terminal amino acid sequence and heterogeneity in glycosylation of rat renal renin, J. Biol. Chem. 266: 7044-7050 (1991).
20. D.J. Campbell, A,J. Valentijin and R. Condron, Purification and amino-terminal sequence of rat kidney renin: evidence for a two-chain structure, J. Hypertens. 9: 29-33 (1991).
21. L.B. Larsen, A. Boisen and T.E. Petersen, Procathepsin D cannot autoactivate to cathepsin D at acid pH FEBS Lett. 319: 54-58 (1993).
22. T. Yamauchi, F. Suzuki, A. Takahashi, I. Tsutsumi, H. Hori, T. Watanabe, Y. Ishizuka, Y. Nakamura and K. Murakami, Expression of rat renin in mammalian cells and its purification,Clin. Exper. Hyper. A14: 377-392 (1992).

HUMAN PROCATHEPSIN D: THREE-DIMENSIONAL MODEL AND ISOLATION

Gerald Koelsch,[1] Peter Metcalf,[2] Vaclav Vetvicka,[3] and Martin Fusek[1*]

[1]Oklahoma Medical Research Foundation
825 NE 13th Street, Oklahoma City, Oklahoma 73104
[2]European Molecular Biology Organization
Mayerhofstrasse 1, D-6900, Heidelberg 1, Germany
[3]Department of Microbiology and Immunology
University of Louisville
Louisville, KY 50292

ABSTRACT

Human procathepsin D was isolated from medium of human breast cancer cell line ZR-75-1 potentiated with estrogen. The isolation involved both immunoaffinity chromatography and ion-exchange chromatography. The affinity chromatography employed polyclonal antibodies raised against a synthetic activation peptide of human cathepsin D. We have started preliminary crystallization trials using the isolated material. A model of human procathepsin D was also built using coordinates of human cathepsin D and pig pepsinogen. The model aids understanding of multiple roles played by activation peptides of aspartic proteinases and will be used as a starting model for molecular replacement.

INTRODUCTION

Cathepsin D (E.C.3.4.23.5) is one of the major lysosomal proteinases[1]. In addition to this function it was found to be related to processes of antigen processing[2], to neurodegenerative changes[3], and in the case of several human breast cancer cell lines procathepsin D is secreted and has an autocrine mitogenic effect[4]. The main route of procathepsin D (PCD) targeting is the mannose-6-phosphate (M6P) targeting mechanism[5]. It is proposed that under physiological conditions, most of the newly synthesized procathepsin D is localized into intracellular acidic compartments by the action of M6P targeting mechanism. Subsequently the activation peptide of 44 amino acids and a surface loop are proteolytically excised[6].

*Address correspondence to Dr. Fusek.

```
     1P         10P        20P           30P
     |          |          |             |
PCD  LVRIPL.HKFTSIRRTMSEVGGSVEDL.IA..K.GPVSKY.SQ...AV..
                                                     *
PGN  LVKVPLVRK.KSLRQNLIK.NGKLKDF.LKTHKHNPASKYFPE.AAALI.

PEP4 .AKV...HKAKIYKHELSD...EMKEVTFEQHLAHLGQKYLTQFEKANPE


                40P
                |   *
PCD  .......PAVTEGPIPEVLKNY

PGN  ..............GDEPLENY
                      *
PEP4 VVFSREHPFFTEGGHDVPLTNY
```

Figure 1. Sequence alignment[25] of procathepsin D (PCD), pig pepsinogen (PGN) and the zymogen of yeast proteinase A (PEP4). Amino acids for PCD are numbered; mature N-termini of activated enzymes are noted with a * above the amino acid. Region of similarity between PCD and PEP4 is underlined.

The three-dimensional structure of PCD comprises multiple functions. First, a signal on the three dimensional structure of PCD is recognized by the phosphotransferase, a key enzyme of M6P targeting mechanism[5]. Second, there are multiple reports on M6P- independent association of PCD with intracellular membranes which seems to be exclusive for PCD[7-10]. Also the alternative (M6P- independent) targeting of PCD was proposed to be associated with the zymogen form[11]. Third, PCD functions as a mitogen in some types of human - derived cell lines[4]. To be able to determine the structural reasons for these functions we initiated studies on the three-dimensional structure of procathepsin D.

Abbreviations used : PCD - procathepsin D; PGN - pepsinogen; M6P - mannose-6-phosphate; SDS - sodium laurylsulfate.

MATERIALS AND METHODS

Modelling of PCD Structure

The crystal structures of human cathepsin D[12] and porcine pepsinogen[13] (PGN) were aligned using the program OVERLAP[14]. The activation peptide and first 14 residues of PGN were spliced to cathepsin D at residue 15. The processed loop (residues 90 - 107) was implemented from a model by Tang et al[15] and the "flap region" (residues 65 - 90) was modeled using coordinates from the pepsinogen structure. Based on the sequence alignment (Figure 1) and on secondary structure prediction we replaced the residues of the activation peptide, the flap, inserted residues 19P and 36P-38P (procathepsin D numbering) and removed residues 29P-31P (pepsinogen numbering). The model was constructed and refined with FRODO[16] on an E&S PS390 computer. The activation peptide (excluding Lys 34P, Tyr 35P and Tyr 10), the processed loop, and the flap region were subjected to *in vacuo* simulated

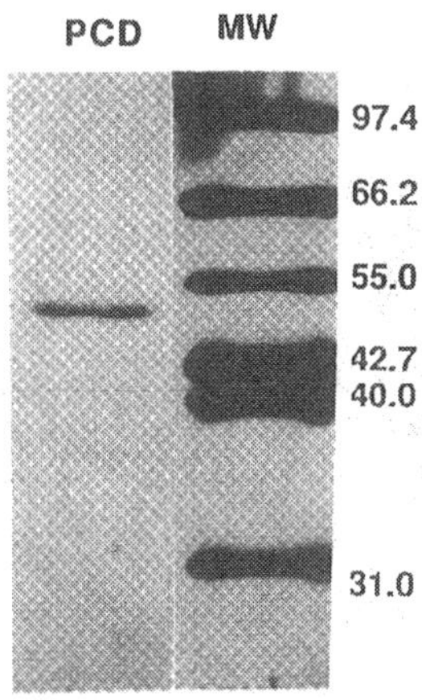

Figure 2. SDS electrophoresis of PCD purified from medium of ZR-75-1 human breast cell line (4% stacking gel, 15% separating gel; stained by silver method).

annealing using XPLOR[17] followed by further refinement of regions structurally different from these of pepsinogen. Static solvent accessibility[18] and ion pairs[19] were assessed.

Isolation of PCD

Human breast cancer cell line ZR-75-1 was grown in roller bottles in RPMI-1640 medium supplemented with 10 % Fetal Clone (Hyclone Laboratories, Logan, UT) and antibiotics. Secretion of procathepsin D was potentiated by the addition of 10^{-8} M b-estradiol (Sigma, St. Louis, MO). The supernatants were collected every 3-4 days and then concentrated by ultrafiltration to 100 ml (20 x concentration). Affinity chromatography. Pig antiserum was raised against the synthetic activation peptide of human PCD[20]. Ten ml of the antiserum was diluted with 20 ml of 20 mM Sodium Phosphate, 150 mM NaCl pH 7.2 (PBS), and then incubated for 1 hour with 2 ml of Protein A Sepharose (Pharmacia, Upsala, Sweden). The resin was washed 10 times with 10 ml of PBS and 5 ml of 0.25 % glutaraldehyde (Sigma) was added and left overnight at 4°C on a rotary shaker. Resin was then washed 10 times with 250 mM Tris-HCl, 150 mM NaCl, pH 7. 5. Ten ml of the concentrated medium from ZR-75-1 cells was diluted with 10 ml of PBS and applied on the 2 ml Protein A - anti-activation peptide-IgG Sepharose column and successively washed with 30 ml of PBS, 30 ml of PBS with 500 mM NaCl, and again with 10 ml of PBS. Bound PCD was eluted by 100 mM glycine buffer pH 3.2 containing 10 mM pepstatin A (Serva, Germany). The pH of the eluent was immediately adjusted to pH 8.0 with 1 M Tris base. FPLC: Combined fractions from elution of PCD from the immunoaffinity chromatography were concentrated using amicon concentrators and dialyzed successively against 25 mM Tris-HCl pH 7.2. The dialysate was applied to a Mono Q (5x5) column and eluted by a gradient of NaCl from 0 to 20 mM during 30 minutes, flow rate 1 ml/ml. The resulting material was homogeneous according to the SDS electrophoresis (Figure 2).

RESULTS

Modelling of PCD Structure

Although the activation peptides of PCD and PGN are the same length the alignment between PCD and PGN (Figure 1) indicated insertions or deletions with a positional

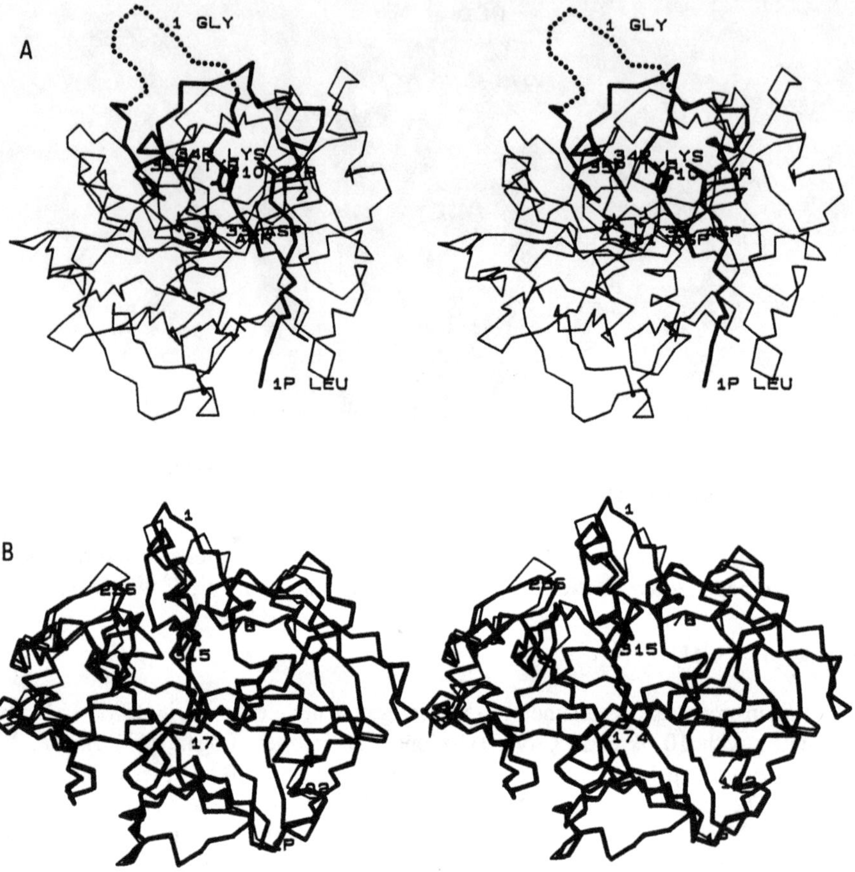

Figure 3. Stereo drawings of the three-dimensional model of PCD. **A**: Alpha carbon trace of PCD. Bold lines are PCD residues 1P - 15; thin lines PCD residues 15-346; dotted lines residues 40P - 5 which are not defined in this model. Residues 33, 231, 34P, 35P and 10 are drawn and labeled to locate the active site; 1P and 1 are labeled also. **B**: PCD model (bold) aligned with PGN. Regions in which the backbone is substantially different from PGN are labeled: 102 (processed region of PCD), 174, 256, and 315. Residues 1P, 1, and 78 (conserved tyrosine of the beta flap) are also labeled.

conservation of Lys34P, Tyr 35P, and Tyr 10 which were expected to have important interactions with the active site. Modelling of PCD residues between Tyr 35P and Tyr 10, namely in the vicinity of the maturation cleavage point, required expansion of the random coil in the PGN backbone.

The starting model was refined using XPLOR to a final RMS of 1.32 Å for 332 common alpha carbon positions of cathepsin D (Figure 3). The simulated annealing process did not alter the secondary structure of the activation peptide from that of the starting model. However, in the vicinity of the cleavage site the random coil retained a conformation accessible to solvent and had not packed to the protein surface. This unexpected result may be due to backbone hindrance as this region is sequentially flanked with three proline residues modelled as *trans*. Therefore further refinement was not expected to alter the conformation of that region and is presented in Fig. 3A with an undefined position.

Ionic interactions of the activation peptide are deemed important to the activation process[13]. Four ionic interactions of the activation peptide of PGN which have conserved titrable sidechains in PCD are conserved in the model. Two additional ion pairs were found in the PCD model which were not present in the PGN crystal structure. The conserved ion pairs occurred in regions which sequentially flank the cleavage site region.

Isolation of PCD

The isolation of PCD from secretions of the ZR-75 produced a contaminant-free PCD and the total yield was 0.3 mg of PCD from 1 l of the starting media. There are two bands corresponding to MW of 52 and 51 kDa respectively present in some preparations. This heterogeneity is caused by different levels of glycosylation (M.Fusek, unpublished data). The isolated PCD had proteolytic activity which was sensitive to the inhibition of pepstatin A (data not shown).

DISCUSSION

Aspartic proteinases are mostly synthesized as zymogens. An activation peptide of approximately 45 amino acids is attached at the amino terminus of mature aspartic proteinases[21]. The function of the activation peptide includes blocking of the active site at high pH, initiation of correct folding of newly synthesized zymogens[22], pH-dependent activation, targeting[18], and possibly other specialized functions. These functions differ among members of the aspartic proteinase family and the specific structure of a zymogen may determine its physiological compartmentalization and function. The three - dimensional structure of zymogens can aid understanding the mechanisms of these functions.

Conservation of ionic interactions in the PCD model and PGN crystal structure indicates titrable residue pairs whose positions should be evolutionarily conserved in order to maintain an acid-activateable function, namely residues Arg 3P and Asp 187, Arg 13P and Asp 11 or Glu 260, and finally Lys 34P and the active site aspartates. Additionally, conservation of residues in pro-aspartic proteinases which interact with the active site aspartates and binding subsites are expected to be conserved; namely Lys 34P, Tyr 35P and Tyr 10.

The region between the activation peptide and the N-terminus of the mature enzyme (mature cleavage site) has high mobility in PGN and is accessible to solvent which is in accord intuitively with requirements for bi-molecular activation. Although the position of the region is not defined in the model it is plausible that it is mobile and solvent accessible and yet very different from pepsinogen. The sequence insertion in that region is justified by the alignment and supported by the conservation of ionic interactions and secondary structure in regions of the polypeptide which sequentially flank the site. Sequence similarity exists in this region between procathepsin D and a zymogen of yeast vacuolar proteinase A (Figure 1). This may be an archetypal motif for interactions with some intracellular or extracellular receptors which is responsible for targeting of proteinase A in yeasts[23] and for association of PCD with mammalian intracellular membranes. On the other hand the model of PCD suggests a conservation of ion pair motifs, active site interactions, and secondary structures of porcine pepsinogen and possibly of other zymogens of the aspartic proteinase family[24].

ACKNOWLEDGMENTS

The authors thank Dr. Jordan Tang and Dr. Steven Foundling for their support.

REFERENCES

1. Barret, A.J. (1970) Biochem. J. 117, 601-607.
2. Factorovich, Y. and Puri, J. (1988) J. Immunol. 141, 3313 - 3317.
3. Whitaker, J.N. and Seyer, J.M. (1979) J. Biol. Chem. 254,6956-6963.
4. Rochefort, H. (1992) Acta Oncol. 31, 125 -130
5. Kornfeld, S. and Mellman, I. (1989) Annu. Rev. Cell Biol. 5, 483-525.
6. Hasilik, A. (1992) Experentia 48, 130-151.
7. Diement, S., Leech, M.S. and Stahl, P.D. (1988) J. Biol. Chem. 263, 6901-6907.
8. McIntyre, G.F. and Erickson, A.H. (1991) J. Biol. Chem. 266, 15438-15445.
9. Rijnbout, S., Aerts, H.M.F.G., Geuze, H.J., Tager, J.M. and Strous, G.J. (1992) J. Biol. Chem. 266, 4862-4868.
10. Wiliams, K.P. and Smith, J.A. (1993) Arch. Biochem. Biophys. 305, 298 - 306.
11. Kornfeld, S. (1987) FASEB J. 1, 462 - 468.
12. Metcalf, P. and Fusek M. (1993) EMBO J. 12, 1293-1302.
13. Hartsuck, J.A., Koesch, G. and Remington, S.J. (1992) Proteins: Structure, Function and Genetics 13, 1-25.
14. Remington, S.J. and Matthews, B.W. (1980) J. Mol. Biol. 140, 77-99.
15. Yonezawa, S., Takahashi, T., Wang, X-J., Wong, R.N.S., Hartsuck, J.A. and Tang, J. (1988) J. Biol. Chem. 263,16504-16511.
16. Jones, T.A. (1978) J. Appl. Crystallogr.11,268-272.
17. Brunger, A.T., Kuriyan, J. and Karpulus, M. (1987) Science 235, 458-460.
18. Lee, B. and Richards, F.M. (1979) J. Mol. Biol. 55, 379-400.
19. Barlow, D.J. and Thornton, J.M. (1983) J. Mol. Biol. 168,867-885
20. Vetvicka, V., Vagner, J., Baudys, M., Tang, J., Foundling, I. S. and Fusek, M., (1993) Biochem. Mol. Biol. Int. 30, 921 - 928.
21. Foltman, B. (1988) Biol. Chem. Hoppe-Seyler 369 Suppl., 311 - 314.
22. van den Hazel, H. B., Kielland-Brandt, M. C. and Winther, J. R. (1993) J. Biol. Chem. 268, 18002 - 18007
23. Klionsky, D. J., Banta, L. M. and Emr, S. D. (1988) Mol. Cell. Biol. 8, 2105 - 2116.
24. Moore, S., Sielecki, A., Tarasive, N., Chernaia, M., Andreeva, N. and James, M., in this issue.
25. Genetics Computer Group (1991), 575 Science Drive, Madison, Wisconsin, USA 53711.

IDENTIFICATION OF FIVE MOLECULAR FORMS OF CATHEPSIN D IN BOVINE MILK

Lotte Bach Larsen and Torben Ellebæk Petersen

Protein Chemistry Laboratory
University of Aarhus
The Science Park, DK-8000 Aarhus, Denmark

INTRODUCTION

Bovine milk contains a wide range of enzymes (1), including zymogens for the proteolytic enzymes plasmin and cathepsin D (2,3). Inside the cell procathepsin D is rapidly processed to the mature cathepsin D found in lysosomes (4). Resulting from the limited availability of procathepsin D the activation process leading to cathepsin D has been less clarified than e.g. the activation pathways of pepsinogen.

Small amounts of procathepsin D have been purified from human cell culture supernatant. It has been observed that the mature human cathepsin D, but not procathepsin D, was able to bind pepstatinyl-Sepharose at pH 5.3, while both enzymes bound at pH 3.5 (5).

Here we show that bovine milk contains five different molecular forms of cathepsin D, of which procathepsin D is the major component. Further, their pH dependent binding to immobilized pepstatin was investigated.

MATERIALS AND METHODS

Ion Exchange and Affinity Chromatography

Cathepsin D forms were purified from a mixture of ammonium sulphate precipitated whey proteins as described earlier (3). The precipitate was redissolved in 50 mM ammonium acetate, pH 5.5, dialysed, and loaded on a column of DEAE-Sepharose equilibrated in the same buffer. Bound proteins were eluted by 50 mM ammonium acetate, pH 5.5, 2 M sodium chloride. Breakthrough fractions were pooled and concentrated to 150-300 ml by ultrafiltration on Amicon using PM10 membranes. A small precipitate was removed by centrifugation. Sodium chloride was added to the supernatant to a final concentration of 0.5 M, and the sample was used for affinity chromatography.

Pepstatinyl-Sepharose affinity chromatography was carried out as described earlier (3) using a 30 minutes batch adsorption at pH 3.5, followed by elution at pH 8.5. Two other sets of pepstatinyl-Sepharose chromatography conditions were also used, methods A and B. All steps were carried out at 4°C.

Method A: The pH of the DEAE-Sepharose pool was adjusted to 3.5. The sample was pumped through a column of pepstatinyl-Sepharose (15 ml), equilibrated in 50 mM sodium acetate, 0.5 M sodium chloride, pH 3.5 (buffer A) with a flow rate of 12 ml/hour. The column was washed with buffer A until the absorbance at 280 nm returned to baseline. A continuous pH gradient was then created using a gradient mixer containing 175 ml buffer A and 175 ml 50 mM Tris, pH 8.5, 0.2 M sodium chloride (buffer B). Finally the column was washed with buffer B.

Method B: This method involves two sequential affinity chromatography steps, and was modified after (5). Both steps were carried out using batch adsorption as described earlier (3), except that the pH during adsorption in the first affinity chromatography was adjusted to 5.4, and the incubation time was 1 hour. The column was packed and the unbound material was used for a second affinity chromatography step, which was carried out using batch adsorption overnight after adjustment to pH 3.5. Bound enzymes were eluted by buffer B.

Electroblotting for Immunovisualization and Sequence Analysis

Proteins (non-reduced) were separated by SDS-PAGE on 10-20% gradient gels, and electroblotted onto PVDF membranes (6). For immunovisualization the membranes were initially washed in water and air dried. Blocking was performed in 50 mM Tris, pH 7.4, 0.5 M sodium chloride, 2 % Tween 20. Washing and incubation buffer was 50 mM Tris, pH 7.4, 0.5 M sodium chloride, 0.1 % Tween 20 (TBS-T). For antibody production bovine spleen cathepsin D (Sigma) was further purified by pepstatinyl-Sepharose affinity chromatography prior to antibody production in a rabbit. Immunoglobulins were purified from immune rabbit serum by protein A-Sepharose, and used as primary antibody (1 μg/ml in TBS-T). Detection was by the alkaline phosphatase system (7) using enzyme conjugated swine immunoglobulins to rabbit immunoglobulins as secondary antibody. Biotinylated and reduced SDS-PAGE marker proteins were used as molecular mass standard, and detected by alkaline phosphatase conjugated streptavidin.

For N-terminal sequence determination the sample was concentrated by ultrafiltration and precipitated by trichloroacetic acid. After electroblotting the PVDF membrane was stained with Coomassie brilliant blue R, and the bands were cut out and placed in the cross-flow reaction cartridge of the Applied Biosystems sequencer.

RESULTS

Identification of Five Molecular Forms of Cathepsin D in Bovine Milk

The breakthrough fractions from the DEAE-Sepharose column were analysed by immunoblotting. Fractions containing cathepsin D immunoreactivity were pooled, and the cathepsin D forms further purified by pepstatin affinity chromatography. The proteins were allowed to bind to the column material at pH 3.5, and eluted at pH 8.5. Immunoblotting of the eluted, and therefore concentrated, material revealed that five molecular forms of cathepsin D were actually present: 46, 45, 43, 39 and 31 kDa (result not shown). We have previously shown that the 45 kDa band is procathepsin D, and that the 43 kDa form corresponds to pseudocathepsin D formed by autoproteolysis (3). The N-terminal sequence of all five molecular forms was determined. The sequence results and the identified cathepsin D forms are shown schematically in figure 1. Both the 46 and 45 kDa bands yielded the procathepsin D N-terminus. The 43 kDa band corresponded to pseudocathepsin D, while the

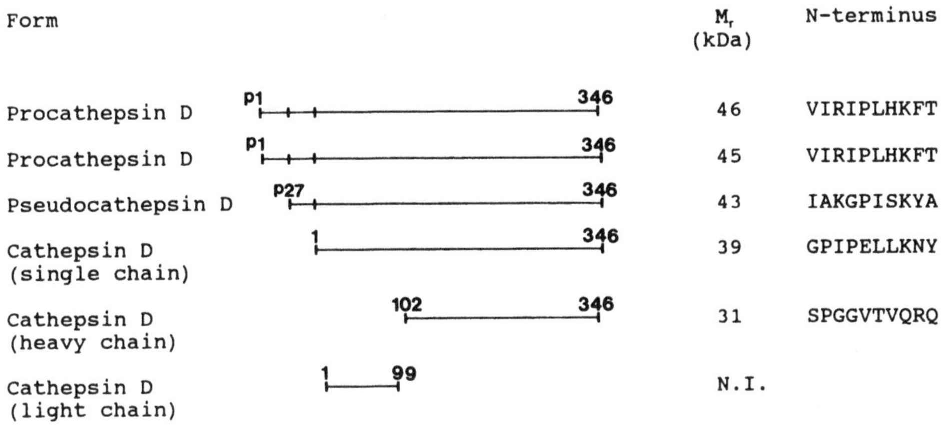

Figure 1. Molecular forms of bovine cathepsin D. The residue length of each form is indicated by numbers above the line (references: 3,10,11,12). The prefix p denotes residues in the propart, which in bovine procathepsin D covers 44 residues. Procathepsin D is found both in a 46 and a 45 kDa form with identical N-termini. Cathepsin D light chain was not identified (N. I.).

39 kDa and 31 kDa bands were mature single chain and heavy chain cathepsin D, respectively. Cathepsin D light chain was not identified.

The elution of the 45 kDa procathepsin D from the DEAE Sepharose column was very sensitive to the ionic strength of the buffer. When either 10 or 20 mM ammonium acetate, pH 5.5, is used for elution, the procathepsin D elutes as a very broad peak, mainly after other non-bound proteins have been washed out. This tailing was overcome using 50 mM ammonium acetate. Elution of mature cathepsin D was not affected by the ionic strength in this way, and always appeared in the breakthrough fraction using 10, 20 or 50 mM ammonium acetate.

pH-Dependent Binding to Pepstatin of Different Cathepsin D Forms

The differential binding to pepstatin of the bovine cathepsin D forms as a function of pH was investigated by two different experiments. The pool from the DEAE-Sepharose column was loaded onto the pepstatinyl-Sepharose column at pH 3.5, and eluted by a pH gradient (method A, materials and methods). The pH of the fractions was measured, and the fractions were analysed for cathepsin D by immunoblotting (figure 2). Procathepsin D (46 and 45 kDa) emerged at a lower pH value, starting in the fraction with pH 6.08, and in a larger amount than cathepsin D (39 and 31 kDa), of which most eluted in the fractions with pH 6.95 or above. Nearly pure procathepsin D is achieved in the pH 6.08 fraction, as seen in lane 3. A small amount of procathepsin D did not bind to the column (lane 1). A distinct band of immunoreactivity with M_r about 20 kDa is also seen in the non-bound fraction. The identity of this band is not known, but it is possible that it represents degradation product(s) of cathepsin D, having lost their affinity for pepstatin.

The binding characteristics of the cathepsin D forms were also investigated by another approach using two sequential affinity chromatography steps at different pH values (method B in materials and methods). The DEAE-Sepharose pool was allowed to bind to the pepstatinyl column at pH 5.4 and eluted at pH 8.5. Most procathepsin D appeared in the non-bound fraction together with the 20 kDa protein band described above, while most mature cathepsin D bound. After a second round of affinity chromatography at pH 3.5 of the

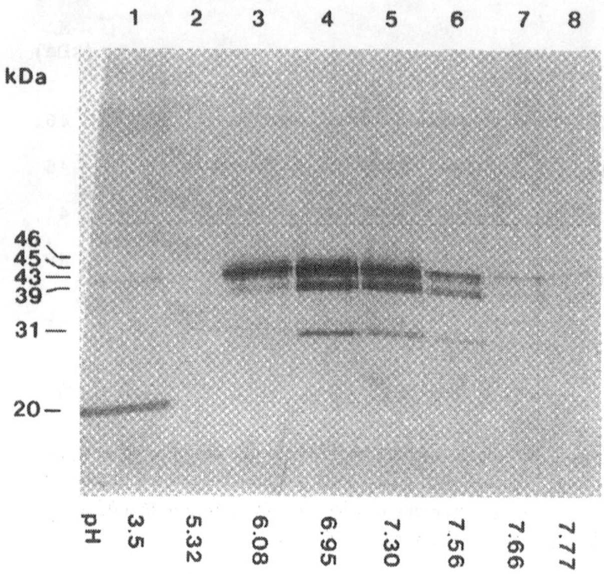

Figure 2. Immunoblot of the five different forms of cathepsin D in bovine milk eluted from a column of pepstatinyl-Sepharose by a pH gradient. The gradient was formed by titration of pH 8.5 buffer in pH 3.5 buffer. Lane 1 is non-bound proteins, lanes 2 to 8 contain samples of eluted protein from successive 6 ml fractions. The pH of each fraction is indicated.

non-bound proteins, the procathepsin D bound the pepstatin. The purity of the obtained procathepsin D was essentially as that seen in figure 2, lane 3.

DISCUSSION

Five molecular forms of cathepsin D were detected in bovine milk: 46 and 45 kDa procathepsin D, 43 kDa pseudocathepsin D, single chained (39 kDa) and two-chained mature cathepsin D (of which only the 31 kDa heavy chain was detected). According to the immunoblot in figure 2 the 45 kDa procathepsin D is the major form. Even though we have not established whether the different forms are recognized equally well by the antibody this interpretation is in accordance with the single 45 kDa band seen by Coomassie brilliant blue staining (3). What distinguishes the 45 and 46 kDa procathepsin D molecule is not known, but carbohydrate variation is a possibility. We observed that procathepsin D, but not cathepsin D, bound weakly to the DEAE-Sepharose column at low ionic strength. The explanation is not straightforward, as the propeptide is expected to have a net positive charge (3).

Bovine lysosomal cathepsin D is a mixture of equal amounts of single-chained and two-chained cathepsin D. It therefore seems likely that the presence of both 39 and the 31 kDa forms in bovine milk is a result of cell lysis and release from lysosomes, while the procathepsin D probably is secreted. The 43 kDa pseudocathepsin D may be formed by autocatalytic proteolysis during purification, as the pH in the milk itself is about 6.7, which seems too high for both intramolecular and intermolecular activation to occur. In this study we did not detect the cathepsin D light chain, which would be expected to be present in equimolar amounts with the heavy chain. The light chain was not seen in immunoblots of commercially available bovine cathepsin D. A likely explanation might be low antigenecity of the light chain, combined with its very poor staining with Coomassie brilliant blue on the PVDF membrane used for sequence analysis.

In the desorption experiment (method A) it is shown that the affinity of procathepsin D for pepstatin declines rapidly between pH 5.32 and 6.08, while the affinity of mature cathepsin D showed a clear decrease from pH 6.08 to 6.95. When adsorption of procathepsin D and cathepsin D to pepstatinyl-Sepharose was studied, we found that most cathepsin D was able to bind at pH 5.4 in contrast to procathepsin D of which most bound at pH 3.5. These adsorption results are in accordance with the gradient elution result in figure 2. The pH induced binding of procathepsin D to pepstatin agrees with a development of an exposed active site of aspartic proteinase zymogens at lower pH values (8,9). The differential affinity of zymogens and mature enzymes for pepstatin has been exploited to obtain a high degree of purification of procathepsin D from the mature forms.

ACKNOWLEDGMENTS

We thank Anni Boisen for technical assistance and Andreas Madsen, MD Foods Research and Development Center, Brabrand, Denmark, for bovine milk samples. This work is part of the plasmin project which is supported by the Danish Ministry of Agriculture and the Danish Dairy Board.

REFERENCES

1. R. M. Whitney, Proteins of milk *in* "Fundamentals of dairy chemistry" p. 81, N. P. Wong (ed.), van Nostrand Reinhold, New York (1988).
2. W. N. Eigel, C. J. Hofmann, B. A. K. Chibber, J. M. Tomich, T. W. Keenan and E. T. Mertz, Plasmin-mediated proteolysis of casein in bovine milk, *Proc. Natl. Acad. Sci. USA* 76:2244 (1979).
3. L. B. Larsen, A. Boisen and T. E. Petersen, Procathepsin D cannot autoactivate to cathepsin D at acid pH, *FEBS Lett.* 319:54 (1993).
4. V. Gieselmann, R. Pohlmann, A. Hasilik and K. von Figura, Biosynthesis and transport of cathepsin D in cultured human fibroblasts, *J. Cell Biol.* 97:1 (1983).
5. G. E. Conner, Isolation of procathepsin D from mature cathepsin D by pepstatin affinity chromatography, *Biochem. J.* 263:601 (1989).
6. P. Matsudaira, Sequence from picomole quantities of proteins electroblotted onto polyvinylidene difluoride membranes, *J. Biol. Chem.* 262:10035 (1987).
7. M. S. Blake, K. H. Johnston, G. J. Russell-Jones and E. C. Gotschlich, A rapid, sensitive method for detection of alkaline phosphatase-conjugated anti-antibody on western blots, *Anal. Biochem.* 136:175 (1984).
8. J. Marciniszyn, J. S. Huang, J. A. Hartsuck and J. Tang, Mechanism of intramolecular activation of pepsinogen, *J. Biol. Chem.* 251:7095 (1976).
9. M. N. G. James and A. R. Sielecki, Molecular structure of an aspartic proteinase zymogen, porcine pepsinogen, at 1.8 Å resolution, *Nature* 319:33 (1986).
10. T. Takahashi and J. Tang, Cathepsin D from porcine and bovine spleen *in* "Meth. enzymol." 80:565, Colowick and Lorand (eds.), Academic Press, New York (1981).
11. S. Yonezawa, T. Takahashi, X. Wang, R. N. S. Wong, J. A. Hartsuck and J. Tang, Structures at the proteolytic processing region of cathepsin D, *J. Biol. Chem.* 263:16504 (1988).
12. P. Metcalf and M. Fusek, Two crystal structures for cathepsin D: the lysosomal targeting signal and active site, *EMBO J.* 12:1293 (1993).

SITE-DIRECTED MUTAGENESIS OF A DISULFIDE BRIDGE IN CATHEPSIN D: EXPRESSION, ACTIVATION, PURIFICATION, AND CHARACTERIZATION

Brian M. Beyer and Ben M. Dunn

Department of Biochemistry and Molecular Biology
University of Florida
JHMHC, Gainesville, FL 32610-0245

INTRODUCTION

The intracellular aspartic proteinase Cathepsin D is found in the tissues of most higher eukaryotes. It is located in both lysosomes and endosomes, and is believed to play an important role in the normal turnover of proteins within the cell, as well as in the activation of other precursor proteins.

Procathepsin D is the intracellular precursor of the mature enzyme. Within the cell, procathepsin D is rapidly processed into the proteolytically active, two chain form of the enzyme. However, *in vitro*, procathepsin D activation via acidification is unusually slow and inefficient as reported by Conner and Udey[1]. This slow autoproteolytic processing may possibly be attributed to some aspects of the production and folding of the recombinant enzyme and is our current subject of investigation. Of particular interest is a disulfide bridge at positions 26 and 93 that is unique to Cathepsin D (with respect to other aspartic proteinases) and that remains present in the autoactivated recombinant protein. Because non-native disulfide bridge formation can impede the correct refolding of a protein, this extra pair of cysteines could be affecting our active recovery which is currently only approximately 1%.

With this in mind, we have targeted these residues for site-directed mutagenesis in hopes of enhancing successful folding and proteolytic activation. Utilizing a degenerative oligonucleotide, we attempted to substitute an Ala, Asp, Tyr and Ser at each one of these positions. Consequently, we have obtained a variety of mutants whose substitutions range in size and properties.

MATERIALS AND METHODS

The variety of substitutions at positions Cys(26) and Cys(93) were generated through a modified version of the overlap extension method of site-directed mutagenesis via PCR[2].

Table 1. Comparison Between Activities of WT Recombinant Cathepsin D and Disulfide Mutants Utilizing the Substrate Lys-Pro-Ile-Glu-Phe*Nph-Arg-Leu

Substitutions at Cys (26) and Cys (93)		Enzyme Activity AU • sec^{-1} •μg $enzyme^{-1}$
Asp	Asp	2.14×10^{-7}
Ala	Asp	2.91×10^{-7}
Ala	Ser	2.48×10^{-6}
Asp	Ala	7.80×10^{-6}
Cys (WT)	Cys	1.55×10^{-5}
Ala	Ala	1.73×10^{-5}
Ala	Tyr	2.58×10^{-5}

Nph = *p*-nitrophenylalanine

Mutants were identified through the sequencing of DNA isolated from transformed HMS174 *E. coli* cells. Primers were designed to target the sequence which corresponded with the targeted areas of mutagenesis.

For the expression of each Cathepsin D mutant and wild type(WT), DNA isolated from HMS174 *E. coli* cells was retransformed into BL21(DE3) *E. coli* expression cells. From this, a single colony was isolated and an overnight culture of M9 media (10 mg/mL thiamine, 0.5% casamino acids, 0.2% glucose and 50 mg/L ampicillin) was grown. The next day, an aliquot of each overnight culture was diluted 1:50 into LB media containing the same amount of ampicillin and grown to an OD_{600} of 0.5.

Cells were then induced with 1mM IPTG from a 1,000X stock (1 M). After the 3 hour induction period, the cells were pelleted at 3,500 x g for 10 minutes and lysed via French Press in TN buffer (50mM TRIS pH 7.4, 150 mM NaCl) with DNAse (80 U/ml). Lysed cells were then applied over a 27% sucrose cushion and centrifuged at 12,000 x g for 30 minutes in order to isolate purified insoluble inclusion bodies that can freely sediment through the cushion[3]. The pelleted inclusions were then resuspended in TN buffer containing 1% Triton X-100 and repelleted as previously described. Purified inclusion bodies were then resuspended in TE pH 7.8 at a concentration of 50 mg/ml and stored in 1 ml aliquots at -70 °C.

Purified inclusion bodies were solubilized at a concentration of approximately 2 mg/ml in 50mM CAPS pH 10.7, 50mM DTT and 8M urea. Following this, the solution was allowed to stand at room temperature for 30 minutes. The solution was then diluted 100-fold by dropwise addition into d-H_20 with constant slow stirring, and the pH was adjusted to the optimal refolding value of 8.7. After two hours of slow stirring, reduced glutathione, oxidized glutathione, and cysteine were added at concentrations of 1mM, 0.1mM and 1mM respectively to facilitate reshuffling of incorrectly formed disulfide bridges. This addition was accomplished by dissolving the three components in d-H_2O, adjusting the pH to 8.7, then adding to the refolding solution. Otherwise, the addition of the glutathione will make the solution very acidic. The solution was stirred for an additional 2 hours and the pH is then adjusted to 3.7 to begin activation of the aspartic proteinase.

Following the activation of the WT and various mutant enzymes, an activity assay was performed daily to compare and contrast the activities of the mutants with respect to the native enzyme. This analysis of catalytic activity was achieved through the utilization of a chromogenic substrate. The assay depends on the shift in absorbance of the *p*-NO_2Phenylalanine chromophore upon cleavage of the peptide bond preceding that amino acid and has been extensively described by Scarborough *et al.*[4] and Dunn *et al.*[5] The peak activities of the various mutants and WT have been reported in Table 1.

DISCUSSION

We have attempted, through site-directed mutagenesis, to simplify the folding and activation pathway of Cathepsin D. In mutating the non-conserved disulfide bridge, we have proven that this fourth disulfide is not essential for the activity of the enzyme. However, some degree of complementarity between these positions does appear to be required for correct refolding, activation and stability. This is evident in the variety of activities exhibited by the mutants. These lower activities indicate that, although refolding has possibly been simplified, it appears that stability has been decreased in some mutants resulting in decreased activity rates. Considering this, it is possible that our attempts to enhance folding and activation were successful but not represented fully in activity increases due to the active enzymes transient life.

REFERENCES

1. G.E. Conner, G. E. and J. A. Udey, *DNA Cell Biology* 9:1-9 (1990).
2. S.N. Ho, H.D. Hunt, R.M. Horton, J.K. Pullen, and L.R. Pease, *Gene* 77:51-59 (1989).
3. G. Taylor, M. Hoare, D.R. Gray, and F.A.O. Marston, *Biotechnology* 4:553-557 (1986).
4. P.E.Scarborough, K. Guruprasad, C. Topham, G.R. Richo, G.E. Conner, T.L. Blundell, and B.M. Dunn, *Protein Science* 2:264-276 (1993).
5. Ben M. Dunn, Paula E. Scarborough, Ruth Davenport, and Wieslaw Swietnicki, "Analysis Of Proteinase Specificity by Studies of Peptide Substrates: The Use of Ultraviolet and Fluorescence Spectroscopy to Quantitate Rates of Enzymatic Cleavage", in Peptide Protocols: Synthesis Purification, and Characterization, Ben M. Dunn and Michael Pennington, eds., (a volume of the METHODS IN MOLECULAR BIOLOGY series, John M. Walker, series editor), Humana Press, Clifton, New Jersey, in press (1994).

EXPRESSION OF RAT CATHEPSIN D cDNA IN *SACCHAROMYCES CEREVISIAE*: INTRACELLULAR SORTING OF CATHEPSIN D TO YEAST VACUOLE

Yukio Nishimura, Hideaki Fujita, Keitaro Kato, and Masaru Himeno

Division of Physiological Chemistry
Faculty of Pharmaceutical Sciences
Kyushu University
Higashi-ku, Fukuoka 812, Japan

INTRODUCTION

Lysosomal cathepsin D is a major aspartic proteinase and is considered to have physiological functions in intracellular protein degradation. Cathepsin D is one of the best characterized lysosomal proteinase for its intracellular protein sorting pathway in mammalian cells. Cathepsin D is synthesized as its preproform on the membrane-bound polysomes of the rough ER membranes, and the enzyme undergoes co-translational cleavage of the NH_2-terminal signal peptide in the rough ER membranes and simultaneously the high-mannose type oligosaccharide chains are added to form the glycosylated proenzyme, then initially follows the secretory pathway. During the pathway from the ER to the Golgi complex, the mannose 6-phosphate residues are formed on their oligosaccharides and the cathepsin D carrying the mannose 6-phosphate signal are recognized in the Golgi complex by a specific receptor (MPR) which plays a critical role in the segregation and targeting of lysosomal enzymes to the endosomes and lysosomes, and the propeptide-portion of cathepsin D is processed and is activated under acidic conditions upon the arrival of the enzyme in the lysosomes (1-4).

In contrast to this lysosomal enzyme sorting machinery, recent evidences suggest that MPR-independent sorting system for lysosomal enzyme would operate in the mammalian cell, since unglycosylated precursor cathepsin D synthesized in the presence of tunicamycin is correctly trnsported to the lysosomes and processed to the mature form enzyme in HepG2 clls (5,6). Rijnboutt et al. have reported that precursor cathepsin D is membrane-associated presumably in the Golgi complex as revealed by celll fractionation experiments in HepG2 cells and this membrane-bound form disappears during the enzyme maturation, therefore, postulating that the MPR-independent transport system would complement the MPR-dependent sorting mechanism in the liver cells (5,6). Membrane-association for procathepsins D and L in acidic microsomal vesicles has also been shown by Mclntyre and Erickson (7).

Aspartic Proteinases: Structure, Function, Biology, and Biomedical Implications
Edited by Kenji Takahashi, Plenum Press, New York, 1995

In the case of yeast, the targeting of vacuolar prtoein to the vacuole is not dependent on the specific N-glycosylation of the precursor protein and the sorting information for the vacuolar enzyme resides within the polypeptide backbone of the precursor protein (8,9).

In the present study, we tried to express the rat cathepsin D cDNA in the yeast *Saccharomyces cerevisiae* to examine whether the expressed cathepsin D is correctly sorted for targeting to the vacuoles. We found the evidence showing that the recombinant rat cathepsin D is expressed in the yeast cells and that the precursor cathepsin D is transported to the vacuoles independent of the N-glycosylation of the polypeptide. The data presented in this paper is also in good accord with our previous finding that the recombinant mouse cathepsin L is translocated to the vacuoles in the yeast (10).

RESULTS

Expression of Rat Cathepsin D cDNA in *Saccharomyces cerevisiae*

To express rat cathepsin D cDNA in *Saccharomyces cerevisiae*, the 1879-bp *Eco* RI fragment of the cat-1 cDNA (11) which contains the complete open reading frame was inserted with *Sal* I site of pYG100 which is located about 120-bp downstream of the TATA box of the GAPDH promoter and the resulting plasmid, pYGcat-1, was used to transform *Saccharomyces cerevisiae*. Yeast transformants carrying the rat cathepsin D cDNA were grown to midlogarithmic phase in SD medium with necessary growth supplements without casamino acids.

To investigate the level of cathepsin D expression, the wild-type cells or the yeast cells producing cathepsin D were lysed in lysis buffer containing 1 % Triton X-100 and the proteinase inhibitors, and then the total cell extract was applied to SDS-PAGE followed by immunoblotting with anti-cathepsin D antibody. The results revealed that the yeast transformants carrying coding regions for rat cathepsin D produced the polypeptide of 44-kDa which appears to be slightly higher than that of mature form cathepsin D of 43-kDa. Wild-type cells were found to exhibit no immunoreactive band. Therefore, the expression of rat cathepsin D cDNA in the yeast cells not normally producing this polypeptide has become possible to produce a stable form of cathepsin D. To investigate the precise localization of the expressed polypeptide, we fractionated the cells containing the cathepsin D polypeptide on discontinuous Ficoll step-gradients. Thus prepared total vacuolar fraction was analyzed for cathepsin D by immunoblotting. The results revealed that the processed form of cathepsin D whose size appears to be corresponded with the intermediate form in the spheroplast lysate was present in the vacuolar fraction.

Biosynthesis of Cathepsin D

To study the biosynthesis of cathepsin D, yeast cells transformed with rat cathepsin D cDNA were subjected to pulse-chase experiments. The cells of yeast strain SEY6210 expressing cathepsin D polypeptide were labeled for 10 min with Tran^{35}S-label at 30°C and chased with nonradioactive amino acids for different times. Cathepsin D or carboxypeptidase Y was immunoprecipitated from the labeled cell extract with each specific antibodies and analyzed by SDS-PAGE and fluorography, respectively. After a 10-min pulse, cathepsin D appears to have two distinct forms of 46 and 47 kDa apparent molecular mass, which after 60 min are chased to one major precursor form 46 kDa, and persisted as its precursor forms after 120 min chase. The higher molecular weight form of 47 kDa is considered to be derived from the differently glycosylatyed forms. Alternatively, it is possible to speculate that the NH_2-terminal 20 amino acids signal peptide for presursor cathepsin D polypeptide may not be cleaved effeciently during biosynthesis, therefore, two distinct precursor forms were detected after short pulse experiment.

When we analyzed the processing kinetics of carboxypeptidase Y protein (CPY) which is a typical vacuolar hydrolase by pulse-chase labeling of the yeast cells, the observed maturation kinetics appear to be consistent with previously reported processing kinetics for CPY (12). In our experimental conditions, post-translational processing events for the vacuolar hydrolase appears to operate correctly in the transformed cells.

Vacuolar Delivery of Cathepsin D in the Presence of Tunicamycin

To determine whether the glycosyl modification of cathepsin D in the yeast cells is required for the sorting of this polypeptide to the vacuole, yeast cells were labeled with Tran^{35}S-label and chased at 30°C for 2 h in the presence of tunicamycin and the vacuolar fraction was prepared from the yeast lysates and then the immunoprecipitation was carried out with cathepsin D- or CPY-specific antibody. As expected, unglycosylated mCPY was observed in the prepared vacuolar fraction, suggesting that the intracellular targeting of CPY to the vacuole is not dependent on the carbohydrate addition. When the biosynthesis of cathepsin D was examined in the presence of tunicamycin, the unglycosylated polypeptide as well as the glycosylated form was demonstrated to be present in the vacuolar fraction of the 2-h chase treated cells, suggesting that the unglycosylated cathepsin D is correctly sorted for targeting to the vacuoles, and supporting the conclusion that like most vacuolar hydrolase cathepsin D would follow the secretory pathway from the ER to the Golgi complex and efficiently transported to the vacuoles independent of the carbohydrate modification.

DISCUSSION

Recent studies on the intracellular sorting of lysosomal hydrolases in mammalian cells indicate that the newly synthesized lysosomal soluble hydrolases are correctly transported to the lysosomes instead of being secreted into the culture medium in the presence of tunicamycin, suggesting that the carbohydrate moiety may not be a prerequisit for the intracellular sorting of lysosomal hydrolase in mammalian cells, and that alternative sorting mechanisms for lysosomal hydrolases operate in the sorting pathway (5,6). To examine the intracellular sorting mechanism for lysosoaml enzymes, we attempted to express rat cathepsin D cDNA in *Saccharomyces cerevisiae.* The present paper have provided the evidence that the recombinant cathepsin D is successfully expressed in *Saccharomyces cerevisiae* and that the processed form of cathepsin D is identified in the isolated vacuolar fraction of the yeast cells by immunoblotting analysis. This result suggests that the newly synthesized cathepsin D enters into the secretory pathway and is efficiently transported to the vacuole and during its intracellular transport pathway precursor cathepsin D would be processed to the mature form. Biosynthesis study have further confirmed that the newly synthesized precursor cathepsin D was N-glycosylated, and then successively transported to the yeast vacuole.

Taken together, these data indicated that cathepsin D is stably expressed in the *Saccharomyces cerevisiae* and that the newly synthesized molecule is targeted to the vacuoles irrespective of carbohydrate addition. The development of expression system for cathepsin D along with our previous work for cathepsin L (10) would provide us a useful system to determine whether the elements responsible for recognizing the lysosomal proteinase is proteinaceous.

REFERENCES

1. M.G. Kornfeld, and I. Mellman, Ann. Rev. Cell Biol.5: 483-525 (1989).
2. M.G. Rosenfeld, G. Kreibich, D. Popov, K. Kato, and D.D. Sabatini, J. Cell Biol. 93:135-143 (1982).
3. Y. Nishimura, M. Higaki, and K. Kato, Biochem. Biophys. Res. Commun. 148: 335- 343 (1987).
4. Y. Nishimura, T. Kawabata, S. Yano, and K. Kato, Acta Histochem. Cytochem. 23: 53-64 (1990).
5. S. Rijnboutt, H. M. F. G. Aerts, H. J. Geuze, J. M. Tager, and G. J. Strous, J. Biol. Chem. 266: 4862-4868 (1991).
6. S. Rijnboutt, A. J. Kal, H. J. Hans, H. Aerts, and G. J. Strous, J. Biol. Chem. 266: 23586-23592 (1991).
7. G. McIntyre, and A. H. Erickson, J. Biol. Chem. 266: 15438-15445 (1991).
8. L.M. Johnson, V. A. Bankaitis, and S. D. Emr, Cell 48: 875-885 (1987).
9. L. A. Valls, C. P. Hunter, J. H. Rothman, and T. H. Stevens, Cell 48:887- 897 (1987).
10. Y. Nishimura, and K. Kato, Arch. Biochem. Biophys. 298: 318-324 (1992).
11. H. Fujita, Y. Tanaka, Y. Noguchi, A. Kono, M. Himeno, and K. Kato, Biochem. Biophys. Res. Commun. 179: 190-196 (1991).
12. A. Hasilik, and W. Tanner, Eur. J. Biochem. 85: 599-608 (1978).

MOLECULAR CLONING AND IMMUNOCYTOCHEMICAL LOCALIZATION OF JASMONIC ACID INDUCIBLE CATHEPSIN D INHIBITORS FROM POTATO

Borut Štrukelj,[1] Maja Ravnikar,[2] Pika Meško,[1] Mateja Poljšak-Prijatelj,[3] Jože Pungerčar,[1] Gregor Kopitar,[1] Igor Kregar,[1] and Vito Turk[1]

[1]Department of Biochemistry and Molecular Biology
Jožef Stefan Institute
Jamova 39, 61000 Ljubljana, Slovenia
[2]Institute of Biology
University of Ljubljana
Karlovška 19, 61000 Ljubljana, Slovenia
[3]Institute of Microbiology, Medical Faculty
University of Ljubljana
Zaloška 2, 61000 Ljubljana, Slovenia

INTRODUCTION

Proteins which have the ability to inhibit proteolytic activity of certain enzymes are found throughout the plant kingdom. They are usually accumulated as storage proteins in seeds and tubers (1). Many inhibitors of serine (2-4), cysteine (5-8) but only a few inhibitors of aspartic (9) proteinases have been isolated from potato tubers. The protein sequences of two closely related isoinhibitors, PDI (10) and NID (11), isolated from potato tubers, were determined.

From potato tubers, we have recently isolated and characterized two cDNA clones, PI-8 (12) and PI-13 (13), and a genomic (PIG) clone (14) encoding aspartic proteinase inhibitor homologues. Comparison of their deduced protein sequences with the amino acid sequences of the isolated isoinhibitors PDI and NID shows more than 70% amino acid identity. Southern blot analysis confirmed that they all belong to one multigene family (15).

The physiological role of proteinase inhibitors in various plants of the *Solanaceae* family has been the subject of research in many laboratories. Results indicate that they are involved in the defensive mechanism of plants (16). It has been demonstrated that defensive genes can be activated in plants by small pectic fragments of cell walls (17), chitosan (18), abscisic acid (19), auxin (20) and jasmonic acid (JA) (21). Farmer and co-workers showed that methyl jasmonate induces the accumulation of serine proteinase inhibitors in wounded and non-wounded tomato leaves (21). In the present experiments, we show two novel deduced amino acid sequences of the mature forms of aspartic proteinase inhibitor homologues (PI-3 and PI-6), obtained by RT-PCR from jasmonic acid-treated potato shoots. Additionally, we demonstrate that aspartic proteinase inhibitors are accumulated in the vacuoles in shoots grown on media supplemented with higher concentrations of JA.

MATERIALS AND METHODS

Plant Material

Stems of healthy potato plants (*Solanum tuberosum* L. cv. Desiree) were cut into segments containing one leaf node and inserted in the Murashige and Skoog medium supplemented with 3% sucrose, 0.7% Bacto agar (Difco, USA) and (±) jasmonic acid (SA Firmenich, Switzerland) in concentrations of 0.1-100 μM. Medium without JA was used as a control medium. Cultures were grown at 25 ± 2°C, with illumination to provide a 16 h-day length.

RNA Isolation and PCR Analysis

Total cellular RNA was isolated from potato shoots according to the procedure of Logemann *et al* (22). Poly(A^+) RNA was isolated by affinity chromatography on a oligo(dT)-cellulose column. Poly(A^+) RNA samples (1 μg) from JA-treated (100 μM JA) leaves were reverse transcribed in a reaction mixture containing 1 mM of each dNTP, 1 X PCR buffer, 1 U/μl RNasin, 0.2 μg of oligo$(dT)_{12-18}$ and 200 U MMLV reverse transcriptase (Perkin-Elmer Cetus, USA). The mixture was incubated for 20 min at 22°C, than 45 min at 42°C and afterwards it underwent heat treatment for 10 min at 95°C. Subsequently, 25 pmol of the 5'-amplimer, 5'-d(GGGAATTCGAATCTCCTGTACCT)-3', 25 pmol of the 3'-amplimer, 5'-d(CTGGGATCCCTAGACTTCCTGGAA)-3', and 3 U of AmpliTaq DNA polymerase (Perkin-Elmer Cetus, USA) were added. PCR reaction was performed for 23 cycles using the thermal cycler (Perkin-Elmer Cetus, USA). After fractionation by gel electrophoresis on 1.2% agarose, the PCR products were excised, purified using a glass milk suspension (Geneclean II kit, BIO 101, USA) and inserted into pUC19. Both DNA strands of inserts were sequenced by the dideoxy chain termination method (23) using T7 DNA polymerase (Pharmacia, Sweden).

Immunohistochemistry

Rabbit antibodies to aspartic proteinase inhibitor (NID) were purified by affinity chromatography on protein A-Sepharose. After four weeks of cultivation, potato apical shoots were fixed for 2 hours in 2% paraformaldehyde, 0.1% glutaraldehyde in cacodylate buffer, pH 7.2. After washing, the plant tissues were dehydrated in a series of ethanol solutions (from 30% to 100%). The tissues were then placed in a mixtures of pure acrylic resin LR White (The London Resin Co) : 100% ethanol of 1:1 and 2:1, each for 45 min, pure LR White for 60 min, and after that, the tissues were finally embedded and polymerized in pure LR White for 3 days at 50°C. After cutting, ultrathin tissue sections were incubated at room temperature for one hour with the antibody solution diluted with PBS-BSA buffer in a ratio of 1:30, and for another hour with protein A-gold, diluted 1:50. After immunolabeling, sections were stained with 1% Reynolds' lead citrate for 10 min, followed by 1% uranyl acetate for 5 min., and inspected by a Jeol JEM 1200 EX II transmission electron microscope at 80 kV.

```
      -30       -20       -10       -1        10        20        30
PI-8  MMKCLFLLCLCLLPIVVFSSTFTSQNLIDLPSESPLPKPVLDTNGKELNPDSSYRIISIGRG
PI-13 ...............L..........P.N.....V...............N...........
PDI                                 ....................N...........
p749  ............F..L..........P.N.....V............K..N.......TFW.
PIG    ...........V...........K.P.N...DAT---....VAD...DSRL......TFW.
NID                                 ....................N...........
PI-3                                                    N...........
PI-6                                                    N...........

            40        50        60        70        80        90
PI-8  ALGGDVYLGKSPNSDAPCPDGVFRYNSDVGPSGTPVRFIPLSG-GIFEDQLLNIQFNIPTV
PI-13 ..........................................T-N................
PDI   ...........................................-..............A..
p749  ...........................................AN................
PIG   .................AN.I.................SHFGQ-....NE......A.S.S
PI-4                                                          .S.S
NID   ...........................................-.................
PI-3  ..........................................T-N................
PI-6  ...........................................-.................

            100       110       120       130       140       150
PI-8  KLCVSYTIWKVGNLNAYFRTMLLETGGTIGQADSSYFKIVKLSNFGYNLLYCPITPPFLC
PI-13 ......RN........HLW......................S.K...........RH...
PDI   .................................N..........................
p749  .............I..HL.......................S.K.........L.RH...
PIG   ...........DYD.SLG..................W.....S.Q.........V.STMS.
PI-4  ...........DYD.SLG...............N................S..F.S-II.
NID   .................................N................S..F.S-II.
PI-3  .................................N.......S.KI.....S..F.S-II.
PI-6  .................................N.......S.KI.....S..F.S-II.

            160       170       180
PI-8  PFCRDDNFCAKVGVVIQNGKRRLALVNENPLDVLFQEV
PI-13 ..............D..........S............
PDI   ......................................
p749  ......................................
PIG   ..SS..Q..L.....H..........KD.....S.KQ.Q
PI-4  LR.PE.Q........H...........G...EC..KQ.Q
NID   LR.PE.Q...............................
PI-3  LR.PE.Q...............................
PI-6  LR.PE.Q...............................
```

Figure 1. Comparison of the deduced amino acid sequences of two newly-identified cDNA clones PI-3 and PI-6 with sequences of other aspartic proteinase inhibitor homologues from potato. The deduced amino acid sequences of PI-8, PI-13, p749 and PIG are from references 12, 13, 24 and 14, respectively. The amino acid sequences of isolated isoinhibitors PDI and NID were taken from references 10 and 11. A dot (.) represents the same amino acid residue as shown for PI-8. Numbering is according to PI-8 cDNA clone.

RESULTS AND DISCUSSION

Previous results have shown (13) that aspartic proteinase inhibitors, similarly to serine potato inhibitors I and II, are accumulated in potato shoots due to exogenous application of jasmonic acid into the growth medium, therefore we decided to find new JA-inducible aspartic proteinase inhibitor cDNA clones and to elucidate their amino acid sequences. The sequences of amplimers corresponded to the N- and C-terminal regions of

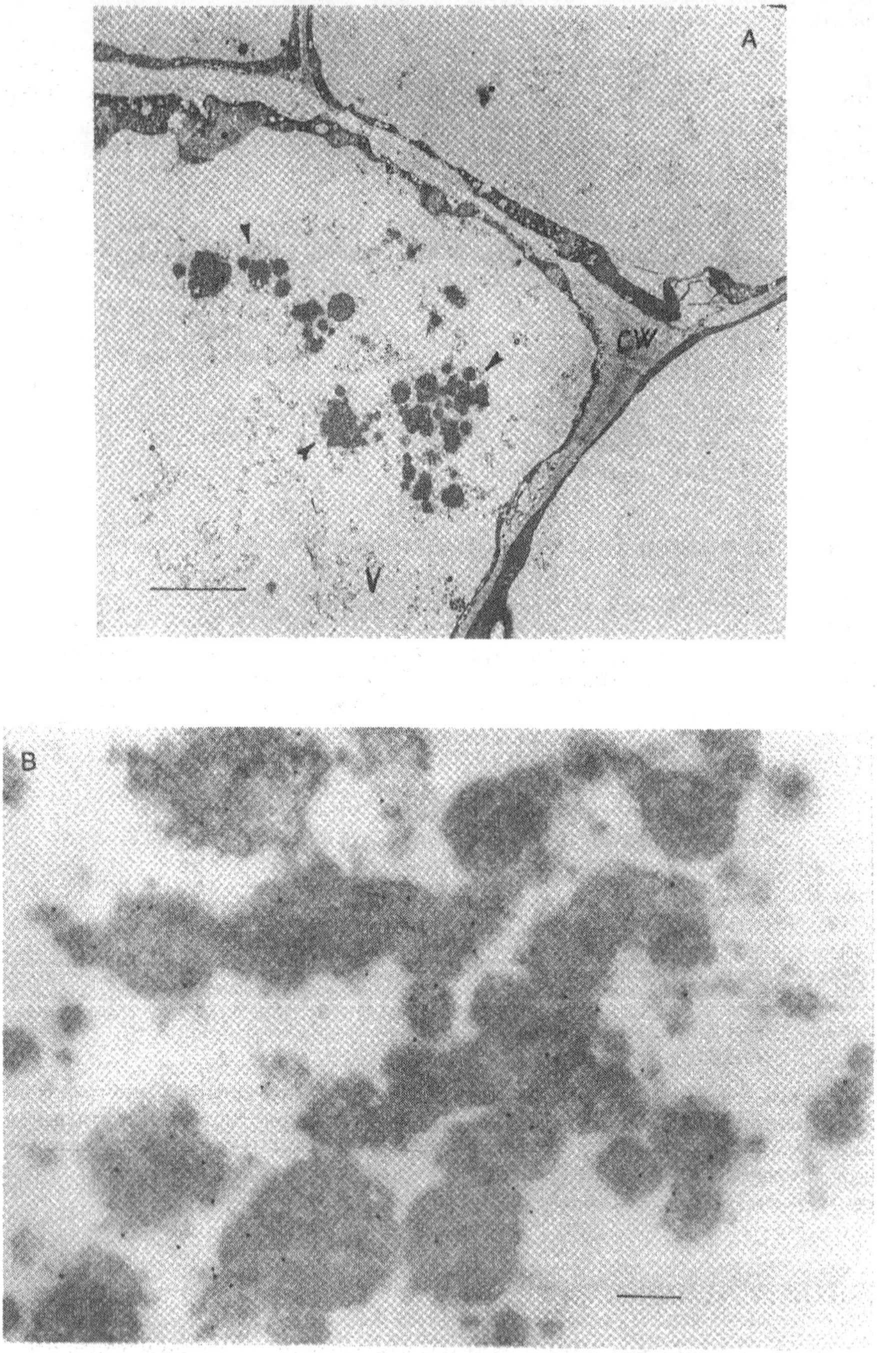

Figure 2. A: Aspartic proteinase inhibitors (arrowheads), located in the vacuole of differentiated shoot apex cell; V, vacuole; CW, cell wall; bar = 2 μm. **B**: Immuno-gold labelling of aspartic proteinase inhibitors associated in protein agregates; bar = 200 nm.

deduced amino acid sequences of the mature PI-8 (12). After 23 cycles of PCR amplification using the first strand cDNA as a template, obtained by reverse transcription of JA-induced poly(A^+) RNA, an intense band of approximately 540 bp could be detected in ethidium bromide-stained agarose gel. The excised band was pooled and cloned. Totally ten PCR clones were sequenced. Four of them were recognized as previously published PI-13 cDNA clone (13), while six clones represented two novel aspartic proteinase inhibitor homologues, which were named PI-3 and PI-6. The deduced amino acid sequences of PI-3 and PI-6 are shown in Fig. 1. Comparison of their deduced amino acid sequences with the protein sequences of known aspartic proteinase inhibitors and deduced sequences from cDNA (12,13, 24) or gene sequences (14) shows that both clones are more closely related to each other than to any of the others. Since we used the N-terminal amplimer constructed on the base of a nucleotide sequence of mature PI-8, the N-termini of PI-3 and PI-6 are incomplete. The newly characterized clones differ in only two amino acid residues in the central part of the molecule, where Thr72 and Asn73 in PI-3 are replaced by Gly72 and Gly73 in PI-6, respectively. Both clones show more than 97% amino acid identity with the protein sequence of NID. The putative active center responsible for trypsin inhibitory activity (Arg67-Phe68) is also present. The N-terminal region of the compared cathepsin D inhibitor sequences (Fig.1) is highly conserved, whereas the C-terminal region is more heterogenous. Therefore, these proteins could be divided into three groups: group I is composed of clones PI-8, PI-13, p749 and isolated inhibitor PDI, group II represents clones PIG and PI-4, and in group III are clones PI-3, PI-6 and isolated inhibitor NID. These features suggest that in potato aspartic proteinase inhibitor multigene family, the differences in a C-terminal part of the proteins might occur due to gene elongation or duplication of an ancestral gene.

The active site of the molecule, responsible for inhibition of cathepsin D is not known yet. Primary structures of cathepsin D inhibitors from potato share a moderate similarity (more than 50% identical residues) to the double-headed Kunitz soybean trypsin inhibitor family (10). Contrary to the trypsin inhibitor family, all the compared sequences of aspartic proteinase inhibitors possess an extra region of seven (PI-3, PI-6 and NID) or eight (PI-4, PIG, PI-8, PI-13 and PDI) amino acid residues at the C-terminal part of the molecule which might be involved in cathepsin D inhibition. Their three-dimensional structure and aspartic proteinase inhibitory site is still a subject of further research.

In the present paper, we have demonstrated that jasmonic acid added to the growth medium in concentrations from 10 to100 μM strongly stimulated synthesis of aspartic proteinase inhibitors in potato shoot apex. They can be found in the form of membraneless protein bodies in the vacuole, especially in differentiated cells with one large vacuole (Fig. 2). In less differentiated meristematic cells with many small vacuoles, they are mainly located on the both sides of vacuole membrane (data not shown). Full-length clones PI-8 and PI-13 contain in their deduced propeptide sequences a stretch of six amino acid residues (N-P-I-N-L-P in PI-13, and N-L-I-D-L-P in PI-8) which is very likely a vacuolar targeting signal (25). The whole prepropeptide with a signal region is apparently post-translationally lost, presumably during transportation to the central vacuole, where the excess of synthesized inhibitors are accumulated. We can only speculate that there they are stored to act as a defense proteins in potential stress response.

ACKNOWLEDGMENTS

This work was supported by the Ministry of Science and Technology of Slovenia, grant number P4-5322.

REFERENCES

1. C.A.Ryan, Proteinase inhibitors. In: *The Biochemistry of Plants* (P.K. Stumpf, E.E. Conn, eds), Academic Press, New York **6**:351 (1981).
2. G. Pearce, G. Sy, C. Russell, C.A. Ryan, G.M. Hass, Isolation and Characterization from Potato Tubers of Two Polypeptide Inhibitors of Serine Proteinases, *Arch. Biochem. Biophys.* **213**:456 (1982).
3. T. Iwasaki, T. Kiyohara, M. Yoshikawa, Amino Acid Sequence of an Active Fragment of Potato Proteinase Inhibitor IIa, *J. Biochem.* **79**:381 (1976).
4. T.A. Walsh, W.P. Twitchell, Two Kunitz-Type Proteinase Inhibitors from Potato Tubers, *Plant Physiol.* **97**:15 (1991).
5. K.V.S. Fernandes, F.A.P. Campos, R.R. DoVal, J. Xavier-Filho, The Expression of Papain Inhibitors during Development of Cowpea Seeds, *Plant Science* **74**:179 (1991).
6. S.G. Suh, J.E. Peterson, W.J. Stiekema, D.J. Hannapel, Purification and Characterization of the 22-Kilodalton Potato Tuber Proteins, *Plant Physiol.* **94**:40 (1990).
7. I. Hirashiku, F. Ogata, N. Yoshida, S. Makisumi, A. Ito, Purification and Complex Formation Analysis of a Cysteine Proteinase Inhibitor (Cystatin) from Seeds of *Wisteria floribunda*, *J. Biochem.* **108**:604 (1990).
8. I. Krišaj, M. Drobnic-Košorok, J. Brzin, R. Jerala, V. Turk, The Primary Structure of Inhibitor of Cysteine Proteinases from Potato, *FEBS Lett.* **333**:15 (1993).
9. H. Keilova, V. Tomašek, Isolation and some Properties of Cathepsin D Inhibitor from Potatoes, *Collect. Czech. Chem. Commun.* **41**:489 (1976).
10. M. Mareš, B. Meloun, M. Pavlik, V. Kostka, M. Baudyš, Primary Structure of Cathepsin D Inhibitor from Potatoes and Its Structure Relationship to Soybean Trypsin Inhibitor Family, *FEBS Lett.* **251**:94 (1989).
11. A. Ritonja, I. Krišaj, P. Meško, M. Kopitar, P. Lučovnik, B. Štrukelj, J. Pungerčar, D.J. Buttle, A.J. Barrett, V. Turk, The Amino Acid Sequence of a Novel Inhibitor of Cathepsin D from Potato, *FEBS Lett.* **267**:13 (1990).
12. B. Štrukelj, J. Pungerčar, A. Ritonja, I. Krišaj, F. Gubenšek, I. Kregar, V. Turk, Nucleotide and Deduced Amino Acid Sequence of an Aspartic Proteinase Inhibitor Homologue from Potato Tubers (*Solanum tuberosum* L.), *Nucleic Acids Res.* **18**:4605 (1990).
13. B. Štrukelj, M. Ravnikar, P. Meško, J. Pungerčar, V. Turk, I. Kregar, Aspartic Proteinase Inhibitors from Potato: Isolation, cDNA Cloning, Induction of Their Expression by Jasmonic Acid and Distribution of Related Genes in Some Other Solanaceous Species. submitted for publication.
14. D. Barlič Maganja, B. Štrukelj, J. Pungerčar, F. Gubenšek, V. Turk, I. Kregar, Isolation and Sequence Analysis of the Genomic DNA Fragment Encoding an Aspartic Proteinase Inhibitor Homologue from Potato (*Solanum tuberosum* L.), *Plant Mol. Biol.* **20**:311 (1992).
15. B. Štrukelj, J. Pungerčar, P. Meško, D. Barlič Maganja, F. Gubenšek, I. Kregar, V. Turk, Characterization of Aspartic Proteinase Inhibitors from Potato at the Gene, cDNA and Protein Levels, *Biol. Chem. Hoppe-Seyler* **373**:477 (1992).
16. C.A. Ryan, Proteinase Inhibitor Gene Families: Strategies for Transformation to Improve Plant Defenses Against Herbivores, *BioEssays* **10**:20 (1989).
17. P. Bishop, K.J. Makus, G. Pearce, C.A. Ryan, Proteinase Inhibitor Inducing Factor Activity in Tomato Leaves Resides in Oligosaccharides Enzymatically Released from Cell Walls, *Proc. Natl. Acad. Sci. USA* **78**:3536 (1981).
18. M. Walker-Simmons, L. Hadwiger, C.A. Ryan, Chitosans and Pectic Polysaccharides Both Induce The Accumulation of The Antifungal Phytoalexin Pisatin in Pea Pods and Antinutrient Proteinase Inhibitors in Tomato Leaves, *Biochem. Byophys. Res. Comm.* **110**:194 (1983).
19. H. Peña-Cortes, L. Willmitzer, J. Sanchez-Serrano, Abscisic Acid Mediates Wound-Induction But Not Developmental Expression of The Proteinase Inhibitor II Gene Family, *Plant Cell* **3**:963 (1991).
20. R.W. Thornburg, X. Li, Auxin Levels Decline in Tobacco Foliage Following Wounding, *Plant Physiol.* **93**:500 (1990).
21. E.E. Farmer, C.A. Ryan, Interplant Communication: Air-Borne Methyl Jasmonate Induces the Synthesis of Proteinase Inhibitor Genes in Plant Leaves, *Proc. Natl. Acad. Sci. USA* **87**:7713 (1990).
22. J. Logemann, J. Schell, L. Willmitzer, Improved Method for The Isolation of RNA From Plant Tissues, *Anal. Biochem.* **163**:16 (1987).
23. F. Sanger, S. Nicklen, A.R. Coulson, DNA Sequencing With Chain-Terminating Inhibitors, *Proc. Natl. Acad. Sci. USA* **74**: 5465 (1977).
24. D.J. Hannapel, Nucleotide and Deduced Amino Acid Sequence of The 22-Kilodalton Cathepsin D Inhibitor Protein of Potato (*Solanum tuberosum* L.), *Plant Physiol.* **101**:703 (1993).
25. M.J. Chrispeels, N.V. Raikhel, Short Peptide Domains Target Proteins to Plant Vacuoles, *Cell* **68**:613 (1992).

PURIFICATION AND CHARACTERIZATION OF AN ACID PROTEINASE FROM *DIROFILARIA IMMITIS* WORMS

Kumiko Sato,[1] Yutaka Nagai,[2] and Mamoru Suzuki[3]

[1]College of Medical Care and Technology
Gunma University
Maebashi 371
[2]Department of Tissue Physiology
Medical Research Institute
Tokyo Medical and Dental University
Tokyo 101
[3]Department of Parasitology
Gunma University School of Medicine
Maebashi 371

ABSTRACT

An acid proteinase of *Dirofilaria immitis* worms was purified 437-fold by gel filtration on Sephadex G-75 followed by pepstatin-Agarose gel affinity chromatography. The enzyme with a molecular weight of 42 kDa was homogeneous as judged by both affinity chromatography and SDS-polyacrylamide gel electrophoresis. Polyacrylamide disc electrophoresis at pH 8.9, however, revealed that the enzyme was composed of five multi-forms, all carrying proteinase activity. Optimum pH of the enzyme was in the range of pH 2.8 to 3.4, and its isoelectric point ranged between 5.8 and 6.4. The purified proteinase showed a potent activity against hemoglobin and myoglobin releasing acid soluble peptides, but not free amino acids. When enzymatic properties of the proteinase was compared with mammalian cathepsin D and pepsin, *D. immitis* proteinase activity was reduced to about 80% of the initial activity by incubating at neutral pH and 50 °C for 5 min, just like cathepsin D, which remained intact. Pepsin activity was completely destroyed under the same condition. An aspartic proteinase inhibitor, 1, 2-epoxy-3-(*p*-nitrophenoxy)propane, which inhibited pepsin by 30% at 37 °C for 10 min, did show little effects on *D. immitis* proteinase and cathepsin D. Inhibitory effect of diazoacetyl-DL-norleucine methyl ester (DAN) on *D. immitis* proteinase was intermediate (50% after 60 min). Immunolocalization of the proteinase in the worm tissue using its monoclonal antibodies revealed that the enzyme was localized in the intestine as well as uterine wall and some small granules of microfilariae in the uterus.

INTRODUCTION

The acid proteinase of parasitic worms is thought to play an important role in the physiological breakdown of host proteins for their nutritional metabolism [1]. Extensive data have been accumulated regarding the enzymes from parasitic helminths and protozoa, i.v. *Angiostrongylus cantonensis*, *Dirofilaria immitis*, *Ascaris suum*, *Trichuris muris*, *Diphyllobothrium erinacei*, *Hymenolepis nana* [2]. The acid proteinase of *D. immitis* worms was first demonstrated in our laboratory [3,4]. Sreedhara Swamy and Jaffe reported the presence of two proteases, designated Fp-I (Mr: 170,000) and Fp-II (Mr: 48,000), in the extract of adult *D. immitis* [5], in which the latter enzyme behaved like cathepsin D on pepstatin-Agarose gel. Recently, we described a new purification method and partial characterization of *D. immitis* proteinase [6]. In this report, we describe the purification procedure of *D. immitis* proteinase and some physicochemical and enzymatic characteristics of the purified enzyme compared with mammalian cathepsin D and pepsin. Immunolocalization of *D. immitis* acid proteinase in the worm tissue is also described by using its monoclonal antibodies.

MATERIALS AND METHODS

Materials

Adult *D. immitis* worms were obtained from the pulmonary arteries and hearts of naturally infected dogs. The collected worms were washed several times with water, lyophilized, and stored at -20°C until use. All reagents used were of analytical grade.

Enzyme Assay

One ml of human hemoglobin (5 mg/ml) in 100 mM citrate-200 mM disodium phosphate buffer (citrate buffer), pH 3.1, was incubated with 500 ul of enzyme in the same buffer at 37°C for 1 hr. After stopping the enzyme reaction with 1.5 ml of chilled 10% trichloroacetic acid, the resulting mixture was kept on ice for 30 min, centrifuged at 2,000 x g at 4°C for 15 min and the absorbance at 280 nm of the supernatant was determined. One unit of proteolytic activity is defined as the amount of enzyme necessary to release protein degradation products at the rate of 1.0 absorbance unit at 280 nm (1.0 A 280 unit)/h under the above conditions.

Preparation of Crude Enzyme

Lyophilized worms (about 12 g) were minced, defatted, suspended in 500 ml of 100 mM acetate buffer, pH 4.0, and refrigerated overnight. The material was homogenized and sonicated for 15 min on ice. The resulting suspension was centrifuged at 10,000 x g for 30 min to obtain proteinase in the supernatant, which was concentrated by ultrafiltration with cellulose tubing 8/32 (Visking Co.). Protein content was determined by the Lowry's method [7] using bovine serum albumin as a standard.

Purification of Acid Proteinase

An aliquot of the concentrated crude enzyme was applied onto a Sephadex G-75 column equilibrated with 100 mM acetate buffer, pH 4.0, and eluted with the same buffer. Fractions containing enzyme activity were pooled, concentrated by ultrafiltration and

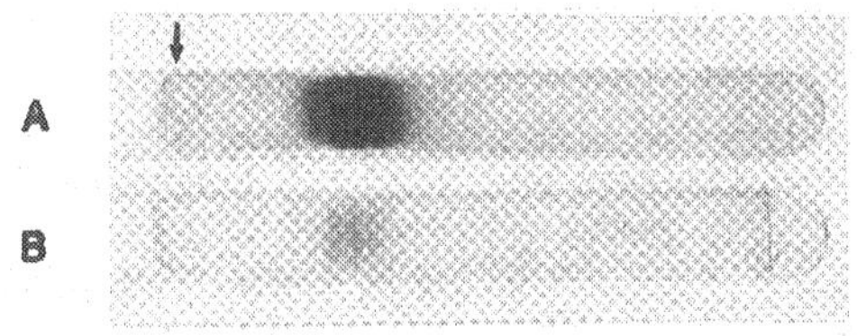

Figure 1. Polyacrylamide gel electrophoresis of the purified *D. immitis* proteinase on a 7% gel/Tris-HCl buffer, pH 8.9. A: Coomassie brilliant blue staining and B: Schiff's reagent staining. The arrow indicates the origin.

rechromatographed on the same column. Enzyme-containing fractions were pooled, and dialyzed against citrate buffer, pH 3.1, containing 500 mM NaCl, and applied onto a pepstatin-Agarose 4B column equilibrated with the same buffer. After washing the column with the same buffer containing 6 M urea, and then with the starting buffer, the bound enzyme was eluted with 100 mM Tris-HCl buffer, pH 8.5, containing 500 mM NaCl. Effluent fractions containing enzyme activity were collected, dialyzed against citrate buffer overnight and rechromatographed on a pepstatin-Agarose gel column as described above, and finally fractions containing enzyme activity were pooled, concentrated and dialyzed against saline.

RESULTS AND DISCUSSION

D. immitis proteinase was purified 437-fold with a specific activity of 384 units/mg protein and some 600 μg of purified enzyme was obtained from 12 g of lyophilized adult worms with a recovery of 16.5% [6].

Characterization of Enzyme Protein

SDS-PAGE of the purified enzyme showed a single protein band with an estimated molecular weight of 42,000, which was consistent with an earlier report by Sreedhara Swamy and Jaffe [5] on *D. immitis* protease, Fp-II, with molecular weight of 48,000. When purified

Table 1. Relative Hydrolysis Rates of Various Substrates with *D. immitis* Acid Proteinase, Cathepsin D and Pepsin

Substrate	Relative hydrolysis rate (%)		
	D. immitis proteinase	Cathepsin D	Pepsin
Human hemoglobin	100	100	100
Bovine hemoglobin	58	80	102
Bovine serum albumin	38	30	76
Human serum albumin	57	37	58
Equine myoglobin	121	83	71
Human γ-globulin	15	15	27
Edestin	13	33	31
Cytochrome C	30	30	101

Incubation: 1 hr

Table 2. Effects of Proteinase Inhibitors on *D. immitis* Acid Proteinase, Cathepsin D and Pepsin

Compound	Concentration (mM)	Inactivation (%)		
		D. immitis proteinase	Cathepsin D	Pepsin
None	-	0	0	0
EDTA	0.5	7	10	0
p-Chloromercuri-benzoate	0.5	50	7	48
Iodoacetic acid	0.5	3	2	6
p-Tosyl-L-lysyl-chloromethyl ketone	0.5	56	7	39
p-Tosylphenylalanine chloromethyl ketone	0.5	3	0	0
1, 2-Epoxy-3-(*p*-nitro-phenoxy)-propane	0.5	5	0	30
Phosphoramidon	0.01	4	0	0
Antipain	0.01	4	3	0
Pepstatin	0.0001	100	100	100
	0.00025	59	100	100
	0.00005	9	24	100
Amastatin	0.01	3	0	0
Leupeptin	0.01	0	5	0
Chymostatin	0.01	0	0	0

D. immitis acid proteinase, cathepsin D and pepsin were incubatedwith various inhibitors at 37 °C for 10 min prior to addition of substrate (human hemoglobin), and assayed for enzyme activity.

enzyme was subjected to classical polyacrylamide gel disc electrophoresis at pH 8.9, however, the enzyme was found to migrate into five separate protein bands consisting of three major and two minor bands [6]. All protein bands showed proteolytic activity with PAS staining positive (Fig. 1), suggesting that distinct migrations of *D. immitis* proteinase on polyacrylamide disc gel are due to different degrees of glycosylation on the enzyme protein, although attempts to treat the enzyme with sialidase failed to modify its migration patterns. Multi-isoforms of *D. immitis* proteinase were further confirmed by demonstrating distinct isoelectric points (pI) ranging 5.8-6.4 by isoelectric-focusing electrophoresis,. The presence of multi-forms was also observed with a proteinase from *Fasciola hepatica* [8], suggesting that the existence of isozymes in parasite proteinase(s) may be a common feature beyond the species.

Enzymatic Characteristics of *D. immitis* Proteinase as Compared with Mammalian Cathepsin D and Pepsin

The pH optimum (2.8-3.4) of *D. immitis* proteinase was rather similar to that (2.8-4.5) of cathepsin D than that of pepsin (1.5-2.5), when human hemoglobin was used as substrate (Fig. 2). This is consistent with a similarity of amino acid compositions of the former two enzymes, in which relative ratio of acidic amino acids to basic amino acids was about 2:1 [6]. When heat stability of the enzymes at neutral pH and 50 °C for

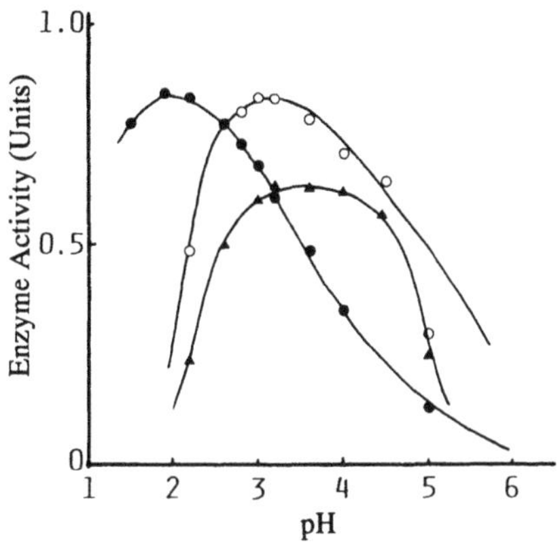

Figure 2. The effect of pH on the enzyme activity of *D. immitis* proteinase, cathepsin D and pepsin. Enzyme activity was assayed using human hemoglobin as substrate at varying pH of 0.1 M citrate-0.2 M Na_2HPO_4 buffer as indicated. Buffer solutions below pH 2.5 were prepared by adjusting with 20% HCl. *D. immitis* proteinase (□), Cathepsin D (▲), Pepsin (•).

5 min was compared, *D. immitis* enzyme activity remained as much as 80%, just like cathepsin D, which was intact, while pepsin showed the complete loss of enzyme activity. After incubation at their optimum pH and 50 °C for 5 min, however, *D. immitis* proteinase and pepsin showed some 50% reduction in their enzyme activities, while cathepsin D remained intact. Table 1 gives the substrate specificity of the three enzymes, which are quite similar to each other, except for cytochrome C, which was a good substrate for pepsin. Among various compounds tested for inhibitory effects on the three enzymes (Table 2), 1, 2-epoxy-3-(*p*-nitrophenoxy)propane (EPNP), a specific inactivator of pepsin and other gastric acid proteases[9], showed little effects on *D. immitis* proteinase and cathepsin D. Under the condition employed, about 30% of pepsin activity was inhibited. Diazoacetyl-DL-norleucine methyl ester (DAN) is known to inactivate pepsin rapidly by esterifying specifically the carboxyl group of aspartic acid at the active site [10]. Reactions of DAN with the three enzymes at individual optimum pH showed that approximately 90% of pepsin activity was inactivated within 10 min at 37°C, while *D. immitis* proteinase was gradually inactivated and approximately 50% of its initial activity was lost after 60 min incubation. Cathepsin D was little affected by DAN even after 60 min incubation at its optimum pH (Fig. 3). This is quite similar to that with *Mucor pusillus*, with which about 40% of the initial activity was inactivated by DAN (Takahashi et al. [11]). These results strongly suggest that *D. immitis* proteinase belongs to the aspartic proteinase family and could be classified as a prototype of mammalian cathepsin D and pepsin.

Immunolocalization of the Proteinase in the Worm

To elucidate tissue distributions of the proteinase in the worm, monoclonal antibodies against the enzyme was generated by immunizing Balb/C mice and fusing the mouse spleen cells with myeloma cells. Immunostainings for the enzyme revealed that the intestine and the inside of uterine wall were well stained as well as some small granules of microfilariae in the uterus. These results led us to the conclusion that the *D. immitis* acid proteinase is secreted into the intestine of the worm. It is well known that proteinases of *Fasciola hepatica* and *Schistosoma mansoni*, which are classified into cysteine proteinase, are synthesized in the intestinal epithelia, and secreted into the lumen [12].

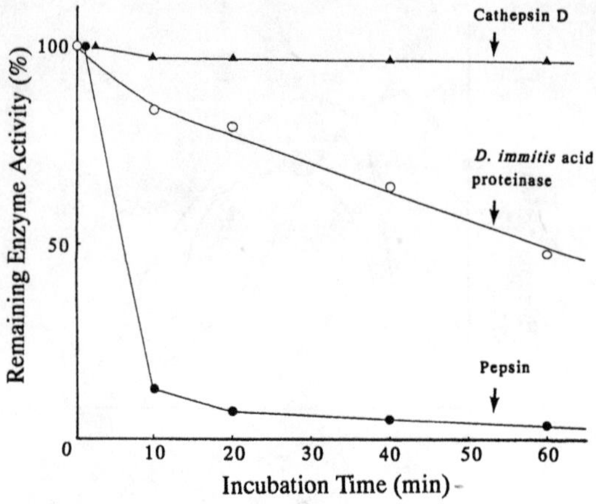

Figure 3. Inactivation curves of *D. immitis* acid proteinase, cathepsin D and pepsin by diazoacetyl-DL-norleucine methyl ester (DAN) as a function of incubation time. Enzyme treatment with DAN was performed as described by Takahashi et al., [11]. The emzymes were incubated with 100-fold molar excess of DAN in the presence of 200-molar excess of cupric acetate in 0.01 M citrate buffer (at optimum pH of them) at 20 °C for 60 min.

ACKNOWLEDGMENT

This study was supported in part by the Japan-US Medical Cooperation Program.

REFERENCES

1. Sauer, M. C. V. and Senft, A. W. (1972) Comp. Biochem. Physiol. 42B, 205-220.
2. Maki, J., Furuhashi, A. and Yanagisawa, T. (1982) Parasitology 84, 137-147.
3. Sato, K., Takahashi, J. and Sawada, T. (1976) Jpn. J. Parasitol. 25, 8-9.
4. Sato, K. and Suzuki, M. (1983) Jpn. J. Parasitol. 32, 577-586.
5. Sreedhara Swamy, K. H. and Jaffe, J. J. (1983) Mol. Biochem. Parasitol. 9, 1-14.
6. Sato, K., Nagai Y. and Suzuki, M. (1993) Mol. Biochem. Parasitol. 58, 293-300.
7. Lowry, O. H., Rosebrough, N. J., Farr, A. L. and Randall, R. J. (1951) J. Biol. Chem. 193, 265-275.
8. Aoki, T. and Oya, H. (1977) Jpn J. Parasitol. 26 (Suppl.), 68.
9. Tang, J., Sepulveda, P., Marciniszyn, J. Jr., Chen, K. C. S., Huang, W-Y., Tao, N., Liu, D. and Lanier, J. P. (1973) Proc. Nat. Acad. Sci. USA 70, 3437-3439.
10. Tang, J. (1973) J. Biol. Chem. 246, 4510-4517.
11. Takahashi, K. Mizobe, F. and Chang W-J. (1972) J. Biochem. 71, 161-164.
12. Yamazaki, H., Aoki, T. and Oya, H. (1989) Jpn. J. Parasitol. 38, 373-384.

INHIBITION AND ENTRAPMENT OF ASPARTIC PROTEINASES BY α_2-MACROGLOBULIN

Senarath B.P. Athauda,[1,2] Hideo Arakawa,[3] Takayuki Takahashi,[1]
Masaaki Nishigai,[1] Eiji Ido,[4] Hiroyuki Kyushiki,[5] Yoshiyuki Yoshinaka,[4]
Atsushi Ikai,[3] Jordan Tang,[6] Masanori Ukai,[7] and Kenji Takahashi[1]

[1]Department of Biophysics and Biochemistry
Faculty of Science
The University of Tokyo, Tokyo 113, Japan
[2]Department of Biochemistry
Faculty of Medicine
University of Peradeniya, Peradeniya, Sri Lanka
[3]Laboratory of Biodynamics
Faculty of Bioscience and Biotechnology
Tokyo Institute of Technology
Yokohama 227, Japan
[4]Research Center for Immunodeficiency Virus
Institute for Virus Research
Kyoto University
Kyoto 606, Japan
[5]Microbiological Research Institute
Otsuka Pharmaceutical Co., Ltd.
Tokushima 771-01, Japan
[6]Protein Studies Program
Oklahoma Medical Research Foundation
Oklahoma City, Oklahoma 73104
[7]Ukai Hospital
Nagoya 453, Japan

Human α_2-macroglobulin(α_2M) is a plasma glycoprotein that inhibits many kinds of proteinases by a molecular trapping mechanism(1,2). It is composed of four identical subunits of M_r~185K, which are linked in pairs by disulfide bonds and two pairs are associated noncovalently to form the native tetramic molecule(3,4). Inhibition of proteinases by α_2M occurs through complex formation initiated by specific limited proteolysis in the bait region of α_2M, located near the middle of each subunit(3,4). Following this cleavage α_2M undergoes a conformational change resulting in entrapment of the proteinases(3).

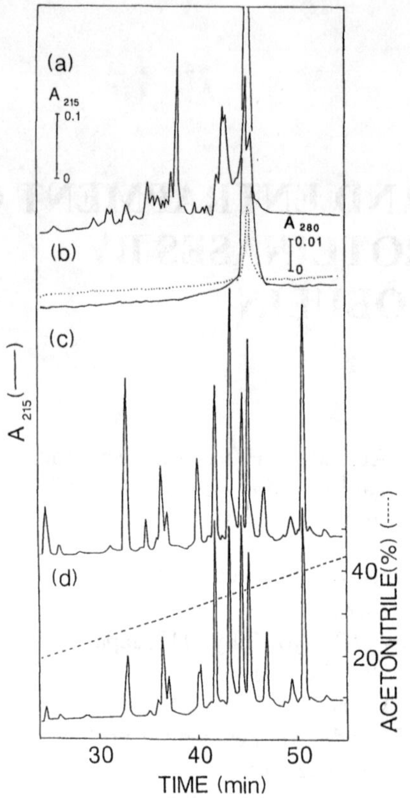

Figure 1. HPLC patterns of 1.5-h digests at pH 5.5 of RCm-ribonuclease A with cathepsin E(a) and α_2M-treated cathepsin E(b) and the B chain of oxidized insulin with cathepsin E(c) and α_2M-treated cathepsin E(d) on a TSKgel ODS-120T column. Solid line, A_{215}; dotted line, A_{280}. A_{280} is shown only for (b).

Inhibition of serine, metallo and thiol proteinases by α_2M was studied extensively. However only a few studies have been reported on intraction of α_2M with aspartic proteinases, because most of the aspartic proteinases are thought to be active mainly in the acidic pH range, in which α_2M loses its proper conformation. Recently, significant proteolytic activities of aspartic proteinases were recognized at neutral pHs(5,6). This finding led us to investigate the intraction of α_2M with aspartic proteinases and the resulting conformational changes of the inhibitor.

MATERIALS AND METHODS

Pepsinogens and procathepsin E were purified from the human gastric mucosa and their mature forms, pepsins and cathepsin E, were obtained by activating the respective zymogen as reported previously(7,8). A chemically synthesized gene of HIV-1 proteinase was overexpressed in *Escherichia coli* and the expressed protein was purified from the inclusion bodies to apparent homogeneity and refolded as describrd previously(9). α_2M was purified from pooled human blood samples as described previously(10). The B chain of oxidized bovine insulin and bovine pancreatic ribonuclease A were purchased from Sigma

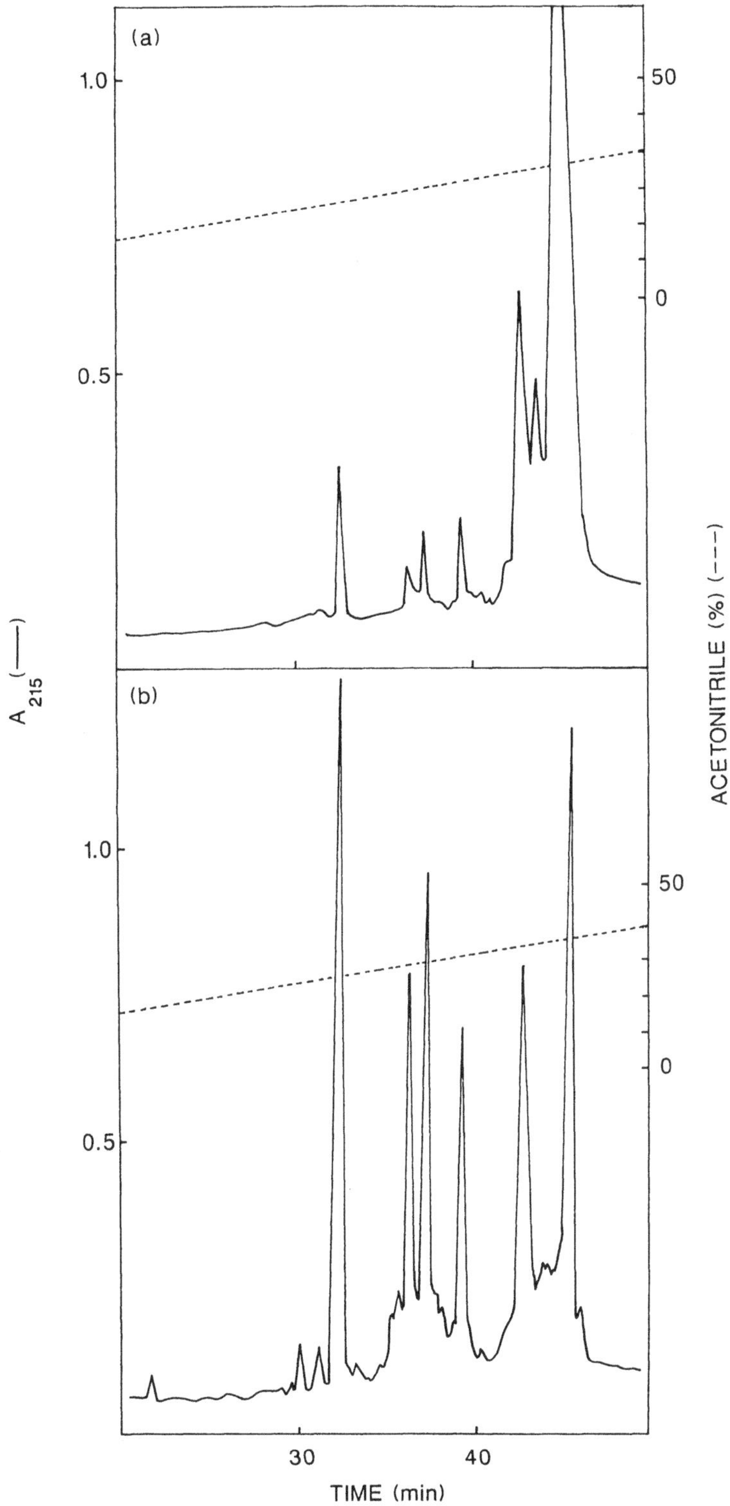

Figure 2. HPLC patterns of 2-h digests at pH 5.5 of RCm-ribonuclease A with α_2M-treated pepsin C(a) and pepsin C(b) on a TSKgel ODS-120T column.

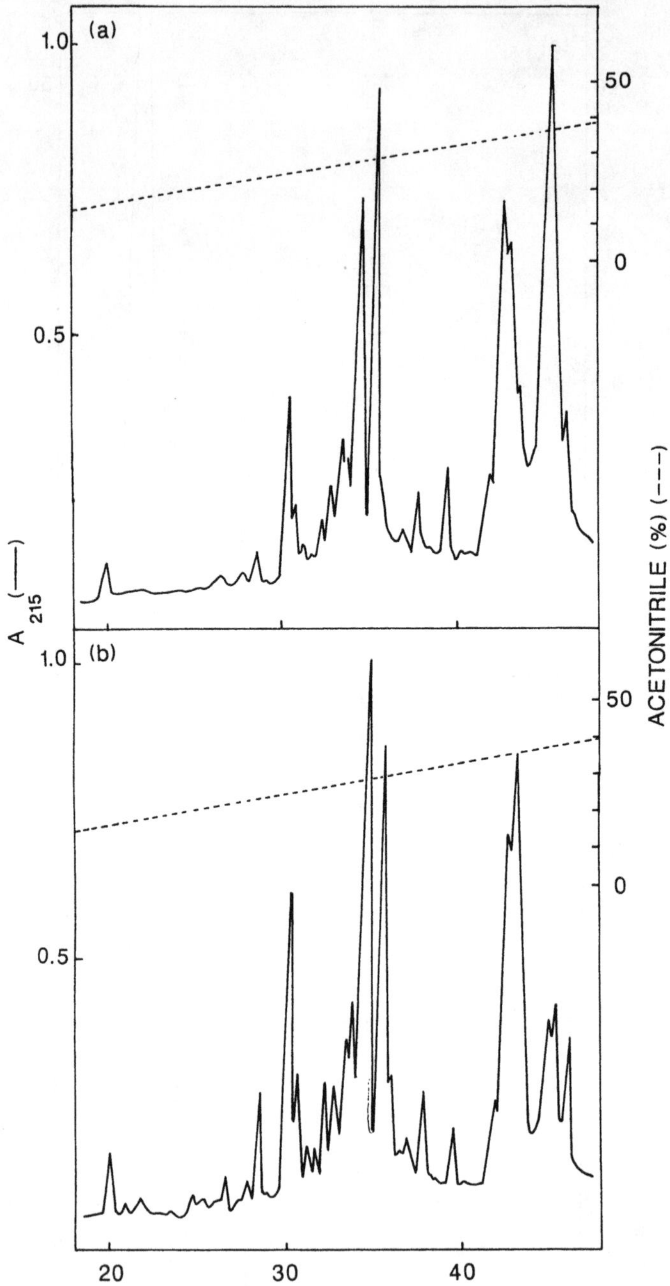

Figure 3. HPLC patterns of 2-h digests at pH 5.5 of RCm-ribonuclease A with α_2M-treated pepsin A-5(a) and pepsin A-5(b) on a TSKgel ODS-120T column.

and Worthington, respectively. RCm-ribonuclease A was prepared according to Crestfield *et al.*(11). The Moloney murine sarcoma virus(MSV) substrate (Pr^{65} gag polyprotein) was prepared as described by Yoshinaka and Luftig(12).

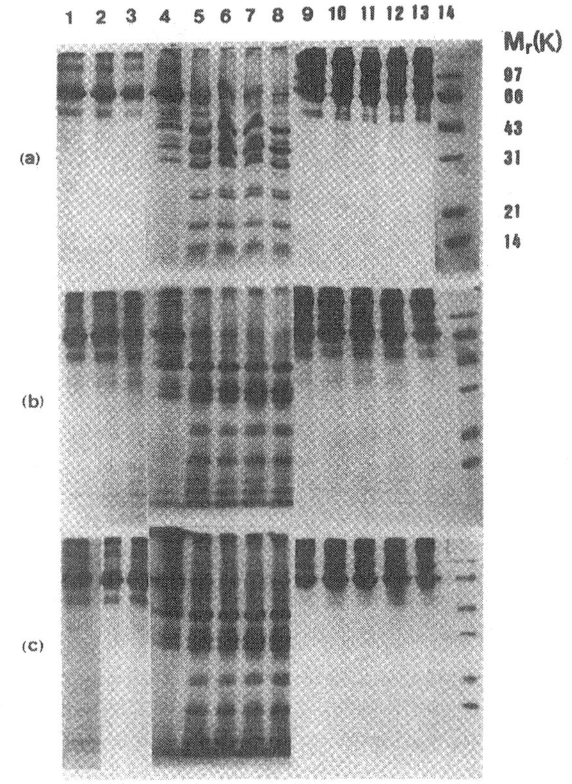

Figure 4. SDS-PAGE analysis of the MSV substrate(lanes 1-3) and its digests at pH 5.5(a), 6.5(b), and 7.4(c) with HIV proteinase(lanes 4-8) or HIV proteinase treated with α_2M at an α_2M: enzyme molar ratio of 1:2(lanes 9-13). Lanes 1,4 and 9; 5 and 10; 2,6 and 11; 7 and 12; 3,8 and 13 represent the aliquots collected at 0, 0.5, 4, 12 and 18 h, respectively, and lane 14 shows molecular weight marker proteins(Pharmacia).

α_2M-treated proteinase was prepared by mixing proteinase and α_2M in a molar ratio of 2:1 in 0.25 M sodium acetate buffer, pH 5.5, and incubating the mixture at 37 °C for 30 min.

RESULTS AND DISCUSSION

Inhibition of Aspartic Proteinases by α_2M

When RCm-ribonuclease A(M_r~13K) was incubated with cathepsin E (500:1 mol/mol) at pH 5.5 for 1.5 h, about 73% of the original substrate protein disappered and several peptide peaks appeared on HPLC (Fig. 1a). On the other hand, only a signle peak of intact RCm-ribonuclease A was observed when the protein was incubated with α_2M-treated cathepsin E at pH 5.5 for 1.5 h (Fig. 1b). This indicates an apparently complete inhibition of proteolytic activity of cathepsin E toward RCm-ribonuclease A by α_2M.

When RCm-ribonuclease A was incubated with pepsin C(100:1 mol/mol) at pH 5.5 for 2 h, about 80% of the original substrate protein disappeared and several peptide peaks which represent the cleavage products of RCm-ribonuclease A appeared on HPLC (Fig. 2b). On the other hand, a higher percentage(80%) of the intact RCm-ribonuclease A with a smaller

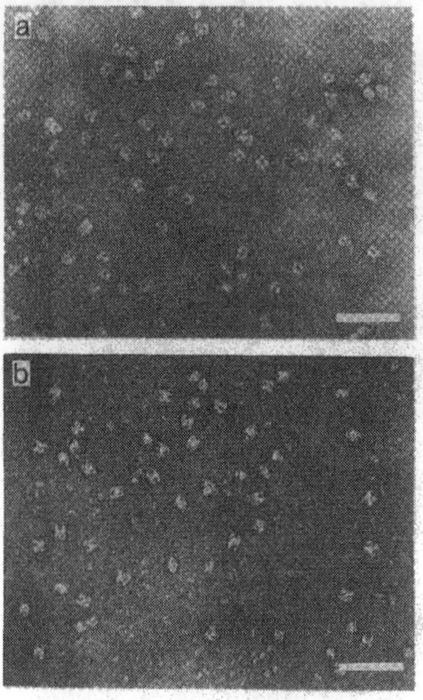

Figure 5. Electron micrographs of α_2M before(a) and after(b) cathepsin E treatment at pH 5.5. Proteins were negatively stained with 3% uranyl acetate without prior fixation. The scale bar represents 1,000 A.

number of peptide peaks were observed when the protein was incubated with α_2M-treated pepsin C at pH 5.5 (Fig. 2a). This indicates a significant inhibition of proteolytic activity of pepsin C toward RCm-ribonuclease A by α_2M.

When RCm-ribonuclease A was incubated with pepsin A-5 (100:1 mol/mol) at pH 5.5 for 2 h about 95% of the original substrate protein disappeared and several peptide peaks, which represent the cleavage products of RCm-ribonuclease A appeared on HPLC (Fig. 3b). On the other hand, a significant amount(30%) of the remaining intact RCm-ribonuclease A with a smaller number of peptide peaks were observed when the protein was incubated with α_2M-treated pepsin A-5 at pH 5.5 (Fig. 3a). This shows a partial inhibition of the proteolytic activity of pepsin A-5 toward RCm-ribonuclease A by α_2M.

When the MSV substrate was incubated with HIV proteinase at pH 5.5, 6.5 and 7.4, the original protein rapidly disappeared (within 30 min) and several low molecular weight polypeptide bands appeared on SDS-PAGE(Fig. 4, lanes 4-8). On the other hand, the polypeptide band of the original protein remained intact on SDS-PAGE when the protein was incubated with α_2M-treated HIV proteinase at pH 5.5-7.4 for 18 h (Fig. 4, lanes 9-13). This clearly demonstrated the complete inhibition of the proteolytic activity of HIV proteinase toward MSV substrate by α_2M.

Similar extents of inhibition by α_2M were observed at proteinase:α_2M molar ratios of 0.25:1 to 2:1 with the above aspartic proteinases (cathepsin E, pepsin C, pepsin A-5 and HIV proteinase). At molar ratios above 2:1, however, the excess enzyme was not inhibited (data not shown). Thus the stochiometry of the inhibition was approximately 2:1 (enzyme:α_2M mol/mol). This stochiometry was further confirmed by analysis using gel filtration and polyacrylamide gel electrophoresis(data not shown). Interstingly this stochiometry did not depend on the size of the aspartic proteinases analyzed (cf. M_r : cathepsin E, 80K ; pepsins A-5 and C, 38K ; and HIV proteinase, 20K).

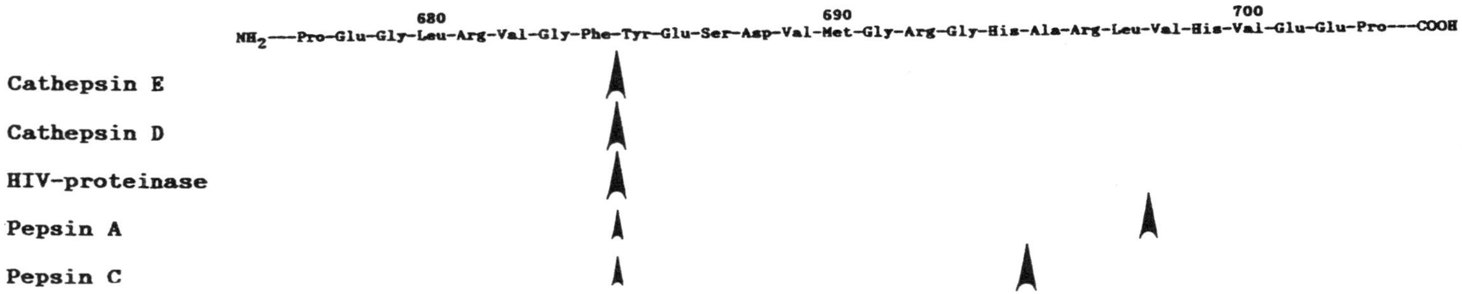

Figure 6. Cleavage sites in the bait region of α_2M by aspartic proteinases at pH 5.5. Large and small arrowheads indicate the major and minor cleavage sites.

On the contrary, similar HPLC elution patterns were obtained with 2-h digests of the B chain of oxidized insulin(M_r~3K) by cathepsin E and α_2M-treated cathepsin E(Fig. 1 c and d). Similar results were also obtained with pepsins A and C, and HIV proteinase. This indicates no inhibition of the activity of the above proteinases toward the B chain of oxidized insulin. Thus the active site of the proteinase in the α_2M-proteinase complex is not freely accessible to high molecular weight substrates(RCm-ribonuclese A for cathepsin E, and pepsins A-5 and C, and MSV substrate for HIV proteinase), although accessible to the low molecular weight substrate, the B chain of oxidized insulin. This indicates the entrapment of these aspartic proteinases by α_2M.

Structural Changes in α_2M upon Complex Formation with Aspartic Proteinases

The electron micrographs of negatively stained native α_2M and cathepsin E treated-α_2M at an α_2M : proteinase molar ratio of 1:2, at pH 5.5 are shown in Fig. 5. The shape of the intact native α_2M resembled a doughnut with a prominent central cavity. On the other hand H-shaped molecules were observed with cathepsin E treated-α_2M. This indicates that a drastic conformational change of α_2M took place upon cathepsin E treatment. A similar drastic conformational change of α_2M was also observed upon treatment at pH 5.5 with pepsins A and C and HIV proteinase. This H-shaped forms of aspartic proteinase-α_2M complexes are similar to those observed for the complexes of α_2M with trypsin(13), chymotrypsin(14) and plasmin(15).

Limited Proteolysis in the Bait Region of α_2M by Aspartic Proteinases

Previously, cleavage sites in the bait region of α_2M by several proteinases belonging to serine, metal and cysteine proteinase families were studied extensively but few studies were reported on those by aspartic proteinases. The cleavage sites in the bait region of α_2M by aspartic proteinases were deduced by determining the NH_2-terminal amino acid sequences of the cleavage products which were separated by SDS-PAGE under reducing conditions. Figure 6 summarizes the cleavage sites in the bait region of α_2M by aspartic proteinases at pH 5.5. Interestingly, the cleavage site by cathepsin E and HIV proteinase (Phe^{684}-Tyr^{685}) is identical with that by chymosin (16). However this was a minor cleavage site for pepsin isozymes, A-5 and C. The major cleavage sites by pepsin A-5 and pepsin C were Leu^{697}-Val^{698} and His^{694}-Ala^{695}, respectively.

Physiological significance of entrapment of these aspartic proteinases by α_2M and the resulting inhibition is not clear at present. However, occurrence of antibodies against HIV proteinase in the plasma of AIDS patients were reported(17), suggesting possible presence of HIV proteinase in their sera. Therefore, if HIV proteinase is released into the plasma of AIDS patients either from the virus infected lymphocytes or from virus particles disrupted by immunological reaction, it will be trapped by α_2M. Further, cathepsin E was isolated from neutrophils and erythrocytes as membrane-bound forms and pepsinogens A and C were found in serum. If these membrane bound forms are released into serum or if serum pepsinogens are activated under certain pathological conditions, α_2M will form complexes with them to hinder any unwanted proteolysis of physiologically important proteins.

ACKNOWLEDGMENTS

We thank Prof. Hikoichi Sakai for the use of an electron microscope, Drs. Hideshi Inoue and Masaru Tanokura in our laboratory and Dr. Masao Ichinose (Faculty of Medicine,

The University of Tokyo) for helpful discussions and Mrs. Yasuko Sakurai for her able technical assistance.

REFERENCES

1. P.M. Starkey and A.J. Barrett, Proteinases in Mammalian Cells and Tissues(Barrett, A.J., ed.), Elsevier North-Holland Biomedical Press, Amsterdam pp. 663-696(1977).
2. A.J. Barrett, Methods Enzymol. 80:737-754(1981).
3. A.J. Barett and P.M. Starkey, Biochem. J. 133:709-724(1973).
4. A.J. Barrett, M.A. Brown and C.A. Sayers, Biochem. J. 181:401-418(1979).
5. S.B.P. Athauda, T. Takahashi, H. Inoue, M. Ichinose and K. Takahashi, FEBS Lett. 292,53-56(1991).
6. S. Billich, M.T. Knoop, J. Hansen, P. Strop, J. Sedláček, R. Mertz and K. Moelling, J. Biol. Chem. 263,17905-17908(1988).
7. S.B.P. Athauda, M. Tanji, T. Kageyama and K. Takahashi, J. Biochem. 106,920-927(1989).
8. S.B.P. Athauda, T. Takahashi, T. Kageyama and K. Takahashi, Biochem. Biophys. Res. Commun. 175:152-158(1991).
9. E. Ido, H. Han, J.F. Kezdy and J. Tang, J. Biol. Chem. 266:24359-24366(1991).
10. H. Arakawa, M. Nishigai and A. Ikai, J. Biol. Chem. 264:2350-2356(1989).
11. A.M. Crestfield, S. Moore and W.H. Stein, J. Biol. Chem. 238:622-627(1963).
12. Y. Yoshinaka and R.B. Luftig, J. Gen. Virol. 48:329-340(1980).
13. J. Tapon-Bretaudiere, A. Bros, E. Couture-Tosi and E. Delain, EMBO J. 4:85-89(1985).
14. M. Nishigai, T. Osada and A. Ikai, Biochim. Biophys. Acta 831:236-241(1985).
15. S.L. Gonias, M.M. Allietta, S.V. Pizzo, F.J. Castellino and T. W. Tillack, J. Biol. Chem. 263:10903-10906(1988).
16. J. Travis and G.S. Salvesen, Annu. Rev. Biochem. 52:655-709(1983).
17. C.A.B. Boucher, M.H. De Jager, C. Debouck, L.G. Epstein, F. De Wolf, T.F.W. Wolfs and J. Goudsmit, J. Clin. Microbiol. 27:1577-1581(1989).

RECOMBINANT HUMAN CATHEPSIN E

Jeffrey Hill,[1] Douglas Montgomery,[2] and John Kay[1]

[1]Department of Biochemistry
University of Wales
Cardiff, CF1 1ST, Wales, UK
[2]Glaxo Group Research Ltd
Greenford, Middlesex, UB6 OHE, UK

INTRODUCTION

Cathepsin E is a non-lysosomal, non-secretory, intracellular aspartic proteinase. Its cellular distribution is variable between species but includes the stomach, spleen, red and white blood cells and the skin. The physiological role of this enzyme is unclear, but it has been implicated in a number of vital processes. Possible functions include, biogenesis of the potent vasoconstrictor peptide, endothelin (1), antigen processing (2) and aiding the infection of cells by the human immunodeficiency virus (3). Cathepsin E, being neither secretory nor lysosomal, is difficult to isolate in reasonable amounts and, due to its physiological importance, is an ideal candidate for cloning and expression in a heterologous host.

RESULTS AND DISCUSSION

Total RNA was isolated from a human gastric adenocarcinoma cell line, which is known to produce large amounts of cathepsin E (4). From this, procathepsin E cDNA was generated as described in (5) using specific primers which were designed to include *Nco*I and *Sal*I restriction sites to facilitate subsequent cloning.

The 1100 bp *Nco*I-*Sal*I fragment, the procathepsin E coding sequence, was subcloned into a series of *E.coli* expression vectors, under the control of different promoters. pET3d contains the T7 promoter and was employed with a view to production of authentic procathepsin E preceded simply by the initiation Met residue. pET22b was assessed in an attempt to secrete procathepsin E as a consequence of expressing the *pel* B leader sequence fused at the N-terminus of the protein. This vector contains a T7*lac* promoter (a T7 promoter immediately followed by the *lac* operator). It also contains the *lac*i gene to provide sufficient amounts of *lac* repressor to offset the basal level expression of the target gene. This is a particularly beneficial feature if the target gene product is toxic. The results of the expression studies carried out are summarised in Table I.

Comparison of the levels of full length procathepsin E expressed from the pET3d and pET22b constructs indicated that pET22b was considerably better. Possible explanations

Aspartic Proteinases: Structure, Function, Biology, and Biomedical Implications
Edited by Kenji Takahashi, Plenum Press, New York, 1995

Table I. Expression Plasmids Constructed to Encompass the Human Procathepsin E Gene

Construct	Promoter	Induced by	Level of Expression (mg/l of culture)
pPLRES	λPL	heat jump 30°C→42°C	ND
pAT*tac*	*tac*	IPTG	ND
pET3d	T7	IPTG	3
pET22b	T7*lac*	IPTG	12

(ND = not detectable)

for this include improved translation initiation with the *pel* B leader sequence and tighter transcriptional control of the T7*lac* promoter compared with the plain T7 promoter. In both cases, the recombinant protein accumulated in the form of insoluble aggregates. The insolubility of the recombinant enzyme may have allowed the detection of its expression, as the formation of inactive aggregates prevented host cell toxicity and / or degradation of the recombinant enzyme.

For larger scale purification, the *E.coli* expression strain, BL21(DE3)pLysS, was transformed with the recombinant pET22b plasmid containing the gene for procathepsin E and a single colony was used to inoculate a 10 ml culture. When the absorbance of the culture at 600 nm was approximately 0.6, the cells were harvested by centrifugation (3,000g for 10 minutes) and the pellet was resuspended in 2 ml of fresh medium. The following day, the resuspended cells were used to inoculate a 4 l culture. When the absorbance of the culture at 600 nm was approximately 0.6, IPTG was added to a final concentration of 1 mM and growth was allowed to continue for a further 2 hours. The cells were separated from the culture medium by centrifugation (3,000g for 10 minutes) and the insoluble material was extracted, washed and solubilised as described previously (5).

A simple procedure was devised to purify the recombinant material in which the majority of contaminating *E.coli* proteins were precipitated out by lowering the pH to 3.1 (5) and then reneutralising. This straightforward purification scheme generated homogeneous material without the need to resort to column chromatography.

A unique N-terminal sequence was determined for the recombinant material through 25 cycles:

Recombinant:	Thr-Glu-Ser-Cys-Met-Asp-Gln-Ser-Ala-Lys~
Naturally-occurring:	Ile-Gln-Phe-Thr-Glu-Ser-Cys-Met-Asp-Gln-Ser-Ala-Lys~
Naturally-occurring:	Thr-Glu-Ser-Cys-Met-Asp-Gln-Ser-Ala-Lys~

Depending on its source of origin (stomach or red blood cells), naturally-occurring mature cathepsin E has been shown to be heterogeneous at its N-terminus (6,7), consisting of two closely similar isozymes (8) which differ by only three residues. Thus, the recombinant material that was isolated by the above purification procedure was the mature form of the enzyme, ie cathepsin E and its sequence was exactly coincident with that predicted for mature cathepsin E from the nucleotide sequence (4). Since the final step of the purification procedure involved acidification, it is perhaps not surprising that procathepsin E had undergone activation to generate the shorter of the two isozymes of mature cathepsin E in a homogeneous form.

The concentration of active enzyme in each preparation was determined by active site titration using isovaleryl pepstatin (9). On this basis, it was estimated that the yield of homogeneous, active enzyme was approximately 1 mg from a 4 litre culture. This amount

Table II. Kinetic Parameters for the Interactions of Substrates and Inhibitors with Recombinant Human Cathepsin E at pH 3.1 and 37°C

Substrate or inhibitor	K_i (nM)	K_m (μM)	k_{cat} (s^{-1})
Pro-Pro-Thr-Ile-Phe*Nph-Arg-Leu	—	65	120
Lys-Pro-Ile-Glu-Phe*Nph-Arg-Leu	—	50	115
Ascaris protein	5	—	—
Isovaleryl-pepstatin	0.3	—	—

is considerably more than can be purified readily from naturally-occurring human cells or tissues.

Kinetic parameters (K_m, k_{cat}) were determined for the cleavage of two chromogenic substrates by the recombinant enzyme (Table II). The values derived were directly comparable to those obtained previously with naturally-occurring cathepsin E from both gastric and erythrocyte sources (8). We have also shown previously that cathepsin E may be distinguished from the lysosomal enzyme, cathepsin D, by its susceptibility to inhibition by a 17 kDa protein inhibitor from the parasitic worm, *Ascaris lumbricoides* (9). The K_i value determined (Table II) for the interaction of this protein with the recombinant enzyme was essentially identical to those determined previously (8) for naturally-occurring cathepsin E from different tissues/cells/species. Isovaleryl pepstatin was a very tight binding inhibitor (Table II), of the aspartic proteinase, cathepsin E (10).

Cathepsin E isolated from natural sources, is notorious for its heterogeneity: (a) at its N-terminus because of the presence of two scissile peptide bonds that are hydrolysed during autoactivation of the precursor at acid pH (7); (b) in being a glycoprotein (10, 6); (c) in having an extra cysteine residue, (residue 4- see sequence above), compared to other aspartic proteinases which forms an interchain disulphide bond so that an equilibrium mixture of monomeric and dimeric enzymes results. These factors, as well as lack of reasonable quantities of naturally occurring material, have not allowed crystallisation and structural determination to be undertaken. However, now that large quantities of homogeneous cathepsin E can be generated in, and purified easily from *E.coli,* these studies can now be carried out.

ACKNOWLEDGMENTS

JH was supported by an SERC-CASE award in collaboration with Glaxo Group Research Ltd. The valuable contributions of Wendy Lees, Pete Tatnell, Ben Dunn, Benne Parten and Benny Chain are gratefully acknowledged.

REFERENCES

1. Lees, W.E., Kalinka, S., Meech, J.D., Capper, S.J., Cook, N.D. and Kay, J. Biochem. J. 273, 99 (1990).
2. Bennet, K., Levine, T., Ellis, J.S., Peanasky, R.J., Samloff, I.M., Kay, J. and Chain, B.M. Eur. J. Immunol. 22, 1519 (1992).
3. Clements, G.J., Price-Jones, M.J., Stephens, P.E., Sutton, C., Schulz, T.F., Clapham, P.R., McKeating, J.A., McClure, M.O., Thompson, S., Marsh, M., Kay, J., Weiss, R.A. and Moore, J.P. AIDS Res. 7, 3 (1991).

4. Azuma, T., Pals, G., Mohandas, T.K., Couvreur, J.M. and Taggart, R.T. J. Biol. Chem. 264, 16748 (1989).
5. Hill, J., Montgomery, D.S. and Kay, J. FEBS Lett. 326, 101 (1993).
6. Athauda, S.B.P., Matsuzaki, O., Kageyama, T. and Takahashi K. Biochem. Biophys. Res. Commun. 168, 878 (1990).
7. Athauda, S.B.P., Takahashi, T., Kageyama, T. and Takahashi K. Biochem. Biophys. Res. Commun. 175, 152 (1991).
8. Jupp, R.A., Richards, A.D., Kay, J., Dunn, B.M., Wyckoff, J.B., Samloff, I.M. and Yamamoto, K. Biochem J. 254, 895 (1988).
9. Valler, M.J., Kay, J., Aoyagi, T. and Dunn, B.M. J. Enzyme Inhib. 1, 77 (1985).
10. Samloff, I.M., Taggart, R.T., Shiraishi, T., Branch, T., Reid, W.A., Heath, R., Lewis, R.W., Valler, M.J. and Kay, J. Gastroenterology 93, 77 (1987).

EXPRESSION OF HUMAN CATHEPSIN E IN METHYLOTROPHIC YEAST, *PICHIA PASTORIS*

Masayuki Yamada,[1] Takao Matsuba,[1] Takeshi Azuma,[2] Hideo Suzuki,[1] Kenji Yamamoto,[3] and Hitoshi Hori[1]

[1]Biotechnology Research Laboratory
Tosoh Corporation
Hayakawa, Ayaseshi, Kanagawa 252, Japan
[2]Second Department of Internal Medicine
Fukui Medical School
Matsuoka, Yoshino-gun, Fukui 910-11, Japan
[3]Department of Pharmacology
Kyushu University, Faculty of Dentistry
Fukuoka 812, Japan

The human gastric cathepsin E (CTSE) was expressed in the methylotrophic yeast *Pichia pastoris* by placing the CTSE cDNA under the control of the methanol-inducible alcohol oxidase promoter. The human CTSE expressed in *P.pastoris* was efficiently secreted into the culture medium as an active enzyme directed by its native signal sequence, whereas CTSE has been shown to be retained in mammalian tissue cells. The recombinant human CTSE was secreted as a 90-kDa molecule and then converted via an 84-kDa intermediate to an 82-kDa molecule. The 90-kDa molecule and the 82-kDa molecule were considered to be the proenzyme and the mature enzyme as dimeric forms, respectively.

INTRODUCTION

Cathepsin E (CTSE) (EC 3.4.23.34) is an intracellular aspartic proteinase consisting of two identical subunits with a molecular mass of around 42-kDa [1-3].

In contrast to a ubiquitous tissue distribution of cathepsin D (CTSD), a lysosomal aspartic proteinase, CTSE has been localized only to certain tissues such as gastric mucosa and several lymphoid associated tissues [3-7].

Although CTSE is thought to play an important physiological role in the cell and have pathological significance, the function of this proteinase is still unclear.

Moreover, CTSE has not been characterized well due to the difficulties in preparing this proteinase in large quantities resulting from its limited tissue distribution and low concentration in tissues. It would be of great help in facilitating studies on CSTE to establish a CTSE production system by use of a genetically engineered host.

The human gastric CTSE cDNA has been cloned and found to code for a 379-residue CTSE proenzyme (calculated molecular weight: 40,883 daltons) that contained 7 cysteine residues and a 17-residue signal peptide [8].

Expressions of human cathepsin E in *E.coli* [9] and CHO cells [10] have been reported.

Here we report the expression of the enzymatically active human gastric CTSE by secretion in the methylotrophic yeast, *Pichia pastoris.* This will also serve as a basis for the characterization of genetically modified CTSE.

MATERIALS AND METHODS

Bacterial and Yeast Strains

E.coli JM103 was used for all plasmid constructions and propagation.

P.pastoris GS115 (his4) [11] was used for expression of human CTSE.

Media

BMGY medium contained the following (per liter): Yeast extract (10 g), peptone (20 g), glycerol (10 mL), biotin (400 μg), yeast nitrogen base with ammonium sulfate (13.4 g) and 1M potassium phosphate buffer (pH6.0) (100 mL). BMMY medium had the same composition as BMGY medium with the exception that 5mL methanol/liter was added in the place of glycerol.

Antibody to Human CTSE

Rabbit polyclonal antiserum against purified human CTSE was prepared as described previously [12].

Construction of Human CTSE Expression Vectors

The 1.7-kb *Eco*RI fragment (bp -36 to 1695) of the human CTSE cDNA in clone 412 [8] was subcloned into the *Eco*RI site of Bluescript II SK$^+$ (Stratagene) to give the pGA412 plasmid. *P.pastoris* expression vector pAO804 contains a unique *Eco*RI site flanked by the 5' and 3' regulatory sequences of the methanol-induced alcohol oxidase gene (AOX1) of *P.pastoris* [13].

The full coding sequence of human CTSE was excised out of pGA412 as a 1.8-kb *Sma*I-*Eco*RI fragment and ligated with pAO804, which had been digested with *Eco*RI and treated with DNA polymerase I Klenow fragment. Plasmid with the insert in proper orientation was designated pHIL-D1CTSEv1. Then the 5' noncoding sequence of 80-bp derived from human CTSE cDNA in pHIL-D1CTSEv1 was replaced with the oligonucleotides that has the same sequence as that of *P.pastoris* AOX1 gene to generate pHIL-D1CTSEv2 (Figure 1).

Expression of Human CTSE in *P.pastoris* Shake Flask Cultures

Cells grown in 100 mL of BMGY at 30°C to an A600 of 15-20 were harvested, resuspended in 100 mL of BMMY (methanol induction) and incubated at 30°C for 100 hours

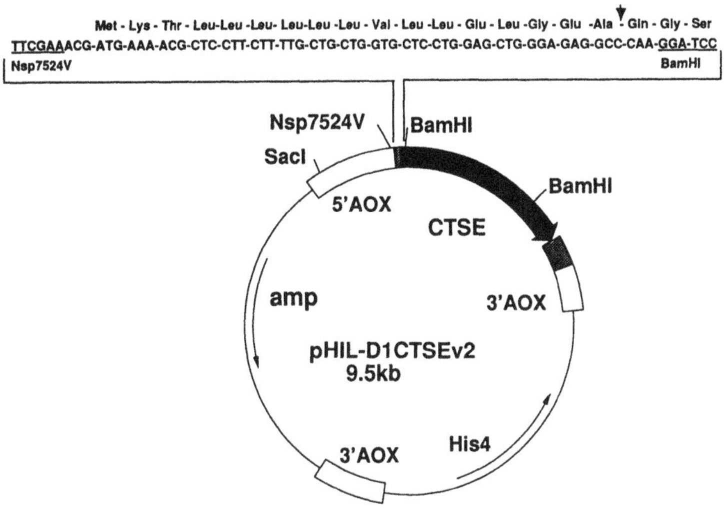

Figure 1. Structure of the human CTSE expression plasmid. The sequence of the synthetic DNA inserted between the Nsp7524V site and the *Bam*HI site to replace the native 5' noncoding region is shown. The predicted amino acid sequence of the N-terminal region of prepro-human CTSE is indicated above the DNA sequence. The arrow head indicates the predicted signal sequence cleavage site.

to induce expression. After methanol induction, methanol concentrations in culture medium were measured by HPLC on a TSKgel ODS120T column and maintained between 0.2 and 0.8%w/v by proper additions of methanol.

Western Blotting Analysis and Hemoglobin Zymography

Proteins in 300 μl culture supernatants were concentrated by TCA precipitation, separated by sodium dodecyl sulfate-polyacrylamide gel electrophoresis (SDS-PAGE) and were transferred to nitrocellulose using the Transblot apparatus (BioRad Laboratories) according to the company's specifications. The transferred proteins were probed with rabbit anti-human CTSE polyclonal antibodies. Goat anti-rabbit IgG coupled to horseradish peroxidase was used as the second antibody followed by development with chloronaphthol or using ECL system (Amersham) according to the company's specifications. For the detection of the proteolytically active band, 800 μl culture supernatants were concentrated by Centricon (Amicon) and subjected to polyacrylamide gel electrophoresis according to the method of Williams and Reinsfeld [14]. Then the gel was first immersed in 0.7% hemoglobin solution (pH3.0), then incubated for 30 min at room temperature, and stained with 0.05% CBB (R250) , followed by destaining with 10% acetic acid.

RESULTS AND DISCUSSION

Spheroplasts of GS115 were transformed by polyethylene glycol method [15] with *Bgl* II-digested pHIL-D1CTSEv2. His$^+$/methanol slow (i.e., impaired growth on media containing methanol as the sole carbon and energy source) transformants, in which the AOX1 structural gene was presumably replaced with the human CTSE expression cassette and His4, were selected as described by Cregg et al. [15]. Spheroplasts were also transformed with pAO804 to obtain a His$^+$/methanol slow tranformant used as a negative control.

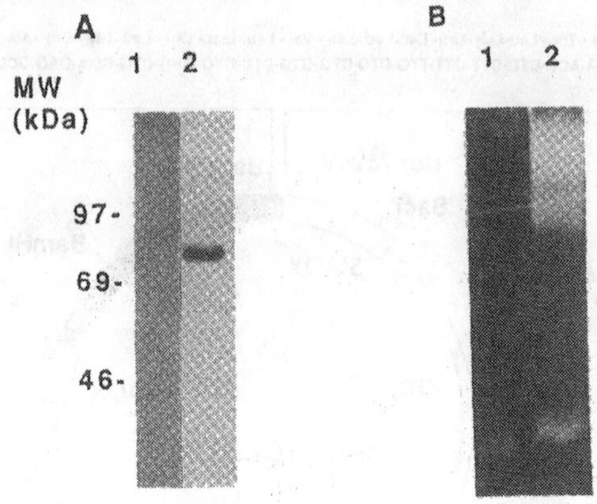

Figure 2. **A**: Western blot analysis of the culture supernatants. Samples of the culture supernatants of 98 hours after methanol induction were analyzed by Western blot as described in materials and methods. Lane 1: GS115/pAO804 (negative control); lane 2: GS115/pHIL-D1CTSEv2. **B**: Hemoglobin zymography of produced protein. Samples of the culture supernatants of 98 hours after methanol induction were loaded on an 8% polyacrylamide gel and proteolitically active bands were detected as described in materials and methods. Lane 1: GS115/pAO804 (negative control); lane 2: GS115/pHIL-D1CTSEv2.

His$^+$/methanol slow transformants of GS115 with the human CTSE expression plasmids were examined in shake flask cultures as described in materials and methods.

GS115/pHIL-D1CTSEv2 were found to secrete an 82-kDa protein reactive with rabbit anti-human CTSE serum in the culture medium (Figure 2A). The result of hemoglobin zymography demonstrated that produced protein was enzymatically active at pH3.0 (Figure 2B). The signals appeared in the lower part of each lane was considered to be the activity of an acid proteinase derived from *P.pastoris* .

The Western blot analysis following SDS-PAGE under a reduced condition showed that the secreted protein was a dimeric form of proteins with the molecular mass expected from the predicted amino acid sequence of human CTSE [8] (Figure 3).

These results indicated that the human CTSE expressed in *P.pastoris* was correctly folded and was recognized as a protein destined for secretion after it transited the ER directed by its native signal sequence, while CTSE has been shown to be retained in mammalian cells [16-20]. The secretion of the human CTSE in *P.patoris* may reflect a difference in the

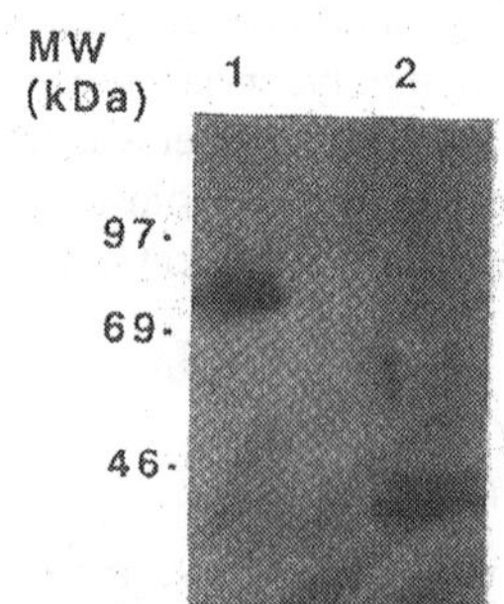

Figure 3. Reduction of the produced protein. Reduced (lane 2) and non-reduced (lane 1) samples of the GS115/pHIL-D1CTSEv2 culture supernatants were subjected to a 10% SDS-polyacrylamide gel electrophoresis and analyzed by Western blot using ECL system (Amersham).

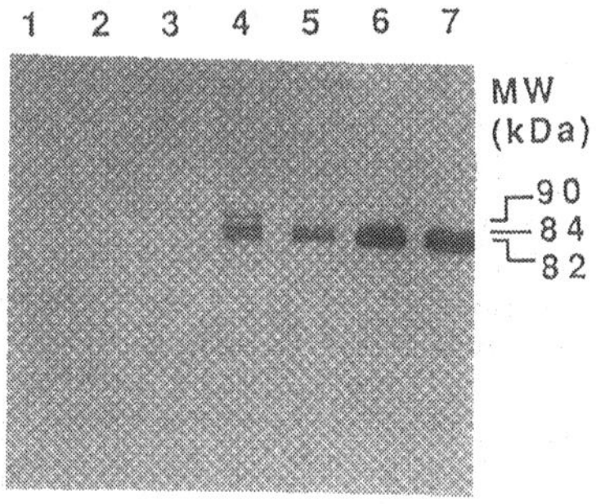

Figure 4. Time course of humanCTSE production. Samples of the GS115/pHIL-D1CTSEv2 culture supernatants were subjected to a 10% SDS-polyacrylamide gel electrophoresis and analyzed by Western blot. Lane 1, 0 hour; lane 2, 16 hours; lane 3, 22 hours; lane 4, 40 hours; lane 5, 46 hours; lane 6, 63 hours; lane 7, 98 hours after methanol induction, respectively.

mechanism of protein compartmentalization or in the glycosylation pattern of proteins between *P.pastoris* and mammalian cell.

The culture medium of GS115/pHIL-D1CTSEv2 was withdrawn at several intervals after methanol induction and was subjected to Western blot analysis. The result showed that the human CTSE was secreted as a 90-kDa protein, converted to an 84-kDa protein and finally accumulated as an 82-kDa protein in the culture medium (Figure 4) , suggesting that the human CTSE was secreted as a dimeric proenzyme and processed to a dimeric mature enzyme via an intermediate form. The molecular weight of the produced human CTSE also indicated that it did not receive hyperglycosylation as reported on several proteins produced in *S.cerevisiae* [21,22].

Whether the conversion of dimeric proenzyme to its mature form was due to self-activating activity remains to be elucidated.

In conclusion, we have shown that the recombinant human CTSE expressed in *P.pastoris* can be secreted by its native signal sequence as an active dimmer enzyme. This human CTSE production system will provide an invaluable tool to characterize CTSE and , moreover, to investigate genetically modified CTSE.

REFERENCES

1. Yamamoto, K., Katsuda, N., and Kato, K. (1978) Eur.J.Biochem. 92, 499-508.
2. Takeda, M., Ueno, E., Kato, Y., and Yamamoto,K. (1986) J.Biochem.(Tokyo) 100, 1269-1277.
3. Kageyama, T., and Takahashi, K.(1980) J.Biochem.(Tokyo) 87, 725-735.
4. Roberts, N.B., and Taylor, W.H.(1978) Biochem.J. 169, 617-624.
5. Samloff, I.M., Taggert, R.T., Shiraishi, T., Branch, T., Reid, W.A., Heath, R., Lewis, R.W., Vallen, M.J., and Kay, J.(1987) Gastroenterology 93, 77-84.
6. Jupp, R.A., Richards, A.D., Kay, J., Dunn, B.M., Wyckoff, J.B., Samloff, I.M., and Yamamoto, K. (1988) Biochem.J. 254, 895-898.
7. Muto, N., Yamamoto, M., Tani, S., and Yonezawa, S. (1988) J.Biochem. (Tokyo) 103, 629-632.
8. Azuma, T., Pals, G., Mohandas, T.K., Couvreur, J.M., and Taggaert, R.T. (1989) J.Biol.Chem. 264, 16748-16753.
9. Hill, J., Montgomery, D.S., and Kay, J. (1993) FEBS lett. 326, 101-104.
10. Tsukuba, T., Hori, H., Azuma, T., Takahashi, T., Taggart, T., Akamine, A., Ezaki, M., Nakanishi, H., Sakai, H., and Yamamoto, K. (1993) J. Biol. Chem. 268, 7276-7282.
11. Cregg, J.M., Barringer, K.J., Heseler, A.Y., and Madden, K.R. (1985) Mol.Cell.Biol. 5, 3376-3385.
12. Saku, T., Sakai, H., Kato, Y., and Yamamoto, K. (1991) J.Biochem.(Tokyo) 110, 946-956.

13. Ellis, S.B., Brust, P.J., Waters, A.F., Harpold, M.M., and Gingeras, T.R.(1985) Mol.Cell.Biol. 5, 1111-1121.
14. Williams, D.F., and Reisfeld, R.A. (1964) Ann.N.Y.Acad.Sci. 121, 373-381.
15. Cregg, J.M., and Madden, K.R. (1987) in Biological Research on Industrial Yeasts(Stewart, G.G., Russell, I., Klein, R.D., and Hiebsch, R.R., Eds.) Vol.II, pp1-18, CRC Press, Boca Raton, FL.
16. Etherington, D.J., and Taylor, W.J.(1972) Clin.Sci. 42, 79-90.
17. Mangla, J.C., Guarasci, G., and Turner, M.D. (1974) Biochem.Med. 10, 83-96.
18. Reichelt, D., Jacobsohn, E., and Haschen, R.J. (1974) Biochim.Biophys.Acta 341, 15-26.
19. Mangla, J.C., Lillibridge, C.B., Navab, F., and Turner, M.D.(1975) Biochem. Med. 13, 184-191.
20. Hirsch-Marie, H., Loisillier, F., Touboul, J.P., and Bartin, P. (1976) Lab.Invest. 34, 623-632.
21. Tarentino, A.L., Plummer, T.H., and Maley, F. (1974) J.Biol.Chem. 249, 818-824.
22. Wiliams, R.S., Ziegler, F.D., Trimble, R.B., and Maley, F. (1988) J.Biol.Chem. 263, 6978-6985.

PURIFICATION AND CHARACTERIZATION OF RECOMBINANT HUMAN CATHEPSIN E

Hiroshi Iida,[1] Takao Matsuba,[1] Masayuki Yamada,[1] Takeshi Azuma,[2] Hideo Suzuki,[1] Kenji Yamamoto,[3] and Hitoshi Hori[1]

[1]Biotechnology Research Laboratory
Tosoh Corporation
Hayakawa, Ayaseshi, Kanagawa 252, Japan
[2]Second Department of Internal Medicine
Fukui Medical School
Matsuoka, Yoshino-gun, Fukui 910-11, Japan
[3]Department of Pharmacology
Kyushu University, Faculty of Dentistry
Fukuoka 812, Japan

The human cathepsin E was purified from the culture supernatant of *Pichia pastoris* strain transformed with a human cathepsin E expression plasmid.

Purification was performed by a three-step procedure, TSKgel Phenyl-5PW, Toyopearl HW55S and TSKgel DEAE-5PW column chromatographies.

The purified recombinant cathepsin E had the molecular mass of around 82-kDa with the amino-terminal sequence started with Ile37 of the predicted amino acid sequence, suggesting the human cathepsin E was accumulated in the culture suparnatant as the mature dimer enzyme. The result of endoglycosidase-H digestion followed by Western blot analysis of the purified recombinant cathepsin E suggested that the human cathepsin E expressed in *Pichia pastoris* received N-linked high-mannose type glycosylation.

INTRODUCTION

Cathepsin E is an intracellular aspartic proteinase that exists in the cell as the dimeric enzyme with a molecular mass of around 82-kDa [1,2].

Although cathepsin E has several properties similar to those of cathepsin D, such as optimum pH, substrate specificity and susceptibility to various protease inhibitors [2, 3], their tissue distributions and intracellular localizations are quite different. In contrast to the wide spread tissue distribution of cathepsin D, cathepsin E has a limited distribution in certain tissues such as gastric mucosa and several lymphoid associated tissues [4,5]. Cathepsin E has been found in the membranous and cytosolic compartments, not found in the lysosome [6,7], whereas cathepsin D is a lysosomal proteinase.

Aspartic Proteinases: Structure, Function, Biology, and Biomedical Implications
Edited by Kenji Takahashi, Plenum Press, New York, 1995

Table 1. Scheme of Purification of Recombinant Human Cathepsin

Step		Volume	A280	Total protein	Total activity	Specific activity	Yield
		ml		*mg*	*units*	*units/mg*	%
1. Culture supernatant		6000	2.198		3.45×10^5		100
2. Phenyl-5PW		240	0.850		1.45×10^5		42
3. Toyopearl HW55S		49	0.367		1.40×10^5		41
4. DEAE-5PW	Pi-1	20	0.156	2.80	1.76×10^4	6.3×10^3	5.1
	Pi-2	12	0.193	1.18	2.39×10^4	2.0×10^4	6.9
	Pi-3	8	0.215	0.88	2.78×10^4	3.2×10^4	8.1

The limited tissue distribution of cathepsin E and its different intracellular localization than cathepsin D suggest that cathepsin E has specialized functions in certain cell types. However, the precise role of cathepsin E is still unclear.

We established the expression system of human cathepsin E using *P.pastoris* . This system made it possible to obtain the glycosylated and enzymatically active human cathepsin E in a large amount. Here we report the purification and characterization of recombinant human cathepsin E.

MATERIALS AND METHODS

Strain

P.pastoris transformant harboring the human cathepsin E expression plasmid, GS115/pHIL-D1CTSEv2 was used for expression of the human cathepsin E. Construction of GS115/pHIL-D1CTSEv2 is described in the chapter by Yamada M. et al. in this proceedings.

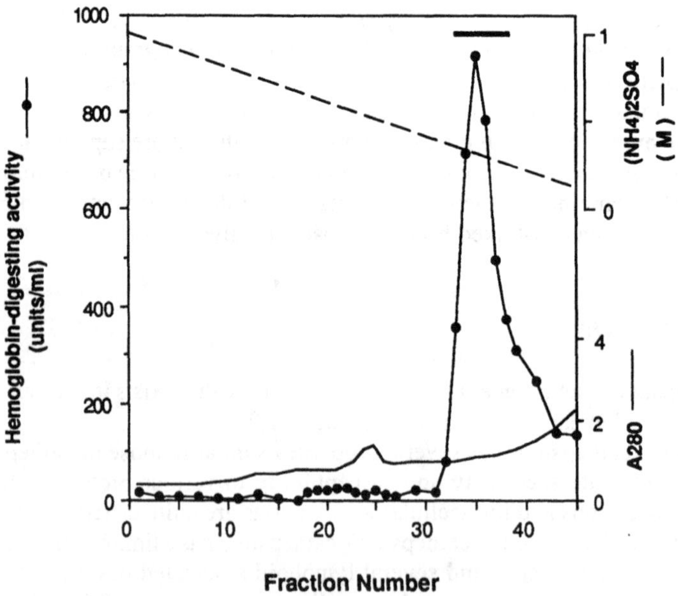

Figure 1. Cathepsin E elution pattern on a phenyl-5PW column after application of 6 liter of the culture supernatant. Each fraction was 4.7 ml. The bar indicates the fractions pooled.

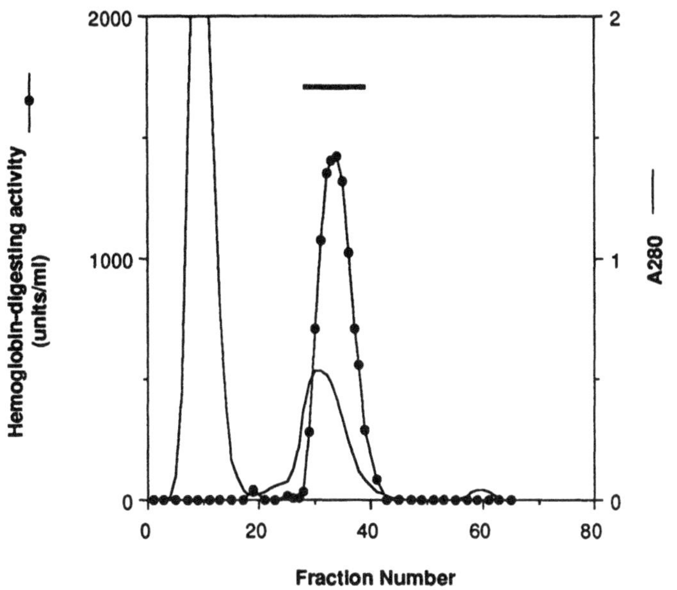

Figure 2. Gel filtration pattern of the pooled sample from the Phenyl-5PW chromatography on a Toyopearl HW55S column. Flow rate was 1.4 ml/min. Each fraction was 4.7 ml. The bar indicates the fractions pooled.

Media Composition

Minimal glycerol medium (MGY) and FM21 medium (for 2.5% carbon source) were described by Hagenson M. et al [8].

Production of Human Cathepsin E in a High Cell Density Fermentor

The production of human cathepsin E was performed in a batch-type fermentation using a 10-L New Brunswick Scientific fermentor equipped with monitors and controls for pH, dissolved oxygen, agitator speed, temperature, air flow and air pressure. Temperature was maintained at 30 °C and pH was maintained at 5.0 by the addition of aqueous ammonia (14% solution). The seed culture was grown in 100 ml of MGY in a 250-ml Erlenmeyer flask at 30 °C. When the available glycerol was exhausted, the carbon source was changed to methanol to induce cathepsin E expression. During the fermentation run, the methanol concentration was kept between 0.2 and 0.8% w/v by the addition of methanol. The culture was maintained for 120 h and the culture supernatant of 6 liters was collected by centrifugation.

Purification of Recombinant Human Cathepsin E

To the culture supernatant, PMSF and EDTA were added to give the final concentrations of 0.5 mM and 5 mM, respectively. And the pH was adjusted to 8.5 by the addition of Tris-base to 0.2 M and 500 ml of 2N NaOH. Debris was removed by filtration using celite.

Ammonium sulphate was added to the supernatant to 20% saturation. The supernatant was left at room temperature for 1 h and then filtrated through an filter paper (Advantech No. 2). Phenyl-5PW column chromatography was performed as the

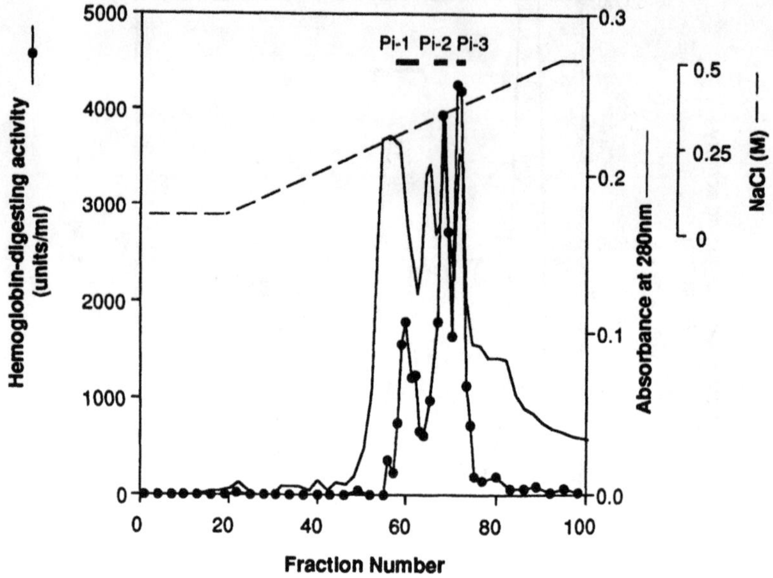

Figure 3. Cathepsin E elution pattern on a DEAE-5PW column. Each fraction was 4 ml. The recombinant cathepsin E was separated into three CTSE fractions, Pi-1, Pi-2, Pi-3. Fractions under the bar were pooled.

first step. The supernatant was applied to the column (5.5 x 20 cm) equilibrated with 10 mM Tris-Cl buffer, pH 8.6, containing 1 M $(NH_4)_2SO_4$, 2 mM EDTA. The column was washed with the same buffer and adsorbed proteins were eluted with a linear gradient of $(NH_4)_2SO_4$ in 10 mM Tris-Cl buffer. The flow rate was 20 ml/min. The enzyme pool obtained in the first step was subjected to gel filtration on a Toyopearl HW55S column performed as the second step. The column was equilibrated with 10 mM Tris-Cl, pH 8.6, containing 0.15 M NaCl. The collection of fractions was not started until 2 h was passed after the sample application. As the third step, DEAE column chromatography was performed. The column was equilibrated with 10 mM Tris-Cl, pH 8.6, containing 50 mM NaCl and proteins were eluted with a linear gradient of NaCl in the same buffer.

Endoglycosidase H Treatment

The purified cathepsin E was boiled for 2 min with 10 μl of 1 % SDS in water. To this sample, 90 μl of 50 mM sodium acetate buffer, pH6.0, containing 10 mU of endoglycosidase H was added. The reaction mixture was incubated for 18 h at 37 °C and analysed by SDS-PAGE under the reduced and non-reduced conditions.

Enzyme Assay

Aspartic proteinase activity was determined at pH3.8, 37 °C using 1.5% acid-denatured hemoglobin as a substrate. The reaction was stopped by the additon of trichloroacetic acid (TCA). 2.5%TCA soluble fractions were taken for assay according to Lowry et al [9] . One unit is defined as the amount of enzyme which gives an absorbance at 660nm equivalent to 1mg of tyrosine in 1min.

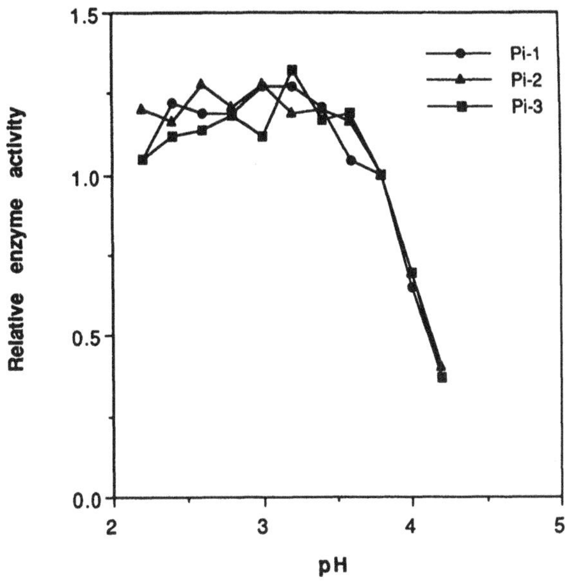

Figure 4. pH profile of the activity of the recombinant human cathepsin E. Hemoglobin (400 μl of 1.5% solution) was incubated at various pHs with 4 units of the purified cathepsin E at 37 °C for 15 min.

RESULTS AND DISCUSSION

The recombinant cathepsin E (CTSE) was highly purified from the culture supernatant of the *Pichia pastoris* transformant harboring the human CTSE expression plasmid by chromatography on Phenyl-5PW, Toyopearl HW55S and DEAE-5PW. Table 1 is the schematic representation of purification procedure.

The chromatographic patterns on the respective column are shown in Fig. 1, Fig. 2 and Fig. 3.

The host cell derived aspartic proteinase(s) passed through the first Phenyl-5PW column and was successfully separated from the CTSEs expressed from the expression plasmid which bound to the column. Fig. 3 shows that the recombinant CTSE was separated into three CTSE fractions, Pi-1, Pi-2, Pi-3, by DEAE-5PW chromatography, indicating the purified CTSE consists of three forms that are different in charge.

Fig. 4 shows that the optimum pH of the purified CTSE resided between 2.0 and 4.0 and the pH effect on the three CTSE fractions were the same as that of the gastric cathepsin E from guinea pigs.

The purified CTSE consisted of two polypeptides with the same molecular mass of 42-kDa (Fig. 5) .

The endoglycosidase H treatment of the purified CTSE resulted in the reduction of molecular mass by around 2-kDa, suggesting that the CTSE expressed in *P.pastoris* is glycosylated with the high mannose-type oligosaccharide chains (Fig. 5). The 42-kDa polypeptide obtained by reduction of the 84-kDa form was found to start with Ile as determined by N-terminal amino acid analysis (data not shown).

These results indicate that the prosequence of the CTSE expressed in *P.pastoris* was processed between Met^{36} and Ile^{37} and secreted as the mature enzyme. This cleavage site is the same as that of human CTSE expressed in CHO cells [10].

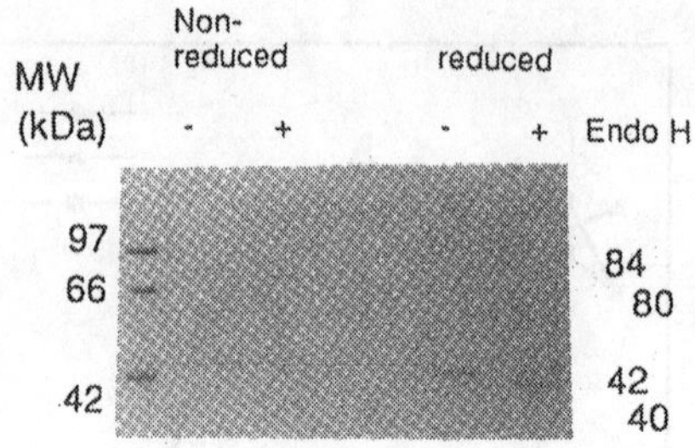

Figure 5. Endoglycosidase H treatment of the purified cathepsin E.

In conclusion, we established the expression system and purification procedure of recombinant human cathepsin E. We believe that this will serve to enhance the study of cathepsinE - for example, the determination of 3D structure of cathepsin E.

REFERENCES

1. Kageyama, T., and Takahashi K. (1980) J. Biochem. (Tokyo) 87, 727-735
2. Yamamoto, K., Katsuda, N., and Kato, K. (1978) Eur. J. Biochem. 92, 499-508
3. Yamamoto, K., Katsuda, N., Himeno M., and Kato K. (1979) Eur. J. Biochem. 95, 459-67
4. Sakai, H., Saku, T., Kato, Y., and Yamamoto K. (1989) Biochim. Biophys. Acta 991, 367-375
5. Muto, N., Yamamoto, M., Tani, S., and Yonezawa, S. (1988) J. Biochem. (Tokyo) 103, 629-632
6. Samloff, I.M., Taggert, R.T., Shiraishi, T., Branch, T., Reid, W.A., Heath, R., Lewis, R.W., Vallen, M.J., and Kay, J. (1987) Gastroenterology 93, 77-84
7. Reid, W.A., Valley, M.J., and Kay, J. (1986) J. Clin. Pathol. 39, 1323-30
8. Hagenson, M.J., Holden, K.A., Parker, K.A., Wood, P.J., Cruze, J.A., Huke, M., Hopkins, T.R., and Stroman, D.W. (1989) Enzyme Microb. Technol. 11, 650-656
9. Lowry, O. H., Rosebrough, N. J., Farr, A.L., and Randall, R.J. (1951) J. Biol. Chem. 193, 265-275
10. Tsukuba, T., Hori, H., Azuma, T., Takahashi, T., Taggart, T., Akamine, A., Ezaki, M., Nakanishi, H., Sakai, H., and Yamamoto, K. (1993) J. Biol. Chem. 268, 7276-7282

COMPARISON OF BIOCHEMICAL PROPERTIES OF NATURAL AND RECOMBINANT CATHEPSIN E

Takayuki Tsukuba,[1] Yamada Masayuki,[2] Hitoshi Hori,[2] Takeshi Azuma,[3] and Kenji Yamamoto[1]

[1]Department of Pharmacology
Kyushu University
Faculty of Dentistry
Fukuoka 812, Japan
[2]Biotechnology Research Laboratory
Tosoh Co., Kanagawa 252, Japan
[3]Second Department of Internal Medicine
Fukui Medical School
Fukui 910-11 Japan

INTRODUCTION

Cathepsin E (CE) (EC 3.4.23.34) is an intracellular aspartic proteinase consisting of two identical subunits with a molecular mass of about 42kDa (1-4). This enzyme is also known as a glycoprotein (1-3) and to have oligosaccharide chains of both the high-mannose-type and the complex-type (5). In contrast to cathepsin D (EC 3.4.23.5), a lysosomal aspartic proteinase, CE has a limited distribution (6, 7) and is not found in the typical lysosomal structures (8). Recent studies provide some information about its physiological significance, e.g., processing of propeptides (9), catabolism of neuropeptides (10) and participation in neuronal cell death (11). However, the precise roles of CE in cellular protein catabolism has not yet been clarified. In addition, little information about the biosynthesis and the processing and activation mechanisms of this enzyme in cells has so far been available. The major problems for these studies are the difficulty in purification of the enzyme due to its low concentration and the lack of a specific substrate that is discriminated only by CE. To solve these problems, we have tried to gain sufficient materials of CE. More recently, we have succeeded in preparing the recombinant CE (rCE) which are expressed in Chinese hamster ovary (CHO) cells at a high rate (4). In this report, we describe the purification and biochemical properties of rCE and compare these properties with those of the natural enzyme purified from human erythrocyte membranes.

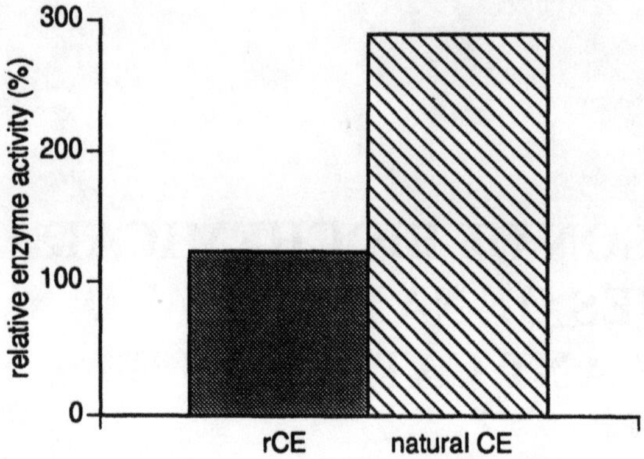

Figure 1. Effect of ATP on the hydrolysis of hemoglobin by the recombinant and natural human CEs at pH 5.5. Aliquots of each enzyme sample in 0.1 M sodium acetate buffer (pH 5.5) were preincubated at 37 °C for 5 min with or without ATP (6.25 mM) and then added to the substrate solution (pH 5.5). The reaction was stopped by addition of 5% trichloroacetic acid and acid-soluble products were determined by the fluorescamine method.

SUBCELLULAR LOCALIZATION AND MOLECULAR FORMS OF rCE IN THE TRANSFECTED CHO CELLS

By immunoelectron microscopic studies, rCE was found to be in two different intracellular fractions; the cytosolic compartment and the vacuolar system. Namely, the immunolabeling for CE was abundant in the cytoplasmic matrix, the endoplasmic reticulum and endocytic vacuoles. Sodium dodecyl sulfate-polyacrylamide gel electrophoresis and immunoblotting revealed that the cells contained three different forms of the enzyme (the 90-kDa proform, the 84-kDa intermediate form and the 82-kDa mature form). The recombinant enzyme was purified by a simple method involving Percoll density fractionation, chromatography on concanavalin A-Sepharose, Mono Q, and TSKgel G2000SW. The N-terminal amino acid sequence analyses showed that the 90-kDa, 84-kDa, and 82-kDa proteins started with Ser^{3}, Lys^{30}, and Ile^{37}, respectively, of the sequence of human gastric CE predicted from its cDNA sequence (12). Upon acid treatment at pH 3.5 and 37 °C for 5 min, the 90- and 84-kDa forms were completely converted to the 82-kDa form. This conversion was rapid and completely inhibited by pepstatin. In addition, the conversion was dependent on the protein concentration, thereby suggesting that the predominance of intermolecular reaction occurs during the process of autocatalytic activation of the precursor proteins.

When these forms of rCE were treated by endoglycosidase H which is specific for high-mannose-type oligosaccharide chains and cleaves between the two proximal *N*-acetylglucosamine residues of asparagine-linked carbohydrate chains, all of the forms resulted in a reduction of about 4-kDa in molecular mass. Therefore, rCE was glycosylated with high-mannose-type of oligosaccharide chain. These results was consistent with the data with the rat spleen and gastric CE, but different from those obtained with the human and rat erythrocyte CE which has the complex-type oligosaccharide chain (5). Thus, it seems likely that the type of oligosaccharide moiety of CE is cell-specific or varies with the cellular location.

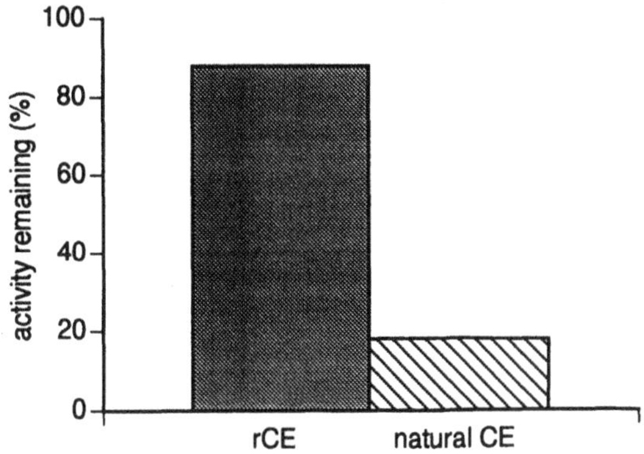

Figure 2. Thermal stability of the recombinant and natural human CEs. Each enzyme sample was preincubated at pH 5.5 and 55 °C for 20 min. Then, the remaining enzyme activity was determined by the hydrolysis of hemoglobin at pH 3.5 and 40 °C for 40 min.

COMPARISON OF BIOCHEMICAL PROPERTIES OF NATURAL AND RECOMBINANT CEs

Next, we compared biochemical properties of rCE with those of natural CE from human erythrocyte membranes. The K_m values for the hydrolysis of protein substrates (e.g., hemoglobin) and synthetic substrates (e.g., Pro-Pro-Thr-Ile-Phe-Phe(4-NO_2)-Arg-Leu and Lys-Pro-Ile-Glu-Phe-Phe(4-NO_2)-Arg-Leu) at pH 3.5 were essentially identical between rCE and natural CE. Also, no significant difference in the K_i values for inhibition of the hydrolysis of hemoglobin at pH 3.5 by pepstatin, Pro-Thr-Glu-Phe(CH_2-NH)Nle-Arg-Leu (H297), and Boc-His-Pro-Phe-His-Sta-Leu-Phe-NH_2 (L363,564) was observed between the two enzymes. Furthermore, with respect to the pH dependency of hydrolysis for these substrates and the substrate preference, there were no significant differences between the recombinant and natural human CEs. Therefore, rCE was found to be catalytically indistinguishable from natural CE.

It has been known that, although CE from human erythrocyte membranes is essentially inactive at pH values above 5.5 by itself, it is stabilized by ATP (13) and RNA (3) in such a way as to maintain its active conformation and exhibits its activity on hemoglobin and the synthetic substrates even at pH 5.8. Therefore, we examined the effect of ATP at pH 5.5 on rCE, as well as the natural CE (Fig. 1). As expected, the natural CE was activated by ATP to apparently exhibit its activity on hemoglobin, which corresponded to 290% of the enzyme activity measured in the absence of ATP. In contrast, rCE was not significantly activated by ATP.

On the other hand, there was a marked difference in the thermal stability between the recombinant and natural human CEs. The natural CE was greatly inactivated by heating at pH 5.5 and 55 °C for 20 min, whereas rCE exhibited the resistance to this treatment to retain about 90% of the initial enzyme activity (Fig. 2). Despite the similarity in catalytic properties, the difference in these physicochemical properties between the recombinant and natural CEs may suggest the difference in their three-dimensional structures mainly due to the difference in post-translational modifications such as glycosylation processing.

REFERENCES

1. K. Yamamoto, N. Katsuda and K. Kato, Affinity Purification and Properties of Cathepsin E-Like Acid Proteinase from Rat Spleen, Eur. J. Biochem. 92: 499 (1978)

2. K. Yamamoto and V.T. Marchesi, Purification and Characterization of Acid Proteinase from Human Erythrocyte Membranes, Biochim. Biophys. Acta 790: 208 (1984)
3. K. Yamamoto, H. Sakai, E. Ueno and Y. Kato, In Structure and Function of the Aspartic Proteases : Genetics, Structures, and Mechanisms (B. M. Dunn, ed.) pp. 297-306, Plenum Press, New York
4. T. Tsukuba, H. Hori, T. Azuma, T. Takahashi, R.T. Taggart, A. Akamine, M. Ezaki, H. Nakanishi, H. Sakai and K. Yamamoto, Isolation and Characterization of Recombinant Human Cathepsin E Expressed in Chinese Hamster Ovary Cells, J. Biol. Chem. 268: 7276 (1993)
5. M. Takeda-Ezaki and K. Yamamoto, Isolation and Biochemical Characterization of Procathepsin E from Human Erythrocyte Membranes, Arch. Biochem. Biophys. 304: 352 (1993)
6. H. Sakai, T. Saku, Y. Kato and K. Yamamoto, Quantitation and Immunohistochemical Localization of Cathepsins E and D in Rat Tissues and Blood Cells, Biochim. Biophys. Acta 991: 367 (1989)
7. N. Muto, M. Yamamoto, S. Tani and S. Yonezawa, Characteristic Distribution of Cathepsin E Which Immunologically Cross-Reacts with the 86-kDa Acid Proteinase from Rat Gastric Mucosa, J. Biochem. 103: 629 (1988)
8. T. Saku, H. Sakai, Y. Shibata, Y. Kato and K. Yamamoto, An Immunocytochemical Study on Distinct Intracellular Localization of Cathepsin E and Cathepsin D in Human Gastric Cells and Various Rat Cells, J. Biochem. 110: 956 (1991)
9. W.E. Lees, S. Kalinska, J. Meech, S.J. Capper, N.D. Cook and J. Kay, Generation of Human Endothelin by Cathepsin E, FEBS Lett. 273: 99 (1990)
10. T. Kageyama, Rabbit Procathepsin E and Cathepsin E: Nucleotide Sequence of cDNA, Hydrolytic Specificity for Biologically Active Peptides and Gene Expression during Development, Eur. J. Biochem. 216: 717 (1993)
11. H. Nakanishi, T. Tsukuba, T. Kondou, T. Tanaka and K. Yamamoto, Transient Forebrain Ischemia Induces Increased Expression and Specific Localization of Cathepsins E and D in Rat Hippocampus and Neostriatum, Exp. Neurol. 121: 215 (1993)
12. T. Azuma, G. Pals, T.K. Mohandas, J.M. Couvreur and R.T. Taggart, Human Gastric Cathepsin E: Predicted Sequence, localization to Chromosome 1, and Sequence Homology with Other Aspartic Proteinases, J. Biol. Chem. 246: 16748 (1989)
13. D.J. Thomas, A.D. Richards, R.A. Jupp, E. Ueno, K. Yamamoto, I.M. Samloff, B.M. Dunn and J. Kay, Stabilisation of Cathepsin E by ATP, FEBS Lett. 243: 145 (1989)

INCREASED EXPRESSION AND SPECIFIC LOCALIZATION OF CATHEPSINS E AND D IN VULNERABLE BRAIN REGIONS OF AGED AND POSTISCHEMIC RATS

Hiroshi Nakanishi, Kazuyoshi Tominaga, and Kenji Yamamoto

Department of Pharmacology
Kyushu University
Faculty of Dentistry
Fukuoka 812, Japan

INTRODUCTION

Most neuronal cell death could be classified into apoptosis and necrosis on the basis of morphological analysis. Apoptosis is initially characterized by condensation and fragmentation of nucleus mainly due to the increased expression of endonucrease (1). Apoptosis is an important biological pathway for programmed cell death which occurs during development. On the other hand, necrosis is primarily characterized by destruction of the cell membrane and the cytosolic cell structures, but lysosomes are rather intact until advanced stage of neuronal degeneration. Necrosis is an accidental neuronal cell death and includes cell death due to aging and ischemia. In the course of necrotic type of neuronal degeneration, a variety of lytic enzymes normally present in the cytoplasm are considered to be activated.

Recently, attention has been paid to proteinase functions in relation to the process of neuronal degeneration leading to cell death. Several recent articles have outlined the deleterious effect of calpain, a calcium-activated cysteine proteinase, on cytoskeleton protein network, which is thought to lead necrosis (2,3). However, little is known about the other cellular proteolytic systems in the degenerative process. Cathepsin E is a relatively newly characterized enzyme and believed to be involved in the non-lysosomal proteolytic system (4,5). There is accumulating evidence that this enzyme has ability to degrade a variety of neuropeptides including tachykinins such as substance P and neurokinin A (6) and big-endothelin (7), suggesting the possible involvement in the catabolism of neuropeptides in the central nervous system. Therefore, it is of interest to determine whether cathepsin E is involved in the process of the necrotic type of neuronal degeneration (8,9).

Aspartic Proteinases: Structure, Function, Biology, and Biomedical Implications
Edited by Kenji Takahashi, Plenum Press, New York, 1995

CATHEPSINS E AND D IN BRAIN TISSUES OF NORMAL YOUNG RATS

Firstly, levels of cathepsin E in various brain regions of normal young rats were determined by activity measurement which combined the hemoglobin-hydrolyzing assay with immunoprecipitation. By this method, cathepsin E was hardly detectable in any brain regions of normal young rats, including the cerebellum, hippocampus, neostriatum, or cerebral cortex. Then we have developed a highly sensitive enzyme immunoassay (EIA) system for cathepsin E. This assay was based on the sandwich method which used polystylene balls coated with anti-cathepsin E IgG and horseradish peroxidase-conjugated Fab' fraction of IgG. The concentrations of cathepsin E in various non-neuronal tissues such as the stomach and lung determined by EIA corresponded with those estimated by activity measurement. In addition, EIA enabled to detect cathepsin E in such tissues as the heart, which contained a very low concentration of the enzyme to be not detectable by activity measurement. Thus, EIA was found to be a sensitive and reliable method to know the precise contents of cathepsin E. Among the brain regions examined, the cerebellum showed the highest concentration of cathepsin E (3.7 ng/mg protein). In contrast, cathepsin D was ubiquitously but unevenly distributed in all of the brain tissues examined, as determined by activity measurement. This distribution of cathepsin E as well as cathepsin D in the brain tissues was further substantiated by immunohistochemistry.

Immunohistochemical localizations of cathepsins E and D in the brain tissues of young rats were examined by using discriminative antibodies specific for each enzyme. Cathepsin E was slightly and diffusely stained only in a small number of relatively large neurons. Purkinje cells in the cerebellum were most intensely stained by anti-cathepsin E antibody and this is consistent with the quantitative data by EIA. In contrast to cathepsin E, cathepsin D was intensely stained as coarse intraplasmic granules in almost all neurons.

CATHEPSINS E AND D IN BRAIN TISSUES OF AGED RATS

As the model of chronic neuronal degeneration, we used aged Fischer 344 rats, since relatively large neurons were found to be degenerated during normal aging process of the brain. Furthermore, these aged Fischer 344 rats are known to exhibit memory deficits seen in aged humans.

As mentioned above, the activity levels of cathepsin E was barely detectable in any brain regions of normal young rats. However, the activity levels of this protein were clearly demonstrated in both the cerebral cortex and neostriatum of aged rat brain. Cathepsin E contents in these aged rat brain tissues were estimated to be 8.6 - 43.0 ng/mg protein by activity measurement. These values were about 10-fold of the normal young rats. Although no significant activity of cathepsin E was detectable in either the hippocampus or the cerebellum of aged rats, the increased immunoreactivity was apparently found in all the brain regions. This may be due to the low neuronal density in these brain regions, as compared with other regions. On the other hand, the activity of cathepsin D in all the brain regions of aged rats also increased by about 2-fold of those of normal young rats.

At the light microscopic level, immunoreactivity of cathepsin E markedly increased in both neurons and glial cells of all brain regions examined from aged rats (30 - 35 months). Large granular immunoreactive products indicating the antigenicity for cathepsin E were found to accumulate in the relatively large neurons. This granular immunostaining pattern in large neurons of aged rat brain was clearly distinct from the diffuse immunostaining pattern in those of the normal young rat brain. On the other hand, the immunoreactivity for

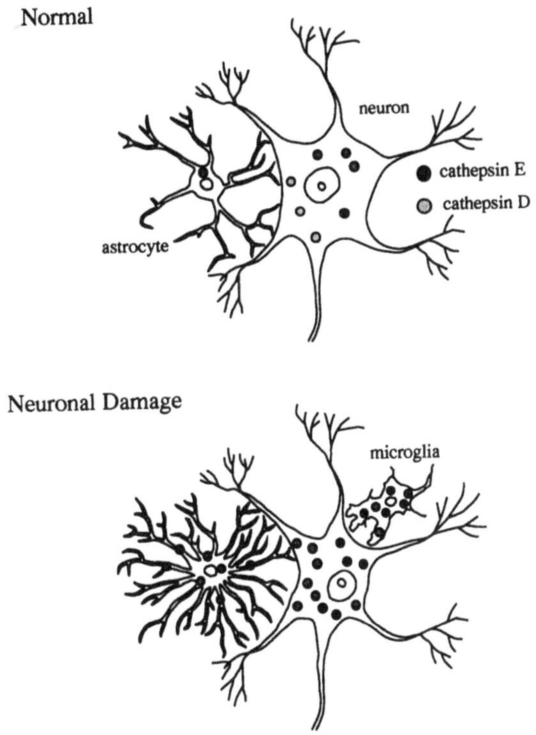

Figure 1. Schematic representation of cathepsins E and D localizations in normal and damaged neuronal tissues.

cathepsin D increased predominantly in glial cells of these brain regions. Reactivation of glial cells is the common response to brain injury and two types of glial cells, astrocyte and microglias, are known to be involved. Reactive astrocytes fill in the space that results from neuronal loss by forming glial scar. On the other hand, reactive microglia are considered to be resident macrophages in the brain and play important roles in phagocytosis and immune response. In order to determine what cathepsins E and D-positive glial cells are, double immunostaining was carried out in the same section by using two different glial cell markers, anti-GFAP antibody as a marker for astrocytes and OX42 as a marker for microglias. The results showed that cathepsin E-positive glial cells exclusively coincided with OX42-positive cells but not with GFAP-positive cells. On the other hand, cathepsin D-positive cells were associated mainly with GFAP-positive cells and partially with OX42-positive cells. Therefore, the increased activity of cathepsin E is partially due to reactivation of microglias or infiltration of macrophages besides neuronal degeneration. The increased activity of cathepsin D is considered to mainly result from glial reactivation.

CATHEPSINS E AND D IN BRAIN TISSUES FROM POSTISCHEMIC RATS

Transient forebrain ischemia in the rat is known to induce neuronal degeneration in particularly vulnerable brain regions including hippocampal CA1 neurons and dorsolateral striatal neurons and produce marked impairment of learning and memory. After cardiac arrest, patients also often show a persistent memory deficit and neurological examination of the brain revealed necrosis of the hippocampus. In the present study, transient forebrain ischemia was induced in the rat by carotid and vertebral arteries occlusion for 15 min according to the method described by Pulsinelli and Brierley (10).

The accumulation and localization of cathepsins E and D in the rat hippocampus and neostriatum during the neurodegenerating process induced by transient forebrain ischemia were investigated by immunoprecipitation and immunohistochemistry. While significant amounts of cathepsin D were found in both tissues of normal rats, cathepsin E was barely detectable in these tissues. At 7 days after ischemia, cathepsin E was markedly increased in both tissues (about 10-fold over the normal values). Cathepsin D in these tissues was also increased at 7 days after ischemia by about 1.5-fold. Increased levels of cathepsins E and D were further confirmed by immunohistochemical stainings. In the normal rat, cathepsin E-positive cells were barely detectable in either the hippocampus or the neostriatum, whereas cathepsin D-positive cells were widely but unevenly distributed in these tissues. At 7 days after ischemia, cathepsin E immunoreactivity was found mainly in reactive glial cells and degenerating neurons of both the hippocampal CA1 subfield and the neostriatal dorsolateral portion. Increased cathepsin D immunoreactivity was also observed in glial cells of these brain regions. By the double-staining technique, it was found that the cathepsin E-positive glial cells were largely OX42-positive reactive microglias, whereas cathepsin D-positive glial cells were associated mainly with GFAP-positive reactive astrocytes. These results are consistent with observations obtained in the aged rat brain.

Figure 1 summarizes our findings on cathepsins E and D in the rat brain in normal and pathological states. In the normal rat brain, a small amount of cathepsin E localized in a small number of relatively large neurons. On the other hand, a large amount of cathepsin D was widely but unevenly distributed in almost all neurons. Cathepsin D was also occasionally found in glial cells. In response to brain injury, both cellular levels of cathepsins E and D markedly increased. The increased cathepsin E is considered to be associated closely with neuronal degeneration and partially with glial reactivation, which is common response to brain injury. The increased cathepsin E besides in the degenerating neurons was found exclusively in reactive microglias, the resident macrophages of the brain, and not in reactive astrocytes. On the other hand, the increased cathepsin D is associated with glial reactivation. Cathepsin D increased mainly in reactive astrocytes and partially in reactive microglias. Therefore, pathological significance of cathepsin E increasingly expressed in the response to brain injury may be also distinct from those of cathepsin D. Further study will be needed for understanding the exact role of cathepsin E in neuronal degeneration.

REFERENCES

1. P.G.H. Clarke, Developmental Cell Death : Morphological Diversity and Multiple Mechanisms, Anat. Embryol. 181:195 (1990).
2. K.S. Lee, S. Frank, P. Vanderklish, A. Arai and G. Lynch, Inhibition of Proteolysis Protect Hippocampal Neurons from Ischemia, Proc. Natl. Acad. Sci. USA 88 :7233 (1991).
3. R. Siman and J. C. Nozek, Excitatory Amino Acids Activate Calpain I and Induce Structural Protein Breakdown in vivo, Neuron 1:279 (1988).
4. H. Sakai, T. Saku, Y. Kato and K. Yamamoto, Quantitation and Immunohistochemical Localization of Cathepsins E and D in Rat Tissues and Blood Cells, Biochim. Biophys. Acta 991:367 (1989).
5. T. Saku, H. Sakai, Y. Shibata, Y. Kato and K. Yamamoto, An Immunocytochemical Study on Distinct Intracellular Localization of Cathepsins E and D in Human Gastric and Various Rat Cells, J. Biochem. 110:956 (1991).
6. T. Kageyama, Rabbit Procathepsin E and Cathepsin E : Nucleotide Sequence of cDNA, Hydrolytic Specificity for Biologically Active Peptides and Gene Expression During Development, Eur. J. Biochem. 216:717 (1993).

7. W.E. Lees, S. Kalinka, J. Meech, S.J. Capper, N.D. Cook and J. Kay, Generation of Human Endothelin by Cathepsin E, FEBS lett. 273:99 (1990).
8. H. Nakanishi, T. Tsukuba, T. Kondou, T. Tanaka and K. Yamamoto, Transient Forebrain Ischemia Increased Expression and Specific Localization of Cathepsins E and D in Rat Hippocampus and Neostriatum, Exp. Neurol. 121:215 (1993).
9. H. Nakanishi, K. Tominaga, T. Amano, I. Hirotsu, T. Inoue and K. Yamamoto, Age-Related Changes in Activities and Localizations of Cathepsins D, E, B and L in the Rat Brain Tissues, Exp. Neurol. (1994) in press.
10. W.A. Pulsinelli and J. B. Brierley, A New Model of Bilateral Hemisheric Ischemia in the Unanesthetized Rat, Stroke 10:267 (1979).

CHARACTERIZATION OF CATHEPSINS E AND D ACCUMULATED AT EARLY STAGES OF NEURONAL DAMAGE IN HIPPOCAMPAL NEURONS OF RATS

Kazuyoshi Tominaga, Hiroshi Nakanishi, Motoyuki Yajima,[1]
and Kenji Yamamoto

Department of Pharmacology
Kyushu University
Faculty of Dentistry
Fukuoka 812, Japan
[1]Central Laboratories
Kaken Pharmaceutical Co. Ltd.
Kyoto 607, Japan

INTRODUCTION

Excitatory amino acid-mediated excitotoxic damage has been implicated in the pathogenesis of a variety of acute and chronic neurodegenerative diseases including stroke, trauma, epilepsy and Alzheimer's disease (for reviews see: 1,2). Despite the clinical importance, the understanding of subcellular mechanism of excitotoxic neuronal cell death is limited. There is increasing evidence that calcium-activated enzymes such as phospholipases, protein kinases, endonucleases or non-lysosomal proteinases play a pivotal role in the excitotoxic neuronal cell death. Among these candidates, the activation of calpain, a calcium-activated cysteine proteinase, has been widely accepted as an important factor contributing to excitotoxic neuronal cell death (3,4). However, little is known about the role of other non-lysosomal proteolytic systems. Recently we have suggested that cathepsin E, a non-lysosomal aspartic proteinase, is associated with the pathological process of ischemia (5) and normal aging (6). Thus it is of special interest whether cathepsin E is involved in the process of excitotoxic neuronal cell death. Here we report evidence indicating that cathepsin E as well as cathepsin D, a lysosomal aspartic proteinase, is closely related to the neuronal cell death in the CA3 subfield of rat hippocampus induced by intraventricular injection of kainate, a potent glutamate agonist.

Aspartic Proteinases: Structure, Function, Biology, and Biomedical Implications
Edited by Kenji Takahashi, Plenum Press, New York, 1995

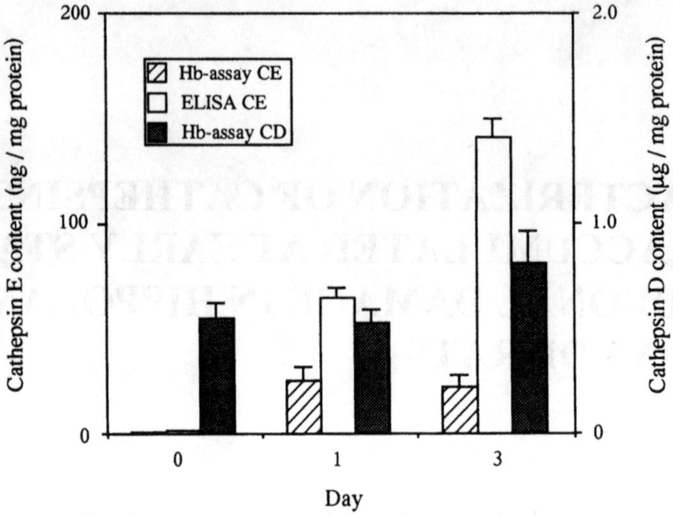

Figure 1. Levels of cathepsins E and D in the hippocampus on various days after kainate injection. Striped and filled columns: cathepsins E and D contents estimated by their enzymatic activities determined by combination of hemoglobin-hydrolyzing assay with immunoprecipitation (Hb-assay), respectively. Knowing the specific activities of purified rat enzymes (cathepsin E: 35000 units/mg, cathepsin D: 8200 units/mg), the amount of each enzyme per milligram protein was calculated from the amount of the enzyme activity immunoprecipitated. Open columns: cathepsin E contents per milligram protein measured by enzyme-linked immunosorbent assay (ELISA) based on the sandwich method which used polystylene balls coated with anti-cathepsin E IgG and horseradish peroxidase-conjugated Fab' fraction of IgG. Each column and vertical bar represent the mean and S. D. of 5 - 7 observations, respectively.

CHARACTERISTICS OF CATHEPSINS E AND D IN RAT HIPPOCAMPUS BEFORE AND AFTER KAINATE INJECTION

After intraventricular injection of kainate, a generalized convulsion was gradually developed after several hours. Changes in functional and morphological aspects of hippocampal neurons following kainate injection were examined by electrophysiology and Nissl staining. At 1 day after injection, the hippocampal CA3 neurons were found to suffer from functional damage as revealed by decrease in amplitude of synaptic potential but to be morphologically rather intact. At 3 days after injection, the hippocampal CA3 neurons were found to die functionally and morphologically since there were no stainings by cresyl violet and no electrical activities in this subfield.

Immunohistochemical localizations of cathepsins E and D in the rat hippocampus before and after kainate injection were examined by using discriminative antibodies specific for each enzyme. In the hippocampus of normal rat, immunoreactivity of cathepsin E was barely detectable. At 1 day after kainate injection, increased cathepsin E immunoreactivity was observed especially in the neurons of the hippocampal CA3 subfield. At 3 days after injection, cathepsin E-positive cells further increased and localized in the entire hippocampal CA3 subfield. On the other hand, the apparent increase in cathepsin D immunoreactivity in this subfield was not demonstrated until 3 days after the injection. The results indicate that the increased expression of cathepsin E occurs at an early stage of neuronal damage, while the cathepsin D contents increase at a more advanced stage of neuronal damage.

In order to determine the amounts of cathepsin E and D in the tissue extract, discriminative antibodies specific each enzyme were employed. The amounts of cathepsin

E were determined by two different methods: one was the activity measurement method which combined hemoglobin-hydrolyzing assay and immunoprecipitation (7) and the other was enzyme immunoassay method which was developed to recognize both the active and inactive forms of the enzyme (8). As shown in Figure 1, the cathepsin E contents in the hippocampus markedly increased after kainate injection. It was also noted that there was a clear difference in the values for cathepsin E contents between the two methods. The cathepsin E contents measured by enzyme assay were 2.6 - 6.5-fold higher than those estimated by activity measurement. The cathepsin D contents in the hippocampus were not significantly changed until 3 days after kainate injection.

The hippocampus extracts before and after kainate injection were analyzed by sodium dodecyl sulfate-polyacrylamide gel electrophoresis under reduced condition followed by Western blotting. No clear immunoreactive band for cathepsin E was found in the normal hippocampal extract. However, the hippocampal extract at 3 days after kainate injection revealed mainly the 54-kDa immunoblotted band, with the faint 42-kDa band. The relative density of the 54-kDa polypeptide in the extract was much greater at 3 days than 1 day after the injection. These results were consistent with the immunohistochemical observations as mentioned above. Although the 42-kDa polypeptide corresponded with the reduced form of the mature cathepsin E, the molecular mass of the 54-kDa polypeptide was much higher than that of procathepsin E (46-kDa). Then, in order to examine whether the 54-kDa protein was a catalytically inactive precursor form of cathepsin E and whether it was able to convert to the mature enzyme, the hippocampal extract at 3 days after the injection was subjected to a brief acid treatment, since procathepsin E is known to completely convert to the catalytically active mature enzyme by an acid treatment at pH 4 and 37 °C for 5 min (9). The results showed that the 54-kDa polypeptide did not convert to the mature form even after acid treatment at pH 4 and 37° C for 40 min. On the other hand, the molecular forms of cathepsin D increased in the hippocampus at 3 days after kainate injection was not distinct from the normal forms of cathepsin D (the 46-kDa proenzyme and the 42-kDa mature enzyme). Therefore, the present study indicates that the intraventricular injection of kainate induces the increased expression and specific accumulation of the seemingly abnormal form of the 54-kDa cathepsin E in the neurons of the CA3 hippocampal subfield, which is likely to lead to the neuronal cell death.

REFERENCES

1. C.W. Choi, Glutamate Neurotoxicity and Diseases of the Nervous System, Neuron 1:623 (1988).
2. S. M. Rothman and J. W. Olney, Excitotoxicity and NMDA Receptor, Trends Neurosci. 10:299 (1987).
3. P. Seubert, J. Larson, M. Oliver, M.W. Jung, M. Braudry and G. Lynch, Stimulation of NMDA Receptor Induces Proteolysis of Spectrin in Hippocampus, Brain Research 458:233 (1988).
4. R. Siman and J. C. Noszek, Excitatory Amino Acids Activate Calpain I and Induce Structural Protein Breakdown in vivo, Neuron 1:279 (1988).
5. H. Nakanishi, T. Tsukuba, T. Kondou, T. Tanaka and K. Yamamoto, Transient Forebrain Ischemia Increased Expression and Specific Localization of Cathepsins E and D in rat Hippocampus and Neostriatum, Exp. Neurol. 121:215 (1993).
6. H. Nakanishi, K. Tominaga, T. Amano, I. Hirotsu, T. Inoue and K. Yamamoto, Age-Related Changes in Activities and Localizations of Cathepsins D, E, B and L in the Rat Brain Tissues, Exp. Neurol. (1994) in press.
7. H. Sakai, T. Saku, Y. Kato and K. Yamamoto, Quantitation and Immunohistochemical Localization of Cathepsins E and D in Rat Tissues and Blood Cells, Biochim. Biochem. Acta 991:367 (1989).
8. E. Ichimaru, H. Sakai, T. Saku, K. Kunimatsu, Y. Kato, I. Kato and K. Yamamoto, Characterization of Hemoglobin-Hydrolyzing Acidic Proteinases in Human and Rat Neutrophils, J. Biochem. 108:1009 (1990).
9. M. Takeda-Ezaki and K. Yamamoto, Isolation and Biochemical Characterization of Procathepsin E from Human Erythrocyte Membrane, Arch. Biochem. Biophys. 304:352 (1993).

FUNCTIONAL ASPECTS OF CATHEPSIN E: IS IT AN EMBRYONIC OR FETAL TYPE OF ASPARTIC PROTEINASE?

Satoshi Yonezawa,[1] Masao Ichinose,[2] Shinko Tsukada,[2] Kazumasa Miki,[2] and Takashi Kageyama[3]

[1]Department of Embryology
Institute for Developmental Research
Aichi Prefectural Cology, Kasugai 480-03, Japan
[2]First Department of Internal Medicine
Faculty of Medicine
University of Tokyo
Tokyo 113, Japan
[3]Department of Cellular and Molecular Biology
Primate Research Institute
Kyoto University
Inuyama 484, Japan

INTRODUCTION

Cathepsin E is now accepted as a unique member of mammalian aspartic proteinases, in that it is non-secretory and non-lysosomal (1). In rodents, the enzyme is distributed in the stomach, bladder, myeloid and lymphoid tissues as well as mature blood cells (2). The functional role of cathepsin E is not yet established, but recent studies have raised a possibility that the enzyme may be involved in the proteolytic processing of bioactive peptides (3,4). We found, however, that the distribution of the enzyme in mammalian blood cells is species-specific (5). Furthermore, our preliminary data showed that cathepsin E was not detectable in cultured human, bovine and rat endothelial cells. We then extended our study to specify its role in the stomach, the main distributional site of the enzyme. We demonstrate here an early appearance of cathepsin E in developing rat stomachs, as revealed by immunochemical and immunohistochemical examinations. Attention was paid for a possible functional association with the developmental differentiation of the gastric gland and for its possible role as a digestive enzyme in the fetal stomach.

Aspartic Proteinases: Structure, Function, Biology, and Biomedical Implications
Edited by Kenji Takahashi, Plenum Press, New York, 1995

MATERIALS AND METHODS

Sampling of Fetal Stomach

Wistar strain rats were mated overnight and pregnancy was confirmed by examination of vaginal smears. The day that sperm was detected was defined as day 0 of gestation.

Biochemical and Immunochemical

The activity of acid proteinases was measured at pH 3.2 with [^{14}C]verdohemoglobin as the substrate. Stomachs at each fetal stage were pooled,homogenized in ten volumes of 20 mM Tris-1 mM EDTA-0.1% Triton X-100, pH 7.5, and centrifuged at 12,000 x g for 20 min. The supernatants (crude extracts) were used as samples for analyses. Polyacrylamide gel electrophoresis (PAGE) of native acid proteinases and identification of acid proteinases on gels were carried out as described (1). SDS-PAGE-immunoblotting was carried out by using monoclonal antibody 3E5, which was raised against rat cathepsin E, as a probe (6). Monoclonal antibody 5B10 was used as a probe for pepsinogen C.

Immunohistochemical

Tissue specimens were fixed in Bouin's solution and embedded in paraffin. Sections (8 μm thick) were cut, mounted on slides and immunostained with monoclonal antibody by the peroxidase/antiperoxidase method. For conventional histological staining, specimens were fixed with Carnoy's solution.

RESULTS

Elevation of Cathepsin E Activity in Developing Fetal Stomachs

The total acid proteinase activity in rat fetal stomachs, when examined in the conventional hemoglobin-hydrolyzing assay system, increased dramatically as development proceeded (Table 1): an approximately 40-fold increase in specific activity during prenatal development from day 15 of gestation to birth. Electrophoretic analysis, combined with immunological tests (1), revealed that the increase was due almost exclusively to increases in the activity of the monomeric and dimeric forms of cathepsin E. Cathepsin D was detectable in all extracts examined, and some bands of activity, though not yet identified, were also found near that of the monomeric form of cathepsin E and at the position of pepsinogen C at late gestational and postnatal stages. However, these activities were relatively weak and, consequently, cathepsin E appeared to be the predominant aspartic proteinase in rat stomachs during fetal and postnatal (1-day-old and 7-day-old) development. In 20-day-old rats, the activity of pepsinogen C was greater than that of cathepsin E. SDS-PAGE immunoblot analysis revealed that cathepsin E in fetal stomachs exists in its pro-form with a molecular mass of 43-kDa.

Immunohistochemical Localization of Cathepsin E in Fetal Stomachs

In developing fetal stomachs, the presence of cathepsin E was first demonstrated in specimens as early as day 16 of gestation. The enzyme was localized in the cytoplasm of the proliferating epithelium in the region destined for pars glandularis, but not for pars proven-

Table 1. Total Acid Proteinase Activity in the Developing Rat Stomach

Gestation (day)	Activity (units/mg of protein)	
15	0.017	(28)[1]
16	0.029	(14)
17	0.054	(11)
18	0.110	(7)
19	0.236	(4)
20	0.648	(5)
21	0.591	(4)
1-day-old[2]	0.706	(3)
20-day-old[2]	1.16	(2)
40-day-old[2]	1.50	(4)

[1] The numbers of pooled stomachs are indicated in parentheses.
[2] The pars glandularis regions were used for the preparation of samples.

tricularis (Fig. 1). At this developmental stage, conventional histological methods could not clearly distinguish the cells destined to differentiate into the simple columnar epithelium of pars glandularis from those destined to differentiate into the stratified squamous epithelium of pars proventricularis. In this respect, cathepsin E could be useful as a molecular marker for identifying the stem cells of the glandular epithelium.

The occurrence of surface mucous cells, as judged from accumulation of periodic acid-Schiff(PAS)-positive material, was first observed in a limited number of cells on day 19 of gestation. Large eosinophilic cells, which were identified as parietal cells, also appeared in specimens on day 19. Interestingly, the epithelial cells at these gestational stages were all immunopositive for cathepsin E. By contrast, no positive signals to indicate the presence of pepsinogen C were observed with monoclonal antibody (5B10) raised against rat pepsinogen C during both pre- and postnatal development.

DISCUSSION

Several possibilities have so far been suggested on the functional significance of cathepsin E, but its role has not yet been established. It would be interesting to note and discuss these possibilities in the light of the present results that indicate an early appearance of the enzyme in fetal stomachs.

In mature gastric glands, cathepsin E has been localized in the surface mucous cells, foveolar epithelial cells and some parietal cells (7,8). This distributional pattern contrasts that of pepsinogen C, which is located in the chief cells and neck mucous cells of the glands. Thus, there is a possibility that cathepsin E is involved in differentiation by acting as a factor that controls rearrangements of intracellular structures or movements of cells. In the present study, we found that all epithelial cells of pars glandularis remained immunopositive for cathepsin E during later stages of gestation and even after birth, and that the time of the appearance of cathepsin E (day 16) was apparently different from that of the parietal cells or the surface mucous cells (day 19). These results do not seem to favor the view that the enzyme is involved directly in the differentiation of these cells. It may, rather, be the case that the enzyme acts to disrupt the association between the epithelial cells and the basal

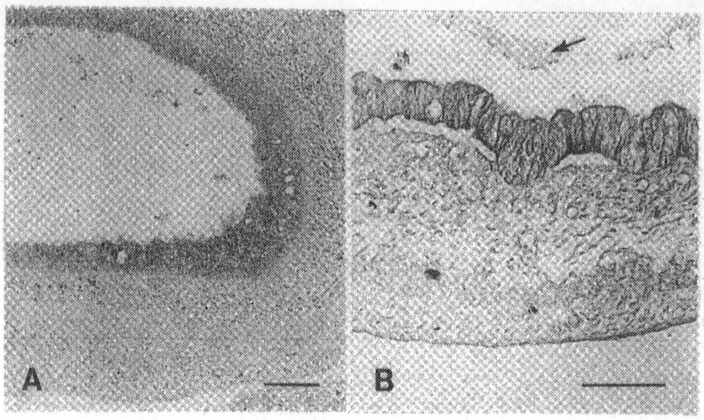

Figure 1. Immunohistochemical localization of cathepsin E in fetal stomachs on day 16 of gestation (A) and day 18 of gestation (B). All epithelial cells in the region of pars glandularis are immunopositive for cathepsin E. The monoclonal antibody 3E5 was used. In specimens on day 17 of gestation or later, the sloughing off of the growing epithelium can be seen (arrow). Bar, 100 μm.

lamina in the falling or sloughing off of cells from the epithelium, since the growing epithelium begins to be sloughed off into the lumen as early as day 17 of gestation, and since cathepsin E is known to be localized in the central and basal regions, but not on the free-surface side, of the mature surface mucous cells (7). These observations further suggest another view that cathepsin E may function as a self-digestive enzyme in the gastric lumen at the fetal and adult stages. Indeed, it might be advantageous if such cells that have extremely short life spans contained a strong self-digestive enzyme.

In this study, no aspartic proteinase exhibiting higher activity than cathepsin E was found in developing rat stomachs, even at neonatal stages. Although cathepsin E has milk-clotting activity *in vitro*, it seems unlikely that "non-secretory" cathepsin E functions *in vivo* as a fetal or neonatal type of digestive enzyme, like prochymosin and pepsinogen F. It remains to be examined whether the minor bands of activity detected in our electrophoretic analyses correspond to such known newborn-specific types of aspartic proteinases.

REFERENCES

1. Yonezawa, S., Fujii, K., Maejima, Y., Tamoto, K., Mori, Y., and Muto, N. (1988) Arch. Biochem. Biophys. 267, 176-183.
2. Muto, N., Yamamoto, M., Tani, S., and Yonezawa, S. (1988) J. Biochem. 103, 629-632.
3. Lees, W. E., Kalinka, S., Meech, J., Capper, S. J., Cook, N. D., and Kay, J. (1990) FEBS Lett. 273, 99-102.
4. Athauda, S. B. P., Takahashi, T., Inoue, H., Ichinose, M., and Takahashi, K. (1991) FEBS Lett. 292, 53-56.
5. Yonezawa, S., and Nakamura, K. (1991) Biochim. Biophys. Acta 1073, 155-160.
6. Yonezawa, S., Maejima, Y., Hagiwara, N., Aratani, T., Shoji, R., Kageyama, T., Tsukada, S., Miki, K., and Ichinose, M. (1993) Develop. Growth Differ. 35, 349-356.
7. Samloff, I. M., Taggart, R. T., Shiraishi, T., Branch, T., Reid, W. A., Heath, R., Lewis, R. W., Valler, M. J., and Kay, J. (1987) Gastroenterol. 93, 77-84.
8. Sakai, H., Saku, T., Kato. S., and Yamamoto, K. (1989) Biochim. Biophys. Acta 991, 367-375.

TISSUE- AND CELL-SPECIFIC CONTROL OF GUINEA PIG CATHEPSIN E GENE EXPRESSION

Shinko Tsukada,[1] Masao Ichinose,[1] Kazumasa Miki,[1] Nobuyuki Kakei,[1] Masashi Matsushima,[1] Naohisa Yahagi,[1] Satoshi Ishihama,[1] Yasuhito Shimizu,[1] Masahiro Kido,[1] Hiroshi Fukamachi,[2] Kiyoshi Kurokawa,[1] Satoshi Yonezawa,[3] Takashi Kageyama,[4] and Kenji Takahashi[5]

[1]First Department of Internal Medicine
Faculty of Medicine
University of Tokyo
7-3-1 Hongo, Bunkyo-ku, Tokyo 113, Japan
[2]Laboratory of Molecular Embryology
Zoological Institute
University of Tokyo
7-3-1 Hongo, Bunkyo-ku, Tokyo 113, Japan
[3]Department of Embryology
Institute for Developmental Research
Aichi Prefectural Colony, Kasugai 480-03, Japan
[4]Department of Cellular and Molecular Biology
Primate Research Institute
Kyoto University
Inuyama 484, Japan
[5]Department of Biophysics and Biochemistry
Faculty of Sciene
University of Tokyo
7-3-1 Hongo Bunkyo-ku, Tokyo 113, Japan

INTRODUCTION

Mammalian stomach mucosa has several types of acid proteinase, namely pepsins, cathepsin D and cathepsin E (CTSE). Compared with cathepsin D, the tissue distributions of pepsins and cathepsin E are relatively limited and they are synthesized mainly in the stomach mucosa. Pepsins are well-known secretory type proteinases, and its precursor forms are pepsinogens (Pg). CTSE is a non-secretory, non-lysozomal intracellular proteinase and its physiological functions are still unclear. Also unclear is the control mechanisms of its

Aspartic Proteinases: Structure, Function, Biology, and Biomedical Implications
Edited by Kenji Takahashi, Plenum Press, New York, 1995

biosynthesis. In order to understand the tissue and cell-specific control of the synthesis of these acid proteinases, we analyzed normal guinea pig tissues and fractions of stomach mucosal cells separated by centrifugal elutriation. The results obtained indicate a highly tissue- and cell-specific control of Pg and CTSE gene expression at the transcriptional level, suggesting that both proteinases are probably involved in specific functions of particular differentiated cells rather than playing a role in general intracellular proteolysis.

MATERIALS AND METHODS

Animals

Adult male guinea pigs (Hartley strain, weighing around 350g) were used in all experiments. Some tissues for biochemical analysis were stored in liquid nitrogen until use, and the remaining for histological studies were fixed with 10% formalin.

Histological Studies

Tissues were embedded in paraffin and were stained with hematoxylin/eosin. Deparaffinized sections were first reacted with rabbit anti-Pg antibodies (1) or rabbit anti-CTSE antibodies (2). The sites of antibody binding were visualized by the avidin-biotin-peroxidase complex method (3) using 3,3'-diaminobenzidine as chromogen (4). As a negative control, preimmune rabbit serum was used instead of the antiserum.

Separation of Dispersed Mucosal Cells of the Stomach

Cell fractions with various contents of Pg or CTSE-producing cells were prepared from guinea pig stomach mucosa using centrifugal-elutriation by the previously described method (5). Cell viability was more than 90% as determined by Trypan blue exclusion test. The percentage of Pg or CTSE-producing cells in each cell fraction was immunohistochemically determined with the anti-Pg and anti-CTSE antibody. Pg and CTSE activity in the crude extract of each fraction was determined as described (6). Protein was measured by the method of Lowry et al. (7).

Isolation and Analysis of DNA and RNA

High-molecular-weight DNA from various guinea pig tissues and from the separated stomach mucosal cells was prepared by the method of Gross-Bellard et al. (8), and total RNA was isolated by the guanidium/cesium chloride method (9). For methylation analysis of DNA, two methylation-sensitive enzymes, *Hpa*II and *Hha*I, were used. Both enzymes recognize the doublet 5'-CG-3' within their recognition sites (*Hpa*II, CCGG; *Hha*I, GCGC) and both fail to cleave DNA if the cytosine is methylated (10). The enzyme *Msp*I provides a control for *Hpa*II since it cleaves CCGG independently of the methylation of the internal cytosine (10). Restriction enzyme digestion, electrophoresis, blotting and detection of DNA were performed by the method of Southern (11). RNA sample was denatured (12) and analyzed by Northern blotting as described elsewhere (13). The cDNA insert for guinea pig Pg in recombinant DNA clone GP (14) was excised by digestion with *Eco*RI and *Bam*HI and the cDNA insert for guinea pig CTSE in recombinant DNA clone GP477 (14) was excised by digestion with *Pst*1, labeled by nick-translation with [α-^{32}P]dCTP (15) and used as the hybridization probe. Both inserts of about 1.07 kb contain almost the entire length of the coding region.

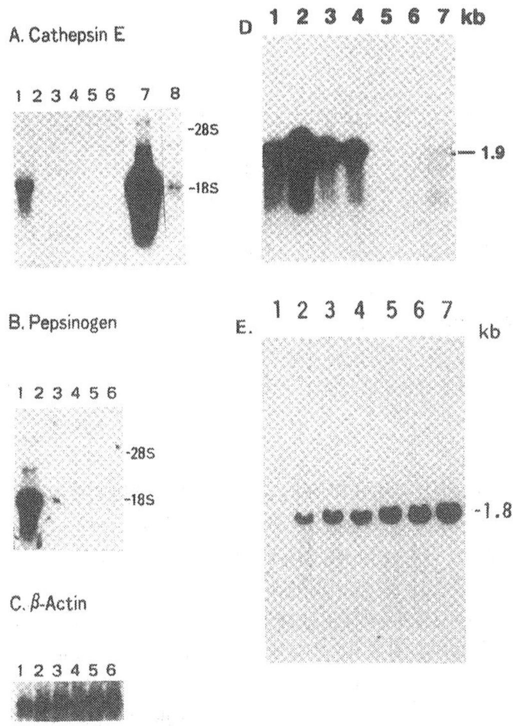

Figure 1. Northern blot hybridization of total RNA from normal guinea pig tissues (A, B, and C) (1, stomach mucosa; 2, small intestine; 3, large intestine; 4, spleen; 5, liver; 6, kidney; 7 and 8, over exposure of lanes 1 and 4, respectively) and from fractions of dispersed elutriation (D,E) (Lanes 1-7 correspond to 1-7 in Table 1) with a cDNA probe for guinea pig cathepsin E (A,D), or for pepsinogen (B,E) or for β-actin (C).

RESULTS AND DISCUSSION

First, we analyzed Pg and CTSE mRNA expression in normal guinea pig tissues by Northern blotting. As shown in Fig. 1A and B, both mRNA expressions were relatively limited to the stomach mucosa. Although the level was very low, the expression of CTSE mRNA was also observed in the spleen. These results were in good agreement with the distribution of the cells producing each proteinase as revealed by immunohistochemistry (Fig. 2A). In the separated fractions of dispersed mucosal cells prepared from the stomach mucosa by centrifugal elutriation, the extents of each mRNA expression were closely correlated with the enrichment of each producing cells (Fig. 1D and E and Table 1). And the results show that pepsinogen activity was highest in fraction 5-7, although CTSE activity was highest in fraction 2 (Fig. 2B). The Pg and CTSE mRNA expression correlated well with the relative enrichment of the producing-cells and with the proteolytic activity of the proteinases in each fraction, suggesting that the mRNA expression is regulated in a cell-specific manner. In other than the stomach mucosa, a much lower level of CTSE mRNA expression was also observed in the spleen, where only a few cells were stained positively with the anti-CTSE antibody (Fig. 2C). Most positive cells were located in the red pulp. The level of CTSE mRNA in the stomach mucosa was about 50-fold higher than that in the spleen. This accords well with previous reports that the spleen of several species contains trace amounts of CTSE (16). It is notable that the stomach and spleen showed different patterns

Table 1. Cell Separation by Elutriator

Fraction	Rpm[a]	Flow[b] rate (mℓ/min)	% of pepsinogen-producing cell	% of cathepsinE-producing cells
1	2000	5	3.0	28.5
2	2000	13	13.0	68.0
3	2000	22	14.8	44.3
4	2000	29	37.2	30.8
5	2000	38	51.0	6.8
6	2000	48	52.3	5.6
7	1450	48	50.0	8.0

a & b The rpm and flow rate at which each fraction was collected.

of hybridized bands: in addition to the major mRNA of approximately 1.9 kb, RNA from the stomach generated additional bands at 4.8 and 2.9 kb, while the spleen generated an additional band at 3.5 kb. Although previous studies indicate that CTSE from the stomach and spleen are enzymologycally and immunologically indistinguishable (17), the molecular nature of CTSE from these tissues remains to be characterized. Thus, further studies are necessary to determine whether the difference results from transcription of different genes or from alternative splicing of a single type of transcript. In other organs, no cell was positively stained for CTSE. RNA from these organs gave only background hybridization signals. Thus, our results demonstrate that the distribution of CTSE-producing cells and its mRNA expression are relatively limited. The enzyme is mainly produced in the stomach mucosa and its biosynthesis is controlled by the cell-specific transcriptional regulation.

Previous studies indicate that transcriptional activity correlates to the extent of methylation in the region of the gene (18,19). So, the methylation statuses of Pg and CTSE gene were analyzed in various normal tissues and cells using methylation sensitive enzymes,

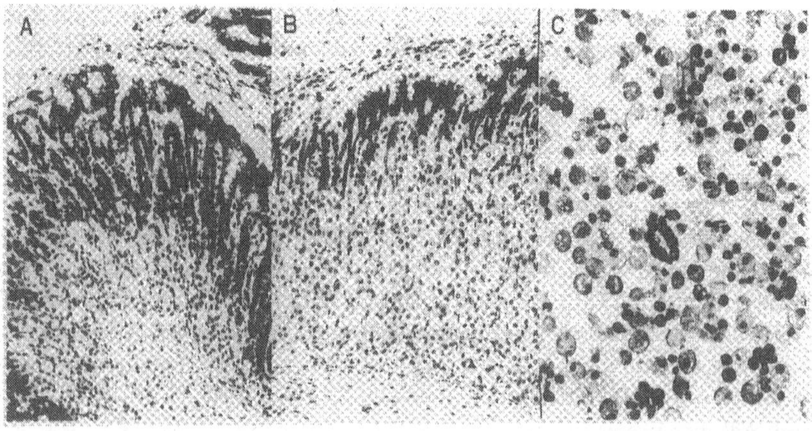

Figure 2. Localization of cells that produce cathepsin E in normal guinea pig tissues and in a separated fraction of dispersed mucosal cells prepared from guinea pig stomach by centrifugal elutriation. A, stomach mucosa (x180); B, a separated mucosal cell fraction corresponding to 2 in Table 1 (x180); C, spleen(x180). Cells strongly stained with anti-guinea pig cathepsin E antibody are seen in both tissues and in the cell fraction.

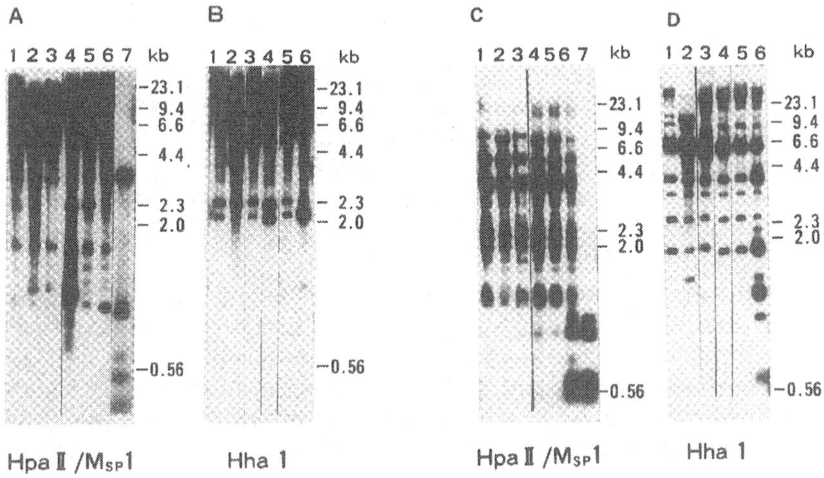

Figure 3. Southern blot hybridization of high-molecular-weight DNAs from normal guinea pig tissues (Lane 1, lung; 2, liver; 3, large intestine; 4, spleen; 5, small intestine; 6 and 7, stomach) with cDNA for guinea pig cathepsin E (A, B), or for pepsinogen (C, D) as a probe. Restriction enzymes used were *Hpa*II (A, C, lanes 1-6), *Msp*I (A, lane 7) and *Hha*I (B,C).

*Hpa*II and *Hha*I. The results of *Msp*1/*Hpa*II analysis are shown in Fig. 3A and 3C. The *Msp*1 digestion patterns of DNAs from various tissues were identical (Fig. 3A and 3C, lane 7). However, when the same DNAs were digested with HpaII, the hybridization patterns were different from those of Msp1 digestion, suggesting methylation of CCGG sites in the gene

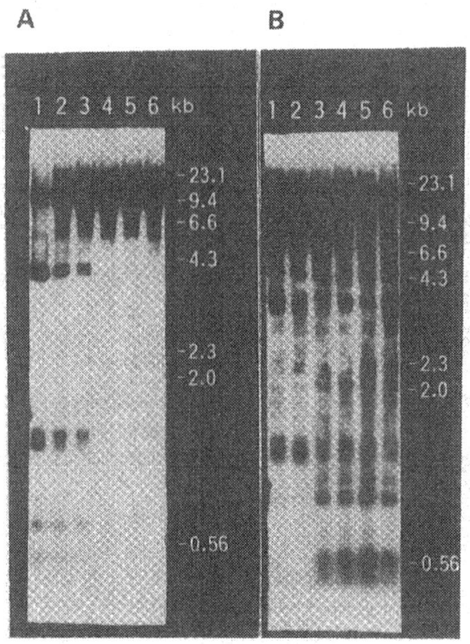

Figure 4. Southern blot hybridization of high-molecular-weight DNAs from fractions of dispersed stomach mucosal cells corresponds to 1-7 in Table 1 with a cDNA probe for guinea pig cathepsin E (A),or for guinea pig pepsinogen (B). Restriction enzymes used were *Hpa*II.

region. CCGG sites in the gene region in the Pg and CTSE-producing tissues were less methylated than those in the non-producing tissues. In CTSE gene of spleen, the methylation pattern was intermediate between that of the stomach mucosa and that of other tissues, as revealed by the appearance of low-molecular-weight bands. Similar methylation analysis with another enzyme, *Hha*I, also revealed that the GCGC sites in Pg or CTSE gene were less methylated in the producing tissues than in other tissues (Fig. 3B and D). The same analysis in the separated fractions of the dispersed mucosal cells of the stomach also revealed the step-wise increase in the hypomethylated bands with the increase in the relative enrichment of the producing cells, suggesting that CCGG sites in Pg and CTSE gene region were selectively hypomethylated (Fig. 4).

These results demonstrate the differential methylation of Pg and CTSE genes not only among tissues but also cell subpopulations within a single tissue. There is an inverse correlation between the extent of methylation and expression of Pg and CTSE genes. In addition, the observed differential metylation of the two genes in the subpopulations of the stomach mucosal cells supports the idea that some mechanisms utilizing DNA methylation are involved in the normal development and differentiation of the stomach mucosa by regulating the expression of the specific phenotype. Demonstration of such mechanisms is a problem for future study.

REFERENCES

1. C.Furihata, D.Saito, H.Fujiki, Y.Kanai, T.Matsushima and T.Sugimura,Purification and characterization of pepsinogens and a unique pepsin from rat stomach, *Eur.J. Biochem.*105:43 (1980).
2. S.Tsukada, M.Ichinose, K.Miki, M.Tatematsu, S.Yonezawa, M.Matsushima, N.Kakei, H.Fukamachi, S.Yasugi, K.Kurokawa, T.Kageyama and K.Takahashi,Tissue- and cell-specific control of guinea pig cathepsin E gene expression, *Biochem.Biophys.Res. Commun.* 187:1401 (1992).
3. S.M.Hsu, L.Raine and H.Fanger,Use of avidin-biotin-peroxidase complex (ABC) in immunoperoxidase techniques: a comparison between ABC and unlabeled antibody (PAP) procedures,*J.Histochem.cytochem.*29:577 (1981).
4. J.C.Adams, Heavy metal intensification of DAB-based HRP reaction product. *J. Histochem.Cytochem.*29:775 (1981)
5. A.H.Soll, The actions of sectetagogues on oxygen uptake by isolated mammalian parietal cells, *J.Clin.Invest.*61:370 (1978).
6. N.Muto, M.Yamamoto, S.Tani and S.Yonezawa,Characteristic distribution of cathepsin E which immunologically cross-reacts with the 86-kDa acid proteinase from rat gastic mucosa, *J.Biochem.*103:629 (1988).
7. O.H.Lowry, N.J.Rosenbrough, A.L.Farr and R.J.Randall, Protein measurement with the Folin-phenol reagent, *J.Biol.Chem.*193:265 (1951).
8. N.Gross-Bellard, P.Oudet and P.Chambon,Isolation of high-molecular-weight DNA from mammalian cells, *Eur.J.Biochem.*36:32 (1973).
9. A.Ullrich, J.Shine, J.Chirgwin, R.Pictet, E.Tischer, W.J.Rutter and H.M.Goodman, Rat insulin genes:construction of plasmids containing the coding sequence, *Science* 196: 1313 (1977).
10. M.McClelland, The effect of sequence specific DNA methylation on restriction endonuclease cleavage, *Nucleic Acids Res.*9:5859 (1981).
11. E.Southern, Detection of specific sequences among DNA separated by gel electrophoresis, *J.Mol.Biol.*98:503 (1975).
12. E.A.Fyrberg, J.W.Mahaffey, B.J.Bond and N.Davidson, Transcripts of the six Drosophila actin genes accumulate in a stage- and tissue-specific manner, *Cell* 33: 115 (1983).
13. D.A.Goldberg, Isolation and partial characterization of the Drosophila alcohol dehydrogenase gene, *Proc.Natl.Acad.Sci.USA* 77:5794 (1980).
14. P.W.Rigby, M.Dieckmann, C.Rhoudes and D.Berg,Labeling deoxyribonucleic acid to high specific activity in vitro by nick translation with DNA polymease, *J.Mol.Biol.* 113: 237 (1977).

15. S.Yonezawa, K.Fujii, Y.Maejima, K.Tamoto, Y.Mori and N.Muto, Further studies on rat cathepsin E: subcellular localization and existence of the active subunit form.*Arch. Biochem.Biophys.* 267:176 (1988).
16. V.Puizdar, C.Lapresle and V.Turk, Evidence for the presence of large amounts of cathepsin E in rat spleen,*FEBS Lett.* 185:236 (1985).
17. N.Muto, M.Yamamoto and S.Tani, Comparative studies of two types of acid proteases from rat gastric mucosa and spleen, *J.Biochem.* 101:1069 (1987).
18. A.P.Bird, DNA-methylation -How important in gene control? *Nature*,321:209 (1984).
19. H.Cedar, DNA methylation and gene activity, *Cell* 53:3 (1988).

CATHEPSIN E IS EXPRESSED IN FETAL RAT GLANDULAR STOMACH EPITHELIAL CELLS IN PRIMARY CULTURE IN THE ABSENCE OF MESENCHYMES

Hiroshi Fukamachi,[1] Masao Ichinose,[2] Satoshi Ishihama,[2]
Shinko Tsukada,[2] Chie Furihata,[3] Satoshi Yonezawa,[4] and Kazumasa Miki[2]

[1]Zoological Institute
Faculty of Science
University of Tokyo
Tokyo 113, Japan
[2]1st Department of Internal Medicine
Faculty of Medicine
University of Tokyo
Tokyo 113, Japan
[3]Department of Molecular Oncology
Institute of Medical Science
University of Tokyo
Tokyo 108, Japan
[4]Department of Embryology
Institute for Developmental Research
Aichi Prefectural Colony, Kasugai 480-03, Japan

INTRODUCTION

The glandular stomach epithelium consists of surface mucous, mucous neck, parietal, chief and endocrine cells, but it remains to be solved how their differentiation is controlled. The importance of epithelial-mesenchymal interaction in inducing differentiation of glandular stomach epithelial cells has been repeatedly reported using avian embryos [1]. We have also found that the differentiation of glandular stomach epithelial cells is affected by the type of combined mesenchymes in fetal mice in organ culture [2], and that disorganization of stroma alters differentiation of glandular stomach epithelial cells in adult mice in vivo [3], demonstrating that tissue interactions play important roles in regulating the differentiation of glandular stomach epithelial cells in mammals.

One way to elucidate the mechanism of the interaction is to culture pure epithelial cells and to examine their differentiation in the absence of mesenchymes. So far, many researchers have reported culture systems for differentiated glandular stomach epithelial

Aspartic Proteinases: Structure, Function, Biology, and Biomedical Implications
Edited by Kenji Takahashi, Plenum Press, New York, 1995

cells, but only a few studies have been done and controversial results have been reported on the differentiation of fetal glandular stomach epithelial cells in culture. Matsuoka et al. [4] showed that cultures of fetal rabbit gastric epithelial cells consisted mainly of surface mucous cells which expressed glucose-6-phosphate dehydrogenase activity, while Kondo et al. [5] reported that epithelial differentiation was not observed in cultures of fetal rat gastric epithelial cells.

The glandular stomach epithelial cells express cell type-specific acid proteases. Lai et al. [6] have reported that pepsinogens are expressed in mucous neck and chief cells while cathepsin E is localized in surface and foveolar epithelial cells in adult rat gastric mucosa. Here we established a primary culture system for the fetal rat glandular stomach epithelial cells, and examined whether the cultured cells expressed acid proteases.

MATERIALS AND METHODS

The method to obtain and culture pure epithelial tissues from 16.5-day fetal rats was a modification of the previously described one [7]. Glandular stomach tissues were obtained from 16.5-day F344/Du rats (Charles River Japan), and epithelial tissues were cultured on collagen gel in Ham's F12 supplemented with 10% horse serum (Handai Biken), bovine pituitary extract (100 μg/ml, Kyokuto Pharmaceutical Co.), epidermal growth factor (10 ng/ml, Collaborative Res.), insulin (5 μg/ml, Sigma) and cholera toxin (0.5 μg/ml, List Biol. Lab.). Acid protease activity was measured as previously described [8]. Protein concentration was determined according to the method of Bradford [9]. For immunohistochemical analysis, adult rat stomach tissues or cultured glandular stomach epithelial cells on collagen gels were processed according to Saint-Marie [10]. Rabbit anti-rat pepsinogen antibodies [8] and rabbit anti-rat cathepsin E antibodies [11] were used to detect pepsinogen and cathepsin E, respectively. Normal rabbit serum instead of anti-pepsinogen or anti-cathepsin E antibodies was used for controls.

To examine the nature of acid protease activity, samples were subjected to electrophoresis on 4 to 15% polyacrylamide gradient gel according to Davis [12]. After electrophoresis, some gels were soaked in a hemoglobin solution, incubated, fixed and stained to indicate the location of potential acid protease activity in the gel (zymogram). Proteins on some gels were electrophoretically transferred to membrane filters by the method of Towbin et al. [13], and pepsinogen and cathepsin E on the filter were detected using specific antibodies (Western blotting). Ferritin (MW = 443,000), lactate dehydrogenase (MW = 139,850), albumin (MW = 66,300) and trypsin inhibitor (MW = 20,100) were used as molecular weight standards.

RESULTS

The developmental change of acid protease activity in fetal rat glandular stomach tissues has been described in detail elsewhere [14]. The activity was very weak (less than 0.3 units/mg protein) in 16.5-day fetal tissues, but it gradually increased to about 0.8 and 10 units/mg protein in newborn and adult rats, respectively. The change of acid protease activity in the glandular stomach epithelial cells in primary culture was examined. Acid protease activity could be detected in cells on day 4 in culture (about 0.7 units/mg protein), gradually increased (about 1.1 and 2 units/mg protein on days 6 and 8, respectively) and then decreased (about 1.3 units/mg protein on day 14).

Two types of acid proteases are known to be expressed in the glandular stomach epithelial cells. By using specific antibodies, we confirmed that pepsinogens were localized

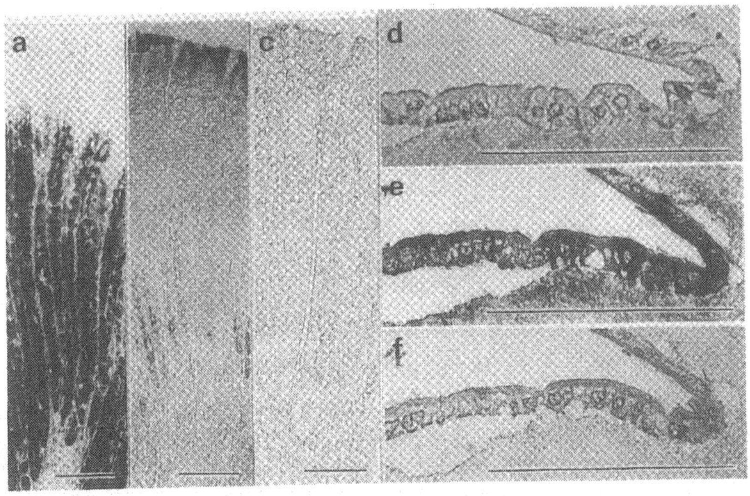

Figure 1. Light micrographs of an adult rat glandular stomach (a-c) and cultured glandular stomach epithelial cells (d-f), stained with anti-pepsinogen antibodies (a,d), anti-cathepsin E antibodies (b,e) and normal rabbit serum (d,f), followed by peroxidase conjugated second antibodies. The localization of antigens was visualized by DAB staining. Note that pepsinogens and cathepsin E are expressed by mucous neck and chief cell, and surface mucous cells, respectively, in the adult rat glandular stomach. A few parietal cells in the body of the glands are also weakly stained with anti-cathepsin E antibodies. Cultured glandular stomach epithelial cells express cathepsin E only. Bars = 100 μm.

in mucous neck and chief cells, while cathepsin E was expressed in surface mucous cells in adult rat glandular stomach (Fig. 1a-c). Immunohistochemical analysis also showed that the epithelial cells in primary culture were strongly stained with anti-cathepsin E antibodies, but were not stained with anti-pepsinogen antibodies (Fig. 1d-f), suggesting that the cells expressed cathepsin E only.

The nature of acid protease activity was biochemically analyzed with zymogram and Western blotting. Analysis with zymogram showed that acid protease activity could be separated into several bands by electrophoresis (Fig. 2). Adult and newborn tissues gave 2 bands, while cultured epithelial cells gave only 1 band with approximate molecular weight of 80,000. Western blotting analysis showed that bands of acid protease activity of adult and newborn tissues were positively stained with both anti-pepsinogen and anti-cathepsin E antibodies while those of the cultured epithelial cells were stained just with anti-cathepsin E antibodies but not with anti-pepsinogen antibodies, indicating that the cells in culture expressed cathepsin E only.

DISCUSSION

In the present study we found that glandular stomach epithelial cells expressed acid protease activity in primary culture. Using specific antibodies, we found that the cells in primary culture were positively stained with anti-cathepsin E antibodies but not with anti-pepsinogen antibodies (Fig. 1). Analysis with zymogram showed that the acid protease activity was localized in a band with approximate molecular weight of 80,000, and Western blotting analysis revealed that the band was positively stained with anti-cathepsin E antibodies but not with anti-pepsinogen antibodies (Fig. 2). Muto et al. [15] estimated that the

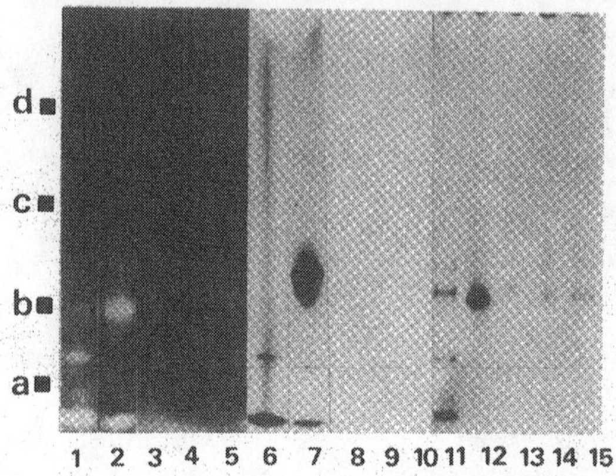

Figure 2. Analysis of acid protease activity with zymogram and Western blotting. Crude extract of adult rat glandular stomachs (lanes 1,6,11), newborn rat glandular stomachs (lanes 2,7,12), and cultured glandular stomach epithelial cells on days 4 (lanes 3,8,13), 8 (lanes 4,9,14) and 15 (lanes 5,10,15) were subjected to gel electrophoresis, and stained for acid protease activity (lanes 1-5), or stained with anti-pepsinogen antibodies (lanes 6-10) or anti-cathepsin E antibodies (lanes 11-15). The marker proteins used were (a) trypsin inhibitor, (b) albumin, (c) lactate dehydrogenase and (d) ferritin.

molecular weight of rat cathepsin E was about 86,000 using gel filtration on a Sephadex G-100 column. This difference in the estimation of molecular weight of cathepsin E might be due to the difference in the method used. Thus we concluded that fetal rat glandular stomach epithelial cells expressed cathepsin E in primary culture in the absence of mesenchymes. Our results are consistent with those of Matsuoka et al. [4], but are more convincing because we demonstrated that the cells expressed a marker protein specific to surface mucous cells.

The superficial portion of gastric mucosa has a rapid rate of cell renewal. Stevens and Leblond [16] estimated that surface mucous cells of adult rats were renewed in about 3 days, and mucous neck cells in about a week. Other 3 types of cells are relatively long lived and are renewed only very slowly. Thus it would be reasonable to think that most epithelial cells produced in the stem cell zone differentiate into surface mucous cells in adult rat gastric mucosa in vivo. Then it is quite natural that most cells differentiated into surface mucous cells when glandular stomach epithelial cells were cultured.

In the present study, we could not find any pepsinogen-producing in primary culture. It is possible that our culture period was too short to induce differentiation of pepsinogen-producing cells. But it is more probable that other factors secreted by mesenchymal cells are necessary for their differentiation because (i) the epithelial cells on day 4 in culture express pepsinogens in a week, when implanted under kidney capsules of isogenic animals in combination with mesenchymal tissues (unpublished observation), and (ii) pepsinogens were expressed when 16.5-day fetal glandular stomach tissues were cultured organotypically for a week (Tsukada et al., manuscript in preparation). It is the problem for the future to find culture conditions which induce the differentiation of pepsinogen-producing cells, and to examine how the expression of cathepsin E and pepsinogen is regulated in the glandular stomach.

REFERENCES

1. S. Yasugi, Role of Epithelial-Mesenchymal Interactions in Differentiation of Epithelium of Vertebrate Digestive Organs, *Dev. Growth Differ.* 35:1 (1993).

2. H. Fukamachi, T. Mizuno, S. Takayama, Epithelial-Mesenchymal Interactions in Differentiation of Stomach Epithelium in Fetal Mice, *Anat. Embryol.* 157:151 (1979).
3. H. Fukamachi, Disorganization of Stroma Alters Epithelial Differentiation of the Glandular Stomach in Adult Mice, *Cell Tissue Res.* 243:65 (1986).
4. K. Matsuoka, M. Tanaka, Y. Mitsui, S. Murota, Cultured Rabbit Gastric Epithelial Cells Producing Prostaglandin I_2, *Gastroenterology* 84:498 (1983).
5. Y. Kondo, G. P. Young, I. Rose, R. H. Whitehead, Organ Specificity of Epithelial Cells Grown in Tissue Culture from Explants Obtained from Various Levels of the Rat Gut, *Exp. Cell Res.* 159:158 (1985).
6. K. Lai, J. B. Wyckoff, I. M. Samloff, Aspartic Proteinases in Gastric Mucosa of the Rat: Absence of Pepsinogen I, Genetic Polymorphism of Pepsinogen II, and Presence of Slow-Moving Proteinase, *Gastroenterology* 95:295 (1988).
7. H. Fukamachi, Proliferation and Differentiation of Fetal Rat Intestinal Epithelial Cells in Primary Serum-free Culture, *J. Cell Sci.* 103:511 (1992).
8. C. Furihata, D. Saito, H. Fujiki, Y. Kanai, T. Matsushima, T. Sugimura, Purification and Characterization of Pepsinogens and a Unique Pepsin from Rat Stomach, *Eur. J. Biochem.* 105:43 (1980).
9. M. M. Bradford, A Rapid and Sensitive Method for the Quantitation of Microgram Quantities of Protein Utilizing the Principle of Protein-Dye Binding, *Anal. Biochem.* 72:248 (1976).
10. G. Sainte-Marie, A Paraffin Embedding Technique for Studies Employing Immunofluorescence, *J. Histochem. Cytochem.* 19:250 (1962).
11. N. Muto, M. Yamamoto, S. Tani, S. Yonezawa, Characteristic Distribution of Cathepsin E which Immunologically Crossreacts with the 86-kDa Acid Protease from Rat Gastric Mucosa, *J. Biochem.* 103:629 (1988).
12. B. J. Davis, Disc Electrophoresis II: Methods and Application to Human Serum Proteins, *Ann. New York Acad. Sci.* 121:404 (1964).
13. H. Towbin, T. Staehelin, J. Gordon, Electrophoretic Transfer of Proteins from Polyacrylamide Gels to Nitrocellulose Sheets: Procedures and Some Applications, *Proc. Natl. Acad. Sci, USA*, 76:4350 (1979).
14. S. Yonezawa, Y. Maejima, N. Higashihara, T. Aratani, R. Shoji, T. Kageyama, S. Tsukada, M. Ichinose, Changes with Development in the Expression of Cathepsin E in the Fetal Rat Stomach, *Dev. Growth Differ.* 35:349 (1933).
15. N. Muto, K. Murayama-Arai, S. Tani, Purification and Properties of a Cathepsin E-like Acid Protease from Rat Gastric Mucosa, *Biochim. Biophys. Acta*, 745:61 (1983).
16. C. E. Stevens, C. D. Leblond, Renewal of the Mucous Cells in the Gastric Mucosa of the Rat, *Anat. Rec.* 115:231 (1953).

CATHEPSIN E EXPRESSED IN PANCREATIC CANCER

Takeshi Azuma,[1*] Masayuki Yamada,[2] Hajime Murakita,[1] Yasuyuki Nishikawa,[1] Yoshihiro Kohli,[1] Kenji Yamamoto,[3] and Hitoshi Hori[2]

[1]Second Department of Internal Medicine, Fukui Medical School
Fukui 910-11, Japan
[2]Biotechnology Research Laboratory, Tosoh Corporation
Hayakawa, Ayaseshi, Kanagawa 252, Japan
[3]Department of Pharmacology
Kyushu University, Faculty of Dentistry
Fukuoka 812, Japan

INTRODUCTION

The aspartic proteinase family, each member of which has two essential aspartyl residues at the active site, includes pepsins, renin, cathepsin D, and cathepsin E (CTSE). Unlike pepsins, CTSE is a non-secretory, intracellular, but non-lysosomal proteinase found in the highest concentration in the superficial epithelial cells of the stomach (1,2). CTSE has been also localized to several lymphoid associated tissues (3,4). Although CTSE is thought to play an important role in the cell, the function of this proteinase is still unclear. Since it is not secreted into the stomach, it does not appear to have a role in digestion of dietary protein. Due to its intracellular location and distribution in lymphoid associated tissue, it has been suggested that the enzyme may have a role in immune function. It would be of great help in facilitating studies on the function of CTSE to determine the biological behavior of the enzyme in pathophysiological conditions. In the present study, we demonstrated the expression of CTSE in pancreatic ductal adenocarcinomas.

* Correspondence: Takeshi Azuma, M.D., Second Department of Internal Medicine, Fukui Medical School, Matsuoka-cho, Yoshida-gun, Fukui 910-11, Japan.

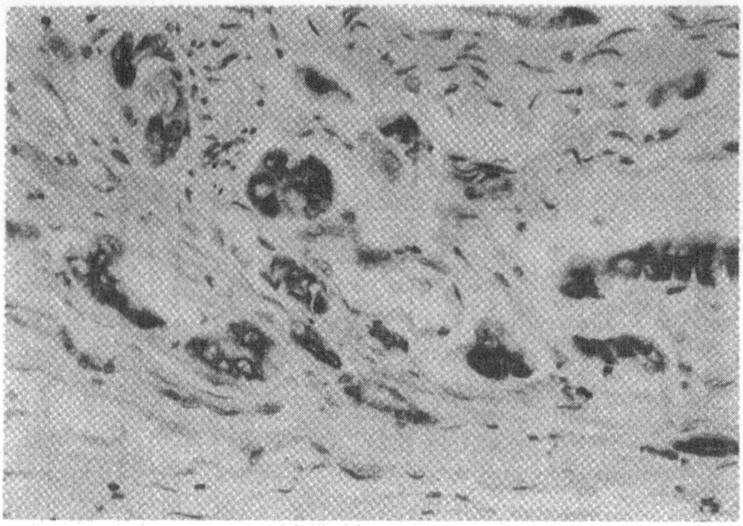

Figure 1. Immunohistochemical staining of CTSE in human pancreatic ductal adenocarcinoma.

MATERIALS AND METHODS

Immunohistochemistry

Ten surgical resected human pancreatic ductal adenocarcinomas were subjected to the present study. CTSE expression was examined by immunohistochemistry. All specimens were fixed in formalin and embedded in paraffin. The antisera against CTSE purified from the human erythrocyte membranes were raised in rabbits as described previously (5). Immunohistochemical staining was performed by the streptoavidin technique (6). Each sections (4 μm thick) were immersed in 0.3% H_2O_2 added methanol to destroy endogenous peroxidase activity. After incubation for 30 min with normal goat serum to block non-specific Ig binding sites, they were incubated overnight with 1:1000 dilution of rabbit polyclonal antiserum specific for CTSE. Washed in Tris buffered saline (TBS), they were incubated with biotinylated goat anti-rabbit IgG. After washed in TBS, they were incubated with peroxidase-conjugated streptoavidin and were developed by 3,3'-diaminobenzidine. Nuclear staining was carried out with hematoxylin. As a control,the first antibodies were replaced by normal rabbit IgG.

Western Blotting Analysis

Pancreatic juice was obtained from patients with pancreatic cancer by endoscopic canulation to pancreatic duct, and subjected to sodium dodecyl sulfate-polyacrylamide gel electrophoresis (SDS-PAGE). For immunoblotting, proteins electrophoresed on SDS-PAGE gels were transfered to nitrocellulose using the transblot apparatus (BioRad Lab.) according to the company's specifications and probed with rabbit anti-human CTSE polyclonal antibodies coupled to horseradish peroxidase.

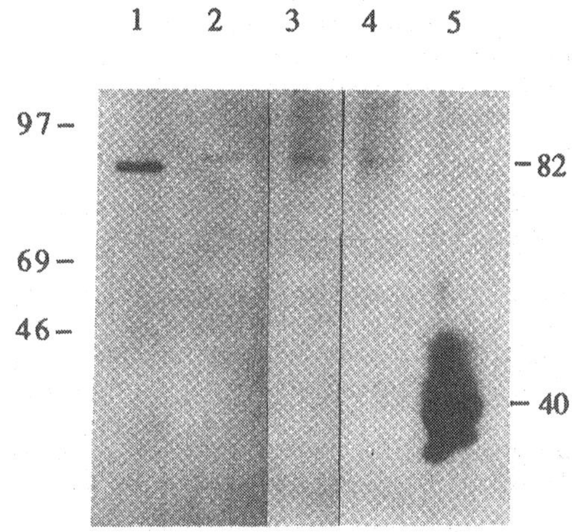

Figure 2. Western blot analysis for CTSE. Lane #1: purified recombinant human CTSE produced by yeast, #2-4: pancreatic juice samples from patients with pancreatic ductal adenocarcinoma, #5: the pancreatic juice from patient with pancreatic cystadenoma.

RESULTS AND DISCUSSION

All human pancreatic ductal adenocarcinomas examined in the present study expressed CTSE. CTSE was found diffusely in the cytoplasm (Figure 1). There were no immunoreactions to CTSE in the normal pancreatic duct. CTSE is an intracellular proteinase found in highest concentration in the superficial epithelial cells of the stomach. Saku et al. examined the biological behavior of CTSE in abnormal gastric tissue in order to understand its pathophysiological roles in gastric epithelia, and reported that chronic inflammation and ulcer did not influence the distribution of this enzyme, that no positive staining was obtained in the incomplete type of intestinal metaplasia, dysplasia, and well differentiated adenocarcinoma, and that tumor cells of signet ring cell carcinoma and poorly differentiated adenocarcinoma cells gave strong diffuse staining for CTSE (7). In contrast, it has been reported that pyloric-type metaplasia was found in pancreatic ducts, that pancreatic tumor cell lines expressed M1, an antigen characteristic of gastric mucin, and suggested that the potential for gastric-type differentiation may be an inherent characteristic of pancreatic stem cells (8). Sessa et al. reported that antigens normally found in gastric and/or intestinal epithelial cells are expressed in most differentiated duct cell carcinomas of the pancreas, and that more than 90% of pancreatic ductal adenocarcinoma expressed M1 and CTSE (8). These findings suggest that the expression of CTSE in pancreatic duct cells is cancer specific. In addition, CTSE, a non-secretory intracellular proteinase, was detected in the pancreatic juice of the patients with pancreatic cancer, and the molecular weight of the enzyme in the pancreatic juice was larger than that of the mature form (82 kDa) (Figure 2). These findings suggest that the intracellular localization and processing of the enzyme expressed in pancreatic ductal adenocarcinomas may be different from those in normal gastric superficial epithelial cells, and that CTSE may be a useful tumor marker for pancreatic ductal adenocarcinoma.

REFERENCES

1. I.M.Samloff, R.T.Taggart, T.Shiraishi, et al. Slow moving proteinase. Isolation, characterization, and immunohistochemical localization in gastric mucosa. Gastroenterology 93:77-84 (1987).

2. K.Yamamoto, N.Katsuda, M.Himeno,K.Kato. Cathepsin D of rat spleen. Affinity purification and properties of two types of cathepsin D. Eur.J.Biochem. 95:459-467 (1979).
3. K.Yamamoto, O.Kamata, N.Katsuda, K.Kato. Immunochemical difference between cathepsin D and cathepsin E-like enzyme from rat spleen. J.Biochem.87:511-516 (1980).
4. N.Muto, M.Yamamoto, S.Tani, S.Yonezawa. Characteristic distribution of cathepsin E which immunologically crossreacts with the 86-kDa acid proteinase from rat gastric mucosa. J.Biochem. 103:629-632 (1988).
5. K.Yamamoto, E.Ueno, H.Uemura, Y.Kato. Biochemical and immunochemical similarity between erythrocyte membrane aspartic proteinase and cathepsin E. Biochem.Biophys.Res.Commun. 148:267-272 (1987).
6. S.M.Hsu, L.Raine, H.Fanger. Use of avidin-biotin-peroxidase complex (ABC) in immunoperoxidase techniques. J.Histochem.Cytochem. 29:577-580 (1981).
7. T.Saku, H.Sakai, N.Tsuda, H.Okabe, Y.Kato, K.Yamamoto. Cathepsin D and E in normal, metaplastic, dysplastic, and carcinomatous gastric tissue: an immunohistochemical study. Gut 31:1250-1255 (1990).
8. F.Sessa, M.Bonato, B.Frigerio, et al. Ductal cancers of the pancreas frequently express markers of gastrointestinal epithelial cells. Gastroenterology 98:1655-1665 (1990).

PLANT ASPARTIC PROTEINASES FROM *CYNARA CARDUNCULUS* SPP. *FLAVESCENS* CV. CARDOON; NUCLEOTIDE SEQUENCE OF A cDNA ENCODING CYPROSIN AND ITS ORGAN-SPECIFIC EXPRESSION

M.C. Cordeiro,[1] Z. -T. Xue,[2] M. Pietrzak,[3] M. S. Pais,[4] and P. E. Brodelius[2]

[1]Departamento de Biologia
Universidade de Evora
Apartado 94, P-7001 Evora Codex, Portugal
[2]Department of Plant Biochemistry
Lund University
P.O.Box 7007, S-22007 Lund, Sweden
[3]Department of Botany
University of Basel
Hebelstrasse 1, CH-4056 Basel, Switzerland
[4]Centro de Biotecnologia Vegetal
Faculdade de Ciências de Lisboa
Bloco C2, Campo Grande, P-1700 Lisboa, Portugal

1. INTRODUCTION

Flowers of the plant *Cynara cardunculus* spp. *flavescens* cv. cardoon have been traditionally used since many centuries to produce ewe's cheese, in Portugal and Spain. One of the first reports on this use by farmers was made by Brotero in 1804 [1]. Today, the Serpa and Serra cheeses are some examples of typical and highly appreciated products in Portugal. The cheeses are made with water extracts of dried flowers.

The clotting activity present in the flowers of cardoon is due to three endo-proteases - cyprosin 1, 2 and 3 (previously designated as cynarases), which have been purified and partly characterised [2-4]. They are aspartic proteinases with a native M_r of around 49 kD; they are composed of two subunits with different molecular weight. Furthermore, these enzymes preferentially cleave peptides bonds between hydrophobic amino acid residues.

Aspartic proteinases from human, animal, parasite, fungal and retroviral origin have been extensively investigated [5,6]. The 3-D structure of several such enzymes has been resolved and they all show some common features [7]. Our knowledge about plant enzymes

```
    1  L  K  K  R  K  V  N  I  L  N  H  P  G  E  H  A  G  S  N  D  A  N  A  R  R  K  Y  G  V  R     30
    1 CCCTTAAAAAGAGGAAAGTCAATATACTGAACCACCCTGGTGAACATGCCGGCTCTAATGATGCGAATGCTAGAAGAAAGTATGGTGTCCGT     92
                   !
   31  G  N  F  R  D  S  D  G  E  L  I  A  L  K  N  Y  M  D  A  Q  Y  F  G  E  I  G  I  G  T  P     60
   93 GGTAATTTCAGGGATTCAGACGGTGAGCTTATTGCTCTAAAGAACTACATGGATGCTCAGTACTTTGGTGAGATTGGTATTGGCACTCCA    182
                               *  *  *
   61  P  Q  K  F  T  V  I  F  D  T  G  S  S  N  L  W  V  P  S  S  K  C  Y  F  S  V  A  C  L  F     90
  183 CCCCAAAAGTTCACAGTGATCTTTGACACTGGAAGTTCTAATCTGTGGGTGCCTTCTTCAAAGTGCTACTTCTCAGTAGCTTGCCTTTTT    272

   91  H  S  K· Y  R  S  T  D  S  T  T  Y  K  K  N  G  K  S  A  A  I  Q  Y  G  T  G  S  I  S  G    120
  273 CACTCGAAGTATAGGTCTACCGATTCAACTACCTATAAGAAAAATGGGAAATCTGCTGCAATTCAATATGGAACTGGCTCTATCTCTGGT    362

  121  F  F  S  Q  D  S  V  K  L  G  D  L  L  V  K  E  Q  D  F  I  E  A  T  K  E  P  G  I  T  F    150
  363 TTCTTTAGCCAAGATTCTGTCAAACTTGGTGATCTTCTTGTTAAAGAGCAGGACTTTATAGAGGCAACCAAAGAGCCTGGCATCACTTTC    452

  151  L  A  A  K  F  D  G  I  L  G  L  G  F  Q  E  I  S  V  G  D  A  V  P  V  W  Y  T  M  L  N    180
  453 TTGGCAGCCAAGTTTGATGGTATCCTTGGCCTTGGGTTTCAAGAGATCTCTGTTGGGGATGCTGTCCCTGTTTGGTACACCATGCTGAAC    542

  181  Q  G  L  V  Q  E  P  V  F  S  F  W  L  N  R  N  A  D  E  Q  E  G  G  E  L  V  F  G  G  V    210
  543 CAAGGTCTTGTTCAAGAACCTGTATTTTCATTTTGGCTTAATCGCAATGCTGATGAGCAGGAAGGGGGTGAACTTGTGTTTGGTGGGGTT    632

  211  D  P  N  H  F  K  G  E  H  T  Y  V  P  V  T  Q  K  G  Y  W  Q  F  E  M  G  D  V  L  I  G    240
  633 GATCCTAATCATTTCAAGGGTGAACACACATATGTCCCTGTGACTCAAAAGGGCTATTGGCAGTTTGAAATGGGTGATGTCCTGATTGGT    722
                                                    *  *  *
  241  D  K  T  T  G  F  C  A  S  G  C  A  A  I  A  D  S  G  T  S  L  L  A  G  T  T  T  I  V  T    270
  723 GACAAAACAACTGGATTTTGCGCCAGTGGTTGTGCAGCAATTGCTGATTCTGGAACCTCTTTGTTGGCTGGTACCACGACGATTGTCACT    812

  271  Q  I  N  Q  A  I  G  A  A  G  V  M  S  Q  Q  C  K  S  L  V  D  Q  Y  G  K  S  M  I  E  M    300
  813 CAAATCAATCAAGCCATTGGGGCTGCGGGGGTTATGAGCCAGCAATGCAAGTCATTGGTTGATCAGTACGGAAAGAGTATGATTGAGATG    902

  301  L  L  S  E  E  Q  P  E  K  I  C  S  Q  M  K  L  C  S  F  D  G  S  H  D  T  S  M  I  I  E    330
  903 CTGTTATCTGAGGAGCAACCTGAGAAAATCTGCTCCCAAATGAAGCTATGTTCTTTTGATGGTTCTCATGATACTAGCATGATAATTGAG    992

  331  S  V  V  D  K  S  K  G  K  S  S  G  L  P  M  R  C  V  P  C  A  R  W  V  V  W  M  Q  N  Q    360
  993 AGTGTGGTTGACAAGAGTAAGGGAAAATCTTCTGGATTACCGATGAGATGTGTACCATGTGCCAGATGGGTTGTTTGGATGCAAAACCAG   1082
                +  +  +
  361  I  R  Q  N  E  T  E  E  N  I  I  N  Y  V  D  K  L  C  E  R  L  P  S  P  M  G  E  S  A  V    390
 1083 ATCAGGCAAAACGAGACTGAAGAAAACATAATTAACTACGTTGATAAGCTGTGTGAGCGGTTGCCAAGTCCAATGGGAGAGTCTGCAGTC   1172
                                                                   +  +  +
  391  D  C  S  S  L  S  S  M  P  N  I  A  F  T  V  G  G  K  T  F  N  L  S  P  E  Q  Y  V  L  K    420
 1173 GACTGTAGCAGTCTTTCCTCCATGCCCAATATTGCCTTCACTGTCGGTGGTAAAACCTTTAACCTCTCCCCCGAACAGTATGTCCTGAAA   1262

  421  V  G  E  G  A  T  A  Q  C  I  S  G  F  T  A  M  D  V  A  P  P  H  G  P  L  W  I  L  G  D    450
 1263 GTTGGGGAAGGAGCAACAGCACAATGCATAAGTGGATTCACCGCCATGGATGTGGCACCTCCTCATGGACCTCTATGGATCTTGGGAGAT   1352

  451  V  F  M  G  Q  Y  H  T  V  F  D  Y  G  N  L  R  V  G  F  A  E  A  A  *                      474
 1353 GTTTTTATGGGTCAATACCATACGGTGTTTGACTATGGTAACTTACGAGTTGGGTTTGCAGAAGCAGCTTGAGATCTCTCTCTCTCTCTC   1442

 1443 TCTCTCATTATCTGCGTATTACTACTATTACTGCTACTTGTTTTCTGTAAGTAATGTGGCAATAAATGTATGGCATTATATGAGTTATGG   1532

 1533 AGCAGGTACATGACTTAAGCTGATGTTTATATATACAAGGGTCTCTTTCATTCCAAAAAAAAAAAAAA                         1600
```

Figure 1. Nucleotide and deduced amino acid sequences for cyprosin from *Cynara cardunculus*. The nucleotide sequence is from cDNA clone *cypro1s*. An internal amino acid sequence obtained from a BrCN fragment is **bold** (amino acids 178 to 186). (!) Putative start of mature protein; (*) active site Asp-Thr/Ser-Gly; (+) putative glycosylation sites. The stop codon (position 1422-1424), the polyadenylation signal (position 1503-1508) and an internal *Pst*1 site (position 1165-1170) are underlined in the nucleic acid sequence.

is, however, very limited. It is not until very recently that investigations on the primary structure of such plant enzymes have been initiated [8,9].

A cDNA clone, *cypro1s*, encoding one of the cyprosin enzymes has been isolated, and sequenced. A comparison of the deduced aa sequence of *cypro1s* to the sequence of other aspartic proteinases is presented here.

2. MATERIALS AND METHODS

Sequencing of *cypro1s* cDNA was performed by the chain termination method of Sanger *et al.* [10]. Purification of single-stranded template DNA from pBluescript SK-cDNA clone and sequencing was carried out as described in the Sequenase 2.0 kit (United States Biochemicals) protocol. Initial DNA sequencing was carried out using reverse primer or M13-40 primer. Oligonucleotides (15-18 mers) were synthesized according to sequence information obtained and used directly as primers for further sequencing.

3. RESULTS AND DISCUSSION

3.1. The Nucleotide Sequence of *Cypro1s*

The nucleotide sequence of *cypro1s* is given in Figure 1. The sequence contains a 1281 bp open reading frame coding for 427 amino acids. Upstream the first ATG codon another 46 aa may be included in this open reading frame. Part of a prosequence frequently observed in aspartic proteases is included within these 46 aa. However, a highly hydrophobic region for the characteristics of a signal peptide could not yet be found in the 5' region of *cypro1s*.

Northern analysis using the ^{32}P labelled *Eco*RI 1.7 kb insert as probe revealed that the size of the mRNA encoding cyprosin in *C. cardunculus* is 1.8 kb and therefore it is highly likely that the *cypro1s* cDNA is near full length and that the translation starts at a ATG codon about 100 bp upstream of the 5' end of this cDNA. This assumption is supported by comparisons to the nucleotide and amino acid sequences of other aspartic proteases. The open reading frame is broken by a TGA stop codon at position 1421 and a 162 bp 3' non-coding region is followed together with a poly(A) tail.

This cDNA has a typical eukaryotic poly(A) addition signal AATAAA at nucleotides 1503 to 1508. The sequence data can confirm the restriction mapping described elswhere [9,11], *i.e.* that there is no *Bam*HI or *Hind*III site but a *Pst*I site at nucleotides 1165 to 1170.

3.2. The Deduced Amino Acid Sequence of Cyprosin

The deduced amino acid sequence of *cypro1s* is also shown in Figure 1. The identity of the cloned cyprosin was confirmed by an internal amino acid sequence (**bold** in Figure 1) obtained by N-terminal microsequencing of a peptide isolated after BrCN-cleavage of the large subunit of cyprosin 2. A high homology was observed to other plant aspartic proteinases: barley (78%) [8] and rice (68%) [12] as shown in Figure 2.

Two putative active site aspartic acid residues (asp35 and asp222 in Figure 2) have been identified. It is interesting to note that the normal Asp-Thr-Gly sequence of catalytic aspartic acid residues has been substituted for Asp-Ser-Gly for one of the residues. Furthermore, two putative glycosylation sites have been found (asn332 and asn379 in Figure 2). It is well established that the cyprosins are glycoproteins [2].

The general consensus sequence (based on the alignment of 30 aspartic proteinases) in Figure 2 shows that the mature proteins contain some highly conserved regions. In particular, the sequences around and at the active site aspartyl residues (in bold): VIF**DTG**SSNLWVPS and AIV**DTG**TSLL are very conserved. An exception to this is observed for the three plant enzymes in which the **T**hreonine of the second active site has been substituted for by **S**erine. However, in spite of this substitution the residues around this active site aspartyl residue are conserved in the plant enzymes. This substitution has been observed in some other aspartic proteinases, *e.g. Saccharomycopsis fibuligera* [13] and *Plasmodium falciparum* [14].

3.3. The Plant Specific Region

A plant specific insert is well identified in the three plant enzymes included in the alignment in Figure 2. The size of these inserts is 102, 104 and 99 aa for *Cynara*, barley and rice aspartic proteinases, respectively. The size of the mature aspartic proteinases of plant origin is, therefore, somewhat larger (440 (putative), 442 and 439 (putative) aa for *Cynara*,

```
                                 -68                                                                                  -1
           Cynara                ........ .......... .......... ......LkKr kvnilnhpge hagsndanar rkyGvrgnfr
           Barley                mgtrglal allaavlllq tvlpaaseae glvriaLkKr pidr.nsrva tglsggeeqp llsGanplrs
             Rice                .makrhll lvttclwals calllhassd gflrvnLnKk rldkedltaa k.laqqgnrl lktG......
  Plant consensus                -------- ---------- ---------- ------L-K- ---------- ---------- ---G------
General consensus                -------- ------L--- ----L----- ------L--- -S-R--L--- G--------- ----------

                   1                                                                                                 100
           Cynara  dsDgElIaLk nYMdaQYFGe IGiGtPPQkF TVIFDTGSSN LWVPSsKCYF SVAClFHSkY rstdSttYKk nGksaaIqYG tGsIsGFFSq DsVklGDllV
           Barley  eeEgDiVaLk nYMnaQYFGe IGvGtPPQkF TVIFDTGSSN LWVPSaKCYF SIACyLHSrY kagaSstYKk nGkpaaIqYG tGsIaGYFSe DsVtvGDlvV
             Rice  ssDsDpVpLv dYLntQYYGv IGlGsPPQnF TVIFDTGSSN LWVPSaKCYF SIACyLHSrY nskkSssYKa dGetckItYG sGaIsGFFSk DnVlvGDqvV
  Plant consensus  --D-D-V-L- -YM--QYFG- IG-G-PPQ-F TVIFDTGSSN LWVPS-KCYF SIAC-LHS-Y ----S--YK- -G----I-YG -G-I-GFFS- D-V--GD--V
General consensus  ----V--PL- NYLD--YFG- I-IGTPPQ-F -VIFDTGSSN LWVPS--C-- S-AC--H--Y -P--SSTY-- -G--FSI-YG TGS--G-L-- DTV-VGGI-V

                   101                                                                                               200
           Cynara  KeQdFIEATk EpgITFlaaK FDGILGLGFq EISVGdAvPV WytMlnQgLv qEpVFSFWLN RnaDeqeGGE lVFGGvDPnH FkGEHTYVPV tqKGYWQFeM
           Barley  KdQeFIEATk EpgITFlvaK FDGILGLGFk EISVGkAvPV WykMieQgLv sDpVFSFWLN RhvDegeGGE iIFGGmDPkH YvGEHTYVPV tqKGYWQFdM
             Rice  KnQkFIEATr EtsVTFiigK FDGILGLGYp EISVGkApPI WqsMqeQeLl aDdVFSFWLN RdpDassGGE lVFGGmDPkH YkGDHTYVPV srKGYWQFnM
  Plant consensus  K-Q-FIEAT- E--ITF---K FDGILGLGF- EISVG-A-PV W--M--Q-L- -D-VFSFWLN R--D---GGE -VFGG-DP-H Y-GEHTYVPV --KGYWQF-M
General consensus  --Q-FG---- EPG--F--A- FDGILGLAYP -I-V----PV FDNM--Q-LV ----FSVYL- -------GG- -VFGGID-S- Y-GSL-YVPV ----YWQI-M

                   201                                                                                               300
           Cynara  GDvLIgdktT GFCAsGCAAI aDSGTSLLAG tTtIVtqINq aIGAaGVmSq qCKslVdqYG ksmieMLlse eqPeKICSQm kLCsFDGshd tSmiIeSVVD
           Barley  GDvLVggksT GFCAgGCAAI aDSGTSLLAG pTaIIteINe kIGAaGVvSq eCKtiVsqYG qqildLLlae tqPkKICSQv gLCtFDGtrg vSagIrSVVD
             Rice  GDlLIdghsT GFCAkGCAAI vDSGTSLLAG pTaIVaqVNh aIGAeGIiSt eCKevVseYG emilnLLiaq tdPqKVCSQv gLCmFDGkrs vSngIeSVVD
  Plant consensus  GD-LI----T GFCA-GCAAI -DSGTSLLAG -T-IV--IN- -IGA-GV-S- -CK--V--YG -----LL--- --P-KICSQ- -LC-FDG--- -S--I-SVVD
General consensus  D---VG---- --C--GC-AI VDTGTSLL-G P-S-I--I-- AIGA------ ---------- ---------- ---------- ---------- ----------

                   301                                                                                               400
           Cynara  kskgkSsgLp m..rCvpCar wVVWmqNQir qNeTeEnIin YvdkLCeRLP SPmGESaVdC sslssMPnia FTVggKtFnL sPEqYVlKvg eGataqCISG
           Barley  depvkSngLr adpmCsaCem aVVWmqNQla qNkTqDlIld YvnqLCnRLP SPmGESaVdC gslgsMPdie FTIggKkFaL kPEeYIlKvg eGaaaqCISG
             Rice  kenlgSdaM. ....CsvCem aVVWieNQlr eNkTkElIln YanqLCeRLP SPnGEStVsC hqiskMPnla FTIanKtFiL tPEqYIvKle qGgqtvCISG
  Plant consensus  -----S--L- ----C--C-- -VVW--NQ-- -N-T-E-I-- Y---LC-RLP SP-GES-V-C -----MP--- FTI--K-F-L -PE-YI-K-- -G----CISG
General consensus  ---------- ---------- ---------- ---------- ---------- ---GEY-V-C -----LPDI- F-I-G--Y-L -P--YIL--- ------C-SG

                   401                                       441
           Cynara  FtAmDVaPPh GPLWILGDVF MGqYHTVFDY GnlRVGFAea A
           Barley  FtAmDIpPPr GPLWILGDVF MGpYHTVFDY GklRIGFAka A
             Rice  FmAfDIpPPr GPLWILGDVF MGaYHTVFDF GkdRIGFAks A
  Plant consensus  F-A-DI-PP- GPLWILGDVF MG-YHTVFDY G--RIGFA-- A
General consensus  F--MDI--P- G-LWILGDVF -R-YY-VFDR -NNRVG-A-A A
```

Figure 2. Alignment of amino acid sequences of *Cynara*, barley [8] and rice aspartic proteinase [12]. The plant concensus sequence represents conserved amino acids in all three plant enzymes and the general concensus sequence represents conserved amino acids in at least 15 of 30 aspartic proteinases (basically those shown in the phylogenetic tree [15]). Points indicate insertions of gaps at positions at which there are no homologous residues. The numbering of the amino acids starts from the putative N-terminus of the mature cyprosin. The prosequences are indicated by negative numbering. The active site aspartyl residues are **bold**. Potential glycosylation sites are underlined. The plant specific insert is in *italics*.

barley and rice aspartic proteinases, respectively) than that of aspartic proteinases from other genera (325-350 aa).

The plant consensus sequence presented in Figure 2 shows 43% homology for the plant specific insert suggesting that this region has similar characteristics. The six cysteine residues present in the plant specific region are conserved in all three enzymes indicating a possible conserved tertiary structure of this region through disulphide bridges. Furthermore, a putative glycosylation site (underlined in Figure 2) is conserved in the plant specific region in all three enzymes. The function of this large insert is at present not known.

4. CONCLUSION

The N- and C-terminal regions of cyprosin show a relatively high homology to other aspartic proteinases. All enzymes are highly conserved around the active site aspartyl residues. It is, however, interesting to note that the normal Asp-Thr-Gly sequence of catalytic aspartyl residues has been changed in one case Asp-Ser-Gly in all three plant enzymes. When comparing cyprosin to the other two plant enzymes the homology in the plant specific insert is considerably lower than for the N- and C-terminal regions. Apparently, the amino acid sequences of the plant specific insert have diverted during evolution to a greater extent than the rest of these plant enzymes. Therefore, the plant specific insert appear to be less important for enzyme structure and function. However, a conserved structure of the plant specific insert may be assumed due to some common features of this region in the three plant enzymes. The function of the plant specific insert is not known.

Proteinases found in other species of the genus *Cynara* may be closely related enzymes showing only slight differences in their primary structures. These differences could be responsible for the differences in clotting activities observed by cheese manufacturers using different mixtures of flowers as source for the clotting enzyme.

5. ACKNOWLEDGMENTS

The participation of M.C Cordeiro in the V International Congress on Aspartic Proteinases was made possible by financial support by the Organizers of the Congress and by Fundação Oriente, Portugal.

6. REFERENCES

1. F. Brotero, *Flora lusitanica*, Olissipone (1804).
2. U. Heimgartner, M. Pietrzak, R. Geertsen, P. Brodelius, A.C. da Silva Figueiredo, and M.S.S. Pais, Purification and partial characterization of milk clotting proteases from flowers of *Cynara cardunculus*. *Phytochemistry* 29:1405 (1990).
3. M. Cordeiro, E. Jacob, Z. Puhan, M.S. Pais, and P. Brodelius, Milk clotting and proteolytic activities of purified cynarases from *Cynara cardunculus*: A comparison to chymosin. *Milchwissenschaft* 47:683 (1992).
4. M. Cordeiro, M.S. Pais, and P.E. Brodelius, Tissue-specific expression of multiple forms of an aspartic proteinase in flowers of *Cynara cardunculus*. *Physiol. Plant*, in press (1994).
5. D.R. Davies, The structure and function of the aspartic proteinases. *Annu. Rev. Biophys. Biophys. Chem.* 19:189 (1990).
6. J. Tang and R.N.S. Wong, Evolution in the structure and function of aspartic proteases. *J. Cell. Biochem.* 33:53 (1987).
7. A. Sali, B. Veerapandian, J.B. Cooper, D.S. Moss, T. Hofmann, and T.L. Blundell, Domain flexibility in aspartic proteinases. *Proteins* 12:158 (1992).

8. P. Runeberg-Roos, K. Törmäkangas, and A. Östman, Primary structure of barley-grain aspartic proteinase. A plant aspartic proteinase resembling mammalian cathepsin D. *Eur. J. Biochem.* 202:1021 (1991).
9. M. Cordeiro, Z.-T. Xue, M. Pietrzak, M.S. Pais, and P.E. Brodelius, Isolation and characterization of a cDNA from flowers of *Cynara cardunculus* encoding cyprosin (an aspartic proteinase) and its use to study the organ-specific expression of cyprosin. *Plant Mol. Biol.*, in press (1994).
10. F. Sanger, S. Nicklen, and A.R. Coulson, DNA sequencing with chain termination inhibitors. *Proc. Natl. Acad. Sci. USA* 74:5463 (1977).
11. M.C. Cordeiro, Milk clotting enzymes from *Cynara cardunculus* spp. *flavescens* cv. Cardoon; Characterization and molecular cloning of the enzymes and studies on their expression in flower tissues. PhD Thesis, University of Lisbon, Lisbon (1993).
12. H. Hashimoto, R. Nishi, H. Uchimiya, and A. Kato, Nucleotide sequence of a cDNA encoding aspartic proteinase in rice. *EMBO Data Library* D12777 (1992).
13. D. Hirata, S. Fukui, and I. Yamashita, Nucleotide sequence of the secretable acid protease gene pep1 in the yeast *Saccharomycopsis fibuligera*. *Agric. Biol. Chem.* 52:2647 (1988).
14. J. Dame, C. Berry, B.M. Dunn, and J. Kay, An aspartic proteinase from the human malaria parasite *Plasmodium falciparum* HB3. *Mol. Biochem. Parasitology*, in press (1994).
15. P.E. Brodelius, M.C. Cordeiro and M.S. Pais, Aspartic proteinases (cyprosins) from *Cynara cardunculus* spp. *flavesens* cv. cardoon; Purification, characterization and tissue-specific expression. In: *Aspartic Proteinases: Structure, Function, Biology, and Biomedical Implications.* (K. Takahashi, ed.) Plenum Press, New York, p. 255 (1994).

CARDOSIN A AND B, ASPARTIC PROTEASES FROM THE FLOWERS OF CARDOON

Carlos Faro,[1] Paula Veríssimo,[1] Yingzhang Lin,[2] Jordan Tang,[2] and Euclides Pires[1]

[1]Departamento de Bioquímica
Universidade de Coimbra
Apartado 3126, 3000 Coimbra, Portugal
[2]Protein Studies Program
Oklahoma Medical Research Foundation
Oklahoma City, OK

INTRODUCTION

Plant aspartic proteinases have been detected and purified from several plant monocotyledoneous, dicotyledoneous and gymnosperms. The physiological role of these aspartic proteinases is not as yet well elucidated, although the involvement in the hydrolysis of exogenous and storage proteins have been reported (for review see reference 1)

Recently, aspartic proteinases were also isolated from the flowers of the cardoon *Cynara* (2,3). The flowers of this genera are traditionally used in Portugal in the manufacturing of artisanal cheeses. These enzymes seem to initiate milk clotting by cleaving the bond Phe105-Met106 in κ-casein (4) which is the most common cleavage site for aspartic proteinases with milk clotting activity (5).

In the present report it is shown that the flowers of the cardoon *Cynara cardunculus* L. contain two different aspartic proteinases instead of just one as previously predicted. The enzymes were named cardosin A and cardosin B which follow the descriptive names for other proteases.

Purification of Cardosins

The cardosins were isolated from flowers of *Cynara cardunculus* L. plants grown from seeds supplied by Botanic Garden, University of Coimbra. The enzymes were extracted from fresh flowers, or liquid nitrogen frozen fresh flowers, of cardoon at pH 3.0, yielding an enzyme preparation with about five times higher specific activity than the one obtained by extraction with 0.1 M Tris-Cl pH 7.6, as determined by using the synthetic peptide Leu-Ser-Phe(NO_2)-Nle-Ala-Leu-OMe as substrate. The acidic extraction eliminates most protein contaminants as shown by SDS-PAGE analysis where only the bands from cardosins were observed. Thus, further purification was required essentialy to eliminate other non-protein contaminants such as phenols and

Aspartic Proteinases: Structure, Function, Biology, and Biomedical Implications
Edited by Kenji Takahashi, Plenum Press, New York, 1995

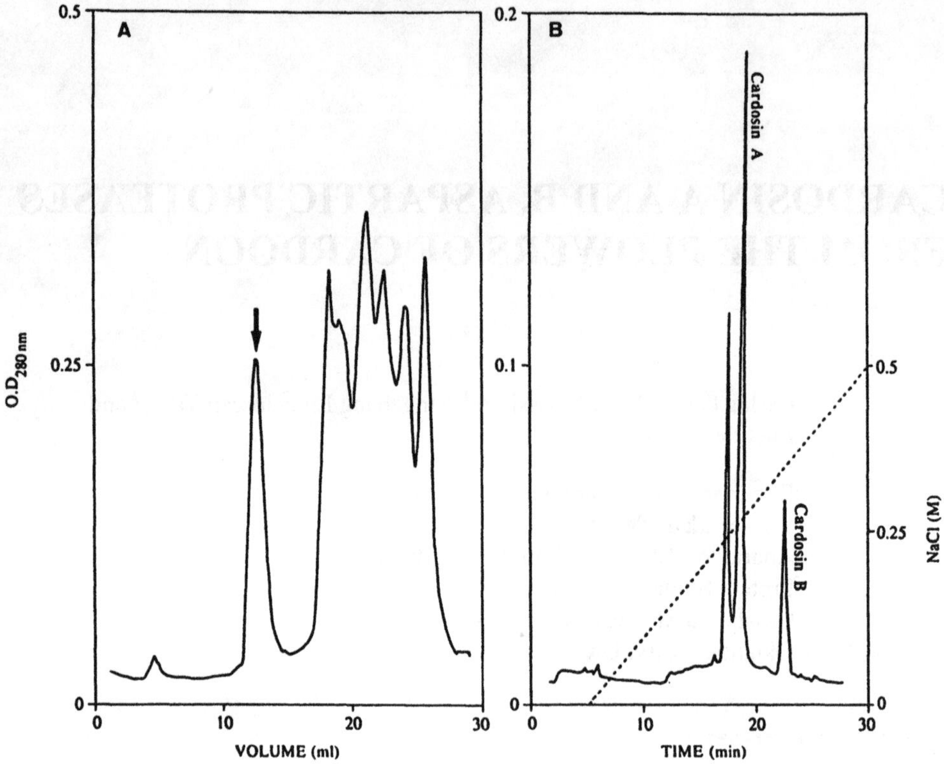

Figure 1. Purification of cardosin A and B. The cardosins were purified by extraction at low pH followed by chromatography on Superdex (A) and Mono Q (B).

pigments which was achieved by gel filtration on Superdex 200 (Fig. 1A). The proteolytic activity was recovered from this chromatography as a single peak with an yield of about 72 % and was further fractionated into two active peaks by ion-exchange chromatography on Mono Q (Fig. 1B). The specific activity of the major peak (cardosin A) was 7.9 I.U./mg protein whereas that of the minor peak (cardosin B) was 91.8 I.U./mg protein. SDS-PAGE of the material from these two peaks revealed that each produced two bands either in the presence or absence of 2-mercaptoethanol with apparent molecular masses of 31 and 15 kDa for cardosin A and 34 and 14 kDa for cardosin B respectively (Fig. 2a). The bands of the 34, 31 and 15 kDa chains were positively stainned by periodate Schiff reagent indicating that they are glycosylated. This two-step purification procedure takes only two hours for overall process and yields pure cardosins which were found to be suitable for sequence and other chemical studies.

Enzymic Properties

Cardosins are active between pH 2 and pH 7 with maximal activities between pH 5 to pH 5.5 and were inhibited by diazoacetyl norleucine methyl ester and pepstatin indicating that they belong unequivocally to the class of aspartic proteases. The Ki for pepstatin inhibition is 3 nM for cardosin A and <1 nM for cardosin B. Both enzymes were active up to 65°C although the stability at high temperatures was relatively low. Furthermore, the cardosins proved to be stable in urea solutions up to 6M and in some selected aqueous-organic biphasic systems (6)

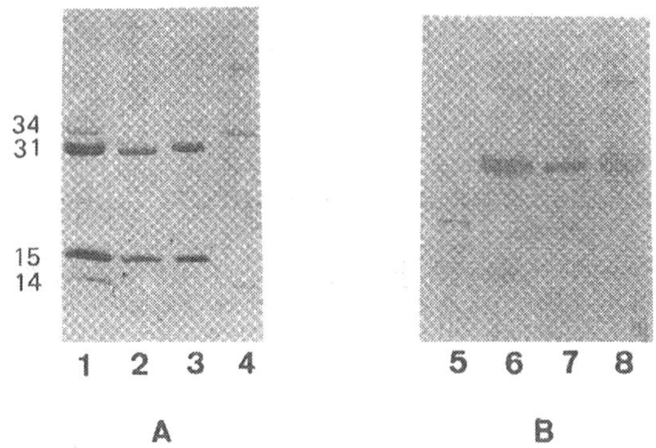

Figure 2. Imunochemical analysis of the relatioship between cardosin A and cardosin B. (A) SDS-PAGE in homogeneous 20 PhastGel developed with silver stainning. (B) Western blotting using an antisera raised against 31 kDa chain of cardosin A. (1,8) Acidic extract; (2,7) Inactive cardosin; (3,6) Cardosin A and (4,5) Cardosin B.

The specificity of cardosins was first investigated by using oxidised insulin B chain. Each cardosin was incubated with the substrate at pH 3.0 for 3 hours and the insulin peptide fragments were separated by HPLC and analysed by N-terminal amino acid sequencing. Cardosin A hydrolyzes oxidised insulin B chain at bonds Leu15-Tyr16, Leu17-Val18 and Phe25-Tyr26. Cleavage by cardosin B was found to occur at Glu13-Ala14, Ala14-Leu15, Leu15-Tyr16, Leu17-Val18, Phe24-Phe25 and Phe25-Tyr26. The additional sites hydrolysed by cardosin B represent indeed genuine differences in specificity as longer incubation of insulin B chain with cardosin A did not generate additional peptides. The results above described indicate therefore that cardosins prefer peptide bonds between residues with large hydrophobic side chains. In this respect, cardosins are similar to the majority of the aspartic proteinases (7). The kinetics parameters Km, kcat and kcat/Km for the hydrolysis of the chromophoric peptide Leu-Ser-Phe(NO_2)-Nle-Ala-Leu-OMe used throughout the work as substrate for enzyme assays are 0.8 mM, 23.6 s^{-1} and 21.87 s^{-1} mM^{-1} for cardosin A and 0.08 mM, 85.5 s^{-1} and 1057 $s^{-1}mM^{-1}$ for cardosin B respectively. The catalytic coefficient for cardosin A is of the same order as the one determined for chymosin whereas the value determined for cardosin B is similar to that reported for pig pepsin (8). Taken together these results suggest that in what it concerns the specificity and kinetics, cardosin A appears to be similar to chymosin whereas cardosin B closely resembles pepsin. It would appear therefore that in terms of the enzymic properties, the components of cardoon rennet are similar to those of calf rennet.

The pH dependence of kinetic parameters produced bell shaped curves and was used to calculated the apparent active-site ionization constants. The values of pKe_1 and pKe_2 are respectively 2.5 (±0.2) and 5.3 (±0.2) for cardosin A and 3.73 (±0.09) and 6.7 (±0.1) for cardosin B. In spite of the differences the pK values for both enzymes fall within the range of the active site ionization constants determined for other aspartic proteinases (9).

Relationship of Cardosins

In order to investigate the relationship of cardosins, structural and immunological studies were carried out. Western blotting analysis using an antisera raised against isolated 31 kDa chain showed that cardosin does not cross react with this antisera (Fig. 2B) indicating that there is no immunoidentity between the two enzymes. This was further confirmed by

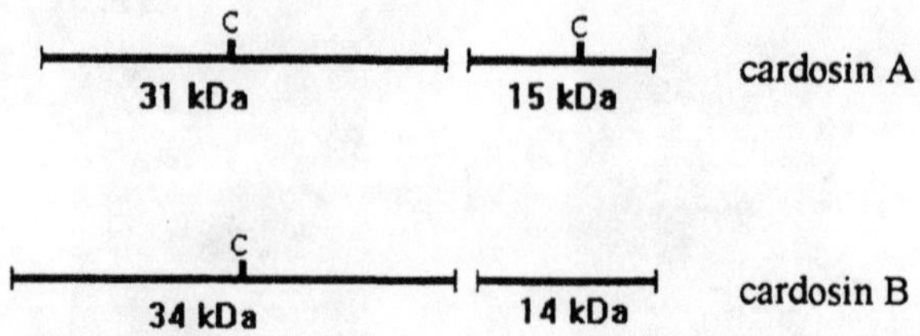

Figure 3. Schematic representation of the chain arrangement of cardosin A and B. The two chain enzymes are generated from single chains by proteolytic processing at the same cleavage site. C denotes the glycosylated nature of the 34, 31 and 15 kDa chains.

peptide mapping with trypsin (10) and partial amino acid sequencing. For amino acid sequence analysis the polypeptide chains of each cardosin were isolated by RP-HPLC upon redution and alkylation with vinylpyridine. The N-terminal amino acid sequence of the 31 kDa chain is NSGSAIVATDNQNTY whereas the 34 kDa chain produced no detectable N-terminally residue suggesting that it is N-terminally blocked. Analysis of 14 kDa and 15 kDa chains produced the following sequences:

SEELQVDNNTLSXMP (15 kDa chain)

SAEDIVNNNGISXMP (14 kDa chain)

Taken together the results above described indicate that cardosins have significant differences throughout their amino acid sequences and although they may derive from the same ancestral protein they are probably the products of different genes.

Chain Arrangement and Putative Proteolytic Processing

The molecular masses of the polypeptide chains were determined by SDS-PAGE upon deglycosilation with TFMS. Deglycosylation of cardosin A yielded polypeptide chains with molecular mass of 28000 and 14000 daltons whereas deglycosylated cardosin B produces polypeptide chains with 30000 and 14000 daltons. This result cleary indicates that the small chains have about the same polypeptide length. The differences in apparent molecular mass are due to the glycosylated nature of the 15 kDa chain.

The two cardosins are probably generated from a single chain by proteolytic cleavage. This is supported by the appearance of a predominant band in SDS-PAGE with an apparent molecular mass of 45 kDa when the flowers are extracted with SDS buffer. The order of the two chains was investigated by using diazoacetyl norleucine methyl ester, an active site directed reagent that esterifies aspartic acid 215 (11). Norleucine was found on the 14 and 15 kDa chains and thus the order of the two chains in each cardosin is from the large to the small chain. These results suggested a chain arrangement that is depicted in Fig. 3. According to the N-terminal amino acid sequences of the small chains, the putative proteolytic cleavage site that generates the two-chain enzymes seems to be the same in both cardosins. In terms of localization of the cleavage site, the processing closely resembles that of barley aspartic proteinase (12) suggesting that it might be a common property of plant heterodimeric aspartic proteinases.

CONCLUDING REMARKS

Cardosins are clearly aspartic proteases. They are inhibited by the general aspartic proteases inhibitors and exhibit proteolytic activity at acid pH. Although both cardosins cleave preferentially between hydrophobic residues, cardosin A has a narrower specificity than cardosin B. The specificity of cardosin A suggests a role in a physiological process mediated by limited proteolysis while that of cardosin B suggests a more general proteolytic role. Cardosins have clearly differences in the amino acid sequence and in spite of some homology they appear to be products of different genes. Like the barley aspartic proteinase, cardosins are heterodimeric proteases that are probably generated from a single chain through proteolytic processing at the C-domain.

REFERENCES

1. J.Kervinen, K.Törmäkangas, P.Runeberg-Roos, K.Guruprasad, T.Blundell and T.H.Teeri. Struture and possible function of aspartic proteinases in barley and other plants. In "*Aspartic Proteinases: Structure, Function, Biology, and Biomedical Implications*" K.Takahashi ed., p. 241, Plenum Press (1995).
2. C.Faro, J.S.Alface and E.Pires. Purification of a protease from the flowers of Cynara cardunculus L. *Cienc. Biol.* 12 (suppl 5A): 201 (1987).
3. U.Meimgartner, M.Pietrzak, R.Geertsen, P.Brodelius, A.C.Figueiredo and M.S.Pais. Purification and characterization of milk clotting enzymes from *Cynara cardunculus. Phytochemistry* 29:1405 (1990).
4. C.Faro, A.J.Moir and E.Pires. Specificity of a milk clotting enzyme extracted from the thistle *Cynara cardunculus* L.: action on oxidised insulin and k-casein. *Biotech. Lett.* 14:841 (1992).
5. H.B.Droshe and B.Foltmann. Specificity of milk clotting enzymes towards bovine kappa-casein. *Biochim. Biophys. Acta* 995: 221 (1989).
6. M.T.Barros, M.G.Carvalho, F.Garcia and E.Pires. Stability performance of the acid *Cynara cardunculus* L. proteases in biphasic aqueous-organic solvents systems. *Biotech.Lett.* 14: 179 (1992).
7. J.Kay and B.Dunn Substrate specificity and inhibitors of aspartic proteinases *Scand. J. Clin. Lab. Invest.* 52 (Suppl.210): 23 (1992).
8. P.Martin (1984) Hydrolysis of the synthetic chromophoric hexapeptide Leu-Ser-Phe(NO_2)-Nle-Ala-Leu-OMe catalyzed by bovine pepsin A. *Biochim. Biophys. Acta* 791: 28 (1984).
9. Y.Lin, M.Fuzek, X.Lin, J.Hartsuck, F.Kezdy and J.Tang. pH dependence of kinetic parameters of pepsin, rhizopuspepsin and their active site hydrogen bonds mutants. *J.Biol. Chem.* 267: 18413 (1992)
10. P.Veríssimo, C.Faro, Y.Lin, J.Tang and E.Pires. Isolation and properties of two cardosins, aspartic proteases from *C.cardunculus* L. (submitted)
11. P. Sepulveda, J. Marciniszyn, D. Liu and J.Tang. Prymary structure of porcine pepsin. *J. Biol. Chem. 250: 5082 (1975).*
12. K.Törmäkangas, P.Runeberg-Roos, A.Östman, C.Tilgmann, P.Sarkkinen, J.Kervinen, L.Mikkola and N.Kalkkinen. Aspartic proteinase from barley seeds is related to animal cathepsin D. In "*Structure and Function of the Aspartic Proteinases*" B.Dunn ed., p.355, Plenum Press, New York (1991).

MECHANISM OF AUTOPROCESSING OF A MINI-PRECURSOR OF THE ASPARTIC PROTEASE OF HUMAN IMMUNODEFICIENCY VIRUS TYPE 1

Edward Co, Gerald Koelsch, Jean A. Hartsuck, and Jordan Tang

Protein Studies Program
Oklahoma Medical Research Foundation and
The Department of Biochemistry and Molecular Biology
University of Oklahoma Health Sciences Center
Oklahoma City, OK 73104

INTRODUCTION

During the expression of human immunodeficiency virus (HIV) genome to make new virions, the 99-residue protease monomer is synthesized as part of the Pr160$^{gag\text{-}pol}$ precursor. After the virions are budded-out from the cell surface, the Pr160$^{gag\text{-}pol}$ polyprotein is autoprocessed at eight different sites to produce mature gag proteins (p17, p24, p7, and p6) and 3 HIV enzymes (protease, reverse transcriptase, and integrase). The autoprocessing is essential for the assembly of the interior of HIV virion and is essential for the infectiousness of the virus.

In spite of the importance of Pr160$^{gag\text{-}pol}$ precursor protein autoprocessing, its mechanism is not well understood. It is assumed that inside the virion, where the precursors are assembled with each p17 oriented toward the lipid membrane, the local concentration of the precursors is increased. The protease monomer units of different precursor molecules must dimerize to generate activity for autoprocessing. However, mechanistic questions have been difficult to answer for two main reasons. First, native Pr160$^{gag\text{-}pol}$ precursor contains too many processing sites for effectively monitoring of cleavage rates. Second, recombinant precursors synthesized in *E. coli* or in yeast spontaneously process into the 99-residue mature HIV protease. The isolation of precursor for mechanism studies has not been possible.

To overcome these difficulties, we have used a designed gene to direct the synthesis of a mini-precursor of HIV protease which contains an extension of 10 native pol residues upstream from the N-terminal position of HIV-1 protease (Fig. 1). Because the designed gene contains *E. coli* preferable codons, the level of expressed mini-precursor is high enough to allow kinetic investigations. We also discovered that a mutation of Ala^{28} to Ser slows the processing rate so the mutant mini-precursor can be isolated (1). (The k_{cat}/K_m value of the mutant enzyme is about 300 fold smaller than that of the native HIV-1 protease.) Using this system, we measured the autoprocessing rate of this mini-precursor; these studies have

```
-26    -20          -10          1          10          20
 .      .            .           .           .           .
MASMTGGQQM GRDQFMDRQG TVSFNFPQIT LWQRPLVTIK IGGQLKEALL

    30         40         50         60         70
    .          .          .          .          .
DTGADDTVLE EMSLPGRWKP KMIGGIGGFI KVRQYDQILI EICGHKAIGT

    80         90         99
    .          .          .
VLVGPTPVNI IGRNLLTQIG CTLNF
```

Figure 1. Sequence of A28S-mini-precursor of HIV-1 protease. The 14 kDa recombinant precursor is synthesized in *E. coli* under the direction of a designed gene (1). The 10 amino acids upstream (bold) from the N-terminus of HIV-1 protease are from the native gag-pol sequence. The expression vector pET-HIVPR contains an N-terminal fusion which adds 16 amino acids at the N-terminus of the precursor.

enabled formulation of a processing mechanism proposal. These results are described in this chapter.

EXPERIMENTAL PROCEDURES

Expression and Preparation of A28S-Mini-Precursor of HIV-1 Protease

The expression vector pET-HIVPR-A28S, which encodes HIV-1 protease mini-precursor gene (Fig. 1) with a residue replacement of Ala^{28} to Ser, has been described previously (1). Radioactive A28S-mini-precursor was synthesized by the addition of 100 μCi of ^{14}C amino acid hydrolysate (Amersham) to a 4 L *E. coli* culture at logarithmic growing phase (A_{600} range 0.4-0.5). The produced mini-precursor in 'inclusion bodies' was recovered, dissolved in 8 M urea, and purified through a DEAE-Sephacel column as previously described (1). A slight modification of

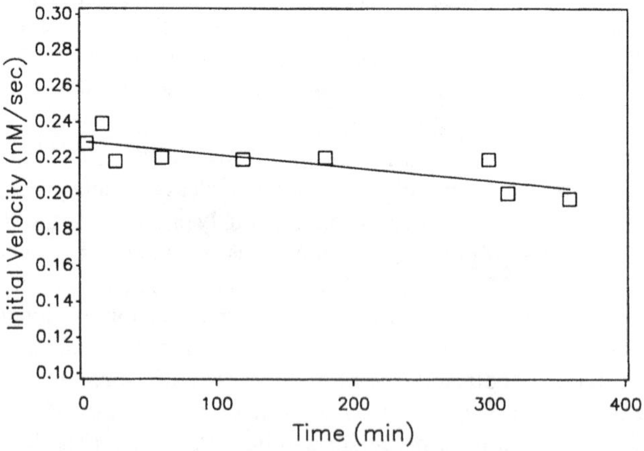

Figure 2. Proteolytic activity of the A28S-mini-precursor during the course of activation. An aliquot of the mini-precursor solution under processing is assayed at different time intervals for activity using a synthetic peptide substrate.

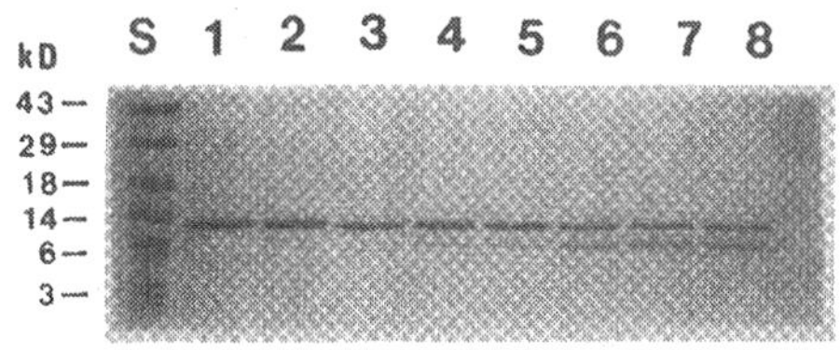

Figure 3. SDS-polyacrylamide gel electrophoresis pattern of the processing of the 14-kDa A28S-mini-precursor to the 11-kDa A28S-HIV-1 protease during the course of the processing reaction. Lane S contains molecular weight standards; lanes 1-8, samples taken at processing time of 0, 0.5, 1, 2, 3, 4, 5, and 6 h.

the refolding conditions involved the change of pH of the breakthrough peak from the DEAE column to pH 3.5 with 10% trifluoroacetic acid and dialyzing for 4 h at 4 °C against 4 changes of 10 mM Na acetate, pH 3.5 with 1 mM dithiothrietol (2).

Assay for Proteolytic Activity

The assay for proteolytic activity was carried out as described by Richards *et al.*, 1990 (3). A typical reaction mixture consisted of 1 ml of 10 mM Na acetate, pH range 3.5 to 6, 1 mM dithiothrietol, and approximately 0.1 mM mini-precursor. The reaction was initiated by the addition of 60 µl of 8.48 mM peptide substrate K-A-R-V-L-Z-E-A-M (Z, p-nitrophenylalanine). The rate of hydrolysis was monitored at A_{300} with a Hewlett Packard 8452A Diode Array Spectrophotometer.

Rate of Autoprocessing of A28S-Mini-Precursor

The conversion of A28S-mini-precursor to the corresponding protease was initiated by the addition of 0.1 ml of 1 M Na acetate, pH 4.5, to a 2-ml solution of mini-precursor dialyzed in 10 mM Na acetate, pH 3.5, and 1 mM DTT. The solution was incubated at 26 °C up to 6 h. Aliquots of 0.1 ml were removed at intervals, mixed with a 10X sample buffer (4% SDS) and subjected to SDS-polyacrylamide (15%) gel electrophoresis. After staining a guide strip, the bands representing the 14-kDa mini-precursor and 11-kDa mature protease were cut out from the gel, solubilized with gel solubilizer (NEN) at 55 °C overnight, and

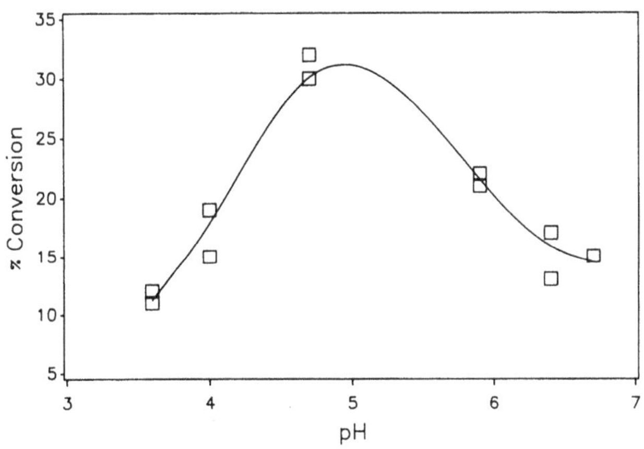

Figure 4. pH dependence of A28S-mini-precursor processing.

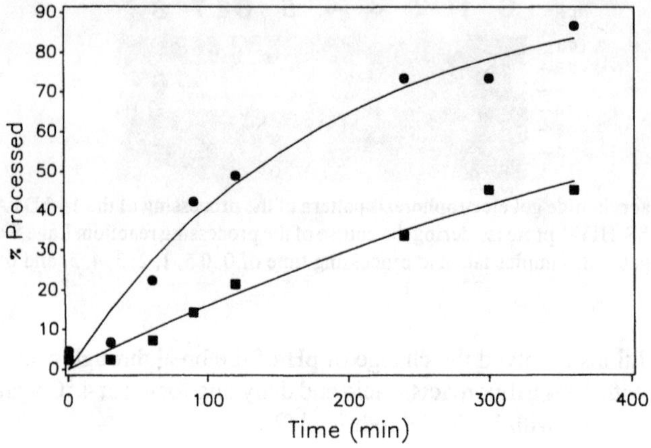

Figure 5. Time course of processing of A28S-mini-precursor in 2 concentrations, 0.03 mg/ml (closed square) and 0.34 mg/ml (closed circle).

counted for radioactivity in 10 ml scintillation fluid. Other details of this procedure will be described separately (2).

RESULTS AND DISCUSSION

Proteolytic Activity of A28S-Mini-Precursor and Its Autoprocessed Product, A28S-HIV-1 Protease

In the processing of A28S-mini-precursor to its corresponding protease, the initial reaction must be catalyzed by the precursor. However, as the reaction proceeds, the catalysis by the protease or its heterodimer with precursor becomes more important. In order to follow the rate of precursor to protease conversion during the course of processing, it is important to know their relative proteolytic activity.

The proteolytic activity of a processing mini-precursor solution was determined at different time points over 6 h, a sufficient time to convert all mini-precursor. Only a slight (about 10%) decrease of activity was observed (Fig. 2) which is attributable to autolysis known to take place over a long period of incubation. When the amounts of active sites at 0 and 6 h of incubation (2) were used to calculate specific activities, the results were about the same (62 U/mg at 0 h and 70 U/mg at 6 h). These results indicate that A28S-mini-precursor and the mature A28S-HIV-1 protease have the same specific proteolytic activity. The pH optimum for A28S-mini-precursor hydrolysis of the above described synthetic peptide substrate is between pH 4.5-5.5.

Conversion of A28S-Mini-Precursor to A28S-HIV Protease

Fig. 3 shows an SDS-gel electrophoresis containing samples taken at different times during the processing reaction. The increase of the 11-kDa protease during the course of incubation can be clearly seen. The decrease of radioactivity in the 14-kDa precursor band is very near that of the increase in the protease band, confirming the precursor-product relationship. At 6 h incubation, where all the protease had been converted, a fraction of the 14 kDa precursor usually still remains. This fraction, which varies with batch of precursor preparation and ranges from 15%-40%, represents precursor molecules that failed to attain

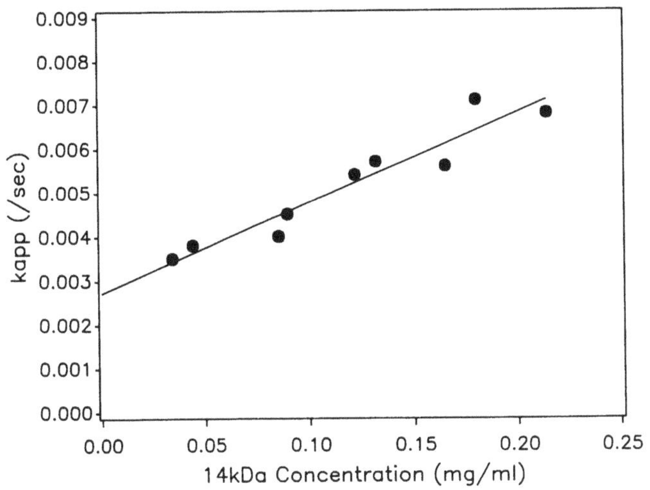

Figure 6. Dependence of the apparent first-order constant, k_{app}, on the concentration of A28S-mini-precursor. This figure represents an example from 5 different experiment sets.

correct conformation during the refolding. The total convertible precursor was therefore estimated from the decrease of the 14-kDa band and the increase of the 11-kDa band at long incubation time. The quantitation method was substantiated by active-site titration using tight-binding inhibitors (2).

The pH dependence of precursor-protease conversion shows an optimum near pH 5 (Fig. 4), as in the case of precursor activity. The time course of precursor-protease conversion is shown in Fig. 5 for two mini-precursor concentrations. The time courses of all experiments could be fitted to a first-order kinetic model, which permitted the calculation of k_{app}, the apparent first-order rate constant. From 5 different precursor preparations, we have observed that the value of k_{app} is dependent on the precursor concentrations. This is illustrated in an example in Fig. 6.

There are three possible dimer species, P•P, P•E and E•E where P represents precursor monomer, and E represents mature protease monomer. The conversion reactions to be considered are

$$\text{First order reaction: } P \xrightarrow{k_1} E \tag{1}$$

$$\text{Second order reaction: } P + X \xrightarrow{k_2} E + X \tag{2}$$

where P is P•P or P•E, E is P•E or E•E and X is any of the three possible species.

The concentration dependence of k_{app} (Fig. 6) suggests the presence of a second-order reaction. However, the value of k_{app} extrapolated to zero concentration of mini-precursor does not approach zero, indicating the presence of a first-order component. The rate for the appearance of protease is described by the following equation:

$$d[E]/dt = \{k_1 + k_2([X])\}\ [P] = k_{app}\ [P] \tag{3}$$

where [X] and [P] are the concentration of X and P. Since [X] represents total protein concentration, $\{k_1, + k_2\ [X]\}$ is constant, and the apparent first order conversion rate is

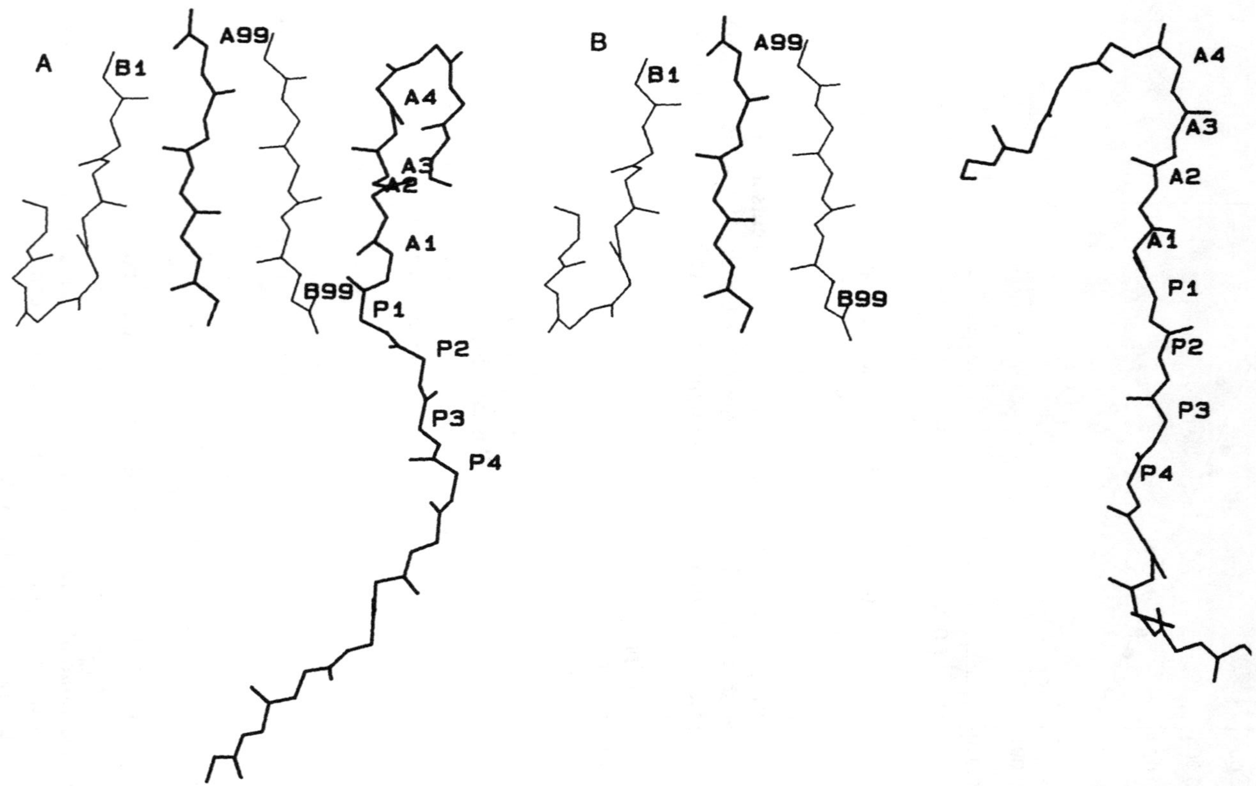

Figure 7. Computer graphic representation of the 4-strand β-structure involving the proteolytic cleavage site at the N-terminal position of the HIV-1 protease. A. The N- and C-terminal positions of HIV protease monomer B (light lines) are marked B1 and B99 respectively. The 4 specificity and recognition-determining residues for cleavage of the N-terminal strand of monomer A (heavy lines) are marked A1, A2, A3, and A4. This monomer is depicted with an upstream extension of the mini-precursor. The 4 residues preceding the mature cleavage site are marked P_4 through P_1 and are also required for specific recognition. It is clear that in an intact 4-strand structure, the peptide containing these 8 subsites can not be placed in the active site of HIV-1 protease for hydrolysis. B. The N-terminal strand of monomer A is detached from the 4-strand β-structure in a position that allows interdimer processing. The figure is based upon the crystal structure of Miller *et al.*, 1989 (8).

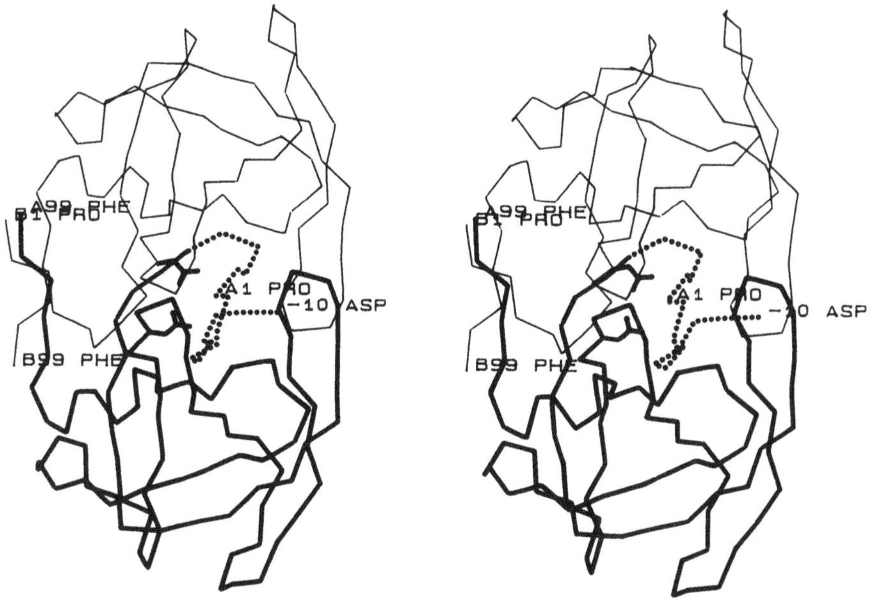

Figure 8. Stereoview of a computer graphic model of the HIV-1 protease mini-precursor engaging in intradimeric processing. Dotted line, region of the protease which is relocated from the N-terminal β-strand, including residue 6 (A6) of monomer A of the homodimer through 10 amino acid residues of the upstream pol sequence. Heavy line, residues A6 through A99; light line, residues B1 through B99 of monomer B. Both catalytic aspartic acids are illustrated and in this model are adjacent to the scissile bond between Pro A1 (labeled) and Phe -1 of the monomer A. The model was based upon the structure of the HIV-1 protease inhibitor complex of Miller *et al.* (1989).

rationalized. In light of unchanging activity during activation, the use of X rather than the individual species concentrations is appropriate. Thus in Fig. 6, the slope is k_2 and the intercept on the y axis is k_1. Values for k_1 and k_2 were calculated from 36 independent processing experiments using 5 different batches of precursor preparations. The average k_1 is 0.003 ± 0.001 min^{-1} and k_2 is 0.03 ± 0.01 min^{-1}(mg/ml)$^{-1}$.

These results indicate that the mini-precursor is capable of processing at the N-terminal position of HIV protease by two mechanisms. The major mechanism is processing a dimer by another dimer molecule, an interdimer mechanism. The second mechanism involves cleaving of the peptide bond in a dimer by the same dimer molecule, an intradimer mechanism. The presence of A28S mutation in the protease active site should have no influence on the overall mechanism observed. In a larger precursor such as the Pr160$^{gag\text{-}pol}$ polyprotein, there might be other tertiary structural factors that could affect these mechanisms. If the pol sequence immediately upstream from the N-terminus of HIV protease is without a definitive tertiary structure, the processing mechanism can be the same as in a mini-precursor. There is insufficient current information on the conformation of a Pr160$^{gag\text{-}pol}$ polyprotein for answering this question at the present.

From the structural viewpoint, either mechanism requires conformational adjustments of the precursor. The N-terminal strand of HIV-1 protease is part of the four-strand β-sheet composed of the N- and C-terminal strands from each monomer. Assuming that the precursor has the same 3-D structure as the protease but, in addition, has an unstructured upstream extension, the specificity subsites P_1', P_2', P_3', and P_4' correspond to the N-terminal 4 residues of the protease. These residues A1, A2, A3 and A4 in Fig. 7A are part of the 4-strand β-sheet. For proteolytic processing, the N-terminal strands, which occupy the outside positions in the sheet, must detach from the four-strand β-structure in order to be fitted into the active-site of HIV protease for hydrolysis (Fig. 7B). In the case of intradimer processing, the detached N-terminal strand of the precursor must be able to fit into the active site of the same dimer for proteolysis. Computer graphic modelling of a mini-precursor processing intermediate indicates that placement of an N-terminal strand of the precursor into its own active site is conformationally possible and does not necessitate other alterations in the structure of the protease moiety (Fig. 8). In this model, the residues in the N-terminal strand and 4 upstream residues are in extended conformation as required for regular interactions with the substrate-binding subsites in the active site (4). Also, the Phe-Pro scissile bond is properly positioned with the catalytic aspartyl residues. It is clear from this model that the intradimer processing does not impose a more severe conformational alteration than that of the interdimer processing.

The potential biological functions of intradimer processing are interesting. It is possible that this mechanism ensures processing when $Pr160^{gag\text{-}pol}$ concentration is low relative to $Pr55^{gag}$ inside of the virions. Another possibility is that $Pr160^{gag\text{-}pol}$ molecules are packed in the virions in such a way that initially the interdimer processing is not topologically feasible until the protease dimer is freed first by intradimer processing. This appears to be the case in the processing of Avian retroviral $Pr76^{gag}$ precursor in which the protease dimer appears at a very early stage (5). It is also interesting to note that the intradimer and interdimer processing mechanism is closely analogous to the intramolecular and intermolecular activation mechanism of pepsinogen (6,7). Since there are many structural and catalytic relationships between the eukaryotic aspartic proteases and the retroviral aspartic proteases, the close relationship in the proteolytic cleavage of precursors perhaps is not entirely surprising.

ACKNOWLEDGMENTS

This work is supported by NIH grant AI 26762. The authors wish to thank Marcus Dehdarani for his technical assistance.

REFERENCES

1. Ido, E., Han, H.P., Kezdy, F.J. and Tang, J. (1991) "Kinetic Studies of Human Immunodeficiency Virus Type 1 Protease and Its Active-site Hydrogen Bond Mutant A28S" *J. Biol. Chem.* **266:**24359-24366.
2. Co, E., Koelsch, G., Lin, Y., Ido, E., Hartsuck, J.A. and Tang, J. (1994) "Proteolytic Processing Mechanisms of a Mini-precursor of the Aspartic Protease of Human Immunodeficiency Virus Type 1" *Biochemistry*, in press.
3. Richards, A.D., Phylip, L.H., Farmerie, W.G., Scarborough, P.E., Alvarez, A., Dunn, B.M., Hirel, P.-H.., Konvalinka, J., Strop, P., Pavlickova, L., Kostka, V. and Kay, J. (1990) "Sensitive, Soluble Chromogenic Substrate for HIV-1 Proteinase", *J. Biol. Chem.* **265:**7733-7736.
4. Jaskolski, M., Tomasselli, A.G., Sawyer, T.K., Staples, D.G., Heinrikson, R.L., Schneider, J., Kent, S.B.H. and Wlodawer, A. (1991) "Structure at 2.5-Å Resolution of Chemically Synthesized Human Immunodeficiency Virus Type 1 Protease Complexed with a Hydroxyethylene-Based Inhibitor" Biochemistry **30:**1600-1609.
5. Dickson, C., Eisenman, R., Fan, H., Hunter, E. and Teich, N. (1984) *RNA Tumor Viruses: Molecular Biology of Tumor Viruses* Vol 1, Second Edition (Weiss, R. Teich, N., Varmus, H. and Coffin, J., eds.) Cold Spring Harbor, New York.
6. Al-Janabi, J., Hartsuck, J.A. and Tang, J. (1972) "Kinetics and Mechanism of Pepsinogen Activation" *J. Biol. Chem.* **247:**4628-4632.
7. Marciniszyn, J., Jr., Huang, J.S., Hartsuck, J.A. and Tang, J. (1976) "Mechanism of Intramolecular Activation of Pepsinogen" *J. Biol. Chem.* **251:**7095-7102.
8. Miller, M., Schneider, J., Sathyanarayana, B.K., Toth, M.V., Marshal, G.R., Clawson, L., Selk, L., Kent, S.B.H. and Wlodawer, A. (1989) "Structure of Complex of Synthetic HIV-1 Protease with a Substrate-Based Inhibitor at 2.3 Å Resolution" *Science* **246:**1149-1152.

MUTANTS OF HIV-1 PROTEASE WITH ENHANCED STABILITY TO AUTODEGRADATION

Alfredo G. Tomasselli, Ana M. Mildner, Donna J. Rothrock,
Jean L. Sarcich, June Lull, Joseph Leone and Robert L. Heinrikson

Upjohn Laboratories
The Upjohn Company
Kalamazoo, Michigan 49001

INTRODUCTION

The essential role of the protease from human immunodeficiency virus (HIV) in the final maturation phase of the viral life cycle makes this enzyme an important target for the design of drugs that might be effective in the treatment of acquired immunodeficiency syndrome (AIDS). The widespread use of the HIV protease as a reagent in kinetic studies, structure-function analysis, and x-ray crystallography research, depends upon the technology to produce milligram quantities of homogeneous enzyme with high specific activity. One difficulty in achieving this goal arises from the fact that the enzyme can undergo autoproteolysis at at least three sites in the molecule: Leu_5-Trp_6, Leu_{33}-Glu_{34}, and Leu_{63}-Ile_{64} [1,2,3]. This means that during the course of enzyme storage and long-term kinetic or structural analysis, the protease is continually generating inactive, or proteolyzed products. It would be useful to have available a protease that is resistant to autodigestion, and yet has the same kinetic properties as the wild-type enzyme. This would facilitate studies alluded to above, and would also provide a useful tool in protein chemistry for excision of peptides and proteins from fusion constructions and in other applications where a stable protease is essential. Rosé *et al.* [2] have described an HIV-1 protease mutant (Q7K) with increased stability over the wild type enzyme, but still vulnerable to autolysis at Leu_{33}-Glu_{34} and Leu_{63}-Ile_{64}. In the present work we have attempted to eliminate these sites of autolysis. This paper deals with the strategy we adopted to select certain mutations at the P_1 and/or P_2' positions to obtain mutants much more resistant to autolysis than the wild type enzyme. Moreover, it analyzes the effects of the mutations on the enzyme stability and activity. A companion paper [4] describes the technology employed for the design and expression of the HIV-1 protease mutants, their purification and the refolding of inactive proteins to active enzymes.

The Material and Methods section is given in the companion paper [4] and is not repeated here.

Aspartic Proteinases: Structure, Function, Biology, and Biomedical Implications
Edited by Kenji Takahashi, Plenum Press, New York, 1995

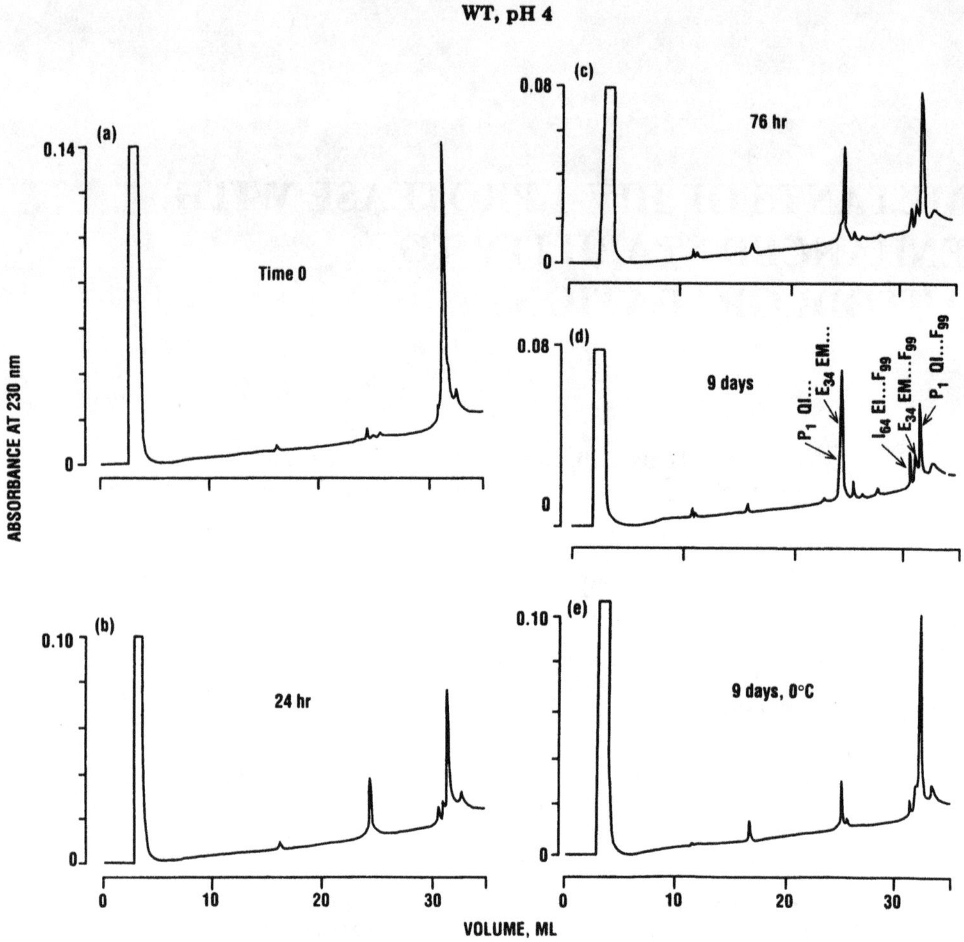

Figure 1. (a) through (d). Course of wild type HIV-1 protease (WT) autocleavage over a 9 day period at 25°C, pH 4.0 in the absence of DTT; (e) autocleavage after a 9 day incubation at 0°C, pH 4.0 no DTT. All the chromatograms described in this figure and in the following figures are Reverse-Phase HPLC elution profiles obtained in the following conditions. A Vydac C_4 column (0.46 x 25 cm) was eluted with a linear gradient of acetonitrile in 0.15% TFA and monitored at a wavelength of 220 nm or 230 nm. Specifically the concentration of acetonitrile was increased from 0 to 70% over a 45 minute period at a flow rate of 1 ml/min.

RESULTS AND DISCUSSION

Recombinant, wild type HIV-1 protease was incubated at pH 4.0, 25°C and peptides generated by autolysis were separated by HPLC (Fig. 1a-1d). Fragments corresponding to the peaks in Fig. 1d were identified by amino acid analysis and protein sequencing. The same experiment was performed at pH 5.5, and the HPLC profiles are shown in Fig. 2a-2c. We, and others [1,2,3] have documented sites of autoprocessing in the HIV-1 protease at positions 5, 33, and 63. (Rosé *et al.* [2] have also reported the sites of autoprocessing in the HIV-2 protease.) These sites for both enzymes, and their surrounding sequences, are shown in the primary structure in Fig. 3. Part of the strategy for mutagenesis of the HIV-1 protease was derived from inspection

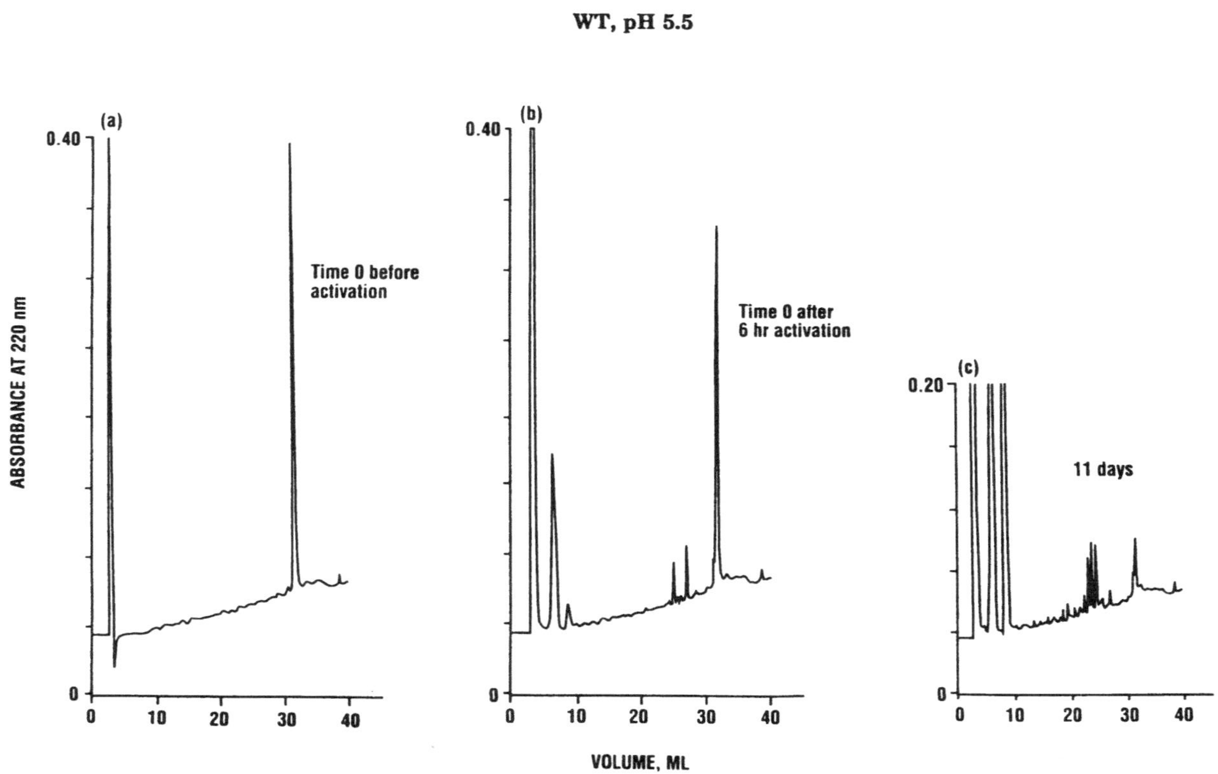

Figure 2. Course of WT HIV-1 protease autocleavage over 11 day period at 25°C, pH 5.5 in the presence of DTT.

```
                      ⇓                        10                                      20
HIV-1   Pro-Gln-Ile-Thr-Leu-Trp-Gln-Arg-Pro-Leu-Val-Thr-Ile-Lys-Ile-Gly-Gly-Gln-Leu-Lys-
                   ⇓                                        ⇓
HIV-2   Pro-Gln-Phe-Ser-Leu-Trp-Lys-Arg-Pro-Val-Val-Thr-Ala-Tyr-Ile-Glu-Gly-Gln-Pro-Val-

                                               30                 ⇓                    40
HIV-1   Glu-Ala-Leu-Leu-Asp-Thr-Gly-Ala-Asp-Asp-Thr-Val-Leu-Glu-Glu-Met-Ser-Leu-Pro-Gly-
                                                                    ⇓
HIV-2   Glu-Val-Leu-Leu-Asp-Thr-Gly-Ala-Asp-Asp-Ser-Ile-Val-Ala-Gly-Ile-Glu-Leu-Gly-Asn-

                                               50                                      60
HIV-1   Arg-Trp-Lys-Pro-Lys-Met-Ile-Gly-Gly-Ile-Gly-Gly-Phe-Ile-Lys-Val-Arg-Gln-Tyr-Asp-
                                                       ⇓
HIV-2   Asn-Tyr-Ser-Pro-Lys-Ile-Val-Gly-Gly-Ile-Gly-Gly-Phe-Ile-Asn-Thr-Lys-Glu-Tyr-Lys-

                  ⇓                            70                                      80
HIV-1   Gln-Ile-Leu-Ile-Glu-Ile-Cys-Gly-His-Lys-Ala-Ile-Gly-Thr-Val-Leu-Val-Gly-Pro-Thr-
                                                                        ⇓
HIV-2   Asn-Val-Glu-Ile-Glu-Val-Leu-Asn-Lys-Lys-Val-Arg-Ala-Thr-Ile-Met-Thr-Gly-Asp-Thr-

                                               90                                  99
HIV-1   Pro-Val-Asn-Ile-Ile-Gly-Arg-Asn-Leu-Leu-Thr-Gln-Ile-Gly-Cys-Thr-Leu-Asn-Phe-COOH
                                              ⇓
HIV-2   Pro-Ile-Asn-Ile-Phe-Gly-Arg-Asn-Ile-Leu-Thr-Ala-Leu-Gly-Met-Ser-Leu-Asn-Leu-COOH
```

Figure 3. Sites of autolysis in the HIV-1 and HIV-2 proteases.

of the primary structure of the HIV-2 enzyme, one which as we mentioned above undergoes autoproteolysis, but at sites distinct from those in the HIV-1 enzyme (Fig. 3). Moreover, analysis of 64 protein substrates of the HIV-1 protease (Table I) indicated the absence of β-branched amino acids in position P_1 and the absence of lysine residues in the segment P_2-P_2', and suggested two important guidelines for abolishing a protease-sensitive site: 1. insert a β-branched amino acid (Val or Ile) at the P_1 position; and 2. insert a Lys residue anywhere from P_2 through P_2'. The same guidelines are also suggested by a series of studies in which peptide substrates modeled after cleavage sites in the viral polyproteins had a variety of amino acids substituted at P_4 through P_4' and showed that substitution of the β-branched amino acids Ile/Val in P_1 and/or Lys in P_2' blocked cleavage by HIV protease [13,14,15]. The following narrative describes a number of HIV-1 protease mutants created on the basis of this strategy.

Mutants Q7K and Q7K/L33V

In a recent paper, Rosé *et al.* [2] described an HIV-1 protease mutant in which the Gln residue at position 7 was changed to a Lys. This change blocks cleavage at the Leu_5-Trp_6 position, in keeping with the guidelines mentioned above; i.e., a Lys cannot be tolerated at P_2'. Moreover, there is a Lys at position 7 in the HIV-2 protease and there is no autodigestion at position 5 in this enzyme (Fig. 3). We also made this Q7K mutant and, although its stability is markedly enhanced over the wild-type enzyme, over a period of 11 days there is about 30% breakdown to products with an average loss of 12% in activity. The mutant Q7K/L33V was not an improvement over Q7K since it still showed degradation, especially at Leu_{63}.

Mutant L33I/L63I

The substitution of Leu residues at positions 33 and 63 by the β-branched amino acid Ile resulted in stabilization of the mutant enzyme over the wild-type protease, but cleavage still occurred at Leu_5, and there was some precipitation of this mutant during the course of incubation.

Table 1. P4 through P4′ Residues in 64 Protein Substrates of the HIV-1 Protease

P_4	P_3	P_2	P_1	↓	P_1'	P_2'	P_3'	P_4'	Protein	P_4	P_3	P_2	P_1	↓	P_1'	P_2'	P_3'	P_4'	Protein
T	Q	I	M	↓	F	E	T	F	Actin	R	K	I	L	↓	F	L	D	G	*pol*
G	Q	V	N	↓	Y	E	E	F	Calmodulin	G	S	H	L	↓	V	E	A	L	Insulin
P	F	I	F	↓	E	E	E	P	pro-IL1-β	G	G	V	Y	↓	A	T	R	S	Vimentin
S	F	N	F	↓	P	Q	I	T	*pol*	F	R	S	G	↓	V	E	T	T	*gag*
D	T	V	L	↓	E	E	M	S	Autolysis	V	E	V	A	↓	E	E	E	E	AAP[a]
A	R	V	L	↓	A	E	A	M	*gag*	L	P	V	N	↓	G	E	F	S	AAP[a]
A	E	E	L	↓	A	E	I	F	Troponin C	E	T	T	A	↓	L	V	C	D	Actin
S	L	N	L	↓	R	E	T	N	Vimentin	H	L	V	E	↓	A	L	Y	L	Insulin[b]
A	T	I	M	↓	M	Q	R	G	*gag*	H	Y	G	F	↓	P	T	Y	G	NF-κB[c]
A	E	C	F	↓	R	I	F	D	Troponin C	D	S	A	D	↓	A	E	E	D	AAP[a]
D	Q	I	L	↓	I	E	I	C	Autolysis	G	W	I	L	↓	G	E	H	G	LDH[d]
D	D	L	F	↓	F	E	A	D	pro-IL1-β	G	W	I	L	↓	A	E	H	G	LDH
Y	E	E	F	↓	V	Q	M	M	Calmodulin	Q	A	I	Y	↓	L	A	L	Q	*pol*[e]
P	I	V	G	↓	A	E	T	F	*pol*	E	K	V	Y	↓	L	A	W	V	*pol*[e]
T	L	N	F	↓	P	I	S	P	*pol*	V	E	I	C	↓	T	E	M	E	*pol*
R	E	A	F	↓	R	V	F	D	Calmodulin	T	Q	D	F	↓	W	E	V	Q	*pol*
A	E	T	F	↓	Y	V	D	K	*pol*	L	W	M	G	↓	Y	E	L	H	*pol*
A	Q	T	F	↓	Y	V	N	L	*pol*	G	D	A	Y	↓	F	S	V	P	*pol*
P	T	L	L	↓	T	E	A	P	Actin	E	L	E	L	↓	A	E	N	R	*pol*
S	F	I	G	↓	M	E	S	A	Actin	S	K	D	L	↓	I	A	E	I	*pol*
D	A	I	N	↓	T	E	F	K	Vimentin	L	E	V	N	↓	I	V	T	D	*pol*
Q	I	T	L	↓	W	Q	R	P	Autolysis	I	I	V	A	↓	C[f]	E	G	N	RNase A
E	L	E	F	↓	P	E	G	G	PE664E	A	A	A	K[g]	↓	F	E	R	Q	RNase A
	A	N	L	↓	A	E	E	A	LysPE40	G	G	N	Y	↓	P	V	Q	H	*gag*[h]
S	Q	N	Y	↓	P	I	V	Q	*gag*	A	R	L	M	↓	A	E	A	L	*gag*
P	G	N	F	↓	L	Q	S	R	*gag*	P	F	A	A	↓	A	Q	Q	R	*gag*
K	L	V	F	↓	F	A	E		AAP[a]	P	R	N	F	↓	P	V	A	Q	*gag*
G	D	A	L	↓	L	E	R	N	LysPE40	G	L	A	A	↓	P	Q	F	S	*gag/pol*
K	E	L	Y	↓	P	L	T	S	*gag*	S	L	N	L	↓	P	V	A	K	*pol*
R	Q	A	N	↓	F	L	G	K	*gag*	A	E	T	F	↓	Y	T	D	G	*pol*
S	R	S	L	↓	Y	A	S	S	Vimentin	R	Q	V	L	↓	F	L	E	K	*pol*
A	E	A	M	↓	S	Q	V	T	*gag*	Q	M	I	F	↓	E	E	H	G	Fibronectin[i]

Table 1 is reproduced from [9] with permission from publisher.
[a]Alzheimer amyloid protein; [b]all entries to this point were referenced in [5]; [c]from [6]; [d]from [7]; [e]from [8]; following *pol* gene entries are from [9]; [f]present as carboxamidomethyl Cys [10]; [g]present as ε-succinyl Lys [10]; [h]the following eight entries are from [11]; [i]from [12].

Mutant Q7K/L33I/L63I

We have incorporated features of two mutants listed above (Q7K and L33I/L63I) in this mutant. Its stability at pH 5.5 over a period of 11 days is shown in Fig. 4a-4b. Interestingly, although we appear to have blocked cleavage at the three major sites of autolysis, a new position at Leu_{89}-Leu_{90} is now evident as a site of minor degradation (seen at 11 days). This site nearly coincides with a natural site of autolysis in the HIV-2 protease (Fig. 3), so there must be some preference of the enzyme for this region of the molecule. In fact, Leu_{89} in the HIV-1 enzyme is an Ile in the HIV-2 protease, which would block cleavage at 89-90. The HIV-2 protease merely "slips" a residue downstream, and cleaves the Leu_{90}-Thr_{91} bond. In any case, mutant Q7K/L33I/L63I represents a highly stabilized form of the HIV-1 protease and one which holds much potential for the kind of applications in protein chemistry which require a stable protease.

The picture of stability at pH 4 is similar (Fig. 4c-4d), but here beside the cleavage at 89 we see some minor cleavage at 33 and 63, positions in which Ile was inserted at P_1 to block hydrolysis. This is in keeping with what we have seen in peptides modelled after these sequences; insertion of the Ile does, indeed decrease cleavage about 100 fold, but does not abolish it entirely. This lends credence to the idea that at pH 4 the protease is in a less

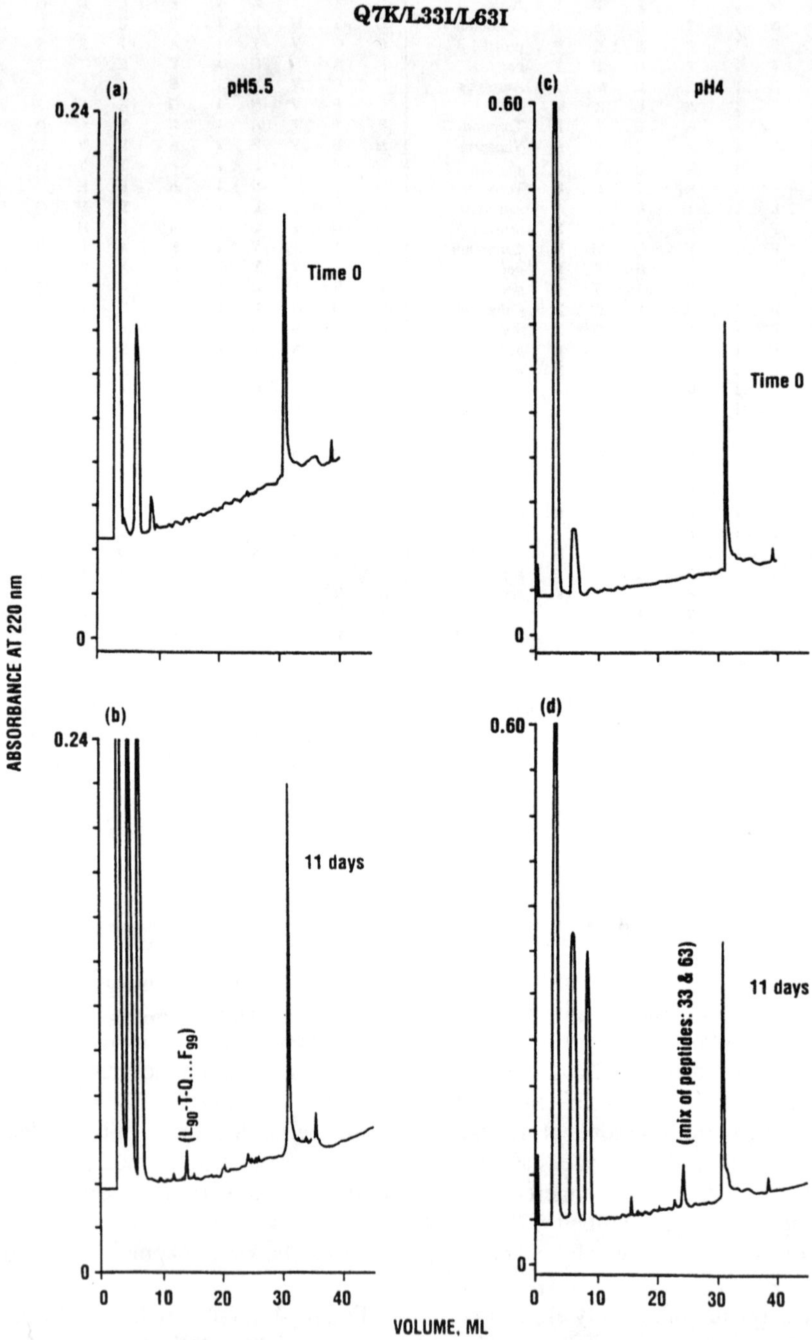

Figure 4. Course of Q7K/L33I/L63I mutant autocleavage over 11 day period at 25°C in the presence of DTT, at pH 5.5 (a and b), and at pH 4.0 (c and d).

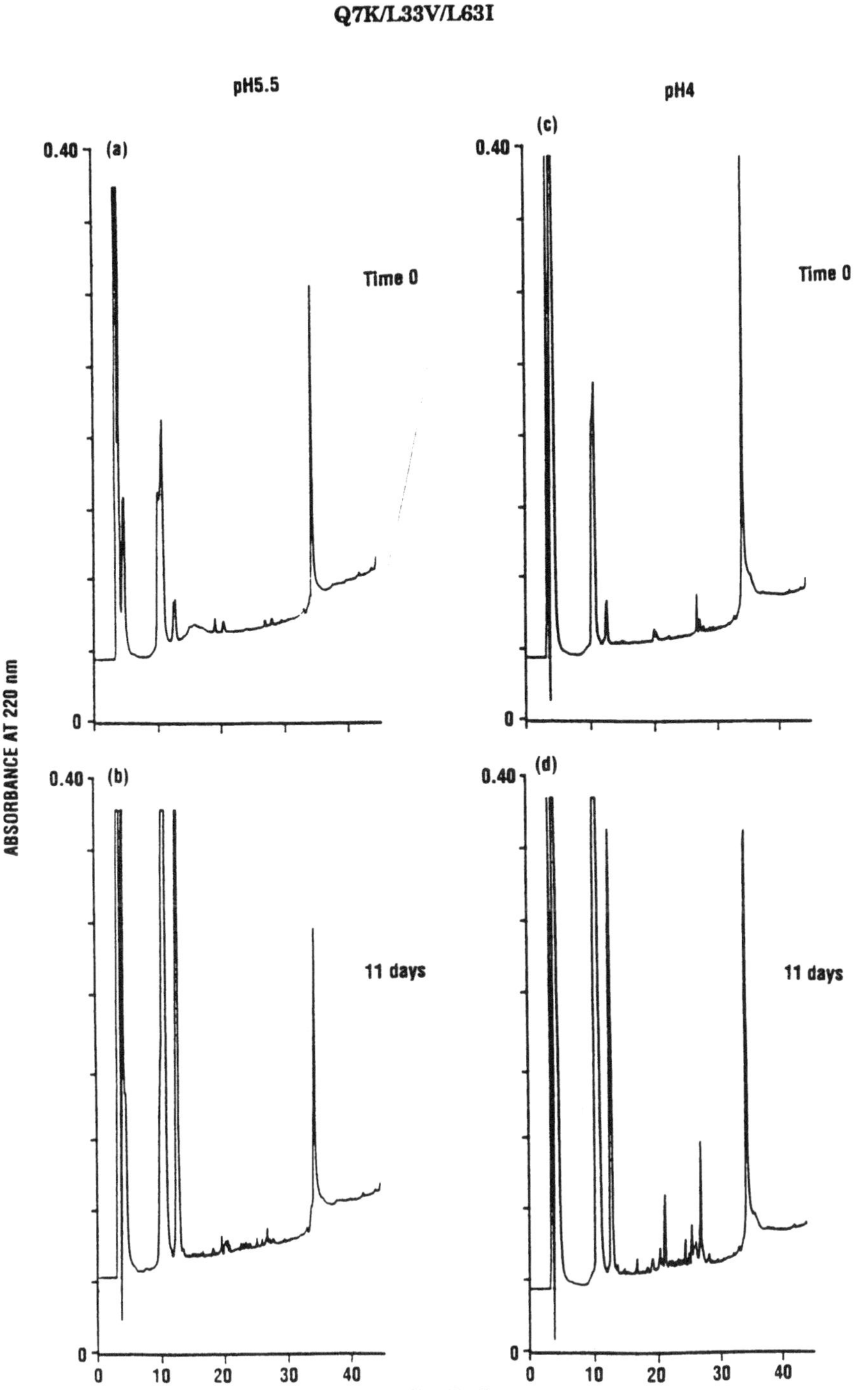

Figure 5. Course of Q7K/L33V/L63I mutant autocleavage over 11 day period at 25°C in the presence of DTT, at pH 5.5 (a and b), and at pH 4.0 (c and d).

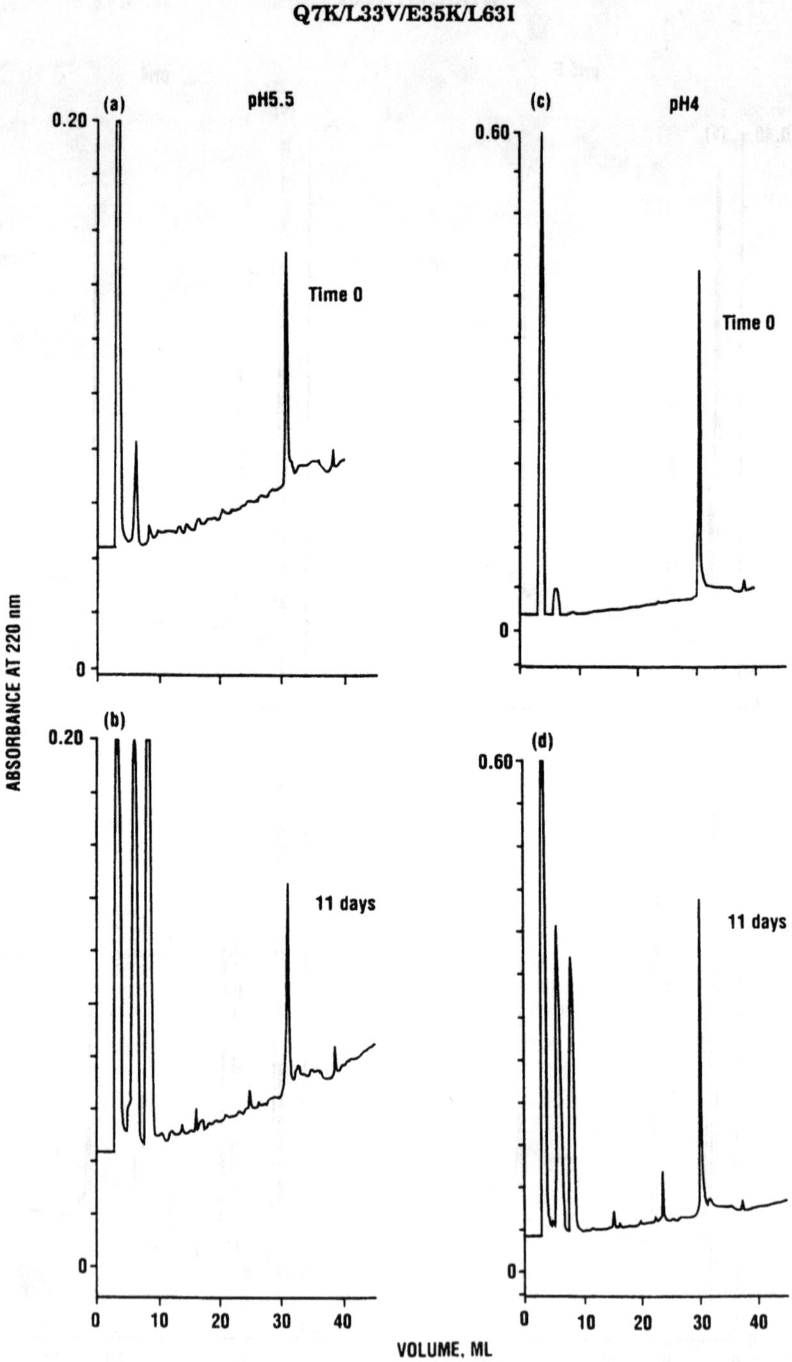

Figure 6. Course of Q7K/L33V/E35K/L63I mutant autocleavage over 11 day period at 25°C in the presence of DTT, at pH 5.5 (a and b), and at pH 4.0 (c and d).

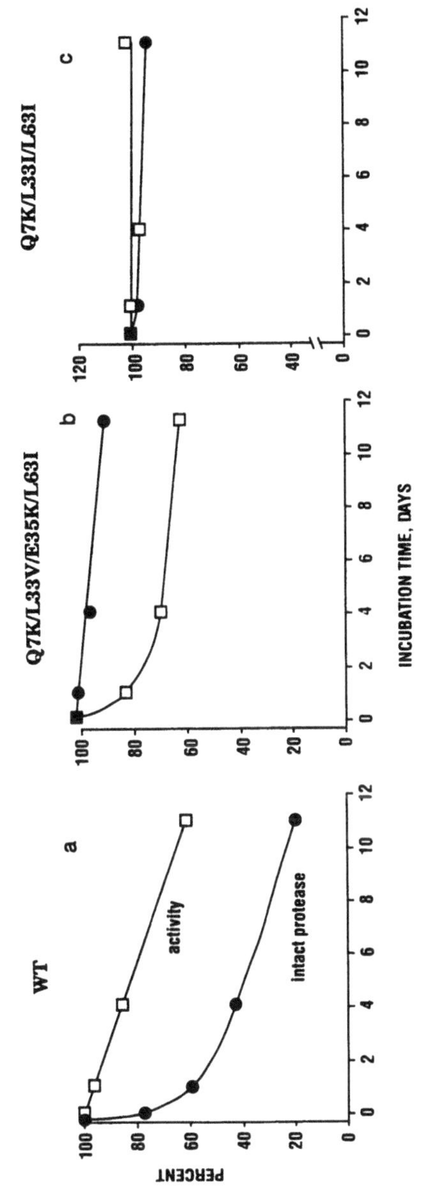

Figure 7. Generated from experiments shown in Figures 4 through 6 measuring the loss of protein over 11 day period and from experiments measuring the loss of proteolytic activity over the same 11 day period.

constricted conformation. Thus, if the mutant enzyme unfolds to reveal these sites, they will be more likely to undergo hydrolysis.

Val vs Ile as a Blocker of Cleavage

In order to test whether Val would work as well as Ile as a P_1 blocker of hydrolysis, we inserted Val at 33 in the same mutant mentioned above. The stability of this Q7K/L33V/L63I mutant was much the same at pH 5.5 (Fig. 5a-5b), but somewhat less so at pH 4 (Fig. 5c-5d). Again, the model peptide containing Val at P_1 is considerably less stable to hydrolysis than that having a P_1 Ile, and if the protein is unfolding more at pH 4, then one would expect greater hydrolysis at Val_{33}.

Lysine as a Blocker of Hydrolysis

Because we saw relatively more cleavage at position 33 when Val was inserted (Fig. 5c-5d), we inserted a Lys at position 35, similar to the change at 7. The stability at pH 5.5 (Fig. 6a-6b) and at pH 4.0 (Fig. 6c-6d) is in keeping with that seen with Q7K/L33I/L63I (Fig. 4). Although we have increased stability with respect to autolysis by insertion of a P_2' Lys, this does not always coincide with enzyme stability, about 30% of the activity is lost at pH 5.5. In the case of all of our Lys mutants involving positions 35 and 65 (P_2' mutations), we saw increased stability to autolysis, but decreased enzyme activity (data not shown). This loss of enzyme activity could be due to precipitation, or to problems with dimerization or dimer stability. Initially we see high activity, but then with time of incubation the loss is more extreme. The Q7K mutation seems to have no adverse effects on enzyme folding or stability, but this is not unexpected due to a Lys present at position 7 in the HIV-2 protease molecule. Replacement of Glu residues at 35 and at 65 by Lys, however, represents charge reversals in these regions, and may not be as well tolerated.

Comparison of Wild Type and Mutant Proteases

In Fig. 7, we compare the stabilities at pH 5.5 of the wild type HIV-1 protease and its Q7K/L33V/E35K/L63I and Q7K/L33I/L63I mutants. The enhanced loss of activity with respect to autolysis of the E35K mutant mentioned above is demonstrated in Fig. 7. Mutant Q7K/L33I/L63I looks the best in terms of activity and stability to autolysis, while wild type HIV-1 protease is the worst. For the latter enzyme the curve representing intact protease drops off quickly at about 20% during the 6 hours of refolding at 0°C, meaning that there is considerable autolysis of the unfolded species (see Fig 2). Another 20% is cleaved during the first 24 hour incubation at 25°C, and again unfolded species are cleaved since the drop in activity is minor. Between 24 hour and 11 day incubations there is a parallel decrease of intact protease and activity, meaning that the properly folded, active protease unfolds gradually, and as a result is digested. Mutant Q7K/L33V/E35K/L63I that is resistant to autolysis as much as Q7K/L33I/L63I loses activity over the time because of protein precipitation.

Does Mutation Alter Kinetic Properties of the Mutant Proteases?

The aim of this study was to make a protease resistant to autolysis, but with no change in kinetic character as exemplified by substrate specificity and kinetic parameters. The mutant proteases are as active as the wild-type enzyme, at least in the few cases tested, e.g. V_{max} is 24.1 and 22.4 μmoles × mg^{-1} × min^{-1} for Q7K/L33I/L63I and wild type protease, respectively. What about the enzyme specificity? We know that the HIV-1 and HIV-2 proteases are similar in specificity in that they will process polyproteins from each others virus. However, we have also shown that they are distinct with respect to particular substrates, including Ca^{++} free calmodulin [16] and troponin C [17]. Clearly, the patterns obtained with the HIV-1 protease and with the Q7K/L33I/L63I mutant are identical and distinct from that generated by the HIV-2 protease (results not shown). Thus, it would appear that the specificity of the HIV-1 protease has been preserved in this mutant enzyme.

CONCLUSIONS

1. Wild type HIV-1 protease undergoes autolysis at three major sites; Leu_5-Trp_6, Leu_{33}-Glu_{34}, and Leu_{63}-Ile_{64}. Particular mutants possess a minor site of cleavage at Leu_{89}-Leu_{90} which becomes apparent upon prolonged incubation periods.

2. We found a way for blocking, or at least reducing dramatically, a susceptible site of hydrolysis by the HIV-1 protease: this is done by inserting an Ile at P_1; and/or inserting a Lys at P_2' (or, presumably, at any position from P_2 through P_2').

3. We performed the substitutions mentioned in 2 in order to create mutant forms of the HIV-1 protease that are resistant to autodegradation. Conservative Leu→Ile changes at P_1 are effective and seem to be well tolerated in terms of folding and dimer stability. Insertion of positively charged Lys residues are highly effective in blocking hydrolysis, but may influence enzyme stability in other ways.

4. Mutant Q7K/L33I/L63I seemed particularly stable to autolysis and had specific activity and specificity for the substrate troponin C, like those of the wild type enzyme. It may find application in kinetics, inhibition and structural studies, and in protein chemistry where a stable protease with defined specificity is required.

REFERENCES

1. J.O. Hui, A.G. Tomasselli, I.M. Reardon, J.M. Lull, D.P. Brunner, C.-S.C. Tomich & R.L. Heinrikson. *J. Protein Chem.* 12:321-325 (1993).
2. J.R. Rosé, R. Salto & C.S. Craik. *J. Biol. Chem.* 268:11939-11945 (1993).
3. J.E. Strickler, J. Gorniak, B. Dayton, T. Meek, M. Moore, V. Magaard, J. Malinowski & C. Debouck. *Proteins* 6:139-154 (1989).
4. A.G. Tomasselli, A.M. Mildner, D.J. Rothrock, J.L. Sarcich, J. Lull, J. Leone & R.L. Heinrikson. Companion Paper (1994).
5. R.A. Poorman, A.G. Tomasselli, R.L. Heinrikson & F.J. Kézdy. *J. Biol. Chem.* 266:14554-14561 (1991).
6. Y. Riviere, V. Blank, P. Kourilsky & A. Israel. *Nature* 350:625-626 (1991).
7. T.A. Tomaszek Jr., M.L. Moore, J.E. Strickler, R.L. Sanchez, J.S. Dixon, B.W. Metcalf, A. Hassell, G.G. Dreyer, I. Brooks, C. Debouck & T.D. Meek. *Biochemistry* 31:10153-10168 (1992).
8. D. Chattopadhyay, D.B. Evans, M.R. Deibel Jr., A.F. Vosters, F.M. Eckenrode, H.M. Einspahr, J.O. Hui, A.G. Tomasselli, H.A. Zürcher-Neely, R.L. Heinrikson & S.K. Sharma. *J. Biol. Chem.* 267:14227-14232 (1992).
9. A.G. Tomasselli, J.L. Sarcich, L.J. Barrett, I.M. Reardon, W.J. Howe, D.B. Evans, S.K. Sharma & R.L. Heinrikson. *Prot. Science* 2:2167-2176 (1993).
10. J.O. Hui, A.G. Tomasselli, H.A. Zürcher-Neely & R.L. Heinrikson. *J. Biol. Chem.* 265:21386-21389 (1990).
11. J. Tozser, I. Blaha, T.D. Copeland, E.M. Wondrak & S. Oroszlan. *FEBS Lett.* 281:77-80 (1991).
12. M. Oswald & K.von der Helm. *FEBS Lett.* 292:298-300 (1991).
13. J. Konvalinka, P. Strop, J. Velek, V. Cerna, V. Kostka, L.H. Phylip, A.D. Richards, B.M. Dunn & J. Kay. *FEBS Lett.* 268:35-38 (1990).
14. J.T. Griffiths, L.H. Phylip, J. Konvalinka, P. Strop, A. Gustchina, A. Wlodawer, R.J. Davenport, R. Briggs, B.M. Dunn & J. Kay. *Biochemistry* 31:5193-5200 (1992).

15. J. Tozser, I.T. Weber, A. Gustchina, I. Blaha, T.D. Copeland, J.M. Louis & S. Oroszlan. *Biochemistry* 31:4793-4800 (1992).
16. A.G. Tomasselli, J.O. Hui, L. Adams, J. Chosay, D. Lowery, B. Greenberg, A. Yem, M.R. Deibel, H. Zürcher-Neely & R.L. Heinrikson. *J. Biol. Chem.* 266:14548-14553 (1991).
17. A.G. Tomasselli, W.J. Howe, J.O. Hui, T.K. Sawyer, I.M. Reardon, D.L. DeCamp & C.S. Craik. *Proteins: Struct. Funct. Genet.* 10:1-9 (1991).

IDENTIFICATION OF AMINO ACID RESIDUES OF THE RETROVIRAL ASPARTIC PROTEINASES IMPORTANT FOR SUBSTRATE SPECIFICITY AND CATALYTIC EFFICIENCY

C. E. Cameron,[1] H. Burstein,[2] D. Bizub-Bender,[2] T. Ridky,[1] I. T. Weber,[3] A. Wlodawer,[4] A. M. Skalka,[2] and J. Leis[1]

[1]Case Western Reserve University
School of Medicine
Cleveland, Ohio 44106-4935
[2]Fox Chase Cancer Center
Institute for Cancer Research
Philadelphia, Pennsylvania 19111
[3]Thomas Jefferson University
Jefferson Cancer Institute
Philadelphia, Pennsylvania 19107
[4]NCI-Frederick Cancer Research and Development Center
ABL-Basic Research Program
Frederick, Maryland 21701 USA

Kinetic analysis of a set of 40 systematically altered peptide substrates based on a Rous sarcoma virus (RSV) polyprotein cleavage site allowed characterization of the 7 individual subsites (S4-S3') of the substrate-binding pocket of the aspartic proteinases (PRs) from RSV and human immunodeficiency virus type 1 (HIV-1) required for recognition of substrates [1,2]. Initial analysis suggested that subsites of the PR substrate-binding pocket acted independently in selection of substrates and implicated RSV amino acids His65 (*Gly48*), Arg105 (*Pro81*) and Gly106 (*Val82*) as primary determinants of specificity in the S1 and S1' subsites of the substrate-binding pocket [1]. The italicized amino acids are derived from HIV-1 PR.

Subsequent interpretation of steady-state kinetic data using models of RSV and HIV-1 PRs built from cystallographic data have allowed us to predict the amino acid residues in the 5 additional subsites of the retroviral PR substrate-binding pocket which might account for differences observed in substrate specificity when the RSV and HIV-1 PRs were compared kinetically [2]. These residues are as follows: Ser 38 (*Thr26*), Ile42 (*Asp30*), Ile44 (*Val 32*), Asn61, Pro62, Gln63, Met73 (*Val56*), Ala100 (*Leu76*), Val104 (*Thr80*) and Ser107 (*Asn83*). The position of these residues in the PR substrate-binding pocket relative to the

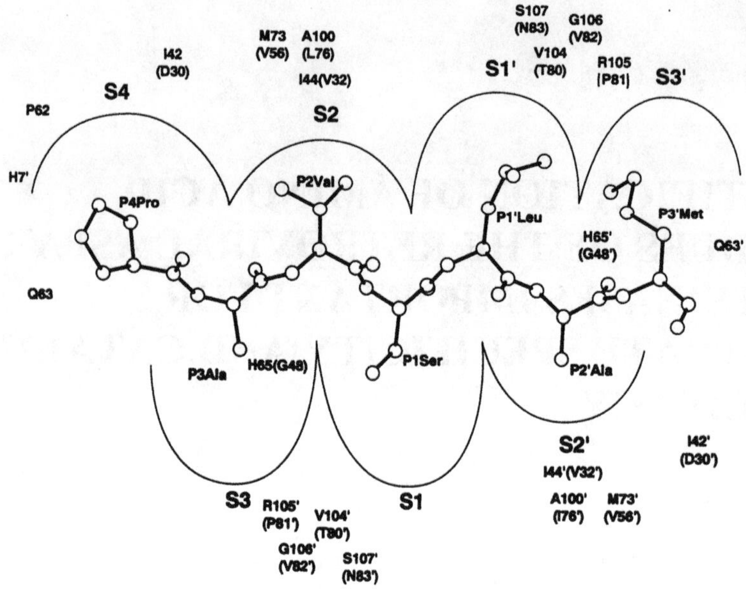

Figure 1. A schematic representation of the RSV NC-PR substrate, PAVSLAM, from P4-P3' in the S4-S3' subsites of PR. The relative size of each subsite is indicated approximately by the area enclosed by the *curved line* around each substrate side chain. Protease residues forming the subsites are shown for those that differ between the RSV and HIV-1 PRs. The RSV PR residue is shown outside the *parentheses*. Most of the residues contribute to more than one adjacent subsite, and this is indicated by the position of the label.

NC-PR cleavage site is shown diagrammatically in Fig. 1 to indicate the specific subsites in which the aforementioned amino acid residues reside.

In this report, we describe results obtained from the analysis of purified, recombinant RSV PRs in which amino acids predicted to be important for substrate selection have been changed individually to those present in the structurally equivalent position of HIV-1 PR using molecular genetics. This analysis has allowed us to identify most, if not all, of the residues in the 7 subsites of the retroviral PR substrate-binding pocket that contribute to the difference in substrate specificity existing between the RSV and HIV-1 PRs. Additionally, these studies have unexpectedly revealed amino acid residues responsible for the increased catalytic efficiency of HIV-1 PR relative to RSV PR. The implications of these results on the design of protease inhibitors to be used as chemotherapeutic agents against retroviral infection are discussed.

RESULTS

RSV PRs with Mutations in the Substrate-Binding Pocket Retain Catalytic Activity

A series of mutations were placed into the RSV PR coding region of the bacterial-expression plasmid pPR [3,4] using site-directed mutagenesis. The resulting amino acid changes substituted RSV residues predicted to be important for substrate specificity for those located in the structurally equivalent positions of HIV-1 PR. These mutant RSV PRs are listed in Table 1. After induction, purification from the inclusion body fraction of *Escherichia*

Table 1. Activity of RSV Proteases with Mutations in the Substrate-binding Pocket

PR mutation[a]	Activity[b] % relative to wild type
S38T	200
I42D	62
I44V	211
Q63M	448
M73V	31
A100L	<1[c]
M73V,A100L	96
V104T	229
S107N	377
ΔN61-Q63	245

[a]Mutations were introduced in the PR coding region of the bacterial-expression plasmid, pPR, using oligo-deoxynucleotide-directed site-specific mutagenesis as described previously [4]. The PR gene of pPR is under the control of the λ PL promoter. The *E. coli* strain used for PR expression is MC1061 containing pRK248*cI*ts which permits constitutive expression of a temperature-sensitive allele of λ cI repressor. Thus, induction of PR expression is effected by a shift from 30 °C to 42 °C. Cells are generally induced for 4 hours once the culture has reached an OD600 of 0.8. PR is purified from the inclusion body fraction obtained from lysozyme-lysed cells. Inclusion bodies are washed several times with a 1% Triton X-100/25% sucrose solution. The final wash contains 6 M urea in the triton-sucrose solution. Inclusion bodies are solubilized in 8 M urea (recrystallized from absolute ethanol) and 150 mM 2-mercaptoethanol. 1 ml of the solubilized inclusion bodies (1 mg/ml) is dialyzed against 2 l of 25 mM sodium phosphate (pH 7.6), 150 mM sodium chloride and 10 mM 2-mercaptoethanol for 12-18 hours at 25 °C. The refolded PR solution is then clarified and the protein concentration determined using the Bio-Rad protein assay with bovine serum albumin as a standard.
[b]PR activity was determined by incubation of 2 pmol of active PR in a 25 µl volume of 0.1 M sodium phosphate (pH 5.9), 2.4 M sodium chloride with 0.1 mM NC-PR peptide substrate (PPAVS-LAMTMRR, the hyphen indicates the site of cleavage) for 15 min at 37 °C. Reactions were stopped by the addition of 0.3 ml of 0.5 M sodium borate (pH 8.5) and followed by the addition of 20 µl of fluorescamine (50 mg/100 ml acetone). Relative fluorescence intensity was determined using a Perkin-Elmer LS-50B fluorimeter with an excitation wavelength of 386 nm (5 nm slit) and an emission wavelength of 477 nm (10 nm slit). The specific activity of the bacterially-expressed wild-type PR is 575 nmol of peptide cleaved/min/mg of enzyme after correction by active-site titration.
[c]The activity of this mutant PR has not yet been corrected by active-site titration.

coli, and refolding as described previously [4], the activity of each RSV PR mutant was determined on a peptide representing the RSV NC-PR cleavage site. The specific activity of each mutant PR is presented in Table 1 relative to the bacterially-expressed, wild-type RSV PR.

Each of the purified RSV PR mutants retained some level of activity on this substrate ranging from approximately 30 percent to 400 percent that of wild type. PR mutants which were less active than wild type contained the substitutions I42*D*, M73*V*, or A100*L*. If the M73*V* and A100*L* substitutions were combined, a PR with activity comparable to that of wild type was obtained (Table 1). Thus we conclude that an interaction between these two residues has to be maintained for activity. All other mutant PRs had activities which were at least 2-fold greater than that of wild type.

Amino Acid Substitutions in the Substrate-Binding Pocket of RSV PR Affect Substrate Selection in Individual Subsites

We previously showed that a library of peptide substrates containing single amino acid substitutions in the RSV NC-PR cleavage site was capable of distinguishing individual subsites of the RSV and HIV-1 PRs [1,2]. Using peptides containing substitutions in the P1 and P1' positions, it was possible to demonstrate that partial HIV-1 substrate specificity could be conferred on RSV PR by replacing non-conserved residues of the RSV PR located primarily in the S1 and S1' subsite with the residues of HIV-1 PR from structurally analogous positions. We now have begun to characterize the specificities of PR(I44*V*) and PR(ΔN61-Q63), which are predicted to have altered S2 and S2' or S4 PR subsites, respectively. The activities of these mutant RSV PRs were determined on NC-PR peptide substrates with substitutions at positions P4, P2, P1 or P2' (see Fig. 1). As indicated in Fig. 2A and Fig. 2B, the I44*V* substitution allowed cleavage of NC-PR peptides with leucine in the P2 or P2' positions, respectively, with a relative efficiency comparable to that of HIV-1 PR. This change in specificity, however, is limited to the S2 and S2' subsites as evidenced by the retention of RSV PR specificity with P4-modified (Fig. 2C) and P1-modified (Fig. 2D) NC-PR peptide substrates. Additionally, the deletion of residues 61-63 produced a PR with HIV-like specificity in the S4 subsite. This enzyme was capable of processing a NC-PR peptide with asparagine in the P4 position as efficiently as HIV-1 PR (Fig. 2C). Again, the change in specificity appeared to be limited to the S4 subsite in that this enzyme recognized the P1-Trp NC-PR analogue with an efficiency comparable to that of the wild-type RSV PR (Fig. 2D). Of the other mutants listed in Table 1, the S38*T* and V104*T* (data not shown) and S107*N* (Fig. 2) mutations did not affect their relative activities on the different test substrates, compared with wild type. Thus, these mutations appear to affect only the catalytic rate of these enzymes.

Amino Acid Substitutions in the Substrate-Binding Pocket of RSV PR Increase Activity on an HIV-1 Cleavage Site Peptide Substrate

To further characterize the specificity of the two mutant RSV PRs discussed above, a series of peptide substrates representing RSV and HIV *gag* and *pol* polyprotein cleavage sites was analyzed as previously described [1,5]. While these mutant RSV PRs retained the ability to cleave each of the RSV-based cleavage site substrates (data not shown), both enzymes exhibited increased activity towards the HIV-1 CA-NCb cleavage site substrate (Fig. 3).

DISCUSSION

Our previous studies of the RSV PR substrate-binding pocket led us to conclude that the individual subsites of retroviral proteases act independently in the selection of substrates [1]. This was based on the observation that the preference of RSV PR for Trp at P1 of the NC-PR peptide substrate could be changed to Leu, the preference of HIV-1 PR. This was accomplished by changing three "key" RSV PR residues (His65, Arg105, and Pro106) at the S1 and S1' subsites to those found in the structurally analogous positions of HIV-1 PR. Furthermore, we hypothesized that recognition of an amino acid sequence as a substrate for

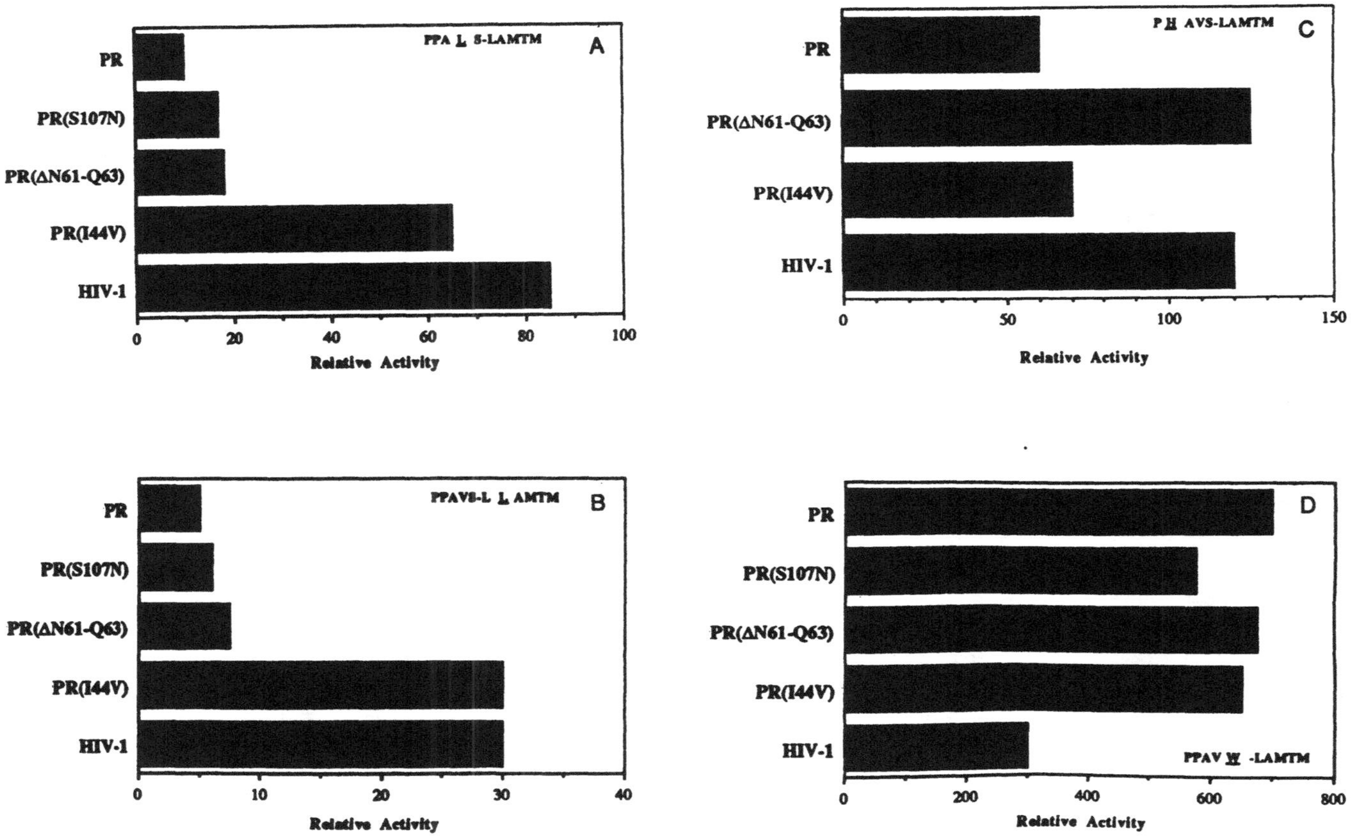

Figure 2. Activities of PR(I44*V*), PR(S107*N*), and PR(ΔN61-Q63) on P2, P2', P4 and P1-modified NC-PR peptide substrates. The activity of each enzyme was determined as described in the legend of Table 1 using NC-PR peptides with the following modifications: (**A**) ValP2Leu; (**B**) AlaP2'Leu; (**C**) ProP4Asn; (**D**) SerP1Trp. The activity of each enzyme on the modified peptide substrates is expressed as percent relative to the activity on the wild-type NC-PR peptide substrate.

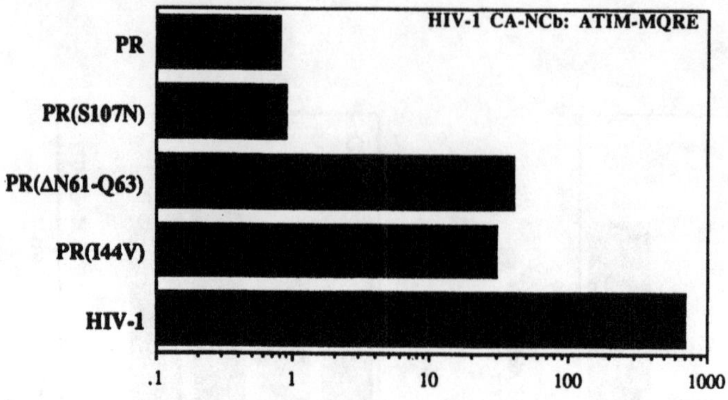

Figure 3. Activity of PR(I44V), PR(S107N), and PR(ΔN61-Q63) on the HIV-1 CA-NCb peptide substrate. The activity of each enzyme was determined as described in the legend of Table 1 using the HIV-1 CA-NCb peptide substrate. The sequence of this peptide is PATIM-MQRETRR; the *hyphen* denotes the site of cleavage. The activity of each enzyme on this peptide substrate is expressed as percent relative to the activity on the wild-type NC-PR peptide substrate.

a retroviral protease would require optimization of interactions between key amino acids in each of the 7 subsites and the side-chains of the substrate. Therefore, these key amino acids *en bloc* would dictate the specificity of a specific PR.

In this report, we have described RSV PR mutants which further define the repertoire of amino acids present in the PR substrate-binding pocket which are key determinants for substrate specificity and catalytic efficiency. It is noteworthy that each of the mutant RSV PRs described is active, and that the sum of the increased activity observed for each of these enzymes can account for the increased catalytic rate of HIV-1 PR relative to that of RSV PR (Table 1). Furthermore, the ability of the I44*V* and ΔN61-Q63 alterations to confer HIV-1-PR-like specificity to the S2 (Fig. 2A) and S2' (Fig. 2B) or S4 (Fig. 2C) subsites, respectively, along with increased activity on yet another HIV-1-based cleavage site substrate (Fig. 3) is consistent with our model for the molecular basis of substrate recognition by the retroviral protease [2]. In addition to those described above, the I42*D*, M73*V* and A100*L* substitutions also affect specificity in the S2 and S2' subsites and permit cleavage of additional HIV-1-based cleavage site substrates(data not shown).

Optimized interactions between key residues of the individual PR subsites and the side-chains of a heptapeptide are necessary and probably sufficient to confer upon a peptide the ability to be specifically and efficiently processed by PR. In this regard, using peptide substrates with single amino acid substitutions in each of the 7 positions required for cleavage has facilitated identification of the key determinants of specificity by leaving 6 of the 7 optimized interactions intact at all times.

Structural analyses indicate that van der Waal's interactions are used primarily in determining the types of amino acids selected by an individual subsite although polar substrate side chains can form ionic or H-bond interactions at some positions. This is best illustrated with the S2 subsite. In RSV PR, an isoleucine is present in this subsite which interacts optimally with a valine in the P2 position of the NC-PR substrate. Changing the valine at P2 of NC-PR to leucine increases the van der Waal's radius at this position, due to the additional methylene group. This causes steric interference with Ile44 present in the RSV PR S2 subsite, which is reflected in the reduced activity of RSV PR observed with this NC-PR analogue (Fig. 2A). The reduction in activity with this modified NC-PR peptide substrate is corrected by changing Ile44 to *valine* which restores the optimal van der Waal's interaction. Thus, the ability of HIV-1 PR to accommodate leucine at P2 of NC-PR is explained by the presence of a valine in the S2 subsite of this enzyme.

The observation that optimization of van der Waal's interactions is critical in determining the stringency of substrate selection provides an explanation for the broad and different substrate recognition by the RSV and HIV-1 PRs. Both enzymes can accommodate multiple amino acids into a given enzyme subsite. However, in our comparison of the 7 subsites of RSV and HIV-1 PRs, we noted that, in many instances, the subsites of HIV-1 PR were larger. For example, in S2 and S2' Ile42 and Ile44 of RSV PR are replaced in HIV-1 PR by *aspartic acid* and *valine*, respectively. In S4, 3 amino acids present in RSV PR have no structurally comparable residues in HIV-1 PR. This means that the HIV-1 PR can sterically accommodate a larger number of amino acids and would thus have a wider substrate range than RSV PR.

Also, we have noted that the active site of the HIV-1 PR appears to be more rigid due to stabilization by intramolecular hydrogen bonds involving *Thr80* and *Asn83*. This accounts in part for the higher catalytic activity of HIV-1 PR. Moreover, it would make the PR less dependent upon side chain-side chain hydrophobic interactions with substrate for catalysis. This idea is supported by the observation that substitution of a Gly residue at most positions of the NC-PR substrate has little affect on catalysis by HIV-1 PR while with the same Gly-substituted substrates elicit a 10 to 100-fold diminution in activity for RSV PR [2].

Implicit in the above model, is that the substrate-binding pocket of HIV-1 PR will be more tolerant to amino acid substitutions at residues of each subsite critical to substrate selection. Furthermore, these changes would be predicted to have negligible effects on *gag* and *pol* cleavage site recognition. Therefore, the emergence of HIV mutants that are resistant to substrate-based inhibitors of PR can occur with little loss of activity towards natural cleavage sites. Such mutations are likely to occur at the "key" residues reported herein and previously [2]. The ease with which residues of HIV-1 PR change to produce viruses resistant to inhibitors has been confirmed by several groups. Otto et al. [6] have found a HIV-1 PR drug resistant variant with *Val82* changed to alanine while the laboratories of John Kay and Ron Swanstrom have isolated resistant HIV variants with changes in *Gly48* and *Val32*, respectively, of HIV-1 PR (personal communications).

ACKNOWLEDGMENTS

This work was supported in part by United States Public Health Service Grants CA38046 and CA52047 (to J. L.), CA47486, CA06927 and RR05539, an appropriation from the Commonwealth of Pennsylvania (to A. M. S.), and National Cancer Institute, Department of Health and Human Services Contract N01 C01 74101 with ABL (to A.W.) C. E. C. was supported by Predoctoral Fellowship GM13628 from the National Institutes of Health. T. R. is a Medical Scientist Trainee supported by Grant GM07250 from the National Institutes of Health.

REFERENCES

1. Grinde, B., C. E. Cameron, J. Leis, I. T. Weber, A. Wlodawer, H. Burstein and A. M. Skalka (1992) Analysis of Substrate Interactions of the Rous Sarcoma Virus Wild Type and Mutant Proteases and Human Immunodeficiency Virus-1 Protease Using a Set of Systematically Altered Peptide Substrates. *J. Biol. Chem.* **267:** 9491-9498.

2. Cameron, C. E., B. Grinde, J. Jentoft, J. Leis, I. T. Weber, T. D. Copeland and A. Wlodawer (1992) Mechanism of Inhibition of the Retroviral Protease by a Rous Sarcoma Virus Peptide Substrate Representing the Cleavage Site between the *gag* p2 and p10 Proteins. *J. Biol. Chem.* **267:** 23735-23741.
3. Kotler, M., R. Katz and A.M. Skalka. (1988) Activity of Avian Retroviral Protease Expressed in *Escherichia coli. J. Virol.* **62:** 2696-2700.
4. Grinde, B., C. E. Cameron, J. Leis, I. T. Weber, A. Wlodawer, H. Burstein, D. Bizub and A. M. Skalka (1992) Mutations That Alter the Activity of the Rous Sarcoma Virus Protease. *J. Biol. Chem.* **267:** 9481-9490.
5. Cameron, C. E., B. Grinde, P. Jacques, J. Jentoft, J. Leis, A. Wlodawer and I. T. Weber (1993) Comparison of the Substrate-binding Pockets of the Rous Sarcoma Virus and Human Immunodeficiency Virus Type 1 Proteases. *J. Biol. Chem.* **268:** 11711-11720.
6. Otto, M., Garber, S., Winslow, D., Reid, C., Aldrich, P., Jadhav, P., Patterson, C., Hodge, C. and Cheng, Y-S. (1993) *In vitro* Isolation and Identification of Human Immunodeficiency Virus (HIV) Variants with Reduced Sensitivity to C-2 Symmetrical Inhibitors of HIV Type 1 Protease. *Proc. Natl. Acad. Sci. USA* **90:** 7543-7547.

INHIBITOR-RESISTANT MUTANTS OF THE HIV-1 ASPARTIC PROTEASE

Bruce D. Korant

DuPont Merck Virus Laboratory
500 South Ridgeway Avenue
Glenolden, PA 19036

INTRODUCTION

There are three approved antiviral drug therapies for individuals infected with human immunodeficiency viruses (HIV). They are all nucleosides, with the most widely used known as AZT, and appear to act by inhibiting essential reactions carried out by the viral reverse transcriptase. Unfortunately, in patients treated for some time with one of these drugs, viral progeny often occur which are much less susceptible to the antiviral action of the inhibitor. The resistant phenotypes are often associated with a characteristic set of changes in the viral RNA sequence coding for the reverse transcriptase (1). As a rule, high-level resistance is accompanied by multiple amino acid changes in the transcriptase. This is not a rare phenomenon in anti-microbial chemotherapies, and occurs with antiviral as well as antibacterial medicines, and can reduce or even eliminate any value of the therapy. If there is an alternative therapy, that can be attempted. Unfortunately, for viruses, there are usually no useful options, and so targeting a second viral target is highly desirable.

With HIV protease as a therapeutic target, there are presently no approved drugs, and so it has not yet been possible to assess the development of resistance in a large population of treated patients. Lufting (2) recently surveyed protease coding sequences, and concluded that variations reported did not lie in relevant domains of the enzyme; namely the catalytic or substrate-binding regions. However, the sequences were from virus populations which presumably had not been exposed to a genuine protease inhibitor. More disturbing then, is the report of a genetic change in a codon for a protease amino acid in a HIV-1 stock which was selected for ability to replicate in presence of a symmetric diol protease inhibitor (3). The change found, valine 82 to alanine, is accompanied by a single base change (thymine to cytosine), and resulted in a new strain of HIV-1 with a 6 fold reduced sensitivity to the inhibitor. Studies with a mutated form of the viral protease (V82A) were consistent with the viral phenotype being a direct result of the protease alteration. Since HIV and other retroviruses are notoriously prone to mutagenic events during transcription, it is predictable that V82A and other mutations will occur and may be selected for in presence of particular antiviral protease inhibitors. From a number of structural studies, it is known that val 82 is a residue involved in binding of inhibitors to the protease. Preliminary studies on the ala 82

Aspartic Proteinases: Structure, Function, Biology, and Biomedical Implications
Edited by Kenji Takahashi, Plenum Press, New York, 1995

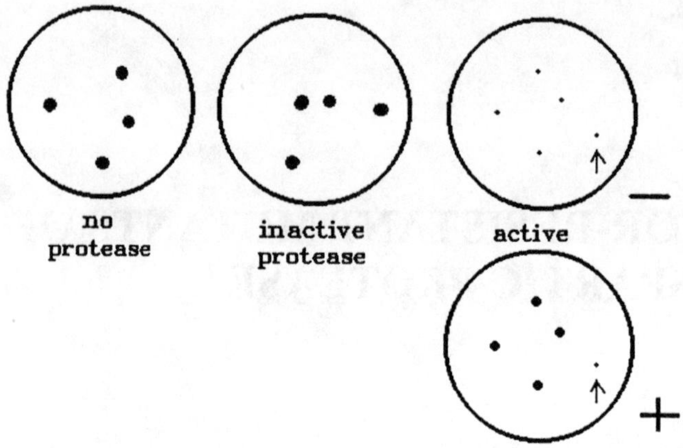

Figure 1. Diagram of the procedure to identify bacterial colonies containing HIV protease mutants which are drug-resistant. *E. coli*, is grown on LB nutrient agar in petri dishes at 37°, overnight. The colony size differs depending on genetic make-up as well as culture conditions. In this experiment cells lacking a gene for HIV protease produce large colonies. Cells containing a gene for an inactive form of the protease (e.g. D25C) also produce large colonies. By comparison, cells containing active forms of the protease produce only small colonies in the absence of an inhibitor, but colony size increases in the presence of a potent protease inhibitor. Those few colonies which demonstrate a small colony phenotype (arrow) in presence (+) of inhibitor are putative mutants which may be resistant. They can be isolated, grown and further characterized to confirm the change in protease qualities.

variant show that although drug-resistant, it is able to process the gag-pol precursor of HIV-1 with high fidelity, and the growth phenotype of the mutant virus is similar to wild-type in lymphocyte cultures (our unpublished results).

In order to begin a wider and more rapid assessment of the potential for development of drug-resistant forms of the HIV protease, we decided to turn to a plasmid-based bacterial expression system for the viral protease we had reported at the 4th Aspartic Proteinase symposium (4). Using a combination of screening of bacterial colonies expressing the HIV protease for drug-resistant phenotypes, as well as site directed mutagenesis of the protease, we have further investigated alterations leading to resistance to a new class of inhibitors, the cyclic ureas (5).

MATERIALS AND METHODS

Construction and mutagenesis of expression plasmids for use in *E. coli* was based on the earlier work of Studier (6), referred to in our studies with HIV protease (4, 7).

PCR-based saturation mutagenesis was done as described by Leung (8).

The assay procedure for HIV protease activity using radiolabelled p55 gag has been previously reported (9).

Isolation and refolding of active HIV-1 protease from bacterial inclusion bodies was essentially as described by Cheng (10).

RESULTS

Our strategy for the use of bacterial colonies to discover protease mutations leading to drug-resistance is outlined in figure 1. A PCR-mutagenized library of HIV-1 protease

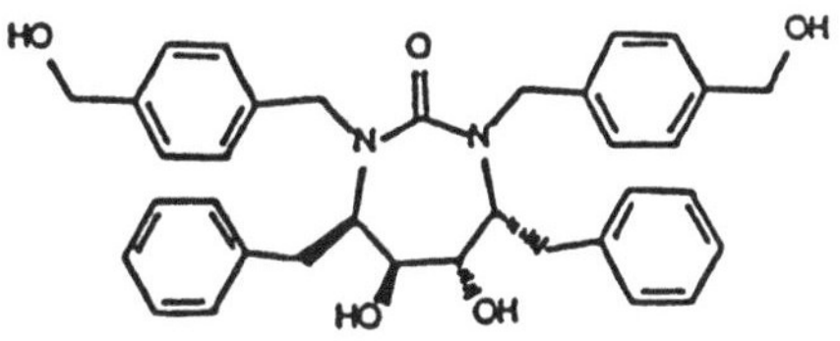

DMP 323

Figure 2. Structure of the cyclic urea, DMP 323.

cDNA is used to transform *E. coli* BL-21 (DE-3). The plasmid used is pET-3, which allows very low-level expression of the protein of interest, in the absence of an inducer of the lac operon. However, this low-level expression of HIV protease is sufficient to reduce bacterial growth, so that a small colony phenotype results. In the presence of a potent HIV protease inhibitor, the toxicity of the protease is blocked, and the colony size is increased. However, drug-resistant forms of the enzyme still result in a small-colony phenotype, and these can be amplified and further characterized to confirm that the phenotype is based on a structural change in the protease.

Based on this simple screening method, and a number of site-directed mutants prepared subsequently, we identified several mutations leading to decreased sensitivity of the HIV protease to a member of the new cyclic urea class of HIV protease inhibitors. The compound used in the selection process was DMP 323, and its structure is provided in figure 2. A potent HIV protease inhibitor (Ki = 0.3 nM), the compound is able to reduce by 90% the replication of HIV-1 in cultured human lymphocytes at a concentration of fifty nanograms/ml (5).

Two mutations we found of particular interest are R8Q and V82I, which both arise as single base pair changes from the parent. They yield enzymes which are similar to wild type in activity level, but show a drug-resistant profile in cell-free assays (see Table 1). Both are able, as single changes, to support virus replication (unpublished results). Crystallography provides a rational explanation for the reduced binding of certain inhibitors to the protease mutants. At position 82, there is potential for interaction with the P1 and P1' functions of the inhibitor; which are benzoyl groups in the case of DMP 323. Compared with the wild-type valine, the bulkier 82 isoleucine occupies too much space, making the fit of DMP 323 problematic. The situation with glutamine at position 8 rather than arginine is less clear. There is at position 8 the possibility of direct interactions with the inhibitor, and also of indirect influences due to shifts of protease amino acids 29 and 30, which form ionic bonds with arg 8. The positioning of aspartics 29 and 30 could be altered coincident with a change in the residue at position 8 leading to altered binding to inhibitor groups P2 and P2'. This will require further analysis.

DISCUSSION

Our results, as well as three other published accounts (3, 11, 12), support the concern that HIV may adapt to replicate in presence of protease inhibitors. Thus far, the degree of resistance is small, and could be compensated by slight increases in dosage. More problematic is whether there will be selection of multiple mutations with continuing exposure to drug, as has been observed with nucleoside therapeutics in clinical settings. For reverse transcriptase, this has led to very high-level resistance, and failure of the therapy.

Table 1. Enzymatic Properties of the Wild Type and Two Mutant Forms of HIV-Protease

Enzyme	Vmax	Km	Ki (nanomolar) for DMP 323
Wild-type	0.035	0.55	0.21
V82I	0.063	0.59	0.87
R8Q	0.083	1.34	0.55

The two mutants reported in this study, R8Q and V82I are both similar to wild-type in their enzymatic properties, and process p55 gag unambiguously (not shown). Also, they are able to permit virus replication in cell cultures (unpublished results). A subsequent study will be to combine these changes into a double mutant, and see whether enzymatic activity is preserved and drug resistance is potentiated. Preliminary results indicate this is the case.

There are other possible scenarios which could lead to protease inhibitor-resistant HIVs. For example, increases in the intracellular level of viral protease up to 20-fold could result from improved efficiency of ribosomal frame shifting. Alternatively, changes in efficiency of transcription from within the viral LTR could select for viruses which overproduce all viral gene products, including protease. These are only theoretical possibilities, which will be looked for as more experience is obtained with the use of anti-HIV therapeutics in clinical situations.

ACKNOWLEDGMENTS

The studies reported in this paper were carried out by C. Rizzo, D. Schilling and Z. Lu. Dr. R. Yates carried out the large-scale fermentations which helped provide enzymes for detailed characterization. Thanks also to Drs. Y.S.E. Cheng and M. Otto for their helpful discussions and for supplying important reagents.

REFERENCES

1. B. Larder, K. Coates and S. Kemp, Zidovudine-Resistant HIV Selected by Passage in Cell Culture, *J. Virol.* 65:5232 (1991).
2. G. Fontenot, K. Johnston, J. Cohen, W. Gallaher, J. Robinson and R. Luftig, PCR Amplification of HIV-1 Proteinase Sequences Directly from Laboratory Isolates Allows Determination of Five Conserved Domains, *Virology* 190:1 (1992).
3. M. Otto, S. Garber, D. Winslow, C. Reid, P. Aldrich, P. Jadhav, C. Patterson, C. Hodge and Y. Cheng, In Vitro Isolation and Identification of HIV Variants with Reduced Sensitivity to C-2 Symmetrical Inhibitors of HIV Protease, *Proc. Natl. Acad. Sci. USA* 90: 7543 (1993).
4. B. Korant and C. Rizzo, Expression of the HIV Aspartic Protease Fused to a Bacterial Phenotypic Marker, in: *Structure and Function of the Aspartic Proteases* (B. Dunn, ed.) p. 429, Plenum Press, N.Y. (1991).
5. P. Lam, P. Jadhav, C. Eyermann, C. Hodge, Y. Ru, L. Bacheler, J. Meek, M. Otto, M. Rayner, Y. Wong, C. Chang, P. Weber, D. Jackson, T. Sharpe and S. Erickson-Viitanen, Cyclic, Non-Peptide HIV Protease Inhibitors: DeNovo Design and Characterization of Potent, Orally-Available Cyclic Ureas, *Science* (in press).
6. F. Studier and B. Moffatt, Use of Phage T7 RNA Polymerase to Direct Selective High-Level Expression of Cloned Genes, *J. Mol. Biol.* 189:113 (1986).
7. B. Korant and C. Rizzo, Assay of HIV Protease Dimerization, *Biol. Chem.* Hoppe-Seyler, 373:517 (1992).

8. D. Leung, E. Chen and D. Goeddel, A Method for Random Mutagenesis of a Defined DNA Segment Using a Modified Polymerase Chain Reaction, *Technique* 1:11 (1989).
9. S. Erickson-Viitanen, J. Manfredi, P. Viitanen, D. Tribe, C. Hutchinson, D. Loeb and R. Swanstrom, A Sensitive Assay for HIV Protease Using Gag Polyprotein Synthesized In Vitro, *AIDS Research and Human Retroviruses* 5:577 (1989).
10. Y. Cheng, M. McGowan, C. Kettner, J. Schloss, S. Erickson-Viitanen and F. Yin, High-Level Expression of Recombinant HIV-1 Protease and the Recovery of Active Enzyme from Inclusion Bodies, *Gene* 87:243 (1990).
11. F. Dianzani, G. Antonelli, O. Turriziani, E. Riva, G. Dong and D. Bellarosa, In Vitro Selection of HIV-1 Resistant to RO31-8959 Proteinase inhibitor, *Antiviral Chem. and Chemother.* 4:329 (1992).
12. J. Craig, L. Whittaker, I. Duncan and N. Roberts, In Vitro Resistance to an Inhibitor of HIV Proteinase (RO31-8959) Relative to Inhibitors of Reverse Transcriptase (AZT and TIBO), *Antiviral Chem. and Chemother.* 4:335 (1993).

DESIGN AND SYNTHESIS OF HIV PROTEASE INHIBITORS CONTAINING ALLOPHENYLNORSTATINE AS A TRANSITION-STATE MIMIC

Yoshiaki Kiso

Department of Medicinal Chemistry
Kyoto Pharmaceutical University
Yamashina-ku, Kyoto 607, Japan

1. INTRODUCTION

The human immunodeficiency virus type-1 (HIV-1), the causative agent of acquired immunodeficiency syndrome (AIDS), codes for a virus-specific aspartic protease responsible for processing the *gag* and *gag-pol* polyproteins and for the proliferation of the retrovirus (Fig. 1). The HIV-1 protease functions as a homodimer and can recognize Phe-Pro and Tyr-Pro sequences as the cleavage site, but mammalian aspartic proteases do not have such specificity. These features provided a basis for the rational design of selective HIV protease-targeted drugs for the treatment of AIDS and related complex.

Based on the substrate transition state (Fig. 2), we designed and synthesized a novel class of HIV protease inhibitors containing allophenylnorstatine [Apns; (2*S*, 3*S*)-3-amino-2-hydroxy-4-phenylbutyric acid]. The critical hydroxyl group as a transition-state mimic interacted with the aspartic acid carboxyl groups of HIV protease active site, and the stereochemistry of the hydroxyl group was significant for the inhibition.

2. DESIGN OF SUBSTRATE-BASED INHIBITORS

We focused on Phe-Pro scissile site, a unique structure for HIV-1 protease, in the design of substrate-based HIV protease inhibitors (Fig. 3 and 4), and incorporated the hydroxymethylcarbonyl (HMC) isostere as a transition-state mimic at P_1 site in a heptapeptide amide, Ser-Phe-Asn-Phe-Pro-Ile-Val-NH_2, similar to the TF/PR and p17/p24 sequences [1].

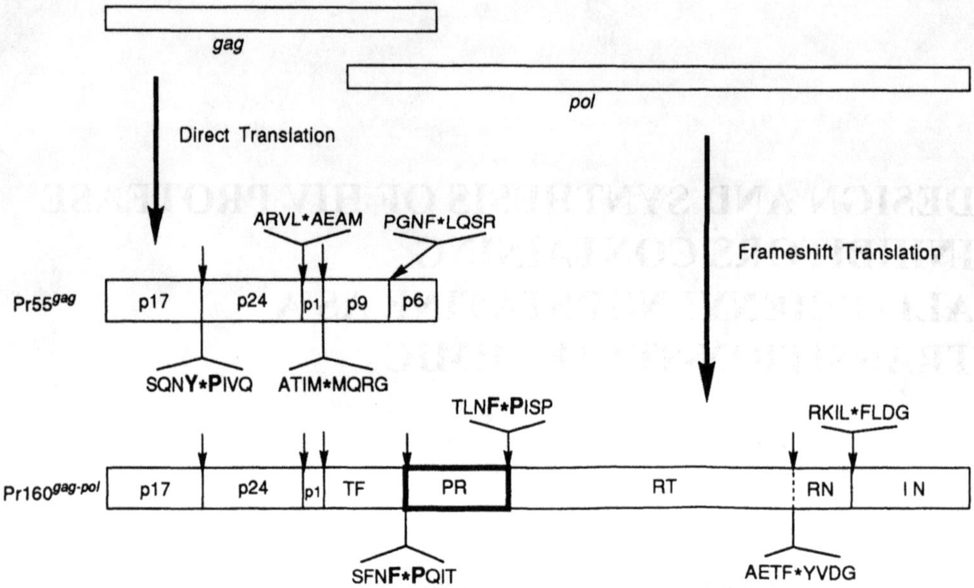

Figure 1. The overlapping *gag/pol* reading frame (above) of HIV-1,their polyprotein translation products Pr55*gag* and Pr160*gag-pol* and the proteolytic cleavage sites. *gag*, group-specific antigen; *pol*, polymerase; TF, transframe protein; PR, protease; RT, reverse transcriptase; RN, ribonuclease H; IN, integrase.

HIV Protease

HIV-1 protease is formed from two identical 99 amino acid peptides by means of hydrophilic and hydrophobic interactions, not by a disulfide bond. Kent et al. [2] synthesized an HIV-1 protease analogue, in which the two Cys residues were replaced by the isosteric L-α-amino-n-butyric acid, whereas we replaced the two Cys residues by a natural amino acid, L-alanine. The [Ala^{67,95}] derivative (amino acid sequence: PQITLWQRPL^{10} VTIKIGGQLK^{20} EALL**DTG**ADD^{30} TVLEEMNLPG^{40} RWKPKMIGGI^{50} GGFIKVRQYD^{60} QILIEI**A**GHK^{70} AIGTVLVGPT^{80} PVNIIGRNLL^{90} TQIG**A**TLNF^{99}) of HIV-1 protease (NY-5 isolate) was synthesized by the solid-phase method based on the Boc-Bzl strategy using PAM resin [1]. Each Boc-amino acid derivative was incorporated by the efficient method [3] and the final deprotection was performed by the HF method. The product was purified by the usual method [2]. The chemically synthesized [Ala^{67,95}]-HIV-1 protease had sufficient cleaving effect on the synthetic nonapeptide substrate, Ac-Arg-Ala-Ser-Gln-Asn-Tyr-Pro-Val-Val-NH_2[4] (pH 5.75, 37°C). We examined the protease inhibitory activities of the peptide compounds containing the HMC structure by this assay system.

KNI-93

Unexpectedly, the (2*S*)-HMC inhibitor, Ser-Phe-Asn-Apns-Pro-Ile-Val-NH_2 (KNI-93; **1S**: *syn* diastereomer) containing Apns was more active against HIV-1 protease than the *anti* diastereomer, Ser-Phe-Asn-Pns-Pro-Ile-Val-NH_2 (KNI-122; **1A**) containing phenylnorstatine [Pns; (2*R*,3*S*)-3-amino-2-hydroxy-4-phenylbutyric acid], in contrast to the case of renin inhibitors [5] which show a preference for the *anti* diastereomer. [e.g., cyclohexylnorstatine (Chns)-containing inhibitor (Fig. 5)].

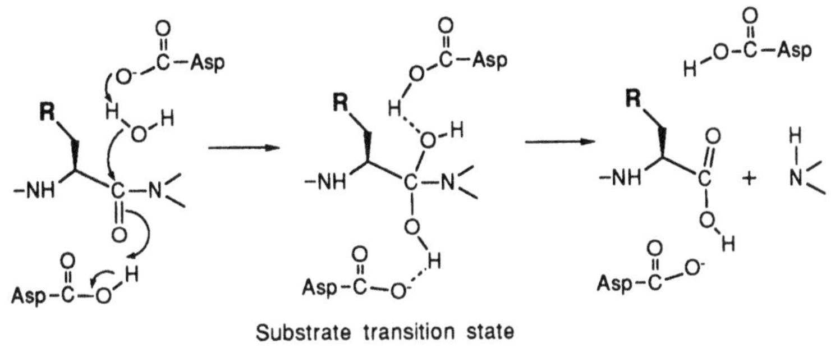

Figure 2. Enzyme reaction mechanism of HIV protease.

3. PROTEASE ACTIVE SITE-TARGETED ANTI-HIV AGENTS

However, a preferred configuration of the hydroxyl group in a series of HMC inhibitors of HIV protease remains to be established. Also, long-chain peptides are unsuitable for a metabolically stable anti-HIV drug, and it is necessary to minimize the number of natural peptide bonds and reduce the molecular weight. We therefore sought the minimum size required for potent inhibition and the preferred hydroxyl group configuration in HMC-inhibitors, and found that a novel tripeptide, Z-Asn-Apns-Pro-NHBu^t (KNI-102; **6S**: *syn* diastereomer)(Z=benzyloxycarbonyl; Bu^t=t-butyl) exhibited a potent, selective inhibitory activity against HIV-1 protease (Fig. 4, Table 1). In contrast, its *anti* diastereomer, Z-Asn-Pns-Pro-NHBu^t (**6A**) exhibited little inhibitory activity [6].

Stereochemistry of the Hydroxyl Group

Thus it became clear that the *syn* configuration of the hydroxyl group was preferred in both cases of low molecular-weight and long-chain HMC-Pro inhibitors of HIV protease. In contrast, symmetric-type inhibitors containing P1'Phe showed a preference for the *anti* diastereomer (Val-Val-Pns-Phe-Val-Val-NH_2; **7A**), in agreement with the case of renin inhibitors [5]. This difference of HMC-Pro inhibitors from symmetric-type inhibitors may be based on the unique conformation of the HMC-Pro moiety.

In the case of hydroxyethyl-Pro-type inhibitors of HIV protease (Fig.6) [7,8], the preference shifted from the *syn* diastereomer in short-chain inhibitors to the *anti* diastereomer in long inhibitors, in contrast to the series of HMC-Pro inhibitors. The discrepancy between the HMC-Pro inhibitors and the hydroxyethyl-Pro inhibitors seems

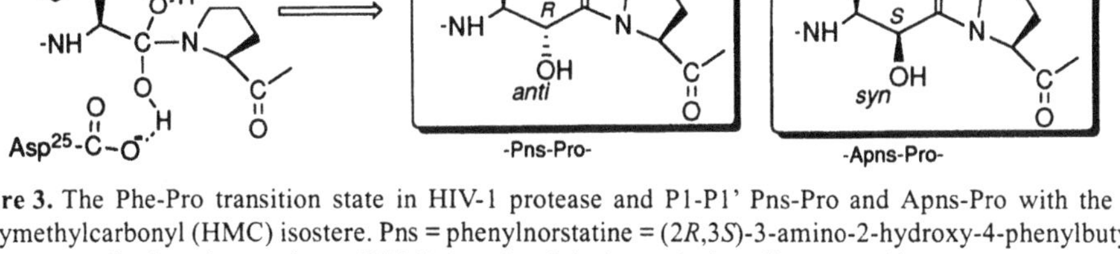

Figure 3. The Phe-Pro transition state in HIV-1 protease and P1-P1' Pns-Pro and Apns-Pro with the hydroxymethylcarbonyl (HMC) isostere. Pns = phenylnorstatine = (2*R*,3*S*)-3-amino-2-hydroxy-4-phenylbutyric acid; Apns = allophenylnorstatine = (2*S*,3*S*)-3-amino-2-hydroxy-4-phenylbutyric acid.

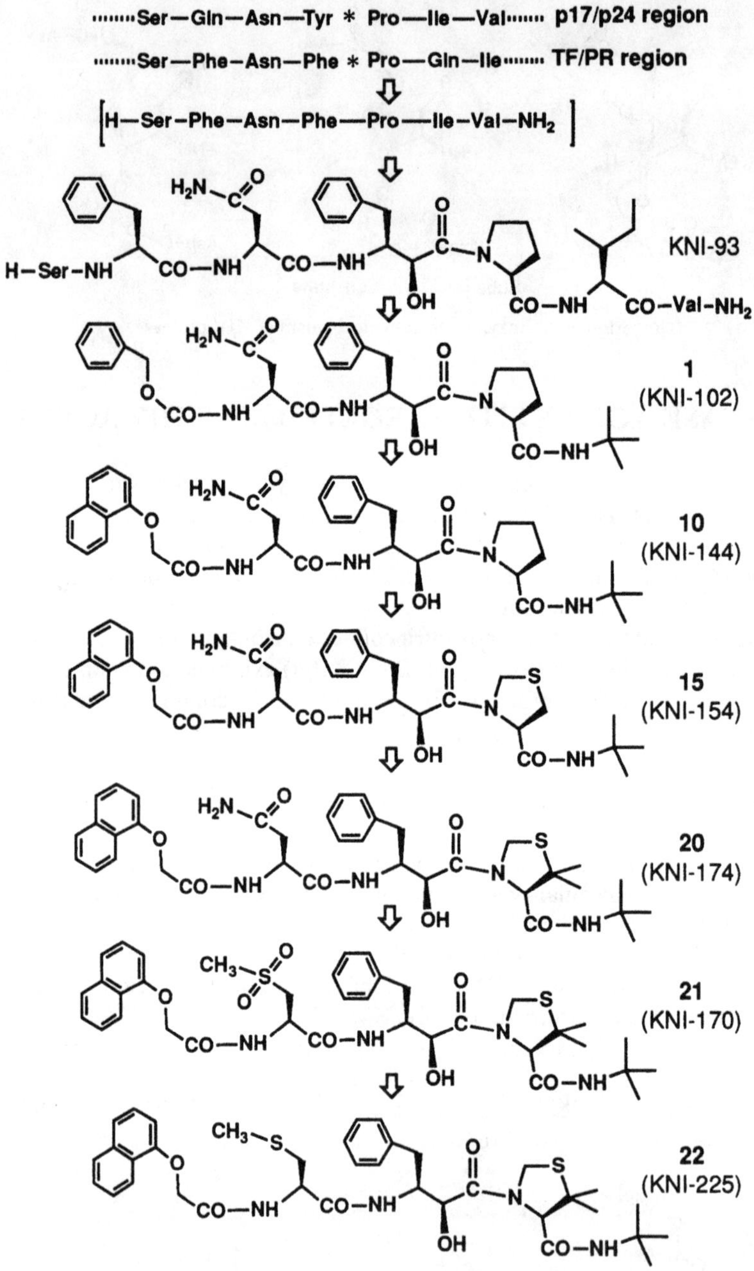

Figure 4. Design of selective and potent HIV protease inhibitors.

to be due to the conformational difference between the constrained peptide bond and the relatively flexible methylene-amine bond. A preference for the *syn* hydroxyl group shown in this series of HMC-Pro inhibitors is exceptional among various inhibitors of aspartic proteases such as HIV protease, renin and pepsin, which implies the uniqueness of the HMC-Pro structure.

Table 1. Protease Inhibitory Activities of HMC Compounds (IC_{50}, nM)

No	P4	P3	P2	P1	P1'	P2'	P3'	HIV Protease	Pepsin
1A (KNI-122)	Ser-	Phe-	Asn-	Pns-	Pro-	Ile-	Val-NH_2	100	>10,000
1S (KNI-93)	Ser-	Phe-	Asn-	Apns-	Pro-	Ile-	Val-NH_2	5	>10,000
2A		Pp-	Asn-	Pns-	Pro-	Ile-	Val-NH_2	3,000	>10,000
2S (KNI-81)		Pp-	Asn-	Apns-	Pro-	Ile-	Val-NH_2	468	>10,000
3A		Pp-	Ser-	Pns-	Pro-	Ile-	Val-NH_2	>10,000	N.D.
3S		Pp-	Ser-	Apns-	Pro-	Ile-	Val-NH_2	1,594	N.D.
4A		Pa-	Ser-	Pns-	Pro-	Ile-	Val-NH_2	>10,000	N.D.
4S		Pa-	Ser-	Apns-	Pro-	Ile-	Val-NH_2	5,041	N.D.
5A		Pp-	Asn-	Chns-	Pro-	Ile-	Val-NH_2	>10,000	N.D.
5S		Pp-	Asn-	Achns-	Pro-	Ile-	Val-NH_2	1,999	N.D.
6A		Z-	Asn-	Pns-	Pro-NHBu^t			>10,000	N.D.
6S (KNI-102)		Z-	Asn-	Apns-	Pro-NHBu^t			89	>100,000
7A		Val-	Val-	Pns-	Phe-	Val-	Val-NH_2	350	4,000
7S		Val-	Val-	Apns-	Phe-	Val-	Val-NH_2	2,600	>10,000

Pp = 3-phenylpropyl
Pa = phenylacetyl
Achns = allocyclohexylnorstatine = (2*S*,3*S*)-3-amino-4-cyclohexyl-2-hydroxybutyric acid
N.D. = not determined.

As shown in Table 1, HMC-Pro inhibitors of HIV protease did not practically inhibit pepsin but the symmetric-type compound containing P1' Phe (**7A**) inhibited pepsin, which showed that Apns-Pro inhibitors of HIV protease were selective. Especially, the tripeptide containing Apns-Pro (**6S**; KNI-102) was a potent and highly selective HIV protease inhibitor, causing no inhibition of pepsin even at a concentration of 80μM [6].

KNI-102

None of four peptide bonds in KNI-102 are natural type, and KNI-102 exhibits substantial anti-HIV activity whereas long-chain peptides have little antiviral activity, which implies that the tripeptide is relatively stable in the cell.

4. LEAD OPTIMIZATION

KNI-174

Having identified the tripeptide derivative KNI-102 as a lead compound, we undertook a study of lead optimization to find a highly selective and potent HIV protease inhibitor, KNI-174 (Noa-Asn-Apns-Dmt-NHBu^t; **20**), with anti-HIV activity [9]. Further structure-activity relationship study considering the penetration across cell membrane and the behaviour *in vivo* resulted in the generation of highly potent protease active site-targeted anti-HIV agents.

Conformationally Constrained Peptides

For the lead optimization of KNI-102, we determined the structural requirements for potent activity at each subsite. The most interesting results were obtained at the P1' residue

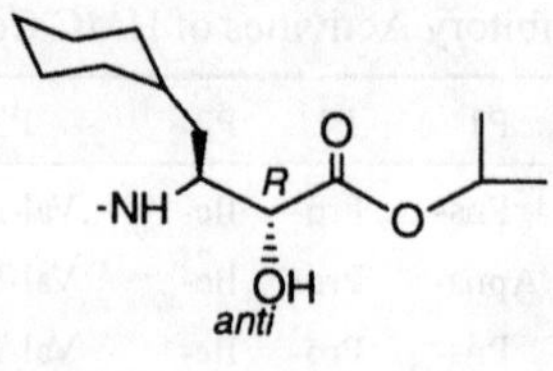

Figure 5. P1-P1' Chns isopropyl ester in renin inhibitors. Chns = cyclohexylnorstatine = (2*R*,3*S*)-3-amino-4-cyclohexyl-2-hydroxybutyric acid.

(Table 2). The pyrrolidine ring of proline was more suitable than the expanded piperidine ring of pipecolinic acid (compound **5**), in contrast to the case of hydroxyethylamine (HEA) type inhibitor [7]. Molecular modeling techniques indicated that the hydroxyl group of HMC isostere played a very important role for interaction with the protease active site (Fig. 7) [10]. The expanded piperidine ring seems to interfere with the interaction of the hydroxyl group of HMC isostere and the protease, which implies a conformational restriction in the HMC isostere-containing peptides, as discussed above. The replacement of the pyrrolidine by the thiazolidine (KNI-125; **7**), dimethylpyrrolidine (compound **8**), or dimethylthiazolidine (KNI-162; **9**) enhanced the protease inhibitory activity.

Especially, the conformationally constrained Apns-Dmt (dimethylthioproline) structure enhanced the HIV protease inhibitory activity. Our preliminary NMR, X-ray crystallography and molecular modeling studies showed that the *trans* conformation of the HMC-peptide bond at the P1-P1' site of KNI compounds was predominant and fitted favorably to HIV-1 protease. Thus, the constrained conformation of these HMC-containing peptides might be responsible for the high activity.

The naphthyl group (KNI-144; **10**) of P_3 residue fitted favorably in the large hydrophobic S_3 subsite, and the aryloxyacetyl type groups were preferred (compounds **10** *vs.* **11**). At the P_2 position, Asn residue could be replaced by L-methanesulfonylalanine (Msa) residue [KNI-151 (**16**) and KNI-170 (**21**)] or L-methylthioalanine (Mta) residue [KNI-217 (**17**) and KNI-225 (**22**)], with a slight improvement in activity, in contrast to the case of HEA type inhibitor [7].

Combinations of each preferred side chain led to highly selective and potent HIV protease inhibitors (Table 2), such as KNI-174 (**20**) (IC_{50} = 2.8 nM, K_i = 0.0068 nM), KNI-170 (**21**) (IC_{50} = 2.6 nM) and KNI-225 (**22**) (IC_{50} = 2.7 nM), with little inhibition of other aspartic proteases such as porcine pepsin (IC_{50} > 10,000 nM for each inhibitor) and human plasma renin (IC_{50} > 100,000 nM for each inhibitor). These compounds exhibited potent antiviral activities against HIV-1 in $CD4^+$ ATH8 cells (Table 3) [11].

KNI-272

The behaviors of compounds *in vivo*, such as penetration across the cell membrane and non-specific adsorption in blood, are important factors for the *in vivo* antiviral activity.

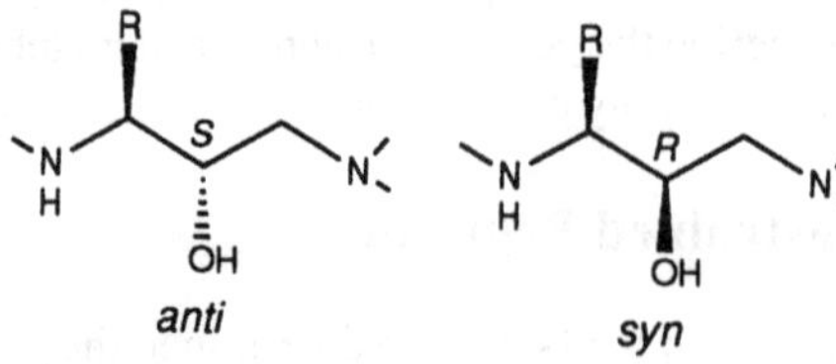

Figure 6. Hydroxyethylamine isostere.

Table 2. HIV-1 Protease Inhibition of Tripeptides[a)]

Number	P_3	P_2	P_1	P_1'	P_2'	IC_{50}(nM)[b)]
1(KNI-102)	Z-	Asn-	Apns-	Pro-	$NHBu^t$	89
2	Z-	Asn-	Apns-	Pro-	OBu^t	868
3	Z-	Asn-	Apns-	Pro-	$NHPr^i$	320
4	Z-	Asn-	Apns-	Pro-	NHCh	572
5	Z-	Asn-	Apns-	Pip-	$NHBu^t$	450
6	Z-	Asn-	Apns-	Tic-	$NHBu^t$	>1,000
7(KNI-125)	Z-	Asn-	Apns-	Thz-	$NHBu^t$	31
8	Z-	Asn-	Apns-	Dmp-	$NHBu^t$	24
9(KNI-162)	Z-	Asn-	Apns-	Dmt-	$NHBu^t$	3.5
10(KNI-144)	Noa-	Asn-	Apns-	Pro-	$NHBu^t$	12
11	Nmoc-	Asn-	Apns-	Pro-	$NHBu^t$	24
12	Fmoc-	Asn-	Apns-	Pro-	$NHBu^t$	45
13	Dcoa-	Asn-	Apns-	Pro-	$NHBu^t$	28
14(KNI-153)	Qc-	Asn-	Apns-	Pro-	$NHBu^t$	20
15(KNI-154)	Noa-	Asn-	Apns-	Thz-	$NHBu^t$	8.8
16(KNI-151)	Noa-	Msa-	Apns-	Thz-	$NHBu^t$	4.0
17(KNI-217)	Noa-	Mta-	Apns-	Thz-	$NHBu^t$	3.2
18(KNI-273)	iQoa-	Msa-	Apns-	Thz-	$NHBu^t$	7.2
19(KNI-272)	iQoa-	Mta-	Apns-	Thz-	$NHBu^t$	6.5
20(KNI-174)	Noa-	Asn-	Apns-	Dmt-	$NHBu^t$	2.8
21(KNI-170)	Noa-	Msa-	Apns-	Dmt-	$NHBu^t$	2.6
22(KNI-225)	Noa-	Mta-	Apns-	Dmt-	$NHBu^t$	2.7
23(KNI-208)	*m*Bpoa-	Asn-	Apns-	Dmt-	$NHBu^t$	2.2
24(KNI-226)	*m*Bpoa-	Mta-	Apns-	Dmt-	$NHBu^t$	2.3
25(KNI-227)	iQoa-	Mta-	Apns-	Dmt-	$NHBu^t$	2.3

[a)]These tripeptide derivatives were synthesized by essentially the same procedure as described previously [6] (for example, see Chart 1). [b)]Protease inhibitory activity was determined using a synthetic [$Ala^{67,95}$]-HIV-1 protease.
Abbreviations: Z = benzyloxycarbonyl, Apns = (2*S*,3*S*)-3-amino-2-hydroxy-4-phenylbutyric acid, Bu^t = t-butyl, Pr^i = isopropyl, Ch =cyclohexyl, Pip = L-pipecolinic acid, Tic = L-tetrahydroisoquinolinecarboxylic acid, Thz = L-thiazolidine-4-carboxylic acid, Dmp = L-3,3-dimethylpyroridine-2-carboxylic acid, Dmt = L-5,5-dimethylthiazolidine-4-carboxylic acid, Noa = 1-naphthoxyacetyl, Nmoc = 1-naphthylmethyloxycarbonyl, Fmoc = 9-fluorenylmethoxycarbonyl, Dcoa = 4,4'-dichlorobenzhydryloxyacetyl, Qc = quinolin-2-ylcarbonyl, Msa = L-methanesulfonylalanine, Mta = L-methylthioalanine, iQoa = 5-isoquinolinyloxyacetyl, *m*Bpoa = *m*-biphenyloxyacetyl.

Therefore, considering the subtle balance of lipophilicity-hydrophilicity and molecular size, we incorporated the 5-isoquinolinyloxyacetyl (iQoa) moiety at the P_3 position and combined it with each preferred side chain. Such modifications resulted in the superactive HIV protease inhibitors, kynostatin (KNI)-227 (iQoa-Mta-Apns-Dmt-$NHBu^t$: **25**) (K_i = 0.0023 nM) and kynostatin (KNI)-272 (iQoa-Mta-Apns-Thz-$NHBu^t$: **19**) (K_i = 0.0055 nM) (Fig. 8).

Both compounds exhibited potent antiviral activities against clinical HIV-1 isolates in phytohemagglutinin-stimulated peripheral blood mononuclear cells assay (IC_{50} = 0.01 μM for both compounds) (Table 4), which might correspond relatively well to the *in vivo* antiviral

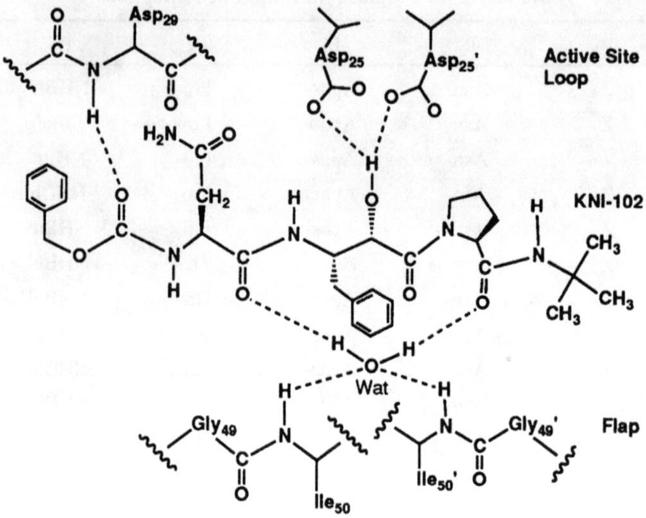

Figure 7. KNI-102 in the active site of HIV protease.

activity. These antiviral activities against clinical HIV-1 isolates were more than 10-fold potent compared to a C2 symmetric protease inhibitor, A-77003 [12]. As expected from the protease active site-targeted antiviral mechanism, these also exhibited potent antiviral activities (IC_{50} = 0.01 μM for both compounds) against AZT-insensitive clinical HIV-1 isolates [11].

Also, KNI-227 (**25**) and KNI-272 (**19**) showed the potent antiviral activities against the infectivity and cytopathic effect of HIV strains, including HIV-1_{LAI}, HIV-1_{RF}, HIV-1_{MN}

Table 3. HIV-1 Protease Inhibition and Antiviral Activities against HIV-1_{LAI} in ATH8 Cells

Compound	Structure P3 Phe	P2 -Asn	P1 -Phe	P1' -Pro	P2' -Ile	IC_{50}[a] (nM)	IC_{50}[b] (μM)	TC_{50}[c] (μM)	TC_{50}/IC_{50}
1(KNI-102)	Z-	Asn-	Apns-	Pro-	NHBu^t	89	1.1	>20	>18
14(KNI-153)	Qc-	Asn-	Apns-	Pro-	NHBu^t	20	1.0	>20	>20
10(KNI-144)	Noa-	Asn-	Apns-	Pro-	NHBu^t	12	0.9	>20	>22
15(KNI-154)	Noa-	Asn-	Apns-	Thz-	NHBu^t	8.8	0.5	>50	>100
20(KNI-174)	Noa-	Asn-	Apns-	Dmt-	NHBu^t	2.8	0.4	30	75
21(KNI-170)	Noa-	Msa-	Apns-	Dmt-	NHBu^t	2.6	0.1	16	160
22(KNI-225)	Noa-	Mta-	Apns-	Dmt-	NHBu^t	2.7	0.2	7	35
25(KNI-227)	iQoa-	Mta-	Apns-	Dmt-	NHBu^t	2.3	0.1	40	400
19(KNI-272)	iQoa-	Mta-	Apns-	Thz-	NHBu^t	6.5	0.1	>50	>500

[a]HIV-1 protease inhibitory activity determined using a synthetic [$Ala^{67,95}$]-HIV-1 protease.
[b]50 percent inhibitory concentration against cytopathic effect of HIV-1.
[c]50 percent toxic concentration of the compound.

Table 4. Antiviral Activity of KNI Compounds against AZT-Sensitive and -Insensitive Clinical HIV-1 Isolates in Target PHA-PBM

Compound	Structure P3 P2 P1 P1' P2' Phe-Asn-Phe- Pro- Ile	TC_{50} /IC_{50}	TC_{50} (μM)[a]	AZT-sensitive strain IC_{50} (μM)[b]	AZT-sensitive strain IC_{90} (μM)	AZT-insensitive strain IC_{50}(μM)[b]
15 (KNI-154)	Noa-Asn-Apns-Thz-NHBu^t	550	77	0.14	0.24	0.16
20 (KNI-174)	Noa-Asn-Apns-Dmt-NHBu^t	246	32	0.13	0.23	0.17
21 (KNI-170)	Noa-Msa-Apns-Dmt-NHBu^t	635	19	0.03	0.05	0.02
22 (KNI-225)	Noa-Mta-Apns-Dmt-NHBu^t	267	8	0.03	0.06	0.03
25 (**KNI-227**)	iQoa-Mta-Apns-Dmt-NHBu^t	4900	49	0.01	0.04	0.01
19 (**KNI-272**)	iQoa-Mta-Apns-Thz-NHBu^t	>8000	>80	0.01	0.04	0.01

[a] TC_{50} values were assessed by [^{3}H]-thymidine incorporation assay.
[b] IC_{50} values were determined based on levels of p24 gag protein production.

and HIV-2_{ROD}, as tested in $CD4^+$ ATH8 cells. The 50% inhibitory concentrations (IC_{50}) of KNI-227 (**25**) against these strains were 0.1, 0.02, 0.03 and 0.13 μM, while those of KNI-272 (**19**) were 0.1, 0.02, 0.04 and 0.14 μM, respectively. From the viewpoint of the action mechanism, the active site-targeted HIV protease inhibitors have reason to exhibit activities against a wide spectrum of HIV strains, including HIV-2.

In terms of the ratio of *in vitro* enzyme inhibition to antiviral activity in cell culture, KNI-272 (**19**) containing thioproline (Thz) seems to have superior cell membrane penetration characteristics to Dmt-containing inhibitors, KNI-174 (**20**), -170 (**21**), -225 (**22**) and -227 (**25**). This can be explained by change of balance of hydrophilicity-hydrophobicity and size reduction due to the removal of the bulky and lipophilic dimethyl group.

Enzyme Selectivity

KNI-272 (**19**) exhibited excellent enzyme selectivity, with practically no inhibition of other aspartic proteases such as human plasma renin (IC_{50} > 100,000nM), porcine pepsin (IC_{50}>80,000nM) and bovine cathepsin D (IC_{50}>100,000nM). On the other hand, IC_{50} values

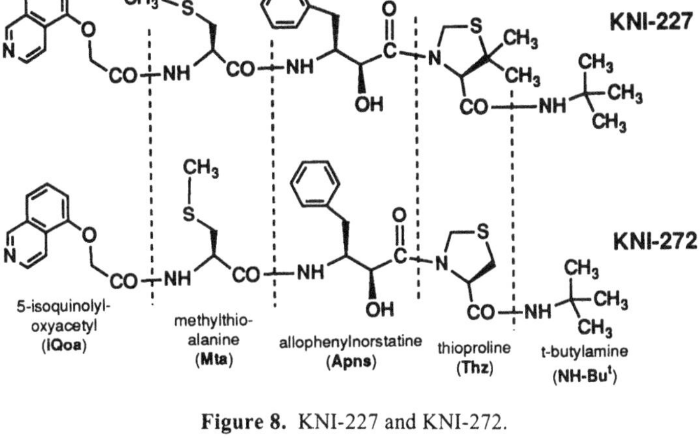

Figure 8. KNI-227 and KNI-272.

of KNI-227 (**25**) were >100,000 nM (renin), 13,000nM (pepsin) and 23,000nM (cathepsin D).

Although the P1' Dmt residue remarkably enhanced the protease inhibitory activity due to the conformational constraint and hydrophobicity of the bulky dimethyl group, enzyme selectivity of KNI-227 (**25**) was a little lower than KNI-272 (**19**) with P1' Thz. This may be explained by that the Dmt residue is regarded as a hybrid-type of Thz and Val, and that Apns-Thz is a closer mimic to the virus specific Phe-Pro sequence than Apns-Dmt.

Stability

When KNI-227 (**25**) and KNI-272 (**19**) were incubated in a hydrochloric acid solution (pH 1) at 37°C, both compounds were found to be stable and no degradation was detected even after 24 h. This was also true when both compounds were incubated in the presence of pepsin A (1 mg/ml) in 0.2 M citrate solution (pH 2.2) at 37°C for 24 h. The control synthetic nonapeptide was completely digested under identical conditions.

Cytotoxicity and Bioavailability

Interestingly, a relatively low-lipophilic and small-sized tripeptide derivative, KNI-272 (**19**), combined with iQoa moiety and Thz residue, exhibited highly potent antiviral activities and low cytotoxicity ($TC_{50} > 80\mu M$). After intraduodenal (i.d.) administration in rats [13,14], bioavailabilities of KNI-174 (**20**), KNI-227 (**25**) and KNI-272 (**19**) were 5.37%, 5.90% and 42.3% respectively. As compared with other potent HIV-protease inhibitors, the bioavailability of KNI-272 (**19**) was improved.

The superiority of cell membrane penetration of KNI-272 (**19**) may be due to appropriate balance of lipophilicity-hydrophilicity and size reduction owing to the removal of lipophilic and bulky dimethyl group from Dmt moiety.

Evaluation and Synthesis of KNI-272

Many protease-targeted anti-HIV agents have been so far reported. Among them, KNI-272 is one of the most potent, selective and compact inhibitors. And KNI-272 is a closer peptide derivative to the virus specific substrate (Phe-Pro) transition state than other potent anti-HIV agents, and therefore seems to satisfy the following requirements for anti-AIDS drugs: 1, *in vitro* enzyme inhibition (potency, mechanism, selectivity and stability); 2, antiviral activity in cells (potency, mechanism, cytotoxicity, stability and permeability); 3, *in vivo* animal tests (toxicity, stability and oral bioavailability).

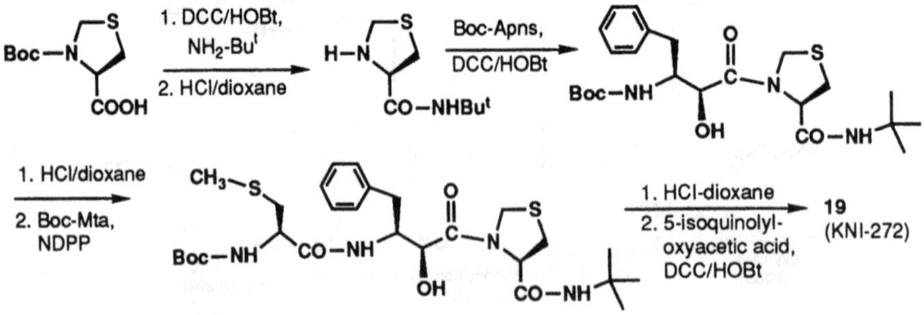

Chart 1. Synthetic scheme of KNI-272 (**19**). Abbreviations: Boc=t-butoxycarbonyl, DCC = dicyclohexylcarbodiimide, HOBt = N-hydroxybenzotriazole, NDPP = norborn-5-ene-2,3-dicarboximido diphenyl phosphate.

It is also noteworthy that the tripeptide derivative KNI-272 was conveniently synthesized by the solution method in a stepwise manner (Chart 1). Pure compound was readily obtained in a high yield and at relatively low cost. This potential advantage is mainly due to the simple synthetic procedure using only materials and reagents which are easily available.

5. CONCLUSION

Ease in the synthetic procedure, the HIV protease-specific inhibition and potent antiviral properties as well as favorable cytotoxicity profile warrant further studies in the direction for clinical application of KNI-272 as an oral anti-HIV drug.

REFERENCES

1. T. Mimoto, J. Imai, S. Tanaka, N. Hattori, O. Takahashi, S. Kisanuki, Y. Nagano, M. Shintani, H. Hayashi, H. Sakikawa, K. Akaji, and Y. Kiso, Chem. Pharm. Bull., 39: 2465 (1991).
2. A. Wlodawer, M. Miller, M. Jaskolski, B. K. Sathyanarayana, E. Baldwin, I. T. Weber, L. M. Selk, L. Clawson, J. Schneider, and S. B. H. Kent, Science, 245: 616 (1989)
3. Y. Kiso, Y. Fujiwara, T. Kimura, A. Nishitani, and K. Akaji, Int. J. Peptide & Protein Res., 40: 308 (1992).
4. M. L. Moore, W. M. Bryan, S. A. Fakhoury, V. W. Magaard, W. F. Huffman, B. D. Dayton, T. D. Meek, L. Hyland, G. B. Dreyer, B. W. Metcalf, J. E. Strickler, J. G. Gorniak, and C. Debouck, Biochem. Biophys. Res. Commun., 159: 420 (1989).
5. K. Iizuka, T. Kamijo, T., H. Harada, K. Akahane, T. Kubota, H. Umeyama, T. Ishida, and Y. Kiso, J. Med. Chem., 33: 2707 (1990).
6. T. Mimoto, J. Imai, S. Tanaka, N. Hattori, S. Kisanuki, K. Akaji, and Y. Kiso, Chem. Pharm. Bull., 39: 3088 (1991).
7. N. A. Roberts, J. A. Martin, D. Kinchington, A. V. Broadhurst, J. C. Craig, I. B. Duncan, S. A. Galpin, B. K. Handa, J. Kay, A. Krohn, R. W. Lambert, J. H. Merrett, J. S. Mills, K. E. B. Parkes, S. Redshaw, A. J. Ritchie, D. L. Taylor, G. J. Thomas, and P. J. Machin, Science, 248: 358 (1990).
8. D. H. Rich, C. -Q. Sun, J. V. N. V. Prasad, A. Pathiasseril, M. V. Toth, G. R. Marshall, M. Clare, R. A. Mueller, and K. Houseman, J. Med. Chem., 34: 1222 (1991).
9. T. Mimoto, J. Imai, S. Kisanuki, S. Tanaka, N. Hattori, O. Takahashi, R. Kato, T. Yumisaki, H. Sakikawa, K. Akaji, and Y. Kiso, In: Peptide Chemistry 1991, edited by A. Suzuki, pp. 395-400. Osaka: Protein Research Foundation (1992).
10. T. Mimoto, J. Imai, S. Kisanuki, H. Enomoto, N. Hattori, K. Akaji, and Y. Kiso, Chem. Pharm. Bull., 40: 2251 (1992).
11. S. Kageyama, T. Mimoto, Y. Murakawa, M. Nomizu, H. Ford, T. Shirasaka, S. Gulnik, J. Erickson, K. Takada, H. Hayashi, S. Broder, Y. Kiso, and H. Mitsuya, Antimicrob. Agents Chemother., 37: 810 (1993).
12. H. Mitsuya, R. Yarchoan, S. Kageyama, and S. Broder, FASEB J., 5: 2369 (1991); S. Kageyama, J. N. Weinstein, T. Shirasaka, D. J. Kempf, D.W. Norbeck, J. J. Plattner, J. Erickson, and H. Mitsuya, Antimicrob. Agents Chemother., 36: 926 (1992).
13. A. Kiriyama, T. Mimoto, Y. Kiso, and K. Takada, Biopharm. Drug. Disp., 14: 199 (1993).
14. A. Kiriyama, T. Mimoto, S. Kisanuki, Y. Kiso, and K. Takada, Biopharm. Drug. Disp., 14: 697 (1993).

BREAKING THE SHACKLES OF THE GENETIC CODE: ENGINEERING RETROVIRAL PROTEASES THROUGH TOTAL CHEMICAL SYNTHESIS

Stephen B.H. Kent, Manuel Baca, John Elder, Maria Miller,[1]
Raymond Milton, Saskia Milton, J. K. M. Rao,[1] and Martina Schnölzer

The Scripps Research Institute, La Jolla, California 92037
[1]NCI-Frederick, Maryland

INTRODUCTION

Since early this century one of the fundamental goals of organic chemistry has been to understand the molecular basis of enzyme action (1). These protein molecules are the machines of the cell, catalyzing essentially all of the chemical reactions in the biological world. The action of enzymes is characterized by extraordinary specificity and by the ability to effect, in aqueous solvent and at normal temperatures, chemical transformations that are otherwise imperceptibly slow. Early approaches to the study of enzymes involved the laborious and painstaking isolation of pure molecules from biological sources, such as tissue homogenates. Only indirect methods, such as selective chemical modification of amino acid side chains, could be used to study the functional imperatives of the protein molecule itself. Consequently, much more was known about the (usually low MW) substrates than about enzymes themselves.

In recent years, the techniques of recombinant DNA molecular biology have added an important new dimension to the study of enzymes. For the first time it has been possible to produce large amounts of high purity enzyme almost at will, for any enzyme. On top of that, it has been possible to routinely produce variant enzymes by techniques such as site-directed mutagenesis. This has made it possible to systematically vary the amino acid sequence of an enzyme molecule in a controlled fashion, and evaluate the effects on catalysis and specificity. Combined with structural techniques such as X-ray crystallography and multidimensional NMR, this has led to a dramatic increase in our knowledge of enzymes (2).

Despite this, we have only a descriptive understanding of even the simplest enzymes. What is lacking is an understanding of the quantitative contribution to enzyme action of individual functional groups, in the context of the enzyme·substrate complex. In this respect, recombinant DNA-based protein engineering has proved to be of limited utility, although extensions of the method now allow the incorporation of a wider range of functionalities (3).

This situation has now changed. With the recent advent of powerful new protein chemistry technologies, the tools are available for a definitive exploration of the molecular basis of enzyme action. These tools include new analytical methods such as the exact measurement of protein molecular weight by mass spectrometry (4), the ability to study native-like protein complexes by mass spectrometry (5), and the rapid protein ladder sequencing method (6) which allows routine determination of post-translational and other modifications to protein structure. But perhaps most important is a new synthetic methodology (7) which for the first time allows the routine total chemical synthesis of enzymes in reproducible fashion and with high purity. In combination, these novel analytical and synthetic methods give complete control over every aspect of the covalent structure of the enzyme molecule. Each atom of both backbone and side chain functionalities can be varied at will, and utterly non-standard elements (such as fixed elements of secondary structure) can be incorporated in any region of the protein molecule. This integrated analytical and synthetic chemistry approach opens the science of enzymes to the whole universe of organic chemistry, and for the first time breaks the shackles of the genetic code.

In this paper, the power of these new protein chemistry technologies is illustrated by their application to the study of key aspects of the action of the aspartyl proteinases from the type 1 human immunodeficiency virus (HIV-1 PR) and from the feline immunodeficiency virus (FIV PR). Four case studies are described; 1. the elucidation of the covalent structure of the Feline Immunodeficiency Virus protease (sic), its preparation by total chemical synthesis, and the elucidation of its action in the specific processing of the Gag and Gag-Pol viral polyproteins; 2. the total chemical synthesis of HIV-1 protease in large amount and high purity by means of "chemical dovetailing", the chemical ligation of large unprotected peptide segments; 3. the elucidation of the high resolution X-ray crystal structure of the mirror image, ligated enzyme D-HIV-1 protease; 4. and, the backbone engineering of L-HIV-1 PR by chemical dovetailing, in order to quantify the contribution of flap-substrate hydrogen bonding to catalysis.

CASE STUDIES

1. Covalent Structure and Processing Activity of the FIV Protease

Domestic cats have been shown to be susceptible to infection by a lentivirus which is closely related to the human immunodeficiency virus. Like HIV-1, the feline immunodeficiency virus (FIV) causes a severe, complex disease state leading to a full-blown AIDS-like syndrome and eventual death. This FIV-induced disease is the best animal model for AIDS. For this reason it is important to understand the life cycle of FIV in relation to the progression of the disease state. To this end the molecular genetics of FIV have been determined in some detail (8) including the complete sequence of the RNA genome.

FIV has a genetic organization that is very similar to that of HIV. In particular, there is an aspartyl proteinase homology unit at the 5' end of the POL open reading frame. We used on line gradient reverse phase HPLC-electrospray mass spectrometry to examine the total proteins of FIV disrupted in 6M guanidine·HCl. Most of the processed Gag proteins and two members of the enzyme cassette from processing of the Pol translation products were identified (9). For the aspartyl proteinase homology unit the MW was determined to be 13,232±2 dalton. In combination with N-terminal sequence analysis by gas phase Edman degradation of the same protein electroblotted from SDS-PAGE, and by alignment with the cDNA-derived amino acid sequence, this exact mass defined the covalent structure of the FIV PR monomer (9). The deduced 116 amino acid residue polypeptide chain was prepared by total chemical synthesis using highly optimized stepwise Boc chemistry solid phase methods (10). The approach used is shown in Figure 1(A). The synthetic 116 amino acid polypeptide chain was purified by preparative reverse phase HPLC, and had a measured MW of 13, 234±2 dalton by electrospray mass spectrometry.

A crude preparation of the folded synthetic protein was used to evaluate the activity of the enzyme on a set of synthetic peptide substrates corresponding to regions overlapping

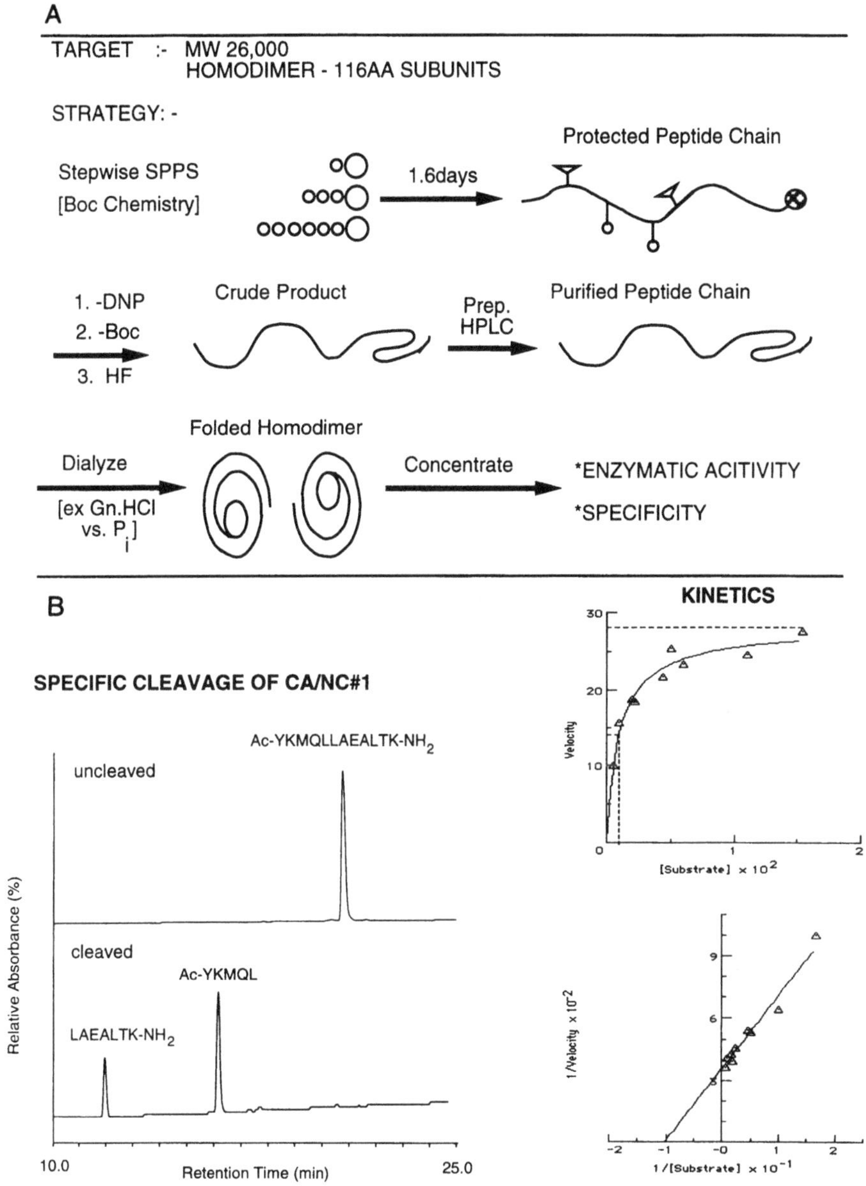

Figure 1. Total chemical synthesis and enzymatic activity of FIV protease. **A.** Schematic of the synthesis. Highly optimized stepwise SPPS was used to give a product of the correct MW. **B.** The partially purified product was folded and assayed for its ability to cleave a synthetic peptide substrate corresponding to the expected processing site between the matrix and capsid proteins. Specific, progressive cleavage was observed by HPLC. Kinetic studies carried out with this HPLC assay showed saturation behavior typical of enzyme catalyzed reactions.

the expected proteolytic processing sites within the FIV Gag and Gag-Pol translation products. The results are illustrated by the data shown in Figure 1(B), using a synthetic peptide corresponding to the first capsid/nucleocapsid cleavage site. Specific cleavage was observed, and the activity showed substrate saturation kinetics typical of an enzyme catalyzed reaction. Similar results were obtained for all eight processing sites. Together with N-terminal sequence analysis of viral proteins and accurate MW data from mass spectrometry, the data obtained from the action of FIV PR prepared by total chemical synthesis on synthetic peptide substrates defined all the Gag-Pol processing events observed in the maturation of the virus (9).

The use of electrospray mass spectrometry and total chemical synthesis has thus defined the covalent structure of the FIV PR, and has defined the activities of the enzyme in virus maturation.

2. Chemical Dovetailing - a General Approach to the Total Chemical Synthesis of Enzymes

Since Emil Fischer articulated the goal more than 80 years ago (1), attempts have been made to develop an approach to the total chemical synthesis of enzymes using the methods of organic chemistry. Despite heroic efforts by a number of research groups and several partial successes*, to this point a routine, reproducible synthesis of enzymes has eluded the organic chemistry research community. The problem has been twofold. In the classical approach to the synthesis of proteins, large peptide segments are condensed in organic solvents by chemical means in a convergent synthetic strategy. In order to prevent side reactions, all side chain functional groups are reversibly masked by protecting groups. Despite being one of the principal focuses of organic chemistry in the decades 1950-1970, this approach proved inadequate. This is because of racemization problems in the condensation of activated peptide segments, and particularly because of the disabling low solubility of fully protected peptide segments which inevitably led to slow and incomplete condensation reactions. Additionally, no good methods exist for purifying and characterizing the large protected peptide segments.

The alternative approach to the chemical synthesis of peptides is stepwise solid phase peptide synthesis (SPPS). SPPS was originally developed by Merrifield with the goal of studying proteins by chemical synthesis (16). This method proved to be a very powerful general approach to the chemical synthesis of peptides, and has revolutionized the use of peptides in biological research. Both the racemization problem and the poor solubility of protected peptides were overcome by SPPS. However, the inability to purify the protected peptides remained because of the resin-bound nature of the intermediate products. This demanded essentially quantitative yields in the stepwise assembly of a target peptide chain, in order to avoid intractable complex mixtures when the peptide products are finally released from the resin support. It is a tribute to the power of the SPPS method that overall yields averaging 99% per amino acid are routinely obtained, allowing the straightforward synthesis of peptides in the 30+ amino acid residue range.

---

* Most notably, highly optimized SPPS was used in our own laboratory to prepare the HIV-1 PR and define its enzymatic properties (11). This enzyme was used to obtain the X-ray structure of the HIV-1 PR molecule (12) and of a series of three cocrystal structures of the enzyme with canonical inhibitors (13-15). These were the first three dimensional molecular structures of any component of the AIDS virus, and were obtained using enzyme prepared by total chemical synthesis. This was also the first crystal structure of a chemically synthesized enzyme, and the first crystal structure of an enzyme containing non-coded amino acids (the four Cys residues in each enzyme molecule were replaced by L-α-amino-n-butyric acid).

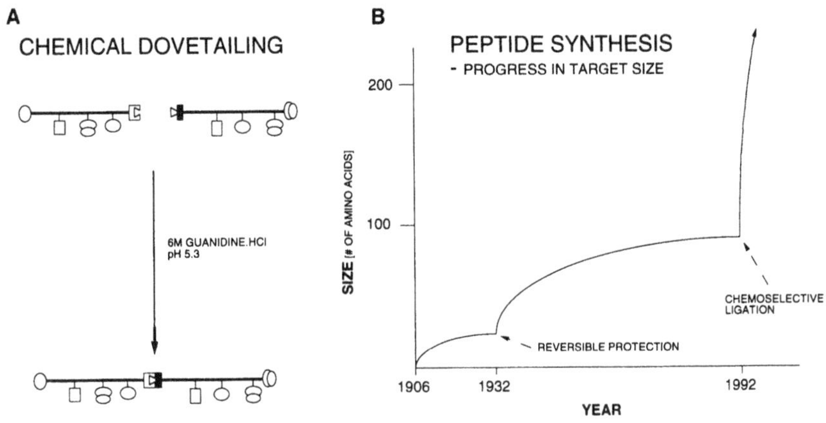

Figure 2. "Chemical Dovetailing", the chemical ligation of unprotected peptide segments by means of unique, mutually reactive functionalities. **A.** Principle of the method, schematically illustrated. Note that a modified, non-peptide bond backbone structure is formed at the ligation site. **B.** The use of chemical dovetailing is a paradigm shift, and removes constraints imposed by the peptide bond as an obligatory synthetic target. This has resulted in a vast increase in accessible size of synthetic targets, and has removed the constraints imposed by the genetic code.

Prolonged efforts, in our own laboratory among others, have led to the optimization of SPPS so that simple synthetic procedures routinely give good yields of peptides in the 50-80+ residue range (10). These unprotected peptide products can be purified in a straightforward fashion using preparative reversed phase HPLC, and can be characterized for purity and covalent structure by the powerful new analytical tool of electrospray mass spectrometry. An obvious approach (17) to the total chemical synthesis of proteins would be to take advantage of this ability to make large unprotected peptides in good yield and purity, if a suitable means could be found to hook them together. Attempts have been made to use enzymes to ligate unprotected peptides, or to use partial reprotection of side chains combined with chemical activation (18). These have had only limited success.

Recently, it occurred to us that there was a simple alternative means to turn our ability to make large unprotected peptides to the task of preparing proteins by total chemical synthesis. In principle, all the difficulties of covalently reacting two peptide segments are imposed by the frequent occurrence of side chain carboxyl and amino groups and the necessity of forming the joining peptide bond from similar functionalities in a specific fashion in the presence of these. A simple way of circumventing this problem would be to form a *non-peptide bond* at the site of reaction between the two peptides. This would allow the use of selective chemical reaction to join two peptide segments, while avoiding the necessity of protecting side chain or other functionalities. We have termed this "chemical dovetailing": the use of chemoselective reaction between unique, mutually reactive groups to covalently ligate unprotected peptides (7). The principle is illustrated in Figure 2A.

We initially applied the chemical dovetailing approach to the total chemical synthesis of the HIV-1 protease (7). Based on the crystal structure of the native backbone enzyme (13), prepared by stepwise SPPS, we designed a chemical synthesis strategy that used a ligation site near the middle of the 99 residue monomer polypeptide but which was not directly involved in the active site of the enzyme. Two unprotected peptides were synthesized: HIV-1 PR(1-51)αCOSH, using a resin developed by Blake and Yamashiro (18); and BrAcetyl(53-99)HIV-1 PR. After purification and characterization, these two fully *un*protected peptide segments were reacted under nucleophilic conditions in 6M guanidine·HCl at pH5.5 to give a near-quantitative yield of a 99 residue polypeptide ligation product. The ligation site was a thioester bond replacing the native peptide bond between residues Gly^{51}-Gly^{52}. Thus, the polypeptide backbone was changed at only one atom,

$$\text{-CO-}\underset{\text{H}}{\underset{|}{\text{N}}}\text{-} \rightarrow \text{-CO-S-}$$

The ligated 99 residue product was purified by reversed phase HPLC, and simple dilution into assay buffer or dialysis from 6M guanidine·HCl gave folded, fully active enzyme with catalytic properties indistinguishable from native-backbone HIV-1 PR (7).

The chemical ligation of large unprotected peptide segments by "chemical dovetailing" represents a true paradigm shift. As illustrated by the above example, chemical dovetailing makes the total chemical synthesis of enzymes a straightforward endeavor. For the first time, we have a general approach that allows the routine, reproducible total chemical synthesis of proteins and their analogues, regardless of size (Figure 2B). The chemical dovetailing principle is a general one. In our laboratory it has been used to synthesize a number of proteins, including: a 4-helix template assembled artificial protein (19); an artificial protein model for the cytoplasmic domains of multichain cell surface receptor molecules (Muir, et al., unpublished); a backbone engineered HIV-1 PR (20; see case Study #4, below); the tenth type 3 fibronectin domain and analogues (Muir, et al., unpublished); basic/helix-loop-helix/zipper transcription factors (Canne, et al., unpublished); and a variety of other synthetic targets.

3. Crystal Structure of a "Mirror Image" Enzyme, Ligated D-HIV-1 Protease

Since the time of Pasteur it has been realized that chirality plays an important role in the biological world. We now know that all proteins found in the biosphere of this planet are made up of amino acids of a single chirality, namely L-amino acids. It has been shown that the ribosomal biosynthetic machinery will not catalyze the incorporation of D-amino acids (3). Not only are the amino acids in ribosomally produced polypeptide chains themselves chiral, but the defined folded form of the polypeptide chain that constitutes a protein is also an inherently chiral structure at a number of levels. First, elements of secondary structure such as helices, antiparallel beta strands, and turns have an inherent handedness (21). Second, the overall relationships of elements within the three dimensional structure of a folded protein are also chiral: even if the polypeptide chain is represented as a featureless "wire" itself devoid of chirality, the *folded structure* found within a globular protein cannot be superimposed on its mirror image. Thus proteins, including enzymes, are chiral. As originally pointed out by Emil Fischer, this is the basis of one of the fundamental characteristics of enzyme action - enzymes are chirally specific. That is, in addition to their exquisite chemical specificity, enzymes invariably act on only one isomer of a stereochemical center in a substrate, and will even accurately discriminate prochiral centers.

Thanks to our ability to prepare proteins by total chemical synthesis, for the first time we are able to experimentally investigate the world of D-enzymes. Recently, we used total chemical synthesis to prepare the enzyme HIV-1 protease from D-amino acids (22). As expected, the all D-amino acid polypeptide chain folded to form an enzymatically active protein, the first mirror image enzyme molecule. This D-HIV-1 PR had the same chemical specificity as HIV-1 protease, but displayed reciprocal chiral specificity: i.e. the D-HIV-1 PR cleaved D-peptide substrates but would not act on L-peptide substrates, whereas the L-enzyme acted only on L-peptide substrates. These results were a vivid demonstration that the folded structure of a polypeptide chain arises solely from the linear sequence of amino acids and that the sequence determines all aspects of the active tertiary and quaternary structure of an enzyme. They definitively rule out the imposition of an external influence on protein folding by biological cofactors (which are not present in the chemical synthesis process, and which in any event would not act on D-amino acid polypeptide chains).

There are a number of interesting aspects of the world of D-proteins. However, perhaps of greatest potential significance is the use of D-proteins in X-ray crystallography in order to improve structure-based drug design. At the present time, the utility of structure-based drug design is limited by uncertainties in the three dimensional structural data obtained from even very high resolution X-ray crystallography of proteins. Much of this lack of precision arises from imperfect phases used in the generation of the electron density map from the original diffraction data. D-proteins can in principle be used to obtain structural data of substantially enhanced accuracy from X-ray diffraction studies. Racemic crystals in centrosymmetric space groups give rise to diffraction data in which all phases are exactly 0° or 180° (the off-diagonal elements cancel each other out) (23). Such crystals formed from the cocrystalization of D- and L- forms of a protein can be used to generate electron density maps of much greater precision (24), which could considerably enhance the utility of such data in drug design.

The HIV-1 protease is a major target of drug design efforts in pharmaceutical research worldwide, because of its critical role in the life cycle of the AIDS virus. It has been demonstrated that inhibition of HIV-1 PR prevents virus replication in tissue culture, and that the virus is much less able to generate viable escape mutants to overcome the effects of HIV-1 PR inhibitors. Using the original X-ray structure data, derived from enzyme prepared by total chemical synthesis (12-15), a number of HIV-1 PR inhibitors have been developed. Several of these have good oral bioavailability and pharmacokinetic properties, and are being taken into clinical trials as candidate AIDs therapeutics. The probable success of this approach to AIDS therapy will make it very important to have more accurate information about the molecular structure of HIV-1 PR, to assist in the systematic development of continuing generations of drugs aimed at this target molecule.

Therefore, we have set out to generate large amounts of D- and L-HIV-1 PR, by means of the chemical dovetailing approach. Using an optimized version of the original approach (7), we were able to generate high purity D- and L-HIV-1 PR in ~50mg amounts from a single synthesis of each (25). The synthesis is summarized in Figure 3A. As with the native backbone chemically synthesized mirror image enzymes, these ligated D- and L-enzymes showed reciprocal chiral specificity and identical kinetics. The UV circular dichroism spectra were also identical but inverted relative to one another.

The three dimensional molecular structure of the ligated D-HIV-1 PR was determined by X-ray crystallography (26). Cocrystals of the D-protease with a D-hexapeptide inhibitor were obtained in the space group $P2_12_12_1$ in a unit cell of dimensions 67.5, 92.8, 29.4Å and the present crystals diffract to a resolution of 2.2Å. The crystal structure has been solved by molecular replacement methods using data collected to 2.4Å. Comparison of the structure with that of the native-backbone synthetic L-amino acid enzyme (13) showed that the

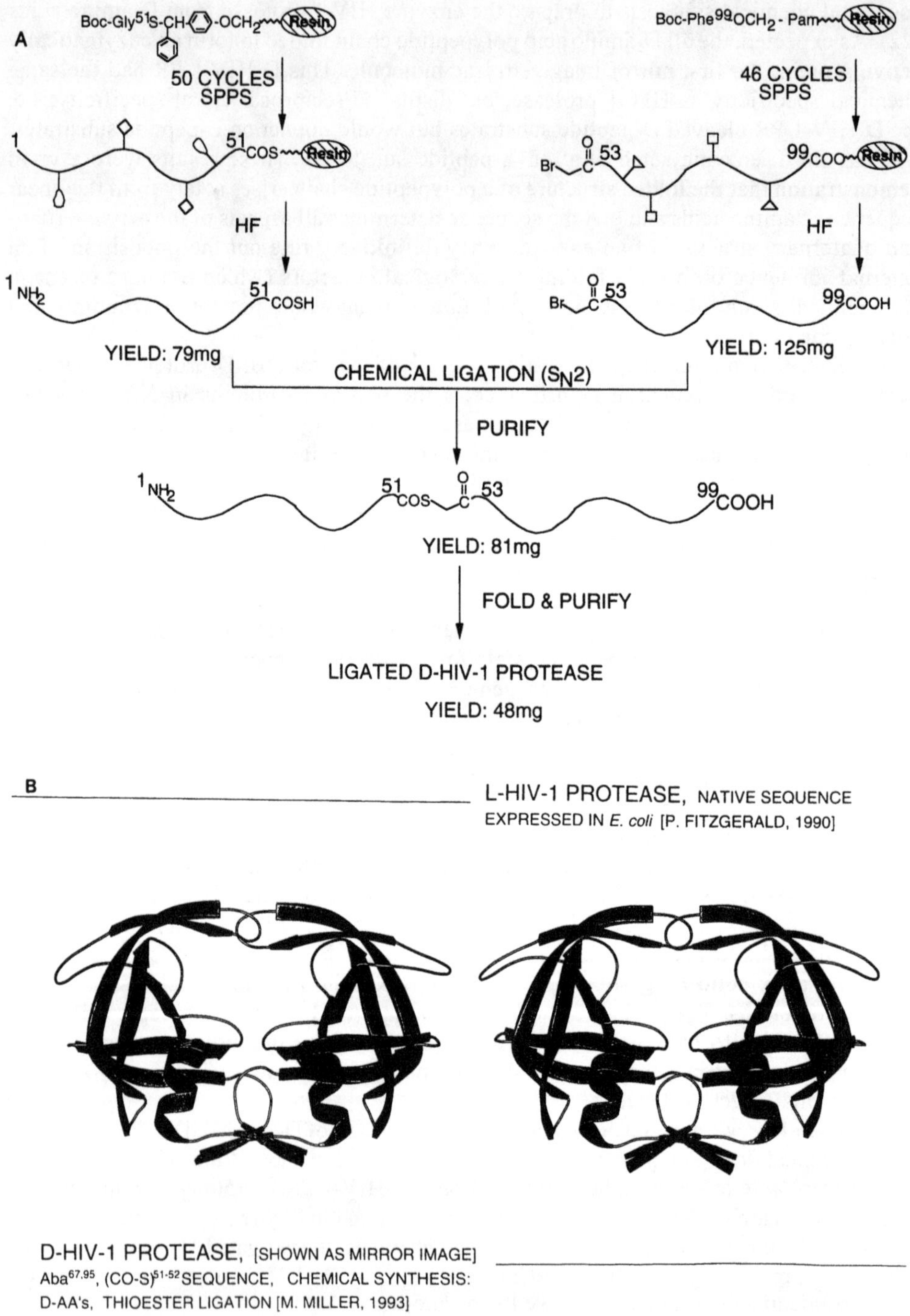

Figure 3. Total chemical synthesis and X-ray structure of ligated D-HIV-1 PR. **A.** Synthetic scheme. The D-enzyme was prepared by chemical dovetailing of the unprotected peptide segments, giving rise to a thioester bond at the ligation site (25). **B.** Crystal structures of ligated D-HIV-1 PR, displayed as the L-structure for comparison purposes, and of HIV-1 PR expressed in *E.coli* by recDNA techniques (27). Only the polypeptide backbone is displayed; secondary strucural elements are shown as ribbons.

D-enzyme was in all essential respects the mirror image of the L-enzyme*. This relationship applied to every aspect of the three-dimensional structure of the homodimeric 22kDa enzyme molecule, and extended to centers of asymmetry not directly determined by the chirality of Cα atoms in the polypeptide backbone. A comparison of the backbone fold of the ligated D-HIV-1 PR with native backbone L-HIV-1 PR expressed in *E. coli* (27) is shown in Figure 3B.

Not only is this the first crystal structure of a D-enzyme but it is also the first crystal structure of a ligated enzyme prepared by total chemical synthesis using chemical dovetailing. This work is currently being extended to the cocrystallization of the enantiomeric forms of the HIV-1 PR with the goal of obtaining racemic crystals and high accuracy structural data to aid in drug design.

4. Contribution of Flap-Substrate H-Bonds to the Activity of HIV-1 PR

Aspartyl proteinases are believed to accelerate the hydrolysis of peptide bonds by a general acid-general base mechanism involving the side chain carboxyl groups of the two essential Asp side chains (28). One feature of this mechanism is the distortion of the scissile peptide bond upon binding of substrate to the enzyme. In crystal structures of cell-encoded aspartyl proteinases, substrate derived peptide inhibitors were observed to bind with an approximately 40° out-of-plane rotation of the bond corresponding to the scissile peptide bond in a substrate. The same phenomenon has been observed in cocrystal structures of substrate-derived inhibitors with the retroviral aspartyl proteinase HIV-1 PR (13-15, 27, 29). This distortion is believed to reflect a "strain" contribution to the mechanism, in which the scissile peptide bond is set up for nucleophilic attack by the out-of-plane rotation, which reduces the double bond character of the C-N bond and increase the electrophilic character of the carbonyl carbon.

However, significant differences have been observed in the way in which the enzyme molecules interact with substrate-derived inhibitors in the crystallographic structures of the cell-encoded and retroviral aspartyl proteinases. The cell-encoded aspartyl proteinases consist of a single polypeptide chain folded into a two domain structure. One of these domains has a flexible "flap" structure, which is made up of two antiparallel β–strands connected by a reverse turn. On binding inhibitor, this single flap closes down with the plane of the reverse turn parallel to the direction of the bound substrate, and with direct H-bonding interactions from the -NH- of the peptide bond at the tip of the flap to the substrate backbone carbonyls on either side of the scissile peptide bond (28,30). The other domain has no flap structure and is not involved in this type of interaction with substrate.

By contrast, the retroviral aspartyl proteinases are homodimeric, with a single enzyme molecule made up of two identical folded polypeptide chains. The substrate binding site is at the interface between the two folded monomers, which correspond to the domains

* The single exception to this identity with the native-backbone L-HIV-1 PR involved the flaps in the region of the thioester bond replacing the Gly^{51}-Gly^{52} peptide bond. The only significant change involved NOT the thioester bond at the exit of the reverse turn at the tip of each flap, but rather the peptide bond at the start of the turn! This carbonyl had been oriented in a manner consistent with H-bonding across the turn to the backbone NH of the n+3 residue. In the ligated enzyme, this NH is a S atom, and the carbonyl at the start of the turn is no longer constrained. It "relaxes" by rotating out of the plane of the turn. Interestingly, separate low occupancy water molecules were observed bound between the rotated NH of this peptide bond in each flap and the P2 and P1' carbonyls respectively of the inhibitor, rather than the single high occupancy water ("water301") tetrahedrally coordinated to both flaps as observed in all other reported structures of complexes of the HIV-1PR with substrate based inhibitors (13-15, 27,29). It should be emphasized that the ligated $[(COS)^{51-52}]$HIV-1 PR displays full enzymatic activity.

of the cell-encoded enzymes. One catalytically essential Asp side chain is contributed from each monomer. Of all the retroviral aspartyl proteinases, the molecular structure of only the HIV-1 PR molecule has been completely solved by X-ray crystallography, both as the empty enzyme (12) and in complexes with a variety of inhibitors (13-15, 27, 29). The HIV-1 PR has *two* flap structures, each with structure essentially identical to the single flap observed in the cell-encoded aspartyl proteinases (28,30). On binding substrate-derived inhibitors,

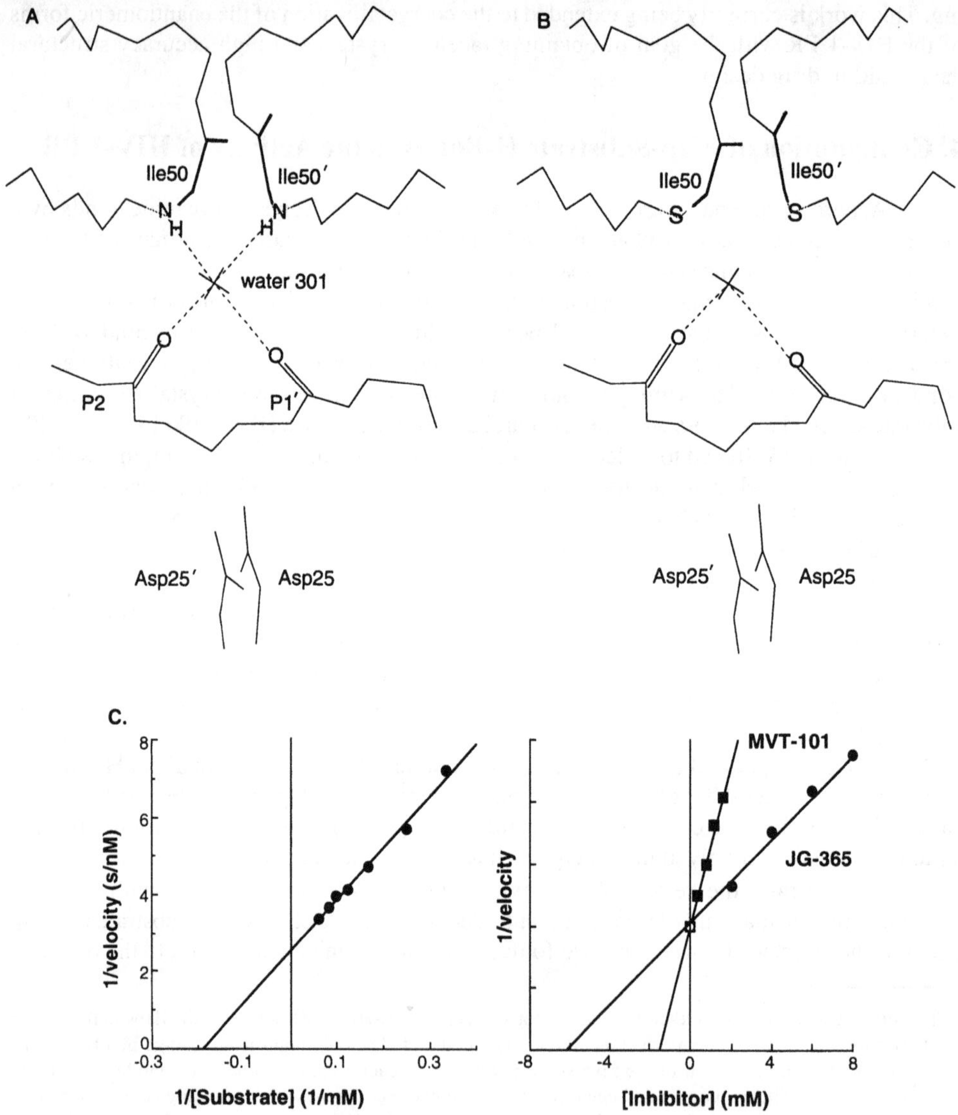

Figure 4. The role of flap-substrate H-bonds in the action of HIV-1 PR. **A.** The H-bonds between the peptide bond at the tip of each flap and the substrate carbonyls on either side of the scissile bond are mediated by a specifc, tetrahedrally coordinated water molecule. [Taken from Miller, et al., (13)] **B.** Design of backbone engineered $[(COS)^{49-50}]_2$HIV-1 PR, in which the H-bond donating ability of the peptide bond at the tip of each flap is precisely and specifically deleted (20). **C.** Kinetic analysis of the action of the backbone enigneered enzyme on a standard substrate (left), and the effects of two different inhibitors on the modified enzyme(right) (20).

both flaps close down over the inhibitor. In contrast to the cell-encoded enzymes, the plane of the reverse turn at the tip of each flap is perpendicular to the direction of the bound peptide chain. That is, the flaps have closed in an "edge on" manner, with respect to substrate.

In all the HIV-1 PR cocrystal structures thus far examined (13-15, 27, 29), a common pattern of H-bonding interactions has been observed between the backbone of the flaps and the bound substrate-derived inhibitor. A single water molecule ("water301") is tetrahedrally coordinated between the -NH- of the peptide bond at the tip of each flap and the substrate backbone carbonyls on either side of the scissile bond (13). This is shown in Figure 4A. Thus, the flap-substrate H-bonds that are part of the strain mechanism are mediated by this specific water molecule. This is a very different situation than the direct flap-substrate H-Bonds found in the cell-encoded enzyme-inhibitor complexes, and could form the basis for the design of inhibitors that would discriminate the cell-encoded enzymes from the retroviral enzymes (14).

Such retroviral-specific inhibitors could be a new class of antiviral agents, and have important applications for a variety of human disease states. For this reason it is important to understand the significance of the flap-substrate H-bonding interactions observed in the enzyme-substrate complexes, and to assess the role of water 301 in the mechanism of the HIV-1 PR.

With this goal in mind, we designed the molecule shown in Figure 4B. In this the backbone engineered version of HIV-1 PR, the -NH- of the peptide bond at the tip of each flap is replaced by a -S- atom in a thioester bond, thus effectively removing the H-bond donating ability. The 99 residue monomer $[(COS)^{49-50}]$HIV-1 PR was prepared by the chemical dovetailing approach described above. Two unprotected peptide segments, HIV-1 PR(1-49)αCOSH and BrIle*(51-99)HIV-1 PR, were reacted under nucleophilic conditions to yield the desired molecule. Note that the nucleophilic ligation reaction proceeds by an S_N2 mechanism, causing an inversion of stereochemistry at the α-carbon. For this reason, we started with the opposite configuration at that center, in order to generate final stereochemistry equivalent to the L-Ile^{50}. The branching of the BrIle side chain at the β-carbon also caused considerable steric hindrance to nucleophilic attack, and alternative ligation reaction conditions had to be worked out. The product 99 residue polypeptide chain was purified and folded to give the dimeric, backbone engineered enzyme $[(COS)^{49-50}]_2$HIV-1 PR in good yield. The synthesis has been described (20).

Backbone engineered $[(COS)^{49-50}]_2$HIV-1 PR displayed the same specificity for cleavage of peptide substrates as the native backbone enzyme. However, detailed kinetic analysis of its action on a standard substrate (results in Figure 4C) showed an unchanged Km, but maximal activity, Vmax, *reduced* 3000-fold relative to the native backbone enzyme (20). This corresponds to a ~5kcal/mole increase in the activation energy for the rate limiting step in catalysis under conditions in which the enzyme is saturated with substrate. This is consistent with the effect of deletion of two stabilizing H-bonds. Investigation of the effects of inhibitors on the modified enzyme gave similarly interesting results (20). A reduced isostere inhibitor, MVT101, had the same Ki for both native backbone and backbone engineered enzymes. However, the hydroxyethylamine inhibitor JG365 which is a potent inhibitor of the native backbone HIV PR had 2,500-fold reduced effect on the backbone engineered enzyme. It has been proposed that the reduced isostere inhibitor is an analogue of the transition sate leading to the formation of the enzyme bound tetrahedral intermediate, which is not rate limiting, and that the hydroxyethylamine inhibitor JG365 is an analogue

*The abbreviation "BrIle" is used to represent 2-bromo-3-methylvaleric acid. The molecule is generated from the corresponding 2-amino-3-methylvaleric acid, i.e. Ile. Note that there are two chiral atoms in this molecule.

of the transition state in the rate-limiting breakdown of the tetrahedral intermediate to form hydrolysis products. The observed effects of the two different types of inhibitors on the backbone engineered enzyme are consistent with this interpretation.

This experiment demonstrates the power of the total chemical synthesis approach to the study of enzyme action. A single backbone atom was precisely replaced, with surgical precision, in each subunit of HIV-1 PR. The effect of deleting these two H-bonding atoms could be quantified, and the properties of the resulting enzyme molecule, changed at only that atom, were consistent with an important functional role for specific H-bonds between the native polypeptide backbone and the substrate molecule.

It should be noted that these results do not directly address the question of the role of Water 301. We have demonstrated the quantitative significance of specific flap-substrate H-bonds, but NOT whether they are water mediated. Two pieces of preliminary evidence suggest that Water301 may be a crystallographic artifact rather than an integral feature of the enzyme mechanism. First, the ligated enzyme $[(COS)^{51\text{-}52}]$HIV-1 PR is fully active and has normal Vmax (7). Yet the crystal structure of this flap-modified enzyme (in the D-form, see above) does not show a single tetrahedrally coordinated water molecule in complex with an inhibitor (26). The same inhibitor complex with native backbone enzyme contains Water 301 (13). Second, we recently reported the first use of electrospray mass spectrometry under native conditions to observe a ternary (homodimeric HIV-1 PR.JG365) protein complex (5). A specifically coordinated water molecule, seen in the X-ray crystallography of this same complex (14), was not observed under conditions in which we did see the specific interaction of the enzyme RNase A with an inorganic phosphate.

Further experimental work will be required to address the role of Water 301 in the action of the HIV-1 PR. We expect the precise modification of enzyme structure by total chemical synthesis to play a critical role in these studies.

DISCUSSION

As the above case studies show, the routine, reproducible preparation of enzymes by total chemical synthesis is now a practical reality. This allows the entire repertoire of organic chemistry to be turned to the study of enzyme action. Precise backbone modifications can be applied to the enzyme molecule, elements with predetermined geometry can be incorporated to dictate the folded structure of the protein, and precise stereochemical engineering can be carried out. This opens up the whole world of D-enzymes to experimental study. Perhaps the most significant use of these new tools will be to allow the construction of protein molecules containing arrays of single-atom-site NMR probe nuclei, designed to address specific aspects of the structural basis of protein function.

We believe that the integrated use of these novel analytical and synthetic protein chemistries will make a major contribution to a definitive exploration of the molecular basis of enzyme action. This will help in the attainment of the long sought after goal of a predictive understanding of enzymes, and will allow the design and construction of catalytically active proteins with precise, predetermined properties.

REFERENCES

1. E. Fischer (1906), quoted in "Molecules and Life", J.S. Fruton, Wiley-Interscience, New York, 1972, pp.112-113, and 114.
2. For example, see "Protein Engineering", R.B. Freedman, R. Wetzel, eds., *Current Opinion in Biotechnology*, **Vol. 3**, Number 4 (August 1992).

3. J.A. Ellman, D. Mendel, P.G. Schultz, Site-specific incorporation of novel backbone structures into proteins, Science **255**, 197-200 (1992).
4. B.T. Chait, S.B.H. Kent, Weighing naked proteins: practical, high-accuracy mass measurement of peptides and proteins, *Science* **257**, 1885-1894 (1992).
5. M. Baca, S.B.H. Kent , Direct observation of a ternary complex between the dimeric enzyme HIV protease and a substrate-based inhibitor, *J.Am.Chem.Soc.* **114**, 3992-3993 (1992).
6. B.T. Chait, R. Wang, R. Beavis, S.B.H. Kent, Protein ladder sequencing, *Science* **262**, 89-92 (1993).
7. M. Schnlzer, S.B.H. Kent, Constructing proteins by dovetailing unprotected synthetic peptides: backbone engineered HIV protease, *Science* **256**, 221-225 (1992).
8. R.L. Talbott, E.E. Sparger, K.M. Lovelace, W.M. Fitch, N.C. Pedersen, P.A. Luciw, J.H. Elder, Nucleotide sequence and genomic organization of feline immunodeficiency virus, *Proc. Natl.Acad. Sci. USA* **86**: 5743-5747 (1989).
9. J. Elder, M. Schnlzer, C.S. Hasselkus-Light, M. Henson, D.A. Lerner, T.R. Philips, P.C. Wagaman, S.B.H. Kent, Identification of proteolytic processing sites within the Gag and Pol polyproteins of feline immunodeficiency virus, *J. Virology* **67**, 1869-1876 (1993).
10. M. Schnlzer, P. Alewood, D. Alewood, S.B.H. Kent, "In situ" neutralization protocols in Boc-chemistry solid phase peptide synthesis: rapid, high yield assembly of difficult sequences, *Int.J.Pept.Prot.Res.* **40**, 180-193 (1992).
11. J. Schneider, S.B.H. Kent, Enzymatic activity of a synthetic 99 residue protein corresponding to the putative HIV-1 protease, *Cell* **54**, 363-368 (1988).
12. A. Wlodawer, M. Miller, M. Jaskolski, B.K. Sathyanarayana, E. Baldwin, I.T. Weber, L. Selk, L. Clawson, J. Schneider, S.B.H. Kent, Crystal structure of synthetic HIV-1 protease: conserved fold in retroviral proteases, *Science* **245**, 616-621 (1989).
13. M. Miller, B.K. Sathyanarayana, M.V. Toth, G.R. Marshall, L. Clawson, L. Selk, J. Schneider, S.B.H. Kent, A. Wlodawer, Structure of a complex of synthetic HIV-1 protease with a substrate-based inhibitor at 2.3Å resolution, *Science* **246**, 1149-1152 (1989).
14. A.L. Swain, M. Miller, J. Green, D.H. Rich, J. Schneider, S.B.H. Kent, A. Wlodawer, X-ray crystallographic structure of a complex between synthetic HIV-1 protease and a substrate-based hydroxyethylamine inhibitor, *Proc. Natl. Acad. Sci. USA* **87**, 8805-8809 (1990).
15. M. Jaskolski, A.G. Tomasselli, T.K. Sawyer, D.G. Staples, R.L. Heinrikson, J. Schneider, S.B.H. Kent, A. Wlodawer, Structure at 2.5Å resolution of chemically synthesized HIV-1 protease complexed with a hydroxyethylene-based inhibitor, *Biochemistry* **30**, 1600-1609 (1991).
16. R.B. Merrifield, Solid phase synthesis, *Science* 232, 341-347 (1986).
17. S.B.H. Kent, Chemical synthesis of peptides and protein, *Ann. Rev. Biochem.* **57**, 957-984 (1988).
18. J. Blake , D. Yamashiro, K. Ramasharma, C.H. Li, Chemical synthesis of α-inhibin-92 by the thiocarboxyl segment coupling method, *Int.J.Peptide Protein Res.* **28**, 468-476 (1986); D. Yamashiro, C.H. Li, New segment synthesis of α-inhibin-92 by the acyl disulfide method, *Int.J.Peptide Protein Res.* **31**, 322-334 (1988).
19. Philip E. Dawson, Stephen B.H. Kent, Convenient total synthesis of a 4-Helix TASP molecule by chemoselective ligation, *J.Am.Chem.Soc.* **115**, 7263-7266 (1993).
20. Manuel Baca, Stephen B.H. Kent, Catalytic contribution of flap-substrate hydrogen bonds in HIV-1 protease explored by chemical synthesis, *Proc. Natl. Acad. Sci. USA.*, **90**, 11638-11642 (1993).
21. J.S. Richardson, D.C. Richardson, Principles and patterns of protein folding, in "Prediction of Protein Structure and the Principles of Protein Conformation", G.D. Fasman, ed., Plenum, New York, 1989, pp. 1-98.
22. R.C deLisle Milton, S.C.F. Milton, S.B.H. Kent, Total chemical synthesis of a D-enzyme: the enantiomers of HIV-1 protease show reciprocal chiral substrate specificity, *Science* **256**, 1445-1448 (1992).
23. A.L. Mackay, Crystal enigma, *Nature* **342**, 133 (1989).
24. L Zawadzke J.M. Berg, The structure of a centrosymmetric protein crystal, *Protein Structure Function and Genetics* **16**, 301-305 (1993)
25. R.C. deLisle Milton, S.C.F. Milton, M. Schnlzer, S.B.H. Kent, Synthesis of proteins by chemical ligation of unprotected peptide segments: mirror-image enzyme molecules, D- and L-HIV protease analogs, *Techniques in Protein Chemistry IV*, Academic Press, New York, pp. 257-267 (1992).
26. M. Miller, J.K.M. Rao, S.C.F. Milton, R.C. deLisle Milton, S.B.H. Kent, Crystal structure of a synthetic D-enzyme prepared by chemical ligation, presented at the meeting of The Protein Society, San Diego, July 1993.
27. P.M.D. Fitzgerald, et al., Crystallographic analysis of a complex between human immunodeficiency virus type 1 protease and acetyl-pepstatin at 2.0 resolution, *J. Biol. Chem.* **265**, 14209-14219 (1990).

28. K. Suguna, E.A. Padlan, C.W. Smith, W.D. Carlson, D.R. Davies, Binding of a reduced bond peptide inhibitor to the aspartyl proteinase from *rhizopus chinensis*: implications for a mechanism of action, *Proc.Natl.Acad.Sci.USA.* **84**, 7009-7013 (1987).
29. A.G. Tomasselli, W.J. Howe, T.K. Sawyer, A. Wlodawer, R.L. Heinrikson, The complexities of AIDS: an assessment of the HIV protease as a therapeutic target, Chimica Oggi pp.6-27 (May, 1991).
30. K.D. Parris, D.J. Hoover, D.R. Davies, Crystal structure of rhizopuspepsin/inhibitor complexes, in "Structure and Function of the Aspartyl Proteinases", B.M. Dunn, ed., Plenum, New York, 1991, pp. 217-231.

X-RAY STRUCTURE OF A TETHERED DIMER FOR HIV-1 PROTEASE

T.N. Bhat, E. T. Baldwin, B. Liu, Y.-S.E. Cheng,[1] and J. W. Erickson

Structural Biochemistry Program
Frederick Biomedical Supercomputing Center
PRI/DynCorp
National Cancer Institute
Frederick Cancer Research and Development Center
Frederick, MD 21702-1201
[1]DuPont-Merck Pharmaceutical Co.
Wilmington, DE 19880

INTRODUCTION

HIV-1 protease (HIV PR) is essential for the replication of human immunodeficiency virus and is responsible for the processing of gag and gag-pol polyproteins. Thus, HIV PR has been the target of intensive efforts to design inhibitors [1-4]. The enzyme is a homodimeric aspartic protease, and its active site is formed by the monomer-monomer interface and exhibits two-fold rotational symmetry. In this respect, HIV PR differs from the single chain, cellular aspartic proteases whose active sites are formed by the interaction of two homologous, but non-identical, domains. Because of its dimeric nature, mutations in HIV PR are expressed in a symmetric pair-wise fashion involving both chains. To facilitate studies of single site mutations, tethered HIV PR dimers have been constructed using recombinant methods [5,6]. In these studies, two identical HIV PR genes were placed in tandem, separated by a short spacer region, such that the C-terminus of subunit 1 would be covalently linked via the spacer to the N-terminus of subunit 2. This design was based on the fact that the N- and C-termini of different subunits are located within several Å of each other near the surface of the native enzyme. While biochemical studies indicated no major differences in the activities of tethered and two-chain HIV PR [5,6], we wished to examine the structure of the single chain enzyme as a basis for further studies on single site mutants. We report here the crystal structure of a recombinant tethered dimer complexed with a C2 symmetry-based inhibitor A-76928. A fuller account of this structure will appear elsewhere[9].

Aspartic Proteinases: Structure, Function, Biology, and Biomedical Implications
Edited by Kenji Takahashi, Plenum Press, New York, 1995

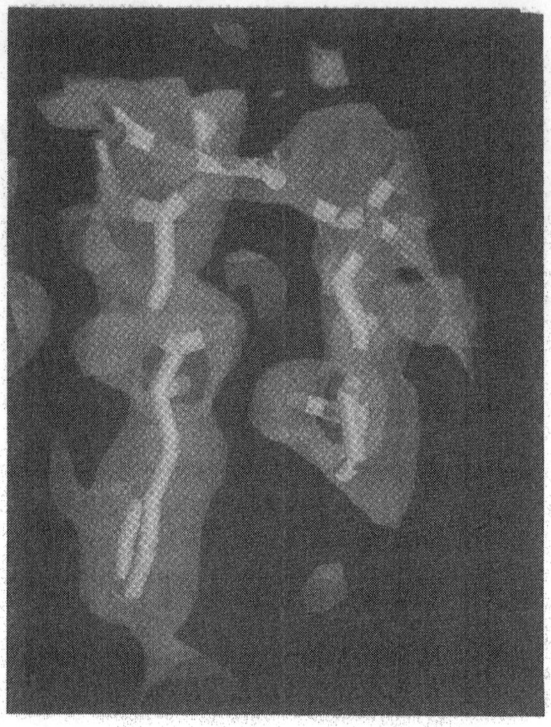

Figure 1. The atomic model for the pentapeptide linker of the HIV-1 protease tethered dimer along with the 2|Fo|-|Fc| electron density surface contoured at 0.5σ is shown. The figure was made using the program GRAMPS kindly supplied by Dr. T.J. O'Donnell.

MATERIALS AND METHODS

Recombinant single chain HIV PR used for the X-ray analysis was produced as previously described [5] by expression of a synthetic gene in *E. coli* in which two identical HIV PR genes are linked by a DNA segment coding for the five residue spacer Gly-Gly-Ser-Ser-Gly. The C2 symmetry-based (S,S)-diol inhibitor, A-76928, was supplied by Dr. Dale Kempf (Abbott Laboratories). Crystals were grown by the hanging drop vapor diffusion method as follows: two microliters of a complex of HIV PR at 6mg/ml with A-76928 at 1:5 molar ratio in 50 mM NaAc buffer containing 10 mM DTT and 10% DMSO at pH 5.5 was mixed with an equal volume of buffer containing 66 mM Na citrate, 132 mM NaPO4, pH6.2 and 34% saturated ammonium sulphate. The crystallization drop was then seeded with microseeds of the corresponding two chain inhibitor complex. Crystals grew at room temperature over 24 hr. Repeated macroseeding was used to increase crystal size. X-ray diffraction data were collected using a Siemens area detector. The space group was $P2_12_12_1$ and the refined unit cell parameters were: a = 52.1; b = 59.8; and c = 62.2Å. Crystallographic coordinates of the wild-type HIV PR/A-76928 complex [7] were used for molecular replacement and for initial refinement using XPLOR. The initial R-factor was 35% and dropped to 23% at the end of the first cycle of refinement. After a few rounds of model building, followed by an additional refinement cycle, the R-factor dropped to 19.4%. The final rms deviation from ideal values for bond lengths and bond angles are 0.018 Å and 3.3°, respectively.

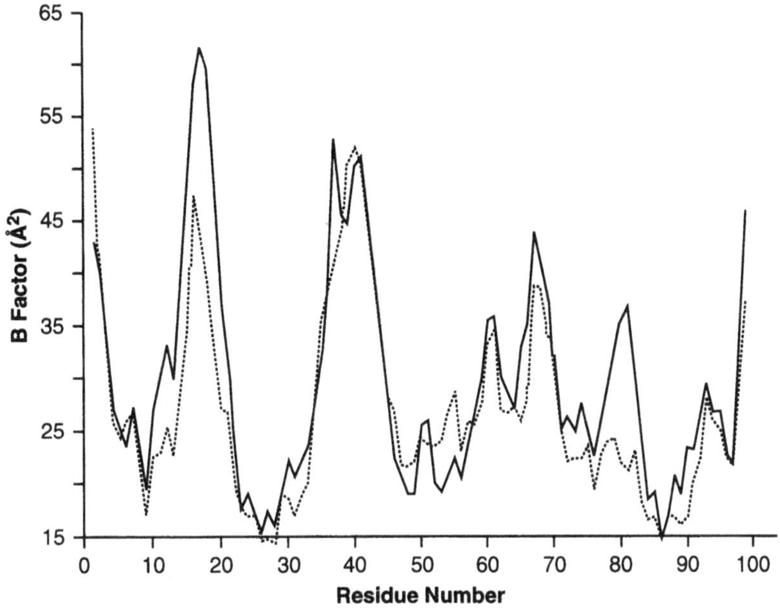

Figure 2. Temperature factors vs residue plot for the tethered dimer of HIV-1 protease. Chain 1 (solid line) and chain 2 (dotted line).

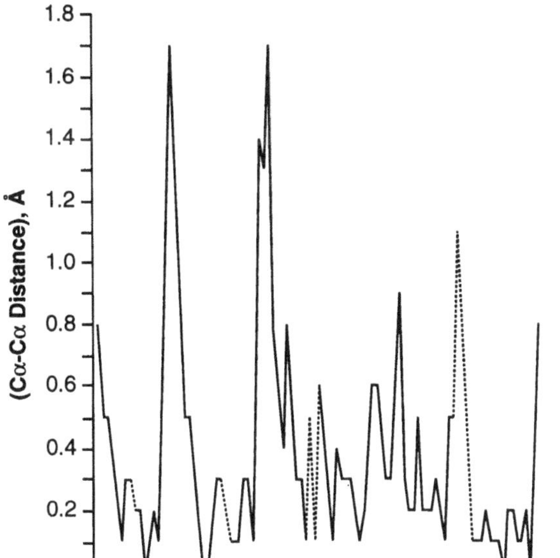

Figure 3. Difference between the Cα atom positions for chain 1 and chain 2 after superposition of all 99 equivalent Cα atom pairs. Residues within 5 Å of the inhibitor are indicated by dotted lines.

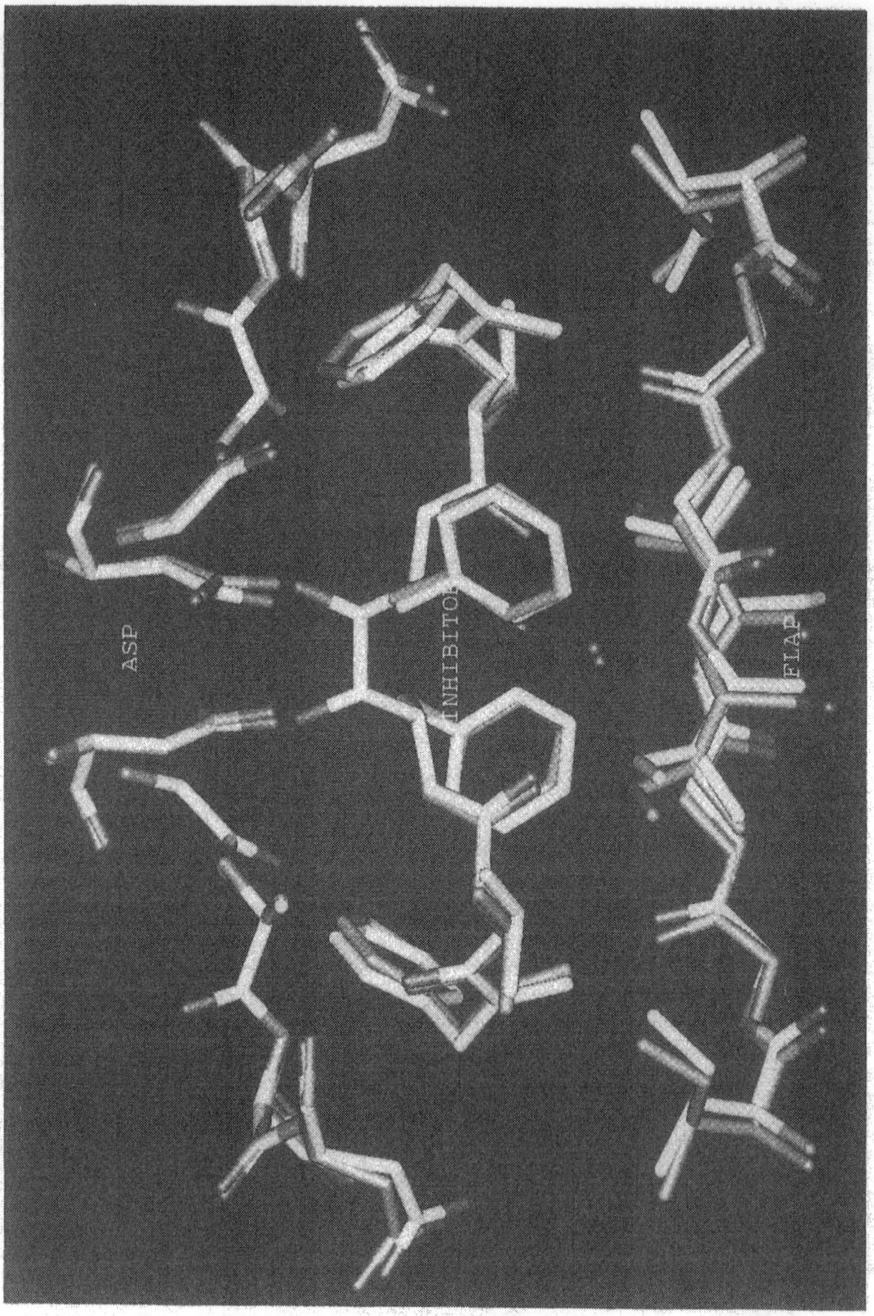

Figure 4. Superposition of the active site region of the A-76928/tethered dimer structure onto that for the two chain complex [7].

RESULTS AND DISCUSSION

Electron density was visible for the Gly-Gly-Ser-Ser-Gly pentapeptide loop that links the two polypeptide chains (Fig. 1).The mean B value for these residues is 75 Å^2 compared to a mean B value of 29 Å^2 for the entire protein. The relatively high B values for the linker residues indicate positional disorder consistent with the fact that this segment is composed of small, flexible amino acids and is exposed to solvent in the crystal structure.

The inhibitor molecule has good electron density and the mean B values over all the inhibitor atoms is 22 Å^2. Active site residues in contact with the inhibitor are also well-defined in the electron density map, and the mean B value for residues that lie within 5 Å of the inhibitor is 23 Å^2. The B value analysis suggests that active site residues and the inhibitor atoms are less mobile on average than the rest of the protease structure.

The mean B values of the two chains differ by about 2 Å^2, 28 Å^2 and 30 Å^2, when averaged over all atoms. The same difference is seen in the B values for the residues that contact the inhibitor 22 Å^2 and 24 Å^2. The two polypeptide chains show a high degree of similarity in the B factor distribution as a function of residue position (Fig. 2). Both chains show large increases in B values on either side of the active site Asp residues.

The two chains (residues 1 - 99 and 101 - 199) are structurally similar (Fig. 3) and the deviations from symmetry between the two chains are similar to those seen in other HIV-1 protease inhibitor complexes belonging to the same space group [8]. Broadly, the two chains differ in three regions comprised of residues 14-20, 35-42, and 78-82. These regions also have relatively high B values which suggests that these segments correspond to more flexible parts of the enzyme. Residues in contact with the inhibitor show smaller differences with the exception of residue 81 which is also involved in a crystal contact.

The crystal structure of the inhibitor complex with the wild type, two-chain form of HIV PR [7] compares well with the inhibitor complex of the tethered dimer. Individual chains of the two structures superpose with an rms difference of 0.18 and 0.20 Å for all Cα atoms. The inhibitor molecule is symmetrically bound as described previously [7] and occupies identical positions in the two crystal structures (Fig. 4).

CONCLUSIONS

The crystal structure of a tethered dimer of HIV PR has been solved in a complex with A-76928. The linker region is visible in electron density maps, and does not appear to significantly affect the crystallization properties of the enzyme or its structure when complexed with inhibitor. These observations suggest that the tethered dimer is a suitable surrogate for mutagenesis studies aimed at elucidating the structure-function effects of single residue, single chain mutations.

REFERENCES

1. J.R.Huff, J.Med.Chem. 34:2305 (1991).
2. T.D.Meek, J.Enzyme Inhib. 6:65 (1992).
3. D.W.Norbeck and D.J.Kempf, Annu.Rep.Med.Chem.26:141(1991).
4. A.G.Tomasselli,W.J.Howe, T.K.Sawer, A.Wlodawer and R.L.Heinrikson, Chem. Oggi. 9:6 (1991).
5. Y.-S. E. Cheng, F.H. Yin, S.Foundling, D. Blomstrom and C.A.Kattner, Proc. Natl. Acad. Sci. USA 87: 9660 (1990).
6. C.L. DiIanni, L.J. Davis, M.K. Holloway, W.K. Herber,P.L.Darke, P.L. Kohl and R.A.F. Dixon, J. Biol.Chem. 265:17348 (1990).

7. M.V.Hosur,T.N.Bhat,D.J.Kempf,E.T.Baldwin,B.Liu,S.Gulnik,N.E.Wideburg,D.W. Norbeck,K.Appelt and J.W.Erickson, J.Am.Chem.Soc. 116:847 (1994).
8. J.W.Erickson,Perspec. Drug Dis. Design, 1:109 (1993).
9. T.N. Bhat, E.T. Baldwin, B. Liu, Y.-S.E. Cheng and J.W. Erickson, Nature Struct. Biol. 1:552 (1994).

STRUCTURE OF HIV-1 PROTEASE WITH KNI-272: A TRANSITION STATE MIMETIC INHIBITOR CONTAINING ALLOPHENYLNORSTATINE

Eric T. Baldwin,[1] T. Narayana Bhat,[1] Sergi Gulnik,[1] Beishan Liu,[1] Yoshiaki Kiso,[2] Hiroaki Mitsuya,[3] and John W. Erickson[1]

[1]NCI-Frederick Cancer Research and Development Center
PRI/DynCorp
Structural Biochemistry Program
Frederick, Maryland 21702
[2]Kyoto Pharmaceutical University
[3]National Cancer Institute
Bethesda, Maryland

INTRODUCTION

Inhibitors of human immunodeficiency virus type 1 protease (HIV PR) are the subject of wide interest for the development of therapeutic drugs against AIDS [1-4]. The crystal structures of numerous HIV PR/inhibitor complexes have been solved to aid the process of inhibitor design and to rationalize the structure-activity relationships (SAR) for various classes of inhibitors [5-7]. Incorporation of allophenylnorstatine (APNS) [(2S,3S)-3-amino-2-hydroxy-4-phenylbutyric acid] into peptidomimetics can lead to highly potent inhibitors of HIV PR [8]. However, no structure of APNS containing inhibitor complexes has been reported. Recently, KNI-272, a conformationally-constrained inhibitor containing APNS (Figure 1a), has been shown to possess good anti-HIV activity [9]. As a step towards understanding the high affinity of this novel class of inhibitors, we determined the three dimensional structure of KNI-272 bound to HIV PR.

METHODS

Crystals of HIV PR complexed with the inhibitor KNI-272 were prepared using the hanging drop method. Recombinant HIV PR from the LAI strain of HIV-1 at 4.1 mg/ml in 50 mM sodium acetate buffer, pH 5.6, 10 mM dithiotreitol was mixed with inhibitor, dissolved in dimethyl sulfoxide (DMSO), in a 1:5 molar ratio. The final DMSO concentration was less than 5%. Crystals were grown at room temperature by mixing 2 μl of 35% saturated

Aspartic Proteinases: Structure, Function, Biology, and Biomedical Implications
Edited by Kenji Takahashi, Plenum Press, New York, 1995

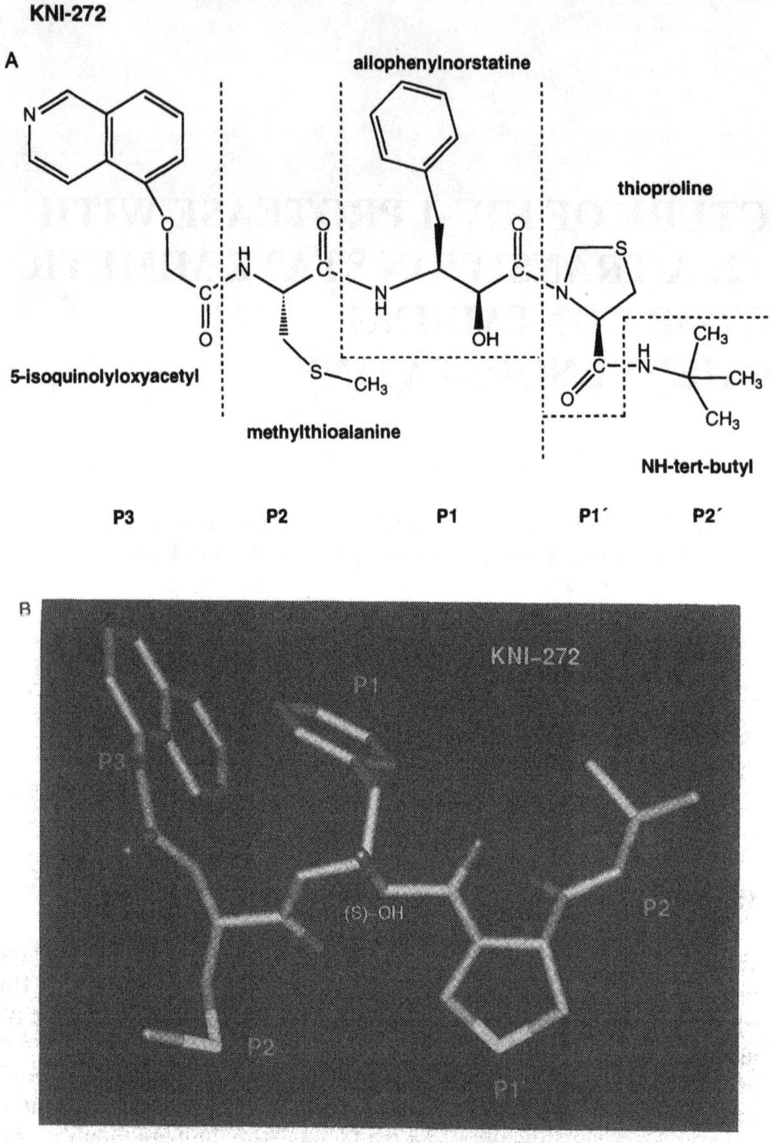

Figure 1. (A) The structure of KNI-272 which shows inhibitor residues extending from P3 through P2'. (B) The conformation of the inhibitor as it is observed in the crystal structure.

ammonium sulfate, 65 mM citric acid, 130 mM sodium phosphate at pH 6.2, (well mix) with an equal volume of protein solution. The drops were microseeded to obtain large, single crystals. X-ray data were collected from a single crystal using an R-axis IIc image plate area detector and an Ru-200 X-ray generator with 0.3 by 3 mm focus, graphite monochromator and 0.3 mm double pinhole collimator operating at 50 KV and 100 mA. Data were processed with software supplied by Molecular Structure Corp. and were 71% complete to 2.0 Å resolution using an intensity/sigma (I/σ) ratio cutoff greater than 2.0 (Table I). Crystals belong to the space group $P2_12_12_1$ with unit cell parameters 51.9, 59.1 and 62.3 Å. Reflections with an $I/\sigma > 2.0$ were used in the structure determination and refinement.

Table 1. Statistics for HIV PR/KNI-272 Complex Data Collection and Scaling

I/σ cutoff	I/σ > 0	I/σ > 1	I/σ > 2
Total Obs	58745	44973	34769
Fulls	42727	33425	26264
Partials	16018	11548	8505
Rejects	204	203	286
Indep. Ref	15593	13119	10242
R-merge	7.72	7.02	6.49
Fulls	7.24	6.60	6.14
Partials	9.07	8.22	7.52
Resolution			
3.0 Å Completeness	99.7	96.3	92.5
2.5 Å Completeness	99.5	93.8	86.1
2.0 Å Completeness	99.2	87.0	71.3
2.25 - 2.0 < I/σ >	2.23	2.98	4.04

The crystal structure was solved by molecular replacement using the protein model of the HIV PR complex with the diol inhibitor, A77003 [10]. The structure was refined with X-PLOR 3.2 on a CRAY YMP from an initial R-factor of 30.4%. The protein structure alone was initially refined with the B-factors reset to 15.0 $Å^2$ and the model adjusted as indicated by inspection of the Fo-Fc and 2Fo-Fc electron density maps. The inhibitor was built using QUANTA and fitted to Fo-Fc and 2Fo-Fc difference maps using phases from the refined protein structure. After several cycles of refinement and adjustment of the model, an R-factor of 16.6% for 9883 reflections with $I/\sigma > 2.0$ was obtained. The model consists of 1676 atoms including 92 waters. All heavy atoms were refined with individual B-factors. The root mean square deviation from ideals values for bond distance restraints, angles, dihedrals and impropers are 0.014 Å, 2.9°, 27.7° and 1.7°, respectively.

RESULTS AND DISCUSSION

The Fo-Fc and 2Fo-Fc electron density maps indicated that the inhibitor is well-ordered in the structure and binds in a unique orientation in an extended conformation (Fig. 1b). The inhibitor makes contacts with the protein in S3, S2, S1, S1' and S2' subsites. There is continuous electron density for all of the inhibitor and the positions of the two sulfur atoms were indicated by the most electron dense regions of the map. The P3 quinoline, P2 methylthioalanine, and P1 phenylalanine groups adopt single orientations. The thioproline residue in P1' occupies a shallow position in the S1' subsite. The terminal t-butyl group is well-resolved in the 2Fo-Fc map and occupies the S2' subsite. The inhibitor makes three hydrogen bonds with the protein mainchain on the P side while it makes no direct mainchain hydrogen bonds on the P' side. The conserved flap water is observed where it makes bridging hydrogen bonds between the backbone NH groups of Ile-50 and Ile-150 and the inhibitor carbonyl groups of the P2 and P1' residues (Fig. 2). Both inhibitor oxygen atoms are accomodated within the active site pocket, whose extent is defined by Gly-27/127 carbonyls [10]. The hydroxyl of the P1 APNS residue is centrally located between the two active site

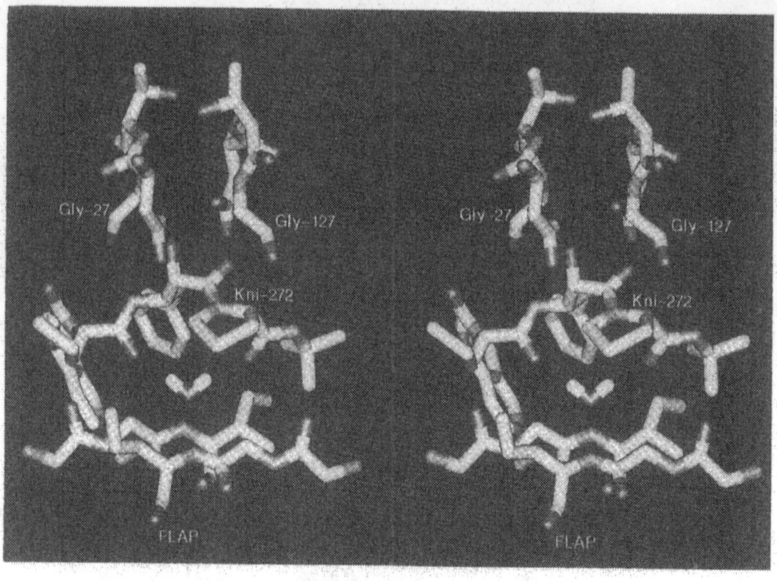

Figure 2. A stereo diagram of KNI-272 bound in the active site pocket. The bridging water between the inhibitor and the flap is also shown. The P3 5-isoquinolyloxyacetyl group is on the left. Carbon atoms are indicated in white, nitrogen atoms in light gray, and oxygen atoms in dark gray.

carboxylates and makes a total of four possible hydrogen bonds to both residues. The APNS carbonyl group is within 2.5 Å of OD2 of Asp-125 which, by implication, must be protonated in this structure (Fig. 2). The isoquinoline group at P3 makes an orthogonal stacking interaction with the APNS phenyl group in P1 and projects towards the flap where it makes some van der Waals interactions. This group also makes weak polar contacts with the Arg-8 sidechain of the enzyme. A similar interaction (planar) was observed for the P3 pyridyl group in the diol A77003/HIV PR complex [10]. Preliminary evidence indicates that the R8Q mutant of HIV PR exhibits decreased sensitivity to KNI-272 (S. Gulnik, unpublished data).

The chirality of the APNS OH group is critical for inhibitor potency. The Ki of the (S) isomer is about 20-fold lower than that of the (R) isomer for KNI-272 [8]. Preliminary attempts to model the (R) isomer into the active site cleft, using the conformation of KNI-272 from the crystal structure, indicate that the (R) hydroxyl points away from the active site aspartic acid residues and would be unable to form hydrogen bonds to Asp-125. Steric contacts in P1' between the flap and the thioproline ring as well as between the thioproline ring and the Gly-127 carbonyl will likely prevent any large adjustments in the (R) OH position. Thus, the increase in Ki can be explained in our model by the loss of optimal interactions with the active site aspartic acid residues. However, the possibility of a different conformation of the inhibitor also needs to be considered. Recent structural studies on the binding modes for C2 symmetric diols [10] and hydroxyethylamines [6] reveal that changing stereochemical configuration about the central OH groups can result in dramatically altered backbone conformations for the adjacent portions of the inhibitor.

CONCLUSION

The crystal structure of KNI-272 complexed with HIV PR has been determined to 2.0 Å resolution. The present structure has allowed us to begin to explain features of the

SAR for this series of inhibitors. The optimal interactions of both the S-OH and the P1' thioproline groups with the active site cannot be achieved simultaneously with the R isomer of KNI-272 due to the rigidity of the P1-P1' amide linkage and the limited ring flexibility of the thioproline. The S-isomer is preferred in all thioproline analogs reported to date [8]. Further analysis of this structure is in progress to determine the structural correlates of tight binding and additional SAR features for this important class of HIV PR inhibitors.

ACKNOWLEDGMENTS

The content of this publication does not necessarily reflect the views or policies of the Department of Health and Human Services (DHHS), nor does mention of trade names, commercial products, or organizations imply endorsement by the U.S. Government. We thank the staff of the Frederick Biomedical Supercomputing Center, Frederick Cancer Research and Development Center, Frederick, MD for assistance.

REFERENCES

1. J.R. Huff, J.Med.Chem. 34, 2305-2314 (1991).
2. T.D. Meek, J.Enzyme Inhib. 6, 65-98 (1992).
3. D.W. Norbeck and D.J. Kempf, Ann.Rev.Med.Chem. 26, 141-149 (1991).
4. A.G. Tomasselli, W.J. Howe, T.K. Sawer, A.Wlodawer, and R.L.Heinrikson, Chim.Oggi. 9, 6-27 (1991).
5. J.W. Erickson, Perspec. Drug Dis. Design, 1, 109-128 (1993).
6. K. Appelt, Perspec. Drug Dis. Design, 1, 23-48 (1993).
7. A. Wlodawer and J.W. Erickson, Ann.Rev.Biochem. 62, 543-585 (1993).
8. T. Mimoto, J. Imai, S. Tanaka, N. Hattori, S. Kisanuki, K. Akaji, and Y. Kiso Chem.Pharm.Bull . 39, 3088-3090 (1991).
9. S. Kageyama, T. Mimoto, Y. Murakaw, M. Nomizu, H. Ford, Jr., T. Shirasaka, S. Gulnik, J. Erickson, K. Takada, H. Hayashi, S. Broder, Y. Kiso, and H. Mitsuya, Antimicrobial Agents and Chemotherapy, April, 810-817 (1993).
10. M.V. Hosur, T.N. Bhat, D.J. Kempf, E.T. Baldwin, B. Liu, S. Gulnik, N.E. Wideburg, D.W. Norbeck, K. Appelt, and J.W. Erickson, J.Am.Chem.Soc., *in press* (1994).

MOLECULAR DYNAMICS OF HIV-1 PROTEASE IN COMPLEX WITH A DIFLUOROKETONE-CONTAINING INHIBITOR: IMPLICATIONS FOR THE CATALYTIC MECHANISM

Abelardo M. Silva,* Raul E. Cachau, Eric T. Baldwin, Sergei Gulnik, Hing L. Sham,† and John W. Erickson

Structural Biochemistry Program
PRI/Dyn Corp
NCI-Frederick Cancer Research and Development Center
Frederick, Maryland 21702

INTRODUCTION

The structures of three aspartic proteases, penicillopepsin, endothiapepsin and rhizopuspepsin, have been determined in complexes with difluorostatone (DFS)-containing peptide-based inhibitors [1-3]. In all three structures, the hydrated core of the inhibitors display similar conformations and interactions with the active site aspartate groups. One of the gem-hydroxyls is positioned between both aspartates, the other one interacts with a single aspartate, and the fluorines do not interact with either of them. Since one of the hydroxyl groups is located near the position of the active site water observed in the native structures, these inhibitors have been proposed as analogs of transition state intermediates of peptide bond hydrolysis.

The structure of HIV-1 protease in complex with a hydrated, or gemdiol, form of the difluoroketone-containing C2 symmetry-based inhibitor, A79285 (Fig. 1) [4,5] was determined using x-ray crystallographic techniques [6]. In contrast to the mode of binding obseved with the DFS analogues [1-3], A79285 is symmetrically bound with its core in a *gauche* conformation and mimics a peptide bond after hybridization collapse and rotation of the C-C-N-C dihedral angle to a *gauche* conformation.

* To whom correspondence should be addressed.
† Abbott Laboratories, Pharmaceutical Discovery Division, Abbott Park, Illinois 60064-3500, USA.

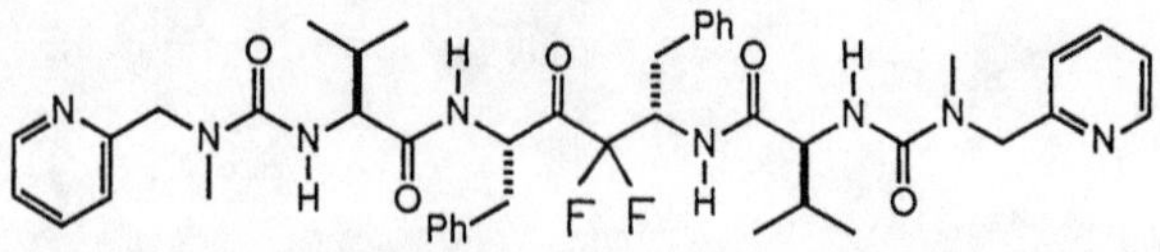

Figure 1. Structure of A79285, a C2 symmetry-based difluoroketone inhibitor of HIV protease.

RESULTS AND DISCUSSION

In order to assess the similarity between the difluorodiol motif of A79285 and a hydrated peptide bond, *ab-initio* Restricted Hartree Fock 6311G** [7,8] and Dmol DNP [9] calculations were performed on an analog of a hydrated peptide bond, CH_3-$C(OH)_2$-NH-CH_3 (Fig. 2). The similarity between the conformations of the peptide bond analog and the difluorodiol motif indicates that a hydrated peptide bond in a *gauche* conformation, homologous to the difluorodiol core, is sterically and energetically feasible at the active site.

To determine whether an *anti-gauche* transition of a hydrated peptide bond could actually occur within the active site pocket of the enzyme, molecular dynamic simulations [10] were performed using the complete protease, the inhibitor with the central motif replaced by a hydrated peptide bond, and 132 water molecules [6]. Only the residues around the inhibitor were allowed full freedom. A shell of atoms of approximately 5 Å was built around the free atoms and a harmonic positional restraint was applied to all atoms within the shell. Atoms beyond this region were fixed at their experimentally determined positions. The *anti* conformation of the hydrated peptide bond was generated in the following manner. First, the complex with the hydrated peptide bond in the difluorodiol conformation was annealed at 500° K for 50 ps and then minimized to obtain a relaxed model. Next, the dihedral angle C-C-N-C was driven towards the *anti* conformation in a 50 ps trajectory at 300° K.

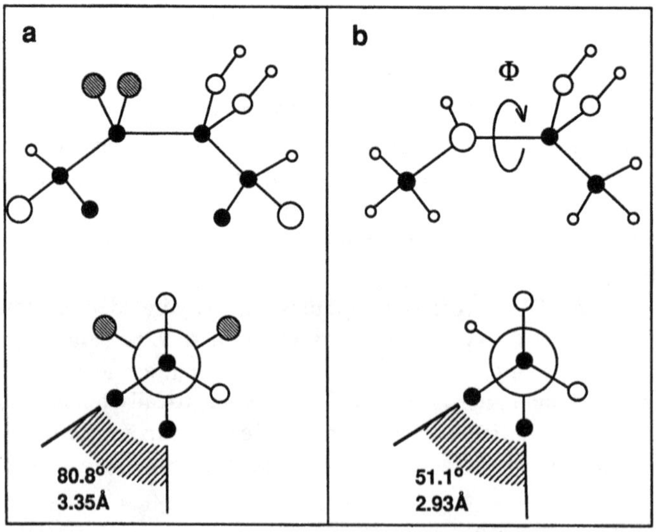

Figure 2. Conformations in terms of the C-C-N-C dihedral angle and C-C distance. a) Difluorodiol core of the inhibitor as determined by x-ray crystallography techniques. b) *Ab-initio* determined hydrated peptide bond in a *gauche* conformation.

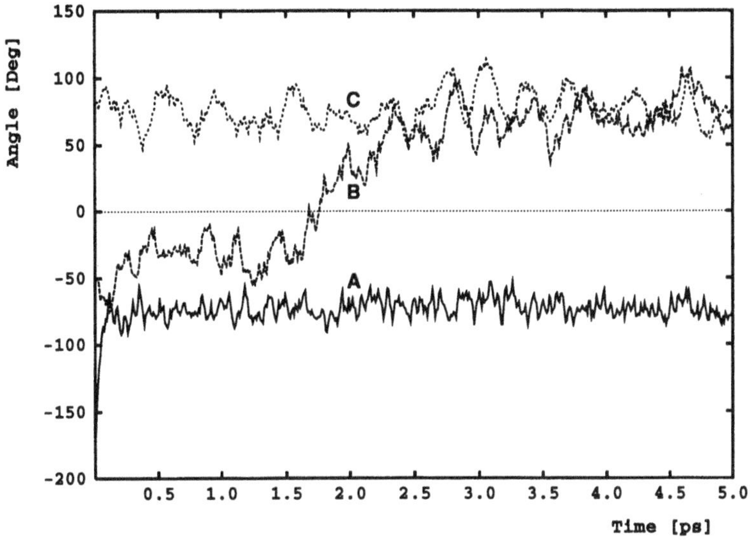

Figure 3. Time evolution of dihedral angles during a molecular dynamics simulation. *A*: Dihedral angle C -C-N-C . The *anti-gauche* transition occurs within the first 15 ps of the trajectory. *B:* Dihedral angle $\chi 2$ of the phenylalanine at the amino end of the hydrated peptide bond. *C:* Dihedral angle $\chi 2$ of the phenylalanine at the diol end of the hydrated peptide bond. In the refined structure of the A79285/protease complex, both $\chi 2$ angles equal 99 [6]. Here only one of them fluctuates around that value.

Finally, at the end of the transition, the hydrated peptide in the *anti* conformation was equilibrated for 50 ps and then minimized. Starting from the *anti* conformation, a total of 15 x 50 ps trajectories were calculated. These trajectories were divided in two segments: 15 ps of equilibration, using a restraint to keep the dihedral angle at the *anti* conformation; and 35 ps without dihedral restraint.

The *anti-gauche* transition can be visualized in terms of the time evolution of the dihedral angle C-C-N-C (Fig. 3). The simulations reveal that the transition usually occurs within 15 ps after equilibration. During the trajectories the diol frequently passes through an intermediate position where a diol hydroxyl lies between both aspartates. This conformation resembles the structures of aspartic proteases with DFS inhibitors. Not only the core but also the whole peptide undergoes conformational changes during the *anti-gauche* transition. The structural change at the core reduces by 0.9 Å the distance between C_α atoms, while the tetrahedral nitrogen is positioned near one of the active site aspartate groups (Fig 4). The time evolution of the conformation of the peptide is complex but deviations from C2 symmetry are clear. In the starting *anti* conformation, the phenylalanine $\chi 2$ dihedral angles differ by about 100 deg but converge to approximately the same value in the symmetric *gauche* conformation (Fig. 3).

The agreement between the simulations and experimental results suggest that the preferred conformation for the hydrated peptide at the active site is *gauche* C2-symmetric. We conclude that the *anti-gauche* transition would readily occur after hydration, is correlated with conformational changes of the peptide and would allow for a proton exchange between the tetrahedral nitrogen and one aspartate group of the active site. The ease of this transition suggests that it should not be the rate-limiting step of the catalytic reaction.

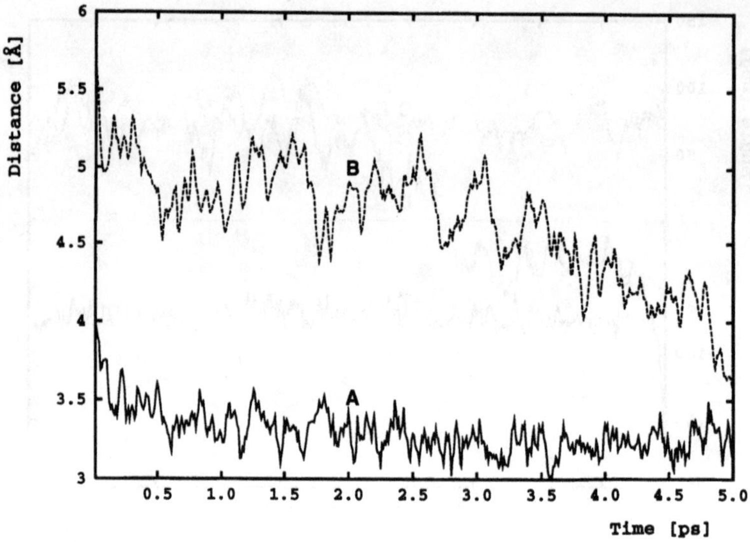

Figure 4. Time evolution of interatomic distances during a molecular dynamics simulation. *A*: C-C distance of the hydrated peptide bond. *B:* Distance between the tetragonal nitrogen of the hydrated peptide bond and an oxygen of one of the active sites aspartates.

REFERENCES

1. K.D. Parris, D.J. Hoover, D.B. Damon and D. R. Davies, *Biochemistry* 31:8125 (1992)
2. M.N.G. James, A. R. Sielicki, K. Hayakawa and M.H. Gelb, *Biochemistry* 31:3872 (1992)
3. Veerapandian B., J.B. Cooper, A. Sali, T.L. Blundell, R.L. Rosati, B. W. Dominy, D.B. Damon and D.J. Hoover, *Protein Science,*1:322 (1992)
4. H.L. Sham, N.E. Wideburg, S.G Spatano, W.E. Kohlbrenner, D.A. Betebenner, D. J. Kempf, D.W. Norbeck, J.J Plattner and J.W. Erickson, *J. Chem. Soc., Chem. Commun.* 110 (1991)
5. H.L. Sham, D.A. Betebenner, N.E. Wideburg, A.C. Saldivar, W.E. Kohlbrenner, S. Vasavanonda, D.J. Kempf, D.W. Norbeck, C. Zhao, J.J. Clement, J.W. Erickson and J. Plattner, *Biochem. Biophys. Res. Comm.* 175:914 (1991)
6. A. M. Silva , R. E. Cachau, I. A. Topol, E. T. Baldwin S. Gulnik, H. L. Sham and J. W. Erickson. Manuscript in preparation.
7. M.J. Frisch, G.W. Trucks, M. Head-Gordon, P.M.W. Gill, M.W. Wong, J.B. Foresman B.G. Johnson, H.B. Schlegel, M.A. Robb, E.S. Replogle, R. Gomperts, J.L. Andres, K. Raghavachari, J.S. Binkley, C. Gonzales, R.L. Martin, D.J. Fox, D.J. Defrees, J. Baker, J.P. Stewart and J.A. Pople, Gaussian 92 Revision A, Gaussian, Inc. Pittsburgh, PA (USA) (1992)
8. J.W. Hehre, L. Radom, P.v.R. Schleyer and J.A. Pople, *Ab Initio Molecular Orbital Theory*, J. Wiley, New York, Chapter 4 (1986)
9. Biosym Technologies, DMol v.2.3, San Diego, CA, (USA)
10. B.R. Brooks, E.R. Bruccoleri, B.D. Olafson, D.F. States, S. Swaminanthan and M. Karplus, *J.Comp.Chem.* 4:187 (1984)

ACTIVATED DYNAMICS OF FLAP OPENING IN HIV-1 PROTEASE

Jack R. Collins, Stanley K. Burt, and J. W. Erickson

Structural Biochemistry Program
Frederick Biomedical Supercomputing Center
PRI/DynCorp
National Cancer Institute
Frederick Cancer Research and Development Center
Frederick, MD 21702-1201

INTRODUCTION

The enzyme active site of HIV protease (HIV PR) is formed by the interface of two dimers capped by two flaps, one from each monomer. The exact role of the flaps in the catalytic mechanism and substrate binding is unknown, but it is generally accepted that the flaps must be flexible to facilitate substrate binding and product release. Numerous crystal structures of HIV PR have been solved in both inhibitor complexed and uncomplexed, or native, forms [1]. Comparison of the MVT-101 inhibited and native structures shows that the Cα positions of the flaps differ by as much as 7 Å (Figure 1a), and that the topographical arrangement of the flaps is quite different, each displaying a different "handedness" (Figure 1b). This handedness precludes conversion between the two conformations via a simple hinge motion without the flaps first changing to a fully open form. Fluorescence data on HIV-PR, suggests that the flaps open on a time-scale of roughly 30 ms, about 6 times faster than the enzyme turnover rate [2], allowing conversion between the different conformations.

Previous molecular dynamics (MD) simulations of the native enzyme both *in vacuo* and including water have resulted in small deviations from the starting crystal structure due to a slight compaction of the flaps toward the active site [3,4]. However, the opening of the flaps, which is envisioned to be necessary for substrate access was not seen. This is not surprising since, large scale, low frequency movements are difficult to simulate in proteins using traditional MD methods. To address this problem, we have applied an "activated" dynamics approach [5] to HIV-PR. We present here, based on our simulations, a study of flap motion that is consistent with the dynamic function of this enzyme. A more detailed account of our study is presented elsewhere [6].

Aspartic Proteinases: Structure, Function, Biology, and Biomedical Implications
Edited by Kenji Takahashi, Plenum Press, New York, 1995

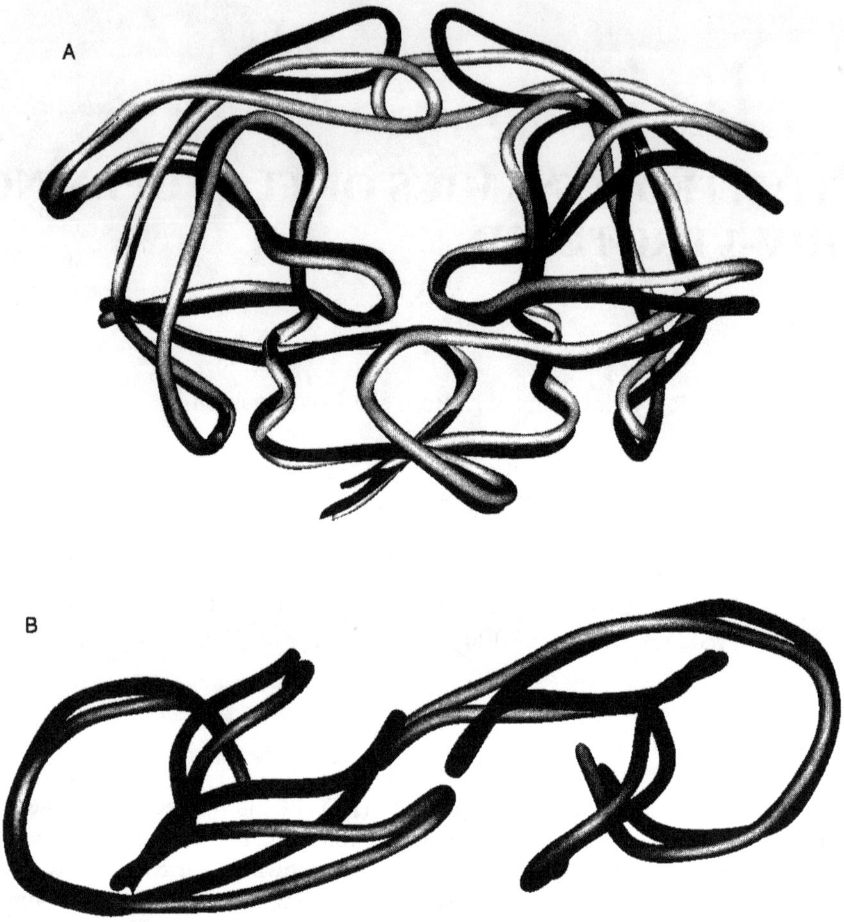

Figure 1. a) View of the "native" (3hvp; dark tubes) and MVT-101 bound (4hvp; light tubes) crystal structures of HIV-1 protease superimposed in the starting coordinate system. b) Top view of Figure 1a emphasizing the "handednes" of the flaps in the two crystal structures. Note that the two flaps would have to physically pass through each other to interconvert between the two structures without the flaps fully opening or melting.

METHODS

Starting from the crystal structures of native (3hvp) or MVT-101 bound (4hvp)HIV-PR [7], molecular mechanics optimizations and MD simulations were performed with the program AMBER[8], explicitly including all hydrogen atoms. Standard AMBER topology files were used to assign charges and add hydrogens. The MD simulations were run for 200ps at a constant temperature of 300K, *in vacuo,* using a distance-dependent dielectric and a non-bonded cutoff of 9 Å.

We adopted an "activated" dynamics approach, similar to template forcing, in our search for a pathway leading to flap opening. Using the Cα atoms of the active site near Asp25 and Asp125, the two crystal structures were superimposed. These atoms were chosen to give the greatest overlap in the "core" region, where the two structures are nearly identical (see Figure 1a). All atoms corresponding to the "non-flap" regions of one structure were

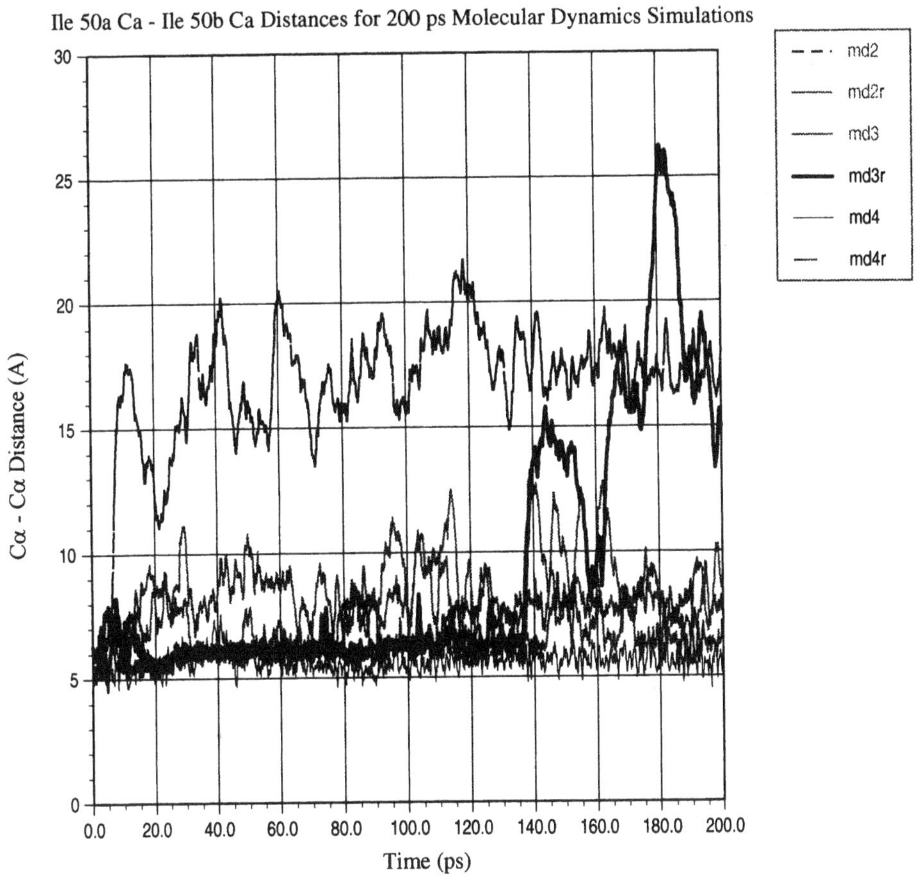

Figure 2. Plot of the distances between the Cα atoms of Ile 50 and 150 for the six different trajectories. Only two, md2 and md3r, separate beyond 15 Å during the course of the simulation.

constrained to move toward the crystal coordinates of the other by applying a harmonic force constant of 1 kcal/$Å^2$ to the former. The assumption underlying these constraints was that a small transition in the "core" atom positions could trigger or more easily allow a large-scale conformational change in the flaps. Effectively, the constraints correlated the atomic motion of the core atoms very early in the trajectories rather than insisting on a random distribution, as is done in most MD simulations. To test our choice of "non-flap" atoms to be constrained, three different sets of "core" atoms were chosen, with each trajectory started from either the inhibited or native coordinates. This combination resulted in six distinct trajectories, three starting from the native conformation, md2, md3, and md4, and three, md2r, md3r, and md4r, starting from the inhibited conformation.

RESULTS AND DISCUSSION

Of the six MD simulations, only two, md2 and md3r, were found to satisfy a distance criteria of 15 Å or more separation between the Cα atoms of Ile50 and 150, which we consider necessary for substrate access (Fig. 2). In trajectory md2, the two flaps fully separated within

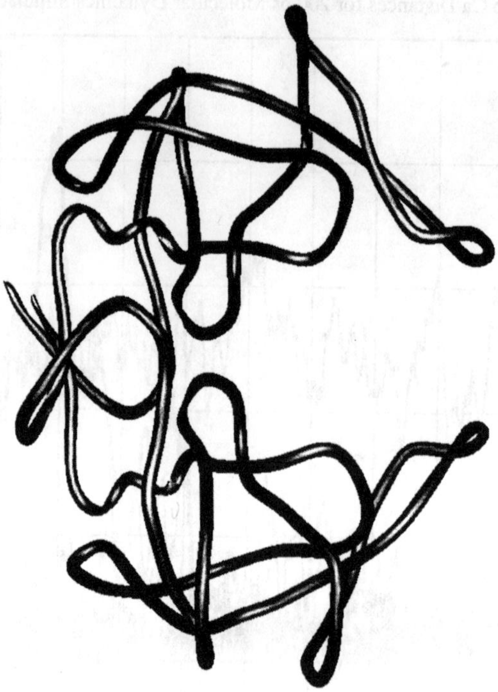

Figure 3. View of the average "open" conformation from trajectory md3r. The average was taken over the structures from 140-200 ps.

20 ps. However, the flaps did not form a structure consistent with substrate binding but changed their conformations to position the Ile50 and 150 sidechains in the active site, effectively blocking access to the catalytic aspartates. In contrast to trajectory md2, the flaps only opened after 140 ps in trajectory md3r, and extended more fully around 180 ps. Visual inspection of this trajectory reveals that the flaps do form a binding cavity that would allow substrate access to the catalytic active site (Fig. 3).

As a first step in examining trajectory md3r, we analyzed the backbone ϕ and φ torsion angles of the non-constrained residues. Residues Met46, Gly52, and Phe53, along with their symmetry counterparts on the other flap, undergo conformational changes in their backbone angles (see Figure 4) at times corresponding to flap opening (Fig. 2), and may serve as "triggers" for flap opening. Thus, flap opening appears to follow a stepwise pathway that requires sequential changes in the ϕ and φ backbone angles for specific residues. The fact that the flaps remained in a conformation intermediate between the MVT101 and native structures for approximately 100 ps prior to the initiation of flap opening suggests that previous simulations may not have been long enough to observe this transition or that there are many stable conformations of the flap that require activation to interconvert.

Our simulations suggest a stepwise pathway for flap opening and provide a complement to experiments attempting to elucidate the possible role of point mutations in flap dynamics. Mutations have been observed at Met46 for HIV-PR obtained from virus grown in the presence of A-77003, an HIV-PR inhibitor [9,10]. Surprisingly, these mutations were found to *not* substantially affect inhibitor binding to HIV-PR. However, mutation of Met46 to Ile appears to restore the wild type growth phenotype to mutant virions that contain potent,

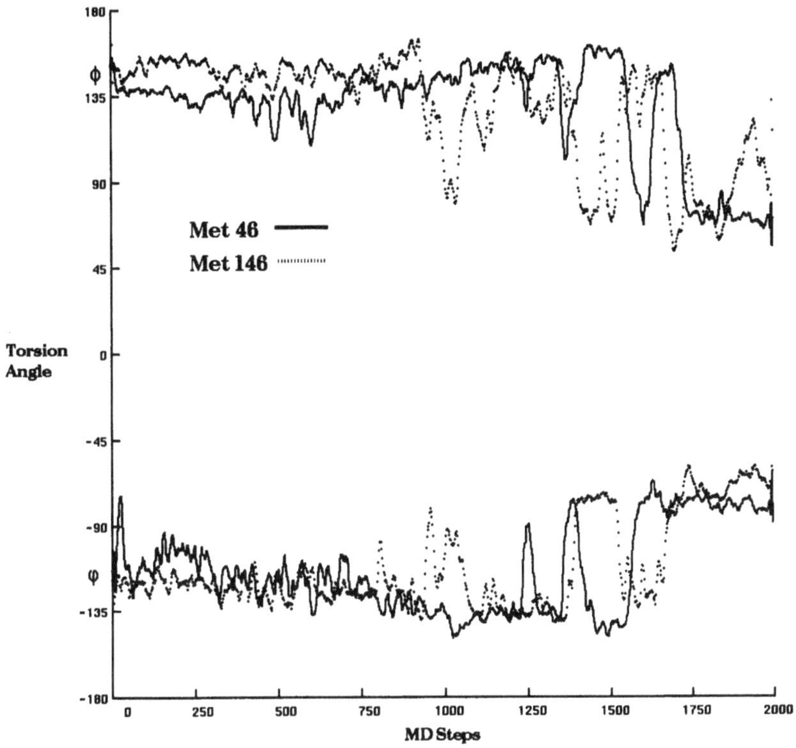

Figure 4. Plot of ϕ/φ backbone torsion angles of methionine 46 and 146 from the md3r trajectory. Upper, ϕ angles; lower φ angles.

drug resistant, mutations of other residues in HIV-PR but that, by themselves, lead to slow growth kinetics. To account for these puzzling observations, it has been proposed that Met46 may influence flap dynamics [9]. Our activated MD simulations lend support for this proposal, and further studies are underway to investigate the effects of mutations at Met46 and other key positions on the flap dynamics observed here.

ACKNOWLEDGMENTS

The content of this publication does not necessarily reflect the views or policies of the Department of Health and Human Services (DHHS), nor does mention of trade names, commercial products, or organizations imply endorsement by the U.S. Government. We thank the staff of the Frederick Biomedical Supercomputing Center, Frederick Cancer Research and Development Center, Frederick, MD for their assistance. Molecular graphics were generated using the MidasPlus software system from the Computer Graphics Laboratory, University of California, San Francisco.

REFERENCES

1. A. Wlodawer and J. Erickson, Ann. Rev. Biochem. 62:543-585 (1993).

2. E.J. Rodriguez, C. Debouck, I.C. Deckman, H. Abu-Soud, F.M. Raushel, and T.D. Meek, Biochemistry 32: 3557-3563, (1993).
3. W.E.Harte, Jr., S. Swaminathan, and D.L. Beveridge, Proteins 13:175-194 (1992)
4. D.M. York, T.A.Darden, L.G.Pedersen, and M.W. Anderson, Biochemistry 32:1443-1453, 1993
5. C.L.Brooks, M.Karplus, and B.M.Pettitt, Proteins: A Theoretical Perspective of Dynamics, Structure, and Thermodynamics (Wiley, New York, 1988).
6. J.R. Collins, S.K. Burt, and J.W. Erickson, Nature (submitted).
7. Brookhaven Protein Databank codes, F.C. Bernstein, J. Mol. Biol. 112:535-542 (1977).
8. D.A. Pearlman, D.A. Case, J.C. Caldwell, G.L. Seibel, U.C. Singh, P. Weiner, and P.A. Kollman (1991), AMBER 4.0, University of California, San Francisco.
9. D.D. Ho, T.Toyoshima, D.J. Kempf, D. Norbeck, C. Chen, N.E. Wideburg, S.K. Burt, J.W.Erickson, and M.K. Singh, J. Virol. (in Press, 1994).
10. A.H. Kaplan, S.F. Michael, R.S. Wehbie, M.F. Knigge, D.A. Paul, L.Everitt, D.J. Kempf, D.W. Norbeck, J.W. Erickson, and R. Swanstrom, Proc. Natl. Acad. Sci. USA (in Press, 1994).

COMPUTER SIMULATION AND ANALYSIS OF THE REACTION PATHWAY FOR THE DECOMPOSITION OF THE HYDRATED PEPTIDE BOND IN ASPARTIC PROTEASES

R. E. Cachau, I. A. Topol, S. K. Burt, A. M. Silva, and J. W. Erickson

Structural Biochemistry Program
PRI/DynCorp
National Cancer Institute
Frederick Cancer Research and Development Center
Frederick, MD 21702-1210

INTRODUCTION

Reaction models based on experimental structures can be used to reveal new details of a reaction, including the exploration of transition states (TS). The TSs usually described in studies of enzyme-inhibitor structures are actually *reaction intermediates* (RI) - short-lived but stable species that can be found during a reaction. Actual TSs are mathematically well-characterized entities, but are inherently unstable species [1]. An enzymatic reaction mechanism can be understood as a recognition process whereby the enzyme *preferentially* recognizes TSs. A very high affinity for substrates, RIs or products will result in a kinetically unfavorable process. On the other hand, enzymatic stabilization of the TS is a requirement for the catalytic process. Thus, a TS mimic is potentially the highest affinity inhibitor that can be designed based on the knowledge of the structure and the details of the mechanism of action of an enzyme. However, the study of reaction paths, which includes the characterization of TSs, is one of the most complex exercises in computational chemistry.

Cleavage of the peptide bond by aspartic proteases is thought to proceed by a two step mechanism [2]: 1) hydration of the peptide bond, followed by 2) proton exchange and bond dissociation. X-ray crystallographic coordinates of a novel difluoroketone-containing inhibitor bound to HIV-1 protease [3] were used in this work to build a model for the study of post hydration reaction steps. A reaction path study for the C-N bond cleavage using this model suggests a concerted mechanism involving simultaneous transfer of two protons between substrate and enzyme.

METHODS

In order to study the reaction path for the peptide bond cleavage, a model RI describing a hydrated peptide bond was built, based on the structure of a difluoroketone inhibitor complexed with HIV-1 protease [2,4]. Subsequently, the aspartic acid side chains of the active site were replaced by two acetic acids, and the hydrated peptide bond by a hydrated N-methyl-acetamide analog, as described elsewhere [4]. Optimization of the model, using AM1 [5] and DGauss [5], resulted in similar geometries (RMSD = 0.12 Å).

An enzymatic reaction is a form of heterogeneous catalysis. Therefore, the treatment of the environment is especially challenging. We have presented a simple protocol [6] for the calculation of effective dielectric constants which is used in this work for the simulation of the average environment. Note that the ionic products of the reaction are too energetically unfavorable from *in-vacuo* calculations, having a positive energy of formation with respect to substrates of about 148Kcal/mol. The reaction field correction, however, reduces the energy of formation of the ionic products to -23Kcal/mol. Even though the exact energy of formation of products will be sequence specific, the calculated energies for the N-methyl-amide reaction exemplifies the importance of including bulk solvation effects.

REACTION PATH CALCULATIONS

A brute force search for a reaction path, starting from the initial geometry of the model, was performed using AM1 and a line search protocol with partial minimizations. At the end of the search the slow changing segments of the energy surface were considered. A putative TS geometry was found, and a restricted search for a TS was performed starting from this point. A geometry whose Hessian matrix has two negative eigenvalues was characterized. By following the energy hypersurface along the direction of the negative eigenvalues, a pictorical representation of the path in the neighborhood of the TS was obtained. The TS shows large displacements along the C-N bond and two dihedrals coordinates which correspond to the C_β-C_γ rotation of the carboxy group near the N, and the carboxy group of the carboxylic acid in formation during the reaction (Figure 2). This extreme point shows the aspartic carboxy groups out of plane, and a *simultaneous* transference of *two* hydrogens taking place. During the path of relaxation of the TS towards substrates and products the carboxy group, proximal to N, rotates 180°. Further refinement of this conformation using DGauss resulted in minor changes in the structure of the TS. The refined TS favors the transfer of a *single* proton from a carboxy group to the N atom. The details of this calculation will be discussed elsewhere [8].

Although the reaction field correction used here is an important contribution to the total energy of the species involved in a reaction, it fails to consider detailed interactions. Therefore, AM1 calculations were performed to include some of the locally important interactions. In particular, the interaction between the aspartic carboxy groups and the nitrogen from the Gly 27/127 were considered. A NH_3, placed in the experimentally determined coordinates of the N, was used to mimic this interaction. A similar TS as the one described in Figure 2 was found for this model. However, if a path is calculated from this TS, the described rotation of one of the Asp carboxy groups is not found. Instead, the carboxy group oscillates 75° about its initial conformation (Figure 2).

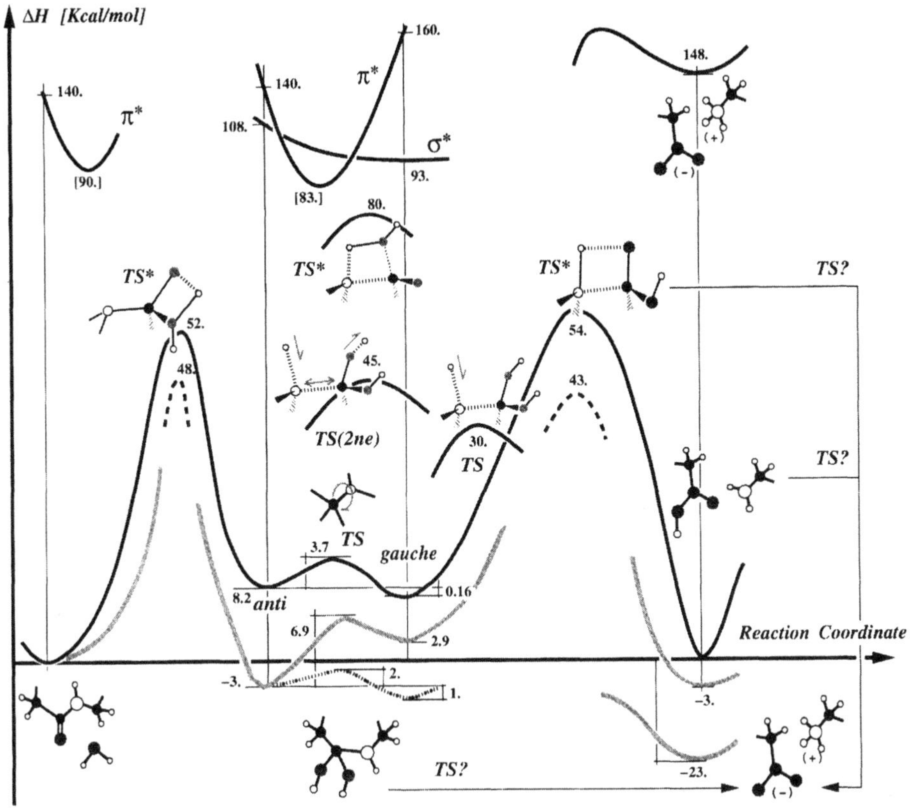

Figure 1. Reaction path calculation. The conformational energy vs. the reaction coordinate is displayed. All energies are referred to an isolated N-methyl-acetamide and 5 water molecules, unless stated otherwise. *TS* are transition states. *TS** are the previously determined TS [3]. *TS(2ne)* is the AM1 TS with two negative eigenvalues of the Hessian matrix. *TS?* are prospective paths of exploration for which no satisfactory TSs have been found yet. In thick black are the *in-vacuo* energies estimated using Gaussian92 (RHF, 6311G** basis set). In gray are the energies corrected by a reaction field approximation using Gaussian92, and an effective dielectric constant (ε=23) for the reaction center simulation. The dotted line shows the energies for the anti-gauche transition estimated from the MD trajectories [4] using the Eyring model [1] and FEPT. This curve was aligned with the RHF calculation for the anti conformation. π^* and σ^* are excited states (triplets and singlets respectively). Dashed lines are the stabilization energies using an external field approximation in Gaussian92. Excited state energies at minima are shown between brackets. The molecular models schematically represent the conformations. The small gray arrows describe the coordinates showing a larger displacement for the negative eigenvector in TS calculations. Oxygen atoms in gray; Nitrogen atoms are large empty circles; Hydrogen atoms are small empty circles; Carbon atoms in black.

In order to assess the feasibility of reaction paths through excited states, a large set of calculations were carried out, with negative results. The lowest energy excited states found are shown in Figure 1.

Reported TSs calculated for small, isolated model peptide systems [7] were analyzed as to their feasibility as putative TSs for the catalytic reaction path. This was done by computing the stabilization energy of each TS as a function of an external electric field. The stabilization energy indicates whether a TS is sensitive to the environment. The enzyme must reduce the activation energy of the process. Two reported TSs were stabilized under an electric field (Figure 1), but not to the extent necessary to considered them for a catalytic path of reaction.

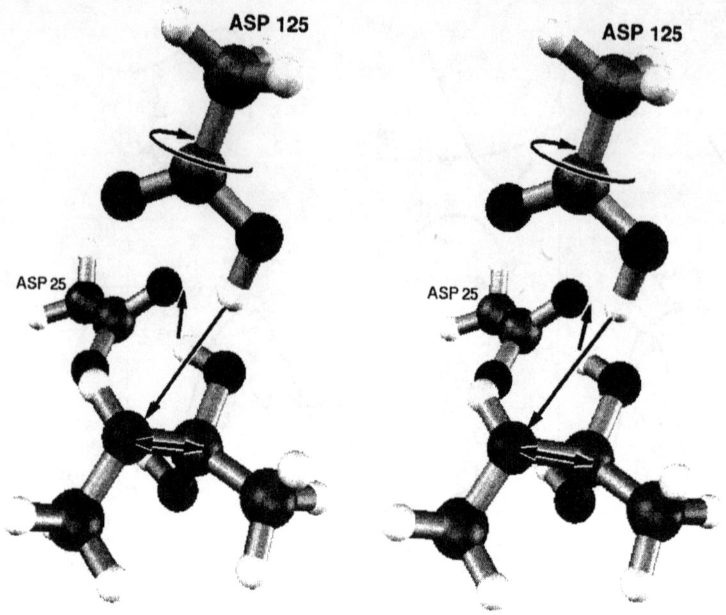

Figure 2. Schematic representation of the TS geometry from the initial geometry (depicted). The larger displacements are on the Cβ-Cγ near the N atom (arrow), the C-C on the forming carboxy group, the C-N bond (arrow), and the transferring H atoms (arrows).

CONCLUSIONS

A simple model for the study of the HIV protease peptide cleavage reaction path was built starting from the experimentally determined structure of a hydrated peptide bond analog. AM1 calculations using this model resulted in a path describing a concerted mechanism where two protons are simultaneously transferred between enzyme and substrate. This is the first reported study of a TS calculation showing the interaction between the hydrated peptide bond and the carboxy groups in an aspartic protease. The role of the environment in these calculations is critical for accuracy. This is exemplified in our calculations by the inclusion of the NH_3-Carboxy groups interaction which modifies the ability of the carboxy groups to freely rotate around the C_β-C_γ bond and therefore affects the reaction path. It is clear from our calculations that a model of an aspartic protease reaction center should include *at least* the carboxy groups of the central aspartic acids, the coordinating Ns, and a reaction field correction to achieve realistic results. Density Functional Theory is the most cost effective approach to simulate these complex systems. However, better mechanisms to transfer information between calculations performed with different Hamiltonians must be developed to optimize the advantage of using different levels of representation of the simulated models.

ACKNOWLEDGMENTS

Research was supported by the National Cancer Institute, DHHS, under contract NOICO-74102 with Program Resources, Incorporated. The contents of this publication do not necessarily reflect the views or policies of the DHHS, nor does mention of trade names, commercial products, or organizations imply endorsement by the U.S. Government.

REFERENCES

1. F.L. Pilar, *Elementary Quantum Chemistry*. McGraw-Hill, New York, 1968.
2. B. Veerapandian, J.B. Cooper, A. Sali, T.L. Blundell, R.L. Rosati, B. W. Dominy, D.B. Damon and D.J. Hoover. *Protein Science,* 1:322 (1992).
3. A. M. Silva, R. E. Cachau, E. T. Baldwin, S. Gulnik, H. L. Sham and J. W. Erickson. Manuscript in preparation.
4. A. M. Silva , R. E. Cachau, E. T. Baldwin S. Gulnik, H. L. Sham and J. W. Erickson. These proceedings.
5. DGauss, Gaussian and MOPAC/AM1 are available as part of the UniChem software from Cray Research, Eagan, MN.
6. I.A.Topol, R.E.Cachau, S.K. Burt and J.W. Erickson. These proceedings.
7. T. Oie, G.H. Loew, S.K. Burt, J.S. Binkley and R.D. MacElroy. J.Am.Chem.Soc. 104:6169 (1982).
8. I.A. Topol, R.E. Cachau, Y. Abashkin, S.K. Burt and J.W. Erickson. Manuscript in preparation.

ACTIVITIES OF PRECURSOR AND TETHERED DIMER FORMS OF HIV PROTEINASE

Lowri H. Phylip,[1] Jonathan T. Griffiths,[1] John S. Mills,[2] Mary C. Graves,[3] Ben M. Dunn,[4] and John Kay[1]

[1]Department of Biochemistry
University of Wales
Cardiff CF1 1ST, Wales, U.K.
[2]Department of Molecular Virology
Roche Research Centre
Welwyn Garden City, Herts, AL7 3AY, U.K.
[3]Department of Biotechnology
Roche Research Center
Nutley, NJ 07110
[4]Department of Biochemistry and Molecular Biology
J. Hillis Miller Health Centre
University of Florida
Gainesville, FL 32610

INTRODUCTION

In the human immunodeficiency viruses (HIV-1 and HIV-2), synthesis of the viral *gag* and *pol* proteins as a *gag-pol* polyprotein necessitates cleavage to release the mature forms of each of the viral proteins. The aspartic proteinase encoded within the polyprotein is responsible for these cleavages, including its own autocatalytic release. Hence the proteinase is initially produced as a molecule much larger than the 11 kDa mature form that has been so well characterised. Since relatively little attention has been paid to these larger forms, it is important to evaluate the nature of the activity when the proteinase component of HIV is present in a precursor form which may be reflective of that in the natural *gag-pol* polyprotein. The synthesis of the proteinase as a zymogen may provide a means of regulating proteolytic activity in a manner reminiscent of the activation of archetypal aspartic proteinase precursors e.g. the activation of pepsinogen at low pH.

The generation of a fully competent aspartic proteinase requires two monomers to combine as each monomer contributes one of the two Asp residues required for catalytic activity. Such dimerisation may occur at the *gag-pol* polyprotein level since the other components of the polyprotein also have an obligate requirement to dimerise i.e. reverse

Aspartic Proteinases: Structure, Function, Biology, and Biomedical Implications
Edited by Kenji Takahashi, Plenum Press, New York, 1995

transcriptase (1) and integrase (2), in addition to the *gag* proteins which naturally oligomerise to form the structural core of the virus. Thus the necessity of the *gag-pol* polyprotein to dimerise to form a functional molecule and the concentration of such species required for dimerisation may serve as an additional level of control over the regulation of the proteolytic activity. In order to examine this aspect further, various tethered homo and heterodimeric forms of HIV-1 and HIV-2 proteinases were constructed whereby the two monomers were covalently linked by a Gly-Gly sequence (3).

The conversion of mammalian and fungal aspartic proteinase precursors into the mature enzymes is often an autocatalytic event involving the removal of the propart region of 40-50 residues located at the N-terminus of the molecules (4). The 65 residues at the start of the *pol* open reading frame immediately upstream of the mature proteinase correspond to the propart found in archetypal aspartic proteinases but no functional role has been suggested for this region. It is well documented that the introduction of a β-branched amino acid into the P_1 position prevents cleavage of both synthetic peptides (5,6) and engineered protein (7) substrates by mature HIV proteinase. Thus, to generate an extended form of HIV-1 proteinase that is not able to undergo self-processing at its N-terminus to liberate the mature proteinase, the N-terminal junction between the 65 residue prosegment and the mature (99 residue) HIV-1 proteinase was mutated from the wild type -Phe*Pro- into an -Ile-Pro- and separately a -Val-Pro- sequence (8). When these mutant constructs were expressed in *E.coli*, a larger form of the proteinase of 18 kDa was detected in each cell extract by Western blotting indicating that the presence of β-branched residues in the P_1 position of the cleavage junction had prevented the autolytic removal of the propart from the mature proteinase resulting in the generation of an extended proteinase. In order to generate sufficient amounts of the extended proteinase for further study, the -Val-Pro- mutant proteinase was placed as an in-frame fusion behind the highly soluble maltose binding protein (MBP 42 kDa) linked through the sequence -IEGRISEF- to the start of the prosegment (8).

Following IPTG-induction of the MBP-Val-Pro- containing mutant, two bands (at ~ 60 and 18 kDa) were detected by staining a Western blot with an anti-proteinase monoclonal antibody. However, only the 60 kDa band which corresponded in size to that anticipated for the fusion between MBP (42 kDa) and the extended proteinase (18 kDa), reacted with an antibody to MBP. The soluble product (approx 10 mg/l of *E.coli* culture) was purified by affinity chromatography on an amylose-agarose column, and the material eluted from the column in the presence of 10 mM maltose was found to contain only two bands of 60 and 18 kDa on SDS-PAGE, followed by Coomassie Blue staining. Both bands were immunoreactive with the anti-proteinase antibody. Affinity chromatography on amylose resin requires the presence of a MBP component, thus the presence in the purified material of the 18 kDa band that was immunoreactive with the anti-proteinase monoclonal antibody may be explained by the formation of heterodimeric molecules through the association of one 60 kDa subunit consisting of full-length MBP-extended proteinase with a second (18 kDa) subunit consisting of the extended proteinase but lacking the MBP moiety. The existence of such a heterodimeric species suggests that correct folding must have been achieved in order for the MBP moiety to bind to the amylose resin and secondly for the extended proteinase components to dimerise.

In order to establish whether the 60 kDa fusion protein might be split into its individual components by autodigestion of the extended proteinase, samples of the column-purified material were incubated in 10 mM Tris-HCl, 100 mM NaCl pH 8.0 at 37°C in the absence and presence of 100 nM Ro31-8959 which is a completely specific inhibitor for HIV-proteinase (9). The conversion of the 60 kDa fusion protein into the 42 kDa MBP and the 18 kDa extended proteinase was initially slow (within the first 24 hours) but accelerated thereafter. However, no conversion was observed in the incubation performed in the presence of Ro31-8959, confirming that the conversion of the 60 kDa fusion protein into its individual components was indeed mediated by the extended HIV proteinase moiety.

In order to characterise the extended proteinase, the 18 kDa band present in the column-purified material was eluted from a preparative SDS-gel. N-terminal sequence analysis by Edman degradation revealed the sequence,

Leu-Gln-Gly-Lys-Ala-Arg-Glu-Phe-Ser-Ser-Glu-Gln-Thr~

Location of this in the propart region indicated that the extended proteinase must have cleaved itself at the sequence -Ala-Phe***Leu-Gln-** in order to generate the 18 kDa band.

Frameshift

```
CAG GCT AAT TTT TTA GGG AAG ATC TGG CCT TCC TAC AAG GGA AGG CCA
Gln Ala Asn Phe Leu Gly Lys Ile Trp Pro Ser Tyr Lys Gly Arg Pro gag
          Phe Phe Arg Glu Asp Leu Ala Phe Leu Gln Gly Lys Ala   pol
          ↑                            ↑
        pol start                  cleavage site
```

Indeed, this sequence matches the consensus cleavage site (5,6,10) mapped for substrate hydrolysis by mature HIV proteinase,

-Ile/Val/Ala-Hydrophobic*Hydrophobic-Glu/Gln-.

To test this, a synthetic peptide

Gly-Ser-Gly-Asp-Leu-Ala-Phe*Leu-Gln-Gly-Lys-Ala-Arg

spanning this putative cleavage site was incubated with homogeneous mature HIV-1 proteinase at pH 4.7. Amino acid analysis of the cleavage products confirmed the location of the scissile peptide bond as being -Phe*Leu-. Values for K_m and k_{cat} of 680 μM and 2 sec^{-1} were derived for the hydrolysis of this peptide. These are comparable to those derived for other peptide substrates containing Ala and Gln in P_2 and P_2' respectively (6,10, 11). All these observations indicate that an additional cleavage junction may well be present in the frameshift region of the *gag-pol* polyprotein.

The pH dependence of the autoconversion of the fusion protein at this newly-identified cleavage site was examined in a further set of incubations performed at pH 5.0, 6.0, 7.0 and 8.0, with the extent of conversion monitored by SDS-PAGE (8). No autolytic release of the MBP from the fusion protein was observed at pH 5.0, whereas at all the higher pH values, conversion did occur. Thus, the extended or precursor form of HIV proteinase (albeit in the form of a heterodimer with one subunit attached to MBP) may not require acidic conditions for activity. If the characteristics of the fusion protein described above (an artificial construct designed for ease of manipulation) should be reflective of the wild type *gag*-PR-*pol* polyprotein fusion, then precursor form(s) with the proteinase still present as part of the polyprotein may be capable of performing the essential cleavages at neutral pH inside the infected cell and budded virion. At the same time, the precursor proteinase mediated processing would result in the release of the mature form of the proteinase. The latter form of the proteinase which is maximally active at pH values of 5.0 and below (5) may remain dormant under such a neutral environment. This may be of advantage to viral replication since it has been demonstrated (12,13,14) that the mature proteinase is capable of digesting various cellular proteins under acidic conditions which could have detrimental consequences on cellular viability.

Table 1. Specificity Constants for the Cleavage of Two Chromogenic Peptides at pH 4.7 by Wild-Type and Tethered Dimer HIV-1 and HIV-2 Proteinases

Substrate 1 = Lys-Ala-Arg-Ile-Nle*Nph-Glu-Ala-Nle-NH_2
Substrate 2 = Ac-Tyr-Arg-Ala-Arg-Val-Phe*Nph-Val-Arg-Ala-Ala-Lys

Substrate	W-T 1	1/1	1/2*	1/2	2/2	W-T 2
			k_{cat}/K_m ($mM^{-1}sec^{-1}$)			
1	2050	2200	920	1400	1080	1050
2	1100	1530	100	100	16	20

It was mentioned earlier that the need for dimerisation to form a competent catalytic proteinase may serve as a regulatory mechanism for viral processing and assembly. It is believed that the close proximity of *gag-pol* polyproteins anchored in the cell membrane by a myristic acid linkage results in the localisation of high concentrations of polyprotein species within a small spatial volume. This may provide the trigger for the association and dimerisation of the proteinase (along with the other polyprotein components). A single chain form of mature HIV-1 proteinase was constructed by insertion of a short linker peptide to connect the N and C termini of two 11 kDa HIV proteinase monomers covalently (15). Insertion of such a tethered dimer proteinase into proviral clones resulted in premature processing of the viral polyprotein and prevented the formation of infectious particles (15). One possible explaination offered for these observations may be that such tethered dimers may be more active than their untethered counterparts.

To further investigate this hypothesis, various tethered dimer form of mature proteinase were produced in which a -Gly-Gly- spacer linked the C terminal residue of the first subunit to the N terminus of the second. In this manner, four plasmids were generated containing single-chain tethered dimer genes for HIV1/1, HIV2/2 (both homodimers), HIV1/2 (heterodimer) and a modified HIV1/2* proteinase (heterodimer) in which the four N terminal (Pro-Gln-Phe-Ser-) and the 5 C terminal (-Met-Ser-Leu-Asn-Leu-) residues of the HIV-2 subunit were replaced by the corresponding residues (Pro-Gln-Ile-Thr- and Cys-Thr-Leu-Asn-Phe, respectively) from HIV-1 proteinase (3).

Expression of each of these constructs in *E.coli* resulted in the production of an immunoreactive proteinase species of 22 kDa for each of the four tethered dimers, as detected by Western blotting. Each of the four tethered dimer proteinases was purified from the *E.coli* lysates (3). In no case was an immunoreactive band of 11 kDa detected, indicating that each of the four tethered dimers was stable and incapable of self-processing.

The hydrolysis of two synthetic chromogenic peptide substrates, each containing a nitrophenylalanine (Nph) residue in the P_1' position as a reporter group (16), by the tethered dimer enzymes was examined and compared with the action of the respective wild-type proteinases from HIV-1 (W-T1) and HIV-2 (W-T2) respectively (Table 1).

It was evident that the activities of the tethered homodimers from HIV-1/1 and HIV-2/2 on the first peptide

Lys-Ala-Arg-Ile-Nle*Nph-Glu-Ala-Nle-NH_2

were directly comparable to those of the wild type HIV-1 and HIV-2 proteinases. The tethered (1/2* and 1/2) heterodimeric proteinases displayed comparable activities to the homodimeric

Table 2. Inhibition Constants (K_i) for the Interaction of Three Synthetic Inhibitors with the Four Tethered Dimeric Proteinases and with Wild-Type HIV-1 and HIV-2 Proteinases

INHIBITOR	W-T 1	1/1	1/2*	1/2	2/2	W-T 2
			K_I (nM)			
1. 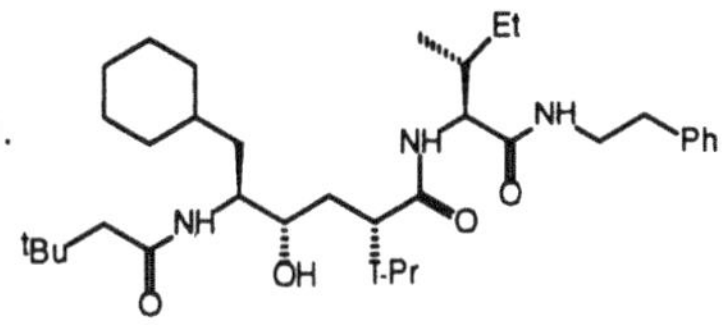	0.8	2	15	10	>500	>500
2. 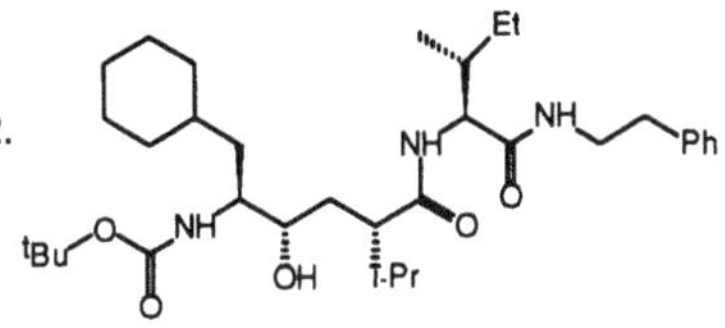	0.4	0.3	3	0.3	200	45
3. 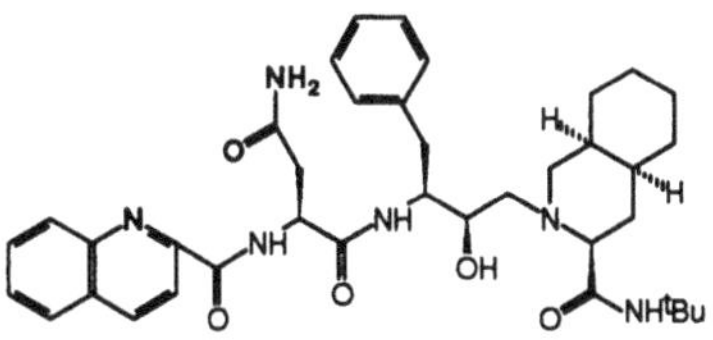	0.3	0.3	0.3	0.2	0.4	0.3

enzymes towards this substrate . Thus it was established that the exchange of non-identical HIV-1 and HIV-2 subunits produce catalytically competent enzymes.

In contrast to substrate 1 which was cleaved only two-fold more effectively by HIV-1 proteinase compared to HIV-2 proteinases (Table 1), the second substrate,

Ac-Tyr-Arg-Ala-Arg-Val-Phe*Nph-Val-Arg-Ala-Ala-Lys

is an excellent substrate for wild type HIV-1 proteinase but is poorly hydrolysed by HIV-2 wild type proteinase. Hence this substrate allows discrimination and detection of subtle differences in the activities of the four tethered proteinases. The results for the tethered dimer enzymes reflect the findings derived for wild type HIV-1 and HIV-2 proteinases (Table 1) in that the tethered HIV-1/1 homodimer cleaved the substrate very efficiently but the HIV-2/2 tethered homodimer displayed a specificity constant (k_{cat}/K_m) that was only 1% of that of the HIV-1/1 homodimer. Thus, it would appear that the covalent linkage of the two subunits to form a single-chain proteinase has no significant effect on intrinsic activity. The kinetic constants determined for the two heterodimeric proteinases were intermediate between those of the two homodimers, so it appears that the two domains of the heterodimers are not equivalent and that the nature of the substrate has considerable influence on the efficiency of cleavage by the family of four tethered enzymes. This effect was maintained when the analyses were performed at a variety of pH values (3).

Further characterisation of the four tethered dimers involved the determination of inhibition constants (K_i) with a series of synthetic inhibitors for comparison with the values determined for wild type HIV-1 and HIV-2 proteinases (3). The specific inhibitor Ro31-8959 (Inhibitor 3-Table 2) was found to be an effective inhibitor of all four tethered dimers. Similarly, K_i values derived for the interaction of a further two inhibitors with the HIV-1/1 and HIV-2/2 tethered homodimers were comparable to the values obtained with their wild type counterparts (3). However, the two inhibitors (1 and 2-Table 2) which were based on hydroxyethylene transition state analogues of the -Hydrophobic*Hydrophobic- type of cleavage junction (11) were more potent against HIV-1 proteinase (wild type and homodimer) but much less effective against HIV-2 proteinases. This is believed to be attributed to the bulky hydrophobic cyclohexylalanine residue in P_1 which is better tolerated by the S_1 subsite of HIV-1 relative to that of HIV-2. The K_i values for these two discriminating inhibitors indicated that the heterodimeric proteinases behaved more like the HIV-1/1 homodimer in terms of binding.

The kinetic constants determined for the two heterodimers indicated that these two enzymes are not simple hybrids. In terms of catalysis, the heterodimers appear to behave more like HIV-2 proteinase, whereas in terms of binding, particularly to certain inhibitors, they resemble HIV-1 proteinase more closely. In the case of such heterodimers, the active site is asymmetrical and it seems that the ligand binding is being selected preferentially to direct certain residues into the HIV-1 domain rather than that contributed by the HIV-2 domain. There are only five different residues in the active sites of HIV-1 and HIV-2 proteinase that contact the ligand, thus the conclusion may be amenable to examination by a mutagenesis approach.

ACKNOWLEDGMENTS

LHP and JTG were supported by the MRC-AIDS Directed Programme. We are very grateful to our many colleagues in Roche for their valuable contributions to this work.

REFERENCES

1. K. Moelling, M. Nawarth, T. Schulze, L. Pavlitzkova, M. Soucek, K.H. Budt, L.H. Pearl, M.T. Knoop & V. Kruft, Retroviral Proteinases, Maturation and Morphogenesis (L.H. Pearl, ed.) Macmillan Press, London pp12-29 (1990).
2. K.S. Jones, J. Coleman, G.W. Merkel, T.M. Laue & A.M. Skalka, J. Biol. Chem. 267:16037 (1992).
3. J.T. Griffiths, L.A. Tomchak, J.S. Mills, M.C. Graves, N.D. Cook, B.M. Dunn & J. Kay, J. Biol. Chem. 269:1 (1994).
4. J. Kay, Aspartic Proteinases and Their Inhibitors (V. Kostka, ed.) Walter de Gruyter, Berlin pp 1-8 (1985).
5. A.D. Richards, L.H. Phylip, W.G. Farmerie, P.E. Scarborough, A. Alvarez, B.M. Dunn, P.H. Hirel, J. Konvalinka, P.Strop, L. Pavlickova, V. Kostka & J. Kay, J. Biol. Chem. 265:7733 (1990).
6. L.H. Phylip, A.D. Richards, J. Kay, J. Konvalinka, P.Strop, V. Kostka, A.J. Ritchie, A.V. Broadhurst, W.G. Farmerie, P.E. Scarborough & B.M Dunn, Biochem. Biophys. Res. Commun. 171:439 (1990).
7. R.A. Jupp, L.H. Phylip, J.S. Mills, S.F.J. Le Grice & J.Kay, FEBS Lett. 283:180 (1991).
8. L.H. Phylip, J.S. Mills, B.F. Parten, B.M. Dunn & J. Kay. FEBS Lett. 314:449 (1992).
9. N.A. Roberts, J.A. Martin, D. Kinchington, A.V. Broadhurst, J.C. Craig, I.B. Duncan, S.A. Galpin, B.K. Handa, J. Kay, A. Krohn, R.W. Lambert, J.H. Merrett, J.S. Mills, Parkes, K.E.B., S. Redshaw, A.J. Ritchie, D.L. Taylor, G.L. Thomas & P.J. Machin, Science 248:358 (1990).
10. R.A. Poorman, A.G. Tomasselli, R.L. Heinrikson & F.J. Zezdy, J. Biol. Chem. 266:14554 (1991).
11. J.T. Griffiths, L.H. Phylip, J. Konvalinka, P.Strop, A. Gustchina, A. Wlodawer, R.J. Davenport, R. Briggs, B.M. Dunn & J. Kay, Biochemistry 31:5193 (1992).
12. M. Wallin, J. Deinum, L. Goodbar & U.H. Danielson, J. Gen. Virol. 71:1985 (1990).
13. A.G. Tomasselli, J.O. Hui, L. Adams, J. Chosay, D. Lowery, B. Greenberg, A. Yem, M.R. Deibel, H. Zurcher-Neely & R.L. Heinrikson, J. Biol. Chem 266: 14548 (1991).
14. B. Honer, R.L. Shoeman & P. Traub, Cell Biol. Int. Rep. 16:603 (1992).
15. Krausslich H.G., Proc. Natl. Acad. Sci. USA 88:3213 (1991).
16. B.M. Dunn, A. Gustchina, A. Wlodawer & J. Kay, Methods Enzymol., in press (1994).

SITE-DIRECTED MUTAGENESIS OF HIV-1 PROTEASE: GENERATION OF MUTANT PROTEASES WITH INCREASED STABILITY TO AUTODIGESTION

Alfredo G. Tomasselli, Ana M. Mildner, Donna J. Rothrock, Jean L. Sarcich, June Lull, Joseph Leone, and Robert Heinrikson

Upjohn Laboratories
The Upjohn Company
Kalamazoo
Michigan 49001

INTRODUCTION

Recently, we reported [1] a single step, large scale purification of recombinant HIV protease, wherein *Escherichia coli* inclusion bodies harboring the enzyme are dissolved in 50% acetic acid and the solution is subjected to size exclusion chromatography in the same solvent on a column of Sephadex G-75. This was an important advance in our studies of the protease because it afforded large quantities of highly pure and active enzyme in a short time. However, we still encountered difficulties because of autodigestion at the three sites (Leu_5-Trp_6, Leu_{33}-Glu_{34} and Leu_{63}-Ile_{64}) [1,2,3] which took place during the concentration process, repeated freeze-thawing, and prolonged storage. Since much of our research had been devoted to understanding the specificity of the HIV-1 and 2 proteases, we knew a good deal about what amino acid substitutions could be made in a peptide sequence in order to destroy its susceptibility to hydrolysis. These ideas were incorporated into site-directed mutagenesis of the HIV-1 protease gene to make a variety of HIV-1 protease mutants that are superior with respect to the wild-type enzyme in terms of stability to autodigestion and yet which maintain the same substrate specificity. This paper deals with the technology employed for the design and expression of HIV-1 protease mutants, their purification, and the refolding of inactive proteins to active enzymes. A companion paper [4] describes in some detail the strategy we used to select certain mutations at the P_1 and/or P_2' position to obtain mutants resistant to autolysis, and the effects of the mutations on the enzyme's stability and activity.

MATERIALS AND METHODS

Materials

Restriction endonucleases, T4 polynucleotide kinase, and T4 ligase were obtained from Boehringer Mannheim or New England Biolaboratories. Taq DNA was from Perkin Elmer and Sequenase version 2.0 from U.S. Biochemicals Media and antibiotics for cell culture were from BRL. All reactions were performed following manufacturer's guidelines. Oligonucleotides were purchased from Genosys. Solvents for protein sequence analysis, and HPLC, and other routine laboratory chemicals were of the highest grade commercially available.

Site-Directed Mutagenesis and Protein Expression

The Polymerase Chain Reaction technique (PCR) was used to synthesize the first mutant (K7/V33). Specifically, two long (165 nucleotides) and two short (36 nucleotides) overlapping oligonucleotides were assembled as detailed in [5]. The PCR method described in [6] was used to generate the other mutants from the K7/V33 gene. For expression, the entire HIV-protease sequence, preceded by a methionine codon and followed by two termination codons, was inserted in a *Cla*I-*Hind*III fragment which was placed behind the phage λ leftward promoter P_L and an AT-rich ribosome binding site (constructed by John Mott, The Upjohn Company) in the pUC-vector background. Recombinant plasmids were constructed and screened in *E. coli* MC1061λ+wildtype λ lysogen) and transformed into *E. coli* 1458 (λ lysogen with tscI857, temperature-sensitive mutation of the repressor cI) to allow the induction of the PL promoter by high temperature. All constructs were sequenced by the dideoxy chain termination method.

Cell cultures containing recombinant plasmids were grown at 30°C with aeration in Luria broth supplemented with 100 μg/mL ampicillin. Expression of protease mutants was induced by dilution of an overnight culture 100-fold into Luria broth with 100 μg/mL ampicillin; growth was continued at 30°C until the A_{550} reached 0.6 to 0.8, at which point the culture was rapidly switched to 40°C for about 4 h. Inclusion bodies production reached a maximum in about 4 h. Inclusion bodies were extracted from the cells by sonication and collected by centrifugation. Sonication and centrifugation were done at 4°C in 10mM Tris-HCL, pH 8.0, containing 1mM EDTA. Expression of the HIV-1 protease and its mutants was analyzed by SDS-PAGE: a prominent band of approximately 11kDa which accounted for greater than 50% of the total protein in the inclusion bodies was routinely found.

Purification

Inclusion bodies from 10mL *E. coli* cultures were solubilized by addition of 1.0 to 1.5 mL of 50% CH_3COOH and the solution was centrifuged 20 min in an Eppendorf benchtop centrifuge. The supernatant was submitted to RP-HPLC on a Vydac C_4 or C_{18} column (0.46 x 25 cm) and eluted at a flow rate of 1.0 mL/min with a linear gradient of acetonitrile (in 0.15% TFA) from 0 to 100% in 60 minutes. A major peak migrating at around 31 minutes was identified by amino acid analysis and protein sequencing as the HIV protease; it was lyophilized. When mg quantities of wild type HIV-1 protease and its mutants were needed, the purification method based on molecular sieving on Sephadex G-75 [1] was used.

Refolding and Activation of HIV-1 PR and Mutants

In a typical experiment 0.3 to 0.5 mg of purified lyophilized HIV-1 protease was dissolved in 100 to 150 µl 50% CH_3COOH and diluted with a 30-fold excess of refolding buffer (0.1 M sodium acetate pH 5.5, containing 5% ethylene glycol, 10% glycerol and 2 to 3 mM DTT). The solution was usually divided in two portions, one of which was adjusted to pH 4.0 and the other to pH 5.5. The enzyme was allowed to fold on ice by gently mixing. The folded and activated enzyme was assayed for catalytic activity. Protein purity was assessed by the following criteria: HPLC on Vydac C_4, SDS gel electrophoresis, amino acid analysis and protein sequencing.

HIV-1 Protease Assay

The synthetic peptide H-Val-Ser-Gln-Asn-Tyr-Pro-Ile-Val-OH (GSP) was used as the substrate for HIV-1 protease activity determination [7]. HIV-1 protease activity was measured at 30°C in 100 to 200 mM sodium acetate, pH 5.5, containing 5% ethylene glycol, 10% glycerol, in a total volume of 50 µL. The reaction was stopped after 30 minutes by the addition of 75 µL of 1% trifluoroacetic acid (TFA) and subjected to HPLC analysis. HPLC was carried out with a Vydac C_4 column (0.46 x 15 cm), eluting with a linear gradient of 0 to 30% acetonitrile over a period of 25 minutes at a flow rate of 1.0 mL/min.

Stability Experiments

HIV-1 protease was incubated at 25 ± 1°C at a concentration of 0.1 to 0.2 mg/mL in 1.0 to 1.5 mL of 0.1 M refolding buffer at pH 5.5 and 4.0 for 11 days. 220 to 250 µL aliquots were withdrawn at 0 hr, 24 hr, 4 days and 11 days: 1 to 5 µL aliquots were used for amino acid analysis and activity assays, and 150 to 180 µL aliquots were submitted to RP-HPLC on a Vydac C_4 column (0.46 x 25 cm) eluted as described in the purification procedure. Peaks generated by proteolysis were collected manually and subjected to amino acid analysis and/or protein sequencing.

Amino Acid Analysis

HIV-1 protease was quantitated by amino acid composition as determined by automated ion-exchange chromatography on a Beckman amino acid analyzer (Model 6300). Samples were hydrolyzed in 6N HCL for 24 hr at 110°C in evacuated sealed tubes. Dried hydrolysates were dissolved in buffer at pH 2.2 (NaS: Beckman) prior to application to the analyzer.

Peptide Sequence Analysis

Automated Edman degradation of protein and peptide samples was performed in an Applied Biosystems sequencer (Model 470) fitted with an on-line HPLC analyzer (Model 120-A) for phenylthiohydantoin amino acids.

Quantitation of the latter was afforded by a PE Nelson Analytical 900 series interface using Turbochrom 3 chromatography software.

SDS-PAGE

SDS-polyacrylamide gel electrophoresis was run in the presence of reducing agents and stained with Coomassie Blue for detection of proteins.

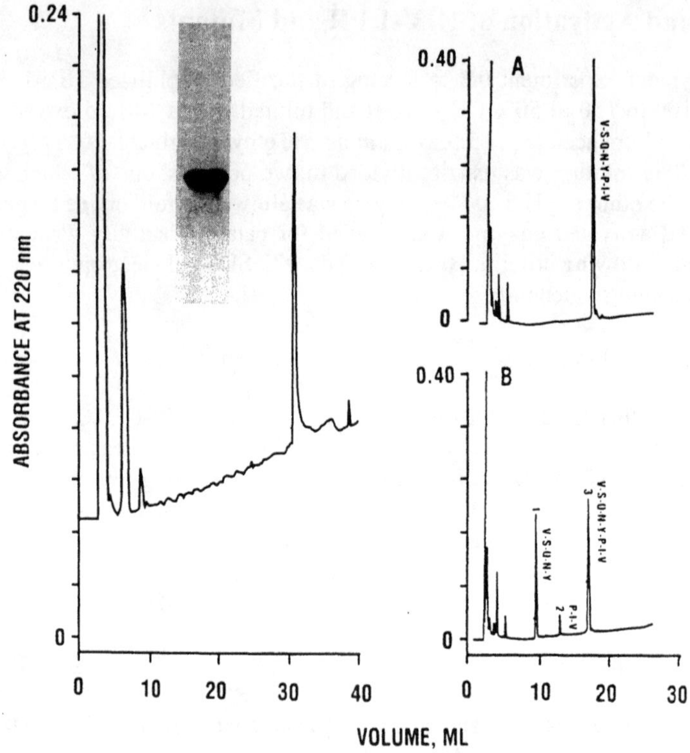

Figure 1. Left panel: elution profile of purified activated Q7K/L33I/L63I mutant from reverse phase HPLC. Right panel: verification of Q7K/L33I/L63I mutant activity. The elution profile of an incubation mixture composed of the octapeptide substrate, buffer and the enzyme is shown at 0 (A) and 30 (B) minute incubation (see details in Methods).

RESULTS AND DISCUSSION

We have performed site-directed mutagenesis of the HIV-1 protease to block autodegradation at the three sites of observed cleavage: Leu_5-Trp_6 , Leu_{33}-Gln_{34}, and Leu_{63}-Ile_{64}. The strategy for mutagenesis, as it will be discussed in the companion paper, has been guided by certain rules that we have learned by examining the specificity of HIV-1 and HIV-2 proteases for viral and non-viral substrates. Specifically, we have introduced Ile/Val at P_1, and/or Lys at P_2' at the three sites of HIV-1 protease autodigestion because, in the various studies of the specificity of HIV-1 protease, these residues are never found in the specified positions [4].

The following mutants were prepared: 1. Q7K/L33V; 2. Q7K; 3.Q7K/L33V/L63I; 4. Q7K/L33V/L63I/E65K; 5. Q7K/L33V/E65K; 6. Q7K/L33I/L63I; 7. Q7K/L33V/E35K/L63I; and 8. Q7K/L33V/E35K/E65K. Expression of these mutant proteases in *Escherichia coli* led to their production in high yield (20 to 50 mg/L cell culture) in inclusion bodies, as is observed for the wild-type enzyme. All preparations were refolded easily following the protocol described in the above Methods and showed excellent degree of purity and high levels of enzyme activity. In addition, the chosen mutations had the desired effect on lowering, or abolishing autocatalysis in the region of amino acid alterations. Of the various mutant HIV-1 proteases that we prepared, mutant (Q7K/L33I/L63I) had minimal alterations in structure with maximal stability against autodigestion. This mutant could be prepared in mg quantities in a highly purified and active form (Figure 1, left and right panels respectively). In the companion paper [4] the stability and other properties of the various mutants is discussed in more detail.

REFERENCES

1. J.O. Hui, A.G. Tomasselli, I.M. Reardon, J.M. Lull, D.P. Brunner, C.-S.C. Tomich and R.L. Heinrikson. *J. Protein Chem.* 12:321-325 (1993).
2. J.R. Rosé, R. Salto and C.S. Craik. *J. Biol. Chem.* 268:11939-11945 (1993).
3. J.E. Strickler, J. Gorniak, B. Dayton, T. Meek, M. Moore, V. Magaard, J. Malinowski and C. Debouck. *Proteins* 6:139-154 (1989).
4. A.G. Tomasselli, A.M. Mildner, D.J. Rothrock, J.L. Sarcich, J. Lull, J. Leone and R.L. Heinrikson. Companion Paper (1994).
5. R.B. Ciccarelli, P. Gunyuzlu, J. Huang, C. Scott and F.T Oakes. *Nucleic Acids Res.* 10: 6007-6013 (1991).
6. R. Higuchi. Recombinant PCR (1990). M.A. Innis, D.H. Gelfand, J.J. Sninsky and T.J. White(ed) PCR Protocols, A Guide to Methods and Applications. Academic Press.
7. A.G. Tomasselli, M.K. Olsen, J.O. Hui, T.K. Sawyer, R.L. Heinrikson and C.-S.C. Tomich. *Biochemistry* 29:264-269 (1990).

MOLECULAR MODELING OF THE STRUCTURE OF FIV PROTEASE

Alla Gustchina

Macromolecular Structure Laboratory
ABL-Basic Research Program
National Cancer Institute
Frederick Cancer Research and Development Center
Frederick, Maryland

INTRODUCTION

Feline immunodeficiency virus (FIV) is a lentivirus isolated from cats with an immune deficiency syndrome (Pedersen et al.,1987). The disease, associated with FIV infection of cats, is very similar to AIDS in human. FIV has an overall life cycle and genomic structure very similar to HIV, and, for example, utilizes a protease in formation of the viral core.

FIV protease (FIV PR) is a virally-encoded aspartyl protease encoded near the 5'-end of the pol open reading frame, and is produced within the gag-pol polyprotein translation product by ribosomal frameshift. The enzyme is responsible for the proteolytic processing of the gag and gag-pol precursor molecules to release the major structural proteins of the virion and the virus-encoded enzymes responsible for virus replications, including the FIV PR itself.

The special advantage of FIV PR in the studies leading to preparation of active compounds with pharmacological properties lies in the fact that they can immediately be tested in an animal model. While the inhibitors of FIV PR would not be necessarily useful in the treatment of human diseases such as AIDS, they may provide good leads in answering questions whether particular chemical groups are compatible with the activity in mammalian cells.

MATERIALS AND METHODS

The complete nucleotide sequence for an infectious molecular clone of the Petaluma strain of FIV was determined by Talbot et al. (1989), and the protease cleavage sites within the gag and pol polyproteins of FIV (Elder et al.,1993) revealed the exact location of the N- and C-termini of the protease.

The three-dimensional model of FIV PR was built on the basis of the crystal structures of Rous sarcoma virus protease (RSV PR) (Jaskolski et al.,1990) and human immunodeficiency virus type I protease (HIV-1 PR) with inhibitor JG-365 (Swain et al.,1990) by using the program FRODO (Jones, 1978) running on an Evans and Sutherland PS390 molecular graphics system.

The conformations for the main chain of particular regions have been chosen in between those in RSV or HIV PRs on the basis of the sequence similarity. One monomer was built in a manner described previously for human immunodeficiency virus type II (HIV-2 PR) (Gustchina and Weber, 1991). The atomic coordinates of the second monomer in the dimer of FIV PR were obtained from the first one by a superposition on the Cα atoms of the second monomer in the crystal structure of HIV-1 PR with inhibitor. Although full energy minimization procedure was not applied to the current model of FIV PR, several cycles of restrained energy minimization have been carried out with the program Discover. The results showed that initial model was reasonable in terms of geometry and prohibited contacts.

RESULTS AND DISCUSSION

FIV PR is a member of a family of homodimeric aspartic proteases, which are commonly found to be active in retroviruses. Crystal structures of four representative members of the family of retroviral aspartic proteases have been determined experimentally. They include RSV PR (Jaskolski et al., 1990), HIV-1 and HIV-2 PRs (Navia et al., 1989; Wlodawer et al., 1989; Lapatto et al., 1989; Mulichak et al., 1993; Tong et al., 1993) and the protease from avian myeloblastosis associated virus (MAV PR) (Ohlendorf et al., 1992). The molecular model of equine infectious anemia virus protease (EIAV PR) was built by Weber et al. (1993), with the overall structure predicted to be close to HIV-1 PR. In addition, a large number of crystal structures of the complexes of HIV-1 PR with different inhibitors have been determined (Wlodawer and Erickson,1993), which provided structural basis for the studies of the specificity of this enzyme and allowed to rationalize drug design to a large extent.

The alignment of the amino acid sequence of FIV PR with the sequences of RSV and HIV-1 PRs, based on careful structural alignment of the latter two proteases, is shown in Table 1. This alignment showed the number of the residues which are of identical or similar type of the residues at the structurally equivalent positions. The results are summarized below:

	identity	similarity
FIV/RSV/HIV	16	18
FIV/RSV	25	21
FIV/HIV	23	20

The number of identical residues in all three proteases is 16, which is rather small, but when they are compared pairwise, the number of identical and similar residues is becoming significantly larger, about 40%. This level of sequence similarity and the fact that all positions in FIV PR sequence, which correspond to the structural template for the aspartic proteases (Andreeva, 1988), are occupied by appropriate residue types, support the idea that the three-dimensional structure for FIV PR is similar, and a conserved core structure is preserved.

Table 1. Structure-Based Alignment of the Amino Acid Sequences of RSV PR and HIV-1 PR, and Predicted Alignment with the Sequence of FIV PR

```
RSV         L A M T M E H K D R P L V R V I L T N T G S H P V K Q R S V Y
            1                10                  20                  30
HIV1        P Q I T L W . . Q R P L V T I R I G . . . . . . . . . . G Q L
            1                    10
FIV   Y N K V G T T T T L E K R P E I L I F V N . . . . . . . . . . G Y P
      1                10                  20

RSV   I T A L L D S G A D I T I I S E E D W P T D W . P V . M E A A N P
                     40                  50                      60
HIV1  K E A L L D T G A D D T V L E E . . M N L . . P G K W K P K M . .
     20                  30                          40
FIV   I K F L L D T G A D I T I L N R R D F Q V K N S I E N G R Q N . .
               30                  40                  50

RSV   Q I H G I G G G I P M R K S R D M I E L G V I N R . . D G S L E R
                   70                  80                      90
HIV1  . I G G I G G F I K V R Q Y . D Q I P V E I C . . . . . . . G H .
             50                    60
FIV   . M I G V G G G K R G T N Y . I N V H L E I R D E N Y K T Q C I F
               60                    70                  80

RSV   P L L L F P A V A M V R G S I L G R D C L Q G L G L R L T N L
                  100                 110                 120
HIV1  K A I G T V L V G P T P V N I I G R N L L T Q I G B T L N F
     70                  80                  90                99
FIV   G N V C V L E D N S L I Q P L L G R D N M I K F N I R L V M
           90                 100                 110
```

The three-dimensional model of FIV PR is shown in Figure 1. The areas of high homology include an extended region in the active site (residues 22-33 in HIV-1 numbering, 27-38 in FIV numbering, helical region and C-terminus (residues 80-99 in HIV-1 PR, 97-116 in FIV PR) and the region around the tip of the flap (residues 47-53 in HIV-1 PR, 57-63 in FIV PR), although exact alignment of this region is difficult, since the flap was not seen in the crystal structure of RSV PR. Predicted structure in other regions of the molecule is less certain. All three sequences differ in their length. The insertions and deletions in all retroviral proteases with known or modeled three-dimensional structure (HIV-1, HIV-2, RSV, MAV, EIAV PR) were accommodated at the regions of the extended surface loops, for example, loops A1 and A2 (using terminology introduced by Andreeva et al., 1984) in HIV and RSV PRs (Weber, 1990). On the basis of the sequence similarity we modeled loop A1 as it is in HIV-1 protease, and loop A2 using as a base the one in RSV PR (Figure 2). Since the insertion in FIV PR in the region of loop A2 is even longer then in RSV PR, the exact conformation of this extended surface loop can not be predicted.

The region 34-40 belongs to the outer wide loop, which has different conformation in all experimental structures of retroviral proteases, thus we are unlikely to come up with the exact structure of this area just by modeling, even though the length of the chain in FIV PR seems to be identical to that in RSV PR. Since this region may determine the area which confers different immunological properties to different proteases, it may be important for the reason (Gustchina and Weber, 1991). N-terminus in our model was simply extended by three residues in comparison with the length of this strand in the structures of HIV and RSV PRs, and the turn following this strand has the length and the conformation of those in RSV PR.

Comparative analysis of the binding sites in FIV PR model to those in the crystal structures of HIV and RSV PRs revealed the important substitutions among the residues forming them, which may be used as a guide for the rational drug design.

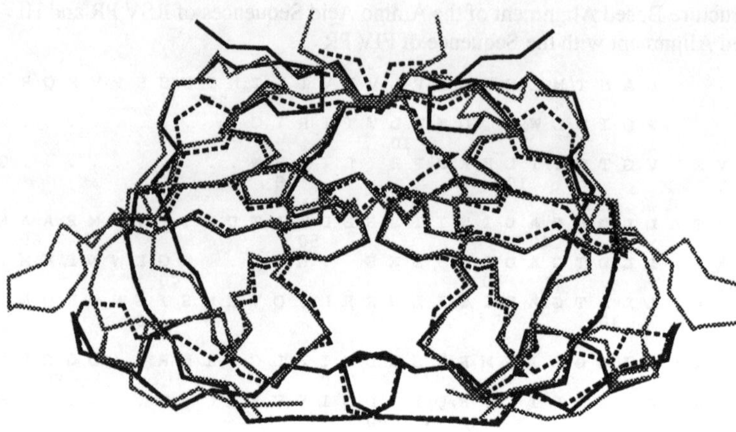

Figure 1. The alignment of crystal structures of HIV-1 PR (dashed lines), RSV PR (grey lines), and the model of FIV PR (solid lines). Only Cα tracing is shown.

Subsites S1 - S1'

The side chains of the residues in the protease comprising these pockets are mostly hydrophobic in HIV PR, and 5 residues were described in Swain et al., (1991) as a main component of the binding pocket in subsites P1/P1': P81, V82, I84, G49 and I50. The substitutions of these residues in all three enzymes are as follows:

HIV	P81	V82	I84	G49	I50	L23
FIV	I98	Q99	L101	G58	V59	L28
RSV	R105	G106	I108	G66	I67	L35

As can be seen, four residues (I84, G49, I50 and L23 in HIV PR) are highly conserved in all three structures, and the other two (P81 and V82) are quite variable. Therefore, Ile98 and Gln99 in FIV PR might be good candidates for the primary specificity of this enzyme.

Subsites S2 - S2'

This subsite has highly hydrophobic nature of the residues forming it in both HIV and RSV PR. The residues which occupy the appropriate positions in the model of FIV PR, also have strictly hydrophobic nature, and even enhance the hydrophobicity of this pocket, since at the position of D30 (the only hydrophilic residue which interact with P2 side chain in the inhibitor MVT-101 (Miller et al., 1989) is isoleucine, as was observed in RSV PR.

HIV	A28	I47	G49	I84	I50	D30	V32
FIV	A33	M56	G58	L101	V59	I35	I37
RSV	A40	I64	G66	I108	I67	I42	I44

There are two candidates for the specific binding in FIV PR for the residues at P2/P2' positions in the substrate, namely Ile35, which could make differences in the specificity of the enzyme towards HIV PR, but would not make any significant differences in comparison with RSV PR, and Met56.

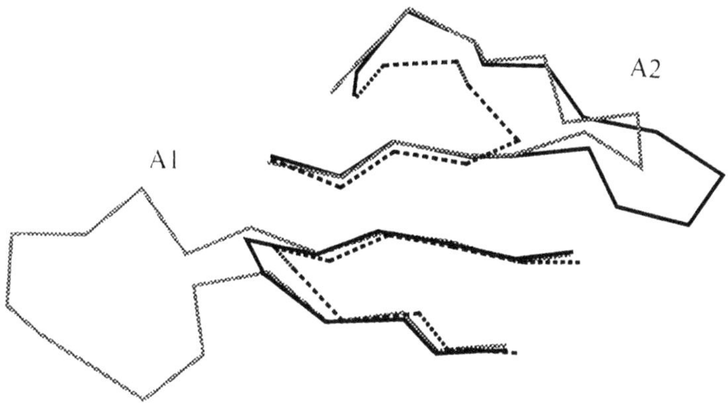

Figure 2. Superposition of the loops A1 (residues 16-26 in FIV PR) and A2 (72-88) in the three proteases, with lines drawn as in Figure 1.

Subsites S3 - S3'

The interactions of different types, aliphatic, polar and ionic, were observed in the crystal structures of HIV PR with the different inhibitors.

HIV	R8	D29	P81	V82
FIV	R13	D34	I98	Q99
RSV	R10	D41	R105	G106

Two residues, R13 and D34, are strongly conserved in all three proteases, and those residues are involved in ionic interactions in the structures of HIV PR with MVT-101 inhibitor. Two other residues are completely different in all three proteases, and may be responsible for the specific interactions for each particular enzyme with P3/P3' side chain of a substrate.

Furthermore, the important substitution is considered to be in the flap region at the position of Gly48 at HIV PR on Ile57 at FIV (according to the current model), because this residue makes important hydrogen bonds in many crystal structures of HIV PR with different inhibitors. At this stage, modeling is not precise enough to predict all details of the subsite specificity, and thus a complete experimental structure is necessary.

REFERENCES

1. Andreeva, N.S., Structural template for aspartic proteases, 1988, *in*: Biosynthesis and Structure of Proteins, Puschino, **3**:97-115.
2. Andreeva, N.S., Zdanov, A.S., Gutschina, A.E., and Fedorov, A.A., 1984, Structure of ethanol-inhibited porcine pepsin at 2-Å resolution and binding of the methyl ester of phenylalanyl-diiodotyrosine to the enzyme, *J. Biol. Chem.* **259**:11353-11365.
3. Elder, J.H., Schnolzer, M., Hasselkus-Light, C.S., Henson, M., Lerner, D.A., Phillips, T.R., Wagman, P.C., and Kent, S.B. H., 1992, Identification of proteolytic processing sites within the gag and pol polyproteins of feline immunodeficiency virus, *J. Virology* **67**:1869-1876.
4. Gustchina, A., and Weber, I.T., 1991, Comparative analysis of the sequences and structures of HIV-1 and HIV-2 proteases, *Proteins: Structure, Function and Genetics* **10**:325-339.
5. Jaskolski, M., Miller, M., Rao, J.K.M., Leis, J., and Wlodawer, A., 1990, Structure of the aspartic protease from Rous sarcoma retrovirus refined at 2-Å resolution, *Biochem.* **29**:5889-5898.

6. Jones, A.T., 1978, A graphics model building and refinement system for macromolecules, *J. Appl. Crystallogr.* **11**:.268-272.
7. Lapatto, R., Blundell, T., Hemmings, A., Overington, J., Wilderspin, A., Wood, S., Merson, J.R., Whittle, P.J., Danley, D.E., Geoghegan, K.R., Hawrylik, S.J., Lee, E.L., Scheld, K.G., and Hobart, P.M., 1989, X-ray analysis of HIV-1 proteinase at 2.7 Å resolution confirms structural homology among retroviral enzymes, *Nature* **342**:299-302.
8. Miller, M., Schneider, J., Sathyanarayana, B.K., Toth, M.V., Marshall, G.R., Clawson, L., Selk, L., Kent, S.B.H., and Wlodawer, A., 1989, Structure of complex of synthetic HIV-1 protease with a substrate-based inhibitor at 2.3 Å resolution, *Science* **246**:1149-1152.
9. Mulichak, A.M., Hui, J.O., Tomasselli, A.G., Heinrikson, R.L., Curry, K.A., Tomich, C., Thaisrivongs, S., Sawyer, T.K., and Watenpaugh, K.D., 1993, The crystallographic structure of the protease from human immunodeficiency virus Type 2 with two synthetic peptidic transition state analog inhibitors, *J. Biol. Chem.* **268**:13103-13109.
10. Navia, M.A., Fitzgerald, P.M.D., McKeever, B.M., Leu, C., Herber, W.K., Sigal, I.S., Darke, P.L., and Springer, J.P., 1989, Three-dimensional structure of aspartyl protease from human immunodeficiency virus HIV-1, *Nature* **337**:615-620.
11. Ohlendorf, D.H., Foundling, S.I., Wendoloski, J.J., Sedlacek, J., Strop, P., and Salemme, F.R., 1992, Structural studies of the retroviral proteinase from avian myeloblastosis associated virus, *Proteins: Structure, Function and Genetics* **14**:382-391.
12. Pederson, N.C., Ho, E.W., Brown, M.L., and Yamamoto, J.K., 1987, Isolation of a T-lymphotrophic virus from domestic cats with an immunodeficiency-like syndrome, *Science* **235**:790-793.
13. Swain, A.L., Miller, M., Green, J., Rich, D.H., Schneider, J., Kent, S.B.H., and Wlodawer, A., 1990, X-ray crystallographic structure of a complex between a synthetic protease of human immunodeficiency virus 1 and a substrate-based hydroxyethylamine inhibitor, *Biochem.* **87**:8805-8809.
14. Talbott, R.T., Sparger, E.E., Lovelace, K.M., Fitch, W.M., Pedersen, N.C., Luciw, P.A., and Elder, J.H., 1989, Nucleotide sequence and genomic organization of feline immunodeficiency virus, *Proc. Natl. Acad. Sci. USA* **86**:5743-5747.
15. Tong, L., Pav, S., Pargellis, C., Do, F., Lamarre, D., and Anderson, P. C., 1993, Crystal structure of human immunodeficiency virus (HIV) Type 2 protease in complex with a reduced amide inhibitor and comparison with HIV-1 protease structures, *Proc. Natl. Acad. Sci. USA*, **90**:8387-8391.
16. Weber, I. T., 1990, Comparison of the crystal structures and intersubunit interactions of human immunodeficiency and rous sarcoma virus proteases, *J. Biol. Chem.* **265**: 10492-10496.
17. Weber, I. T., Tozser, J., Wu, J., Friedman, D., and Oroszlan, S., 1993, Molecular model of equine infectious anemia virus proteinase and kinetic measurements for peptide substrates with single amino acid substitutions, *Biochem.*, **32**:3354-3362.
18. Wlodawer, A., and Erickson, J.W., 1993, Structure-based inhibitors of HIV-1 protease, *Annu. Rev. Biochem.* **62**:543-585.
19. Wlodawer, A., Miller, M., Jaskolski, M., Sathyanarayana, B.K., Baldwin, E., Weber, I.T., Selk, L.M., Clawson, L., Schneider, J., and Kent, S.B.H., 1989, Conserved folding in retroviral proteases: Crystal structure of a synthetic HIV-1 protease, *Science* **245**:616-621.

PROCESSING, PURIFICATION, AND KINETIC CHARACTERIZATION OF THE GAG-POL ENCODED RETROVIRAL PROTEINASE OF MYELOBLASTOSIS ASSOCIATED VIRUS EXPRESSED IN *E. COLI*

Jiří Brynda, Milan Fábry, Petr J. Tichý, Magda Hořejší, and Juraj Sedláček

Department of Gene Manipulation
Institute of Molecular Genetics
Academy of Sciences of the Czech Republic
166 37 Prague 6, Czech Republic

INTRODUCTION

Gag-Pol frameshift translational products of avian retroviruses (e. g. myeloblastosis associated virus, MAV, or Rous sarcoma virus, RSV) contain a putative proteinase species that maps between the NC/PR and PR/RT processing sites. Such Gag-Pol proteinase deffers from the *gag*-encoded proteinase (Gag PR, 124 amino acids length) by a carboxy-terminal extension of 7 amino acids (numbered 125 to 131), i. e. -IGRATVL (see Refs. 1, 2, 3, for review). While the main species of the processing proteinase of avian retroviruses, Gag PR (known also as $p15^{gag}$) was characterized to considerable details (4, 5, 6, 7), direct description of avian retroviral Gag-Pol proteinases remained very limited (3) and substantial knowledge was missing. We attempted to analyze in detail recombinant *gag-pol* encoded species of the MAV avian retroviral proteinase (identical to that of RSV, except for a conservative substitution at amino acid residue 84, Ref. 8) using a truncated, *E. coli* expressed in-frame precursor that retains the natural NC/PR cleavage site and is translationally terminated at the position of the PR/RT site. This work showed that under conditions favourable for processing, an unexpected cleavage occurs between amino acid residues 126 and 127, and that this cleavage bears characteristics of a regular self-processing step. The cleavage product, termed PR(+IleGly), represents an enzyme with catalytic parameters not much different but distinguishable from those of the *gag*-encoded proteinase.

EXPERIMENTAL AND DISCUSSION

Plasmid pMG45, which has been previously used for P_{tac}-controlled massive expression of the MAV PR precursor in *E. coli* (8, 9) was reduced at its NC part by a *Sma*I-*Stu*I delation (preserving only 15 NC amino acid residues upstream of the NC/PR boundary). The plasmid was then altered for in-frame expression of the *gag-pol* encoded sequence by replacing the MAV PR (p15^{gag}) carboxy-terminal sequence between restriction site *Pst*I (PR amio acid 114) and the downstream *Hind*III, with a synthetic oligonucleotide 52/60 duplex, encoding amino acid sequence

```
          gag                     pol
L Q G L G L R L T N L I G R A T V L
  .         .         .         .
 115       120       125       130
```

plus a translational termination signal. In another plasmid construct, the oligonucleotide replacement comprised an amino acid substitution, $Arg_{127} \rightarrow Ser$.

Bacterial fermentations were carried out as described previously (8, 9). Pelleted and washed *E. coli* inclusions were solubilized with 8 M urea and dialysed at room temperature against a buffer containing 50 mM sodium phosphate (pH 6.0), 0.5 M NaCl, 1 mM EDTA, 0.05 % mercaptoethanol, 0.1 mM phenylmethylsulfonyl fluoride and 20 % glycerol, using procedures detailed in (8, 9). Further preparation procedures included a batchwise use of QAE-Sephadex A25 (8), followed by SE-Sephadex column chromatography (10). An illustration of the purification steps is given in Fig. 1. The purified proteinase corresponded to a cleavage product where proteolysis had occurred between amino acid residues 126 and 127, i. e. at the carboxy-terminal sequence which extends downstream of the *gag*-encoded 124 amino acids. Determination of its carboxy terminal structure, i.e. the time course for sequential release of amino acids observed after digestion with carboxypeptidase Y correlated with the presence of a carboxy terminal glycine and penultimate isoleucine residue. Purified PR(+IleGly) differs from Gag PR in its profile of elution from a C18 (Vydac) HPLC column: in a gradient of 60 to 90 % methanol (in water), its elution peak corresponded to a 72.5 % methanol concentration, while that of Gag PR was 71 %.

Product of single cleavage at the Gag-Pol precursor NC/PR site, i. e. the putative proteinase species PR(+IleGlyArgAlaThrValLeu), is detected in the *E. coli* inclusion bodies prior dialysis (Fig. 1, Lane 3). It can also be detected when dialysis is performed at conditions

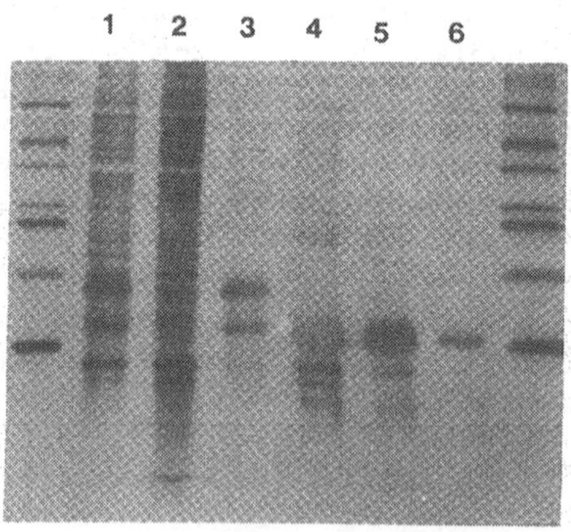

Figure 1. Purification of MAV PR(+IleGly) from the *E. coli* expression product. SDS PAGE, 16.5% gel, silver staining. Lanes: 1, whole cells; 2, cytoplasmic fraction; 3, pelleted and washed inclusions; 4, supernatant after dialysis of solubilized inclusions; 5, effluent from QAE-Sephadex A25 at pH 8.2 (50 mM sodium tetraborate, 4 mM EDTA, 0.05% mercaptoethanol); 6, PR(+IleGly) preparate after SE-Sephadex column (Ref. 10) and final FPLC Mono S chromatography at pH 5.0 using a salt gradient. Molecular size markers were, from bottom, 14.2, 20.1, 24, 29 kDa, etc.

Table 1. Kinetic Parameters for Cleavage of Peptide Substrates by the Retroviral Proteinases of MAV. (Gag-Pol and Gag-Pol* denote PR(+IleGly) species generated, respectively, from the wild-type and mutant expression products. N.D. stands for not determined. All substrates were peptide amides.)

substrate	K_m [μM]			k_{cat} [1/sec]			k_{cat}/K_m [1/sec.1/mM]		
	Gag	Gag-Pol	Gag-Pol*	Gag	Gag-Pol	Gag-Pol*	Gag	Gag-Pol	Gag-Pol*
PAVS*LAMT NC/PR	327 (± 50)	347 (± 84)	261 (± 60)	0.90 (± .07)	0.34 (± .04)	0.26 (± .03)	2.7	0.97	1.0
GRATVL*TVALHL PR/RT	90 (± 16)	93 (± 7)	96 (± 19)	0.044 (± .002)	0.012 (± .001)	0.013 (± .001)	0.49	0.13	0.14
FQAY*PLRE RT/IN	1130 (± 410)	1670 (± 324)	2650 (± 330)	12.4 (± 1.9)	5.5 (± 0.5)	7.4 (± 0.5)	11	3.3	2.8
TLIG*RAVTL P4-modified 126/127	650 (± 143)	645 (± 214)	N.D.	1.26 (± .16)	0.37 (± .04)	N.D.	1.94	0.57	N.D.
LTNL*IGRA, 124/125	------------ not cleaved ----------								

less favourable for enzyme activity, e. g. without NaCl and at 4 °C. However, when such obtained mixtures of the Gag-Pol precursor and the putative MAV PR (+IleGlyArgAlaThrValLeu) product were further incubated at 30 °C with addition of 1 M NaCl (final concentration), gradual conversion of both protein species proceeded, leading to a complete formation of PR(+IleGly) in 4 hours (gels not shown). It is interesting that this conversion took place also with the mutant species of MAV PR(+IleGly*Ser*AlaThrValLeu) and its precursor, i.e. that peptide bonds Gly*Arg and Gly*Ser were cleaved with equal ease.

Cleavage of peptide substrated derived from structures of several MAV processing sites, including the internal Gag-Pol PR 126/127 and 124/125 boundaries, is illustrated in Table I. Reactions were carried out at 37 °C, essentially as in (4, 5); buffer used was 40 mM sodium phosphate (pH 6.0) containing 2 M NaCl and 1 mM EDTA. The active enzyme concentration (E) was determined by active site titration as described elsewhere (4), and adjusted in its sub-micromolar range so that the substrate conversion was typically around 10 % and did not exceed 25 % for the various times of reaction (up to 30 minutes). Substrate concentrations were calculated from amino acid analyses, and varied between 6 μM and 2mM. For solubility reasons, substrate TLIG*RATVL which differs from TNLIG*RATVL (126/127) in relatively unimportant P4 position was used: both substrates had equal conversion rates at 0.1 mM peptide and 0.19μ M enzyme. Cleavage in individual reaction mixtures (total volume 100 μl) was terminated by addition of 5 1 of 20 % trifluoroacetic acid. The reaction rate was determined from the amount of the cleavage product, as monitored spectrophotometrically in RP-HPLC using a C18 (Vydac) column and the methanol-H_2O system. Cleavage products were also collected and identified with amino acid analyses. The values of K_m and $k_{cat} = V_{max}/E$, and their standard deviations were derived by computer fit of the data using the Enzfitter program.

Several conclusions can be drawn from the kinetic data: The presence of Ile-Gly residues in PR(+IleGly) did not alter the specificity of the enzyme as judged fron K_m values. A moderate impairment of catalysis was observed in the velocity parameters, k_{cat}, which were systematically lower by a factor of 3 than those for Gag PR. This phenomenon can be probably attributed to interference of the protruding carboxy ends with subunit dimerization.

Capacity of both Gag PR and PR (+IleGly) to cleave efficiently the peptide bond Gly*Arg in the context of the carboxy terminal amino acid sequence of the putative Gag-Pol PR was convincingly demonstrated. In contrast, a peptide relfecting Gag-Pol PR 124/125 amino acid boundary was not cleaved: this made any processing reaction that would produce the genuine Gag PR structure from *gag-pol* encoded viral polyproteins very unlikely. Taken together, the results suggest that PR(+IleGly) forms ordinarily from the Gag-Pol polyproteins due to additional proteolytic cleavage that removes 5 amino acid "spacer" peptide between the PR 126/127 and PR/RT sites. Kinetic data do not indicate any exceptional properties for MAV PR (+IleGly), and thus do not support any special function of this enzyme in the retroviral processing and assembly.

We wish to thank Dr. Hana Večerková for oligonucleotide synthesis and Mr. J. Zbrožek and Dr. Ivo Bláha from the Institute of Organic Chemistry and Biochemistry for amino acid analyses and peptide synthesis, respectively.

REFERENCES

1. Oroszlan, S., and R. B. Lufting. 1990. Retroviral Proteinases. p. 153-185. In R. Swanstrom and P. K. Vogt (ed.) Current Topics in Microbiology and Immunology. Springer-Verlag, Heidelberg.
2. Jacks, T., H. D. Madhani, F. R. Masiarz, and H. E. Varmus. 1988. Signals for ribosomal frameshifting in the Rous sarcoma virus gag-pol region. Cell 55 : 447-458
3. Bennet, R. P., S. Rhee, R. C. Craven, E. Hunter, and J. W. Wills. 1991. Amino acids encoded downstream of gag are not required by Rous sarcoma virus protease during Gag-mediated assembly. J. Virol. 65 : 272-280
4. Štrop, P., J. Konvalinka, D. Štys, L. Pavlíčkova, I. Bláha, J. Velek, M. Trávníček, V. Kostka, and J. Sedláček. 1991. Specificity studies on retroviral proteinase from myeloblastosis associated virus. Biochemistry 30 : 3437-3443
5. Konvalinka, J., I. Bláha, R. Škrabana, J. Sedláček, I. Pichová, F. Kaprálek, V. Kostka, and P. Štrop. 1991. Subsite specificity of the proteinase from myeloblastosis associated virus. FEBS Letters 282 : 73-76
6. Ohlendorf, D. H., S. E. Foundling, J. I. Wendoloski, J. Sedláček, P. Štrop, and F. R. Salemme. 1992. Structural studies of the retroviral proteinase from avian myeloblastosis associated virus. Proteins. Struct. Funct. Genet. 14 : 382-391
7. Jaskolski, M., M. Miller, J. K. M. Rao, J. Leis, and A. Wlodaver. 1990. Structure of the aspartic protease from Rous sarcoma retrovirus refined at 2 Å solution. Biochemistry 29 : 5889-5898
8. Pichová, I., P. Štrop, J. Sedláček, F. Kaprálek, V. Beneš, M. Trávníček, L. Pavlíčková, M. Souček, V. Kostka and S. Foundling. 1991. Isolation, biochemical characterization, and crystallization of the p15gag proteinase of myeloblastosis associated virus expressed in *E. coli*. Int. J. Biochem. 24 : 235-242
9. Sedláček, J., P. Štrop, F. Kaprálek, V. Pečenka, V. Kostka, M. Trávníček, and J. Říman. 1988. Processed enzymatically active protease ($p15^{gag}$) of avian retrovirus obtained in an *E. coli* system expressing a recombinant precursor ($Pr25^{lac\text{-}\Delta gag}$). FEBS Letters 237 : 187-190
10. Andreánsky, M., O. Hrušková-heidingsfeldová, J. Sedláček, J. Konvalinka, I. Bláha, P. Ječmen, M. Hořejší, P. Štrop and M. Fábry. 1991. High-level expression of enzymatically active bovine leukemia virus proteinase in *E. coli*. FEBS Letters 87 : 129-132

EXTRACELLULAR ASPARTIC PROTEINASES FROM *CANDIDA* YEASTS

Martin Fusek, Elizabeth Smith, and Stephen I. Foundling

Laboratory of Protein Crystallography
Protein Studies Program
Oklahoma Medical Research Foundation
825 N.E. 13th Street, Oklahoma City, OK 73104

INTRODUCTION

Infections caused by *Candida* opportunistic yeasts are becoming one of the major threats for immune compromised patients (Douglas, 1988).

The World Health Organization (WHO) data suggest that about 1.2 million cases of AIDS have occurred worldwide in the past ten years and that ten million people could be infected (Grmek, 1990). This last figure may rise to 30-40 million by the end of the century. Although great advances have been made in understanding the molecular biology of the generally accepted causative agent, the human immunodeficiency viruses, HIV (Broder and Gallo, 1984; Popovic *et al.*, 1984), and some palliative agents are available clinically (Grmek, 1990; Mitsuya *et al.*, 1987), AIDS is still an incurable disease. The impaired immune defense mechanisms of these patients increase their susceptibility to opportunistic infection. Opportunistic infections are responsible for nearly 90% of the mortality of AIDS (Mills and Masur, 1990).

One of the major causes of opportunistic infection in AIDS patients are *Candida* yeasts, which are single-cell dimorphic fungi. They account for from 56% to 81% of the infections in AIDS patients (Holmberg and Myer, 1986). Particularly important in AIDS related infections are three *Candida* strains, *C. albicans, C. tropicalis*, and *C. parapsilosis*, which cause localized infections and persistent fungemia that leads to systemic infections of virtually any organ, especially liver, kidney, lung, and brain (Holmberg and Myer, 1986).

Current therapy for the treatment of systemic fungal infections is largely limited to two classes of compounds, the polyene antibiotics and the azoles (Vanden Bossche *et al.*, 1987; Walsh *et al.*, 1988). Amphotericin B, a polyene antibiotic is a powerful fungicidal agent but its use is limited by the necessity for parenteral administration and by its substantial toxicity. Ketoconazole is an antifungal azole drug. Ketoconazole disrupts a demethylase in

* Abbreviations used: FPLC - Fast Protein Liquid Chromatography, SDS - sodium dodecyl sulfate, CAAP - *Candida albicans* aspartic proteinase, CTAP - *Candida tropicalis* aspartic proteinase, CPAP - *Candida parapsilosis* aspartic proteinase, PP - pig pepsinogen, HCD - human cathepsin D.

Aspartic Proteinases: Structure, Function, Biology, and Biomedical Implications
Edited by Kenji Takahashi, Plenum Press, New York, 1995

the synthesis of ergosterol, a component of the fungal membrane. Treatment with this drug also results in toxemic side effects relative to its limited selectivity for the fungal demethylase (Ator *et al.*, 1992). Thus, the need for effective drugs against invasive fungemia is acute.

Aspartic proteinases produced by *Candida* species have been shown to be virulence factors. *Candida* aspartic proteinases are secreted almost exclusively by pathogenic strains of *Candida* either in culture or at the lesions of the infection, but not by nonpathogenic strains (Ray and Wupper, 1976; ibid, 1978; Ray and Payne, 1990a; Ruchel, 1983; Kondoh *et al.*, 1987). The circulating antibodies against *Candida* acid proteinases are also found in patients with *Candida* infections (MacDonald and Odds, 1980a, 1980b; Ray and Payne, 1987; Ruchel *et al.*, 1987). Additionally, the levels of proteinase secreted correlate with the increased lethality of the strains (Staib, 1969; Schreiber *et al.*, 1985). Mutations of the virulent strains which block the proteinase secretion also become non-virulent (MacDonald and Odds, 1983; Kwon-Chung *et al.*, 1985; Ross *et al.*, 1990). Convincingly, the revertant strains which regain acid protease secretion also regained virulence (Kwon-Chung *et al.*, 1985). The function of the proteinases in these infections is the hydrolysis of proteins of connective tissue, such as collagens and keratins, thus increasing the invasiveness of the organisms.

Purified *Candida* secreted proteinases are active in the pH range of 2.2 to 5.0. The activity in acid as well as their inhibition by pepstatin suggest that they belong to the pepsin family of aspartic proteinases (Remold *et al.*, 1968; Ruchel, 1981; Negi *et al.*, 1984; Ray and Payne, 1990b; Ruchel *et al.*, 1986). However, the *Candida* proteinases differ from all other aspartic proteinases in that they hydrolyze collagen and keratin (Negi *et al.*, 1984; Ray and Payne, 1990b; Hattori *et al.*, 1984; Kaminishi *et al.*, 1986). The aspartic proteinase classification is further supported by the results obtained from the isolation and partial characterization of enzymes from the yeasts *Candida albicans* (CAAP) (Remold *et al.*, 1968), *Candida tropicalis* (CTAP) and *Candida parapsilosis* (CPAP) (Ruchel *et al.*, 1983; MacDonald, 1984). The amino acid sequences of these aspartic proteinases were deduced from the corresponding genomic DNA sequences (Hube *et al.*, 1991 (CAAP); Togni *et al.*, 1991 (CTAP); de Viragh *et al.*, 1993 (CPAP)). The amino acid sequences of these proteinases from three virulent *Candida* strains are homologous (most importantly near the two active site aspartic residues) with other eukaryotic aspartic proteinases (Figure 1).

Based upon the invasive role that *Candida* aspartic proteinases are suggested to play in the development of severe systemic *Candida* infection and disseminated candidamycoses (recently summarized by Rüchel *et al.*, 1992), these proteinases offer a new target for the synthetic inhibitor design. A wealth of information exists for the inhibition of several other aspartic proteinases which has been gained during the design of inhibitors for human renin (Thaisrivongs, 1988), and for the retroviral proteinases of the AIDS virus (Debouck and Metcalf, 1990). These studies show that substrate specificity in the aspartic proteinase family is complex and composite. A minor change in the primary amino acid sequence and so the

---

Figure 1. A multiple sequence alignment of three *Candida* yeast aspartic proteinases with sequences of two architypic aspartic proteinases representative of this family. Numbering of the sequences shown is for the CAAP sequence. Positions of sequence identity are marked by an asterisk. **D**TG/**D***G define the following features of these enzymes. **D**, the active site aspartates, TG/*G the residues of the 'fireman's grip'. Y shows the position of the conserved tyrosine of the 'flap'. The sequences of the three proteinases that were determined by amino terminal sequencing are shown in **underlined bold** typeset. Multiple sequence alignments were performed using the program CLUSTALV (Higgins et al., 1992) on a VAX 4600 computer. The following sequence datasets were used for the respective proteinases, CAAP - sw:car3_canal, CTAP - sw:carp_cantr, CPAP - pir3:S20705, pig pepsinogen (PP) - sw:pepa_pig, human cathepsin D (HCD) - sw:catd_human.

```
                       p1        |         |        1
CAAP       -------------SPGFVTLDFDVIKTPVNATGQEGKVKR.QA-----------------IPV
CTAP       -------------TDKVVSLDFTVIRKPFNATAHRLIQKR.SD-----------------VPT
CPAP       DD-----------NPGFVALDFDVLRKPLNLTEALLREKR.DS-----------------ISL
PP         L-VKVPLVRKKSLRQNLIK---NGKLKDFLKT-----HKH-NPASKYFPEAAALI.G-D---E
HCATD      L-VRIPLHKFTSIRRTMSEV--GGSVEDLIAK---------GPVSKYSQAVPAVT-E.GPIPE

                |         |         |         |         |50       |
CAAP       TLNN-ELVSYAADITIGSNKQKFNVIVDTGSSDLWVPDASVTCDKPRPGQSADFCKGKGI
CTAP       TLIN-EGPSYAADIVVGSNQQKQTVVIDTGSSDLWVVDTDAECQVTYSGQTNNFCKQEGT
CPAP       SLIN-EGPSYASKVSVGSNKQQQTVIIDTGSSDFWVVDSNAQCGKGVD------CKSSGT
PP         PLENYLDTEYFGTIGIGTPAQDFTVIFDTGSSNLWV--PSVYCSSL--A-----CSDHNQ
HCATD      VLKNYMDAQYYGEIGIGTPPQCFTVVFDTGSSNLWV--PSIHCKLLDIA-----CWIHHK
            * *     *      *   *   *  DTG**  **      *           *

                |         |          |         |100                 |
CAAP       YTPKSSTTSQNL-GSPFYIGYGDGSSSQGTLYKDTVGF-----------GGASITKQVF-
CTAP       FDPSSSSSAQNL-NQDFSIEYGDLTSSQGSFYKDTVGF-----------GGISIKNQQF-
CPAP       FTPSSSSSYKNL-GAAFTIRYGDGSTSQGTWGKDTVTI-----------NGVSITGQQI-
PP         FNPDDSSTFEA-TSQELSITYGTGSMT-GILGYDTVQV-----------GGISDTNQIFG
HCATD      YNSDKSSTYVK-NGTSFDIHYGSGSLS-GYLSQDTVSVPCQSASSASALGGVKVERQVFG
                *            * Y       *    ***              *     *

                         |         |                   |        |150
CAAP       ------ADITKTSIPQGILGIGYKTNEAAGD---------YDNVPVTLKNQGVIAKNAYS
CTAP       ------ADVTTTSVDQGIMGIGFTAVEAGYNL--------YSNVPVTLKKQGIINKNAYS
CPAP       ------ADVTQTSVDQGILGIGYTSNEAVYDTSGRQTTPNYDNVPVTLKKQGKIRTNAYS
PP         LSETEPGSFLYYAPFDGILGLAYP------SISASGATPVFDN----LWDQGLVSQDLFS
HCATD      EATKQPGITFIAAKFDGILGMAYP------RISVNNVLPVFDN----LMQQKLVDQNIFS
                           ** *                      *    *  *        *

              |             |         |         |         |200        |
CAAP       LYLNSPNAAT----GQIIFGGVDKAKYSGSLIAVPVTSDRELRITLNSLKAVGKNI--NG
CTAP       CDLNSEDAST----GKIIFGGVDNAKYTGTLTALPVTSSVELRVHLGSINFDGTSV--ST
CPAP       LYLNSPSAET----GTIIFGGVDNAKYSGKLVAEQVTLSQPLTISLASVNLKGSSF--SF
PP         VYLSS-NDDS---GSVVLLGGIDSSYYTGSLNWVPVSVEGYWQITLDSITMDGETIACSG
HCATD      FYLSR-DPDAQP-GGELMLGGTDSKYYKGSLSYLNVTRKAYWQVHLDQVEVASGLTLCKE
             *                ** *   * * *    *         *

                    |         |          |         |250         |
CAAP       NIDVLLDSGTTITYLQQDVAQ-DIIDAFQAELKLDGQGHTFYVTDC--QTS-GTVDFNFD
CTAP       NADVVLDSGTTITYFSQSTAD-KFARIVGATW--DSRNEIYRLPSC--DLS-GDAVVNFD
CPAP       GDGALLDSGTTLTYFPSDFAA-QLADKAGARLVQVARDQYLYFIDCNTDTS-GTTVFNFG
PP         GCQAIVDTGTSLLTGPTSAIAINIQSDIGASENSDGE----MVISCSSIDSLPDIVFTIN
HCATD      GCEAIVDTGTSLMVGPVDEVR-ELQKAIGAVPLIQGE----YMIPCEKVSTLPAITLKLG
                 D*G                    *               *

              |         |         |         |300          |          |
CAAP       NNAKISVPASEFTAPLSYANGQPYPKCQLLLGISDA----NILGDNFLRSAY-LVYDLDD
CTAP       QGVKITVPLSE----LILKDSDS-SICYFGISRNDA----NILGDNFLRRAY-IVYDLDD
CPAP       NGAKITVPNTEY----VYQNGDG--TCLWGIQPSDD----TILGDNFLRHAYYLLYNLDA
PP         -GVQYPLSPSAYILQDDDS----CTSGFEGMDVPTSSGELWILGDVFIRQ-YYTVFDRAN
HCATD      -GKGYKLSPEDYTLKVSQAGKTLCLSGFMGMDIPPPSGPLWILGDVFIGR-YYTVFDRDN
                                                    **** *    *

                   |         |
CAAP       DKISLAQVKYTSASNIAALT CAAP  Candida albicans
CTAP       KTISLAQVKYTSSSDISAL- CTAP  Candida tropicalis
CPAP       NTISIAQVKYTTDSSISAV- CPAP  Candida parapsilosis
PP         NKVGLAPV---------A-- PP    Porcine pepsinogen
HCATD      NRVGFAEA---------ARL HCATD Human Cathepsin D
                *           *
```

tertiary protein structures of these enzymes can lead to markedly different substrate specificities (reviewed by Davies, 1990, ibid).

It is our ambition to isolate the secreted forms of these yeast proteinases in sufficient quantity to further characterize their biochemical properties. This characterization will extend to the analysis of substrate specificity and eventually to crystallographic structural studies. We hope that through this type of analysis we will identify the subtle chemical differences that can be used to discriminate between the different yeast enzymes in this group. Eventually we hope to design novel substrates which can be used for the rapid identification of the infectious organism and design tight binding inhibitors that may eventually lead to new therapeutic drugs. These drugs may play an important role in the treatments of invasive fungemia in the immune compromised patient.

EXPERIMENTAL METHODS

Yeast Cell Culture

The strains of *Candida* yeasts used for cell growth and enzyme production in these analysis are the strains C-74 of *Candida albicans*, ATCC-75 of *Candida tropicalis* and E-18 of *Candida parapsilosis*. Yeast cell cultures are maintained and replated every 6-8 weeks on Sabouraud agar plates (Difco). *Candida* yeast are grown in liquid culture containing 0.2 % bovine serum albumin (BSA) and 1.2 % YCB Difco medium, adjusted to pH 4.0. 50 ml seed cultures are inoculated with several colonies of *Candida* yeast from agar plates. This culture is grown overnight at 30°C in a shaker (200 rpm). 0.5 ml of this overnight culture is used for the inoculation of 500 mls of the same culture media. The cultures are grown for 3 days at a temperature of 30°C, with shaking at 200 rpm. The cells are then centrifuged (30 min. at 7000g) and the resulting supernatant is sterile filtered to remove any infectious cellular material. This sterile supernatant is used for all subsequent steps in the enzyme isolation. The sterile filtered supernatant can be stored at -70°C for several weeks prior to use.

Isolation of Extracellular Aspartic Proteinases

The isolation of four proteolytic enzymes from the three strains of *Candida* yeast differs for each strain of yeast grown. The isolations employ standard methods of salting out or batch adsorption ion exchange chromatography, low pressure column chromatography and FPLC as the final purification step. The chart that follows describes the general steps used for the purification of CAAP, CTAP and CPAP#1 and CPAP#2 respectively. The purification is decribed in detail elsewhere (Fusek *et al.*, 1993a (CPAP#1, CPAP#2); Fusek *et al.*, *submitted* (CAAP,CTAP, CPAP#1); ibid). The relative concentrations of the proteinases were determined using the BCA protein assay (BioRad, Hercules,CA). The four isolated proteins are free of contaminants as determined by SDS poylacrylamide electrophoresis and by N-terminal sequence analysis (Fusek *et al.*, 1993a, Fusek *et al.*, *submitted*).

Activity Determination

The proteolytic activity in chromatographed fractions was followed using the general photometric method described here. For fast screening of enzyme activity in eluted fractions we use a solution of 1% bovine hemoglobin (Sigma, USA) dissolved in 100 mM sodium citrate buffer, pH 3.4. Aliquots (5 μl) of the eluted fractions are incubated with 1 ml of the hemoglobin solution at 37°C for 30 min. The reaction is stopped by the addition of 200 μl of 10 % trichloracetic acid. Precipitate is removed by centrifugation and the absorbance of

General steps used in the purification of *Candida* yeast aspartic proteinases.

CAAP	CTAP	CPAP#1/CPAP#2
Sterile supernatant	Sterile supernatant	Sterile supernatant
↓	↓	↓
Batch Adsorption chromatography DEAE Trisacryl	Salt precipitation centrifugation	Batch Adsorption chromatography SP-Trisacryl
↓	↓	↓
Dialysis	Dialysis	Dialysis
↓	↓	↓
FPLC MONO Q	Low pressure column chromatography DEAE Trisacryl	FPLC MONO S
↓	↓	↓
Dialysis	Dialysis	Dialysis
↓	↓	↓
yield CAAP 4.2 mg/L	FPLC MONO Q	yield CPAP#1 4.4 mg/L CPAP#2 1.0 mg/L
	↓	
	Dialysis	
	↓	
	yield CTAP 5.1 mg/L	

(DEAE Trisacryl, SP-Trisacryl are products of IBF, France; MONO Q and MONO S are products of Pharmacia LKB, Sweden).

the supernatant is read at 280nm. This assay is linear in the range of 0.01 to 1.0 AU. One unit of specific activity is defined as the change in absorbance at A_{280nm} of 1 AU by 1 mg of proteinase in incubation time of 30 mins.

Kinetic Characterization Using Synthetic Chromogenic Substrates

To determine specificity of the three major isolated extracellular proteinases (CAAP,CTAP,CPAP#1) from *Candida* yeast, we used a library of synthetic chromogenic peptides with the general structure X_5-X_4-X_3-X_2-X_1-Phe(*p*-NO_2)-X_2'-X_3'.

The hydrolysis of these substrates is monitored by the cleavage of the bond between X_1-↓-Phe(NO_2). This cleavage results in a decrease in absorbance in the range of 284-324 nm. The library of 60 substrates is derived by the sequential modification of the primary amino acid sequence at positions X_5 to X_2 and X_2' to X_3' in the substrates. Positions X_5 to X_2 and X_2' to X_3' are occupied by either of the following amino acid residues, Ala, Asp, Arg, Ser or Leu. The parent substrates used in this screening are shown in Figure 2. This type of substrate has been widely used for the typification of substrate specificity in other aspartic proteinases (Dunn et al., 1986; Pohl and Dunn, 1988; Scarborough *et al.*, 1993).

The preliminary screening for the three enzymes were carried out under fixed substrate concentrations [*S*] of 50 mM, fixed concentration of enzyme 0.5 μg in a constant 250 μl reaction volume. The initial velocities (ΔA_{300nm} / sec x 10^6) were determined. This allowed a general ranking of initial velocities (ν) for the three enzymes against the full library of 60 chromogenic peptide substrates. Based on this initial ranking of initial velocities we chose a set of peptides for a more detailed kinetic evaluation (Fusek *et al.*, *submitted*). This choice focused on the substitutions in the parent sequences which gave the most dramatic changes in the initial cleavage rates. Kinetic measurements were performed on a Hewlett-Packard 8452A diode array spectrophotometer equipped with a multi-cell transport and thermostated at 37°C following the methods described in Scarborough *et al.*, 1993; ibid. Kinetic parameters were calculated using the ENZYME KINETICS program package supplied by Hewlett-Packard.

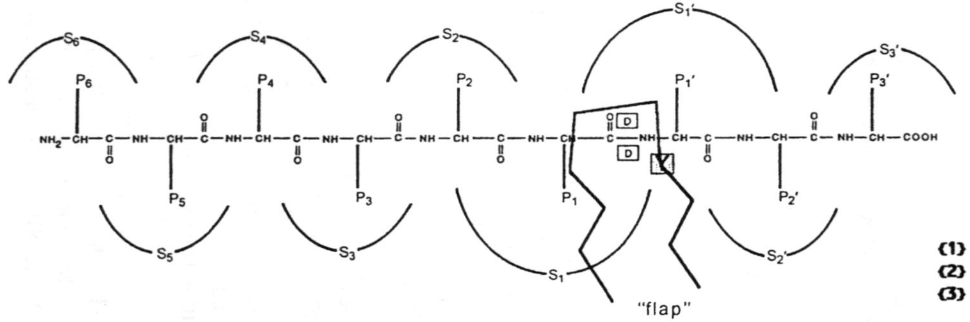

Schematic diagram of enzyme substrate binding sites S_6 to S_3'.

	P_5	P_4	P_3	P_2	P_1		P_1'	P_2' P_3'
(1)	Lys-	Pro-	Ala-	Lys-	Phe-	↓	-Phe(p-NO$_2$)-	Arg-Leu
(2)	Lys-	Pro-	Ile-	Glu-	Phe-	↓	-Phe(p-NO$_2$)-	Arg-Leu
(3)	Lys-	Pro-	Ala-	Glu-	Phe-	↓	-Phe(p-NO$_2$)-	Ala-Leu

Amino acid sequence of parent chromogenic peptide substrates.

	S_5	S_4	S_3	S_2	S_1	S_1'	S_2'	S_3'
CAAP	NS	Pro Asp	Phob Polar	Phob Lys	Phob	Phob	Arg	NS
CTAP	Arg	Pro	Phob Polar	Phob Arg	Phob	Phob	Arg	NS
CPAP#1	NS	Pro	Phob Polar	Phob Glu	Phob	Phob	Phob Arg	NS
PP	NS	Pro	Phob Polar	Phob Lys	Phob	Phob	Phob Arg	Phob Polar

Figure 2. General substrate preferences and specificity of the *Candida* yeast aspartic proteinases compared to porcine pepsin. PP, porcine pepsin. NS - non specific. Phob - hydrophobic amino acids. Polar - polar or neutral amino acids. (Hydrophobic - Ala, Val, Ile, Leu, Phe, Tyr, Trp, Cys, Met. Polar or neutral - Gly, Ser, Thr, Asn, Gln. Acidic - Asp, Glu. Basic -His, Lys, Arg.)

Inhibition by Pepstatin A

The inhibition of enzyme activity by pepstatin A was studied by two methods. In the first inhibition assay we studied the effect of pepstatin A on cell growth. We included pepstatin A in culture media at concentrations of 10 nm, 100nm and 1000nm. Cultures were inoculated with a fixed density of cells and grown up over a period of 120 hours. Samples were drawn from these cultures at regular intervals to determine the viable cell density by measuring the apparent absorbance at 600 nm. This experiment was done at pH 4.0 and 7.0. For the determination of the inhibitory constant K_i of pepstatin A with the three purified enzymes CAAP, CTAP and CPAP#1 peptide **{3}** (Figure 2) was used as a substrate. Values for K_i for pepstatin A were determined by the Linweaver-Burke plot. The pepstatin used in these analyses was purchased from Serva, Germany.

Crystallization of *Candida* Yeast Aspartic Proteinases

Purified fractions of each enzyme were used in crystallization experiments after centrifugal ultrafiltration/concentration in AMICON Centricon cells. Protein concentrations were estimated using the method of Lowry with BSA, IgG and porcine pepsin as standards (Lowry *et al.*, 1957). The strategies that we used to explore crystallization were the SAGS technique (Systematic Automated Grid Searching; Cox and Weber, 1987; 1988), and a modification of the 50 buffer system (Jancarik and Kim, 1991) set up at two temperatures 6°C and 20°C. In these techniques we employ a robotics pipetting system which automates the hanging drop vapour diffusion technique for crystallization (Wlodawer and Hodgson, 1976). Initial crystals obtained from these preliminary searches were generally improved in a systematic manner by careful modification of the variables which give rise to the crystallization event. The details of the crystallizations of CAAP, CTAP, CPAP#1 and CPAP#2 will be reported elsewhere (Foundling and Fusek, *manuscript in preparation*).

RESULTS AND DISCUSSION

Yeast Cell Growth and Enzyme Purification

Four yeast aspartic proteinases have been purified to homogeneity from three strains of *Candida* yeast which include *C.albicans* (CAAP, strain C-74), *C.tropicalis* (CTAP, strain ATCC-75) and *C.parapsilosis* (CPAP#1,CPAP#2, strain E-18). Using standard chromographic techniques we obtained enzyme yields of 4.2 mg (CAAP), 5.1 mg (CTAP), 4.4 mg (CPAP#1) and 1.0 mg (CPAP#2) per litre of culture media. Homogeneity of these purified enzymes was monitored by SDS-polyacrylamide gel electrophoresis and by N-terminal sequencing (Fusek *et al.*, 1993a; Fusek *et al.*, *submitted*). The biophysical properties of these enzymes are tabulated below.

These *Candida* yeast aspartic proteinases were purified in sufficient quantities for the analysis of their substrate specificity using a library of 60 synthetic chromogenic substrates. The purified enzymes were also used extensively in crystallization trials.

Kinetic Characterization Using a Library of Synthetic Chromogenic Substrates

We undertook the study of enzyme specificity in the attempt to derive substrates that could be used to discriminate between the individual *Candida* yeast enzymes. We also

Table 1. Biophysical Properties of *Candida* Aspartic Proteinases CAAP, CTAP, CPAP#1 and CPAP#2

	No. of Residues	Calc. Mol. wt.	Apparent Mol. wt.	pI
CAAP	342*aa*	36.5Kd	42Kd	3.2
CTAP	334*aa*	35.8Kd	40Kd	3.6
CPAP#1	340*aa*	36.1Kd	38Kd	5.6
CPAP#2	334*aa*	35.5Kd	38Kd	6.4

```
        1   .   10   .   20   .   30   .
CAAP    QAIPVTLNNELVSYAADITIGSNKQKFNVIVDTGSSDLW
CTAP    SDVPTTLINEGPSYAADIVVGSNQQKQTVVIDTGSSDLW
CPAP#1  DSISLSLINEGPSYASKVSVGSNKQQQTVIIDTGSSDFW
CPAP#2  SSPSSPLYFEGPSYGIRVSVGSNKQEQQVVLDTGSSDFW
```

N-Terminal sequences of the four *Candida* aspartic proteinases (single letter code). The amino acids shown in ***bold italics*** correspond to the sequences determined by N-terminal sequencing of the isolated enzymes CAAP, CTAP, CPAP#1 and CPAP#2 respectivley.

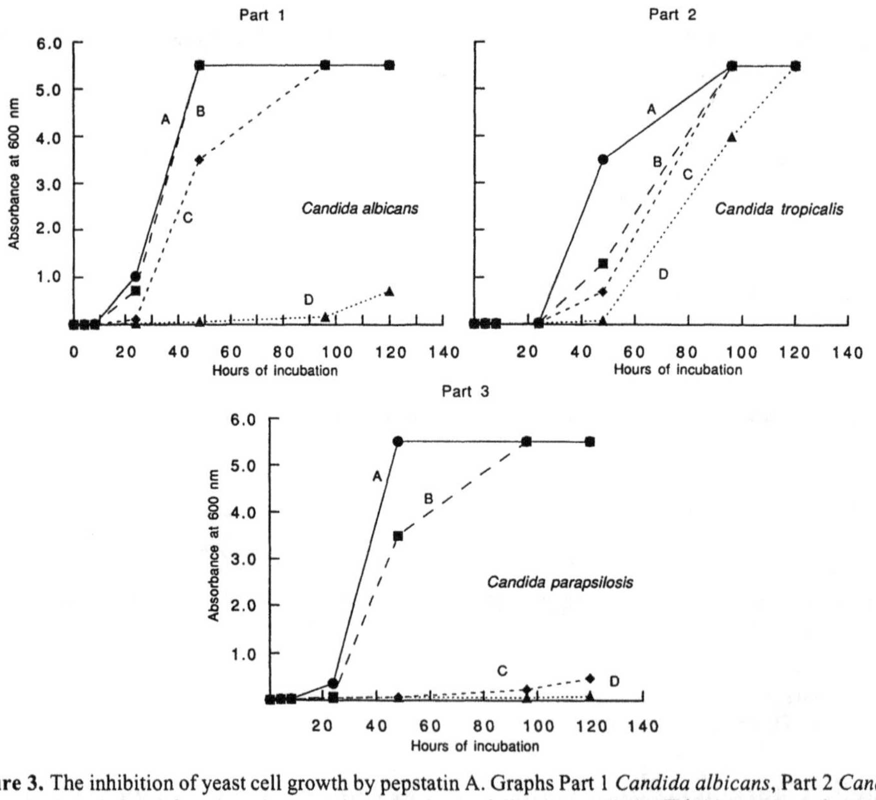

Figure 3. The inhibition of yeast cell growth by pepstatin A. Graphs Part 1 *Candida albicans*, Part 2 *Candida tropicalis*, Part 3 *Candida parapsilosis*. In these graphs the following symbols (■◆▲) represent data points taken at identical times from cell cultures grown in the presence of different concentrations of pepstatin A. Line A (●) no pepstain A. Line B (■) 10 nM pepstatin A. Line C (◆) 100 nM pepstatin A. Line D (▲) 1000 nM pepstatin A. The absorbance of samples that measured 2.0 AU or higher at 600 nm were determined by 1:10 dilution in culture media.

mapped the substrate specificity at the various binding pockets S_6 to S_3' in the enzymes. Substrate and inhibitor binding studies with other aspartic proteinases indicate that the most important residues involved in recognition and specificity are formed by interactions at residue positions P_4 to P_3' with specificity pockets S_4 to S_3' of the enzyme (nomenclature of Berger and Schecter, 1970). The most common primary specificity for aspartic proteinases is toward substrates with hydrophobic residues at positions P_1 and P_1' (Fruton, 1976., reviewed by Davies, 1990). We generally found that the *Candida* enzymes behaved like gastric pepsins, having a broad substrate specificity. However the pockets S_4 and S_2 in the active site cleft stand out as being remarkably restrictive. The scope of this report is limited and we report the detailed substrate typification of the enzymes elsewhere (Fusek *et. al. submitted*). Figure 2 describes the general specificity that we have defined in the substrate specificity mapping. It is evident from our data that specificity differences do exist for each of the yeast enzymes. This type of data may be useful in designing substrates which could be used to discriminate between the individual enzymes in the clinical laboratory.

Inhibition by Pepstatin A

The spatial arrangement of the amino acids that make up the binding cleft and subsites are responsible for the discriminatory substrate specificity of each enzyme. Very subtle changes in the enzyme's amino acid sequence can lead to large differences in the substrates recognized by these enzymes. It was interesting to discover that the classic aspartic proteinase inhibitor Pepstatin A, which has a predominant hydrophobic chemical structure and inhibits the archetypic aspartic proteinases like pepsin and cathepsin D at K_i values in the sub-nanomolar range (Umezawa H., 1972; Marciniszyn et al., 1976), shows some variation in K_i with the *Candida* proteinases. K_i values determined for the *Candida* enzymes were 0.6 nM, 6.0 nM and 12.0 nM for CPAP#1, CAAP and CTAP respectively. CAAP and CTAP are inhibited to a similar extent by pepstatin A and CPAP is at sub-nanomolar K_i.

We also used pepstatin A as a general antifungal in *in vivo* growth inhibition assays. We tested pepstatin A for inhibitory activity at acidic and neutral pH. Figure 3 shows that pepstatin A is moderately effective in retarding cell growth in cultures of CPAP, less effective for CAAP and only marginally effective even at concentrations of 1μM for CTAP. The general antifungal behavior of pepstatin A appears to be correlated with the affinity of pepstatin A for the *Candida* enzymes. No inhibition of growth was found when 1μM pepstatin A was used in cultures grown at pH 7.0 (Fusek and Foundling, unpublished results). This observation indicates that growth inhibition is not a result of inhibition of the intracellular vacuolar aspartic proteinase.

Crystallization of *Candida* Yeast Aspartic Proteinases

The crystal structures of aspartic proteinases are important in explaining specificity and in designing inhibitors for this group of enzymes. These structures and additional crystal structures of enzyme-transition state inhibitor complexes have revealed that the peptide bond hydrolysis is catalyzed by two aspartic acids which are hydrogen bonded to a water molecule thought to be the nucleophile in the hydrolysis (Davies, 1990). The juxtapositions of the transition-state inhibitors to the active site aspartates are virtually identical in all these enzymes. Thus, the structural information suggests that the catalytic mechanisms and transition state conformations of aspartic proteinases are nearly identical.

The *Candida* yeast proteinases CAAP, CTAP, CPAP#1 and CPAP#2 have been purified in sufficient quantity for extensive crystallization trials. Preliminary crystalization of all *Candida* enzymes has been achieved by using a combination of screening methodologies. The first of these uses a modified crystal screen which evalutes 48 different combina-

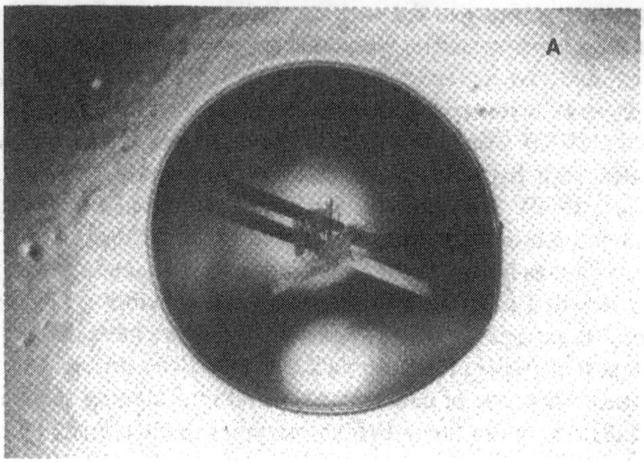

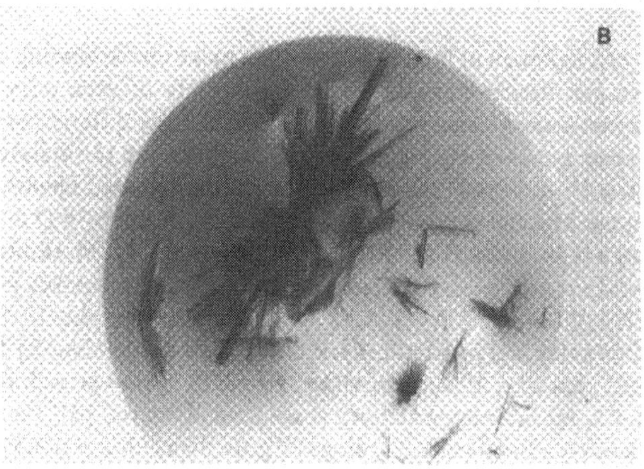

Figure 4. Part A. The secreted *Candida tropicalis* aspartic proteinase (CTAP) grows in two predominant crystal habits shown here in this picture. The crystals grow as large orthorhombic rods of mm dimensions and clusters of thin needles which are unsuitable for diffraction experiments. Part B. The recombinant *Canditropsin* aspartic proteinase grows in several crystal habits which include long needles and thin plates. These morphologies are often obtained in the same droplet as is shown in picture 4, part B.

tions of pH, additive and precipitant for their ability to promote crystal growth. Results from this screening combined with observations of solubility of these proteins derived from a second crystallization screening technique, systematic automated grid searching, has provided sufficient information to optimize the crystallization for several of the yeast aspartic proteinases. The secreted CTAP is crystallized reproducibly from 10 mg ml^{-1} protein with ethanol as precipitant. Preliminary diffraction experiments indicate that these crystals diffract to 2.2 Angstrom resolution and that the space group is orthorhombic. Further characterization of these crystals is underway in our laboratory using film precession photography. The recombinant *Canditropsin* is also crystallized in our laboratory (Lin *et al.*, 1993), these crystals grown from 6 mg ml^{-1} protein with ammonium sulphate as the precipitant. These crystals also diffract to high resolution and are currently being analysed in our laboratory. Crystals of the secreted CTAP and recombinant *Canditropsin* are shown in Figure 4. The full details of the crystallizations of CTAP, rec.*Canditropsin* and the other

Candida proteinases CPAP#1, CPAP#2 and CAAP are being reported elsewhere (Foundling *et al.*, manuscript in preparation).

Different aspartic proteinases have different substrate and inhibitor specificities. Both kinetics (Fruton, 1976; Dunn *et al.*, 1987) and crystal structures suggest that the aspartic proteinases can accommodate about eight residues in their substrate binding sites. However, the residue preferences in these eight subsites is highly related, but not identical, to the side chain preferences in the best inhibitors. This argues that the information of substrate specificity and crystal structure of the enzymes are both important in the design of inhibitors. The understanding of subsite specificity is important for the rational design of first-generation inhibitors, but additional crystal structures on proteinase-inhibitor complexes are important for the designing of subsequent generations of better inhibitors. This general approach, which has been successfully employed to produce potent and specific inhibitors for renin (for review, see Szelke, 1985; Greenlee, 1990; Thaisrivongs, 1991) and HIV proteinase (for reviews, see Debouck and Metcalf, 1990; Petteway *et al.*, 1991; Graves, 1991), will also be employed to design inhibitors for *Candida* proteinases.

ACKNOWLEDGEMENTS

We thank Dr. Ben Dunn and his research staff of the Department of Biochemistry, University of Florida, USA, for their help in the typification of substrate specificity for the *Candida* enzymes. We thank Dr. Michel Monod from the Laboratoire de Mycologie, Center Hopsitalier Universitaire Vaudois, Switzerland, for the *Candida* yeast strains used in our work. We acknowledge the financial support of the Oklahoma Center of Applied Science and Technology (OCAST) for the Scientific Recruitment grant number HS2-012 as well as the National Institutes of Health, National Institute of Allergy and Infectious Diseases (NIAID) for the First Independent Research Support Award, grant number R29 AI 34308 which supports Dr. S.I. Foundling in this research.

REFERENCES

1. Ator, M.A., Schmidt, S.J., Adams, J.L., Dolle, R.E., Kruse, L.I., Frey, C.L. and Barone, J.M. (1992) *J. Med. Chem.* Vol.35. pp.100-106.
2. Berger, A. and Schecter, I. (1970) *Phil. Trans. Roy. Soc.* B257. pp.249-264.
3. Broder, S. and Gallo, R.C.N. (1984) *Engl. J. Med.* Vol.311. pp.1292-1297.
4. Cox, M.J. and Weber, P.C. (1987) *J. of Applied Crystallography* Vol.20. pp.366-373.
5. Cox, M.J. and Weber, P.C. (1988) *J. of Crystal Growth* Vol.90. pp.318-324.
6. Davies, D. (1990) *Ann. Rev. of Biophysics and Biophysical Chemistry* Vol.19. pp.189-215.
7. Debouck, C. and Metcalf B.W. (1990) *Drug Development Research* Vol.21. pp.1-17.
8. de Viragh, P., A., Sanglard, D., Togni, G., Falchetto, R. and Monod, M. (1993) *J. Gen. Microbiol.* Vol. 139. pp.335-342.
9. Douglas, L.J. (1988) *Crit. Rev. Biotechnol.* Vol.8. No.2. pp.121-129.
10. Dunn, B. M., Jimenez, M., Parten, B. F., Valler, M. J., Rolph, C. E. and Kay, J. (1986) *Biochem. J.* Vol.237. pp.899-906.
11. Dunn, B.M., Valler, M.J., Rolph, C.E., Foundling, S.I., Jimenez, M. and Kay, J. (1987) *Biochim. Biophys. Acta* Vol.913. pp.122-130.
12. Foundling, S.I. and Fusek M. *J. Mol. Biol. Manuscript in preparation.*
13. Fruton. J. (1976). *Adv. Enzymology Relat. Areas Mol. Biol.* Vol.44. pp.1-33.
14. Fusek, M., Smith, E.A., Monod, M. and Foundling, S. I. (1993a) *FEBS Lett.* Vol.327. pp.108-112.
15. Fusek, M., Smith, E.A., Monod, M., Dunn B.M. and Foundling, S. I. *Biochemistry - submitted.*
16. Greenlee, W.J. (1990) *Med. Res. Rev.* Vol.10. pp.173-186.
17. Graves, M.C. (1991) in: *Structure and Function of the Aspartic Proteinases*, (Dunn, B.M., ed.) Plenum Press, New York, pp.395-405.

18. Grmek, M.D. (1990) in: *History of AIDS. Emergence and Origin of a Modern Pandemic*. Princeton University Press, Princeton.
19. Hattori, M., Yoshiura, K., Negi, M. and Ogawa, H. (1984) *Sabouraudia* Vol.22. pp.175-183.
20. Holmberg, K. and Myer, R.D. (1986) *Scand. J. Infect. Dis.* Vol.18. pp.179-192.
21. Hube, B., Turver, C. J., Odds, F. C., Eiffert, H., Boulnois, G. J., Kochel, H. and Rüchel, R. (1991) *J. Med. Vet. Mycology*, Vol.29. pp.129-132.
22. Jancarik, J. and Kim, S.H. (1991) *J. Appl. Cryst.* Vol.24. pp.409-411.
23. Kaminishi, H., Hagihara, Y., Hayashi, S. and Cho, T. (1986) *Infect. Immun.* Vol.53. pp.312-316.
24. Kondoh, Y., Shimizu, K. and Tanaka, K. (1987) *Microbiol. Immunol.* Vol.31. pp.1061-1069.
25. Krempl-Lamprecht, L. (1991) "*Factors of the Pathogenicity of Candida albicans - A review*", in *Candida and Candidamycosis* (Tumbay,E. , Seelinger, H.P.R. and Ang, O., eds.), Plenum Press, New York and London, pp.49-54.
26. Kwon-Chung, K., J., Lehman, D., Good, C. and Magee, P.T. (1985) *Infect. Immun.* Vol.49. pp.571-575.
27. Lin, X.L., Tang, J., Koelsch, G., Monod, M. and Foundling, S.I. (1993) *J. Biol. Chem.* Vol.268. pp.20143-20147.
28. Lowry, O.H., Roseburg, N.J., Farr, A.L. and Randall, R.J. (1951) *J. Biol. Chem.* Vol.193. pp. 265-275.
29. MacDonald, F. and Odds, F.C. (1980a) *J. Med. Microbiol.* Vol.13. pp.423-435.
30. MacDonald, F. and Odds, F.C. (1980b) *J. Am. Med. Assoc.* Vol.243. pp.2409-2411.
31. MacDonald, F. and Odds, F.C. (1983) *J. Gen. Microbiol.* Vol.129. pp.431-438.
32. MacDonald, F. (1984) *Sarbouraudia* Vol.22. pp.79-82.
33. Marciniszyn J., Hartsuck J.A. and Tang J. (1976) *J. Biol. Chem.* Vol.251. pp.7088-7094.
34. Mills, J. and Masur, H. (1990) *Scientific American*, August. pp.50-57.
35. Mitsuya, H. and Broder, S. (1987) *Nature* Vol.325. pp.773-778.
36. Negi, M., Tsuboi, R., Matsui, T. and Ogawa, H. (1984) *J. Invest. Dermatol.* Vol.83. pp.32-36.
37. Petteway, S.R. Jr., Dreyer, G.B., Meek, T.D., Metcalf, B.W. and Lamber, D.M. (1991) *AIDS Res. Rev.* Vol.1. pp.267-288.
38. Pohl, J. and Dunn, B. M. (1988) *Biochemistry* Vol.27. pp.4827-4834.
39. Popovic, M., Sarngadharan, M.G., Read, E. and Gallo, R.C.N. (1984) *Science* Vol.224. pp.497-500.
40. Ray, T.L. and Wuepper, K.D. (1976) *J. Invest. Dermatol.* Vol.66. pp.29-33.
41. Ray, T.L. and Wuepper, K.D. (1978) *Arch. Dermatol.* Vol.114. pp.539-543.
42. Ray, T.L. and Payne, C.D. (1987) *Clin. Res.* Vol.35. pp.711A.
43. Ray, T.L. and Payne, C.D. (1990a) *Clin. Res.* Vol.38. pp.678A.
44. Ray, T.L. and Payne, C.D. (1990b) *Infect. Immun.* Vol.58. pp.508-514.
45. Remold, H., Fasold, H. and Staib, F. (1968) *Biochim. Biophys. Acta* Vol.167. pp.399-406.
46. Ross, I.K., De Bernards, F., Emerson, G.W., Cassone, A. and Sullivan, P.A. (1990) *J. Gen. Microbiol.* Vol.136. pp.687-694.
47. Rüchel, R. (1981) *Biochim. Biophys. Acta* Vol.659. pp.99-113.
48. Rüchel, R., Uhlemann, K. and Boning, B. (1983) *Zentralbl. Bakteriol. Mikrobiol. Hyg. I Abt. Orig. A*, Vol.255. pp.537-548.
49. Rüchel, R., Boning, B. and Borg, M. (1986) *Infection and Immunity* Vol.53. pp.411-419.
50. Rüchel, R., Boning-Stutzer, B. and Mari, A. (1987) *Mycoses* Vol.31. pp.87-106.
51. Rüchel, R., De Bernardis, F., Ray, T., L., Sullivan, P. A. and Cole, G. T. (1992) *J. Med. Vet. Mycology* Vol.30. pp.123-132.
52. Scarborough, P., E., Guruprasad, K., Topham, C., Richo, G. R., Conner, G. E., Blundell, T. L. and Dunn, B. M. (1993) *Protein Sci.* Vol.2. pp.264-276.
53. Schreiber, B., Lyman, C.A., Gurevich, J., and Needham, C.A. (1985) *Diagn. Microbiol. Infect. Dis.* Vol.3. pp.1-5.
54. Stiab, F. (1969) *Mycopathol. Mycol. Appl.* Vol.37. pp.345-348.
55. Szelke M. (1985) in: *Aspartic Proteinases and their Inhibitors* (Kostka, V., ed.) pp.421-441. Walter de Gruyter and Co., Berlin.
56. Thaisrivongs, S. (1988) *Drug News and Perspectives* Vol.1. pp.11-16.
57. Thaisrivongs, S., Pals, D.T., Ducharme, D.W., Turner, S.R., Degraaf, G.L., Lawson, J.A., Couch, S.J. and Williams, M.V. (1991) *J. Med. Chem.* Vol.34. pp.633-642.
58. Togni, G., Sanglard, D., Falchetto, R. and Monod, M. (1991) *FEBS Lett.* Vol.286. pp.181-185.
59. Umezawa H. (1972) *Enzyme Inhibitors of Microbial Origin*, University Park Press, Baltimore.
60. Vanden Bossche, H., Willemsens, G. and Marichal, P. (1987) *Crit. Rev. Microbiol.* Vol.15. pp.57-72.
61. Walsh, T.J. and Pizzo, A. (1988) *Eur. J. Clin. Microbiol. Infect. Dis.* Vol.7. pp.460-475.
62. Wlodawer, A. and Hodgson, K.O. (1975) *Proc. Natl. Acad. Sci., USA* Vol.72. pp.398-399.

TYROSINE 75 ON THE FLAP CONTRIBUTES TO ENHANCE CATALYTIC EFFICIENCY OF A FUNGAL ASPARTIC PROTEINASE, *MUCOR PUSILLUS* PEPSIN

Teruhiko Beppu, Young-Nam Park, Jun-ichi Aikawa, Makoto Nishiyama, and Suheharu Horinouchi

Department of Agricultural Chemistry
Faculty of Agriculture
The University of Tokyo
Yayoi 1-1-1, Bunkyo-ku, Tokyo 113, Japan

INTRODUCTION

Aspartic proteinases have been isolated from a wide variety of organisms such as vertebrates, fungi, plants, and more recently retroviruses, which show marked structural similarity in spite of relatively lower homology in amino acid sequence.[1] Their X-ray crystallographic analyses revealed a bilobal structure accommodating an elongated substrate binding cleft.[2,3,4] The two aspartate residues, Asp32 and Asp215 according to the pepsin numbering, are identified as the catalytic residues, which are located at the bottom of the cleft. Two members of the aspartic proteinases, chymosin obtained from calf stomach and *Mucor pusillus* pepsin, MPP, produced by a filamentous fungus *Mucor pusillus*,[5] are characterized by their high milk-clotting activity due to their selective cleavage of κ-casein along with slower nonspecific proteolysis, and thus they are widely used as milk coagulants in cheese production. We have developed an expression system for prochymosin cDNA in *Escherichia coli* to produce the zymogen as inclusion bodies,[6] which are subsequently converted to chymosin through successive refolding and self-processing steps.[7,8] By using this system, site-directed mutagenesis of chymosin was carried out to generate mutant enzymes with a variety of amino acid exchanges. Among them, replacement of a well conserved residue, Tyr75, on the flap, the long stretch from residue 72 to 85 forming a β-hairpin loop extending over the active site, by Phe caused distinct change in kinetic parameters, k_{cat} as well as K_m, for synthetic oligopeptide substrates. This result suggested the involvement of Tyr75 in the catalytic function.[9]

On the other hand, we have also developed an expression system for MPP using *Saccharomyces cerevisiae* as a host, in which the proMPP secreted into the medium is converted to the mature protease through the self-processing activity as well as by the host-dependent cleavage.[10,11] This system possesses an inherent advantage, which enables us to obtain even inactive mutant enzymes in the mature form. Although MPP excreted from the yeast is highly

Aspartic Proteinases: Structure, Function, Biology, and Biomedical Implications
Edited by Kenji Takahashi, Plenum Press, New York, 1995

glycosylated at Asn72 and Asn172, which causes modulation of the catalytic properties of MPP,[12] most part of the glycosyl moieties can be removed by treatment with endo-β-*N*-glucosaminidase H (endo H).[13] By using this yeast expression system, we conducted site-directed mutagenesis at position 75 of MPP (82nd from the NH_2-terminus in MPP).

MATERIALS AND METHODS

Strains and Plasmids

The MPP gene encodes preproMPP, which consists of the pre-sequence of 22 amino acids for secretion, the pro-sequence of 44 residues and the mature enzyme of 361 residues.[14] Multi-copy plasmid, pJP1,[10] and integration vector, pFR14, containing the MPP gene downstream of the *GAL7* promoter were introduced into *Saccharomyces cerevisiae* MC16 (*a, leu2, his4, ade2*) and *Saccharomyces cerevisiae* AH22 (*a*, *leu*2, *his*4, *can*1) as a host to produce the non-mutated and mutated MPPs. *Escherichia coli* TG1[(Δ*lac pro) thi strA endA sbcB15 hsdR4* F'*traD36 proAB lacI*^{q} *LacZ*Δ*M15*] from Amersham Co. was used for site-directed mutagenesis.

Mutagenesis

Site-directed mutagenesis was carried out by using oligonucleotides by Kunkel's method by use of a BioRad MutaGene kit.[15] All the mutations were checked by nucleotide sequencing by using the M13 dideoxy chain termination method.[16]

Culture Conditions

YPD medium containing 2% Bacto-peptone(Difco), 1% Bacto-yeast extract (Difco), and 2% glucose, and YPGal medium, containing 3% galactose instead of glucose in YPD medium, were used for cultivation of yeast cells carrying plasmid pJP1 and pFR14. For purification of MPP, the yeast cells were precultured in YPD medium at 30°C for 24 h. The cells were then harvested and resuspended in 100 volumes of YPGal medium containing 3% galactose instead of glucose in YPD medium to induce the *GAL7* promoter. The culture was continued at the same temperature for additional 3 days.

Purification of Non-Mutated and Mutated MPPs

The yeast expression system allowed the secretion of highly glycosylated MPP in the mature form in a high yield (>200 mg/l). The glycosylated enzyme was purified by the method principally the same as that described previously with the miner modification of substitution of 50 mM phosphate buffer (pH 5.5) for 50 mM acetate buffer (pH 5.5). The purified preparation (1 mg) was incubated with 0.05U endo H (Seikagaku Kogyo,Tokyo,Japan) in 1 ml of 50 mM of phosphate buffer (pH 5.5) at 37°C for 24 h. After removal of Asn-linked oligosaccharides, deglycosylated MPP was purified as described previously and used as enzyme sample.

Assay of Milk-Clotting and Proteolytic Activities

Milk-clotting activity was measured and expressed according to Iwasaki *et al.*[17] Proteolytic activity was measured by modified method of Anson *et al.* using acid-denatured hemoglobin as the substrate.[18]

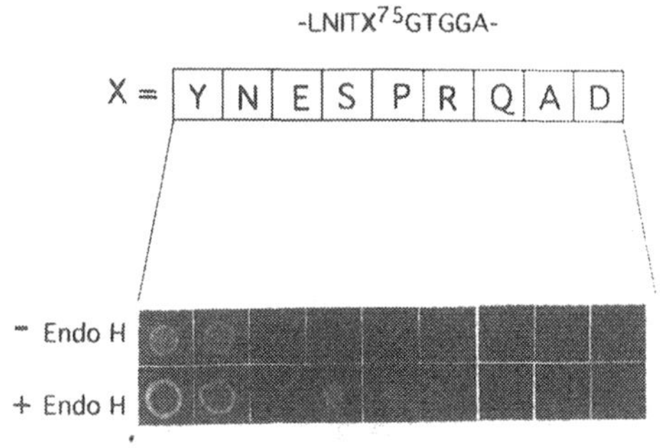

Figure 1. Halo formation on skim-milk agar plate by non-mutated and mutated MPPs carrying substitution of Tyr75 with or without endo H treatment . -LNITX^{75}GTGGA- indicates amino acid sequence of MPP. -Endo H:enzyme was not treated with endo H, +Endo H:enzyme was treated with endo H.

Circular Dichroism and Tryptophan Fluorescence Spectra

Circular dichroism(CD) and tryptophan fluorescence spectra were measured with non-mutated MPP, Tyr75Pro, Tyr75Asp and Tyr75Glu, respectively.

Analysis of Pepstatin-Binding Ability by Differential UV Spectrum

Each enzyme preparation (400 mg/ml) was incubated with various concentrations of pepstatin (Sigma, St. Louis,USA) at 37°C for 10 min. Differential UV spectra were obtained by subtracting the spectrum of the control solution containing each enzyme from that of the above mixture containing the enzyme and pepstatin, according to the methods of Satoi[19] and Nakatani[20].

RESULTS

Site-Directed Mutagenesis at Position 75 of MPP

We constructed 17 mutated MPP genes, in which the codons for Tyr75 residue was exchanged to one for 17 amino acid residues excluding only those for Gly and His. *Saccharomyces cerevisiae* was transformed with pJP1 derivatives carrying these mutated genes, and transformants were cultured aerobically in YPGal medium for 3 days. Activities of the mutant enzymes were monitored by spotting 2 μl of the culture supernatants on milk-agarose plates to examine the halo formation due to milk-clotting activity (Fig. 1). Tyr75Phe and Tyr75Trp mutants, possessing the aromatic amino acid residues as in the non-mutated enzyme, showed milk-clotting activity. In addition, however, Tyr75Asn mutant was found to show a distinct enzyme activity, while all the other mutated MPPs had too weak activity to be detected. The amount and molecular weight of each mutated MPP in culture supernatant was the same when assayed by Coomassie-brilliant-blue staining after sodium dodecyl sulfate (SDS)-polyacrylamide gel electrophoresis. In addition, these mutants showed almost identical circular dichroism spectra as well as tryptophan fluorescence spectra (Fig. 2), suggesting the substitution of Tyr75 did not affect the enzyme activity through distortion of the global conformation. These results indicated that Tyr, Phe, Trp, or Asn at position 75 of MPP has some functional role to express distinct enzyme activity.

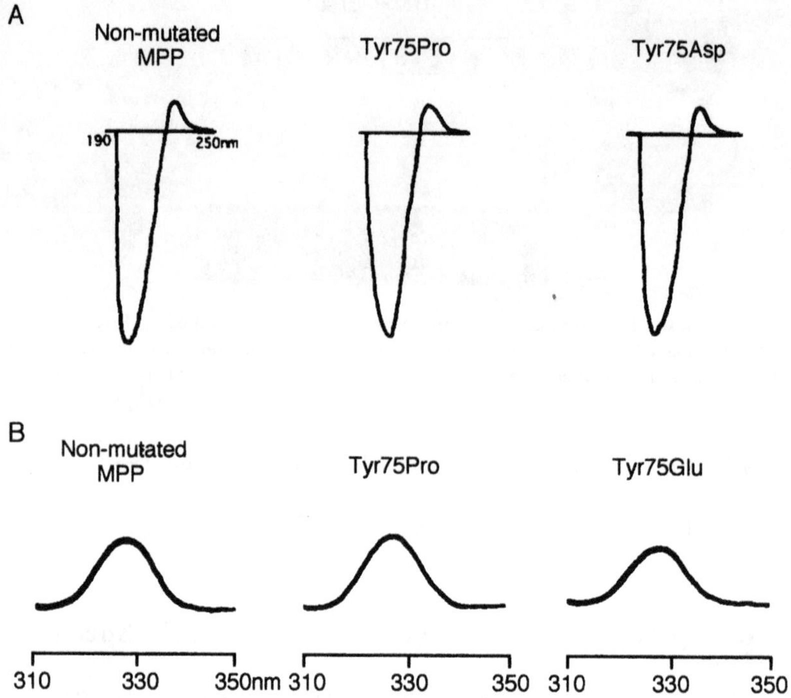

Figure 2. CD and Trp fluorescence spectra of mutants at position 75. A:CD spectra at pH 5.5; B:Trp fluorescence spectra; exciting wavelength, 280 nm; enzyme concentration, 1 μM; 50mM phosphate buffer, pH 5.5, at 25°C.

Catalytic Activities of the Non-Mutated and Mutated MPPs

In order to examine the effect of the replacement of Tyr75 by other amino acid residues on enzyme activity precisely, the catalytic activities of these enzymes were measured. We had observed that MPP produced by the recombinant yeast was highly *N*-glycosylated at Asn72 and Asn172 and the glycosylation caused alteration of both the milk-clotting and proteolytic activities of this protease.[12] Therefore, the enzyme preparations were first treated with endo H to remove most of the Asn-linked glycosyl moieties prior to kinetic study. Although the mutants with amino acid residues except for Tyr, Phe, Trp, and Asn showed only negligible activity in glycosylated forms, treatment with endo H caused appearance of the milk-clotting activity which was detected by halo-formation on milk-agar plates (Fig. 1). This result indicates that the glycosylation causes inhibitory effect on the catalytic activity of MPP. Determination of the proteolytic activities of these deglycosylated mutants using hemoglobin as a substrate revealed that the mutants with very low activity possessed proteolytic activity over 1000-fold less than that of non-mutated enzyme (data not shown). We measured both milk-clotting (C) and proteolytic (P) activities of mutants with Phe and Asn at position 75 (Table 1). The mutant Tyr75Phe was found to exhibit a significant decrease in milk-clotting activity along with the proteolytic activity, in which larger decrease was observed in the clotting activity. On the other hand, Tyr75Asn mutation affected to cause a larger decrease in proteolytic activity, compared to a decrease in milk-clotting activity. The ratio (C/P) of milk-clotting activity to proteolytic activity is a key parameter of milk-clotting enzyme to be used as a milk-coagulant in the cheese industry. Tyr75Phe showed a 3-fold

Table 1. Enzyme Activities and Kinetic Parameters of *Mucor pusilllus* Pepsin and Its Mutants

Enzyme	Milk-Clotting Activity(C) (U/μg)	Proteolytic Activity(P) (U/μg)	C/P	Peptide I			Peptide II		
				K_m (μM)	k_{cat} (s^{-1})	k_{cat}/K_m ($\mu M^{-1}\cdot s^{-1}$)	K_m (μM)	k_{cat} (s^{-1})	k_{cat}/K_m ($\mu M^{-1}\cdot s^{-1}$)
Non-mutated	6.9	1.5	4.6	87.5	8.1	0.09	21.4	46.2	2.2
Tyr75Phe	0.6	0.4	1.5	161.0	0.15	0.001	60.6	55.2	0.9
Tyr75Asn	2.1	0.2	10.5	115.0	0.93	0.008	55.0	56.2	1.0

decrease in C/P ratio, while the mutation of Tyr75Asn resulted in a 2.2-fold increase in the ratio.

Kinetic Analyses of Tyr75Asn and Tyr75Phe Mutants Using Synthetic Substrates

Kinetic analyses were carried out with the mutants Tyr75Phe and Tyr75Asn by using synthetic oligopeptides as substrates. When peptide 1 (Leu-Ser-Phe(NO_2)-Nle-Ala-Leu-OMe) was used, both mutations Tyr75Phe and Tyr75Asn were found to exhibit significant decreases in the k_{cat} values, while such a large change in the K_m values was not caused by the mutations (Table 1). When peptide 2 (Lys-Pro-Ile-Glu-Phe-Phe(NO_2)-Arg-Leu) was used, these mutations affected the K_m value, while no such significant change in the k_{cat} value was observed.

Analysis of Inhibitor-Binding Ability of Non-Mutated and Mutated MPPs

The replacement of Tyr at position 75 by one of 14 amino acids except for Phe, Trp and Asn caused a marked decrease in enzyme activity. In order to clarify the reason for the large decrease in the activity by these mutations, we examined their binding ability to pepstatin, a specific inhibitor for aspartic proteinase, by using differential spectra at UV range (Fig. 3). Non-mutated MPP showed differential spectra upon addition of pepstatin with three peaks at 280, 287, and 293 nm. When differential absorbance at each of three wavelengths was plotted against the concentrations of pepstatin, constant absorbance was obtained by addition of the inhibitor to the enzyme solution at the molar ratio of 1:1, indicating the formation of an enzyme-inhibitor complex. Similar analysis was carried out with other mutants. Almost identical spectra with three peaks were observed with Tyr75Asn mutant. On the other hand, other mutants with extremely low activity showed only two peaks in the differential UV spectra (Fig. 3). Maximal change in absorbance upon addition of pepstatin was different between these mutants, which may reflect the difference in substrate-binding mode between the enzyme-inhibitor complexes.

Stability of the Active Mutant Enzymes

We next analyzed the effects of the mutations Tyr75Phe and Tyr75Asn on the stability of MPP. After incubation at 55°C for 0, 10, 60, and 180 min, their milk-clotting activities were measured. The results are shown in Fig. 4A. Tyr75Phe showed stability similar to that

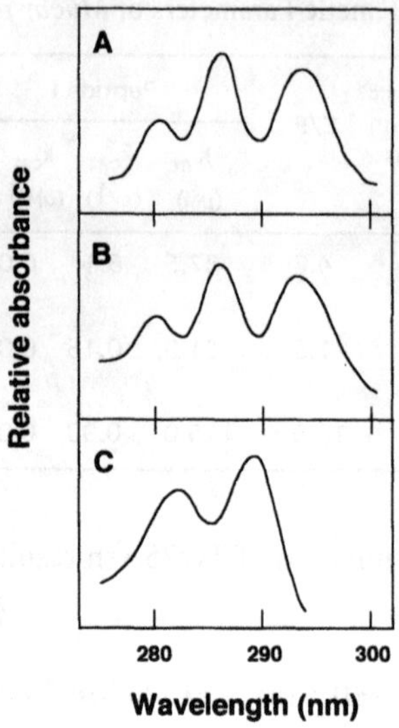

Figure 3. Differential UV spectra of non-mutated and mutated MPPs upon addition of pepstatin A. A: Differential UV spectrum with non-mutated *Mucor pusillus* pepsin. B: Differential UV spectrum with Tyr75Asn mutant. C: Differential UV spectrum with Tyr75Ser mutant. Mixtures of the enzymes and pepstatin were incubated at 37°C for 10min.

of the non-mutated enzyme, retaining 90% residual activity after incubation for 180 min. On the other hand, Tyr75Asn showed distinctly lower stability, and only 20% residual activity was detected after the heat-treatment for 180 min. We also determined the Tm values of the non-mutated and Tyr75Asn enzymes by monitoring the change in the CD spectrum (Fig. 4B). However, only a slight decrease in Tm value was observed with Tyr75Asn (64°C) in comparison with the value of the non-mutated enzyme (67°C). When the residual amounts of the enzyme protein in the incubated mixture were analyzed by SDS-polyacrylamide gel electrophoresis, the non-mutated enzyme showed no significant change in the amount during the heat treatment, whereas Tyr75Asn showed a decrease in its amount upon heating (Fig. 4C). The amounts of residual MPP measured by densitometry of the gels coincided well with the residual activities of both enzymes after the heat treatment. This may suggest that the marked decrease in the residual activity of Tyr75Asn on heating is caused by self-digestion.

DISCUSSION

Site-directed mutagenesis of MPP was conducted to reveal the functional role of Tyr75 on the flap. The yeast expression system for the MPP gene allows us to isolate even an inactive mutant enzyme in the mature form due to processing by the proteolytic activity of the host cells. In fact, all the mutant enzymes were confirmed to be produced in the mature form. Therefore, the extremely low enzyme activities of most of the mutants were not due to the failure of processing but to the direct decrease in the activity caused by the substitutions.

On the other hand, a few mutants, especially Tyr75Phe and Tyr75Asn, were found to possess distinct catalytic activities. They caused a change in kinetic parameters similar to that observed with the Tyr75Phe mutant of chymosin[9]; *i.e.*, a marked drop of k_{cat} but small change of K_m for peptide 1, while a marked increase in K_m but no change of k_{cat} for peptide

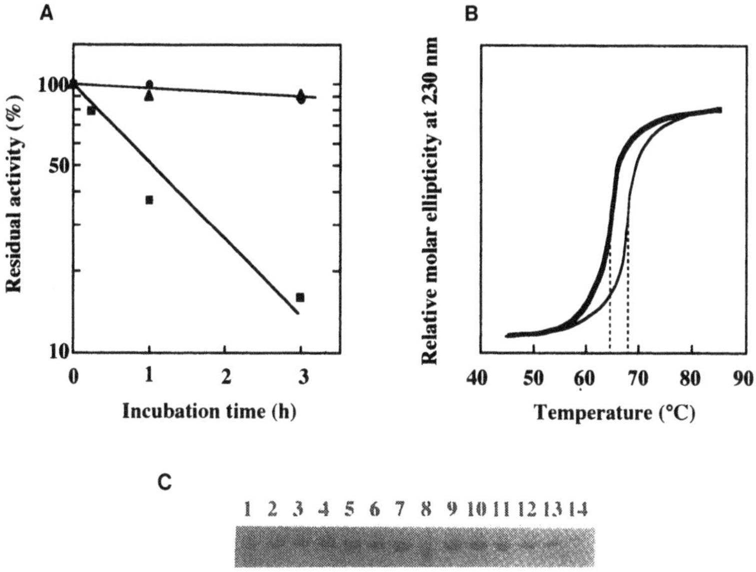

Figure 4. Stability of non-mutated, Tyr75Phe and Tyr75Asn *Mucor pusillus* pepsins. A: Residual activity of non-mutated, Tyr75Phe and Tyr75Asn *Mucor pusillus* pepsins after heating at 55°C. B: CD analysis of the structural degradation of wild-type (thin line) and Tyr75Asn (thick line) *Mucor pusillus* pepsins on heating. C: SDS-polyacrylamide gel electrophoresis of non-mutated and Tyr75Asn *Mucor pusillus* pepsins. Lanes 1 and 8: molecular marker (ovalbumin: 43 kDa), lanes 2-7: non-mutated enzyme, lanes 9-14: Tyr75Asn mutant. Lanes 2 and 9: no heat-treatment, lanes 3 and 10: enzymes heated at 55°C for 12min, lanes 4 and 11: for 30min, lanes 5 and 12: for 1h, lanes 6 and 13: for 2h, lanes 7 and 14: for 3h.

2. The changes in both k_{cat} and K_m in the cleavage of these oligopeptides suggests the involvement of Tyr75 in catalytic function as well as in substrate recognition. Tyr75Phe and Tyr75Asn also showed a significant decrease in k_{cat}/K_m for the synthetic peptides, which may correspond to a decrease in the stability of a transition-state intermediate.[21] Furthermore, by measurement of catalytic activity for deglycosylated preparation, most of the substitutions at position 75 were found to cause remarkable decreases in catalytic activity but do not lead to complete loss of activity. From these results, we speculate that basal activity mediated by two catalytic aspartate residues might be greatly enhanced by possible function of amino acid residue at the 75th, for example, to stabilize a transition state intermediate during catalysis. According to the crystallographic analysis of the co-crystals of endothia pepsin with a statine-containing inhibitor, Blundell *et al.* proposed that the aromatic ring of Tyr75 is involved in stabilizing the transition state intermediate by a possible interaction between the hydrogen atoms of the aromatic ring and one of the OH-groups of the tetrahedral intermediate.[2] In subtilisin, a serine protease, the amide group of the side chain of an Asn residue is positioned to form a hydrogen bond with tetrahedral oxyanion intermediate bound to subtilisin to stabilize the intermediate,[22] and in fact the exchange of the Asn residue caused a 100-fold decrease in catalytic activity.[23] The fact that Tyr75Asn mutant showed a distinct enzyme activity may emphasize the putative catalytic mechanism involving the 75th residue to stabilize the intermediate.

The observation that all the mutants examined showed 1:1 complex with pepstatin confirms that the basic conformation of the mutant enzymes was maintained in the mutants. However, it should be noted that both the non-mutated and Tyr75Asn MPPs with higher enzyme activity gave three signals in the difference spectra upon addition of the inhibitor, whereas only two signals were observed with the other mutants with extremely low activity. The absence of the 3rd signal in the low-activity mutants cannot be attributed to the absence of Tyr75, since the signal was observed with Tyr75Asn. We assume that the abnormality in the spectrum of mutant with low activity reflects a slightly different conformation of the enzyme-inhibitor complex in this mutant probably due to binding of the inhibitor in a non-productive mode which cannot stabilize the possible tetrahedral intermediate.

In this study we found that the Tyr75Asn mutation caused a rapid decrease in the enzyme activity by the heat-treatment, which was accompanied by disappearance of the enzyme protein probably due to self-digestion. We assume that substitution of Tyr75Asn may cause an increased flexibility of the flap, which in turn facilitates self-digestion. Chymosin, MPP and their related enzymes are widely used as a milk coagulant in cheese industry. In comparison with chymosin, however, MPP possesses a lower ratio of milk-clotting activity to non-specific proteolytic activity and higher thermostability, the latter of which causes survival of the enzyme activity after cooking of curd and may cause ill flavor in cheese during the long aging period. Tyr75Asn showed high sensitivity to the heat-treatment and an increase in C/P ratio, which make the mutant a candidate as an improved milk coagulant.

REFERENCES

1. D.R. Davis, The structure and function of the aspartic proteinases, *Annu. Rev. Biophys. Biophys. Chem.* **19**:189-215 (1990)
2. T.L. Blundell, J. Cooper, S.I. Foundling, D.M. Jones, B. Attrash, and M. Szelke, On the rational design of renin inhibitors:X-ray studies of aspartic proteinases complexed with transition-state analogues, *Biochemistry* **26**:5585-5590 (1987)
3. J. Cooper, S. Foundling, A. Hemmings, T.L. Blundell, D.M. Jones, A. Hallett, and M. Szelke, The structure of a synthetic pepsin inhibitor complexed with endothiapepsin, *Eur. J. Biochem.* **169**:215-221 (1987)
4. A. Sali, B. Veerapandian, J.B. Cooper, S.I. Foundling, D.J. Hoover, and T.L. Blundell, High-resolution X-ray diffraction study of the complex between endothiapepsin and an oligopeptide inhibitor:the analysis of the inhibitor binding and description of the rigid body shift in the enzyme, *EMBO J.* **8**:2179-2188 (1989)
5. K. Arima, S. Iwasaki, and G. Tamura, Milk clotting enzyme from Microorganisms, *Agric. Biol. Chem.* **31**:540-545 (1967)
6. Y. Kawaguchi, N. Kosugi, K. Sasaki, T. Uozumi, and T. Beppu, Production of chymosin in *Escherichia coli* cells and its enzymatic properties, *Agric. Biol. Chem.* **51**:1871-1877 (1987)
7. Y. Kawaguchi, N. Shimizu, K. Nishimori, T. Uozumi, and T. Beppu, Renaturation and activation of calf prochymosin produced in an insoluble form in *Escherichia coli*, *J. Biotechnol.* **1**:307-315 (1984)
8. M. Yonezawa, J. Suzuki, M. Nishiyama, S. Horinouchi, and T. Beppu, Role of the amino-terminal amino acid sequences determining the *in vitro* refolding process of prochymosin polypeptide, *J. Biotechnol.* **28**:85-97 (1993)
9. J. Suzuki, K. Sasaki, Y. Sasao, A. Hamu, H. Kawasaki, M. Nishiyama, S. Horinouchi, and T. Beppu, Alteration of catalytic properties of chymosin by site-directed mutagenesis, *Prot. Eng.* **2**:563-569 (1989)
10. T. Yamashita, N. Tonouchi, T. Uozumi, and T. Beppu, Secretion of *Mucor* rennin, a fungal aspartic protease of *Mucor pusillus*, by recombinant yeast cells, *Mol. Gen. Genet.* **210**:462-467 (1987)
11. R. Hiramatsu, J. Aijawa, S. Horinouchi, and T. Beppu, Secretion by yeast of the zymogen from of *Mucor* rennin, an aspartic proteinase of *Mucor pusillus*, and its conversion to the mature form, *J. Biol. Chem.* **264**:16862-16866 (1989)
12. J. Aikawa, T. Yamashita, M. Nishiyama, Suheharu Horinouchi, and T. Beppu, Effects of glycosylation on the secretion and enzyme activity of *Mucor* rennin, an aspartic proteinase of *Mucor pusillus*, produced by recombinant yeast, *J. Biol. Chem.* **265**:13955-13959 (1990)
13. A.L. Tarentino, T.H. Plummer Jr. , and F. Maley, The release of intact oligosaccharide from specific glycoprotein by endo-β-*N*acetylglucosamidase H, *J. Biol. Chem.* **249**:818-824 (1974)
14. N. Tonouchi, H. Shoun, T. Uozumi, and T. Beppu, Cloning and sequencing of a gene for *Mucor* rennin, an aspartate protease from *Mucor pusillus*, *Nucleic Acids Res.* **14**:7557-7568 (1986)

15. T.A. Kunkel, Rapid and efficient site-specific mutagenesis without phenotype selection, *Proc. Natl. Acad. Sci. U.S.A.* **82:**488-492 (1985)
16. F. Sanger, A.R. Coulson, B.G. Barrelle, A.J.M. Smith, and B.A. Ros, Bacteriophage as an aid to rapid DNA sequencing, *J. Mol. Biol.* **143:**161-178 (1980)
17. S. Iwasaki, G. Tamura, and K. Arima, Milk clotting enzyme from microorganisms. Part II. The enzyme production and the properties of crude enzyme, *Agric. Biol. Chem.* **31:**546-551 (1967)
18. M.I. Anson, The estimation of pepsin, trypsin, papain, and cathepsin with hemoglobin, *J. Gen. Physiol.* **22**:79-89 (1938)
19. S. Satoi, and S. Murao, Inhibition of acid proteases by a pesin inhibitor (S-PI), *Agric. Biol. Chem.* **37**:2579-2587 (1973)
20. H. Nakatani, K. Kitagishi, and K. Hiromi, Spectroscopic studies of pepsin and its complex with *Streptomyces* pepsin inhibitor, *Biochim. Biophys. Acta* **452**:521-524 (1976)
21. A.J. Wilkinson, A.R. Fersht, D.M. Blow, and G. Winter, Site-directed mutagenesis as a probe of enzyme structure and catalysis: tyrosyl-tRNA synthetase cysteine-35 to glycine-35 mutation, *Biochemistry* **22**:3581-3586 (1983)
22. P. Carter, and J.A. Wells, Dissecting the catalytic traid of a serine protease, *Nature* **332**:564-568 (1988)
23. P. Bryan, M.W. Pantoliano, S.G. Quile, H.Y. Hsiao, and T. Poulous, Site-directed mutagenesis and the role of the oxyanion hole in subtilisin, *Proc. Natl. Acad. Sci. USA* **83**:3743-3745 (1986)

ASPARTIC PROTEINASES FROM THE HUMAN MALARIA PARASITE *PLASMODIUM FALCIPARUM*

Colin Berry,[1] John B. Dame,[2] Ben M. Dunn,[3] and John Kay[1]

[1]Department of Biochemistry
University of Wales College of Cardiff
PO Box 903, Cardiff CF1 1ST, UK
[2]Department of Infectious Diseases
College of Veterinary Medicine
University of Florida
Gainesville, FL 32610
[3]Department of Biochemistry and Molecular Biology
College of Medicine
University of Florida
Gainesville, FL 32610

INTRODUCTION

The widespread occurrence of drug resistant parasites and insecticide resistant mosquitoes has resulted in a growing public health emergency in malaria control. In species of Plasmodium, aspartic proteinases have been implicated in processes such as release of merozoites from the late schizont stage (1) in invasion of the erythrocyte (2) and digestion of haemoglobin therein (3-5). Following ingestion, haemoglobin is degraded in the food vacuole, a specialised proteolytic organelle (3,4,6). The aspartic proteinase activity demonstrated to be present within this organelle was resolved into three peaks during partial purification (5), although full characterisation of these has yet to be accomplished. The key role of an aspartic proteinase in initiating haemoglobin digestion in the vacuole has been defined (6) and one aspartic proteinase (PFAPG) has been purified to apparent homogeneity from the food vacuoles of *P. falciparum*. The sequence of 22 residues at the N-terminus of the mature enzyme was determined, and its ability to initiate the degradation of haemoglobin by selective cleavage of the -Phe33 -Leu34 peptide bond located in the hinge region of the alpha-chain was demonstrated (6). Inhibition of haemoglobin catabolism or other essential functions catalysed by aspartic proteinases in the parasite offer attractive targets for therapeutic intervention.

Aspartic Proteinases: Structure, Function, Biology, and Biomedical Implications
Edited by Kenji Takahashi, Plenum Press, New York, 1995

```
         .         .         .         .         .         .         .         .        90
TGAACCAAATGGATATTACAGTAAGAGAACATGATTTTAAACATGGCTTTATCAAAAGCAATTCAACATTTGATGGATTAAACATTGACA
   N  Q  M  D  I  T  V  R  E  H  D  F  K  H  G  F  I  K  S  N  S  T  F  D  G  L  N  I  D

         .         .         .         .         .         .         .         .       180
ATTCAAAGAACAAAAAAAAAAATACAGAAAGGATTTCAAATACTATATGTACTTCTCTTTTGTAGTGTAATGTGTGGTTTATTTTATTATG
N  S  K  N  K  K  K  I  Q  K  G  F  Q  I  L  Y  V  L  L  F  C  S  V  M  C  G  L  F  Y  Y

         .         .         .         .         .         .         .         .       270
TGTATGAAAATGTATGGCTTCAAAGAGATAATGAAATGAATGAAATTTTAAAAAATTCCGAACATTTAACTATTGGATTTAAAGTTGAAA
V  Y  E  N  V  W  L  Q  R  D  N  E  M  N  E  I  L  K  N  S  E  H  L  T  I  G  F  K  V  E

         .         .         .         .         .         .         .         .       360
ATGCACATGATAGAATTTTGAAAACTATAAAAACACATAAATTAAAAAATTACATTAAAGAATCTGTCAATTTTCTTAATTCAGGACTTA
N  A  H  D  R  I  L  K  T  I  K  T  H  K  L  K  N  Y  I  K  E  S  V  N  F  L  N  S  G  L

         .         .         .         .         .         .         .         .       450
CTAAAACAAATTATTTAGGTAGTTCAAATGATAATATCGAATTAGTAGATTTCCAAAATATAATGTTTTATGGTGATGCAGAAGTTGGAG
T  K  T  N  Y  L  G  S  S  N  D  N  I  E  L  V  D  F  Q  N  I  M  F  Y  G  D  A  E  V  G
            ↑

         .         .         .         .         .         .         .         .       540
ATAACCAACAACCATTTACATTTATTCTTGATACAGGATCTGCTAATTTATGGGTCCCAAGTGTTAAATGTACAACTGCAGGATGTTTAA
D  N  Q  Q  P  F  T  F  I  L  D  T  G  S  A  N  L  W  V  P  S  V  K  C  T  T  A  G  C  L
                        -------------
                              *

         .         .         .         .         .         .         .         .       630
CTAAACATCTATATGATTCATCTAAATCAAGAACATATGAAAAAGATGGAACCAAAGTAGAAATGAATTATGTGTCAGGAACTGTTAGTG
T  K  H  L  Y  D  S  S  K  S  R  T  Y  E  K  D  G  T  K  V  E  M  N  Y  V  S  G  T  V  S
                                                               -------
                                                                  *

         .         .         .         .         .         .         .         .       720
GATTTTTCAGTAAAGATTTAGTAACTGTTGGTAATTTATCTCTTCCATATAAATTTATTGAAGTAATAGATACTAATGGATTCGAACCAA
G  F  F  S  K  D  L  V  T  V  G  N  L  S  L  P  Y  K  F  I  E  V  I  D  T  N  G  F  E  P
                                 -------
                                 †

         .         .         .         .         .         .         .         .       810
CTTATACTGCTTCAACATTTGATGGTATCCTTGGTTTAGGATGGAAAGATTTATCAATAGGTTCAGTAGATCCAATTGTTGTTGAATTAA
T  Y  T  A  S  T  F  D  G  I  L  G  L  G  W  K  D  L  S  I  G  S  V  D  P  I  V  V  E  L
```

```
         .         .         .         .         .         .         .         .       900
AAAACCAAAACAAAATTGAAAATGCTCTTTTCACCTTTTACTTACCTGTACATGATAAACATACAGGATTCTTAACCATTGGTGGTATTG
K  N  Q  N  K  I  E  N  A  L  F  T  F  Y  L  P  V  H  D  K  H  T  G  F  L  T  I  G  G  I

         .         .         .         .         .         .         .         .       990
AAGAAAGATTTTATGAAGGACCACTAACTTACGAAAAATTAAACCACGATTTATATTGGCAAATAACTTTAGATGCACACGTTGGAAATA
E  E  R  F  Y  E  G  P  L  T  Y  E  K  L  N  H  D  L  Y  W  Q  I  T  L  D  A  H  V  G  N

         .         .         .         .         .         .         .         .      1080
TAATGTTAGAAAAAGCAAACTGTATTGTAGATAGTGGTACTAGTGCCATTACTGTACCAACTGACTTTTTAAATAAAATGTTGCAGAATT
I  M  L  E  K  A  N  C  I  V  D  S  G  T  S  A  I  T  V  P  T  D  F  L  N  K  M  L  Q  N
                        -------------
                              *

         .         .         .         .         .         .         .         .      1170
TAGATGTTATCAAAGTCCCATTCTTACCTTTCTATGTAACTCTTTGTAACAACAGCAAATTACCAACTTTTGAATTTACCTCAGAAAATG
L  D  V  I  K  V  P  F  L  P  F  Y  V  T  L  C  N  N  S  K  L  P  T  F  E  F  T  S  E  N
                                                ---------
                                                †

         .         .         .         .         .         .         .         .      1260
GTAAATACACATTAGAACCTGAATACTACCTTCAACACATAGAAGATGTTGGTCCAGGATTATGTATGCTTAATATCATAGGATTAGATT
G  K  Y  T  L  E  P  E  Y  Y  L  Q  H  I  E  D  V  G  P  G  L  C  M  L  N  I  I  G  L  D

         .         .         .         .         .         .         .         .      1350
TTCCAGTACCAACCTTTATTCTAGGTGACCCATTCATGAGAAAATATTTTACCGTCTTTGATTATGATAATCACAGTGTTGGTATTGCTC
F  P  V  P  T  F  I  L  G  D  P  F  M  R  K  Y  F  T  V  F  D  Y  D  N  H  S  V  G  I  A
                                                                     ---------
                                                                     †

         .         .         .         .         .     1413
TTGCTAAAAAGAATTTATAAAAACGAAAATTTGGAAACCAAAGGAAAAACATAATTATATAAG
L  A  K  K  N  L  ·
```

Figure 1. Nucleotide and deduced amino acid sequences of the *P. falciparum* aspartic proteinase PFAPD. The active site sequences are marked as $\overline{\ \ *\ \ }$ and the predicted N-terminus of the mature enzyme is depicted by ↑. Possible glycosylation signals are marked by $\overline{\ \ \dagger\ \ }$.

```
             1       10        20        30        40          50        60
                      .         .         .         .           .         .
PPD: ----LGSSNDNIELVDFQNIMFYGDAEVGDNQQPFTFILDTGSANLWVPSVKCT--TAGCLTKHLYDSSKSRTYE
PPG: ----  NAGDSVT N VA V Y  E QI   K K A  F           AQ N-- I  K  N    N  K
D:   -----TEGPIPEV KNYMDAQY  EIGI TPP C  VVF    S      IH KLLDIA WIH K N D  SY V
E:   TESCSMDQSAKEP INYLDMEYF TISI SPP N  V F    S       Y  --SPA K HSRFQPSQ S  S

        70   75   80        90                   100          110      120
         .    .    .         .                    .            .        .
PPD: KDGTKVEMNYVSGTVSGFFSKDLVTV-----------GNLSLPYK-FIEVIDTNGFEPTYTASTFDGILGLGWKD
PPG:                       I  I-----------A   F   -    T       A  LGQ    V
D:    N  SFDIH G  SL  YL Q T S PCQSASSASAL GVKVERQV G ATKQP I-- FI AK      MAYPR
E:   QP QSFSIQ GT SL  IIGA Q S -----------EG TVVGQQ G SVTEP Q-- FVDAE        YPS

     130       140       150         160       170  175  180       190       200
       .         .         .     .     .         .    .    .         .         .
PPD: LSIGSVDPIVVELKNQNKIENALFTFYLPVHDKHT-GF-LTIGGIEERFYEGPLTYEKLNHDLYWQITLDA-HV-
PPG:         V           Q V      FD   K- Y-       D     Q             VD L-  F-
D:   I VNN·L VFDN MQ KLVDQNI S   SRDPDAQP GE ML  TDSKY K S S LNVTRKA   VH  QVE A
E:    AV G T VFDNMMA  LVDLPM SV MSSNPEGGA SE IF  YDHSHFS S NWVPVTKQA    A  NIQ G

            210  215  220        230         240       250       260       270
              .    .    .          .           .         .         .         .
PPD: GNIMLEKANC--IVDSGTSAITVPTDFLNKMLQNLDVIKVPFLPF--YVTLCNN-SKLPTFEFTSENGKYTLEPE
PPG:   LTV   TA--       S  A  E    FFEG   V I    L-- I T   -P    L  R ATNV
D:   SGLT C EG EA   T  LMVG V EV-RE  KA-IGA  LIQ-GE MIP EKV T  AITLKLGGKG  K S
E:    TV FCSEG QA   T  L  G S KI- Q   A-IGAA --VDGE A E A LNVM DVT  INGVP    S T

           280           290       300       310       320
             .             .         .         .         .
PPD: YY-LQHIEDVGPGLCMLNIIGLDFPVPT---FILGDPFMRKYFTVFDYDNHSVGIALAKKNL
PPG:   -  Q F F IS   VS  PV LNKN ---                    T  F     K
D:   D T K-VSQA KT  LSGFM M I P SGPLW    V IGR Y V  R  NR  F E ARL-
E:   A T LDFV - MQF SSGFQ   IHP AGPLW    V I QFYSV  RG NR  L P VP--
```

Figure 2. Alignment of the sequence of the putative mature *Plasmodium* aspartic proteinase PFAPD with PFAPG, human cathepsin D (D) and cathepsin E (E). Where no amino acid is shown, this residue is identical to that of PFAPD in the top line of the alignment. - indicates gaps introduced to maximize the alignment. Numbering is based on that for pig pepsin.

RESULTS AND DISCUSSION

A clone encoding an aspartic proteinase (PFAPD) was identified by sequence homology in a *Plasmodium falciparum* strain HB3 gene library (7).

The sequence of the 1413 bp insert was determined in both strands by dideoxy chain-termination sequencing and is shown in Fig. 1. The insert contains a 1368 bp open reading frame extending from base 3-1370. A further 43 bp of 3'-untranslated sequence was also established (8).

The translated sequence designated *Plasmodium falciparum* Aspartic Proteinase D (PFAPD) was aligned with those of human procathepsin D (9) and procathepsin E (10). Homology was virtually absent in the N-terminal one third of the sequence (encoded by bases 3-374) corresponding to the prosegment of this putative zymogen, but extensive homology was apparent between the C-terminal two thirds of PFAPD and the mature proteinase regions of the human aspartic proteinases, cathepsin D and cathepsin E (Fig. 2). Also included in the alignment in Fig. 2 is the sequence of a second aspartic proteinase (called PFAPG) translated from its recently published cDNA sequence (11). PFAPG does contain at its N-terminus, an identical sequence to the 22 residues that were identified previously by Edman degradation of the protein isolated from food vacuoles (6). The

```
                 10        20        30        40        50
                  .         .         .         .         .
PFAPD:   MDITVREHDFKHGFIKSNSTFDGLNIDNSKNKKKIQKGFQILYVLLFCSV
PFAPG:    ALSIK -  SSA A NE AVNSSTFN NMKTW    R      FF LLI

                 60        70        80        90       100
                  .         .         .         .         .
PFAPD:   MCGLFYYVYENVWLQRDNEMNEILKNSEHLTIGFKVENAHDRILKTIKTH
PFAPG:   TGA    LID  LFPKNKKI   MNT K VI   SI  S    M  V Q

                110       120       130
                  .         .         .
PFAPD:   KLKNYIKESVNFLNSGLTKTNYLGSSNDNI
PFAPG:   R        LK FKT   QKPH  NAG SV
```

Figure 3. Alignment of the prosegments of *Plasmodium* aspartic proteinases D and G. Where no amino acid is shown, this residue is identical to that of PFAPD in the top line of the alignment.

sequences of PFAPD and PFAPG in the N-terminal beta-strand region may be clearly distinguished, indicating that *Plasmodium falciparum* encodes two aspartic proteinases. Overall however, the two proteins PFAPD and PFAPG show a high degree of identity, 87% in the mature proteinases (Fig. 2) and 45% in the putative prosegments (Fig. 3).

For ease of alignment, the sequences of the enzymes depicted in Fig. 2 are begun at residues which coincide with the N-terminal residue of pig pepsin (Ile 1). In turn, this has enabled adoption of the conventional numbering system for aspartic proteinases which is based on that of pig pepsin. However, the N-terminal region of aspartic proteinases is of variable length and composition, reflecting the different mechanisms of activation of these proteinases by autolytic processes or requiring involvement of an extrinsic proteinase in some cases. On this basis, a putative self-processing site can be identified in the sequence of PFAPD such that mature PFAPD would begin with the Leu residue encoded at base 375 in Fig. 1 (amino acid 123, Fig. 3) and would thus have a calculated molecular weight of 37,094.

Sequence identities at the amino acid level between PFAPD and the human aspartic proteinases were 33% (renin (12) and pepsin) and 34% (cathepsin D and cathepsin E). Allowing for conservative amino acid substitutions, similarities rise to >50%. Two active site hydrophobic-hydrophobic-Asp-Thr/Ser-Gly sequences of PFAPD (Figure 1; bases 473-487 and 1013-1027) were readily aligned with their counter parts within the mature human enzymes (Figure 2; centering around Asp at residues 32 and 215 respectively). The second catalytic Asp residue is located in an Asp-Ser-Gly sequence in PFAPD. This Ser for Thr replacement is relatively uncommon in two-domain aspartic proteinases but has been observed previously in the cathepsin D-like enzyme from barley (13) and in the dimeric retroviral aspartic proteinases from Rous Sarcoma Virus (RSV) and Myeloblastosis Associated Virus (MAV) (14). The hydroxyl group of the Thr/Ser residue must be present as a component contributing to the "fireman's grip" which maintains the spatial arrangement of the two catalytic carboxyl groups supplied by the Asp residues in this type of proteinase (15). Another feature seen in all archetypal aspartic proteinases described to date is the location of a Tyr residue (Tyr 75) in a beta hairpin loop or "flap" which partially occludes the active site and which contributes to substrate binding/catalysis (16). This motif can also be observed (Fig. 2) in PFAPD although the surrounding sequence is atypical in having a Val in position 76 where Gly is found in many previously described members of this family of proteinases.

On the basis of the extensive homology of PFAPD to the aspartic proteinase family of enzymes, a model of the proteinase was constructed utilising a customised database of aspartic proteinase structures(8). The PFAPD model shows a number of structural features in common with aspartic proteinases that have been characterised previously by X-ray crystallography. Its overall globular shape consists of two domains of well-defined beta-sheet structures with an obvious cleft dividing the N-terminal from the C-terminal domain and a prominent "flap" or beta-hairpin loop containing the Tyr75 residue, that overhangs the active site cleft.

The two Asp residues (32 and 215) in their respective -Asp-Thr-Gly and -Asp-Ser-Gly- sequences are located close to each other mid-way along the length of the cleft in positions closely comparable to the equivalent active site residues of other aspartic proteinases. The sequence -Ile-Leu-Gly which is necessary to complete the psi loops with each catalytic Asp-Thr/Ser-Gly motif (15) is found to occur twice in conserved locations in PFAPD at residues 120-122 and 300-302 (Fig. 2).

A number of other features contribute to the overall compactness in the folded PFAPD molecule. In general, the hydrophilic residues are found predominantly on the solvent-accessible surface while the hydrophobic residues are concentrated in the central core of the protein. The conserved disulphide bond between residues 45 and 50 is found in all members of the aspartic proteinase family, but with some slight variation in the size of the loop formed. Cysteine residues are present at positions 45 and 50 in the PFAPD sequence (Fig. 2) so it is assumed that these will take the form of a disulphide bond. A second conserved disulphide bond is between Cys 249 and Cys 282. Again, these residues in these locations are conserved in PFAPD (Fig. 2). Aspartic proteinases of mammalian origin have a third conserved disulphide bond between residues 206 and 210 which is not found in fungal enzymes such as the *Rhizopus chinensis* proteinase (17). PFAPD is unique in having only a single cysteine residue at position 210, which may exist in the reduced SH state. It is possible that two enzyme molecules might dimerise through formation of an interchain disulphide bond in a similar manner to cathepsin E (10).

In contrast, the PFAPG enzyme from *Plasmodium falciparum* does not contain this unpaired cysteine residue (Fig. 2). A reduction in size of PFAPD relative to other members of the aspartic proteinase family in the region of 295-300 (Fig. 2) results in a loop structure in the C-terminal domain, above the active site cleft which is shorter than those present in mammalian enzymes. However, the equivalent is shorter still in *Rhizopus chinensis* proteinase without causing any adverse effect on proteinase activity (18). Thus, numerous characteristic structural features of the aspartic proteinase superfamily are readily identified in the modelled PFAPD structure and where minor deviations do occur, they are less severe than those observed in the *Rhizopus chinensis* enzyme.

The protein product of the PFAPD gene was produced in *E. coli* by recloning the insert into the expression vector pGEX2T to produce a fusion protein with glutathione S-transferase. Western blots of extracts of transformed *E. coli* after induction of protein synthesis with IPTG were probed with a polyclonal antiserum raised against the synthetic peptide FDYDNHSVGIALAKKNL, which corresponds to the C-terminal sequence of the proteinase (Fig. 2). Expression of the PFAPD gene was successful but relatively little of the fusion protein (anticipated mol. wt. approx. 70 kDa) accumulated. Instead, immunoreactive bands were detected at 45 and 28 kDa. The former coincides with the molecular weight anticipated for precursors of aspartic proteinases.

The immunoreactive material expressed was insoluble in *E. coli.* When solubilised and refolded by the method of Hill et al.(20), activity was assayed using as substrates both haemoglobin and the peptide ALERTF*LSFPT, based on the sequence around the primary cleavage site identified in the alpha chain of haemoglobin (6). This peptide was hydrolysed at the -Phe*Leu bond and cleavage was completely inhibited by inclusion of pepstatin (1

Table 1. Hydrolysis of Synthetic Chromogenic Peptides of the Type Lys-Pro-P_3-P_2-Phe*Nph-P_2'-Leu by PFAPD Expressed in *E. coli*

P3	P2	P2'	Relative Rate
Ile	Val	Arg	1.0
Ile	Ile	Arg	0.8
Ile	Nle	Arg	0.6
Ile	Leu	Arg	0.2
Ile	Ala	Arg	0.1
Ile	Glu	Arg	0.1
Ala	Leu	Arg	0.4
Ala	Ser	Arg	0.1
Ala	Arg	Arg	0.1
Ala	Lys	Asp	0.2
Ala	Lys	Leu	0.1

Measurements were made at pH 4.7 and the hydrolysis of each peptide was monitored by the absorbance change at 300 nm. Relative rates of hydrolysis are displayed

μM) in the incubations. Cleavage of the peptide at a range of pH values showed that the optimum pH for this hydrolysis was in the region of 4.7. Haemoglobin was also digested more efficiently at pH 4.7 than at pH 3.1 and this hydrolysis was also inhibited totally by 100 nM pepstatin (8). PFAPD also shows significant activity at pH 5.0, the estimated pH of the parasite food vacuole (21,22). The acid pH optimum is commensurate with the acidic environment required for activity by most aspartic proteinases, with the exception of renin.

Final purification of the activated proteinase was achieved by affinity chromatography on pepstatin agarose. Edman degradation of the resultant material yielded the N-terminal sequence, LNSGLTKTNYLGSS corresponding to residue 113 onwards of PFAPD. This would indicate a self processing site for PFAPD between Phe 112 and Leu 113 (Fig. 3), a typical aspartic proteinase cleavage site.

The ability of recombinant material to hydrolyse a series of synthetic substrates with systematic substitutions in the P_3, P2, and P2' positions was assessed spectrophotometrically (Table 1) by means of the nitrophenylalanine reporter group positioned in P_1' (See the first chapter by Dunn et al., this volume). At pH 4.7, it is evident from the relative rates of cleavage that bulky, hydrophobic residues were preferred in both P_3 and P_2 positions whereas an Arg residue is acceptable in P_2'. From this, it is apparent that the Plasmodium aspartic proteinase does not have a restricted specificity targeted towards the specific sequence of haemoglobin (such as that displayed by renin towards angiotensinogen, for example). Instead, this enzyme appears to manifest broader tolerance of a variety of residues in a substrate, as is found in pepsin, cathepsin D or cathepsin E although its preference for hydrophobic residues (Table 1) is reminiscent of that observed with cathepsin D (23).

The features described for PFAPD clearly identify this protein as a member of the aspartic proteinase family. These include sequence homology to, and a modelled structure closely resembling that of other aspartic proteinases and proteolytic activity inhibitable by pepstatin. PFAPD shows a high degree of identity (87%) with another aspartic proteinase from *Plasmodium falciparum*, PFAPG, identified by Goldberg and coworkers (11). The extensive homology between these two proteins strongly suggests very similar or overlapping roles for the proteinases in the parasite.

Goldberg et al. (6) demonstrated that the rate limiting first step in haemoglobin breakdown by intra-erythrocytic malaria parasites can be inhibited by pepstatin. This clearly indicates the critical role of an aspartic proteinase at this step in the degradation of virtually the sole nutrient for the parasite at this stage of its life cycle. The potential for the use of selective inhibitors of the parasite aspartic proteinases is clear and it is hoped that such an approach may prove effective in combating a disease manifesting 100 million clinical cases per year.

ACKNOWLEDGMENTS

Supported by awards from The Royal Society, World Health Organisation and a grant from the UF Malaria Genome Project from the Division of Sponsored Research at the University of Florida, Gainesville. The invaluable contributions made by G. Roman Reddy, Charles A. Yowell, Betsy Leveritt, Benne Parten and Rachel Granger are gratefully acknowledged.

REFERENCES

1. Sherman, I and Tanigoshi, L. *Mol. Biochem. Parasitol.* 8:207 (1983).
2. Banyal, H.S., Misra, G.C., Gupta, C.M., and Dutta, G.P. *J. Parasitol.* 67:623 (1981).
3. Goldberg, D.E., Slater, A.F.G., Cerami, A., and Henderson, G.B. *Proc. Natl. Acad. Sci. USA* 87:2931 (1990).
4. Bailly, E., Savel, J., Mahouy, G., and Jaureguiberry, G. *Exp Parasitol.* 72:278 (1991).
5. Vander Jagt, D.L., Hunsaker, L.A., Campos, N.M., and Scaletti, J.V. *Biochim. Biophys. Acta* 1122:256 (1992).
6. Goldberg, D.E., Slater, A.F.G., Beavis, R., Chait, B., Cerami, A., and Henderson, G.B. *J. Exp. Med.* 173:961 (1991).
7. Reddy, G.R., Chakrabarti, D., Schuster, S.M., Ferl, R.J., Almira, E.C., and Dame, J.B. *Proc. Natl. Acad. Sci. USA* 90:9867 (1993).
8. Dame, J.B., Reddy, G.R., Yowell, C.A., Dunn, B.M., Kay, J., and Berry, C. *Mol. Biochem. Parasitol.* In press, (1994).
9. Blum, J.S., Fiani, M.L., and Stahl, P.D. *Adv. Exp. Med. Biol.* 306:281 (1991)
10. Azuma, T., Pals, G., Mohandas, T.K., Couvreur, J.M., and Taggart, R.T. *J. Biol. Chem.* 264:16748 (1989).
11. Francis, E.F., Gluzman, I.Y., Oksman, A., Knickerbocker, A., Mueller, R., Bryant, M.L., Sherman, D.R., Russel, D.G., and Goldberg, D.E. *EMBO J.* 13:306 (1994).
12. Hobart, P.M., Fogliano, M., O'Connor, B.A., Schaefer, I.M., and Chirgwin, J.M. *Proc. Natl. Acad. Sci. USA* 81:5026 (1984).
13. Runeberg-Roos, P., Tormakangas, K., and Ostman, A. *Eur. J. Biochem.* 202:1021 (1991).
14. Ohlendorf, D.H., Foundling, S.I., Wendoloski, J.J., Sedláček, J., Strop, P., and Salemme, F.R. *Proteins* 14:382 (1992).
15. Pearl, L.H., and Blundell, T.L. *FEBS Lett.* 174:96 (1984).
16. Dunn, B.M., Valler, M.J., Rolph, C.E., Foundling, S.I., Jimenez, M., and Kay, J. *Biochim. Biophys. Acta* 913:122 (1987).
17. Takahashi, K. *J. Biol. Chem.* 262:1468 (1987).
18. Suguna, K., Bott, R.R., Padlan, E.A., Subramanian, E., Sheriff, S., Cohen, G.H., and Davies, D.R. *J. Mol. Biol.* 196:877 (1987).
19. [Reference missing].
20. Hill, J., Montgomery, D.S., and Kay, J. *FEBS Lett.* 326:101 (1993).
21. Yayon, A., Cabantchik, Z.I., and Ginsburg, H. *EMBO J.* 3:2695 (1984).
22. Krogstadt, D.J., Schlesinger, P.H., and Gluzman, I.Y. *J. Cell Biol.* 101:2302 (1985).
23. Scarborough, P.E., Guruprasad, K., Topham, C., Richo, G.R., Conner, G.E., Blundell, T.L., and Dunn, B.M. *Protein Sci.* 2:264 (1993).

YEAST AND MAMMALIAN BASIC RESIDUE-SPECIFIC ASPARTIC PROTEASES IN PROHORMONE CONVERSION*

Y. Peng Loh,[1,2] Niamh X. Cawley,[1,2] Theodore C. Friedman,[1] and Le-Ping Pu[1]

[1]Section on Cellular Neurobiology
Laboratory of Developmental Neurobiology
National Institute of Child Health and Human Development
National Institutes of Health
Bethesda, MD 20892
[2]Department of Biochemistry
Uniformed Services University of the Health Sciences
Bethesda, MD 20814

INTRODUCTION

Various proteins and peptides such as albumin, peptide hormones, neuropeptides, complement C3, transforming growth factor, and yeast α-mating factor, are synthesized as precursors which are processed to their biologically active forms by limited proteolysis, generally at paired and to a lesser extent at single basic residues [1,2]. The protein and peptide precursors are usually processed intracellularly and the products secreted from the cells. Limited proteolysis likely occurs in the trans-Golgi network (TGN) for proteins that are secreted constitutively [2], while most prohormones and pro-neuropeptides are processed within secretory granules, although processing can begin at the TGN, and their products are secreted in a regulated manner [3]. Besides limited proteolysis, these pro-proteins may undergo other post-translational modifications such as glycosylation, sulfation, and phosphorylation which, in some cases, can influence the efficacy of the proteolytic processing step. Efforts have been made in the last decade to elucidate the enzymes involved in the processing of the various pro-proteins. A number of enzymes have been found that specifically cleave paired and/or single basic residues of prohormones and pro-neuropeptides to generate the biologically active peptides [for review, see 1]. Enzymes different from those

* The opinions or assertions contained herein are the private ones of the authors and are not to be construed as official or reflecting the views of the Department of Defense or the Uniformed Services University of the Health Sciences.

Aspartic Proteinases: Structure, Function, Biology, and Biomedical Implications
Edited by Kenji Takahashi, Plenum Press, New York, 1995

that cleave prohormones appear to be involved in cleaving pro-proteins secreted *via* the constitutive pathway, such as pro-albumin and the growth factors [2].

Cleavage of pro-proteins occurs in between and/or, on the carboxy side of paired basic residues as evidenced by the presence of a N-terminally extended basic residue on some naturally occurring peptides, such as cholecystokinin 33, substance P and Arg^{o}-[met]-enkephalin. The resulting C-terminally or N-terminally extended basic residues on the peptide are removed by carboxypeptidase H [4] or an aminopeptidase B-like enzyme, respectively [5].

Two major classes of enzymes, serine and aspartic proteases are involved in pro-protein processing at paired and mono- basic residues. The serine protease class consists of the subtilisin-like enzymes which have been found in yeast and mammals. The yeast enzyme kexin, encoded by the KEX-2 gene is a member of this class and it processes pro-α-mating factor [6]. There is a family of mammalian enzymes which is homologous to kexin and these are known as pro-protein convertases (PCs). So far six members of this family have been identified by molecular cloning. They are PC1/3 [7-10], PC2 [11,12] PC4 [13, 14], PC5/PC6 [15,16], furin [17] and PACE 4 [18]. PC1/3 and PC2 appear to be distributed mainly in the endocrine/neuroendocrine systems while PC4 is exclusively found in the testis. The others appear to be more ubiquitous. Thus, PC1/3 and PC2 may be involved primarily in prohormone/proneuropeptide processing, while furin, PACE 4 and PC5/6 may function to process growth factors and other constitutively secreted proteins. PC1/3 and PC2 have been shown to process the ACTH/endorphin prohormone, pro-opiomelanocortin (POMC) by co-transfection experiments in mammalian cells [19,20] as well as *in vitro* using recombinant PC1/3 (Friedman, Loh and Birch, submitted for publication).

The aspartic protease class of prohormone processing enzymes includes the yeast aspartic protease 3 (YAP3) encoded by the YAP3 gene [21,22] and its mammalian homologue, pro-opiomelanocortin converting enzyme (PCE, EC.3.4.23.17) found in the pituitary [23,24]. In addition, a somatostatin-28 generating, single basic residue specific enzyme from anglerfish islets has been purified and characterized as a cathepsin D-like aspartic protease [25]. A cathepsin D-like enzyme has also been reported in chromaffin granules which processes pro-tackykinin [26]. These unique aspartic proteases cleave specifically at paired and/or single basic residue sites of prohormones, unlike all other aspartic proteases which generally cleave at hydrophobic residues [27]. In the following sections, we shall focus on discussing the properties of two members of this novel class of aspartic proteases, YAP3 and PCE.

YEAST ASPARTIC PROTEASE 3 (YAP3)

Physical Properties

Yeast aspartic protease 3 encoded by the non-essential YAP3 gene was expressed in KEX-2 deficient mutants when pro-α-mating factor was overexpressed in these mutants, and it processed this pheromone precursor to yield active α-mating factor [21]. The enzyme structure of YAP3 derived from the nucleotide sequence shows that it is a 569 amino acid protein with a 21 amino acid signal peptide, a pro-region, three potential transmembrane domains, a serine/threonine rich region and 10 potential N-linked glycosylation sites. The structure of YAP3 in comparison to other aspartic proteases is discussed in the chapter by Azaryan *et al.* (this volume).

Recently we have over-expressed the YAP3 gene in yeast using a plasmid which is under the galactose-inducible promoter. The induced recombinant enzyme was found to be in the cellular extract and growth medium. YAP3 from transformed yeast cells and the growth medium has been characterized. A ~70 kDa YAP3 glycoprotein was purified from the soluble extract of induced yeast cells [28]. In contrast, secreted YAP3 consisted of several hetero-

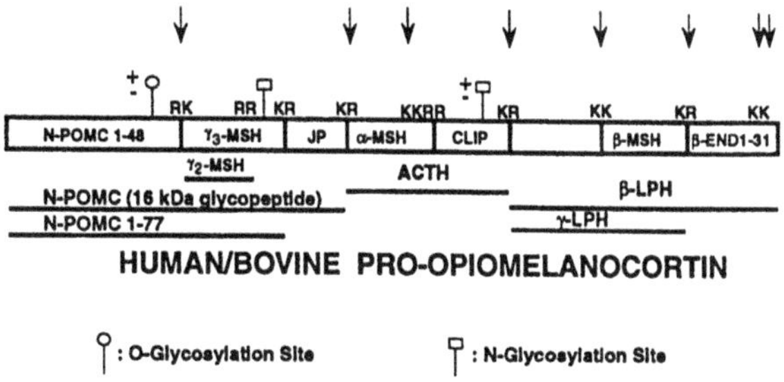

Figure 1. Diagramatic representation of pro-opiomelanocortin (POMC) showing the paired basic residue sites processed by YAP3 purified from yeast cell extracts. The ability of YAP3 to cleave the γ_3-MSH-JP junction has not been determined. MSH=melanocyte stimulating hormone; JP=joining peptide; CLIP=corticotropin-like intermediate lobe peptide; LPH=lipotropin; END=endorphin; ACTH=adrenocorticotropin.

genous forms ranging in molecular weight from 180-200 kDa in size. These various secreted forms are due to hyper-glycosylation, since upon treatment with Endoglycosidase H, they migrated at ~60 kDa on a SDS gel, approximating the size of the protein backbone of YAP3 (Cawley, *et al.*, in preparation).

Specificity of YAP3

The pH optimum, inhibitor profile and specificity of YAP3 purified from induced yeast cells, for various paired basic residues of the prohormone, pro-opiomelanocortin (POMC) is discussed in the chapter of Azaryan *et al* (this volume) and summarized in Figure 1 [28]. The secreted YAP3 activity has been shown to also have an acidic pH optimum, is inhibited by pepsatin A and has a specificity for paired basic residues of POMC, similar to the cellular YAP3. Like the cellular activity, secreted YAP3 also cleaved in between the two Lys residues at the tetrabasic site of $ACTH^{1-39}$ (see Figure 1) to yield $ACTH^{1-15}$, as well as the synthetic substrate Boc-Arg-Val-Arg-Arg-MCA to yield Arg-MCA. The K_m of YAP3 for this synthetic substrate was approximately ten-fold greater than for $ACTH^{1-39}$, suggesting that this enzyme prefers larger substrates (Cawley, *et al.*, in preparation).

The ability of secreted YAP3 to cleave anglerfish pro-somatostatin I and II (see Figure 2) was also investigated. YAP3 cleaved pro-somatostatin I at Arg^{81}-Lys^{82} to yield somatostatin-14 and the Lys extended form (Lys^{-1} SS-14), but did not cleave the single basic residue site, Arg^{68} to yield somatostatin-28. In contrast, YAP3 cleaved the single basic residue Arg^{73} of pro-somatostatin II to yield somatostatin-28, but not the paired basic residues, Arg^{86}-Lys^{87} [29]. This is consistent with the endogenous processing pattern observed in anglerfish pancreatic islet cells [30]. These findings show that YAP3 can cleave at single basic residues and that it is highly selective for a particular paired and single basic residue cleavage site in these two prohormones, respectively. This selectivity is most likely dictated by the conformation of the two anglerfish pro-somatostatin prohormones. Interestingly, pro-somatostatin II has an Arg in the -6 position upstream from the single Arg cleavage site which is substituted for a histidine in pro-somatostatin I. An Arg at -4 or -6 has been suggested as a requirement for single basic residue cleavage to occur [1,31,32].

Immunocytochemical Localization of YAP3

An antibody has been generated against YAP3 and shown to cross-react with the enzyme by Western blot. It was also used for immunocytochemistry to determine the cellular

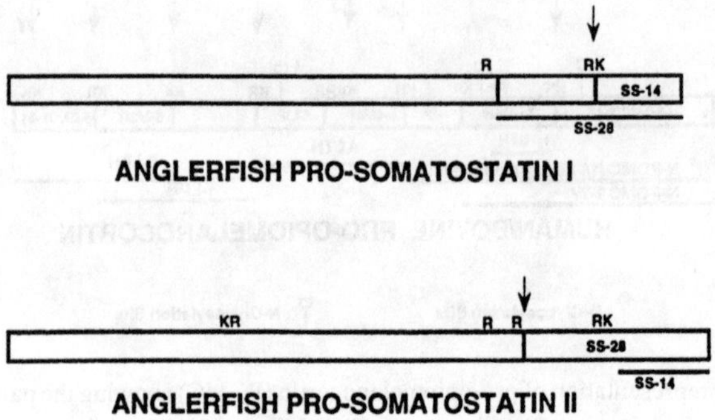

Figure 2. Diagramatic representation of anglerfish pro-somatostatin I and II showing the single and paired basic residue sites cleaved by YAP3 (see arrows). Note the other basic residue sites that are not cleaved by the enzyme *in vitro* or *in vivo* in pancreatic islet cells.

localization of YAP3 in transfected/induced yeast cells. Figure 3 shows that anti-YAP3 staining was most evident surrounding the periphery of the yeast cell, suggesting localization primarily within the periplasm. It is possible that processing of pro-α-mating factor in the KEX-2 deficient yeast mutants may occur in the periplasm which has an acidic pH appropriate for YAP3 function.

PRO-OPIOMELANOCORTIN CONVERTING ENZYME (PCE, EC 3.4.23.17)

Physical Properties

The physical properties of purified PCE from bovine intermediate and neural lobe has previously been described [23,24]. It is a ~70 kDa molecular weight glycoprotein. Western blot analysis of partially purified PCE using an antibody against YAP3 showed a

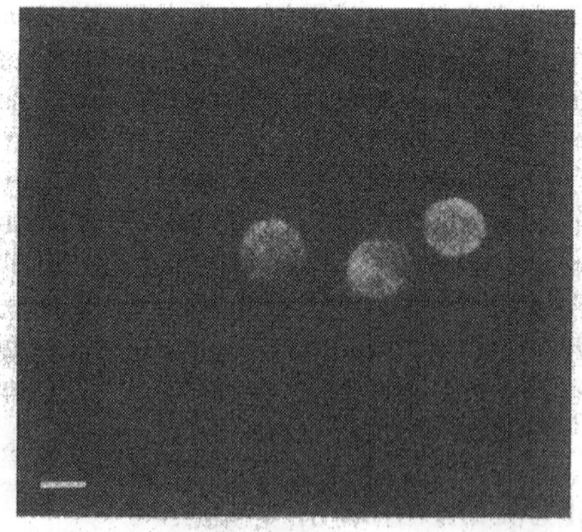

Figure 3. Immunofluorescence of YAP3 in transformed/induced yeast cells treated with an antiserum to YAP3. Note that YAP3 appears to be primarily located in the periplasm of yeast cells. No immunostaining was found in yeast treated with pre-immune serum. Bar = 1 μm.

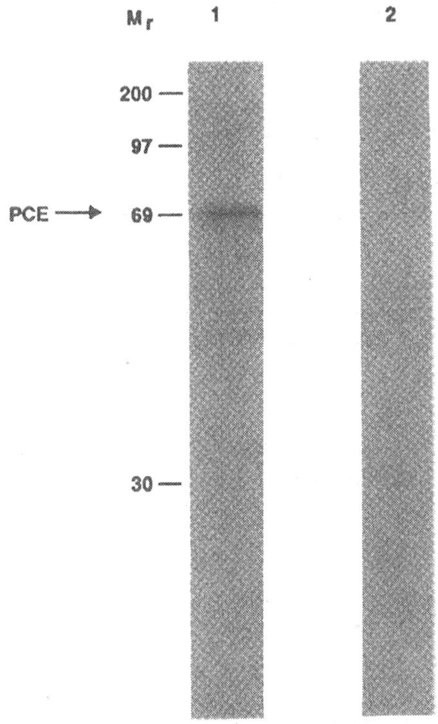

Figure 4. Lane 1. Western blot analysis showing the cross-reactivity of the YAP3 antibody with neural lobe PCE as indicated by the presence of the ~70 kD immunoreactive band (see arrow). Lane 2 shows the lack of immunostaining using pre-immune antiserum.

~70 kDa immunoreactive band (Fig. 4), indicating cross-reactivity of PCE with anti-YAP3. This suggests a structural relationship between these two enzymes.

Specificity of PCE

PCE has been shown to cleave several prohormones at paired basic residues (see Fig. 5). It has a pH optimum of 4.5-5.5 and is inhibited by pepstatin A. In comparison to YAP3 [28], PCE exhibited a more restricted specificity for the paired basic residues of POMC (compare Fig. 1 and 5). PCE did not cleave the Lys^{28}-Lys^{29} of β-endorphin or the tetrabasic residues, Lys^{15}-Lys^{16}-Arg^{17}-Arg^{18} of ACTH [23,33]. Cleavage at the Arg^{49}-Lys^{50} residues of POMC to yield Lys-γ-MSH and γ-MSH (Fig. 5) also depended on the glycosylation state of Thr^{45} of POMC which is an 'O'-linked glycosylation site. Processing at this site by PCE occurred only when Thr^{45} was not glycosylated, or only partially glycosylated, and the sugar chains lacked the terminating sialic acid residues [34]. We speculate that the sialic acid forms a salt bridge with the Arg^{49} of POMC providing steric hinderance and thereby preventing cleavage by PCE. Thus, this is another demonstration of the importance of the secondary structure of the substrate in influencing the specificity of the aspartic protease family of processing enzymes.

PCE cleaved pro-vasopressin at the Lys-Arg pair to yield vasopressin-Lys-Arg (Fig. 5) [24] and pro-insulin at the paired basic residues to generate the A and B chains (Fig. 5) [23]. However, other substrates, such as anglerfish pro-somatostatin I, was poorly cleaved

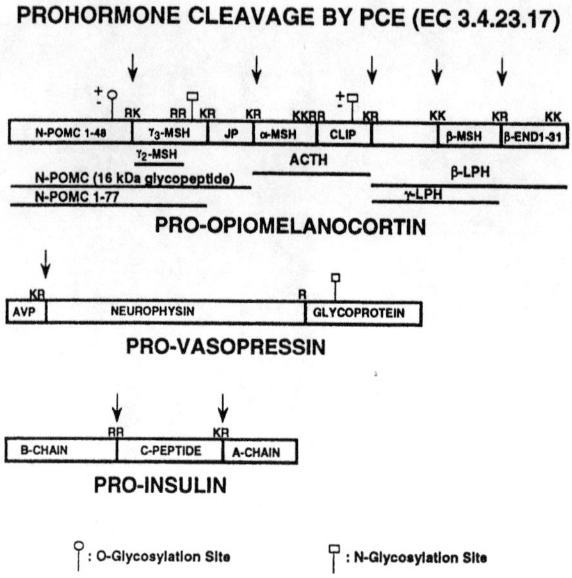

Figure 5. Diagramatic representation of pro-opiomelanocortin, pro-vasopressin and pro-insulin showing the paired basic residue sites cleaved by PCE (see arrows).

at the Arg^{81}-Lys^{82} by PCE to yield somatostatin-14, unlike YAP3 (Cawley *et al.*, unpublished data). Pro-oxytocin was also not cleaved by PCE (Loh, unpublished data). These findings suggest that the primary structure is insufficient and that conformational matching between substrate and enzyme is essential for cleavage at the appropriate pairs of basic residues to occur.

Subcellular Localization of PCE

PCE has been found in purified secretory granules of pituitary intermediate, neural and anterior lobes [23,24,35]. This is consistent with the site of processing of POMC and pro-vasopressin being in the secretory granules [3]. There is both a soluble and membrane associated-form of PCE in the secretory granules and the enzyme is co-ordinately secreted with the hormones from the pituitary intermediate lobe in a regulated manner [36,37]. Since POMC within the granules is initially membrane-associated [31], the first cleavage of the prohormone between ACTH and β-LPH may be carried out by the membrane-associated form. The intermediates formed are non-membrane-associated [38] and may subsequently be cleaved by the soluble form of PCE.

CONCLUSIONS

YAP3 and PCE appear to be novel aspartic proteases that are highly specific for paired- and mono-basic residue cleavage sites of many prohormones tested. Table I summarizes their properties. The similarity of the specificity, pH optimum and cross-reactivity with YAP3 antibodies strongly indicate that PCE and YAP3 are homologous enzymes. Future analysis of the molecular structure, modeling and X-ray crystal data on the three dimensional folding of PCE and YAP3 will shed light on the mechanism by which these two proteases

Table 1. Comparison of PCE and YAP3 Aspartic Proteases

PCE	YAP3
70 kDa glycoprotein	~70 kDa (cellular) and 180-200 kDa (secreted) forms of glycoprotein
pH optimum 4.5-5.5	pH optimum 4.0-5.0
Cleaves paired basic residues	Cleaves paired basic residues
	Cleaves the Arg^{73} monobasic residue of aPSSII
YAP3 antiserum cross-reacted with PCE	
Localized in secretory granules of pituitary	Localized in the secretory pathway and in the periplasm of induced yeast cells.
These data indicate homology in structure and enzymatic activity between PCE (E.C. 3.4.23.17) and YAP3.	

exhibit such unique specificity for paired/single basic residues of prohormones in comparison to other aspartic proteases.

REFERENCES

1. Y.P. Loh, M.C. Beinfeld, and N.P. Birch, Proteolytic Processing of Prohormones and Pro-Neuropeptides, in: Mechanisms of Intracellular Trafficking and Processing of Proproteins, Y.P. Loh, ed., pp. 179-224, CRC Press: Boca Raton, FL. (1993).
2. P.A. Bresnahan, J.S. Hayflick, S.S. Molloy, and G. Thomas, Endoproteolysis of growth factors and other nonendocrine precursor proteins, in: Mechanisms of Intracellular Trafficking and Processing of Proproteins, Y.P. Loh, ed., pp. 225-250, CRC Press: Boca Raton, FL. (1993).
3. H. Gainer, Intracellular protein trafficking and proprotein processing: An overview, in: Mechanisms of Intracellular Trafficking and Processing of Proproteins, Y.P. Loh, ed., pp. 1-17, CRC Press: Boca Raton, FL. (1993).
4. L.D. Fricker, Peptide processing exopeptidases: amino- and carboxypeptidases involved with peptide biosynthesis, in: Peptide Biosynthesis and Processing, L.D. Fricker, ed., pp. 199-229 CRC Press: Boca Raton, FL (1991).
5. H. Gainer, J.T. Russell, and Y.P. Loh, An aminopeptidase activity in bovine pituitary secretory vesicles that cleaves the N-terminal arginine from β-lipotropin$_{60-65}$. *FEBS Lett.* 175:135 (1984).
6. R.S. Fuller, R.E. Sterne and J. Thorner, Enzymes required for yeast prohormone processing. *Annu. Rev. Physiol.* 50:345 (1988).
7. N. Seidah, L. Gaspar, P. Mion, M. Marcinkiewicz, M. Mbikay, and M. Chretien, cDNA Sequence of Two Distinct Pituitary Proteins Homologous to KEX-2 and Furin Gene Products: Tissue-Specific mRNAs Encoding Candidates for Pro-Hormone Processing Proteinases. *DNA and Cell Biol.* 9:415 (1990).
8. S. Smeekens, A. Avruch, J. LaMendola, S. Chan, and D. Steiner, Identification of a cDNA Encoding a Second Putative Prohormone Convertase Related to PC2 in AtT20 Cells and Islets of Langerhans. *Proc. Natl. Acad. Sci. U.S.A.*, 88:340 (1991).
9. N. Seidah, M. Marcinkiewicz, S. Benjannet, L. Gaspar, G. Beaubien, M. Mattei, C. Lazure, M. Mbikay, and M. Chretien, Cloning and Primary Sequence of a Mouse Candidate Prohormone Convertase PC1 Homologous to PC2, Furin, and KEX-2: Distinct Chromosomal Localization and Messenger RNA Distribution in Brain and Pituitary Compared to PC2. *Mol. Endocrinol.*, 5:111 (1991).
10. D.J. Hakes, N.P. Birch, A. Mezey, and J.E. Dixon, Isolation of Two cDNA Clones from a Rat Insulinoma Cell Line Based on Similarities to KEX-2 and Furin Sequences and the Specific Localization of Each Transcript to Endocrine and Neuroendocrine Tissues in Rat, *Endocrinology* 129:3053 (1991).

11. K. Nakayama, M. Hosaka, K. Hatsuzawa, and K. Murakami, Cloning and Functional Expression of a Novel Endoprotease Involved in Prohormone Processing at Dibasic Sites. *J. Biochem.*, 109:803 (1991).
12. S. Smeekens, and D. Steiner, Identification of a Human Insulinoma cDNA Encoding a Novel Mammalian Protein Structurally Related to the Yeast Dibasic Processing Protease KEX-2. *J. Biol. Chem.*, 265:2997 (1990).
13. K. Nakayama, W. Kim, S. Torii, M. Hosaka, T. Nakagawa, J. Ikemizu, T. Baba, and K. Murakami, Identification of the Fourth Member of the Mammalian Endoprotease Family Homologous to the Yeast KEX-2 Protease. *J. Biol. Chem.*, 267:5897 (1992).
14. N.G. Seidah, R. Day, J. Hamelin, A. Gaspar, M.W. Collard and M. Chretien, Testicular Expression of PC4 in the Rat: Molecular Diversity of a Novel Germ Cell-Specific KEX-2/Subtilisin-Like Proprotein Convertase. *Mol. Endocrinol.* 6:1559 (1992).
15. J. Lusson, D. Vieau, J. Hamelin, R. Day, M. Chretien, and N.G. Seidah, cDNA Structure of the Mouse and Rat Subtilisin/kexin-like PC5: A Candidate Proprotein Convertase Expressed in Endocrine and Nonendocrine Cells. *Proc. Natl. Acad. Sci. U.S.A.*, 90:6691 (1993).
16. T. Nakagawa, M. Hosaka, S. Torii, T. Watanabe, K. Murakami and K. Nakayama, Identification and Functional Expression of a New Member of the Mammalian KEX-2-like Processing Endoprotease Family: Its Striking Structural Similarity to PACE4, *J. Biochem.* 113:132 (1993).
17. A. van den Ouweland, H. van Duijnhoven, G. Keizer, L. Dorssers, and W. Van de Ven, Structural Homology Between the Human Fur Gene Product and the Subtilisin-like Protease Encoded by Yeast KEX-2. *Nucleic Acids Res.*, 18:664 (1990).
18. M.C. Kiefer, M.C., J.E. Tucker, R. Joh, K.E. Landsberg, D. Saltman, and P.J. Barr, Identification of a Second Human Subtilisin-Like Protease Gene in the fes/fps Region of Chromosome 15. *DNA and Cell Biol.*, 10:757 (1991).
19. Thomas, L., R. Leduc, B. Thorne, S. Smeekens, D. Steiner, and G. Thomas, KEX-2-like Endoproteases PC2 and PC3 Accurately Cleave a Model Prohormone in Mammalian Cells: Evidence for a Common Core of Neuroendocrine Processing Enzymes. *Proc. Natl. Acad. Sci. U.S.A.*, 88:5297 (1991).
20. S. Benjannet, N. Rondeau, R. Day, M. Chretien, and N.G. Seidah, PC1 and PC2 are Proprotein Convertases Capable of Cleaving Proopiomelanocortin at Distinct Pairs of Basic Residues. *Proc. Natl. Acad. Sci. U.S.A.*, 88:3564 (1991).
21. M. Egel-Mitani, H.P. Flygenring, and M.T. Hansen, A Novel Aspartyl Protease Allowing KEX-2-Independent MFalpha Propheromone Processing in Yeast. *Yeast,* 6:127 (1990).
22. Y. Bourbonnais, J. Ash, M. Daigle, and D.Y. Thomas, Isolation and Characterization of S. Cerevisiae Mutants Defective in Somatostatin Expression: Cloning and Functional Role of a Yeast Gene Encoding an Aspartyl Protease in Precursor Processing of Monobasic Cleavage Sites. *EMBO J.* 12:285 (1993).
23. Y.P. Loh, D.C. Parish, and R. Tuteja, Purification and Characterization of a Paired Basic Residue-specific Pro-opiomelanocortin Converting Enzyme from Bovine Pituitary Intermediate Lobe Secretory Vesicles. *J. Biol. Chem.*, 260:7194 (1985).
24. D.C. Parish, R. Tuteja, M. Altstein, H. Gainer, and Y.P. Loh, Purification and Characterization of a Paired Basic Residue-specific Prohormone-converting Enzyme from Bovine Pituitary Neural Lobe Secretory Vesicles. *J. Biol. Chem.*, 261:14392 (1986).
25. R.B. Macklin, Noe, B.D. and Spiess, J., The Anglerfish Somatostatin-28-generating Pro-peptide Converting Enzyme is an Aspartyl Protease. *Endocrinology,* 129:1951 (1991).
26. T.J. Krieger and Y.H. Hook, Purification and Characterization of Cathepsin D Protease from Bovine Chromaffin Granules, *Biochemistry* 31:4223 (1992).
27. B.M. Dunn, M. Jimenez, B.F. Parten, M.J. Valler, C.E. Rolph, and J.K. Kay, A Systematic Series of Synthetic Chromophoric Substrates for Aspartic Proteinases, *Biochem. J.* 237:899 (1986).
28. A.V. Azaryan, M. Wong, T.C. Friedman, N.X. Cawley, F.E. Estivariz, H.-C. Chen, and Y.P. Loh, Purification and Characterization of a Yeast Paired Basic Residue-specific Aspartic Protease Encoded by the YAP3 Gene: Similarity to the Mammalian Pro-opiomelanocortin Converting Enzyme, *J. Biol. Chem.* 268:11968 (1993).
29. N.X. Cawley, B.D. Noe, and Y.P. Loh, Purified Yeast Aspartic Protease 3 Cleaves Anglerfish Pro-somatostatin I and II at Di- and Monobasic Sites to Generate Somatostatin-14 and -28, *FEBS Lett.* 332:273 (1993).
30. B.D. Noe, and J. Spiess, Evidence for Biosynthesis and Differential Posttranslational Proteolytic Processing of Different (Pre) Prosomatostatins in Pancreatic Islets. *J. Biol Chem.*, 258:1121 (1983).

31. K. Nakayama, T. Watanbe, T. Nakagawa, W.-S. Kim, M. Nagahama, M. Hosaka, K. Hatsuzawa, K. Kondah-Hashiba, and K. Murakami. Consensus Sequence for Precursor Processing at Mono-arginyl Sites. *J. Biol. Chem.* 267:16335 (1992).
32. L. Devi, Consensus Sequence for Processing of Peptide Precursors at Monobasic Sites, *FEBS Lett.* 280:189 (1991).
33. Y.P. Loh, Kinetic Studies on the Processing of β-lipotropin by Bovine Pituitary Intermediate Lobe Pro-opiomelanocortin Converting Enzyme, *J. Biol. Chem.* 261:11949 (1986).
34. N.P. Birch, F.E. Estivariz, H.P.J. Bennett, and Y.P. Loh, Differential Glycosylation of N-POMC$^{1-77}$ Regulates the Production of γ^3-MSH by Purified Pro-opiomelanocortin Converting Enzyme. *FEBS Lett.*, 290:191 (1991).
35. T.L. Chang and Y.P. Loh, Characterization of Pro-opiocortin Activity in Rat Anterior Pitutiary Granules. *Endocrinology,* 112:1832 (1983).
36. M.G. Castro, N.P. Birch, and Y.P. Loh, Regulated Secretion of Pro-opiomelanocortin Converting Enzyme and an Aminopeptidase B-like Enzyme from Dispersed Bovine Intermediate Lobe Pituitary Cells. *J. Neurochem.*, 52:1619 (1989).
37. T.-L. Chang and Y.P. Loh, *In Vitro* Processing of Pro-opiocortin by Membrane-associated and Soluble Converting Activities from Rat Intermediate Lobe Secretory Granules, *Endocrinology* 114:2092 (1984).
38. Y.P. Loh, and W.W.H. Tam, Association of Newly Synthesized Pro-opiomelanocortin with Secretory Granule Membranes in Pituitary Pars Intermedia Cells. *FEBS Lett.* 184:40 (1985).

PEPSTATIN-INSENSITIVE CARBOXYL PROTEINASES

K. Oda,[1] S. Takahashi,[1] T. Shin,[2] and S. Murao[2]

[1]Department of Applied Biology, Faculty of Textile Science
Kyoto Institute of Technology, Kyoto 606, Japan
[2]Department of Applied Microbial Technology
Kumamoto Institute of Technology, Kumamoto 860, Japan

INTRODUCTION

As reported previously [1, 2], we succeeded in isolating *Scytalidium lignicolum* in 1972 [3], which produced S-PI (Pepstatin Ac) [4]-insensitive carboxyl proteinases. This strain produced four distinct carboxyl proteinases: A-1, A-2, B, and C [5-7]. None of them were inactivated by S-PI, diazoacetyl-DL-norleucine methylester (DAN) [8], and 1,2-epoxy-3-(*p*-nitrophenoxy)propane (EPNP) [9]. However, as an exception, the carboxyl proteinase B was inactivated by EPNP. They had unique substrate specificities [10-14] in addition to a unique behavior against inhibitors. The complete amino acid sequence of carboxyl proteinase B [15] was quite different from those of other enzymes. Furthermore, it was found that unlike other carboxyl proteinases, one of the catalytic residues of the enzyme is glutamic acid [16, 17]. To our knowledge, it was the first demonstration of a glutamic proteinase. We have further demonstrated that enzymes having properties similar to those of *Scytalidium* were widely distributed among fungi [18-22], bacteria [23, 24] and thermophilic bacteria [25]. On the basis of the results obtained so far, we proposed that carboxyl proteinases should be classified into two groups: pepstatin-sensitive carboxyl proteinases (aspartic proteinase) and pepstatin-insensitive carboxyl proteinases (*Scytalidium* type) [1, 2].

In this paper, we introduce the background of our research on pepstatin-insensitive carboxyl proteinase which is classified into a new subclass. Later we describe pepstatin-insensitive carboxyl proteinases isolated from *Pseudomonas* sp. [23, 26] and thermophilic *Bacillus* novosp. [25, 27]. The former is the first carboxyl proteinase isolated and characterized from a prokaryote, and the latter is the first carboxyl proteinase isolated from thermophilic bacteria.

DISCOVERY OF PEPSTATIN-INSENSITIVE CARBOXYL PROTEINASES

Pepstatin-insensitive carboxyl proteinases described here were found by testing with carboxyl proteinase inhibitor, S-PI isolated by us in 1970 [28]. The structure of S-PI is

Aspartic Proteinases: Structure, Function, Biology, and Biomedical Implications
Edited by Kenji Takahashi, Plenum Press, New York, 1995

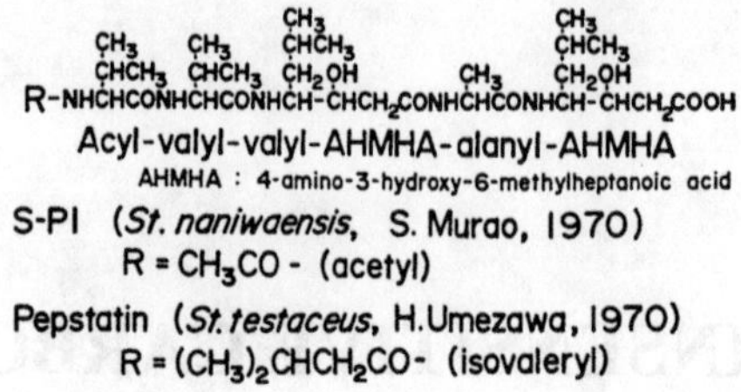

Figure 1. Chemical structures of S-PI (Pepstatin Ac) and Pepstatin.

acetyl-valyl-valyl-AHMHA-Ala-AHMHA as shown in Fig. 1 [4]. In the same year, Professor Umezawa [29] isolated a pepstatin, containing isovaleryl as the acyl residue. Since 1970, we have paid attention to the inhibitory spectrum of S-PI which inhibits all the carboxyl proteinases available. We considered that an S-PI-insensitive carboxyl proteinase, if there was any, should be distinct in the structure of the active site from any other known carboxyl proteinases.

From such a standpoint, we started screening for a particular microorganism producing an S-PI-insensitive carboxyl proteinase and succeeded in isolating *Scytalidium lignicolum* in 1972 [3, 5]. This strain produces four distinct carboxyl proteinases: A-l, A-2, B and C (Table 1) [5-7]. None of them are inactivated by S-PI, DAN or EPNP, with the exception of carboxyl proteinase B, which is exceptionally inactivated by EPNP.

In addtion to this unique behavior against inhibitors, they have unique substrate specificities [10-14]. When Z-tetrapeptides [12, 13], such as Z-X-Leu-Ala-Ala and Z-Phe-Y-Ala-Ala are used as substrates, carboxyl proteinases A-l and A-2 cleaved these peptides at either Leu-Ala or Y-Ala bond. The k_{cat}/K_m values are between 11,000 and 8,000. This is different from S-PI-sensitive carboxyl proteinases. Such S-PI-sensitive carboxyl proteinases as pepsin cleaved the X-Leu and Phe-Y bonds. The k_{cat}/K_m values are between 1,000 and 200. Carboxyl proteinase B is even different from A-l and A-2 because it cleaved the carboxyl terminal bonds. The activities are very low compared with those of A-l and A-2 enzymes. Thus the substrate specificities of carboxyl proteinases from *S. lignicolum* are distinguish-

Table 1. Some Properties of *Scytalidium lignicolum* Carboxyl Proteinases

	S. lignicolum A-1 (A-2)	*S. lignicolum* B	*S. lignicolum* C	Pepsin	*R. chinensis*
Optimum pH	3.0~3.5	2.2	2.0	1.8~2.0	2.9~3.3
Molecular weight	40,000	21,969	360,000	34,500	35,000
pI	3.6 (3.8)	3.2	2.3	1.0	5.0~5.2
Amino acid composition	His; none	His; none	His; yes	Ser; rich	His; none
Carbohydrate content	= 10%	0	= 30%	0	0
Inhibition with S-PI*	−	−	−	+	+
DAN	−	−	−	+	+
EPNP	−	+	−	+	+
Catalytic residues	-COOH**	-COOH** Glu 53 Asp 98	-COOH**	Asp 32 Asp 215	Asp 32 Asp 215

S-PI*; Pepstatin Ac, -COOH**; Zn(II)-PAD[31, 32] and Kinetics.[33]

Table 2. Classification of Carboxyl Proteinase

Type	Inhibitor Pepstatin	DAN	EPNP	Active center (-COOH)	Carboxyl proteinase
Pepstatin-sensitive (Pepsin)	+	+	+	Asp^{32} Asp^{215}	Pepsin, Gastricsin, Chymosin, Cathepsin D, Renin, Microbial carboxyl proteinases(Penicillium, Rhizopus, Aspergillus, Mucor, Rhodotorula, Cladosporium, Saccharomyces, etc.)
	-	-	+		S. lignicolum B--Glu^{53}, Asp^{98}
Pepstatin-insensitive (Scytalidium)	-	-	-	Glu or Asp	S. lignicolum A and C, Lentinus edodes, Ganoderma lucidum, Pleurotus ostreatus, Flammulina velutipes, *Pseudomonas sp. No. 101, Xanthomonas sp. No. T-22, *Bacillus novosp. MN-32(65°C), Asp. niger var. macrosporus A, Irpex lacteus, etc.

DAN :Diazoacetyl-DL-norleucine methylester, EPNP :1,2-epoxy-3-(p-nitrophenoxy)propane.

able from those of other microbial carboxyl proteinases which are sensitive to pepstatin, DAN and EPNP.

Later we determined the amino acid sequence of carboxyl proteinase B (Fig. 2) [15]. The enzyme is a single polypeptide composed of 204 amino acid residues with a molecular weight of 21,969. The amino acid sequence of carboxyl proteinase B is quite different from those pepstatin-sensitive carboxyl proteinases such as pepsin and penicillopepsin. Furthermore, unlike the other carboxyl proteinases, one of the catalytic residues of the enzyme is a glutamic acid. The active amino acid residue modified with EPNP was found to be Glu-53 [16, 17]. The other catalytic residue is found to be Asp-98 by using a new inhibitor, l-diazo-3-phenyl-2-propanone (DPP) [30].

The amino acid sequence around Glu-53 shows high homology with those around the active site Asp-215 residue of porcine pepsin and other carboxyl proteinases (Fig. 3). While the amino acid sequence around Asp-98 also shows high homology with those around the active site Asp-32, there is an insertion of serine residue between Thr and Gly. This was the first demonstration of a glutamic proteinase.

In our subseqent studies, we found that such enzymes having properties resembling *Scytalidium* enzymes were widely distributed among fungi [18, 19], basidiomycetes [20-22], bacteria [23, 24], and even in thermophilic bacteria [25] as shown in Table 2. On the basis of the results described here and others (studies on catalytic residues by means of Zn (II)-PAD [31, 32] and kinetics [33]), we proposed that carboxyl proteinase should be classified into two groups: pepstatin-sensitive carboxyl proteinases and pepstatin-insensitive carboxyl proteinases.

PEPSTATIN-INSENSITIVE CARBOXYL PROTEINASE FROM *PSEUDOMONAS* SP.

In 1986, we started screening microorganisms, especially, bacteria, for carboxyl proteinases. We were interested in looking for pepstatin-insensitive carboxyl proteinases, but at that time, there were not any reports on the occurrence of carboxyl proteinase from

```
1                                       10                                      20
Thr-Val-Glu-Ser-Asn-Trp-Gly-Gly-Ala-Ile-Leu-Ile-Gly-Ser-Asp-Phe-Asp-Thr-Val-Ser-
21                                      30                                      40
Ala-Thr-Ala-Asn-Val-Pro-Ser-Ala-Thr-Gly-Ala-Ser-Gly-Gly-Ser-Ser-Ala-Ala-Trp-Val-
41                                      50          *                           60
Gly-Ile-Asp-Gly-Asp-Thr-Cys-Gln-Thr-Ala-Ile-Leu-Glu-Thr-Gly-Phe-Asp-Trp-Tyr-Gly-
61                                      70                                      80
Asp-Gly-Thr-Tyr-Asp-Ala-Trp-Tyr-Glu-Trp-Tyr-Pro-Glu-Val-Ser-Asp-Asp-Phe-Ser-Gly-
81                                      90                              *      100
Ile-Thr-Ile-Ser-Glu-Gly-Asp-Ser-Ile-Gln-Met-Ser-Val-Thr-Ala-Thr-Ser-Asp-Thr-Ser-
101                                     110                                     120
Gly-Ser-Ala-Thr-Leu-Glu-Asn-Leu-Thr-Thr-Gly-Gln-Lys-Val-Ser-Lys-Ser-Phe-Ser-Asn-
121                                     130                                     140
Glu-Ser-Ser-Gly-Leu-Cys-Arg-Thr-Asn-Ala-Glu-Phe-Ile-Ile-Glu-Asp-Phe-Glu-Glu-Cys-
141                                     150                                     160
Asn-Ser-Asp-Gly-Ser-Asp-Glu-Phe-Val-Pro-Phe-Ala-Ser-Phe-Ser-Pro-Ala-Val-Glu-Phe-
161                                     170                                     180
Thr-Asp-Cys-Ser-Val-Thr-Ser-Asp-Gly-Glu-Ser-Val-Ser-Leu-Asp-Asp-Ala-Gln-Ile-Thr-
181                                     190                                     200
Gln-Val-Ile-Ile-Asn-Asn-Gln-Asp-Val-Thr-Asp-Cys-Ser-Val-Ser-Gly-Thr-Thr-Val-Ser-
201         204                         210                                     220
Cys-Ser-Tyr-Val
```

Figure 2. Amino acid sequence of *Scytalidium lignicolum* carboxyl proteinase B. Disulfide bonds connect Cys47-Cys126, Cys140-Cys163 and Cys192-Cys201. Catalytic residues are Glu53 and Asp98.

bacteria, regardless of pepstatin-sensitivity. In 1987, we succeeded in isolating two strains, *Pseudomonas* sp. No. 101 [23] and *Xanthomonas* sp. No. T-22 [24].

1) General Properties of Carboxyl Proteinase from *Pseudomonas* sp. No. 101

The production of pepstatin-insensitive carboxyl proteinases by this strain was carried out by using a 30-liter Jar Fermentor containing 20 liter medium (1% glucose, 0.5%

```
                                                    53
Carboxyl proteinase B    -Cys-Gln-Thr-Ala-Ile-Leu-Glu*-Thr-Gly-Phe-

                                                    215
Porcine pepsin           -Cys-Gln-----Ala-Ile-Val-Asp*-Thr-Gly-Thr-
Calf chymosin            -Cys-Gln-----Ala-Ile-Leu-Asp*-Thr-Gly-Thr-
Penicillopepsin          -----Ser-----Gly-Ile-Ala-Asp*-Thr-Gly-Thr-
Rhizopuspepsin           -Ser-Phe-Asp-Gly-Ile-Leu-Asp*-Thr-Gly-Thr-

                                 98
Carboxyl proteinase B    -Thr-Ser-Asp*-Thr-Ser-Gly-Ser-Ala-Thr-Leu-

                                 32
Porcine pepsin           -Ile-Phe-Asp*-Thr-----Gly-Ser-Ser-Asn-Leu-
Calf chymosin            -Leu-Phe-Asp*-Thr-----Gly-Ser-Ser-Asp-Leu-
Penicillopepsin          -Asn-Phe-Asp*-Thr-----Gly-Ser-Ser-Asp-Phe-
Rhizopuspepsin           -Asp-Phe-Asp*-Thr-----Gly-Ser-Ala-Asp-Leu-
```

Figure 3. Comparison of amino acid sequences around Glu53 and Asp98 of *Scytalidium lignicolum* carboxyl proteinase B with those of active site aspartic acid residues of some other Pepstatin-sensitive carboxyl proteinases.

Table 3. Some Properties of *Pseudomonas* sp. No. 101 Carboxyl Proteinase

Optimum pH (casein, 37 °C, 15 min)	3.0
pH stability (37 °C, 15 hr)	stable :pH 4.0 ~ 5.5
Inhibition with Pepstatin	—
S-PI(Pepstatin Ac)	—
DAN	—
EPNP	—
Tyrostatin*	+ (Ki=2.5nM)
Molecular weight (SDS-PAGE)	40,000
(amino acid seq.)	38,000
Isoelectric point	3.2
Amino acid sequence**	Ala(1)-His(372)

* Tyrostatin: Isovaleryl-tyrosyl-leucyl-tyrosinal.[34]
** No apparent similarity to carboxyl proteinases reported so far.
No active aspartic acid sequence, Asp-Thr-Gly. S-S bridge: 1.

polypeptone, and 0.5% meat extract, pH 5; agitation at 300 rpm; aeration at 20 liter/min). The carboxyl proteinases began to accumulate at approx. 15 h after the beginning of cultivation and reached a maximum at 25 h. The culture fluid at 25h was used for purification of the carboxyl proteinase.

The purified enzyme represents approx. 190-fold purification over the original culture filtrate, with a 53% recovery. From a 21 liter culture filtrate, about 105 mg of purified enzyme was obtained. We would emphasize that this enzyme was secreted in the culture fluid from gram-negative bacterium which has a double layer membrane around the cells.

The optimum pH for the action toward casein was around 3.0 and its molecular weight 40,000 (Table 3) [23]. This enzyme is an acidic protein having an isoelectric point at pH 3.2. From amino acid analysis, this enzyme contains one methionine residue. The absence of free thiol groups and the presence of 2 mol half-cystine indicates the presence of one disufide bridge/mole enzyme.

S-PI, DAN and EPNP were examined for their effects on this enzyme with porcine pepsin as a control [23]. Pepsin was completely inactivated by S-PI at a 1: 1 ratio. On the other hand, the *Pseudomonas* carboxyl proteinase was not inhibited by S-PI even at a concentration of 200 μg/ml. Also, DAN and EPNP did not inhibit this enzyme. Tyrostatin [34], which was isolated by us in 1989 to use as probe for elucidating the catalytic residues of the carboxyl proteinase from *Pseudomonas* sp. No. 101, strongly inhibited this enzyme. The inhibition constant of tyrostatin was 2.6 nM [26].

2) Substrate Specificity

We examined the enzyme for substrate specificities using a new generation substrate. They have a general structure P_5-P_4-P_3-P_2-P_1*(NO$_2$)Phe-Arg-Leu (P_5, P_4, P_3, P_2, P_1 : a variety of amino acids; (NO$_2$)Phe is *p*-nitro-L-phenylalanine). These peptides were designed and synthesized by Dr. B. M. Dunn. All these peptides are readily soluble in water.

The nature of the amino acid residue in the P_1 position was first investigated (Table 4) [26]. The substrate having Ala at P_1 position is not favorable. The substrate having Leu at P_1 position was hydrolysed rapidly. Phe in P_1 position yield a slightly better substrate than Leu: the value of k_{cat}/K_m increased by about 1.7-times. The substrate having Val at P_1 position

Table 4. Substrate Specificity of *Pseudomonas* sp. No. 101 Carboxyl Proteinase

Substrate P_5	P_4	P_3	P_2	P_1	P_1'	P_2'	P_3'	Km (mM)	kcat (sec^{-1})	kcat/Km (mM^{-1} sec^{-1})
K	P	A	K	F	Nph	R	L	0.095	14.5	153
K	P	A	K	A	Nph	R	L		n.d.*	
K	P	A	K	V	Nph	R	L		0	
K	P	A	K	L	Nph	R	L	0.060	5.5	92
K	P	A	D	F	Nph	R	L	0.013	6.8	523
K	P	I	E	F	Nph	R	L	0.003	6.9	2330
K	P	N	Q	F	Nph	R	L	0.052	5.5	106
K	P	V	S	Y	Nph	R	L	0.019	5.2	274
K	S	A	K	F	Nph	R	L	0.123	2.2	18
K	D	A	K	F	Nph	R	L	0.122	1.0	8
K	R	A	K	F	Nph	R	L	0.077	0.9	12
K	A	A	K	F	Nph	R	L	0.120	11.4	95
K	L	A	K	F	Nph	R	L	0.030	3.8	127
S	P	A	K	F	Nph	R	L	0.055	5.5	100
D	P	A	K	F	Nph	R	L	0.098	5.9	60
R	P	A	K	F	Nph	R	L	0.125	5.2	42
A	P	A	K	F	Nph	R	L	0.100	6.2	62
L	P	A	K	F	Nph	R	L	0.043	4.0	93
K	P	A	K	F	Nph	D	L	0.068	6.9	101

All reactions were carried out at pH 3.5 and 37°C. n.d * : not determined (very weak). Nph; p-nitro-L-phenylalanine.

was not cleaved at all, which is due to unfavorable steric interactions arising from the beta-branched Val side chain in the S_1 pocket.

The next series of experiments was designed to provide an insight into subsite interactions at other positions with a series of substrates. These results are also summarized in Table 4 [26].

Pseudomonas carboxyl proteinase has preference for aromatic amino acid residue at P_1 position, acidic amino acid residue at P_2 position (Asp or Glu), hydrophobic and bulky amino acid residue at P_3 position (Ile), and Pro or Leu residue at P_4 position. Replacement of Lys by other amino acid at P_5 position has not shown such difference in the kinetic parameters as observed in the replacement at the P_4 position. These data may suggest that the length of the active site in the direction of N-terminus of the substrates is nearly the same as those of the enzymes such as porcine pepsin which are occupied by four amino acid residues (P_4-P_1). Replacement of Arg by Asp at P_2' position has not yielded as substantial an effect as observed in the replacement of Lys by Asp at P_2 position.

The kinetic constants obtained from the best substrate tested, were compared with those of other carboxyl proteinases (Table 5) [26]. Compared with porcine pepsin, the k_{cat} and K_m values of the enzyme were very low, but the value of k_{cat}/K_m is about the same level as porcine pepsin.

In summary, it can be said that P_2, P_3 and P_4 subsites are the locations of a great deal of secondary specificity exhibited by *Pseudomonas* sp. No. 101 carboxyl proteinase, and that S_2 subsite of the enzyme may be occupied by basic amino acid residue(s), and that S_4 subsite may be occupied by acidic amino acid residue(s).

Table 5. Kinetic Constants for the Hydrolysis of K-P-I-E-F-Nph-R-L by Carboxyl Proteinases

Proteinase	Km (mM)	kcat (sec^{-1})	kcat/Km (mM^{-1} sec^{-1})
Pig pepsin*	0.05	106	2120
Human gastricsin*	0.56	8	14
Human cathepsin D*	0.64	124	195
Bovine chymosin*	0.36	190	530
Yeast proteinase A*	0.29	54	185
Mucor pusillus*	0.043	19	440
Rhizopus chinensis*	0.001	3	3000
Endothia parasitica*	0.003	16	5330
Pseudomonas sp. No. 101	0.003	6.9	2300

* : B. M. Dunn, et al., Biochem. J. (1986) 237, 899-906.

These initial investigations prompted us to carry out more detailed systematic studies of structure-function relationships of *Pseudomonas* sp. No. 101 carboxyl proteinase.

3) Amino Acid Sequence

We attempted to determine the amino acid sequence of the enzyme by a chemical method and compared the active site structure with that of pepstatin-sensitive ones. The complete amino acid sequence of this enzyme was clarified as shown in Fig. 4 [35]. The enzyme is a single polypeptide composed of 372 amino acid residues, containing one disulfide bridge.

Interestingly, this enzyme has no apparent sequence, similar to either pepstatin-sensitive carboxyl proteinases reported so far, or to the pepstatin-insensitive carboxyl proteinase

```
1                                     20                                    40
A A G T A K G H N P T E F P T I Y D A S S A P T A A N T T V G I I T I G G V S Q
41                                    60                                    80
T L Q D L Q Q F T S A N G L A S V N T Q T I Q T G S S N G D Y S D D Q Q G Q G E
81                                    100                                  120
W D L D S Q S I V G S A G G A V Q Q L L F Y M A D Q S A S G N T G L T Q A F N Q
121                                   140                                  160
A V S D N V A K V I N V S L G W C E A D A N A D G T L Q A E D R I F A T A A A Q
161                                   180                                  200
G Q T F S V S S G D E G V Y E C N N R G Y P D G S T Y S V S W P A S S P N V I A
201                                   220                                  240
V G G T T L Y T T S A G A Y S N E T V W N E G L D S N G K L W A T G G G Y S V Y
241                                   260                                  280
E S K P S W Q S V V S G T P G R R L L P D I S F D A A Q G T G A L I Y N Y G Q L
281                                   300                                  320
Q Q I G G T S L A S P I F V G L W A R L Q S A N S N S L G F P A A S F Y S A I S
321                                   340                                  360
S T P S L V H D V K S G N N G Y G G Y G Y N A G T G W D Y P T G W G S L D I A K
361               370
L S A Y I R S N G F G H
```

Figure 4. Amino acid sequence of *Pseudomonas* sp. No. 101 carboxyl proteinase. Disulfide bonds connect Cys137-Cys176.

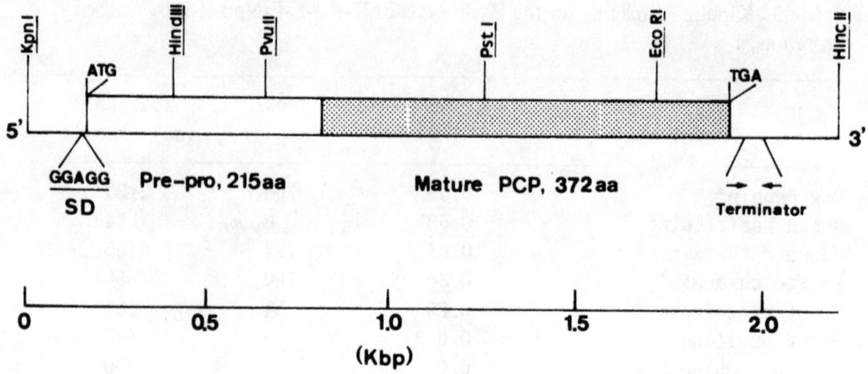

Figure 5. Organization and partial restriction map of *Pseudomonas* sp. No. 101 carboxyl proteinase gene. Open box and shaded box indicate the prepro-peptide and mature *Pseudomonas* sp. No. 101 carboxyl proteinase, respectively. Initiation methionine codon, ATG, and termination codon, TGA, are also indicated. SD, Shine-Dalgarno sequence.

B of *Scytalidium lignicolum* [15]. Also the characteristic active site aspartic acid sequence, -Asp-Thr-Gly- of pepstatin-sensitive carboxyl proteinases (aspartic proteinase) was absent.

4) Cloning of the Gene and Its Analysis

Recently, we have cloned and sequenced *Pseudomonas* No. 101 carboxyl proteinase gene. The nucleotide sequence of the gene involves an open reading frame of 1761

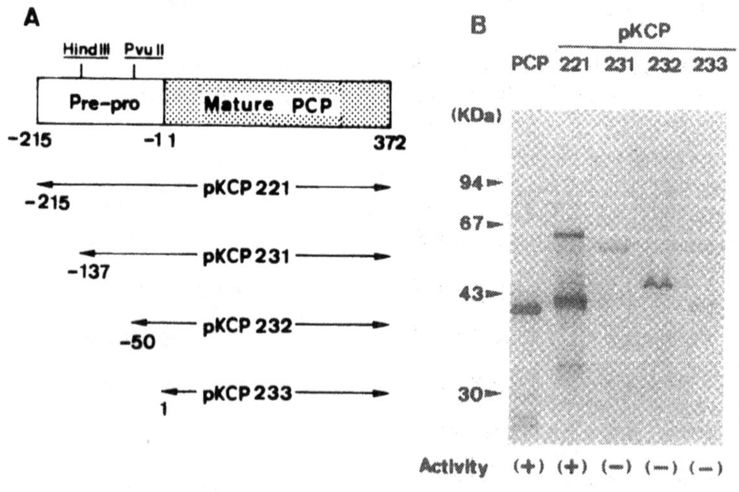

Figure 6. Mutant plasmids construction and Western blot analysis of the expressed proteins in transformed *E. coli*. (A) In the upper panel, the open box represents the prepro- sequence and the shaded box represents mature *Pseudomonas* sp. No. 101 carboxyl proteinase. The number at the left-hand side of each plasmid is the residue number from the amino terminus of mature *Pseudomonas* sp. No. 101 carboxyl proteinase. (B) Total cellular proteins from 0.2 ml culture were electrophoresed in a slab gel of 11.25% polyacrylamide and analyzed by Western blotting. PCP, 50 ng of authentic *Pseudomonas* sp. No. 101 carboxyl proteinase. Phosphorylase b (94 KDa), bovine serum albumin (67 KDa), ovalbumin (43 KDa), and carbonic anhydrase (30 KDa) were used as molecular weight standards.

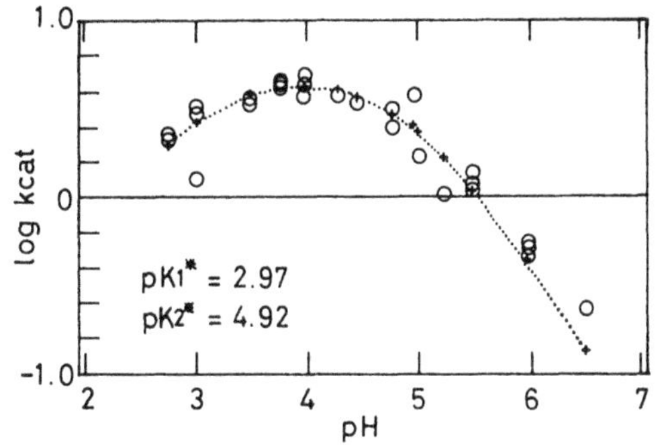

Figure 7. pH-dependence of log k_{cat} for *Pseudomonas* sp. No. 101 carboxyl proteinase catalysed hydrolysis of Ser-Pro-Ala-Lys-Phe* (NO_2)Phe-Arg-Leu at 37°C. All reactions were carried out at 37°C in 0.1 M sodium citrate buffer at each pH. The dotted line is the theoretical line computed from the equation [26].

nucleotides coding for a protein of 587 amino acids as shown in Fig. 5 [36]. The pre-pro part consisted of 215 amino acids and the mature enzyme consisted of 372 amino acid residues. A Shine-Dalgarno like sequence was observed 5 to 9 base pair upstream of the initiation methionine codon. An inverted repeat was also found downstream of the termination codon, TGA.

To investigate the function of pre-pro part of the enzyme precursor, several pre-pro part deletion mutants were constructed and expressed in *E. coli*. We constructed wild type and three pre-pro part deletion mutants as shown in Fig. 6. When the wild type pre-pro carboxyl proteinase was cloned into expression vector pKK223-3 (pKCP221), active mature carboxyl proteinase was synthesized and secreted into periplasmic space. When these deletion mutants were cloned into expression vector pKK233-2, products of mutants were detected by Western blot analysis, although no proteinase activity was observed.

These results indicate that the pre-pro part of *Pseudomonas* carboxyl proteinase precursor plays an important role in the formation of mature carboxyl proteinase.

5) Catalytic Residues

In the course of elucidating the structure and function relationships, kinetic studies were made to reveal the essential groups for catalytic action. The pH-dependence of hydrolysis was studied by using Ser-Pro-Ala-Lys-Phe* (NO_2)Phe-Arg-Leu as substrate.

As shown in Fig. 7 [26], the pH-profile of log k_{cat} is bell-shaped. The theoretical curve fits well with the experimental values. Thus, it was found that two residues with apparent dissociation constants at pH 2.97 or 4.92 are involved in the catalytic action. The results suggest to us the involvement of carboxyl or imidazole residues in the catalytic actions.

Additional experiments for elucidating the catalytic residues by means of a zinc (II)-PAD complex as a probe were performed (data not shown).

Coupled with these results, it is reasonable to conclude that *Pseudomonas* sp. No. 101 carboxyl proteinase has two active carboxyl residues participating in the action, as do pepstatin-sensitive carboxyl proteinases. It remains unclear whether the catalytic residues are aspartic or glutamic acid. To identify the catalytic residues of this enzyme, further investigations are necessary.

As reported previously, we isolated a *Kitasatosporia*, which produced a specific inhibitor, subsequently named as tyrostatin [34]. The chemical structure of tyrostatin is

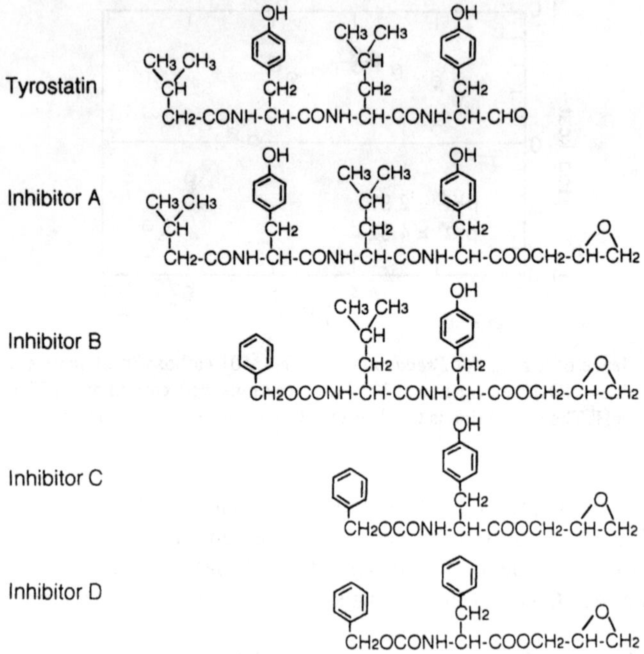

Figure 8. Chemical structures of Tyrostatin and its derivatives.

N-isovaleryl-tyrosyl-leucyl-tyrosinal. The carboxyl terminal tyrosinal residue of tyrostatin is essential for inhibition. Tyrostatin was elucidated to be a competitive inhibitor of the enzyme with a K_i value of 2.6 nM [26].

Chemical synthesis of tyrostatin derivatives capable of binding covalently with enzyme was done, to enable them to be used as probes for elucidating the catalytic residues of the enzyme (Fig. 8). Among them, inhibitor D [N-benzyloxycarbonyl-L-tyrosine-2,3-epoxypropyl ester (ZTEP)] showed inhibitory activity as shown in Fig. 9. The inhibition rate was reduced under the presence of hemoglobin or inhibitor B. This result indicates that ZTEP is a competitive type inhibitor against this enzyme. We are now in a position to identify the catalytic residues of *Pseudomonas* sp. No. 101 enzyme.

PEPSTATIN-INSENSITIVE CARBOXYL PROTEINASE FROM THERMOPHILIC *BACILLUS* NOVOSP.

Our results mentioned here promotes us to screen more extensively for new types of carboxyl proteinases. As the target, we chose thermostable enzymes. One reason is that, there is no report on the occurrence of thermostable carboxyl proteinases, regardless of pepstatin sensitivity. Second reason is that thermostable enzymes are applicable to industry. In 1988, we succeeded in isolating a thermophilic strain [25] from acidic hot springs in Japan, which produces pepstatin-insensitive carboxyl proteinase in the culture filtrate. The carboxyl proteinase was named Kumamolysin [27].

Optimum temperature and pH range for the growth of the strain are 60°C and pH 3.0 to 5.0, respectively. The morphological and physiological characteristics resemble those of *Bacillus acidocaldrius*. However, such characteristics as utilization of citric acid, the ability

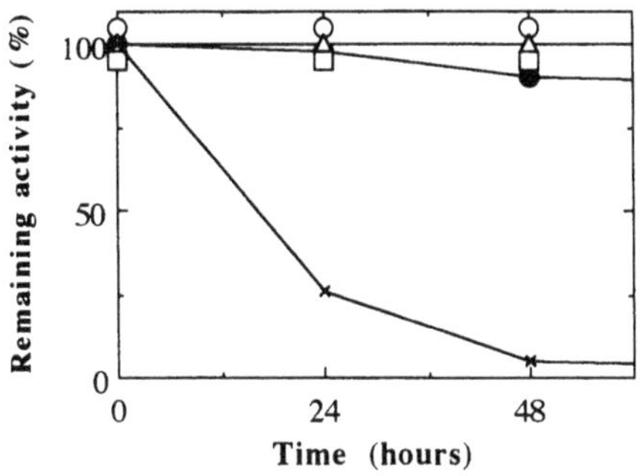

Figure 9. Effect of Tyrostatin derivatives on proteinase activity of *Pseudomonas* sp. No. 101 carboxyl proteinase. Inhibitors A, B, C or D (4 mg/80 μl in methanol) were mixed with 1820 μl 50 mM sodium acetate buffer, pH 4.4 and 100 μl of *Pseudomonas* sp. No. 101 carboxyl proteinase (1 mg/ml in 50 mM sodium acetete buffer, pH 5.0). The mixture was shaken at 70 strokes/min and 25°C. A control was run without inhibitor. At the time indicated, aliquotes were taken for testing the remaining activities. □-□: control, •-•: Inhibitor A, Δ-Δ: Inhibitor B, □-□: Inhibitor C, ×-×: Inhibitor D.

to reduce nitric acid, and so on were different. Therefore we tentatively named this strain as *Bacillus* novosp. [25, 27].

For the production of Kumamolysin, this strain was aerobically cultured at 60°C and pH 3.5, using 200 liter Jar Fermentor. The culture filtrate obtained after 46 h cultivation was used for this purification. After 6 purification steps, about 20 mg of purified enzyme was obtained with 13.8% recovery from 120 liter culture filtrate. The homogeneity of the enzyme was proven by electrophoresis and HPLC.

The optimum pH and temperature for the action of Kumamolysin was 3.2 and 70°C, respectively (Table 6) [25, 27]. The enzyme is fairly acid stable and thermostable, retaining 100% of initial activity between pH 2 to 5 under 24 h incubation at 50°C, and 60% activity under 10 min incubation at pH 4 and 80°C. The effects of various proteinase inhibitors on this enzyme were examined. Kumamolysin was not inhibited by serine-, metallo-, or thiol-proteinase inhibitors. Also the carboxyl proteinase inhibitors such as pepstatin, S-PI, DAN or EPNP did not show any inhibitory activities against this enzyme. The molecular weight of Kumamolysin was found to be 41,000 by SDS-PAGE. The amino acid composition

Table 6. Some Properties of Kumamolysin

Microorganism: Bacillus novosp. MN-32(65 °C)	
Optimum pH (casein, 70 °C, 30 min)	3.0
pH stability (50 °C, 24 hr)	stable:pH 2.0~5.0
Thermal stability(pH 4, 10 min)	up to 70 °C
Inhibition with Pepstatin	—
S-PI(Pepstatin Ac)	—
DAN	—
EPNP	—
Molecular weight (SDS-PAGE)	41,000
(Gel filtration)	40,000
Isoelectric point	3.2
Amino acid composition Gly + Ala + Pro	35%

S. Murao et al., J. Biol. Chem., 268, 349(1993).

Table 7. Substrate Specificity of Kumamolysin

	Substrate P_5 P_4 P_3 P_2 P_1 P_1' P_2' P_3'	Km (mM)	kcat (s^{-1})	kcat/Km (M^{-1} s^{-1})
Ins	---Leu-Val-Glu-Ala-Leu-Tyr-Leu-Val--	0.09	71	7.9×10^5
Ins(12-17)	Val-Glu-Ala-Leu-Tyr-Leu	0.203	20.8	1.03×10^5
Ins(13-17)	Glu-Ala-Leu-Tyr-Leu	0.321	0.159	4.95×10^2
Ins(14-17)	Ala-Leu-Tyr-Leu		no hydrolysis	
Ins(15-17)	Leu-Tyr-Leu		no hydrolysis	
Ins(12-16)	Val-Glu-Ala-Leu-Tyr		no hydrolysis	
Ins(12-R^{14}-17)	Val-Glu-Arg-Leu-Tyr-Leu	0.280	0.771	2.75×10^3
Ins(12-K^{14}-17)	Val-Glu-Lys-Leu-Tyr-Leu	0.514	0.442	8.61×10^2
Ins(12-H^{14}-17)	Val-Glu-His-Leu-Tyr-Leu	0.539	0.596	3.21×10^3

All reactions were carried out at pH 3.0 and 30°C. The sequence of the oxidized insulin B-chain and predominant cleavage site by Kumamolysin are

5 10 15↓
Phe-Val-Asn-Gln-His-Leu-$CySO_3H$-Gly-Ser-His-Leu-Val-Glu-Ala-Leu-Tyr-Leu-Val-$CySO_3H$-

20 25 30
Gly-Glu-Arg-Gly-Phe-Phe-Tyr-Thr-Pro-Lys-Ala

shows high hydrophobic amino acid content. That is, mole per cent of Gly, Ala and Pro residues are 35% of the total residues.

We studied the substrate specificity of Kumamolysin by using oxidized insulin B-chain as a substrate. As reported previously, the Leu15-Tyr16 bond of insulin B-chain was specifically cleaved. Upon overnight incubation, a minor cleavage product at Phe25-Tyr26 bond was observed. An apparent pH optimum is around pH 2.6. At this pH and 30°C, k_{cat}/K_m value was obtained to be 3.0×10^5 $s^{-1}M^{-1}$ [27].

To get more information about the substrate specificity of Kumamolysin, we synthesized some peptide substrates corresponding to scissile sequences in oxidized insulin B-chain and obtained kinetic parameters for these substrates. As shown in Table 7, Kumamolysin cleaved insulin B-chain at Leu-Tyr bond very rapidly. The K_m, k_{cat}, and k_{cat}/K_m values are 0.09, 71, and 7.9×10^5, respectively. Kumamolysin cleaved synthesized peptide, namely insulin (12-17) at reasonable rate and the value of k_{cat}/K_m is 13% of control. In the case of insulin (13-17) lacking P_4 amino acid residue, the kinetic parameters changed drastically. The value of k_{cat}/K_m is 0.06% of control. Synthetic insulin peptide lacking P_3 or P_2' amino acid residue was not cleaved.

From these results, it can be said that Kumamolysin has a narrow specificity favoring large and hydrophobic residues on both sides of the scissile bond, and that the length of the active site of Kumamolysin corresponds to 6 amino acid residues.

These initial investigations prompted us to carry out more detailed systematic studies of structure-function relationships of Kumamolysin. Cloning of this enzyme is now under way.

In this review, we highlighted on the pepstatin-insensitive carboxyl proteinases from *Pseudomonas* sp. No.101 and thermophilic bacteria. The *Pseudomonas* carboxyl proteinase is the first carboxyl proteinase from a prokaryote. Kumamolysin obtained from thermophilic bacteria is also from a prokaryote. Therefore, we believe that sudies on structure and function relationships of both enzymes will be useful to understand the evolution of carboxyl proteinases.

REFERENCES

1. K. Oda and S. Murao, Pepstatin-Insensitive Carboxyl Proteinases, in: "Structure and Function of The Aspartic Proteinases Genetics, Structures, and Mechanisms," B. M. Dunn ed., Plenum Press New York and London (1991) pp.185-201.
2. S. Murao and K. Oda, Pepstatin-Insensitive Acid Proteinases, in: "Aspartic Proteinases and Their Inhibitors," V. Kostka ed., Walter de Gruyter & Co., Berlin, New York (1985) pp.379-399.
3. S. Murao, K. Oda and Y. Matsushita, New Acid Proteases from *Scytalidium lignicolum* M-133, Agric. Biol. Chem., 36: 1647 (1972).
4. M. Fukumura, S. Satoi, N. Kuwana and S. Murao, Structure Elucidation of New Pepsin Inhibitor (S-PI), Agric. Biol. Chem., 35: 1310 (1971).
5. S. Murao, K. Oda and Y. Matsushita, Isolation and Identification of a Microorganism which Produces Non *Streptomyces* Pepsin Inhibitor and N-Diazoacetyl-DL-norleucine Methylester Sensitive Acid Proteases, Agric. Biol. Chem., 37: 1417 (1973).
6. K. Oda and S. Murao, Purification and Some Enzymatic Properties of Acid Protease A and B of *Scytalidium lignicolum* ATCC 24568, Agric. Biol. Chem., 38: 2435 (1974).
7. K. Oda, H. Torishima and S. Murao, Purification and Characterization of Acid Proteinase C of *Scytalidium lignicolum* ATCC 24568, Agric. Biol. Chem., 50: 651 (1986).
8. T. G. Rajagopalan, W. H. Stein and S. Moore, The Inactivation of Pepsin by Diazoacetyl-norleucine Methyl Ester, J. Biol. Chem., 241: 4295 (1966).
9. J. Tang, Specific and Irreversible Inactivation of Pepsin by Substrate-like Epoxide, J. Biol. Chem., 246: 4510 (1971).
10. K. Oda, S. Murao, T. Oka and K. Morihara, Some Physicochemical Properties and Substrate Specificities of Acid Proteases A-l and A-2 of *Scytalidium lignicolum* ATCC 24568, Agric. Biol. Chem., 40: 859 (1976).
11. K. Oda, S. Murao, T. Oka and K. Morihara, Some Physicochemical Properties and Substrate Specificity of Acid Protease B of *Scytalidium lignicolum* ATCC 24568, Agric. Biol. Chem., 39: 477 (1975).
12. K. Morihara, H. Tsuzuki, S. Murao and K. Oda, Pepstatin-Insensitive Acid Proteases from *Scytalidium lignicolum* - Kinetic Study with Synthetic Peptides, J. Biochem., 85: 661 (1979).
13. K. Morihara, Comparative Specificity of Microbial Acid Proteinases, in: "Proteinases and Their Inhibitors," V. Turk and Lj. Vitale, ed., Mladinska Knjiga-Pergamon Press, Ljubljana, Oxford (1981) pp.213-222.
14. K. Oda and S. Murao, Action of *Scytalidium lignicolum* Acid Proteases on Insulin B-chain, Agric. Biol. Chem., 40: 1221 (1976).
15. T. Maita, S. Nagata, G. Matsuda, S. Maruta, K. Oda, S. Murao and D. Tsuru, Complete Amino Acid Sequence of *Scytalidium lignicolum* Acid Protease B, J. Biochem., 95: 465 (1984).
16. D. Tsuru, S. Shimada, S. Maruta, T. Yoshimoto, K. Oda, S. Murao, T. Miyata and S. Iwanaga, Isolation and Amino Acid Sequence of a Peptide Containing an Epoxide-Reactive Residue from the Thermolysin-Digest of *Scytalidium lignicolum* Acid Protease B, J. Biochem., 99: 1537 (1986).
17. D. Tsuru, A. Naotsuka, R. Kobayashi, T. Yoshimoto, K. Oda and S. Murao, Inactivation of *Scytalidium lignicolum* Acid Protease B with 1,2-Epoxy-3-(4'-azido-2'-nitrophenoxy) propane, Agric. Biol. Chem., 53: 2751 (1989).
18. W.H. Chang, S. Horiuchi, K. Takahashi, M. Yamasaki and Y. Yamada, The Structure and Function of Acid Proteases, VI. Effect of Acid Protease-Specific Inhibitors on the Acid Proteases from *Aspergillus niger* var. *macrosporus*, J. Biochem., 80: 975 (1976).
19. H. Kobayashi, I. Kusakabe and K. Murakami, Purification and Characterization of Pepstatin-insensitive Carboxyl Proteinase from *Polyporus tulipiferae* (*Irpex lacteus*), Agric. Biol. Chem., 49: 2393 (1985).
20. K. Oda, T. Terashita, M. Kono and S. Murao, Occurrence of *Streptomyces* Pepsin Inhibitor-insensitive Carboxyl Proteinase in Basidiomycetes, Agric. Biol. Chem., 45: 2339 (1981).
21. T. Terashita, K. Oda, M. Kono and S. Murao, *Streptomyces* Pepsin Inhibitor-Insensitive Carboxyl Proteinase from *Lentinus edodes*, Agric. Biol. Chem., 45: 1937 (1981).
22. T. Terashita, K. Oda, M. Kono and S. Murao, *Streptomyces* Pepsin Inhibitor-Insensitive Carboxyl Proteinase from *Ganoderma lucidum*, Agric. Biol. Chem., 48: 1029 (1984).
23. K. Oda, M. Sugitani, K. Fukuhara and S. Murao, Purification and Properties of a Pepstatin-insensitive Carboxyl Proteinase from A Gram-negative Bacterium, Biochim. Biophys. Acta, 923: 463 (1987).

24. K. Oda, T. Nakazima, T. Terashita, K. Suzuki and S. Murao, Purification and Properties of an S-PI(Pepstatin Ac)-insensitive Carboxyl Proteinase from a *Xanthomonas* sp. Bacterium, Agric. Biol. Chem., 51: 3073 (1987).
25. S. Murao, K. Ohkuni, M. Nagao, K. Oda and T. Shin, A Novel Thermostable, S-PI (Pepstatin Ac)-insensitive Acid Proteinase from Thermophilic *Bacillus* novosp. Strain MN-32 Agric. Biol. Chem., 52: 1629 (1988).
26. K. Oda, H. Nakatani and B. M. Dunn, Substrate Specificity and Kinetic Properties of Pepstatin-insensitive Carboxyl Proteinase from *Pseudomonas* sp. No. 101, Biochim. Biophys. Acta, 1120: 208 (1992).
27. S. Murao, K. Ohkuni, M. Nagao, K. Hirayama, K. Fukuhara, K. Oda, H. Oyama, and T. Shin, Purification and Characterization of Kumamolysin, a Novel Thermostable Pepstatin-insensitive Carboxyl Proteinase from *Bacillus* novosp. MN-32, J. Biol. Chem., 268: 349 (1993).
28. S. Murao and S. Satoi, New Pepsin Inhibitors (S-PI) from *Streptomyces* EF-44-201, Agric. Biol. Chem., 34: 1265 (1970).
29. H. Umezawa, T. Aoyagi, H. Morishima, M. Matsuzaki, M. Hamada and T. Takeuchi, Pepstatin, A New Pepsin Inhibitor Produced by Actinomycetes, J. Antibiot., 23: 259 (1970).
30. D. Tsuru, R. Kobayashi, N. Nakagawa and T. Yoshimoto, Inhibition of *Scytalidium lignicolum* Acid Protease B by 1-Diazo-3-phenyl-2-propane, Agric. Biol. Chem., 53: 1305 (1989).
31. H. Nakatani, K. Hiromi, S. Satoi, K. Oda, S. Murao, and E. Ichishima, Studies on the Interaction between *Streptomyces* Pepsin Inhibitor and Several Acid Proteinases by Means of A Zinc(II)-Dye Complex as A Probe, Biochim. Biophys. Acta, 391: 415 (1975).
32. K. Oda and S. Murao, Additional Evidence for the Identity of *Scytalidium lignicolum* Acid Proteinases with the Carboxyl Proteinase Group: The Interaction between Angiotensin I and S-PI-Insensitive Acid Proteinases by Means of A Zinc(II)-Dye Complex as a Probe, Agric. Biol. Chem., 50: 1995 (1986).
33. K. Oda and S. Murao, Kinetic Studies on S-PI(Pepstatin Ac)-insensitive Acid Proteinases of *Scytalidium lignicolum* ATCC 24568. Evidence Identifying Them with a Carboxyl Proteinase Group, Agric. Biol. Chem., 50: 659 (1986).
34. K. Oda, Y. Fukuda, S. Murao, K. Uchida and M. Kainosho, A Novel Proteinase Inhibitor, Tyrostatin, Inhibiting Some Pepstatin-insensitive Carboxyl Proteinases, Agric. Biol. Chem., 53: 405 (1989).
35. K. Oda, et al., in preparation.
36. K. Oda, et al., in preparation.

THE THREE-DIMENSIONAL X-RAY CRYSTAL STRUCTURE OF THE ASPARTIC PROTEINASE NATIVE TO *TRICHODERMA REESEI* COMPLEXED WITH A RENIN INHIBITOR CP-80794

Jim E. Pitts,[1*] Mark D. Crawford,[1] Phil G. Nugent,[1] Ron T. Wester,[2] Jon B. Cooper,[1] Arja Mantyla,[3] Richard Fagerstrom,[3] and Helena Nevalainen[3]

[1]Laboratory of Molecular Biology
Department of Crystallography
Birkbeck College
University of London
Malet Street, London WC1E 7HX
United Kingdom
[2]Department of Medicinal Chemistry
Central Research Division
Pfizer Inc.
Groton, Connecticut 06340
[3]Research Laboratories
ALKO Ltd.
Helsinki
Finland

INTRODUCTION

Aspartic proteinases comprise a family of proteolytic enzymes found in a diverse range of species, from vertebrates, to lower eukaryotes and retroviruses. They are active at neutral or low pH. They are all characterised by the presence of two aspartic acid residues at the active site and are inhibited by pepstatin A. The aspartic proteinases can be broadly divided into two main groups: the retroviral and pepsin-like enzymes. We shall, however only consider the pepsin-like aspartic proteinases.

The pepsin-like enzymes are monomers and have molecular weights of approximately 35 kD. These enzymes have between 320 and 360 residues and have a variety of functions. The fungal pepsins for example are involved in protein digestion, while cathepsins D and E are involved in protein degradation. The aspartic proteinase renin has pharmacologi-

Aspartic Proteinases: Structure, Function, Biology, and Biomedical Implications
Edited by Kenji Takahashi, Plenum Press, New York, 1995

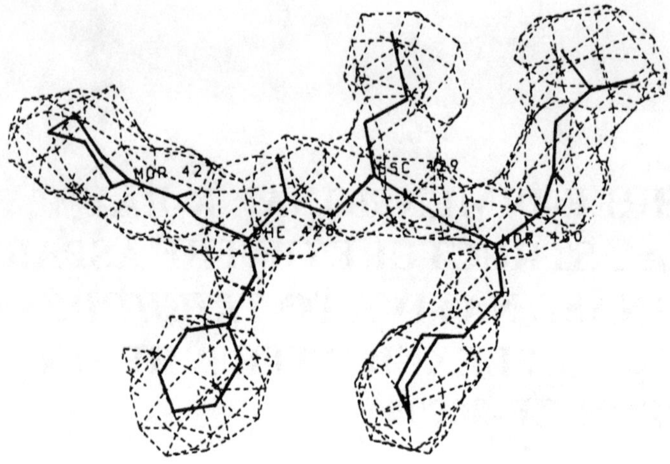

Figure 1. The final electron density map calculated with the $2|F_o|-|F_c|$ coefficients for the CP-80794 inhibitor in the active site of trichodermapepsin.

cal interest and is involved in homeostatic regulation. The structures of the pepsin-like enzymes which have been determined include rhizopuspepsin[1], porcine pepsin[2], mucor pepsin[3], endothiapepsin[4], calf chymosin[5] and human renin[6]. The similarity of the overall fold of human renin and the other aspartic proteinases[7,8] means that co-crystallization of renin inhibitors with fungal aspartic proteinases may offer insight into the design of improved drugs for renin inhibition. For this reason we have complexed a renin inhibitor with the aspartic proteinase native to *Trichoderma reesei.*

METHODS

The trichodermapepsin was crystallized with a norstatine containing renin inhibitor CP-80794. CP-80794 is a tetrapeptide extending from P_4 to P_1' with the sequence MOR-PHE-CSC-NOR where MOR is morpholino, CSC a methylated cysteine and NOR is norstatine. To crystallize the complex trichodermapepsin was dissolved in 0.15M sodium acetate and 0.1M sodium chloride (pH 5.5) and a ten-fold molar excess of the inhibitor was added. This was dissolved overnight and the resulting liquor Millipore filtered to remove any precipitant. Crystallization was achieved by centrifugation[9] of the filtrate through a 10,000 MW cutoff membrane using a Centricon-10 device until crystallization occurred. The procedure is believed to create a protein concentration gradient resulting in supersaturation and crystallization at the membrane/solute interface.

X-ray data (2.4Å resolution) was collected on the synchrotron at Daresbury at station 9.5 using a MAR image plate using 0.89Å radiation. The oscillation range per frame was 1.5° and 87° of data was collected. The data was processed using MOSFLM and the space group determined to be tetragonal with $a=b=76.1$Å and $c=162.2$Å. The program LATTICE[10] indicated that the space group was either $P4_32_12$ or $P4_12_12$, while consideration of solvent content indicated that most probably there was only one molecule per asymmetric unit. The batches were subsequently scaled and merged using ROTAVATA and AGROVATA[11,12]. The cumulative R-factor of the 11 896 reflections was 8.9%. The R-factor had risen to 18.5% in the 2.6Å shell so all data at higher resolution was rejected. The intensities were modified

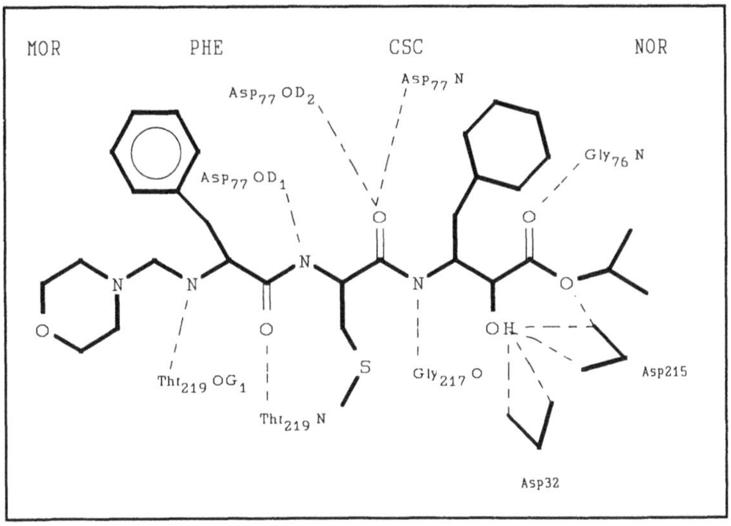

Figure 2. A schematic showing the hydrogen-bonding between trichodermapepsin and CP-80794. Hydrogen bonds of the inhibitor to the enzyme are shown schematically.

using Bayesian statistics by means of the program TRUNCATE[13] that treated negative and weak reflections.

In the absence of the primary structure for trichodermapepsin, a variety of search models were used to solve the structure by molecular replacement. The molecular replacement suite of programs AMORE[14] was used. On the basis of the peak heights from the rotation function endothiapepsin was chosen as the best search model with a signal to noise ratio of 9.9 as opposed to approximately 6.5 for both penicillopepsin and rhizopuspepsin. To resolve the space group ambiguity the translation function was undertaken in both $P4_32_12$ and $P4_12_12$. The analysis indicated that the preferred space group was $P4_32_12$ with an R-factor of 45%, correlation coefficient 60% and for $P4_12_12$ an R-factor of 55%, correlation coefficient of 38%. The correctly positioned model underwent least squares refinement using the program RESTRAIN[15] for 135 cycles using the 2.6Å resolution image plate data. After each run of least squares refinement electron density maps with coefficients $2m|F_o|-D|F_c|$ were used to check the model using the program FRODO. The structure of the aspartic proteinase native to *Trichoderma reesei* complexed with CP-80794 is at present refined to an R-factor of 18.5% at 2.6Å resolution. Figure 1 shows the final $2|F_o|-|F_c|$ map for the inhibitor. The error in the atomic coordinates is approximately 0.15 to 0.20Å as estimated using the method of Luzzati[16]. The r.m.s. deviations from ideality for the stereochemistry are within acceptable limits. In all 2,366 protein atoms and 100 inhibitor atoms have been included in the refinement, as well as 20 water molecules.

RESULTS

The electron density extends along the active site from subsites S_4 to S_1' and defines all the side-chains of CP-80794 (Figure 1). The inhibitor binds in an extended conformation as is typical of other aspartic proteinase inhibitor complexes and almost identically to the endothiapepsin CP-80794 complex[17]. The norstatine mimics a dipeptide analogue with the isopropyloxy group occupying the S_1' instead of residue P_1'. It can be seen from Figure 2

Table 1. The Specificity Pockets for the Human Renin, Trichodermapepsin and Endothiapepsin Inhibitor Complexes Defined by Those Residues in Contact with the Inhibitor

	HUMAN RENIN COMPLEX	TRICHODERMAPEPSIN COMPLEX	ENDOTHIAPEPSIN COMPLEX
S4	T77, L144	T219, F289	D12, T219, L220
S3	Q13, T77, P111, L114, A115, F117, A218, S219	A12, Q13, D77, D114, Q116, S218, T219	I7, D12, A13, D77, D114, G217, T218, T219
S2	S76, T77, G217, S222, M289	Y75, G76, D77, G217, S218, G289	Y75, G76, D77, G217, T218, I297
S1	D32, Y75, F112, F117, V120, D215, G217	N30, D32, Y75, Q116, L120, I123, G217, F111, D215	D30, D32, Y75, D77, S79, L120, D215, G217, T218
S1'	G34, Y75, S76, L213, D215, I291	G34, Y75, G76, F195, I213, D215	G34, Y75, G76, F189, D215

The cut-off distance which defined contacts was 4Å. Single letter codes identify the amino acid and pepsin numbering is used.

that norstatine makes many hydrogen bonds with the enzyme, four of these involve the five oxygen atoms of the catalytic aspartic acid side-chains and the hydroxyl group of the inhibitor. The hydroxyl oxygen occupies essentially the same position as it does in the mouse renin, human renin and endothiapepsin complexes[17]. In common with the human renin complex, there is a hydrogen bond between the ester oxygen of the norstatine and the outer carboxyl oxygen of Asp_{215}. In addition the carboxyl oxygen of norstatine bonds to the amide nitrogen of Gly_{76}. Finally, carboxyl oxygen of Gly_{217} binds to the amide nitrogen of norstatine. All of these bonds are conserved in the endothiapepsin complex[17], although in human renin Ser_{76} [2] fails to make a hydrogen bond with the carboxyl oxygen.

The methylated cysteine at position P_2 has three hydrogen bonds to the enzyme all of which are to the loop which covers the active site. The carboxyl oxygen of the methylated cysteine has a weak bifurcated hydrogen bond going to the Oδ2 and amide nitrogen of Asp_{77}. There is also hydrogen bond to the ODδ1 of Asp_{77}. In the endothiapepsin inhibitor complex there was no evidence of a bond from Oδ2 of Asp_{77} to the carboxyl oxygen. Instead the amide nitrogen of Gly_{76} binds to the carboxyl oxygen of CSC. Human renin hydrogen bonds in a similar way to endothiapepsin at the P_2 position. There is also some evidence that CP-80794 complexed with endothiapepsin is bifurcated[17] with final occupancies approximately 40:60.

The phenylalanine at position P_3 has two hydrogen bonds; both of these come from the main-chain atoms of P_3 to Thr_{219}. The first hydrogen bond is from the amide nitrogen of Thr_{219} to the carboxyl oxygen of P_3. The second bond is from the amide nitrogen of P_3 to the Oγ1 of the Thr_{219}. All of these hydrogen bonds are conserved in the endothiapepsin inhibitor complex. However, in the human renin complex an extra hydrogen bond forms from the Oγ of Thr_{77}.

In summary the subsites of trichodermapepsin bear closest resemblance to endothiapepsin as indicated in Table 1. This analysis provides further details of substrate/enzyme interactions in the aspartic proteinase family and will be of assistance in the design of engineered enzymes in the future.

REFERENCES

1. K. Suguna, R.R. Bott, E.A. Padlan, E. Subramanian, S. Sheriff, G.H. Cohen and D.R. Davies, Structure and refinement at 1.8Å of the aspartic proteinase from *Rhizopus chinensis*, *J. Mol. Biol.* **196**:877 (1987).
2. J.B. Cooper, G. Khan, G. Taylor, I.J. Tickle and T.L. Blundell, X-ray analyses of aspartic proteinase. II. Three-dimensional structure of the hexagonal crystal form of porcine pepsin at 2.3Å resolution, *J. Mol. Biol.* **214**:199 (1990).
3. M. Newman, F. Watson, P. Roychowdhury, H. Jones, M. Badasso, A. Cleasby, S.P. Wood, I.J. Tickle and T.L. Blundell, X-ray analyses of aspartic proteinases. V. Structure and refinement at 2.0Å of the Aspartic Proteinase from *Mucor pusillus*, *J. Mol. Biol.* **230**:260 (1993).
4. T.L. Blundell, J.A. Jenkins, B.T. Sewell, L.H. Pearl, J.B. Cooper, I.J. Tickle, B. Veerapandian and S.P. Wood, X-ray analyses of aspartic proteinase. The three-dimensional structure at 2.1Å resolution of endothiapepsin, *J. Mol. Biol.* **211**:919 (1990).
5. M. Newman, M. Safro, C. Frazeo, G. Khan, A. Zdanov, I.J. Tickle, T.L. Blundell and N. Andreeva, X-ray analyses of aspartic proteinases. VI. Structure and refinement at 2.2Å of the Aspartic Proteinase of bovine chymosin, *J. Mol. Biol.* **221**:1295 (1991).
6. V. Dhanaraj, C.G. DeAlwis, C. Frazeo, M. Badasso, B.L. Sibanda, I.J. Tickle J.B. Cooper, H.P.C. Driessen, M. Newman, C. Aguilar, S.P. Wood, T.L. Blundell, P.M. Hobart, K.F. Geoghegan, M.J. Ammirati, D.E. Danley, B.A. O'Connor and D.J. Hoover, X-ray analyses of peptide-inhibitor complexes define the structural basis of specificity for human and mouse renins, *Nature* **357**:466 (1992).
7. T.L. Blundell, J.B. Cooper, A. Sali and Zhan-yang Zhu, Comparisons of the sequences, 3-D Structures and mechanisms of pepsin-like and retroviral aspartic proteinases, in: *Structure and function of the aspartic proteinases*, (ed. B.M. Dunn), Plenum Press 433 (1991).
8. A. Sali, B. Veerapandian, J.B. Cooper, S.I. Foundling, D.J. Hoover and T.L. Blundell, High resolution X-ray diffraction study of the complex between Endothiapepsin and an oligopeptide inhibitor: the analysis of the inhibitor binding and description of the rigid body shift in the enzyme, *EMBO J.* **8**:2179 (1989).
9. J.E. Pitts, Crystallization by centrifugation, *Nature* **355**:117 (1992).
10. G. Taylor unpublished program.
11. J.M. Smith and A.J. Wonacott, *CCP4 User Documentation*, Daresbury Laboratory, Warrington, UK (1985).
12. G.C. Fox and K.C. Holmes, An alternative method of solving the layer scaling equations of Hamilton, Rollet and Sparks, *Acta Crystallogr.* **20**:886 (1966).
13. S. French and K. Wilson, On the treatment of negative intensity observations, *Acta Crystallogr. A* **34**:517 (1978).
14. J. Navaza, On the fast rotation function, *Acta Crystallogr. A* **43**:645.
15. I. Haneef, D.S. Moss, M.J. Stanford and N. Borkakoti, Restrained structure-factor least-squares refinement of protein structures using a vector processing computer, *Acta Crystallogr. A* **41**:426 (1985).
16. V. Luzzati, Treatment of statistical errors in the determination of crystal structures, *Acta Crystallogr.* **5**:802 (1952).
17. M.D. Crawford, R.T. Wester and J.B. Cooper, unpublished results.

REFERENCES

A QUANTUM MECHANICAL MODEL OF THE HYDRATION AND ACIDITY OF THE ACTIVE SITE IN ASPARTIC PROTEASES

I. A. Topol, R. E. Cachau, S. K. Burt, and J. W. Erickson

Structural Biochemistry Program
Frederick Biomedical Supercomputing Center
PRI/DynCorp
National Cancer Institute - Frederick Cancer Research and Development Center
Frederick, MD 21702-1201

INTRODUCTION

The exact catalytic mechanism of aspartic proteases (APs) is still not clearly understood. Most of the postulated mechanisms are based on a general acid-general base mechanism. Proposed catalytic mechanisms [1-4] for peptide bond hydrolysis in the APs are based on the assumption that the substrate initially binds to an active site in which one of the two active site aspartates is neutral, the other is ionized, and a nearby water molecule acts as a nucleophile.Thus, detailed information about the protonation of the active site and about the hydrogen bonds responsible for binding the water molecule(s) to the active site is required to accurately describe the catalytic mechanism of APs. Indirect evidence on the relative ionization potentials of the two aspartic acids can be obtained by investigating the pK_a changes in the APs induced by mutations of neighboring residues. Site-directed mutagenesis studies in which Ser35 and Thr218 were mutated to Ala in pepsin suggested that the higher pK_a is normally associated with Asp32, such that Asp32 may be protonated when Asp215 is ionized. Several attempts have been made to use quantum mechanical calculations [6-12] to determine the position of the active site protons. However, the protonation site preferences obtained by these various quantum mechanical calculations have been shown to depend on the size of the theoretical active site model used and differ with respect to their prediction of the proton location on the active site aspartates, Asp32 or Asp215 (pepsin residue numbering).

The aims of the present work were to use quantum mechanical calculations to: a) reliably determine the protonation state of the active site of the APs by utilizing a large molecular model that includes several crystallographically determined water molecules ; b) investigate the influence of Ser35Ala and Thr218Ala mutations on the pK_a's of the active site aspartates; and c) investigate the influence of the inclusion of a continuum solvation model on protonation preferences.

a

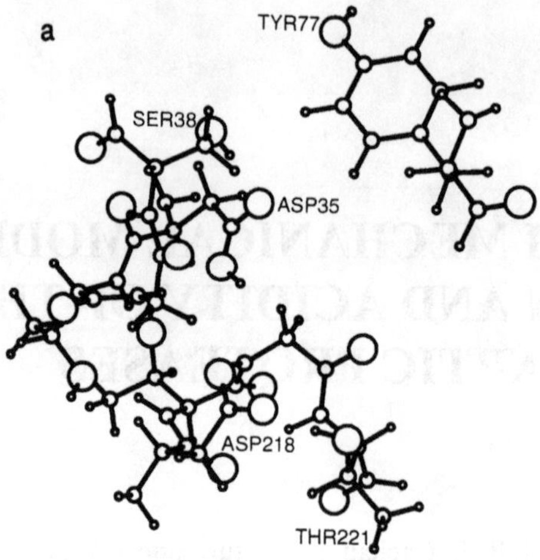

b

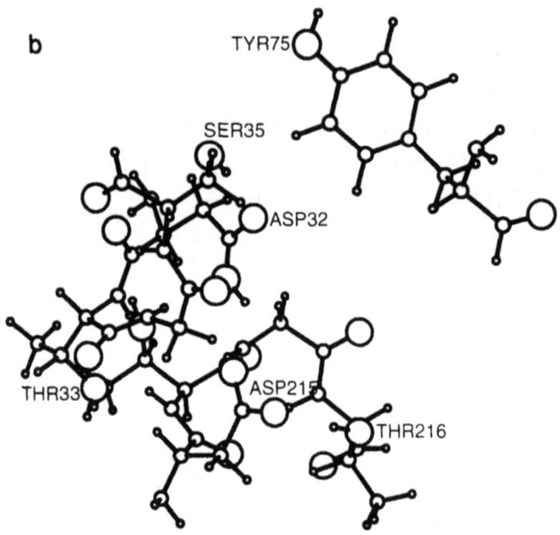

Figure 1. Models of the active site of (a) rhizopuspepsin and (b) pepsin. The models included the folllowing active site residues: a) Asp32, Thr33, Gly34, Ser35, Tyr77, Asp215, Thr216, Gly217, Thr218;. b) Asp35, Thr36, Gly37, Ser38, Tyr77, Asp218, Thr219, Gly220, Thr221. Only the two central aspartates, their immediate neighbors, and the tyrosine residues are numbered.

MOLECULAR MODELS AND METHODS

Molecular models of the active site were built based on the reported crystallographic coordinates of two APs: rhizopuspepsin [4] and pepsin [13]. The models included 9 residues and up to six crystallographically determined active site water molecules. Both models included all atoms of the residues; hydrogens were added to complete the terminal residues (Figure 1). The side chain atoms of Ser35 and Thr218 in pepsin were replaced by a C_β methyl group to model the Ala mutants. Quantum mechanical calculations were performed using the semiempirical MNDO/H method [14-16]. All single and double protonation configurations were considered as well as the double ionized configuration for the two active site carboxylates. Different hydration schemes were considered to include 0,1 and 6 crystallographically determined water molecules. The positions of all protons and water atoms were

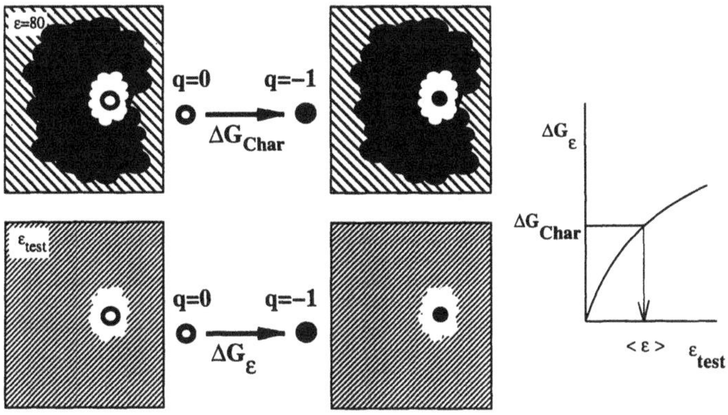

Figure 2. The effective dielectric constant of the environment surrounding the reaction center model was estimated by first calculating the energy necessary for growing a charge (q=0 to q=-1) inside a cavity the size of the reaction center placed inside the complete enzyme in which the entire system is surrounded by a box a water molecules. An all atom representation was used for this calculation. Next, the environment was replaced by a continuum and the calculation repeated using different values of the dielectric constant. The dielectric constant that best reproduced the initial energy for growing the charge was used in subsequent calculations. All calculations were carried out using Delphi [17].

fully optimized using MNDO/H, while the non-hydrogen atoms coordinates were restrained to their crystal structure coordinates. A reaction field model, employing the program Delphi [17] and an effective dielectric constant of 40 (Figure 2), was used to evaluate the effect of full solvation on the calculated protonation energies. The optimized coordinates of the active site models from the MNDO/H calculations were used to estimate the ΔΔG of solvation, $\Delta\Delta G_{Sol}$, between the different protonation schemes in the reaction field calculations; $\Delta\Delta G_{Sol}$ values were added to the calculated protonation energies, ΔH_{Form}, to determine $\Delta\Delta G_{Tot}$ (Table I).

RESULTS AND DISCUSSION

Although it is difficult to determine precisely the stabilization energy of the various charged states, heat of formation calculations show that the energetic preference is: -1>neutral>-2. Our calculations agree with assumptions based on proposed mechanisms for catalysis [1-4], and we assumed a -1 charge state of the active site for all subsequent calculations on protonation site preferences. The relative enthalpies of formation, ΔH_{Form}, for the lowest energy conformation for each potential protonation site were calculated for different hydration schemes (Table I). All energies are expressed with reference to site $O_{\delta 4}$(Asp32) (Figure 3). In the absence of water, the most favorable protonation site is $O_{\delta 6}$(Asp218) in rhizopuspepsin and $O_{\delta 8}$(Asp215) in pepsin. However, when the water molecule between the two active site Asp residues is included in the calculation, the protonation preference shifts to Asp 32 ($O_{\delta 4}$ in rhizopuspepsin, and $O_{\delta 3}$ in pepsin). When the hydration scheme is fully expanded to include the six optimized water molecules, the preferred proton position remains at Asp32 for both pepsin and rhizopuspepsin. The optimized orientation of the innermost water molecule differs for rhizopuspepsin and pepsin (Figure 3) The bifurcated hydrogen bonded configuration seen with the pepsin water was also found to be the most stable structure from ab initio calculations using smaller pepsin active site models [9].

Table I. Relative Heats of Formation (in kcal/mol) for Single Protonation States

	ΔH_{Form}	$\Delta\Delta G_{Sol}$	$\Delta\Delta G_{Tot}$		ΔH_{Form}	$\Delta\Delta G_{Sol}$	$\Delta\Delta G_{Tot}$
Rhiz. Pepsin 0 waters				**Pepsin 0 waters**			
$O_{\delta6}$	-1.70	-1.11	-2.81	$O_{\delta6}$	4.85	-2.75	2.10
$O_{\delta8}$	-1.05	0.48	-0.57	$O_{\delta8}$	-1.55	-0.82	-2.36
$O_{\delta3}$	2.54	-1.68	0.86	$O_{\delta3}$	0.02	-2.54	-2.51
Rhiz. Pepsin +1 waters				**Pepsin +1 waters**			
$O_{\delta6}$	0.38	-1.29	-0.92	$O_{\delta6}$	3.36	-2.39	0.97
$O_{\delta8}$	1.95	0.34	2.29	$O_{\delta8}$	2.04	-1.20	0.84
$O_{\delta3}$	16.29	-1.52	14.76	$O_{\delta3}$	-0.33	-2.42	-2.75
Rhiz. Pepsin +6 waters				**Pepsin +6 waters**			
$O_{\delta6}$	22.41	-1.11	21.30	$O_{\delta6}$	4.08	-1.81	2.28
$O_{\delta8}$	22.91	-0.58	22.33	$O_{\delta8}$	3.54	-0.61	2.93
$O_{\delta3}$	30.35	-0.71	29.64	$O_{\delta3}$	-0.88	-2.31	-3.19

Protonation site $O_{\delta4}$ was used as the zero-point reference energy in all calculations. Positions of protonation sites $O_{\delta3}$, $O_{\delta4}$, $O_{\delta6}$, $O_{\delta8}$ are shown in Figure 3.

The effect of mutations at Ser35 and Thr218 to Ala on relative protonation affinities at Asp32 and Asp215 were calculated using a "gas phase acidity" approach [10] and compared to experimental data [5] (Table II). For each mutation the hydrogen atom positions and the water molecules were optimized while the heavy atom positions were held fixed. We calculated proton affinities of the most stable monoanion (ΔH_1) and dianion (ΔH_2) structures either with or without explicit solvent. As shown in Table II, our results agree with the trend of the experimentally observed pKa shifts [5]. Mutation of Thr218 to Ala results in a large increase in ΔH_1 and pK_1, and a small decrease in ΔH_2 and pK_2. When Ser35 is mutated to Ala, there is a large increse in ΔH_2 and pK_2, and a small increse in ΔH_1 and pK_1. The close agreement between the theoretical and experimental results was only obtained when the six optimized water molecules were included in the calculations. Our calculations indicate that the lower acidity of the mutants is a direct result of the loss of hydrogen bonding between Ser35, Thr218 and the active site carboxylates. This hydrogen bonding network apparently helps to stabilize the active site charge.

Reaction field calculations using different hydration schemes show that the contribution of ΔΔGSol to protonation energies produces values in the range of 0.5 to 3.0 kcal/mol (Table I). Even though the ΔΔGSol is a relatively small contribution compared to the

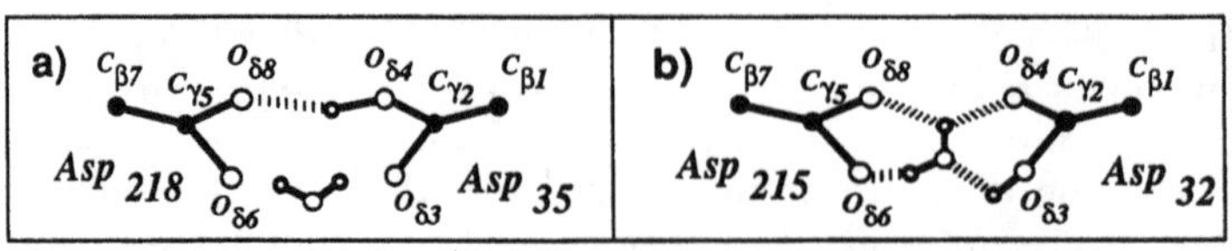

Figure 3. Orientation of the innermost water molecule and numbering of atoms in the active sites of (a) rhizopuspepsin and (b) pepsin.

Table 2. Theoretical Proton Affinities (in Kcal/mol) for Monoanion (DH1) and Dianion (DH_2) States of Native and Mutant Active Site Structures of Pepsin Compared with Experimental pK_a Values [5]

	ΔH_1(6w)	ΔH_2(6w)	ΔH_1(0w)	ΔH_2(0w)	pK_1	pK_2
pepsin	323.1	428.2	341.0	429.3	1.50	3.95
Thr218->Ala	328.0	425.1	340.1	442.4	2.26	3.79
Ser35->Ala	328.9	437.4	346.4	446.0	2.04	4.79
Ser35,Thr218 ->Ala,Ala	345.1	445.6	343.2	447.2		

Calculations were performed using either 6 (6w) or no (0w) water molecules. Values of ΔH were determined by using a gas phase approach [10].

difference in DHForm, it can become an important term in certain cases. The most remarkable example is pepsin with no water molecules, where $O_{\delta 8}$ is the most favorable protonation site in vacuo and $O_{\delta 3}$ and $O_{\delta 4}$ are almost degenerate in energy. However, when the solvation effect is included, $O_{\delta 3}$ and $O_{\delta 8}$ show typical prototropic equilibrium. Inclusion of the reaction field calculations with the hydrated model calculations led to the same conclusion that support Asp32 as the prefered protonation site.

CONCLUSION

Our quantum mechanical calculations agree with recent experimental observations that establish Asp32 as having the higher pKa in the active site of APs. The number of water molecules and their hydrogen bonded patterns influences the calculation of protonation site preference.Our results also show that active site acidity changes for mutant APs will result from distortions of the hydrogen bonding network that involves the active site carboxylates. The judicious use of a reaction field, solvent continuum model can yield more accurate results in certain cases. In conclusion, quantum chemical calculations can be a powerful tool for the study of ionization states and protonation site prefernces when combined with high resolution structural data as long as one is careful to assess and to include essential environment effects.

ACKNOWLEDGMENTS

The content of this publication does not necessarily reflect the views or policies of the Department of Health and Human Services (DHHS), nor does mention of trade names, commercial products, or organizations imply endorsement by the U.S. Government. We thank the staff of the Frederick Biomedical Supercomputing Center, Frederick Cancer Research and Development Center, Frederick, MD for assistance.

REFERENCES

1. B. Veerapandian, J.C. Cooper, A. Sali, T.L. Blundell, R.L. Rosati, B.W. Dominy, D.B. Damon, & D.J. Hoover, Protein Sci. 1:332 (1992).

2. M.N.G. James, A.R. Sielecki, K. Hayakawa, & M.N. Gelb, Biochemistry 31:3872 (1992).
3. L. Pearl, & T. Blundell, FEBS Lett 174:96 (1984).
4. K. Suguna, E.A. Padlam, W.S. Clark, W.D. Carlson, & D.R. Davies, Proc. Natl. Acad. Sci. U.S.A. 84:7009 (1987).
5. X.L. Lin, M. Fusek, Z.Chen, G. Koelsch, H.P. Han, J.A. Hartsuck, & J. Tang, in Structure and Function of the Aspartic Proteinases (Dunn, B.M., Ed.), Plenum Press, New York, pp. 1-8 (1991).
6. D. Hadzi, M. Hodoscek, V. Harb, & D. Turk, J. Mol. Struct. (Theochem) 150:241 (1987).
7. D. Hadzi, M. Hodoscek, & D. Turk, J. Mol. Struct. (Theochem) 181:71 (1988).
8. D. Hadzi, Pure Appl. Chem. 61:2175 (1989).
9. A.J. Beveridge, & G.C. Heywood, Biochemistry 32:3325 (1993).
10. A. Goldblum, Biochemistry 27:1653 (1988).
11. A. Goldblum, in Comput-Assisted Modeling of Receptor-Ligand Interactions: Theoretical Aspects and Applications to Drug Design , Alan R. Liss Inc,pp. 445-454 (1989).
12. L. Turi, & G. Naray-Szabo, Int. J. Quantum Chem. 42:1537 (1992).
13. A.R. Sielecki, A.A. Fedorov, A. Boodhoo, N.S. Andreeva, & M.N.G. James,J. Mol. Biol. 214:43 (1990).
14. K.Ya. Burstein, & A.N. Isaev, Theor. Chim. Acta 64:397 (1984).
15. A. Goldblum, J. Comput. Chem. 8:835 (1987).
16. The MNDO91 code implemented in UniChem1.1.1 version was used. UniChem software environment is available from Cray Research Inc, Eagan, MN, USA.
17. Delphi is available from Biosym Technologies Inc., San Diego, CA, USA.

SITE-DIRECTED MUTAGENESIS OF RHIZOPUSPEPSIN: AN ANALYSIS OF UNIQUE SPECIFICITY

W. Todd Lowther and Ben M. Dunn

Department of Biochemistry and Molecular Biology
University of Florida, JHMHC
Gainesville, FL 32610-0245

Rhizopuspepsin is an aspartic proteinase from the fungus *Rhizopus chinensis*. Interest in rhizopuspepsin as a model system for studying active site interactions stems principally from the need to understand interactions which may be exploited in rational drug design. Targets for these therapeutics are renin, which plays a role in hypertension, the HIV proteinase, essential to the HIV life cycle, and a variety of related, secreted *Candida* yeast aspartic proteinases[1, 2].

Despite primary sequence differences (identity to porcine pepsin and human cathepsin D 39% and 33%, respectively) rhizopuspepsin exhibits a high degree of secondary and tertiary structural homology, particularly in the region of the active site, to the other members of this enzyme family as shown by high resolution x-ray crystallographic analysis[3-6]. Kinetic analyses, utilizing synthetic chromogenic peptides with systematic substitutions, varying hydrophobicity, size, charge and hydrogen bonding potential, in the P_5-P_1 and the P_2'-P_3' positions (Schecter and Berger nomenclature), based on the parent peptide, Lys-Pro-Ala-Lys-Phe*Nph-Arg-Leu (Nph = *p*-nitrophenylalanine), have lead to the observation of interactions which suggest that each member of the aspartic proteinase family has a unique specificity generated by the enzyme residues which create and line the individual binding pockets[7-9]. The mammalian enzymes, porcine pepsin in particular, tend to cleave substrates readily with weaker binding characteristics (higher K_m). Rhizopuspepsin, in contrast, consistently binds a large variety of substrates with lower K_m values, resulting in specificity constant (k_{cat}/K_m) values 2-10 fold higher than those seen for the corresponding substrates cleaved by porcine pepsin.

Two residues which may generate the unique specificity of rhizopuspepsin can be seen by sequence and structural comparisons. Besides the two catalytic aspartic acids, rhizopuspepsin has two *additional* aspartic acids at positions 30 and 77 (pepsin numbering). Asp30 is located between the S_3 and S_1 subsites. Asp77 is located on the flap and points downward into the active site. In order to analyze the contributions of Asp30 and 77 to specificity, mutations were made to the corresponding residues in porcine pepsin: Asp30Ile, Asp77Thr and Asp30Ile/Asp77Thr.

Aspartic Proteinases: Structure, Function, Biology, and Biomedical Implications
Edited by Kenji Takahashi, Plenum Press, New York, 1995

Table 1. Kinetic Comparison of the Wild-Type and Mutant Rhizopuspepsins to Porcine Pepsin with the Substrate Lys-Pro-Asp-Lys-Phe*Nph-Arg-Leu

Enzyme	k_{cat}	K_m	k_{cat}/K_m
	s^{-1}	μM	$M^{-1}s^{-1}$
pI 6	15 ± 2	21 ± 1	690 ± 90 x 10^3
WT-REC	15 ± 1	20 ± 2	650 ± 90 x 10^3
D30I	11 ± 1	68 ± 8	160 ± 20 x 10^3
D77T	08 ± 0.1	71 ± 11	12 ± 3 x 10^3
D30I/D77T	0.4 ± 0.03	127 ± 14	2.9 ± 0.4 x 10^3
POR PEP	10 ± 2	85 ± 19	120 ± 30 x 10^3

pI6 = rhizopuspepsin isozyme pI 6

WT-REC = wild-type recombinant rhizopuspepsin

POR PEP = porcine pepsin

MATERIALS AND METHODS

The zymogen forms of the wild-type and the mutant enzymes were overexpressed in *E. coli* as inclusion bodies[10,11]. The inclusion bodies were denatured in 8 M urea and the proteins refolded by dialysis. After activation by lowering the pH to 3.0, the processed proteins were purified to homogeneity using a Pharmacia Mono S column. The naturally occurring isozyme pI 6 and the recombinant rhizopuspepsins were analyzed conformationally by monitoring intrinsic fluorescence changes induced by guanidinium hydrochloride denaturation[12]. Kinetic comparisons were also made using substrates of the form Lys-Pro-X-Lys-Phe*Nph-Arg-Leu where X = Asp, Arg, Ser and Leu and cleavage occurs specifically between the Phe-Nph bond (*) by porcine pepsin[7-9,13].

RESULTS AND DISCUSSION

The guanidinium hydrochloride denaturation of the rhizopuspepsins was followed by the decrease in fluorescence at 350nm. The resulting curves showed single transitions with inflections representing 50% unfolded protein ($Gdm\text{-}HCl_{1/2}$) around 3.3 M. The derived ΔG values ($-RT\ln K_{eq}$) in the range ± 1.5 kcal mol^{-1} were used to linearly extrapolate to the ΔG_{H2O} value (intercept at 0 M GDM-HCl) for each protein. The $Gdm\text{-}HCl_{1/2}$ and the average ΔG_{H2O} values (11.6 kcal/mol) indicate that the wild-type enzymes are directly comparable and that the mutations, within experimental error, did not alter the protein structures significantly.

Table 1 shows the kinetic parameters of the rhizopuspepsins and porcine pepsin for one of the substrates in the series, Lys-Pro-Asp-Lys-Phe*Nph-Arg-Leu. Similar to the results seen for the structural analysis, the wild-type recombinant enzyme (WT-REC) is kinetically indistiguishable from the native isozyme. In the comparison of WT-REC to porcine pepsin, a large difference can be seen in the K_m and the k_{cat}/K_m parameters, as shown previously[7-9]. Rhizopuspepsin appparently binds substrates more tightly. Mutation of Asp30 to Ile results in an enzyme which behaves more like pepsin. K_m values increased while the k_{cat} values remained the same. The average loss of transition state binding energy for the entire substrate series determined by the equation $\Delta\Delta G = -RT\ln[(k_{cat}/K_{m2})/(k_{cat}/K_{m1})]$ for Asp30Ile was 0.9 kcal mol^{-1}. Mutation of Asp77 to Thr results in an enzyme which exhibits depressed substrate turnover (lower k_{cat} values) and binding (higher K_m values). The average loss of transition state binding energy was 2.6 kcal mol^{-1}. The double mutant exhibited consistently lower affinity for substrates than either of the single mutants. The average loss of transition state binding energy was 3.5 kcal mol^{-1}. Analysis of the binding energy to the transition state for each mutation by double mutant cycle analysis was shown to be additive indicating that the residues at positions 30 and 77 function independently[14]. This study has shown that Asp77 is the predominant residue to confer the unique specificity of rhizopuspepsin.

ACKNOWLEDGEMENTS

This work was supported by NIH grant AM18865. We would like to specially thank the following for their contributions to this work: Xin-li Lin and Jordan Tang at the Oklahoma Medical Research Foundation for their contribution of the rhizopuspepsinogen clone and their help and numerous suggestions toward the production of the recombinant rhizopuspepsins, Bill Farmerie for his expertise in molecular biology cloning and expression systems, Wilhelm deWeerd for his help in the construction, sequencing and expression of the D77T mutant. We would also like to thank the University of Florida Protein Chemistry and DNA Synthesis Core Facilities; in particular, Ruth Davenport, Benne Parten, Joe Geber and Robin Proveaux.

REFERENCES

1. D. R. Davies, Ann. Rev. Biophys. Chem., 19: 189-215 (1990).
2. R. J. Wright, A. Carne, A. D. Hieber, I. L. Lamont, G. W. Emerson and P. A. Sullivan, J. Bacteriol., 174: 7848-7853 (1992).
3. K. Suguna, E. A. Padlan, C. W. Smith, W. D. Carlson and D. R. Davies, Proc. Natl. Acad. Sci., 84: 7009-7013 (1987).
4. K. D. Parris, D. J. Hoover and D. R. Davies, in *Structure and Function of the Aspartic Proteinases: Genetic, Structures and Mechanisms* (B. M. Dunn, Ed.) Plenum Press, New York, pp. 217-231 (1991).
5. J. Cooper, W. Quail, C. Frazao, S. I. Foundling, T. L. Blundell, C. Humblet, E. A. Lunney, W. T. Lowther and B. M. Dunn, Biochemistry, 31: 8142-8150 (1992).
6. M. Jaskolski, A. G. Tomasselli, T. K. Sawyer, D. HG. Staples, R. L. Heinrickson, J. Schneider, S. B. H. Kent and A. Wlodawer, Biochemistry, 30: 1600-1609 (1991).
7. W. T. Lowther, Z. Chen, X. Lin, J. Tang and B. M. Dunn, in *Structure and Function of the Aspartic Proteinases: Genetic, Structures and Mechanisms* (B. M. Dunn, Ed.) Plenum Press, New York, pp. 275-279 (1991).
8. C. Rao, P. E. Scarborough, W. T. Lowther, J. Kay, B. Batley, S. Rapundalo, S. Klutchko, M. D. Taylor and B. M. Dunn in *Structure and Function of the Aspartic Proteinases: Genetic, Structures and Mechanisms* (B. M. Dunn, Ed.) Plenum Press, New York, pp. 143-147 (1991).
9. P. E. Scarborough, K. Guruprasad, C. Topam, G. R. Richo, G. E. Conner, T. L. Blundell and B. M. Dunn, Protein Science, 2: 264-276 (1993).

10. Z. Chen, G. Koelsch, H. Han, X. Wang, X. Lin, J. Hartsuck and J. Tang, J. Biol. Chem., 266: 11718-11725 (1991).
11. X. Lin, M. Fusek, Z. Chen, G. Koelsch, H. P. Han, J. A. Hartsuck and J. Tang, in *Structure and Function of the Aspartic Proteinases: Genetic, Structures and Mechanisms* (B. M. Dunn, Ed.) Plenum Press, New York, pp. 1-8 (1991).
12. C. N. Pace, Methods in Enzymology, 131: 266-280 (1986).
13. B. M. Dunn, P. E. Scarborough, R. Davenport and W. Swietnicki, "Analysis of Proteinase Specificity by Studies of Peptide Substrates: The Use of Ultraviolet and Fluorescence Spectroscopy to Quantitate Rates of Enzymatic Cleavage", in Peptide Protocols: Synthesis Purification, and Characterization, Ben M. Dunn and Michael Pennington, eds., (a volume of the METHODS IN MOLECULAR BIOLOGY series, John M. Walker, series editor), Humana Press, Clifton, New Jersey, in press (1994).
14. J. A. Wells, Biochemistry, 29: 8509-8517 (1990).

SITE-DIRECTED MUTAGENESIS REVEALED ROLE OF SUBSITE RESIDUES OF *MUCOR PUSILLUS* PEPSIN IN CATALYTIC FUNCTION

Young-Nam Park, Makoto Nishiyama, Sueharu Horinouchi, and Teruhiko Beppu

Department of Agricultural Chemistry
Faculty of Agriculture
The University of Tokyo Yayoi 1-1-1, Bunkyo-ku
Tokyo 113, Japan

INTRODUCTION

A few aspartic proteinases, chymosin obtained from calf stomach[1] and *Mucor pusillus* pepsin (MPP) produced by a filamentous fungus *Mucor pusillus*,[2] have been used as the milk-clotting enzymes for industrial production of cheese. Milk-clotting by the aspartic proteases is caused by the specific cleavage of κ-casein, which functions as a stabilizer of milk micelle, at the peptide bond of Phe105-Met106. Chymosin and MPP are discriminated from other members of the aspartic proteinases by high selectivity to the peptide bond of Phe105-Met106 causing destabilization of milk micelles, along with very weak activity of non-specific proteolysis, which assure the production of the clotted materials in a high yield. In the last several years we have constructed an expression system for MPP using *Saccharomyces cerevisiae* as a host,[3] in which the proMPP secreted into the medium is converted to the mature protease through the self-processing activity as well as by the host-dependent cleavage.[4] This system possesses an inherent advantage, which enables us to obtain even inactive mutant enzymes in mature forms. By using this yeast expression system, we conducted site-directed mutagenesis of MPP, especially on the amino acid residues involved in substrate binding.

MATERIALS AND METHODS

Purification of Yeast MPP

Saccharomyces cerevisiae MC16 (α, *leu*2, *his*4, *ade*2) carrying pJP1 or its derivative which contains the preproMPP gene[5] under the control of the yeast *GAL*7 promoter was cultured aerobically at 30°C in YPD medium.[6] The growing cells were transferred into

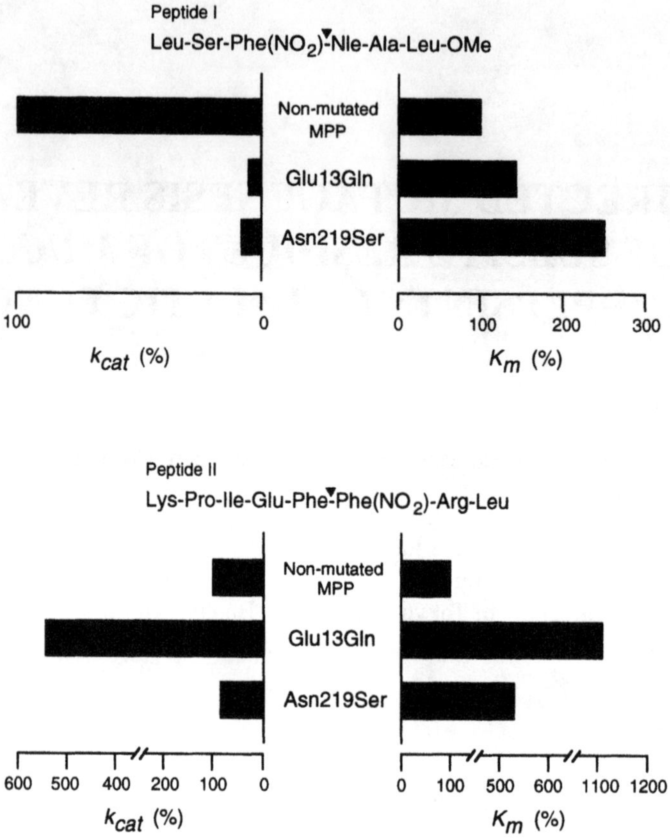

Figure 1. Schematic presentation of kinetic parameters of subsite mutants.

YPGal6 medium containing galactose in place of glucose to induce the *GAL7* promoter and cultured further aerobically. Yeast MPP was purified from the supernatant of the cells in an early stationary phase as described previously,[7] and digested with endo-β-*N*-acetylglucosaminidase H (endo H)[8] to remove two Asn-linked glycosyl moieties at Asn72 and Asn172.[7] The endo H-digested yeast MPP was applied to Mono Q and Superose 12 chromatographies on a FPLC system (Pharmacia Fine Chemicals, Uppsala, Sweden), as described previously,[7] and used as the purified yeast MPP preparation

Mutagenesis

Site-directed mutagenesis of Phe220Leu (S4 subsite), Phe220Glu (S4 subsite), Asn219Ser (S3, S4 subsite), Glu13Gln (S3 subsite), Glu12Thr (S3 subsite), Thr218Ser (S2, S3 subsite), Ile222Thr (S2 subsite), Ile120Thr (S1 subsite), Ile120Asn (S1 subsite), Phe189Ser (S1' subsite), Phe189Asn(S1' subsite), Trp190Phe (S2' subsite), and Trp190Leu (S2' subsite) was carried out using synthetic oligonucleotides by Kunkel's method[9] using a Bio-Rad MutaGene kit. All the mutations were verified by nucleotide sequencing using the M13-dideoxy chain termination method.[10]

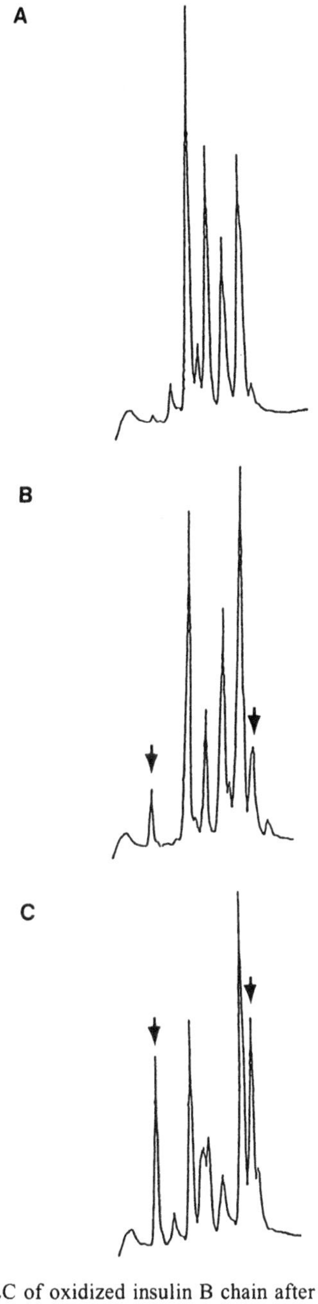

Figure 2. Elution patterns on HPLC of oxidized insulin B chain after reaction with *Mucor pusillus* pepsin, Glu13Gln and Asn219Ser. Samples (50μl) of the trifluoroacetic acid-soluble products of the oxidized insulin B chain were subjected to ODS 1151-N column.(Senshu, Yokyo, Japan) They were eluted with a linear gradient (0-60%) of CH_3CN containing 0.1%(*w*/*v*) trifluoroacetic acid for min. The flow rate was 1.0ml/min and elution was monitored by following absorbance 220nm. Corresponding absorbance peaks from HPLC separations were pooled for peptides analysis (data not shown).

Assay of Enzyme Properties

Milk-clotting activities were determined according to Iwasaki et al.[11] Two chromogenic synthetic oligopeptides, Leu-Ser-Phe(NO_2)-Nle-Ala-Leu-OMe (peptide I) and Lys-Pro-Ile-Glu-Phe-Phe(NO2)-Arg-Leu-OH (peptide II) were used as substrates to determine kinetic parameters, as described previously[12], by Spectronic 3000 array spectrophotometer (Milton Roy Co.,New York,USA). Scissile bond specificity was examined with oxidized insulin B chain[13] and the thermostability was determined by measuring residual milk-clotting activity after incubation at 55°C for an appropriate time.

RESULTS AND DISCUSSION

Kinetic analyses for the mutants were carried out by using the synthetic oligopeptides as substrates. Marked changes in the kinetic parameters were caused by the mutations, Glu13Gln (S3 subsite) and Asn219Ser (S3, S4 subsite) (Fig. 1). When peptide I was used, these two mutants exhibited marked decreases in the k_{cat} values, while such a large change in the K_m values was not caused by the mutations, although a smaller increase in the K_m value was observed with Asn219Ser. On the other hand, when peptide II was used, Asn219 mutation was found to cause a marked increase in the K_m value but have no effects on the K_m value. Glu13Gln mutation caused great effects both on the K_m and k_{cat} values. These results suggest that the amino acid residues apart from the catalytic center contribute to the catalytic function as well as the substrate binding. Scissile bond specificities of the mutants were analyzed with oxidized bovine insulin B chain. After digestion of the B chain with enzymes, the peptides generated were separated by HPLC (Fig. 2). Non-mutated MPP and most of the mutants cleaved the B chain into 4 major peptides. On the other hand, Glu13Gln and Asn219Ser showed distinct change in the scissile bond specificity. These observations suggest that the amino acid residues at position 13 and 219 are residues specifically determining the substrate specificity.

We next analyzed the effect of the mutations on the stability of MPP. After incubation at 55°C for 0, 10, 30, 60, 120, and 180 min, their milk-clotting activities were measured. Most of the mutant enzymes showed stability similar to that of the non-mutated MPP, retaining 80-90% residual activity after incubation for 180 min (data not shown). On the other hand, Trp190Phe, especially, showed distinctly lower stability, and only 1% residual activities were detected after the heat-treatment for 180 min. When the residual amount of the enzyme protein during the heat treatment was analyzed by SDS-polyacrylamide gel electrophoresis, the non-mutated MPP showed no significant change in the amount upon heating, whereas the amount of Trp190Phe was markedly decreased upon heating (data not shown). The amounts of residual MPP measured by densitometry of the gels coincided well with the residual activities of all enzymes after the heat treatment. This suggests that decreases in the residual activities of Trp190Phe on heating were caused by self-digestion. The position of Trp190 is assigned in the hydrophobic core inside the central part of the enzyme. Therefore we speculate that a decrease in the hydrophobic interaction which may stabilize the structure is the cause of the decrease in stability.

REFERENCES

1. A.G. Mackinlay, and R.G. Wake, κ-Casein and its attack by rennin (chymosin), in "Milk Proteins" vol. 2, H.A. Mckenzie, ed., Academic press, New York, pp. 175-215 (1971)

2. K. Arima, S. Iwasaki, and G. Tamura, Milk clotting enzyme from microorganisms, Agric. Biol. Chem. 31:540-545 (1967)
3. T. Yamashita, N. Tonouchi, T. Uozumi, and T. Beppu, Secretion of Mucor rennin, a fungal aspartic protease of *Mucor pusillus*, by recombinant yeast cells, Mol. Gen. Genet. 210:462-467 (1987)
4. R. Hiramatsu, J. Aijawa, S. Horinouchi, and T. Beppu, Secretion by yeast of the zymogen from of Mucor rennin, an aspartic proteinase of *Mucor pusillus*, and its conversion to the mature form, J. Biol. Chem. 264:16862-16866 (1989)
5. N. Tonouchi, H. Shoun, T. Uozumi, and T. Beppu, Cloning and sequencing of a gene for Mucor rennin, an aspartate protease from *Mucor pusillus*, Nucleic Acids Res. 14:7557-7568 (1986)
6. J. Aikawa, R. Hiramatsu, M. Nishiyama, S. Horinouchi, and T. Beppu, A yeast expression system and site-directed mutagenesis of a fungal aspartic proteinase, Mucor rennin, in "Structure and Function of the Aspartic Proteinases" B. M. Dunn, ed., Plenum Press, New York, pp. 233-242 (1991)
7. J. Aikawa, T. Yamashita, M. Nishiyama, Suheharu Horinouchi, and T. Beppu, Effects of glycosylation on the secretion and enzyme activity of Mucor rennin, an aspartic proteinase of *Mucor pusillus*, produced by recombinant yeast, J. Biol. Chem. 265:13955-13959 (1990)
8. A.L. Tarentino, T.H. Plummer Jr., and F. Maley, The release of intact oligosaccharide from specific glycoprotein by endo-β-*N*-acetylglucosamidase H, J. Biol. Chem. 249:818-824 (1974)
9. T.A. Kunkel, Rapid and efficient site-specific mutagenesis without phenotype selection, Proc. Natl. Acad. Sci. U.S.A. 82:488-492 (1985)
10. F. Sanger, A.R. Coulson, B.G. Barrelle, A.J.M. Smith, and B.A. Ros, Bacteriophage as an aid to rapid DNA sequencing, J. Mol. Biol. 143:161-178 (1980)
11. S. Iwasaki, G. Tamura, and K. Arima, Milk clotting enzyme from microorganisms. Part II. The enzyme production and the properties of crude enzyme, Agric Biol. Chem. 31:546-551 (1967)
12. J. Suzuki, K. Sasaki, M. Nishiyama, S. Horinouchi, and T. Beppu, Alteration of catalyutic properties of chymosin by site-directed mutagenesis, Protein Eng. 2:563-569 (1989)
13. P.E. Butler, M.J. McKay, and J.S. Bond, Characterization of meprin, a membrane-bound metalloendopeptidase from mouse kidney, Biochem. J. 241:229-235 (1987)

A NOVEL INTRACELLULAR ACID PROTEINASE FROM THE PLASMODIA OF A TRUE SLIME MOLD, *PHYSARUM POLYCEPHALUM*

Kimiko Murakami-Murofushi,[1] Takayuki Takahashi,[2] Hiromu Murofushi,[2] and Kenji Takahashi[2]

[1]Department of Biology
Ochanomizu University
Bunkyo-ku, Tokyo 113 Japan
[2]Department of Biophysics and Biochemistry
Faculty of Science
The University of Tokyo
Bunkyo-ku, Tokyo, 113 Japan

INTRODUCTION

The plasmodia of a true slime mold, *Physarum polycephalum*, contain intracellular acid proteinases that have highly acidic pH optimum (1,2). The enzyme activity was detected in the particulate fraction and was found to be effectively extracted with a non-ionic detergent, indicating the possibility of the distribution in the lysosome or lysosome-like organelle (2). To investigate its molecular and enzymatic properties, we purified this enzyme to apparent homogeneity from the plasmodia (3). The results are reported in this communication.

MATERIALS AND METHODS

The plasmodia of *P. polycephalum* were cultured as microplasmodia in a sterile semi-defined culture medium according to Daniel and Rusch (4). Microplasmodia from 3-day cultures were collected by centrifugation, washed with distilled water to remove extracellular slime polysaccharide and used for the enzyme preparation.

The enzyme was purified to apparent homogeneity by a combination of detergent extraction, acid precipitation, and column chromatographies of DEAE-Sephadex, hydroxylapatite, CM-Sephadex, and Sephadex G-100 (3).

The acid proteinase activity was assayed at pH 1.7 with hemoglobin as a substrate (2).

Aspartic Proteinases: Structure, Function, Biology, and Biomedical Implications
Edited by Kenji Takahashi, Plenum Press, New York, 1995

Table I. Purification of the Proteinase

Step	Total protein	Total activity	Specific activity	Yield
	mg	unit	units/ mg	%
Homogenate	4433	1955	0.45	100
Crude extract	1266	1708	1.35	85.7
DEAE-Sephadex A-50	38.7	590	15.2	29.6
Acid precipitation	16.0	312	19.5	15.6
Hydroxylapatite	4.46	132	29.6	6.6
CM-Sephadex C-50	2.25	78	34.6	3.9
Sephadex G-100	0.75	48	64.0	2.4

Cleavage site determination in a synthetic peptide, Lys-Pro-Ile-Glu-Phe-Phe(4-NO_2)-Arg-Leu (5), and oxidized insulin B chain was performed as described previously (6) by analyzing the peptides produced upon incubation with the acid proteinase.

RESULTS AND DISCUSSION

Table I summarizes the results of purification of the *Physarum* acid proteinase. An approximately 142-fold increase in specific activity was attained with an overall yield of 2.4% (3). The low yield of the enzyme is primarily due to the presence of a significant amount of other acid proteinase(s) in the crude extract.

Analyses of the purified enzyme by SDS-PAGE and Sephadex G-100 column showed that it was composed of two polypeptide chains (a 31-kDa heavy chain and a 23-kDa light chain) cross-linked by disulfide bond(s). The heavy chain, but not the light chain, of the enzyme was shown to be glycosylated. Direct carbohydrate analysis revealed that one molecule of the native enzyme contained about 7 residues of mannose, 5 residues of glucosamine, and 1 residue each of fucose and glucose (3). The relatively high content of glucosamine suggests the presence of a few N-linked carbohydrate chains in the heavy chain.

The NH_2-terminal amino acid sequence of the heavy chain was determined to be Ala-Gly-Val-Asp-Gly-Tyr-Ile-Val-Pro-Tyr-Val-Ile-Phe-Asp-Leu-Tyr-Gly-Ile-Pro-Tyr and that of the light chain to be Ala-Glu-Pro-Pro-Ile. As for the light chain, the sequence was blocked after five cycles of Edman degradation for unknouwn reasons. No significant homogogy of the sequences of the heavy and light chains with those of other proteinases was found in the protein sequence data base of the Protein Identification Resource.

The enzyme was optimally active at pH 1.7 toward hemoglobin as a substrate. The enzyme was stable at 25°C in the pH range of 3.0-6.0. However, it became less stable at higher pH values. Incubation at 37°C and pH 8.0 for 15 min resulted in 30% activity loss.

The effects of various proteinase inhibitors on the enzyme activity were examined and the results are shown in Table II. A strong inhibition by diazoacetyl-D,L-norleucine methyl ester(DAN)(7) was observed in the presence of cupric ions. Norleucine content of a DAN-modified enzyme was determined by amino acid analysis after acid hydrolysis. An enzyme sample inactivated about 80% by the reagent contained 1.6 mol of norleucine/mol of protein. To identify the modified aspartic acid residue(s), the inhibited enzyme was digested with porcine pepsin at 37°C for 18 h and the digest was subjected to HPLC. A norleucine-containing peptide composed of about 25 amino acid residues was obtained. Since the peptide contained 3-4 aspartic acids, we attempted to further digest it with other proteases, such as trypsin, chymotrypsin, thermolysin, and V8 protease, to generate smaller

Table II. Effects of Various Reagents on the Proteinase Activity

Reagent	Concentration	Relative activity
		%
None		100
DAN	42 µg/ml	5
EPNP	500 µg/ml	99
Pepstatin A	1 mM	97
Leupeptin	1 mM	95
Soybean trypsin inhibitor	10 µg/ml	104
PMSF	1 mM	89
	10 mM	79
TPCK	1 mM	81
	10 mM	77
TLCK	1 mM	93
	10 mM	82
p-Hydroxymercuribenzoic acid	2 mM	93
N-Ethylmaleimide	2 mM	103
EDTA	2 mM	109
$CaCl_2$	1 mM	107
$MgCl_2$	1 mM	92

The enzyme activity was assayed using hemoglobin as substrate.

fragments. However, the peptide was found to be resistant to hydrolysis by the proteases. Thus, we could not identify the modified amino acid residue(s). Other inhibitors tested including 1,2-epoxy-3-(*p*-nitrophenoxy)propane (EPNP)(8) and pepstatin A (9), typical aspartic proteinase inhibitors, showed no significant inhibitory effects.

A synthetic peptide, Lys-Pro-Ile-Glu-Phe-Phe(4-NO_2)-Arg-Leu (6), was cleaved at the Phe-Phe(4-NO_2) bond by the enzyme, but another synthetic pepsin-substrate, *N*-acetyl-L-phenylalanyl-3,5-diiodo-L-tyrosine (10), was not hydrolyzed at all. The enzyme showed a unique substrate specificity toward oxidized insulin B chain. The major cleavage sites were the bonds Gly^{8}-Ser^{9}, Leu^{11}-Val^{12}, Cya^{19}-Gly^{20}, and Phe^{24}-Phe^{25}, and the Gly^{8}-Ser^{9} bond was most susceptible. In addition, the Gly^{20}-Glu^{21} bond was also cleaved to a marked extent. To our knowledge (6,11), the cleavages of these Gly-containing bonds by acid proteinases have never been reported.

In conclusion, the *Physarum* enzyme was demonstrated to be a pepstatin-insensitive acid proteinase, having a two-chain structure with a molecular weight of about 54,000, and exhibited a unique substrate specificity. These results indicate that this enzyme is a novel type of intracellular acid proteinase distinct in several properties from ordinary aspartic proteinases including cathepsin D.

REFERENCES

1. M.M. Polanshek, J.C. Blomquist, T.E. Evans, and H.P. Rusch, Aminopeptidases of *Physarum polycephalum* during Growth and Differentiation, Arch. Biochem. Biophys. 190:261 (1978).
2. K. Murakami-Murofushi, A. Hiratsuka, and J. Ohta, A Novel Acid Protease from Haploid Amoebae of *Physarum polycephalum*, and Its Changes during Mating and Subsequent Differentiation into Diploid Plasmodia, Cell Struct. Funct. 9:311 (1984).
3. K. Murakami-Murofushi, T. Takahashi, Y. Minowa, S. Iino, T. Takeuchi, H. Kitagaki-Ogawa, H. Murofushi, and K. Takahashi, Purification and Characterization of a Novel Intracellular Acid Proteinase from the Plasmodia of a True Slime Mold, *Physarum polycephalum*, J. Biol. Chem. 265:19898 (1990).

4. J.W. Daniel, and H.P. Rusch, The Pure Culture of *Physarum polycephalum* on a Partially Defined Soluble Medium, J. Gen. Microbiol., 25:47 (1961).
5. B.M. Dunn, M. Jimenez, B.F. Parten, M.J. Valler, C.E. Rolph, and J. Kay, A Systematic Series of Synthetic Chromophoric Substrates for Aspartic Proteinases, Biochem. J. 237:899 (1986).
6. O. Matsuzaki, and K. Takahashi, Improved Purification of Slow Moving Protease from Gastric Mucosa and its Action on the B Chain of Oxidized Bovine Insulin, Biomed. Res. 9:515 (1988).
7. T.G. Rajagopalan, W.H. Stein, and S. Moore, The Inactivation of Pepsin by Diazoacetyl-norleucine Methyl Ester, J. Biol. Chem. 241:4295 (1966).
8. J. Tang, Specific and Irreversible Inactivation of Pepsin by Substrate-like Epoxides, J. Biol. Chem. 246:4510 (1971).
9. T. Aoyagi, S. Kunimoto, H. Morishima, T. Takeuchi, and H. Umezawa, Effect of Pepstatin on Acid Proteases, J. Antibiot. (Tokyo) 24:687 (1971).
10. J. Tang, Gastricsin and Pepsin, Methods Enzymol. 19:406 (1970).
11. A.J. Barrett, Cathepsin D and Other Carboxyl Proteinases, in: Proteinases in Mammalian Cells and Tissues (Barrett, A, J., ed) pp. 209, North-Holland Publishing Co., Amsterdam (1977).

CHARACTERISTICS OF YAP3, A NEW PROHOHORMONE PROCESSING ASPARTIC PROTEASE FROM *S. CEREVISIAE*[*]

A.V. Azaryan,[1,2] T. C. Friedman,[1] N. X. Cawley,[1,2] and Y. P. Loh[1,2]

[1]Section on Cellular Neurobiology
Laboratory of Developmental Neurobiology
National Institute of Child Health and Human Development
National Institutes of Health
Bethesda, MD 20892
[2]Department of Biochemistry
Uniformed Services University of the Health Sciences
Bethesda, MD 20814

INTRODUCTION

Neuropeptides and peptide hormones are synthesized as large precursors that are cleaved at paired, or less frequently at single basic amino acid residues by endoproteases to release biologically active peptides [1,2]. In the yeast *Saccharomyces cerevisiae*, processing of the pheromone precursor, pro-α-mating factor, is carried out by kexin, a subtilisin-related proprotein convertase that is the product of the KEX2 gene [3,4]. In 1990, a *S. cerevisiae* aspartic protease gene (YAP3) which was induced when pro-α-mating factor was over-expressed in a KEX2-deficient mutant was described [5]. The YAP3 gene product was able to catalyze the processing of the α-mating factor precursor on the COOH-terminal side of paired basic residues to yield the active pheromone in KEX2-deficient mutants, suggesting that YAP3 may be an alternative pro-α-mating factor processing protease. YAP3 has sequence similarity to the classical two-domain active site region of other aspartic proteinases such as pepsin, renin, cathepsin D and penicillopepsin that exhibit specificity for hydrophobic residues [6]. Despite this sequence similarity, YAP3 exhibits novel specificity for basic residues.

To characterize the YAP3 protein and its activity, the YAP3 gene was over-expressed in *S. cerevisiae* strain BJ 3501 by transformation with pEMBLyex-YAP3 construct [7]. The enzyme was purified and characterized with respect to pH optimum, protease

[*]The opinions or assertions contained herein are the private ones of the authors and are not to be construed as official or reflecting the views of the Department of Defense or the Uniformed Services University of the Health Sciences.

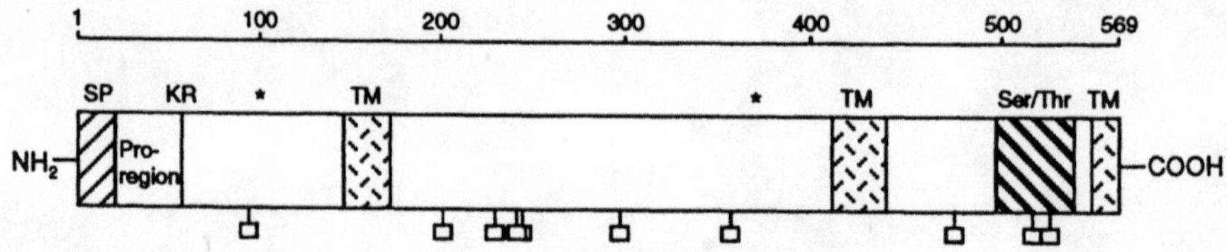

Figure 1. Diagrammatic representation of yeast aspartic protease 3 (YAP3) showing the various structural domains. TM=Potential transmembrane domain; Ser/Thr=Serine/Threonine rich domain; (aa503-545); ■ = Potential N-linked glycosylation sites; KR = Potential zymogen activation site (Lys^{66}-Arg^{67}); SP = Signal peptide (aa1-21); * = Active site aspartic residues (Asp^{101}, Asp^{371}).

inhibitor profile and specificity for paired basic residues of pro-opiomelanocortin (POMC) [7]. In this paper, we summarize the physico-chemical properties of YAP3 and its substrate specificity requirements, as well as discuss the predicted domain structure of this proteinase.

STRUCTURAL DOMAINS WITHIN YAP3

Analysis of the YAP3 gene sequence [5] reveals that it encodes a 569-amino acid protein that has significant homology to a variety of aspartic proteinases and appears to be organized into several domains. There is a predicted amino-terminal signal peptide sequence of 21 amino acids, followed by a putative 45-amino acid proregion (Fig. 1). This putative pro-region indicates the possible existence of a zymogen form of the enzyme that probably

Table 1. Properties of Yeast Aspartic Protease 3 (YAP3)

Molecular weight	~70,000
pH optimum (β-LPH as substrate)	4.0-4.5
Subunit structure	none
Subcellular localization	periplasmic*
Affinity to concanavalin A-sepharose	high (glycoprotein)
Effect of Ca^{2+} (5 mM)	40% activation
Effect of DTT (1 mM)	33% activation
Effect of protease inhibitors	
pepstatin (0.1 mM)	85% inhibition
p-chloromercuribenzoate (1 mM)	no effect
iodoacetamide (1 mM)	no effect
phenylmethylsulfonyl fluoride (1 mM)	no effect
TLCK (0.2 mM)	no effect
leupeptin (1 mM)	5% inhibition
captopril (0.1 mM)	no effect
o-phenanthroline (0.1 mM)	no effect
EDTA (0.1 mM)	8% inhibition

*Cawley, N.X., Wong, M., Pu, L.-P., and Loh, Y.P., in preparation.

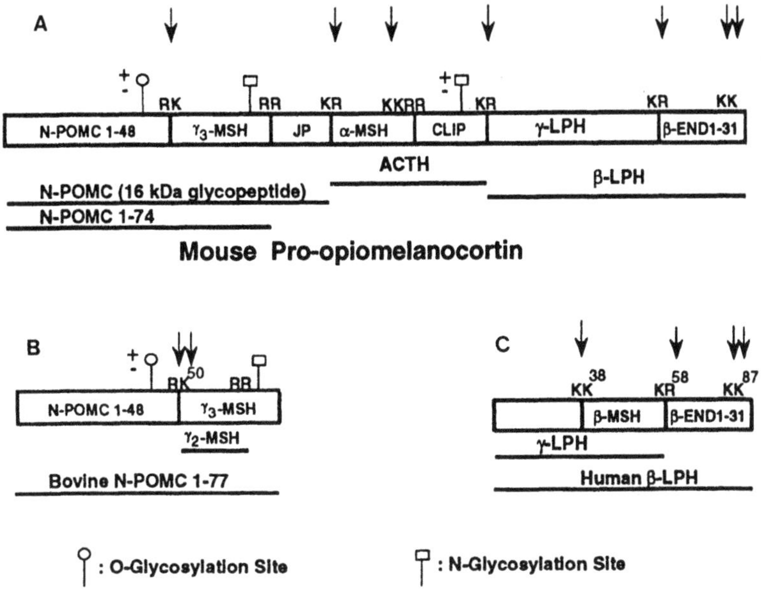

Figure 2. Diagrammatic representation of (A) mouse pro-opiomelanocortin (POMC), (B) bovine N-$POMC^{1-77}$ and (C) human β-lipotropin showing the paired basic residue sites cleaved by YAP3 purified from yeast cell extracts (see arrows). MSH = melanocyte stimulating hormone; JP = joining peptide; CLIP = corticotropin-like intermediate lobe peptide; LPH = lipotropin; END = endorphin; ACTH = adrenocorticotropin.

undergoes activation by cleavage at the Lys^{66}-Arg^{67} site (Fig. 1). The structure of YAP3 also contains a COOH-terminal Ser/Thr-rich domain (aa 503-545), three potential transmembrane anchors (aa 149-174, 414-443, 553-569) and ten potential N-linked glycosylation sites (Fig. 1). The presence of N-linked carbohydrate moieties on YAP3 is evidenced by its affinity to concanavalin A [7]. By alignment to other aspartic proteinases, the two active site aspartyls in YAP3 are Asp^{101} and Asp^{371}; the sequence around Asp^{101} is typical for classical aspartic proteinases Asp-Thr-Gly (DTG), but in the case of Asp^{371}, the sequence is Asp-Ser-Gly (DSG).

The YAP3 sequence exhibits the highest degree of homology to the yeast BAR1-encoded 'barrier' proteinase [8,9], an extracellular aspartic protease involved in the degradation of α-mating factor [9]. This homology is great throughout the NH_2-terminal catalytic domain and both YAP3 and BAR1 proteinase have DSG, instead of the DTG triad, around the second active site aspartyl residue [5,9]. In the COOH-terminal region there is no homology between YAP3 and BAR1 proteinase, as BAR1 does not contain either Ser/Thr-rich or hydrophobic transmembrane domains. The COOH-terminal part of BAR1 proteinase does not affect its catalytic activity [8], but is probably required for secretion of this extracellular enzyme [9]. KEX2 protease also shares some topological features with YAP3, for example a Ser/Thr-rich domain and a transmembrane domain in the COOH-terminal region of the enzyme [3]. The transmembrane domain may be necessary for membrane association and targeting to the secretory pathway. The decrease in production of α-mating factor in yeast cells containing the COOH-terminally deleted forms of YAP3 described previously [5], may well be due to improper cellular localization of those truncated forms, rather than the loss of catalytic potency, although the latter cannot be excluded.

CHARACTERIZATION OF YAP3 ACTIVITY

YAP3 has a molecular weight of ~70,000 and is a glycoprotein, based on its affinity to concanavalin A (Table 1). The pH optimum of the enzymatic activity assayed with β-lipotropin (LPH), is 4.0-4.5. YAP3 activity is inhibited by the selective aspartic protease inhibitor, pepstatin A, but not by serine (TLCK, PMSF) or cysteine (p-CMB, iodoacetamide) protease inhibitors (Table 1). Thus, not only the deduced amino acid sequence of YAP3, but the characteristics of the purified enzymatic activity indicate that YAP3 is a member of the aspartic proteinase family. The activity of YAP3 is increased 1.4-fold in the presence of 5 mM Ca^{2+} and 1.3-fold by 1 mM DTT.

The specificity of purified YAP3 has been studied using the mammalian prohormone, pro-opiomelanocortin (POMC), and its fragments as substrates [7]. YAP3 cleaved either between or on the carboxyl side of paired basic residues of mouse POMC to yield adrenocorticotropin (ACTH) and β-LPH (Fig. 2A), human β-LPH to yield β-melanocyte-stimulating hormone (β-MSH) and β-endorphin$^{1-31,1-29,1-28}$ (Fig. 2C) and bovine N-POMC$^{1-77}$ to yield γ_3-MSH (Fig. 2B). YAP3 also cleaved ACTH$^{1-39}$ at the tetrabasic site Lys^{15}-Lys^{16}-Arg^{17}-Arg^{18} primarily between Lys^{15}-Lys^{16} to generate ACTH$^{1-15}$ and Lys^{16}-Arg^{17}-CLIP (Fig. 2A). Pepstatin A (10 μM) inhibited the processing of POMC, N-POMC$^{1-77}$ and ACTH by YAP3 by 85-95%. When the substrate specificity of YAP3 was assessed with short model peptides, substrates with a single basic residue, e.g. Z-Gln-Gly-Arg-MCA, were not cleaved. However, a substrate with the R-X-R-R motif, Z-Arg-Val-Arg-Arg-MCA, was cleaved by YAP3 at the peptide bond between the two arginine residues [7].

In conclusion, YAP3 was demonstrated to be another aspartic protease, besides the mammalian pro-opiomelanocortin converting enzyme (E.C.3.4.23.17) (see Loh *et al.*, this volume), that has unique specificity for paired basic residues. YAP3 may be a potentially very useful industrial enzyme for proprotein conversion.

REFERENCES

1. K. Docherty, and D.F. Steiner, Posttranslational proteolysis in polypeptide hormone biosynthesis, *Ann. Rev. Physiol.* **44**:625 (1982).
2. Y.P. Loh, M. Brownstein, and H. Gainer, Proteolysis in neuropeptide processing and other neural functions, *Ann. Rev. Neurosci.* **7**:189 (1984).
3. R.S. Fuller, R.E. Sterne, and J. Thorner, Enzymes required for yeast prohormone processing, *Ann. Rev. Physiol.* **50**:345 (1988).
4. R.S. Fuller, A. Brake, and J. Thorner, Yeast prohormone processing enzyme (KEX2 gene product) is a Ca^{2+}-dependent serine protease, *Proc. Natl. Acad. Sci., USA* **86**:1434 (1989).
5. M. Egel-Mitani, H.P. Flygenring, and M.T. Hansen, A novel aspartyl protease allowing KEX2-independent MFα prohormone processing in yeast, *Yeast*, **6**:127 (1990).
6. B.M. Dunn, M. Jimenez, B.F. Parten, M.J. Valler, C.E. Rolph, and J. Kay, A systematic series of synthetic chromophoric substrates for aspartic proteinases, *Biochem. J.* 237:899 (1986).
7. A.V. Azaryan, M. Wong, T.C. Friedman, N.X. Cawley, F.E. Estivariz, H.-C. Chen, and Y.P. Loh, Purification and characterization of a yeast paired basic residue-specific aspartic protease encoded by the YAP3 gene: Similarity to the mammalian pro-opiomelanocortin converting enzyme, *J. Biol. Chem.* **268**:11968 (1993).
8. V.L. MacKay, S.K. Welch, M.I. Insley, T.R. Manney, J. Holly, G.C. Saari, and M.L. Parker, The *Saccharomyces cerevisiae* BAR1 gene encodes an exported protein with homology to pepsin, *Proc. Natl. Acad. Sci., USA*, **85**:55 (1988).
9. V.L. MacKay, J. Armstrong, C. Yip, S. Welch, K. Walker, S. Osborn, P. Sheppard, and J. Forstrom, Characterization of the BAR proteinase, an extracellular enzyme from the yeast *Saccharomyces cerevisiae*, in: Structure and Function of the Aspartic Proteinases, B. Dunn, ed., pp 161-172, Plenum Press, New York (1991).

CANDIDA ALBICANS ASPARTIC PROTEINASE: cDNA CLONING AND COMPARISON AMONG STRAINS

Somay Yamagata Murayama,[1] Osamu Takeda,[2] Hiroyuki Mukai,[2] Kazutoh Takesako,[2] Eiko Sono,[2] Ikunoshin Kato,[2] and Hideyo Yamaguchi[1,3]

[1]Department of Microbiology and Immunology
Teikyo University School of Medicine
2-11-1 Kaga, Itabashi-ku, Tokyo 173, Japan
[2]Biotechnology Research Laboratories
Takara Shuzo Co., Ltd.
3-4-1 Seta, Ohtsu, Shiga 520-21, Japan
[3]Teikyo University Research Center for Medical Mycology
359 Ohtsuka, Hachioji, Tokyo 192-03, Japan

INTRODUCTION

The increasing incidence of fungal infections in immunocompromised individuals has focused attention on the importance of *Candida* species, particularly *Candida albicans* as the major opportunistic fungal pathogen. Among several putative virulence factors of *C. albicans* SAP (secreted aspartic proteinase) produced and excreted by this yeast has been considered primarily responsible for its pathogenicity. Although this possibility is supported by experimental data reported from several different laboratories, the actual pathogenetic role of *C. albicans* aspartic proteinase in the development of candidiasis remains unclear. In the course of studies which were attempted to answer this question, we initially undertook the cDNA cloning of this enzyme.

RESULTS AND DISCUSSION

Yamamoto *et al.*[1], our coworker reported that all strains of *C. albicans* tested secrete proteinase extracellularly, and they sequenced the first 23 amino acids of the N-terminal of the SAP from the *C. albicans* serotype A, TIMM2726 (No. 114) strain. Based on the results, we constructed mixed primers, and using the PCR walking method, we cloned the complete gene of *C. albicans SAP*[2]. DNA sequencing of this *SAP* from TIMM2726 revealed that it includes 167 bp of the 3' flanking sequences in addition to 1,194 bp of coding DNA (Figure 1)[2,3].

```
   1  ATG TTT TTA AAG AAT ATT TTC ATT GGC CTT GCT ATT GCT TTA TTA GTC GAT GCT ACT CCA ACA ACA ACC AAA AGA TCA GCT GGT TTC GTT GCT   93
 -56   M   F   L   K   N   I   F   I   G   L   A   I   A   L   L   V   D   A   T   P   T   T   T   K   R   S   A   G   F   V   A  -26
                                           ---------------------

  91  TTA GAT TTC AGT GTT GTG AAA ACT CCT AAA GCA TCC CCA GTT ACT AAT GGC CAA GAA GGT AAA ACT TCC AAA AGA CAA GCT GTC CCA GTG ACT  183
 -25   L   D   F   S   V   V   K   T   P   K   A   S   P   V   T   N   G   Q   E   G   K   T   S   K   R   Q   A   V   P   V   T    6

 181  TTA CAC AAT GAA CAA GTC ACT TAT GCT GCT GAT ATT ACC GTT GGA TCC AAT AAT CAA AAA CTT AAT GTT ATT GTT GAT ACT GGA TCA TCT GAT  273
   7   L   H   N   E   Q   V   T   Y   A   A   D   I   T   V   G   S   N   N   Q   K   L   N   V   I   V   D   T   G   S   S   D   37
                                                                                                           *

 271  TTA TGG GTT CCT GAT GTT AAT GTT GAT TGT CAA GTC ACT TAT AGT GAT CAA ACT GCA GAT TTC TGT AAA CAA AAG GGG ACA TAT GAT CCA AGT  363
  38   L   W   V   P   D   V   N   V   D   C   Q   V   T   Y   S   D   Q   T   A   D   F   C   K   Q   K   G   T   Y   D   P   S   68

 361  GGT TCA TCA GCT TCA CAA GAT TTG AAT ACT CCA TTC AAA ATT GGT TAT GGT GAT GGA TCT TCA TCT CAA GGT ACT TTA TAT AAG GAT ACC GTT  453
  69   G   S   S   A   S   Q   D   L   N   T   P   F   K   I   G   Y   G   D   G   S   S   S   Q   G   T   L   Y   K   D   T   V   99

 451  GGA TTT GGT GGT GTT TCG ATT AAA AAT CAA GTT TTA GCT GAT GTT GAT TCT ACT TCA ATT GAT CAA GGT ATT TTA GGG GTT GGT TAT AAA ACC  543
 100   G   F   G   G   V   S   I   K   N   Q   V   L   A   D   V   D   S   T   S   I   D   Q   G   I   L   G   V   G   Y   K   T  130

 541  AAT GAA GCC GGT GGT AGT TAT GAT AAT GTC CCT GTC ACT TTA AAA AAA CAA GGA GTC ATT GCT AAG AAT GCT TAT TCA CTT TAT CTT AAT TCT  633
 131   N   E   A   G   G   S   Y   D   N   V   P   V   T   L   K   K   Q   G   V   I   A   K   N   A   Y   S   L   Y   L   N   S  161

 631  CCA GAT GCT GCC ACG GGA CAA ATA ATT TTC GGT GGG GTT GAT AAT GCT AAA TAT AGT GGT TCA TTA ATT GCA TTA CCA GTT ACT TCT GAT CGT  723
 162   P   D   A   A   T   G   Q   I   I   F   G   G   V   D   N   A   K   Y   S   G   S   L   I   A   L   P   V   T   S   D   R  192

 721  GAA TTA AGA ATT AGT TTG GGT TCA GTT GAA GTT TCT GGT AAA ACC ATC AAT ACT GAT AAT GTC GAT GTT CTT TTG GAT TCA GGT ACC ACC ATT  813
 193   E   L   R   I   S   L   G   S   V   E   V   S   G   K   T   I   N   T   D   N   V   D   V   L   L   D   S   G   T   T   I  223
                                                                                                       *

 811  ACT TAT TTG CAA CAA GAT CTT GCT GAT CAA ATC ATT AAA GCT TTC AAT GGT AAA TTA ACT CAA GAT TCC AAT GGT AAT TCA TTC TAT GAA GTT  903
 224   T   Y   L   Q   Q   D   L   A   D   Q   I   I   K   A   F   N   G   K   L   T   Q   D   S   N   G   N   S   F   Y   E   V  254

 901  GAT TGT AAT TTG TCA GGG GAT GTT GTA TTC AAT TTT AGT AAA AAT GCT AAA ATT TCC GTT CCA GCT TCC GAA TTT GCT GCT TCT TTA CAA GGT  993
 255   D   C   N   L   S   G   D   V   V   F   N   F   S   K   N   A   K   I   S   V   P   A   S   E   F   A   A   S   L   Q   G  285

 991  GAT GAT GGT CAA CCA TAT GAT AAA TGT CAA TTA CTT TTC GAT GTT AAT GAT GCT AAC ATT CTT GGT GAT AAC TTT TTG AGA TCA GCT TAT ATT 1083
 286   D   D   G   Q   P   Y   D   K   C   Q   L   L   F   D   V   N   D   A   N   I   L   G   D   N   F   L   R   S   A   Y   I  316

1081  GTT TAT GAT TTG GAT GAT AAT GAA ATT TCT TTG GCT CAA GTC AAA TAT ACT TCT GCT TCC AGT ATT TCT GCC TTG ACC 1194
 317   V   Y   D   L   D   D   N   E   I   S   L   A   Q   V   K   Y   T   S   A   S   S   I   S   A   L   T  342
```

Figure 1. Nucleotide sequence of a cDNA for aspartic proteinase secreted by *C. albicans* TIMM2726. The putative signal peptidase recognition site is underlined. * Conserved active-site aspartic acid residue.

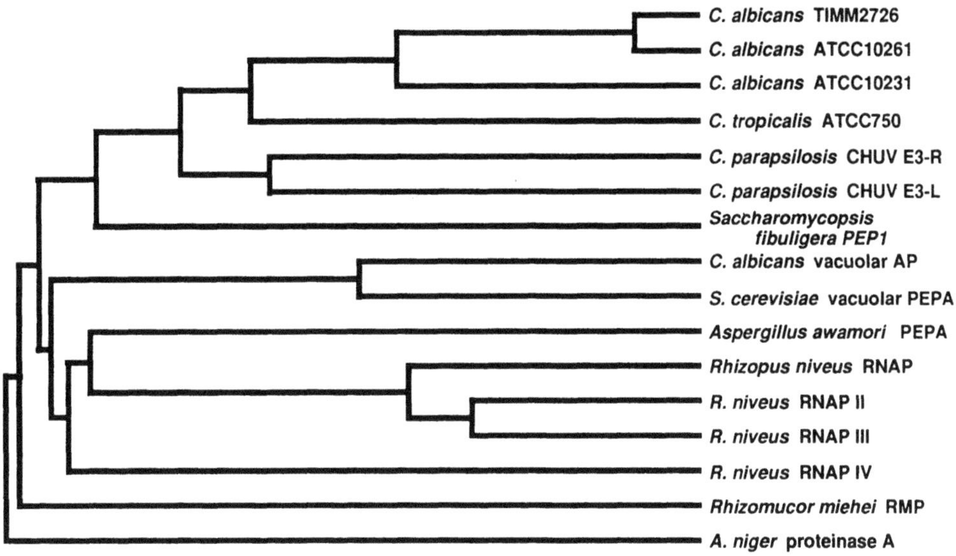

Figure 2. Dendrogram of fungal aspartic proteinases.

To judge from the open reading frame, a mature enzyme is 342 amino acid residues long and has a molecular weight of 37,000. Two active aspartic acid sites, Asp-32 and Asp-219, and the amino acids near these sites were conserved as exemplified in pepsin and cathepsin D.

In Figure 2 the dendrogram of all 16 fungal aspartic proteinases cloned is shown. Mature aspartic proteinase secreted by TIMM2726 had only 12.6 % and 22.6 % homology with vacuolar aspartyl proteinase of *C. albicans* and *Saccharomyces cerevisiae* proteinase A (vacuolar enzyme). Such a low sequence homology of the former enzyme with the latter two may be due to a difference in the cellular localization of the enzyme.

Hube *et al.*[4] and Wright *et al.*[5] very recently reported the DNA sequences of SAPs of two different *C. albicans* strains, ATCC10231 (*SAP1*) and ATCC10261 (*SAP2*), respectively. We compared their sequences with our cloned sequence, and the result showed 77% and 99% DNA sequence homology with *SAP1* or *SAP2*. The DNA sequence of *SAP2* shows a close similarity to cDNA described here, but the difference between the DNA sequences of *SAP1* and the *SAP* of TIMM2726 seems to be greater as a usual enzyme secreted by the same species. Therefore we examined whether TIMM2726 had the same DNA sequence of *SAP1*. We constructed the specific primers for PCR on the basis of sequence data from *SAP1*. Genome DNA of TIMM2726 was used as the PCR template. In this PCR, the amplification product gave a discrete band of about 230 bp. The amplified product was cloned and the result of sequencing of this clone revealed that TIMM2726 had the same nucleotide sequence of a part of *SAP1*. Next, PCR was performed with genome DNA of ATCC10231 as the template, and specific primers to TIMM2726. The result showed that ATCC10231 also had the same nucleotide sequence of a part of *SAP* of TIMM2726.

Next we examined the N-terminal amino acid-sequences of actually secreted aspartic proteinases of ATCC10231 and ATCC10261. The N-terminal amino acid-sequences of both strains are the same as that of TIMM2726. The strain-specific differences between secretory proteinase of *C. albicans* might depend on culture conditions. We did not examine the culture conditions but have concluded in spite of the existence of two genes on *C. albicans* genome, that the gene expressed in normal condition is only one that was cloned by us with TIMM2726. Recently, Morrow[6] have shown that a gene differentially regulated in cells undergoing the phenotypic white-opaque transition encodes aspartyl proteinase. They cloned this opaque specific mRNA from *C. albicans* WO1 strain and the gene is apparently identical to *SAP1*. Although the mRNA for this gene is abundant in opaque cells in WO1 strain, no mRNA is apparent in either white cells in WO1 strain or nonswitchig cells of another *C. albicans* strain which are producing abundant proteinase extracellularly grown with protein as the sole nitrogen source. According to their finding and our results the two SAPs may have different functions and two *SAP* genes may express in different growth conditions.

C. albicans is said that they would invade by attachment to the host cells via cell-surface adhesive molecules and once attached, proliferate and release SAP, which aids them to invade the integument of the host. There seem to be a relationship between the production of SAP by *C. albicans* and its virulence but evidence for SAP as virulence factor is mostly indirect. A number of reports on the role of SAP have been published. The hydrolytic activity of the SAP against important serum proteins, such as immunoglobulins and hemoglobin, keratin and collagen, association of tissue invasion, resistance of phagocytosis, activation of inflammatory mediators and enhancement of vascular permeability[7], relationship to DIC (Disseminated intravascular coagulation) and allergen have been demonstrated. Even if the secreted, inducible, aspartic proteinase of *C. albicans* is a putative virulence factor of the fungus, its actual function is still uncertain. Its precise role could be clarified by disruption of this gene based on the sequence information shown here.

REFERENCES

1. T. Yamamoto, K. Nohara, K. Uchida and H. Yamaguchi, Purification and characterization of secretory proteinase of *Candida albicans*, Microbiol. Immunol. 36 : 637-641 (1992)
2. H. Mukai, O. Takeda, K. Asada , I. Kato , S. Y. Murayama and H. Yamaguchi, cDNA cloning of an aspartic proteinase secreted by *Candida albicans*, Microbiol. Immunol. 36 : 1207-1216, (1992)
3. S. Y. Murayama, cDNA cloning of *Candida albicans* aspartic proteinase and its diagnostic application. In "Molecular Biology and its Application to Medical Mycology" (B. Maresca, G. S. Kobayashi and H. Yamaguchi eds), Springer-Verlag Berlin Heiderberg, pp. 149-158, 1993.
4. B. Hube, C. J. Turver, F. C. Odds, H. Eiffert, G. J. Boilnois, H. Kuhel and R. Ruchel, Sequence of the *Candida albicans* gene encoding the secretory aspartate proteinase, J. Med. Vet. Mycol. 29 : 129-132 (1991)
5. R. J. Wright, A. Carne, A. D. Hieber, I. L. Lamont, G. W. Emerson and P. A. Sullivan, A second gene for a secreted aspartate proteinase in *Candida albicans*, J. Bacteriol. 174 : 7848-7853 (1992)
6. B. Morrow, T. Srikantha and D. R. Soll, Transcription of the gene for a pepsinogen, PEP1, is regulated by white-opaque switching in *Candida albicans*, Mol. Cell. Biol. 12 : 2997-3005 (1992)
7. H. Kaminishi, Y. Hagihara, S. Hayashi and T. Cho, Isolation and characteristics of collagenolytic enzyme produced by *Candida albicans*, Infect. Immun. 53 : 312-316 (1986)

MOLECULAR CLONING AND SEQUENCE ANALYSIS OF A GENE ENCODING AN ASPARTIC PROTEINASE FROM *ASPERGILLUS ORYZAE*

M. Takeuchi, K. Ogura, T. Hamamoto, and Y. Kobayashi

Laboratory of Molecular Biology and Microbial Chemistry
Tokyo University of Agriculture and Technology
Fuchu, Tokyo 183, Japan

INTRODUCTION

Aspergillus oryzae is one of the most important filamentous fungi used as koji in traditional Japanese fermentation industries such as sake, shoyu and miso manufacturing industries.[1] *A. oryzae* secretes a number of hydrolases such as proteinases, amylases and lipases.[2] The koji is used as enzyme source and starters of the fermented foods. In the proteinaceous fermented foods, aspartic proteinases play an important role, since the fermentation is carried out at acidic conditions. Accordingly, many studies on the physico-chemical and enzymatic properties of the proteinases purified from *A. oryzae* have been done.[3] Several forms of the aspartic proteinases have been purified and characterized from *A. oryzae*[4] and it has been proved to have trypsinogen-kinase activity which is capable of activating trypsinogen and chymotrypsinogen.[5] But limited information is available on the structure, function and gene organization of the enzymes. In order to clarify their diversity, structure and function at the molecular level, cloning and analysis of aspartic proteinase genes were carried out.

RESULTS AND DISCUSSION

To facilitate the studies on the aspartic proteinase from *A. oryzae*, we screened large quantities of the aspartic proteinase producing strain. Among 23 strains of *A. oryzae* screened for the aspartic proteinase, *A. oryzae* var. *globosus* M-9 showed the highest enzyme production.

Three aspartic proteinases produced by *A. oryzae* var. *globosus* M-9 were purified to homogeneous state from the extract of solid bran culture. They were designated aspartic proteinase I (Mr 39,000), aspartic proteinase II-1 (Mr 48,000) and aspartic proteinase II-2 (Mr 48,000), respectively. When the purified enzymes were electrophoresed in SDS-PAGE

Aspartic Proteinases: Structure, Function, Biology, and Biomedical Implications
Edited by Kenji Takahashi, Plenum Press, New York, 1995

with or without mercaptoethanol, a single band was obtained. The results suggested that the enzyme is composed of a single chain monomer. These enzymes did not interact with Con A Sepharose and the mobility of the enzymes in SDS-PAGE was not changed by the endo F treatment. These results suggested that the enzyme was not glycosylated. They were completely inhibited by pepstatin. Among them, properties of the aspartic proteinases II-1 and II-2 were the same, except for the elution position from ion-exchange chromatography. But the aspartic proteinase I was different in the molecular weight, immunological properties and N-terminal amino acid sequence. Ouchterlony double diffusion technique using antibody raised against the aspartic proteinase II-1 protein demonstrated good cross reactions between aspartic proteinases II-1 and II-2. On the other hand, no precipitin line was found with aspartic proteinase I. These data indicated that aspartic proteinases II-1 and II-2 were immunologically similar, whereas aspartic proteinase I had different immunogenicity.

We have tried to isolate the aspartic proteinase II gene from a lambda phage cDNA library of *A. oryzae* var. *globosus* M-9 by immunoscreening using the antibody raised against aspartic proteinase II-1 protein. Only one positive clone from 150,000 independent clones was obtained. The inserted DNA was 850 bp. The sequence of the 850 bp fragment did not contain an initiation codon at the 5'-end. In order to obtain a clone containing the initiator Met, the cDNA library was rescreened with digoxigenin labeled 850 bp fragment. One of the clones obtained included a 1.48 kb insert which had an *Eco*RI site. The fragment was divided into 850 bp and 650 bp fragments by *Eco*RI digestion. The 650 bp fragment hybridized to the oligonucleotide corresponding to the N-terminal amino acid sequence of aspartic proteinase II-1 and 850 bp fragment hybridized to the oligonucleotide corresponding to the consensus sequence of the second active center of aspartic proteinases. The amino acid sequence deduced from the nucleotide sequence of the 1.48 kb contained the amino acid sequence identical to that of N-terminal and active site of the enzyme. These results indicated that the 1.48 kb cDNA carried aspartic proteinase II-1 gene. Genomic Southern hybridization was done using a digoxigenin labeled probe derived from the 1.48 kb cDNA. Single band was observed in DNA digested with several restriction enzymes and no other signal was detected. These results indicated that only one copy of aspartic proteinase II gene was present in *A. oryzae* M-9. Using size-fractionated *Hind*III fragments, a genomic library was constructed with λgt11 vector. A digoxigenin labeled probe derived from the 1.48 kb cDNA was used as probe for plaque hybridization of the genomic library. Only one positive clone was obtained from 50,000 plaques screened and it contained a 2.7 kb fragment.

Nucleotide sequencing analysis revealed that a part of the 2.7 kb genomic DNA had the same sequence to that of the 1.48 kb cDNA and the cloned gene encoded a protein of 390 amino acid residues (Fig. 1). High homology(70%) was found in amino acid sequences of aspartic proteinase II-1, *A. awamori* aspartic proteinase[6] and *A. oryzae* 24 (Ref. 7) and RIB 40 (Ref. 8) aspartic proteinases. The aspartic proteinase II-1 had no cysteine residue and its gene had two introns. These were different from other *Aspergillus* aspartic proteinases and their genes which had several cysteine residues and three introns, respectively. In the deduced amino acid sequence of aspartic proteinase II-1, the amino acid sequence corresponding to the N-terminal of aspartic proteinase I was not found. This result suggested that aspartic proteinase I was not a degradation product of aspartic proteinase II-1 and strain M-9 had more than one aspartic proteinase genes. Comparison of the N-terminal amino acid sequences determined from the purified aspartic proteinase II-1 and that deduced from the cloned gene revealed that the aspartic proteinase II-1 consisted of a 67-amino acid prepropeptide and 323-amino acid mature protein. Using *Saccharomyces cerevisiae* expression vector pG-3 (Ref. 9), active M-9 aspartic proteinase was produced and secreted in the medium. But, during the early stage of the cultivation, inactive pro-enzyme was also detected by the Western analysis. Although the mature enzyme had a potential N-linked glycosylation site, the mobility of the enzyme in SDS-PAGE was not changed by the endo F treatment.

```
                                                                                              ↓
AOPEPO2  1 : MVNTSLLAALTAYAV-AVSAAPTAPQ———VKGFSVNQVAVPKGVYRHPAAQLAKAY———GKYHATVPTQV-AAAAAA
AOPEPA   1 : ··IL·KV··VAVGLSTVA··L··G·SHSPHARR··TI··ITRQTARVGPKT·SFPAI·SRALA··GG···AHLKS·V·SG
AOPEPO   1 : ··IL·KV··VAVGLSTVA··L··G·SHSPHARR··TI··ITRQTARVGPKT·SFPAI·SRALA··GG···AHLKS·V·SG
AAPEPA   1 : ··VF·KT···VLGLSS······–··T———R···TI··I·R·ANK—TRTIN·PGM·ARSLA·FGG···QS·KE·–·S–

                                      •    ▼
        81 : TGSVTTNPTSNDEEYITQVTVGDDTLGLDFDTGSADLWVFSSQTPSSERSGHDYYTPGSSAQKIDGATWSISYGDGSSAS
        81 : H·T·V·S·EP··I··L·P·NI·GT··N··············EEL·K··QT···V·K·SGN·S··A··S·D·········
        81 : H·T·V·S·EP··I··L·P·NI·GT··N··············EEL·K··QT···V·K·SGN·S··A··S·D·········
        81 : K··AV·T·QN·····L·P····KS··H··············DEL····QT···L···S···T·LS·Y··D·········

       161 : GDVYKDKVTVGGVSYDSQAVESAEKVSSEFTQDTANDGLLGLAFSSINTVQPTPQKTFFDNVKSSLSEPIFAVALKHNAP
       161 : ····Q·T······TAQG····A·S·I·DQ·V··KN··············K·K··T····T··DQ·DA·L···T··YH··
       161 : ····Q·T······TAQG····A·S·I·DQ·V··KN··············K·K··T····T··DQ·DA·L···T··YH··
       161 : ····R·T······TTNK····A·S·I····V·N·················KA·T····T···Q·DS·L···Q···D··
                                                               ▲
                                                                 •▼
       241 : GVYDFGYTDSSKYTGSITYTDVDNSQGFWGFTADGYSIGS–DSSSDSITGIADTGTTLLLLDDSIVDAYYEQVNGASYDS
       241 : ·S····FI·K··F··ELA·A···D·····Q·······V·KG·AQKAP···········VM···E······K··Q··KN·A
       241 : ·S····FI·K··F··ELA·A···D·····Q·······V·KG·AQKAP···········VM···E······K··Q··KN·A
       241 : ·······I·D···········A·S···Y···ST······DGS···SGFSA········I····E··S······S···GET

       321 : SQGGYVFPSSASLPDFSVTIGDYTATVPGEYISFADVGNGQ–T–FGGIQSNSGIGFSIFGDVFLKSQYVVFDASGPRLGF
       321 : ·A······CETE··E·TAV··S·N·VI··KH·NY·PLQE·SS·CV········L·L··L·············SQ······
       321 : ·A······CETE··E·T·V··S·N·VI··KH·NY·PLQE·SS·CV········L·L··L·············SQ······
       321 : EA·····SC·TNP···T·V····K·V···K··NY·PIST·SS·C·········L·L··L············NSE··K···

       401 : AAQA
       401 : ····
       401 : ····
       401 : ····
```

Figure 1. Comparison of the amino acid sequences of the *A. oryzae* var. *globosus* M-9 and other *Aspergillus* aspartic proteinases. The amino acid sequence of the *A. oryzae* M-9 aspartic proteinasc (AOPEPO2) is compared with those of *A. oryzae* aspartic proteinases (AOPEPA 8 and AOPEPO 7) and *A. awamori* aspartic proteinase (AAPEPA 6). Gaps(-) are introduced to obtain maximum homology. The amino acid sequences are aligned for maximum matches using the GENETYX program(Software Development Co., Japan). Identical amino acid residues in *A. oryzae* M-9 aspartic proteinase and other *Aspergillus* aspartic proteinases are shown by dots. Active site aspartic acids are marked with asterisks. Presumed cleavage sites of the propeptide of the aspartic proteinases are shown by arrow. Closed triangles indicate the position of introns. Cysteine residues and asparagine residues at potential N-linked glycosylation site are underlined.

This result indicated that the enzyme was not glycosylated in *S. cerevisiae*, as not in M-9 strain.

ACKNOWLEDGMENTS

We thank Prof. K.R.Yamamoto, University of California, for the kind gift of *Saccharomyces cerevisiae* expression vector, pG-3. This work was supported in part by a Grant-in-Aid for Scientific Research from the Ministry of Education, Science and Culture Japan.

REFERENCES

1. Hara,S., Kitamoto,K., and Gomi,K., 1992, New developments in fermented beverage and foods with *Aspergillus*, in :"*Aspergillus*: Biology and industrial applications", J.W.Bennet and M.A.Klich, eds., Butterworth-Heineman, Inc., pp.133-153.

2. Berka,R.M., Dunn-Coleman,N., and Ward,M., 1992, Industrial enzymes from *Aspergillus* species, in:"*Aspergillus*: Biology and industrial applications", J.W.Bennet and M.A.Klich, eds., Butterworth-Heineman, Inc., pp.155-202.
3. Sakaguchi,K., Takagi,M., Horiuchi,H., and Gomi,K., 1992, Fungal enzymes used in oriental food and beverage industries, in :"Applied molecular genetics of filamentous fungi", J.R. Kinghorn and G.Turner, ed., Blackie Academic and Professional, pp.54-99.
4. Tujita,Y., and Endo,A., 1976, Purification and characterization of the two molecular forms of *Aspergillus oryzae* acid protease, Biochim. Biophys. Acta, 445:194-204.
5. Davidson,R., Gertler,A., and Hofmann,T., 1975, *Aspergillus oryzae* acid proteinase: purification and properties, and formation of π-chymotrypsin, Biochem. J., 147: 45-53.
6. Berka,R.M., Ward,M., Wilson,L.J., Hayenga,K.J., Kodama,K.H., Carloomagno,L.P., and Thompson, S.A., 1990, Molecular cloning and deletion of the gene encoding aspergillopepsin A from *Aspergillus awamori*, Gene, 86: 153-162.
7. Berka,R.M., Carmona,C.L., Hayenga,K.J., Thompson,S.A., and Ward,M., 1993, Isolation and characterization of the *Aspergillus oryzae* gene encoding aspergillopepsin O, Gene, 125:195-198.
8. Gomi,K., Arikawa,K., Komiya,N., Kitamoto,K., and Kumagai,C., 1993, Cloning and nucleotide sequence of the acid protease-encoding gene(pepA) from *Aspergillus oryzae*, Biosci. Biotech. Biochem., 57: 1095-1100.
9. Schena,M., Picard,D., and Yamamoto,K.R., 1991, Vectors for constitutive and inducible gene expression in yeast, Methods in Enzymol., 194: 389-398.

ASPERGILLUS NIGER VAR. *MACROSPORUS* PROTEINASE B. cDNA CLONING, EXPRESSION, AND ACTIVATION OF THE PROENZYME

Hideshi Inoue,[1] Jing-Fang Lu,[1] Senarath B. P. Athauda,[1] Kwang-Hoon Kong,[1] Takaharu Hayashi,[1] Takao Kimura,[2] Osamu Makabe,[2] and Kenji Takahashi[1]

[1]Department of Biophysics and Biochemistry
Faculty of Science
University of Tokyo
Bunkyo-ku, Tokyo 113, Japan
[2]Meiji Seika, Ltd.
Kohoku-ku, Yokohama 222, Japan

Filamentous fungus *Aspergillus niger* var. *macrosporus* produces two types of extracellular acid endoproteinases, proteinases A and B, commercially called proctases A and B, respectively (1,2). Proteinase A is a non-pepsin-type acid proteinase (1-7), whereas proteinase B has properties of pepsin-type aspartic proteinase. Proteinase B has a molecular mass of about 35 kDa and is inhibited by specific inhibitors of aspartic proteinases such as pepstatin, diazoacetyl-DL-norleucine methyl ester in the presence of cupric ions, and 1,2-epoxy-3-(*p*-nitrophenoxy)propane (4,8).

The extracellular aspartic proteinases from mammals such as pepsin are excreted as zymogens which are transformed into active forms by autoprocessing under acidic conditions. Similarly, the extracellular aspartic proteinases from fungi have been thought to be transformed into active forms autocatalytically from inactive proforms under acidic conditions. No native proforms of the fungal aspartic proteinases, however, have been isolated so far. The presence of the proforms is suggested from the structures of the cloned genes or cDNAs encoding aspartic proteinases from *Mucor pusilus* (9), *Rhizopus niveus* (10), and so on. The putative proforms of aspartic proteinases from *M. pusillus* and from *R. chinensis* were heterologously produced by *Saccharomyces cerevisiae* and *E. coli*, respectively (11,12). The conversion of the recombinant zymogens to the mature forms occurred by autocatalytic mechanism. However, the details of the characters and the activation mechanism of the zymogens are unknown.

In this paper, we report the primary structure of proteinase B deduced from the cDNA sequence, the expression in *E. coli* and autoactivation of its proform, and the results of site-directed mutagenesis on a lysyl residue in its prosegment that is highly conserved in the aspartic proteinase family.

Aspartic Proteinases: Structure, Function, Biology, and Biomedical Implications
Edited by Kenji Takahashi, Plenum Press, New York, 1995

MATERIALS AND METHODS

Enzymatic Cleavage of Proteinase B and Purification of Peptide Fragments

A. niger var. *macrosporus* proteinase B (1.7 mg, Seikagaku Kogyo Co., Tokyo) was digested with TPCK-treated trypsin (30 μg) in 0.2 ml of *N*-ethylmorpholine acetate buffer (50 mM, pH 8.0) for 6 hr at 37°C. The tryptic digest was directly submitted to HPLC using a Hitachi 655A-11 system on a column (0.46 x 25 cm) of TSKgel ODS-120T (Tosoh Co.). Peptides were eluted with a linear gradient of acetonitrile (0 - 50% in 60 min) in 0.1% trifluoroacetic acid. The peptide peak fractions were pooled and lyophilized. The amino acid compositions and sequences were analyzed by using an automated derivatizer-analyzer 420A/130A (Applied Biosystems) and a Protein Sequencer 477A/120A (Applied Biosystems), respectively.

cDNA Cloning

The DNA probe used for screening the cDNA encoding proteinase B was prepared by polymerase chain reaction (PCR) (13). The primers for PCR were designed based on the amino acid sequences of the tryptic fragments, synthesized by use of a DNA synthesizer 391 (Applied Biosystems), and purified through electrophoresis on 12% polyacrylamide gel containing 8 M urea. The reaction mixture for PCR contained 10 ng of total cDNA from *A. niger* var. *macrosporus*, 1 nmol each of the sense and antisense primers, and 2.5 units of *Taq* DNA polymerase in 10 mM Tris-HCl buffer (pH 8.3) containing 50 mM KCl, 1.5 mM $MgCl_2$, 0.2 mM dNTPs, and 0.001% (w/v) gelatin in a total volume of 0.1 ml. PCR, isolation of the products, and labeling of the probe were carried out as described elsewhere (6). Cloning of cDNA was performed according to the standard method (14).

Construction of Expression Plasmid pAR-ANB

The nucleotide sequence encoding putative proproteinase B (amino acid residues 21-394 in the preproenzyme) was amplified by PCR and subcloned into T7 expression vector pAR2113 (15). The resulting expression plasmid pAR-ANB was confirmed to contain the correct sequence encoding proproteinase B by DNA sequencing.

Expression of Proproteinase B in *E. coli*

Expression of proproteinase B was performed essentially according to the method as described (15). *E. coli* BL21(DE3) cells harboring plasmid pAR-ANB were cultured in M9ZB medium for 12 hr at 37°C with shaking (preculture). One liter of M9ZB medium with ampicillin (0.1 mg/ml) was inoculated with 10 ml of the preculture and shaken at 37°C until Abs_{600} reached 0.6. IPTG was then added to a concentration of 1 mM and shaking was continued for 2.5 hr.

Purification and Refolding of Recombinant Proproteinase B

The cells were harvested by centrifugation and resuspended in a lysis buffer (50 mM Tris-HCl, pH 7.5, 0.1 M NaCl and 1 mM EDTA) and 0.3 mg/ml of lysozyme was added. After this suspension stood on ice for 10 min, 0.4 mg/ml of sodium deoxycholate and 1 mM of PMSF were added. After standing on ice for 10 min, the suspension was stirred at 37°C

```
A. niger var. macrosporus    SKGSAVTTPQNNDEEYLTPVTVGKSTLHLDFDTGSADLWVF
A. niger var. awamori        .........................................
P. janthinellum            AAS.V.TN..TA.....I....I.GT..N.N.....S.....

SDELPSSEQTGHDLYTPSSSATKLSGYTWDISYGDGSSASGDVYRDTVTVGGVTTNKQAVEAASK
.................................................................
.T...A.Q.S..SV.N..ATGKE......S...........N.FT.S.......AHG...Q..QQ

ISSEFVQDTANDGLLGLAFSSINTVQPKAQTTFFDTVKSQLDSPLFAVQLKHDAPGVYDFGYIDD
.......N.........................................................
..AQ.Q...N.................QS..........S.AQ.....A.Q.QQ.......F..S

SKYTGSITYTDADSSQGYWGFSTDGYSIGDGSSSSSGFSAIADTGTTLILLDDEIVSAYYEQVSG
.................................................................
......L...GV.N...F.S.NV.S.TA.--.Q.GD...G........L....SV..Q..S....

AQESEEAGGYVFSCSTNPPDFTVVIGDYKAVVPGRYINYAPISTGSSTCFGGIQSNSGLGLSILG
.SGET.........................K..................................
..QDSN......D....L...S.S.SG.T.T...SL...G.SGD..-..L........I.F..F.

DVFLKSQYVVFNSEGPKLGFAAQA
........................
.I.........D.D..Q....P..
```

Figure 1. The amino acid sequence of *A. niger* var. *macrosporus* proteinase B deduced from the cDNA sequence and its comparison with the sequences of *A. niger awamori* pepsin and penicillopepsin. Underlines 1-4 indicate the sequences determined by Edman degradation. Underline 1, the sequence of the N-terminal region of mature proteinase B from the culture medium of *A. niger* var. *macrosporus*. Underlines 2-4, tryptic fragments. The sequences used for designing PCR primers are boxed. -, Deletion. ., Residue identical with that in *A. niger* var. *macrosporus* proteinase B.

until the viscosity was increased, then was sonicated. After centrifugation at 5,000*g* for 15 min, the pellet was dissolved in 1 l of 50 mM Tris-HCl (pH 8.0) containing 8 M urea, 30 mM 2-mercaptoethanol and 1 mM EDTA, and centrifuged at 9,000*g* for 30 min. The supernatant was dialyzed against 50 mM Tris-HCl (pH 8.0) firstly with and secondly without 30 mM 2-mercaptoethanol at 4°C. The resulting solution was applied to a DE52 column (1 cm x 1.5 cm) equilibrated in 50 mM Tris-HCl (pH 8.0). The bound proteins were eluted with 50 mM Tris-HCl (pH 8.0) containing 1 M NaCl. Fractions containing proteolytic activity toward denatured hemoglobin at pH 2.0 were pooled and purified by gel filtration chromatography on a Superose HR column equilibrated with 50 mM Tris-HCl (pH 8.0), containing 0.2 M NaCl, 1 mM EDTA, and 0.1 mM PMSF. Fractions containing proteolytic activity were pooled and further purified by FPLC on a Mono Q column, which had been previously equilibrated with 20 mM triethanolamine buffer (pH 7.3) containing 1 mM EDTA and 0.1 mM PMSF. The bound proteins were eluted with a linear gradient of NaCl (0 - 1.0 M) in the same buffer.

RESULTS AND DISCUSSION

Partial Amino Acid Sequences of Proteinase B

The N-terminal sequence of proteinase B isolated from the culture medium of *A. niger* var. *macrosporus* was determined (Fig. 1, underline 1). Further, the tryptic fragments were fractionated by HPLC, and the amino acid compositions and sequences of three peptides were analyzed (Fig. 1, underlines 2-4). The homologous sequences to them were found in *A. awamori* pepsin (16) and penicillopepsin (17). The location of these sequences in proteinase B were presumed based upon the homology to penicillopepsin, since the complete amino acid sequence of *A. awamori* pepsin deduced from the nucleotide sequence of the genomic DNA (18) was not available at the time when we had done this work.

cDNA Cloning

The DNA probe for screening of cDNA clone for proteinase B was prepared through specific amplification by PCR using total cDNA from *A. niger* var. *macrosporus* as a template. The primers used for PCR were designed based upon the partial amino acid seqences detremined. The sense primers were designed based on the sequence Gln-Asn-Asn-Asp-Glu-Glu which was homologous to the sequence of the region from residues 10 to 15 of penicillopepsin. All the possible nucleotide sequences (17-mer) were synthesized as two groups of sixteen oligonucleotides. The antisense primers were designed based on the sequence Tyr-Asp-Phe-Gly-Tyr-Ile which was homologous to the region from residues 165 to 170 of penicillopepsin. All the possible sequences (17-mer) were synthesized also as four groups of sixteen oligonucleotides. One of the eight combinations of primers gave the largest amount of PCR products which was approximately 440-bp long as predicted by the homology with penicillopepsin. The screening of the cDNA library by using the amplified DNA as a probe gave five positive clones out of 14,000 clones. One of the clones that contained the longest cDNA was analyzed.

The Deduced Amino Acid Sequence from cDNA of Preproproteinase B

The sequence analysis of the longest clone showed that the inserted cDNA was 1,467-bp long and had an open reading frame of 1,182 bp encoding 394 amino acid residues. The amino acid sequences of the mature form and the putative prosegment deduced from cDNA are shown in Fig. 1 and Fig. 3, respectively. The encoded protein contained all of the amino acid sequences determined at the protein level by Edman degradation (Fig. 1, underlines 1-4). The amino acid sequence deduced from cDNA (Fig. 1) was almost identical with that of *A. awamori* pepsin (18). There are only six amino acid residue differences between them. Proteinase B was also homologous (69% identity) to penicillopepsin (Fig. 1).

Construction of Expression Plasmid pAR-ANB

The region of cDNA encoding the putative proproteinase B was amplified by PCR and inserted into pAR2113 vector, which contains T7 promoter. It was confirmed by DNA sequencing that the cloned expression plasmid contains the correct DNA sequence for proproteinase B.

Purification and Refolding of Recombinant Proproteinase B

Proproteinase B was expressed in *E. coli* under the control of T7 promoter. Expression of >100 mg of polypeptide of proproteinase B per one liter of culture in the form of inclusion bodies was detected by Western blotting using antiserum for proteinase B. The inclusion bodies were isolated and suspended into a buffer containing 8 M urea and 100 mM 2-mercaptoethanol. The dissolved polypeptides were refolded through dialysis to remove urea and 2-mercaptoethanol. After the dialysis, proteinase activity was detected by assay with denatured hemoglobin as a substrate under acidic conditions. The proteinase activity was detected also by activity-staining after nondenaturing PAGE.

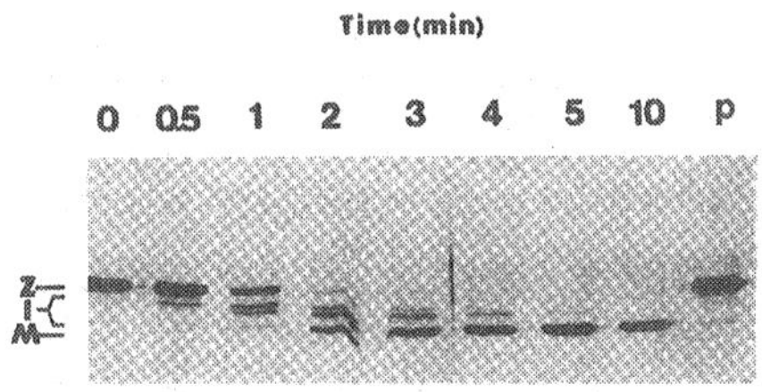

Figure 2. Autoprocessing of recombinant proproteinase B at pH 2.0 and 14°C as analyzed by SDS-PAGE. Z, I and M indicate proform, intermediate forms and mature form, respectively. The lane P shows inhibition of autoprocessing by pepstatin.

The recombinant proproteinase B was purified by column chromatography on DE52, Superose HR and FPLC MonoQ. The purified proproteinase B gave a single band on SDS-PAGE and the specific activity was about 13 units/mg at pH 2.0, which was almost equivalent to the specific activity of the purified proteinase B mature form from *A. niger* var. *macrosporus*. The N-terminal sequence of the recombinant proproteinase B was Ala-Pro-Ala-Pro, which was consistent with the designed sequence except for elimination of the N-terminal methionine residue.

Activation of Proproteinase B

The purified recombinant proproteinase B had potential acid proteinase activity. The pH-dependence of the potential catalytic activity toward denatured hemoglobin was similar to the activity of the native mature form. However, at pH 5.0, the recombinant proproteinase B was inactive while the mature enzyme was active. SDS-PAGE showed that the treatment of the purified proproteinase B under acidic conditions converted the proform to a processed form (Fig. 2) whose migration rate was almost equivalent with that of the wild-type mature form. The N-terminal sequence of the processed form at pH 2.0 was Glu-Ala-Ala-Ser-Lys-Gly-Ser-Ala, which was three residues elongated from the N-terminus of the mature form. This processing was inhibited by pepstatin (Fig. 2). The processed protein was considered to be formed by autocleavage of proproteinase B under acidic conditions as in the cases of other aspartic proteinases such as mammalian and fungal pepsins. The time course of the autoprocessing at pH 2.0 and 14°C is shown in Fig. 2 and two intermediates were observed. The rates of autoprocessing was faster at lower pH in the range of pH 2-4 and the intermediates were also found at pH 3.0 and 4.0 (data not shown). However, no processed form was found at pH 5.0 in 60 min.

Site-directed Mutagenesis of Lys36p

The prosequence of proproteinase B is compared with those of other aspartic proteinases (Fig. 3). There are some more conserved residues, such as those of Pro-Leu and Lys-Tyr, in the prosequences of other aspartic proteinases than in that of proproteinase B. Thus, the prosequence of proproteinase B was less homologous to those of other aspartic proteinases. Only the Lys36p-Phe37p sequence seems to be a counterpart of the Lys-Tyr sequence that is conserved in the proform of almost all the known extracellular or lysosomal aspartic proteinases. The lysine residues of Lys-Tyr and Lys-Phe are supposed to be important for interaction with the active site aspartic residues as indicated in porcine pepsinogen (19). In order to confirm the importance of Lys36p in proproteinase B, the lysine residue was

```
                          1p        10p       20p       30p       40p
A.niger B                 APAPTRKGFTINQIARPANKTRTINLPGMYARSLAKFGGTVPQSVKEAA
R. chinensis pepsin  AVNAAPGEKKISIPLAKNPNYKPSA----------KNAIQKAIAKYNKHKINTSTGGIVPD
M. pusilus rennin          SRRPVSKQSDADDKL---------LALPLTSVNRKYSQTKHGQQAAEKLGGIKAF
Human pepsin A          ECIMYKVPLIRKKSLRRTLSERGLLKDFLKKHNLNPARKYFPQWEAPTL
Human pepsin C           AVVKVPLKKFKSIRETMKEKGLLGEFLRTHKYDPAWKYRFGDL
Calf chymosin           AEITRIPLYKGKSLRKALKEHGLLEDFLQKQQYGISSKYSGF
Human cathepsin D          LVRIPLHKFTSIRRTMSEVGGSVEDLIAK--GPVSKYSQAVPAVTE
Human cathepsin E       QGSLHRVPLRRHPSLKKKLRARSQLSEFWKSHNLDMIQFTES
Rat renin           TFSLPT-GTTFERIPLKKMPSVREILEERGVDMTRLSAEWDVFTKR
```

Figure 3. Comparison of the prosequences of aspartic proteinases. -, Deletion inserted to maximize homology. Residues commom to most enzymes are shadowed.

replaced with asparagine by site-directed mutagenesis to generate a mutant Lys36pAsn. The Lys36pAsn polypeptide was expressed in *E. coli* to form inclusion bodies. The mutant polypeptide was isolated and submitted to refolding process as done for the recombinant wild type. At least under such conditions, however, no proteinase activity was detected by assay with denatured hemoglobin. Further, the denatured Lys36pAsn was partially purified by DE52 chromatography in the presence of 8M urea, and then was allowed to refold at various pHs between 5.0 and 8.0. But no significant activity was detected. The replacement of Lys36p in proproteinase B with asparagine might prevent the correct folding. Lys36p seems to be a counterpart of the conserved lysyl residue in the prosequence of the aspartic proteinase family which is considered to interact with the active site residues. Further investigation is now in progress.

REFERENCES

1. Y. Koaze, H. Goi, K. Ezawa, Y. Yamada and T. Hara, *Agric. Biol. Chem.* 28: 216 (1964). Fungal proteolytic enzymes. Part I. Isolation of two kinds of acid-proteases excreted by *Aspergillus niger* var. *macrosporus*.
2. S. Horiuchi, M. Honjo, M. Yamasaki and Y. Yamada, *Sci. Pap. Coll. Gen. Educ. Univ. Tokyo (Biol. Part)* 19:127 (1969). Studies on acid proteases. I. Homogeneity and molecular weight of acid proteases from *Aspergillus niger* var. *macrosporus* (proctase A and proctase B).
3. S. Horiuchi, M. Yamasaki and Y. Yamada, *Sci. Pap. Coll. Gen. Educ. Univ. Tokyo (Biol. Part)* 19:141 (1969). Studies on acid proteases. II. Amino acid composition of acid proteases from *Aspergillus niger* var. *macrosporus* ("proctase A and B").
4. W.-J. Chang, S. Horiuchi, K. Takahashi, M. Yamasaki and Y. Yamada, *J. Biochem. (Tokyo)* 80:975 (1976). The structure and function of acid proteases. VI. Effects of acid protease-specific inhibitors on the acid proteases from *Aspergillus niger* var. *macrosporus*.
5. K. Takahashi, H. Inoue, K. Sakai, T. Kohama, S. Kitahara, K. Takishima, M. Tanji, S.B.P. Athauda, T. Takahashi, M. Akanuma, G. Mamiya and M. Yamasaki, *J. Bil. Chem.* 266:19480 (1991). The primary structure of *Aspergillus niger* acid proteinase A.
6. H. Inoue, T. Kimura, O. Makabe and K. Takahashi, *J. Biol. Chem.* 266:19484 (1991). The gene and deduced protein sequences of the zymogen of *Aspergillus niger* acid proteinase A.
7. K. Takahashi, M. Tanokura, H. Inoue, M. Kojima, Y. Muto, M. Yamasaki, O. Makabe, T. Kimura, T. Takizawa, T. Hamaya, E. Suzuki and H. Miyano, *Structure and Function of Aspartic Proteinases. Genetics, Structure, and Mechanism*, Plenum Press, N.Y., 203 (1991). Structure and function of a pepstatin-insensitive acid proteinase from *Aspergillus niger* var. *macrosporus*.
8. K. Iio and M. Yamasaki, *Biochim. Biophys. Acta* 429:912 (1976). Specificity of acid proteinase A from *Aspergillus niger* var. *macrosporus* towards B-chain of performic acid oxidized bovine insulin.
9. N. Tonouchi, H. Shoun, T. Uozumi and T. Beppu, *Nucl. Acids Res.* 14:7557 (1986). Cloning and sequencing of a gene for Mucor rennin, an aspartic protease from *Mucor pusillus*.
10. H. Horiuchi, K. Yanai, T. Okazaki, M. Takagi and K. Yano, *J. Bacteriol.* 170:272 (1988). Isolation and sequencing of a genomic clone encoding aspartic proteinase of *Rhizopus niveus*.

11. R. Hiramatsu, J. Aikawa, S. Horiuchi and T. Beppu, *J. Biol. Chem.* 264:16862 (1989). Secretion by yeast of the zymogen form of Mucor rennin, an aspartic proteinase of Mucor pusillus, and its conversion to the mature form.
12. Z. Chen, G. Koelsch, H. Han, X.-J. Wang, X. Lin, J.A. Hartsuck and J. Tang, *J. Biol. Chem.* 266:11718 (1991). Recombinant rhizopuspepsinogen. Expression, purification, and activation properties of recombinant rhizopuspepsinogens.
13. R.K. Saiki, S. Scharf, F. Faloona, K.B. Mullis, G.T. Horn, H.A. Erlich and N. Arnheim, *Science* 230:1350 (1985). Enzymatic amplification of β-globin genomic sequences and restriction site analysis for diagnosis of sickle cell anemia.
14. J. Sambrook, E.F. Fritsh and T. Maniatis, *Molecular Cloning: A Laboratory Mannual, 2nd ed.* Cold Spring Harbor Laboratory Press, N.Y. (1989).
15. F.W. Studier, A.M. Rosenberg, J.J. Dunn and J.W. Dubendorff, *Methods in Enzymol.* 185:60 (1990). Use of T7 RNA polymerase to direct expression of cloned genes.
16. V.M. Stepanov, *Aspartic Proteinases and Their Inhibitors*. Ed. by V. Kostka, Walter de Gruyter, N.Y., p. 27 (1985). Fungal aspartic proteinases.
17. I.-N. Hsu, L.T.J. Delbaere and M.N.G. James, *Nature (London)* 266:140 (1977). Penicillopepsin from Penicillium janthiellum crystal structure at 2.8 Å and sequence homology with porcine pepsin.
18. R.M. Berka, M. Ward, L.J. Wilson, K.J. Hayenga, K.H. Kodma, L.P. Carlomagno and S.A. Thompson, *Gene* 86:153 (1990). Molecular cloning and deletion of the gene encoding aspergillopepsin A from *Aspergillus awamori*.
19. A.R. Sielecki, M. Fujinaga, R.J. Read and M.N.G. James, *J. Mol. Biol.* 219:671 (1991). Refined structure of porcine pepsinogen at 1.8 Å resolution.

EXPRESSION AND SECRETION OF RECOMBINANT ASPARTIC PROTEINASES BY *BACILLUS BREVIS*

Takaharu Hayashi,[1] Hideshi Inoue,[1] Masashi Kato,[2] Shigezo Udaka,[2] and Kenji Takahashi[1]

[1]Department of Biophysics and Biochemistry
Faculty of Science
The University of Tokyo
Hongo, Bunkyo-ku, Tokyo 113, Japan
[2]Department of Applied Biological Sciences
Faculty of Agriculture
Nagoya University
Chikusa-ku, Nagoya 464, Japan

INTRODUCTION

Aspartic proteinases are synthesized as precursors, and their activation takes place upon acidification releasing the NH_2-terminal prosegments. For examples, mammalian gastric aspartic proteinases such as porcine pepsin A and bovine chymosin were shown to be secreted as proforms, and converted into active forms autocatalytically. On the other hand, extracellular aspartic proteinases from fungi such as *Mucor miehei* (1), *Rhizopus chinensis* (2) and *Aspergillus niger* var. *macrosporus* (Inoue *et al.* in preparation) are also suggested to be synthesized as preproforms by analysis of their gene structures, but no evidence for secretion as proforms has been obtained.

To study the mechanism of activation of aspartic proteinase precursors and the relationship between the structure and function of their prosegments, a mass production system of the precursors is essential. *Bacillus brevis* is known to secrete a large amount of proteins but very little protease into the medium (3). Under the optimal conditions, the amount of proteins that accumulates in the medium reaches about 30 g per liter. This bacterium has been shown to produce several foreign proteins including human epidermal growth factor (4) extracellularly in large quantities when the expression vectors developed for mass production are used. By using this system, we have tried to express some acid proteinase precursors, *i.e.*, human pepsinogen A, human procathepsin E, and *Aspergillus niger* var. *macrosporus* proproteinases A and B. In this paper, we report mainly the results of expression of putative proproteinase B from *A. niger* var. *macrosporus* by *B. brevis* HPD 31.

Aspartic Proteinases: Structure, Function, Biology, and Biomedical Implications
Edited by Kenji Takahashi, Plenum Press, New York, 1995

A)

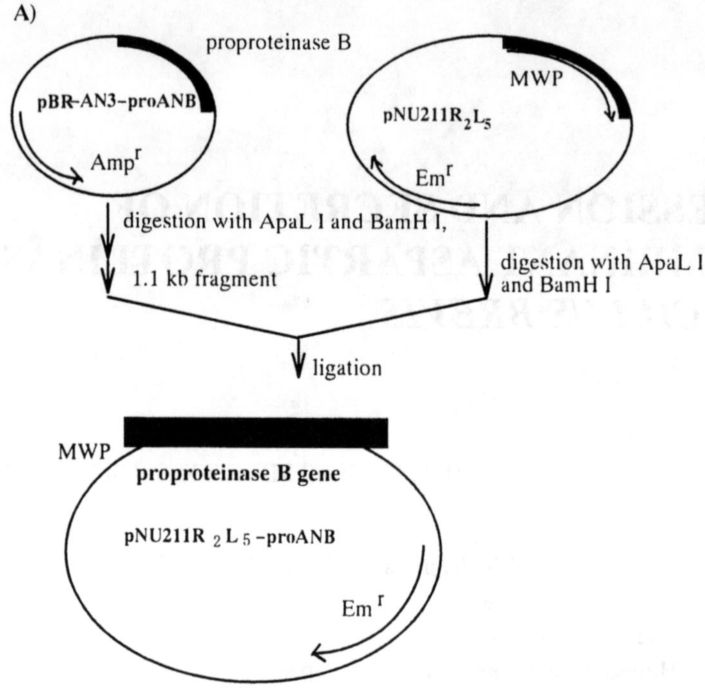

B)

MWP signal peptide

ATGAAAAAG**CGAAGG**GTCGTTAACAGTGTATTG**CTTCTGCTACTA**GCTAGTGCA
MetLysLys**ArgArg**ValValAsnSerValLeu**LeuLeuLeuLeu**AlaSerAla

→ proproteinase B gene
CTCGCACTTACTGTTGCTCCCATGGCTTTTGCG | GCGCCGGCTCCT
LeuAlaLeuTyrValAlaProMetAlaPheAla AlaProAlaPro

↑ cleavage site

Figure 1. A) Construction of the proproteinase B expression vector. Proproteinase B gene was inserted downstream the MWP signal peptide. Amp^r, β-lactamase gene; Em^r, *ermC* gene (erythromycin resistance gene); MWP, middle wall protein. B) The nucleotide and amino acid sequences of the junction region of the fugion gene. The peptide sequences inserted into the native MWP signal peptide and the corresponding nucleotide sequences are shown in bold-face letters.The cleavage site of the MWP signal peptide is shown by a vertical arrow below the amino acid sequences.

MATERIALS AND METHODS

Bacterial Strains, Plasmids, and Media

B. brevis HPD 31 and 47-5Q were grown in T2 medium as described (5). T2M medium contained 0.01% $MgSO_4·7H_2O$, 0.001% $FeSO_4·7H_2O$, 0.001% $MnSO_4·4H_2O$, and 0.0001% $ZnSO_4·7H_2O$ in T2 medium. TF2 medium contained 0.5% polypepton, 0.2% yeast extract, 0.5% glucose, and 0.001% $MnCl_2$, and its pH was adjusted to 6.8 with NaOH. 5'PY medium used for the expression of proproteinase B by *B. brevis* HPD 31 contained 4% polypepton P1, 0.5% yeast extract, 2% glucose, 0.01% $MgSO_4·7H_2O$, 0.001% $FeSO_4·7H_2O$,

0.001% $MnSO_4{\cdot}4H_2O$, and 0.0001% $ZnSO_4{\cdot}7H_2O$ in 0.1 M HEPES (pH 7.5), and after cultivating for 24 hr, glucose was added to 5%. pBR-AN3 was described previously (6). Expression and secretion vector for *B. brevis* used was pNU211R_2L_5, which was constructed from pNU211 (7) that had been developed to express a foreign gene utilizing a promoter of middle wall protein (MWP) gene, having insertions of Arg-Arg and Leu-Leu-Leu-Leu peptides in the signal sequence of MWP for improving the membrane transfer ability (Fig. 1).

Construction of Expression Plasmid and Screening of Transformants

The region encoding the putative proform in the cDNA for *A. niger* var. *macrosporus* proteinase B (Inoue *et al.* in preparation) was amplified by polymerase chain reaction (PCR) by using a sense and an antisense primers which include additive *Nco*I and *Bam*HI sites, respectively. Oligonucleotide primers were synthesized by using a DNA sequencer model 391A (Applied Biosystems). The sequences of the primers were as follows; sense primer, 5'-AAACCATGGCTTTTGCGGCGCCGGCTCCTAC-3'; antisense primer, 5'-ATATGGATCCTTACTAAGCCTGAGCGGCGAATCC-3'. A 1.1 kb PCR product was digested with *Nco*I and *Bam*HI, and ligated with the *NcoI/Bam*H I sites of pBR-AN3. Then the 1.1 kb *Apa*LI/*Bam*HI fragment of the resultant plasmid was subcloned into the *Apa*LI/*Bam*HI sites of pNU211R_2L_5 to obtain pNU211R_2L_5-proANB (Fig. 1). *B. brevis* HPD 31 was transformed with pNU211R_2L_5-proANB by the method of Takagi *et al.* (8). Transformants generating recombinant proproteinase B were screened by colony immunoassay as described (9). The clone used for the further study was confirmed by DNA sequencing to have the correct sequence.

Assay of Aspartic Proteinase Activity

After the cultivation of the transformant in 5'PY medium containing 10 μg/ml erythromycin, cells were removed by centrifugation and the potential activity of acid proteinase in the culture supernatant toward acid-denatured hemoglobin was determined by the method of Anson *et al.* (10). One unit of activity was defined as the amount of enzyme that gives an increase of absorbance at 280 nm by 1.0 per minute.

Activation of Proproteinase B to the Mature Form

Activation of proproteinase B was carried out by acidification to pH 2.2 with 2 N HCl and incubation for 10 min at 37°C. Acid-denatured proteins were removed by centrifugation.

Isolation of Recombinant Proproteinase B for NH_2-Terminal Amino Acid Sequence Determination

Ammonium sulfate was added to the culture supernatant to 60% saturation. After removing the precipitate by centrifugation at 10,000 × *g* for 20 min, ammonium sulfate was added to the supernatant to 80% saturation. The precipitate was collected by centrifugation at 10,000 × *g* for 25 min and dissolved in a small volume of 50 mM Tris/HCl (pH 7.5) containing 1 μM pepstatin. Then this solution was acidified to pH 2.2 and incubated at 37°C for 10 min, and acid-denatured proteins were removed by centrifugation. The supernatant was concentrated by ethanol precipitation and applied to SDS-PAGE. The bands were blotted onto a PVDF transfer membrane and the band to be sequenced was cut off.

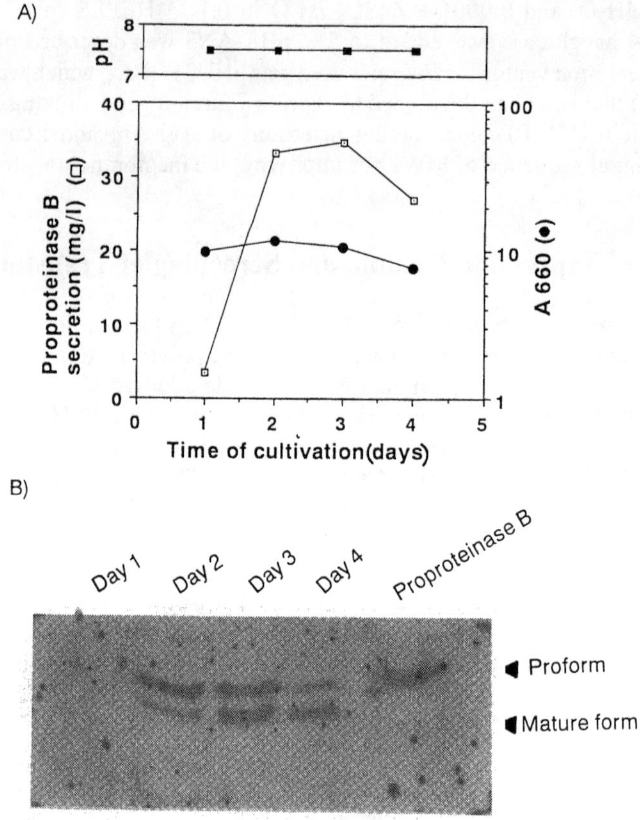

Figure 2. Time course of proproteinase B production. A) The amount of proproteinase B secreted into the medium estimated from its potential acid proteinase activity, cell growth (absorbance at 660 nm) and pH of the medium are shown. *B. brevis* carrying pNU211R_2L_5-proANB was grown in 5'PY medium containing 10 μM erythromycin at 30°C. B) Immunoblot analysis of the culture supernatant. 20 μl of the culture supernatant was applied to SDS-PAGE, blotted onto a PVDF transfer membrane, and detected with the anti-proteinase B antiserum. In the rightmost lane, proproteinase B prepared in the *E. coli* expression system was applied.

Isolation of the Activated Form for NH_2-Terminal Amino Acid Sequence Determination

50 ml of the culture supernatant (1 mg of proproteinase B) was acidified to pH 2.2 with 2 N HCl and incubated at 37°C for 10 min to activate proproteinase B. After removal of acid-denatured proteins by centrifugation at 42,000 rpm (Beckman, 50.2 Ti rotor) for 90 min, 1/30 volume of 1 M sodium acetate (pH 3.5) and 1/10 volume of 5M NaCl were added to the supernatant and the mixture was immediately applied to a pepstatin-agarose column (1.0 × 1.0 cm) (11) equilibrated with 30 mM sodium acetate (pH 3.5) containing 0.5M NaCl. Proteinase B was eluted with 2 ml of 0.1 M Tris/HCl (pH 7.5) containing 6 M guanidine hydrochloride. The eluted fraction was concentrated by ethanol precipitation and applied to SDS-PAGE. The bands were blotted onto a PVDF transfer membrane and the band to be sequenced was cut off.

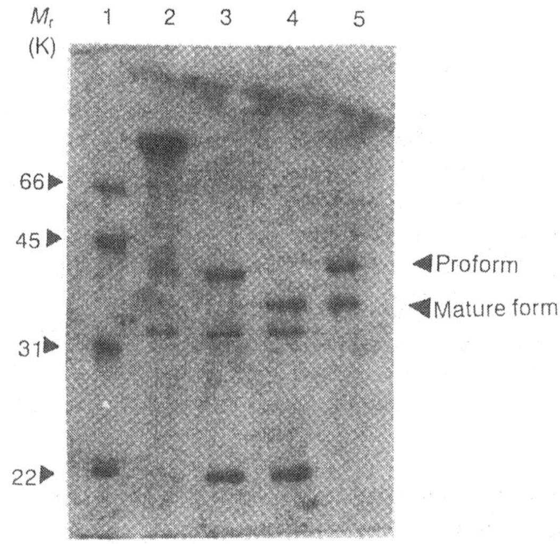

Figure 3. Separation of proproteinase B from the mature form by ammonium sulfate precipitation. The culture supernatant was precipitated between the following ranges of ammonium sulfate saturation. Lane 1, 40-50%; lane 2, 50-60%; lane 3, 60-70%; lane 4, 70-80%.

Determination of NH_2-Terminal Amino Acid Sequence

The NH_2-terminal amino acid sequence was determined by using a protein sequencer model 477A/120A (Applied Biosystems).

Other Procedures

DNA sequencing was performed by the dideoxynucleotide method using a DNA sequencer model 370 (Applied Biosystems). SDS-PAGE was performed by the method of Laemmli (12). Standard DNA manipulation was carried out as described elsewhere (13).

RESULTS AND DISCUSSION

Accumulation of Recombinant Proproteinase B in the Culture Medium

The time course of proproteinase B accumulation in the culture medium of the transformant is shown in Fig. 2. After cultivation for 2 days at 30°C, the recombinant proproteinase B was found to accumulate extracellularly in a soluble form by immunoblotting, and potential acid proteinase activity toward acid-denatured hemoglobin was also detected. More than 30 mg per liter of culture of potentially active proproteinase B was estimated to accumulate on the basis of the specific activity of authentic proteinase B. Immunoblot analysis showed that two protein bands reacted with the anti-proteinase B antiserum. The molecular sizes of the two proteins appeared to correspond to those of the proform and the mature form. The smaller protein might have been produced by proteolysis of the proform by other proteinase(s) during a long time incubation at 30°C, although its NH_2-terminal amino acid sequence was not analyzed. Intensity of the two bands was not altered by addition of 1 μM pepstatin to the culture medium. The possiblity of autoprocess-

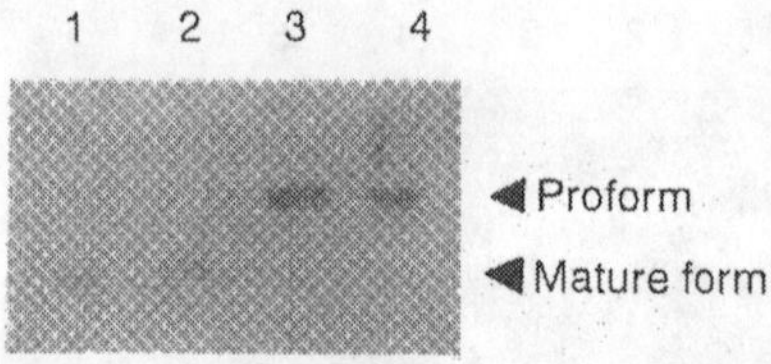

Figure 4. Conversion of proproteinase B to the mature form by acidification and its inhibition by pepstatin. Partially purified proproteinase B, precipitated between 60 and 80% ammonium sulfate saturation, was acidified to pH 2.2 by addition of 2 N HCl with or without pepstatin, and then applied to SDS-PAGE followed by staining with Coomassie brilliant blue G-250. Lane 1, molecular weight standard (bovine serum albumin, ovalbumin, carbonic anhydrase, and soybean trypsin inhibitor); lane 2, partially purified proproteinase B by ammonium sulfate precipitation; lanes 3 and 4, partially purified proproteinase B acidified with and without 1 μM pepstatin, respectively; lane 5, authentic proproteinase B (upper band) and proteinase B (lower band).

ing, however, cannot be ruled out, because the affinity of pepstatin for proproteinase B at neutral pH seemed too low to inhibit the autoprocessing.

Separation of Proproteinase B from Mature Proteinase B

To investigate the properties and the activation mechanism of proproteinase B, separation of proproteinase B from the mature form was essential. As shown in Fig. 3, this separation was performed easily and very efficiently by ammonium sulfate precipitation. Up to 60% saturation of ammonium sulfate, only the mature form was precipitated. The proform was precipitated above 60% saturation.

NH_2-Terminal Sequence Analysis of Expressed Proproteinase

After the ammonium sulfate fractionation, proproteinase B was isolated as described in "Materials and Methods". The NH_2-terminal sequence was determined to be Ala-Pro-Ala-Pro-Thr. This result indicates that the signal peptide of the recombinant fusion protein was cleaved at the expected site by signal peptidase in *B. brevis* (Fig. 1B).

Conversion of Proproteinase B into the Mature Form by Acidification

Figure 4 shows the processing of proproteinase B under acidic conditions and its inhibition by pepstatin. Partially purified proproteinase B, separated from the mature form by ammonium sulfate precipitation, was acidified to pH 2.2 and applied to SDS-PAGE. In the absence of pepstatin, proproteinase B was converted into the mature form and this conversion was inhibited by 1 μM pepstatin. These results are consistent with those obtained with recombinant proproteinase B prepared by an *E. coli* expression system (14).

NH_2-Terminal Sequence Analysis of Activated Proproteinase B

After activation by acidification of the culture supernatant to pH 2.2, the processed proteinase B was isolated as described in "Materials and Methods". The NH_2-terminal sequence was determined to be Glu-Ala-Ala-Ser-Lys which was extended by three residues in comparison with the mature form isolated from the culture medium of *A. niger* var. *macrosporus* but was in agreement with that of the activated proteinase B prepared by an *E. coli* expression system (14). The reason for the difference in the cleavage sites between recombinant and native proteinase B is not clear. This difference might be attributed to proteolysis of the native proform by other proteinase(s) of *A. niger* var. *macrosporus* during intracellular transport and/or after secretion into medium, or to change in the cleavage site at different pH. The cleavage site of recombinant proproteinase B at pH 2.2 was not altered by the addition of proteinase A, one of the major extracellular proteases of *A. niger* var. *macrosporus* (Hayashi *et al.*, unpublished data).

Expression of Other Aspartic Proteinases by *B. brevis*

Expression of a few other acid proteinases was also tried. Human pepsinogen A was expressed by *B. brevis* 47-5Q in T2M medium. After cultivating for 6 days at 30°C, the recombinant human pepsinogen A was found to accumulate extracellularly by immunoblotting, and potential acid proteinase activity was also detected by activity staining assay (data not shown). The amount of the zymogen secreted was lower than that in the case of proproteinase B and was about 1 mg per liter of culture. In a previous study, swine pepsinogen was expressed by *B. brevis* HPD 31 to an extent of about 11 mg per liter of culture (15). In this case, swine pepsinogen was expressed in a form of a fusion protein with a part of MWP. Expression of a larger quantities of human pepsinogen A may be possible by improving the NH_2-terminal sequence. Expression of proproteinase A from *A. niger* var. *macrosporus* by *B. brevis* HPD 31 was not detected by assay with acid-denatured hemoglobin or by immunoblotting partly because of its low sensitivity to both of these analyses. Neither was the expression of human procathepsin E detected.

The production system described here is expected to provide a sufficient amount of pro-proteinase B for site-directed mutagenesis studies, which will be useful to elucidate the structure and function relationship of fungal as well as other aspartic proteinases.

ACKNOWLEDGMENTS

We wish to thank Mrs. Yasuko Sakurai for her assistance in the NH_2-terminal amino acid sequencing.

REFERENCES

1. E. Boel, A.-M. Bech, K. Randrup, B. Draeger, N. P. Fül, and B. Foltmann, Primary structure of a precursor to the aspartic proteinase from *Rhizomucor miehei* shows that the enzyme is synthesized as a zymigen. *Proteins* 1, 363-369 (1986).
2. R. Delaney, R. N. S. Wong, G. Meng, N. Wu, and J. Tang, Amino acid sequence of *Rhizopus* pepsin isoform pI5. *J. Biol. Chem.* 262, 1461-1467 (1987).
3. S. Miyashiro, H. Enei, Y. Hirose, and S. Udaka, Effect of glycine and L-isoleucine on protein production by *Bacillus brevis* no. 47. *Agric. Biol. Chem.* 44, 105-112 (1980).
4. H. Yamagata, K. Nakahara, Y. Suzuki, A. Kakinuma, N. Tsukagoshi, and S. Udaka, Use of *Bacillus brevis* for efficient synthesis and secrtetion of human epidermal growth factor. *Proc. Natl. Acad. Sci. U.S.A.* 86, 3589-3593 (1989).
5. S. Udaka, Extracellular production of protreins by microorganisms. Part I. Secreening for protein-producing bacteria. *Agric. Biol. Chem.* 40, 523-528 (1976).
6. S. Udaka, and H. Yamagata, High-level secretion of heterologous proteins by *Bacillus brevis*. *Methods in Enzymol.* 217, 23-33 (1993).
7. H. Tamura, K. Tameishi, H. Yamagata, S. Udaka, T. Kobayashi, M. Tomita, and H. Ikezawa, Mass production of sphyngomyelinase of *Bacillus cereus* by a protein-hyperproducing strain, *Bacillus brevis* 47, and its purification. *J. Biochem.* 112, 488-491 (1992).

8. H. Takagi, S. Kagiyama, K. Kadowaki, N. Tsukagoshi, and S. Udaka, Screening and characterization of protein-hyperproducing bacteria without detectable exoprotease activity. *Agric. Biol. Chem.* 53, 3099-3100 (1989).
9. V. Nagarajan, Synthesis for secretion of heterologous proteins in *Bacillus subtilis*. *Methods in Enzymol. 185, 214-223 (1990).*
10. M. L. Anson, The estimation of pepsin, trypsin, papain, and cathepsin with hemoglobin. *J. Gen. Physiol.* 22, 79-89 (1939).
11. J. Pohl, M. Zoaral, J. Jr. Jindra, and V. Kostka, Purification of pepsins and cathepsin D by affinity chromatography on Sepharose 4B with an immobilized synthetic inhibitor. *Anal. Biochem.* 139, 265-271 (1984).
12. U. K. Laemmli, Cleavage of structural proteins during the assembly of the head of bacteriophage T4. *Nature* 227, 680-685 (1970).
13. J. Sambrook, E. F. Fritsch, and T. Maniatis, *Molecular Cloning: A Laboratory Manual 2nd ed*, Cold Spring Harbor Laboratory Press, NY (1989).
14. H. Inoue, J. F. Lu, S. B. P. Athauda, K.-H. Kong, T. Hayashi, T. Kimura, O. Makabe, and K. Takahashi, in this issue. *Aspergillus niger* var. *macrosporus* proteinase B.
15. M. Takao, T. Morioka, H. Yamagata, N. Tsukagoshi, and S. Udaka, Production of swine pepsinogen by protein-producing *Bacillus brevis* carrying swine pepsinogen cDNA. *Appl. Microbiol. Biotechnol.* 30, 75-80 (1989).

EXPRESSION IN *E.COLI* OF *ASPERGILLUS NIGER* VAR. *MACROSPORUS* PROTEINASE A, A NON-PEPSIN TYPE ACID PROTEINASE

Naofumi Kagami,[1] Hideshi Inoue,[1] Takao Kimura,[2] Osamu Makabe,[2] and Kenji Takahashi[1]

[1]Department of Biophysics and Biochemistry
Faculty of Science
The University of Tokyo
Tokyo 113, Japan
[2]Meiji Seika, Ltd.
Kohoku-ku, Yokohama 222, Japan

In addition to the extensively studied pepsin type aspartic proteinases, there exists a distinct family of acid proteinases that differ widely in both functional and structural aspects. Functionally, they are insensitive to more than one of the specific inhibitors for the pepsin type aspartic proteinases; i.e., pepstatin, 1,2-epoxy-3-(*p*-nitrophenoxy)propane and diazoacetyl-DL-norleucine methyl ester. Structurally, no homology is observed in the primary structures, even in the Asp-Thr/Ser-Gly sequence that is conserved in the active sites of the aspartic proteinases. Because of these fundamental differences, this family is discriminated by the name, "non-pepsin type acid proteinase".

Primary structures have been reported for four non-pepsin type acid proteinases; i.e., *Scytalidium lignicolum* proteinase B (SLB) (1), *Pseudomonas* sp. No. 101 carboxyl proteinase (2), thermopsin from *Sulfolobus acidocaldarius* (3) and *Aspergillus niger* proteinase A (ANA) (4), which was used in this study. SLB and ANA share 50% amino acid identity, while the other two enzymes have no homology to each other or to any other known proteins. As for ANA, the mature enzyme consists of two polypeptide chains (a 39-residue light chain and a 173-residue heavy chain). However, analysis of the cDNA indicates that it is initially translated as a single-chain preproform (282 residues) (5).

In contrast to the detailed information available on the structure and function relationships of the pepsin type aspartic proteinases, that on the non-pepsin types is severely limited. The insensitivity to specific inhibitors for the aspartic proteinases and the lack of homology suggest that the active site structures of the non-pepsin types differ largely from those of the pepsin-type aspartic proteinases. Thus, studies in this direction could potentially have important implications, not only on the aspartic proteinases, but also on the mechanisms of proteinase function in general. Previously, chemical modification experiments on SLB suggested that the active site residues are a glutamic acid and an aspartic acid (6). However, these residues are not conserved in ANA despite of the 50% identity among these two enzymes, and the residues at the corresponding positions in the latter are a glutamine and a lysine, respectively (4). This strongly argues against the above proposal, and therefore the identity of the catalytic residues is still very uncertain. In this study, we aim to obtain more definite results using the highly specific method of site-directed mutagenesis. As a first step, an expression and purification system for ANA was constructed with *E.coli* as the host. About 20 mg of the active enzyme was obtained after purification, and comparisons of several characteristics indicate that the recombinant protein is sufficiently similar to the natural one. We are now preparing various mutant enzymes to search for the active sites, and by combining this approach with the X-ray crystallographic studies that are also underway in our laboratory, we hope to reach a detailed description of the active site structure.

EXPERIMENTAL PROCEDURES

Hosts and Plasmids

The expression system used consisted of a pAR2113 expression vector containing a T7 promoter and an expression host *E.coli* strain: BL21(DE3) (7). The cDNA for ANA, which had previously been cloned in our laboratory (5), was modified for expression of the putative proform by PCR and inserted into pAR2113 to form pAR-ANA (Fig. 1). The final construction was such that the translated product was the putative proform of ANA with an additional methionine at the N-terminal corresponding to the start codon. pAR-ANA was introduced into the host BL21(DE3) and glycerol stocks were prepared.

Expression and Isolation of the Inclusion Bodies

Expression was performed by the following method. As a preculture, 100 μl of a glycerol stock was first seeded into 10ml of M9ZB broth (0.1% NH_4Cl, 0.3% KH_2PO_4, 1% NZ amine, 0.5% Bacto yeast extract, 0.5% NaCl, 0.4% glucose, 1 mM $MgSO_4$), containing 200 μg/ml ampicillin. This was incubated at 37 °C for a few hours until the culture became slightly turbid. At this point the culture was transferred to 1 l of M9ZB broth, containing 50 μg/ml of ampicillin, and grown at 37 °C in shaking flasks. When the absorbance at 600 nm reached 0.6~1.0, expression of the ANA gene was induced by the addition of 1 mM isopropyl-β-D-thiogalactopyranoside into the culture medium. The culture was further incubated at 37 °C for 150 min.

The cells were harvested by centrifugation and suspended in 50 ml of a lysis buffer (50 mM Tris-HCl pH 8.0, 100 mM NaCl, 1mM EDTA, 1mM PMSF). This was frozen in liquid nitrogen and kept at -30 °C overnight. The cells were thawed and lysozyme was added to a final concentration of 1 mg/ml. This mixture was left on ice for 30 min with occasional gentle mixing. Then the solution was subjected to sonication (100 W, 20 sec x 4 cycles) to completely lyse the cells and to lower the viscosity of the mixture.

The inclusion bodies were extracted from the cell lysate by centrifugation at 12,000 rpm for 10 min. After removing the supernatant, the pellet was dissolved in 50 ml of a denaturation buffer (8 M urea, 100 mM ß–mercaptoethanol, 50 mM sodium phosphate pH 6.0, 50 mM NaCl, 1 mM EDTA). The remaining insoluble materials were removed by centrifugation at 28,000 rpm for 30 min. The insoluble materials did not contain pro-ANA as judged by SDS-PAGE (data not shown).

Partial purification of the solubilized inclusion bodies was performed on the anion-exchanger DE-52. After applying the sample and washing with about 100 ml of the denaturation buffer, pro-ANA was eluted with 50 ml of the denaturation buffer, containing an additional 350 mM of NaCl.

```
1                             10                            20
Met-Lys-Phe-Ser-Thr-Ile-Leu-Thr-Gly-Ser-Leu-Phe-Ala-Thr-Ala-Ala-Leu-Ala-Ala-Pro-
                        signal peptide
21                            30                            40
Leu-Thr-Glu-Lys-Arg-Arg-Ala-Arg-Lys-Glu-Ala-Arg-Ala-Ala-Gly-Lys-Arg-His-Ser-Asn-
                        pro peptide
41                            50                            60
Pro-Pro-Tyr-Ile-Pro-Gly-Ser-Asp-Lys-Glu-Ile-Leu-Lys-Leu-Asn-Gly-Thr-Thr-Asn-Glu-

61                            70                            80
Glu-Tyr-Ser-Ser-Asn-Trp-Ala-Gly-Ala-Val-Leu-Ile-Gly-Asp-Gly-Tyr-Thr-Lys-Val-Thr-
                        light chain
81                            90                            100
Gly-Glu-Phe-Thr-Val-Pro-Ser-Val-Ser-Ala-Gly-Ser-Ser-Gly-Ser-Ser-Gly-Tyr-Gly-Gly-

101                           110                           120
Gly-Tyr-Gly-Tyr-Trp-Lys-Asn-Lys-Arg-Gln-Ser-Glu-Glu-Tyr-Cys-Ala-Ser-Ala-Trp-Val-
 intervening sequence
120                           130                           140
Gly-Ile-Asp-Gly-Asp-Thr-Cys-Glu-Thr-Ala-Ile-Leu-Gln-Thr-Gly-Val-Asp-Phe-Cys-Tyr-

141                           150                           160
Glu-Asp-Gly-Gln-Thr-Ser-Tyr-Asp-Ala-Trp-Tyr-Glu-Trp-Tyr-Pro-Asp-Tyr-Ala-Tyr-Asp-
                        heavy chain
161                           170                           180
Phe-Ser-Asp-Ile-Thr-Ile-Ser-Glu-Gly-Asp-Ser-Ile-Lys-Val-Thr-Val-Glu-Ala-Thr-Ser-

181                           190                           200
Lys-Ser-Ser-Gly-Ser-Ala-Thr-Val-Glu-Asn-Leu-Thr-Thr-Gly-Gln-Ser-Val-Thr-His-Thr-

201                           210                           220
Phe-Ser-Gly-Asn-Val-Glu-Gly-Asp-Leu-Cys-Glu-The-Asn-Ala-Glu-Trp-Ile-Val-Glu-Asp-

221                           230                           240
Phe-Glu-Ser-Gly-Asp-Ser-Leu-Val-Ala-Phe-Ala-Asp-Phe-Gly-Ser-Val-The-Phe-Thr-Asn-

241                           250                           260
Ala-Glu-Ala-The-Ser-Gly-Gly-Ser-The-Val-Gly-Pro-Ser-Asp-Ala-The-Val-Met-Asp-Ile-

261                           270                           280
Glu-Gln-Asp-Gly-Ser-Val-Leu-Thr-Glu-The-Ser-Val-Ser-Gly-Asp-Ser-Val-The-Val-Thr-

281
Tyr-Val
```

Figure 1. The amino acid sequence of the putative precursor of ANA(5). The amino acids that are present in the natural mature enzyme obtained from *Aspergillus niger* are in inverted letters. Those that are predicted to compose the propeptide and the intervening sequence are in white squares. The putative signal sequence is underlined. The proform was thus predicted to be composed of the inverted and squared sequences (propeptide + light chain + intervening sequence + heavy chain).

Folding of ANA

Folding was performed to obtain the functional enzyme. The partially purified sample was diluted into 10 volumes of a refolding buffer (50 mM sodium acetate, pH 5.5) at 4 °C and left overnight.

Processing

During refolding, pro-ANA was autocatalytically cleaved to various extents, resulting in a mixture of fragments of multiple sizes. To fully convert the various intermediates to the mature form, autocatalysis was accelerated by acidification. This was performed by dialyzing the solution against a processing buffer (20 mM sodium citrate, pH 3.5) overnight at 4 °C.

Purification of ANA

The processed sample was dialyzed against buffer A (20 mM sodium citrate, pH 4.5) and applied to a DEAE-Toyopearl column. Elution was performed with a linear gradient from buffer A to buffer B (20 mM sodium citrate, pH 2.5).

RESULTS

Expression

The proform of ANA was successfully translated as confirmed by immunoblotting (data not shown).

Partial Purification

The partial purification procedure in which the urea denatured sample was submitted to DE-52 anion-exchange chromatography, proved to be very efficient, and most contaminants were removed at this step (Fig. 2). This is thought to be due to the highly acidic nature of proteinase A. Since the acidic property is shared with many other acid proteinases, this method may be widely applicable. It has also proved to be efficient for the partial purification of a pepsin type aspartic proteinase also found in the culture filtrate of *Aspergillus niger*, with slight modifications.

Refolding

Several refolding conditions were surveyed to optimize the yield, and the pH of the dilution buffer was found to have a strong effect (Fig. 3). At the optimal pH of about 5.5, potential acid proteinase activity equal to about 20mg of ANA was obtained per liter culture. The refolding efficiency was estimated to be about 30%. However, when the pH was shifted as little as 1 pH unit in either direction, a large decrease in refolding efficiency was observed. The pH of the buffer used in the subsequent dialysis had no effect in the range of pH 3 to 7, and therefore, this was normally set at pH 3.5 to allow simultaneous processing to the mature form.

Under any conditions where the potentially active enzyme was obtained, the proteolysis of pro-ANA also occurred. Since this proteolysis was not observed in mutants without activity, it was thought to proceed autocatalytically (data not shown). Although the activity of ANA is very low under these conditions, the unfolded forms appeared to be particularly easily cleaved.

Purification

After processing at pH 3.5, only one band, which corresponded to the heavy chain of ANA, was visible on SDS-PAGE by Coomassie brilliant blue G-250 staining. The light chain

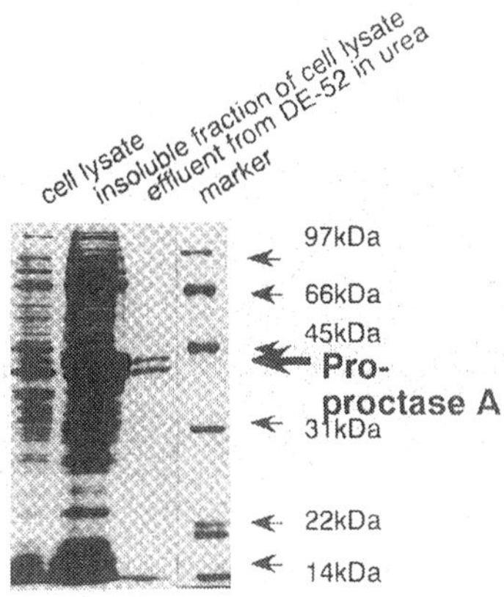

Figure 2. SDS-PAGE of the samples obtained at the steps of extraction and partial purification. The cell lysate (lane 1) was centrifuged and the pellet was dissolved in the denaturation buffer (lane 2). This sample was partially purified by anion-exchange chromatography on DE-52 (lane 3).

was not visible due to its small size. However, this sample was still contaminated with large amounts of misfolded proteins and the cleaved propeptides. They were removed by column chromatography on DEAE-Toyopearl using a pH gradient of 4.5 to 3.0. The elution pattern showed only one peak, which was accompanied by acid proteinase activity and showed only one band on SDS-PAGE. The specific activity was almost identical (~90%) to that of natural ANA purified from *Aspergillus niger* (data not shown). We therefore concluded that mature recombinant ANA had been purified. The yield at this purification step was about 80%.

Processing by Acidification

The refolded sample consisted of the proform and its several autolysis products. When this sample was acidified, it was converted to a form indistinguishable on SDS-PAGE from the natural mature enzyme (Fig. 4). This indicates that, like the proform of the aspartic proteinases, the proform of ANA is also processed to the mature form by autocatalysis under acidic conditions.

We investigated the optimal pH for processing using a sample refolded under conditions where the majority of ANA remained in the proform. Samples were acidified and

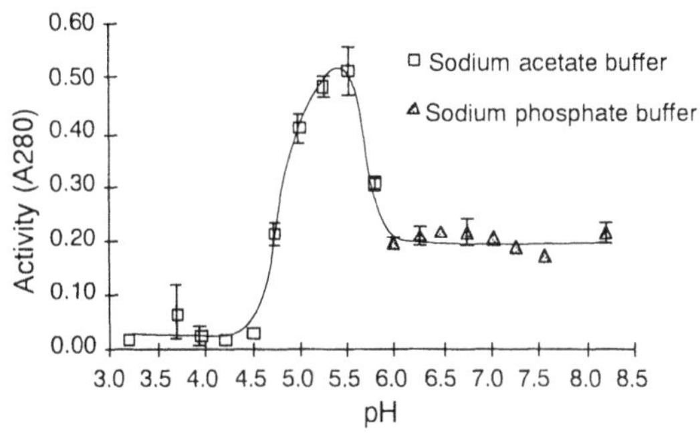

Figure 3. The effect of pH on refolding efficiency at the dilution step. The potential proteolytic activities toward hemoglobin after refolding at various pH were measured and considered to represent the refolding efficiencies.

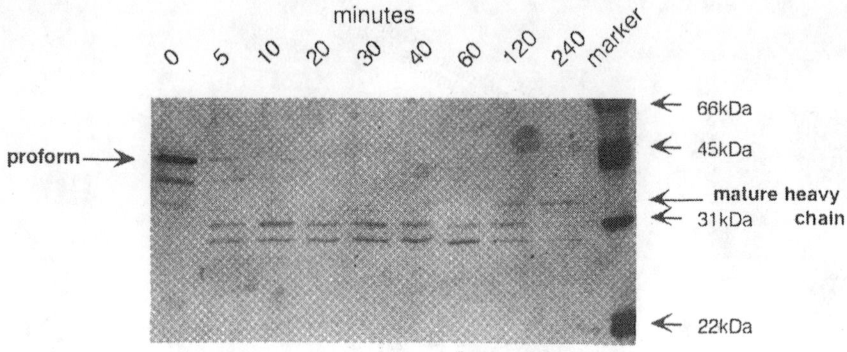

Figure 4. SDS-PAGE showing the time course of autocatalytic processing at pH 3. The sample was acidified to pH 3 by the addition of an appropriate amount of 0.2N HCl. The sample was incubated at 37°C. At the indicated times, small aliquots were withdrawn and the processing reaction was terminated by boiling in SDS-PAGE sample buffer.

incubated at 37°C, then subjected to SDS-PAGE. The optimal pH for processing was in the range of pH 3 to 4. At above pH 6, no processing was observed within the duration of investigation (data not shown).

N-terminal Sequencing

The N-terminal sequences of the light and heavy chains of the recombinant mature enzyme were determined. The light chain of the recombinant mature enzyme was identical to that of natural ANA from *Aspergillus niger*. On the other hand, the heavy chain of the recombinant was two residues longer to the N-terminal direction. The cleavage sites were Thr-Asn-/-Glu-Glu (between residues 59-60 in Fig. 1) and Lys-Asn-/-Lys-Arg (between residues 107-108 in Fig. 1), respectively. These are both compatible with the cleavage specificity of ANA (Takahashi, K., unpublished results).

Temperature Dependency of Activity

The temperature dependencies of activity of recombinant and natural ANA were compared by performing the activity assay at different temperatures. Although the recombinant enzyme appeared to have slightly lower activity at high temperatures, both enzymes displayed maximal activity at about 55 °C, and showed a large decrease in activity at about 60 to 65°C (data not shown).

pH Dependency of Activity

The pH dependencies of activity were compared by performing the assay at various pH, and were shown to be indistinguishable between recombinant and natural ANA (data not shown).

DISCUSSION

Processing of ANA

Analyses of the cDNAs of several non-pepsin type acid proteinases have indicated that they are translated as a preproform (2, 3, 5). However, the proenzymes have never been isolated, and nothing was previously known about their processing. The present study demonstrates that, at least for ANA, the processing can also proceed by autocatalysis. Like the aspartic proteinases, this requires acidic conditions. The processing of ANA generates two new N-termini on the light and heavy chains. That of the light chain was identical between the autocatalytically processed enzyme and the natural one purified from *Aspergillus niger*. On the other hand, that of the heavy chain was different between them; the autocatalytically processed one was elongated by two residues to the N-terminal direction. This suggests that *in vivo*, the N-terminal of the light-chain is generated by autocatalysis, while that of the heavy chain is generated by the action of another proteinase. The unusually long time required for the autocatalytic generation of the heavy chain also supports this view. Since the natural cleavage site is the carboxyl side of two basic residues (Lys-Arg-/-Gln), this enzyme may be a Kex2-like intracellular proteinase. Alternatively, an aspartic proteinase that is present in the culture medium of *Aspergillus niger* may be responsible since it is also known to cleave the carboxyl side of basic residues (Inoue *et al.*, in this issue). It should be noted that SLB is a single-chain enzyme and does not have an intervening sequence as in ANA.

Applicability of the Present Expression System

The objective of this study is to gain decisive information on the structure and function of ANA. Therefore, an expression system must provide a sufficient amount of the active and pure protein for enzymatic and structural studies. Furthermore, the structure and function of the recombinant enzyme must be sufficiently similar to the natural one. The results given above indicate that these requirements have been satisfied in the present expression system. We are currently preparing several site-specific mutants to identify the catalytic residues.

REFERENCES

1. Maita, S., Nagata, S., Matsuda, G., Maruta, S., Oda, K. and Tsuru, D. (1984) Complete amino sequence of *Scytalidium lignicolum* acid protease B, *J. Biochem.*, **95**, 465-475.
2. Oda, K. and Murao, S.(1991) Pepstatin-insensitive Carboxyl Proteinases, *in* "Structure and Function of the Aspartic Proteinases", ed. by B. M. Dunn, Plenum Press, New York, London, 1991, p.185.
3. Lin, X. and Tang, J.(1990) Purification, Characterization, and Gene Cloning of Thermopsin, a Thermostable Acid Proteinase from *Sulfolobus acidocaldarius*, *J. Biol. Chem.*, **265**, 1490-1495.
4. Takahashi, K., Inoue, H., Sakai, K., Kohama, T., Kitahara, S., Takishima, K., Tanji, M., Athauda, S.B.P., Takahashi, T., Akanuma, H., Mamiya, G. and Yamasaki, M. (1991) The Primary Structure of *Aspergillus niger* Acid Proteinase A,*J. Biol. Chem.*, **266**, 19480-19483.
5. Inoue, H., Kimura, T., Makabe, O., Takahashi, K. (1991) The Gene and Deduced Protein Sequences of the Zymogen of *Aspergillus niger* Acid Proteinase A, *J. Biol. Chem.*, **266**, 19484-19489.
6. Murao, S. and Oda, K.(1985) Pepstatin-insensitive Acid Proteinase, *in* "Aspartic Proteinases and Their Inhibitors," ed. by V. Kostka, Walter de Gruyter and Co., Berlin, New York, pp. 379-399.
7. Studier, F.W., Rosenberg, A.H., Dunn, J.J. and Dubendorff, J.W. (1990) Use of T7 RNA Polymerase to Direct Expression of Cloned Genes, *Methods in Enzymol.*, **185**, 60-88.

X-RAY CRYSTALLOGRAPHIC STUDY OF A NON-PEPSIN-TYPE ACID PROTEINASE, *ASPERGILLUS NIGER* PROTEINASE A

Hiroshi Sasaki,[1] Masaru Tanokura,[1] Tomonari Muramatsu,[1] Atsushi Nakagawa,[2] So Iwata,[2] Toru Hamaya,[3] Toshio Takizawa,[3] Toshiaki Kono,[3] and Kenji Takahashi[1]

[1]Department of Biophysics and Biochemistry
Faculty of Science
University of Tokyo
Bunkyo-ku, Tokyo 113, Japan
[2]Photon Factory
National Laboratory for High Energy Physics
Tsukuba, Ibaraki 305, Japan
[3]Bioscience Laboratories
Meiji Seika Kaisha, Ltd.
Sakado, Saitama 350-02, Japan

INTRODUCTION

Acid proteinase A, secreted by the fungus *Aspergillus niger* var. *macrosporus*, is a non-pepsin-type acid proteinase as judged from its distinct differences in various properties from the ordinary pepsin-type aspartic proteinase family [1,2]. It is insensitive to pepstatin, diazoacetyl-DL-norleucine methyl ester (DAN), and 1,2-epoxy-3-(*p*-nitrophenoxy)propane (EPNP) [3], and shows fairly different substrate specificity [4]. The enzyme is composed of two chains, and the relative molecular mass is 22,300, about half the size of the ordinary pepsin-type proteinases. No protein in the GenBank database is homologous in the primary structure with proteinase A except for *Scytalidium lignicolum* proteinase B [5] that shows about 50% identity [2]. Proteinase A has no consensus sequence, Asp-Thr/Ser-Gly, including the catalytic aspartic acid residue, present in the family of pepsin-type proteinases. It, thus, remains to be elucidated which residues participate in the catalysis and how the mechanism operates. Moreover, it is of interest to elucidate how the three-dimensional structure differs from pepsin-type enzymes. Comparison of the crystal structure of proteinase A with those of pepsin-type enzymes will enable us to explain the unusual properties of this enzyme. To approach these queries, we have started the crystallographic study of the enzyme, and have obtained an isomorphous heavy atom derivative of the crystals.

Aspartic Proteinases: Structure, Function, Biology, and Biomedical Implications
Edited by Kenji Takahashi, Plenum Press, New York, 1995

Table 1. Crystal Data of *A. niger* Proteinase A Crystals

	Crystal I	Crystal II	Crystal III
Crystallization conditions			
$[(NH_4)_2SO_4]$/M	1.35-1.45	1.95-2.20	1.30
pH	1.0-3.0	2.5-3.75	2.5-2.75
[DMSO]/%	3-7	9	5
Temp./°C	25, 4	25	4
	2 days	3 months	3 months
Space group	$P2_12_12_1$	$P2_12_12_1$	$P2_12_12_1$
Cell dimensions/Å	a=55.0, b=70.6, c=38.3	a=69.8, b=87.5, c=60.8	a=67.8, b=78.5, c=62.8
Unit cell volume/Å^3	1.49 x 105	3.71 x 105	3.34 x 105
n/asymmetric unit	1	2	2
V_m /Å^3·dalton^{-1}	1.67	2.08	1.87
Density/g·cm^{-3}	1.29	—	—
V_{sol}/%	26	41	34
Resolution/Å	1.5	1.6	1.3

Symbols used are: n, number of molecules; V_m, volume to unit protein mass ratio; V_{sol}, solvent content calculated by assuming the specific volume of the crystals to be 0.74 cm^3/g [12].

MATERIALS AND METHODS

Proteinase A was purified from the crude enzyme mixture obtained from the culture medium of *Aspergillus niger* var. *macrosporus*. For crystallization, the hanging-drop vapor diffusion method was employed at 4°C and 25°C, at various concentrations of ammonium sulfate as a precipitant in the presence of dimethylsulfoxide [6,7,8]. Preparation of heavy atom derivatives of crystals were made by soaking the crystals in heavy-atom solutions.

Preliminary x-ray diffraction experiments were performed using a Huber precession camera with Ni-filtered CuKα radiation of rotating anode x-ray generator (40 kV, 70 mA, focus size 0.2 × 2 mm) and a Rigaku AFC-5 four circle diffractometer with CuKα radiation of sealed tube x-ray generator (40 kV, 30 mA, focus size 0.4 × 8 mm). Intensity data were collected by using a data collection system at the BL6A2 station at the Photon Factory at National Laboratory for High Energy Physics, by combining the Weissenberg camera for macromolecular crystallography, an imaging plate, a Fuji image reader BA100 and a data reduction program WEIS with a synchrotron radiation [9,10]. The wavelength of x-rays used was 1.00 Å and the temperature was 10°C.

Calculation of difference Patterson maps (isomorphous and anomalous differences) was performed on the program package Protein [11]. Wilson plot and calculation of R_F value ($\Sigma| |F_{PH}| - |F_P| |/\Sigma |F_P|$) were performed on the BOSS system (Oxford Univ.).

RESULTS AND DISCUSSION

So far, three types of proteinase A crystals were obtained under different conditions. Although the crystals were rather small (typical type I crystals, 0.05 × 0.05 × 1 mm^3; average type 2 crystals, 0.2 × 0.3 × 0.4 mm^3; and average type 3 crystals, 0.1 × 0.2 × 0.2 mm^3),

Table 2. Search of Heavy Atom Derivatives of Type I Crystals of Proteinase A with the Soaking Method

Compound	Soaking duration (Conc. of reagent)	R_{merge}	No. of independent reflections	R_F
$Yb(OAc)_3$	6 days (Sat.)	5.80	15798	8.52
$Lu(OAc)_3$	6 days (Sat.)	6.49	15203	8.47
WO_3	8 days (Sat.)	6.52	15353	7.39
K_2PtCl_4	6 days (Sat.)	6.52	13566	22.26
K_2HgI_4	4 days (1/50 sat.)	5.51	16253	9.25
$(NH_4)_2Hg(SCN)_4$	4 days (1/10 sat.)	5.38	18194	6.63
$HgI_2{\cdot}2KI$	4 days (1/50 sat.)	5.35	15902	8.83
Hg(OAc)	1 day (Sat.)	6.58	12457	4.58
PCMB	4 days (Sat.)	5.89	16026	7.02

All the heavy metal reagents were dissolved in the solution containing 50mM glycine, 1.45 M ammonium sulfate, and 5% DMSO.

$R_{merge} = \Sigma|I - \langle I \rangle| / \Sigma \langle I \rangle$; $R_F = \Sigma|F_{derivative} - F_{native}| / \Sigma|F_{native}|$

PCMB, p-chloromercuribenzoic acid.

diffraction data were collected up to a high resolution, using the Weissenberg camera for macromolecular crystallography with synchrotron radiation. Their crystallization conditions and crystallographic data are summarized in Table 1. It should be noted that the solvent content (V_{sol}) of proteinase A type I crystals was 26%, among the smallest values for protein crystals [6,8]. For the data of type I crystals collected to 1.3 Å resolution, 16,820 of the total 52,542 observed reflections were independent of one another, and the merging R factor (R_{merge}) was 0.051.

Type I crystals are grown more rapidly under less strict conditions than the other two types of crystals. Heavy atom derivatives have, therefore, been searched for the type I crystals prepared by the soaking method. The details of the soaking conditions are listed in Table 2. Some of the heavy atom reagents made crystals cracked in their saturated solutions. In such a case, the concentration of the heavy atom was lowered (Table 2). Although type I crystals are prepared at an extremely low pH and the solvent content is very low, a K_2PtCl_4 derivative was successfully obtained by the soaking method as judged from isomorphous differences and anomalous differences Patterson maps (Fig 1), and R_F value (Table 2). The search of another heavy atom derivative of type I crystals of the enzyme is in progress to determine the three-dimensional structure of proteinase A with the multiple isomorphous replacement method.

ACKNOWLEDGMENTS

We thank Prof. N. Sakabe (Photon Factory, National Laboratory for High Energy Physics) for his encouragement and interest during the experiments. A part of this work was performed using the Station BL6A2 at the Photon Factory, National Laboratory for High Energy Physics under the approval of the Photon Factory Program Advisory Committee (Proposal Nos. 91-037 and 93-G061). This work was supported in part by Grants-in-Aid for Scientific Research from the Ministry of Education, Science, and Culture of Japan.

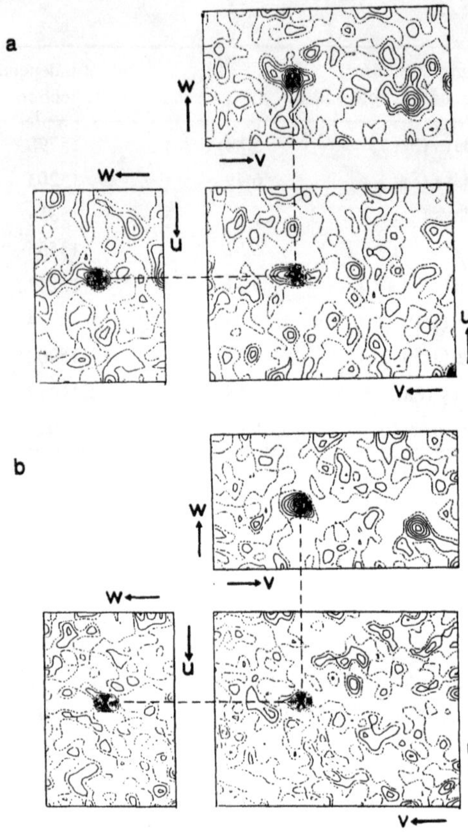

Figure 1. Difference Patterson maps (resolution range of 10.0 to 2.0 Å) for the K_2PtCl_4 derivative of *A. niger* proteinase A: (a) isomorphous differences and (b) anomalous differences.

REFERENCES

1. S. Murao and K, Oda, Pepstatin-Insensitive Acid Proteinases, *in*: Aspartic Proteinases and Their Inhibitors, V. Kostka, ed., pp. 379-399, Walter de Gruyter, Berlin (1985).
2. K. Takahashi, M. Tanokura, H. Inoue, M. Kojima, Y. Muto, M. Yamasaki, O. Makabe, T. Kimura, T. Takizawa, T. Hamaya, E. Suzuki, and H. Miyano, Structure and Function of Pepstatin-Insensitive Acid Proteinase from *Aspergillus niger* var. *macrosporus*. *in*: Structure and Function of the Aspartic Proteinases: Genetics, Structures, and Mechanisms, B. Dunn, ed., pp. 203-211, Plenum Press, New York (1991).
3. W.-J. Chang, S. Horiuchi, K. Takahashi, M. Yamasaki, and Y. Yamada, The Structure and Function of Acid Proteases. VI. Effects of Acid Protease-specific Inhibitors on the Acid Proteases from *Aspergillus niger* var. *macrosporus*, *J. Biochem. (Tokyo)*, **80**: 975-981 (1976).
4. E. Ido, T. Saito, and M. Yamasaki, Substrate Specificity of Acid Proteinase A from *Aspergillus niger* var. *macrosporus*, *Agric. Biol. Chem.* **51**: 2855-2856 (1987).
5. S. Maita, S. Nagata, G. Matsuda, S. Maruta, K. Oda, and D. Tsuru, Complete Amino Acid Sequence of *Scytalidium lignicolum* Acid Protease B, *J. Biochem. (Tokyo)* **95**: 465-475 (1984).
6. M. Tanokura, H., Matsuzaki, S. Iwata, A. Nakagawa, T. Hamaya, T. Takizawa, and K. Takahashi, Crystallization and Preliminary X-Ray Investigation of Proteinase A, a Non-Pepsin-type Acid Proteinase from *Aspergillus niger* var. *macrosporus*, *J. Mol. Biol.* **223**: 373-375 (1992).

7. H. Matsuzaki, S. Iwata, T. Hamaya, T. Takizawa, M. Tanokura, and K. Takahashi, Effect of Dimethylsulfoxide on the Crystallization of *Aspergillus niger* Proteinase A, *Proc. Japan Acad.* **67B**: 209-212 (1991).
8. M. Tanokura, H. Sasaki, T Muramatsu, S. Iwata, T. Hamaya, T. Takizawa, and K. Takahashi, A New Crystal Form of Proteinase A, a Non-Pepsin-type Acid Proteinase from *Aspergillus niger* var. *macrosporus*, *J. Biochem.* **114**: 457-458 (1993).
9. N. Sakabe, X-Ray Diffraction Data Collection System for Modern Protein Crystallography with a Weissenberg Camera and an Imaging Plate Using Synchrotron Radiation, *Nucl. Instr. Meth.* **A303**: 448-463 (1991).
10. T. Higashi, The Processing of Diffraction Data Taken on a Screenless Weissenberg Camera for Macromolecular Crystallography, *J. Appl. Cryst.* **22**: 9-18 (1989).
11. W.Steigemann, Die Entwicklung und Anwendung von Rechenverfahren und Rechenprogrammen zur Strukturanalyse von Proteinen am Beispiel des Trypsin-Trypsininhibitor Komplexes, des Freien Inhibitors und der L-Asparaginase, Ph. D. Thesis, Technical University Munich (1974).
12. B. W. Matthews, Solvent Content of Protein Crystals, *J. Mol. Biol.* **33**: 491-497 (1968).

CONFORMATION ANALYSIS OF NON-PEPSIN-TYPE ACID PROTEINASE A FROM THE FUNGUS *ASPERGILLUS NIGER* BY NMR

Masaki Kojima,[1] Masaru Tanokura,[1] Yutaka Muto,[1] Hiroshi Miyano,[2] Ei-ichiro Suzuki,[2] Toru Hamaya,[3] Toshio Takizawa,[3] Toshiaki Kono,[3] and Kenji Takahashi[1]

[1]Department of Biophysics and Biochemistry
Faculty of Science
University of Tokyo
Tokyo, Japan
[2]Central Research Laboratories
Ajinomoto Co., Inc.
Kawasaki, Japan
[3]Bioscience Laboratories
Meiji Seika Kaisha, Ltd.
Sakado-shi, Saitama, Japan

INTRODUCTION

Proteinase A [EC 3.4.23.6] secreted by the fungus *Aspergillus niger* var. *macrosporus* is a non-pepsin-type acid proteinase.[1] While the enzyme has its optimum at an extremely acidic pH, it is insensitive to pepstatin and also almost insensitive to diazoacetyl-DL-norleucine methyl ester (DAN) and 1,2-epoxy-3-(*p*-nitrophenoxy)propane (EPNP) that are the inhibitors specific and common to pepsin-type aspartic proteinases.[2] Proteinase A shows substrate specificity fairly different from that of pepsin-type aspartic proteinases.[3,4] These results indicate that the enzyme belongs to a group of non-pepsin-type acid proteinases. It consists of two polypeptide chains, a light chain of 39 residues and a heavy chain of 173 residues.[5,6] They are bound non-covalently to each other. The two chains are dissociated above neutral pH with concomitant loss of activity. In order to elucidate the characteristics of the three-dimensional structure of the enzyme, the conformations of the enzyme and its light chain were studied by 2D- and 3D-NMR spectroscopy. There have been known no specific inhibitors for proteinase A, and the catalytic and the substrate-binding sites of the enzyme are not clear. N-acetyl-phenylalanyl-diiodotyrosine (APDT), originally a substrate for pepsin, inhibits weakly the proteinase A activity against peptide substrate. Using APDT as a substrate analog, the enzyme-substrate interaction was also studied by NMR.

MATERIALS AND METHODS

Proteinase A was purified from the crude extract powder prepared from the broth culture medium of *Aspergillus niger* var. *macrosporus* by two-step column chromatography on DEAE-Toyopearl 650S. The light chain was dissociated from the heavy chain at pH 8.0 and purified by Sephacryl S-200 gel filtration. The detailed procedure will be described elsewhere. APDT was purchased from Sigma.

All homonuclear 2D and 3D spectra were recorded with Bruker AM400 and AMX600 spectrometers in the phase-sensitive mode using the TPPI method.[7] 2D TOCSY (total correlation spectroscopy),[8] NOESY (nuclear Overhauser effect spectroscopy), and DQF-COSY (double-quantum filtered correlated spectroscopy)[9] spectra of the light chain were measured at 25 °C at a concentration of 7 mM in dimethyl sulfoxide-d_6 (DMSO-d_6). 2D spectra of intact proteinase A were measured at 45 °C and pH 4.6 at a concentration of 3 mM (in D_2O) or 5 mM (in 90% H_2O/10% D_2O). The solvent water signal was suppressed by selective irradiation during the relaxation delay. To investigate the enzyme-substrate analog interaction, NOESY spectra were measured in 10% DMSO-d_6/90% D_2O. All 2D spectra were recorded with 512 × 2048 data points and spectral width of 7,000 Hz (400 MHz) or 10,000 Hz (600 MHz). Prior to Fourier transformation, the data were zero-filled twice in t_1 time domains and the window function was applied for both dimensions. Homonuclear 3D NOESY-TOCSY spectra[10] were measured at 37 °C and pH 4.5 at a concentration of 10 mM in 90% H_2O/10% D_2O. The experimental data set of 128 × 256 × 512 points was processed on an IRIS Indigo Elan workstation (Silicon Graphics, Inc.) by the software Felix (Biosym Technologies, Inc.).

^{15}N-^{1}H and ^{13}C-^{1}H HMQC spectra[11] of intact proteinase A were recorded with a GE OMEGA 500 spectrometer in the phase sensitive mode as described by States et al.[12] The spectra were measured at 37 °C and pH 4.5 at a concentration of 10 mM.

All calculations were carried out on an IRIS 4D/35 and Indigo Elan workstations (Silicon Graphics, Inc.). NOE peak intensities were measured using the software NMRZ (New Methods Research, Inc.). Distance geometry calculations were performed with the program system EMBOSS[13] (Protein Engineering Research Institute). Restrained molecular dynamics calculations were carried out by the software NMRgraf (Molecular Simulations, Inc.) using a default setting routine.

RESULTS AND DISCUSSION

Structure of the Light Chain of Proteinase A

In order to determine the three-dimensional structure of the light chain alone, 2D NMR spectra were measured in DMSO-d_6 since the chain was well soluble in DMSO, but not in H_2O. In the fingerprint region of the TOCSY spectrum, all 38 cross peaks except for 1 proline residue appeared separately (Figure 1). In addition, NOE peaks between amide protons were observed. These results indicate that the light chain has a regular structure in DMSO. Using interresidue NOE peaks, sequence-specific resonance assignments were carried out, and all the cross peaks in the spectra were identified. Figure 2 shows the short-range, and medium-range NOE connectivities. Vicinal coupling constants between amide proton and a proton larger than 8 Hz are also indicated. This suggests that the light chain is rich in β-strands. The portion including Pro-27 is supposed to form a turn structure. Since NOEs were not observed in the C-terminus, this portion may form no regular structure and fluctuate in the solvent.

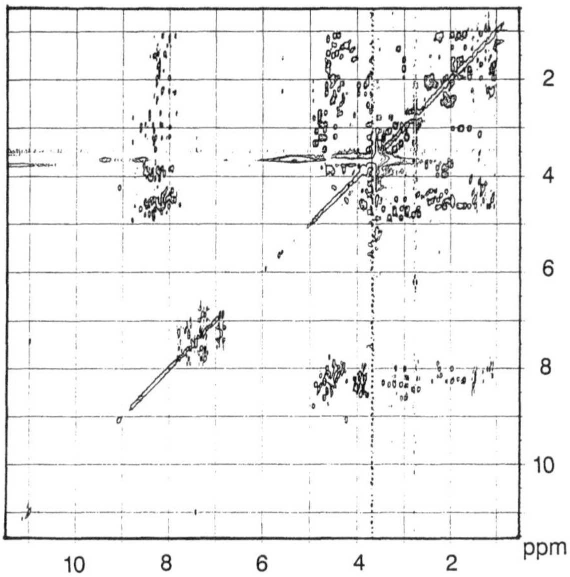

Figure 1. TOCSY spectra of the light chain of proteinase A with the mixing time of 44 ms.

Distance geometry calculation was performed using interresidue NOEs and vicinal coupling constants. Among 400 structures calculated, 84 converged structures were obtained. Then, molecular dynamics and energy minimization were carried out by using these structures as initial structures. The structures thus obtained were similar to one another, and were rich in β-strands.

NMR Spectra of the Intact Enzyme

As for the intact enzyme, 2D-NMR spectra (TOCSY, NOESY, and DQF-COSY) were measured. As shown previously,[14] the distribution of cross peaks in the fingerprint region indicated that the enzyme is rich in β-sheet and contains a tightly packed core structure in the molecule. Since proteinase A is quite soluble in an aqueous solution (up to 10 mM), ^{15}N-^{1}H and ^{13}C-^{1}H HMQC (natural abundance) and homonuclear 3D-NOESY-TOCSY spectra were measured. During the experiment, neither autolysis nor denaturation occurred. In the ^{15}N-^{1}H HMQC spectra, about 130 cross peaks were observed.

The spectral analysis and resonance assignments are in progress. In addition, heteronuclear 3D- and 4D-experiments are now under way using isotope-labeled samples.

Interaction between Enzyme and Substrate Analog

APDT inhibited the proteinase A activity against the bait region peptide of α_2-macroglobulin as a substrate. In order to get information on the mode of interaction between the enzyme and the substrate, NOESY spectra of APDT were measured in the presence and absence of proteinase A. While no cross peaks were observed in the spectra of APDT alone, negative NOEs between protons of APDT were detected in the complex with the enzyme. These results indicate that APDT takes intrinsically a random-coil structure, but has a certain regular structure in the complex. Considering that the molecular weight of APDT is less than 1,000, the results also suggest the existence of chemical exchange between the free state and

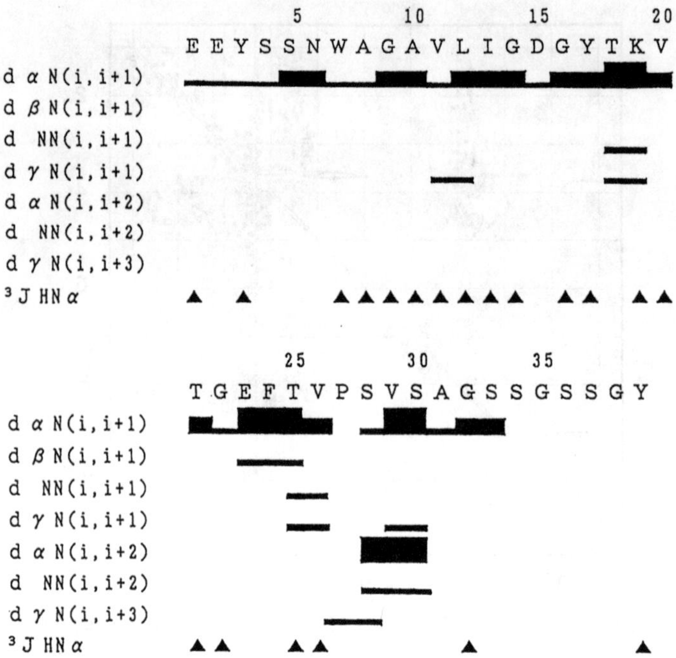

Figure 2. Amino acid sequence of the light chain and a summary of the short-range NOEs. The NOE intensities are shown in the order of the height of the bars. Residues with $^{3}J_{NH\alpha} > 8$ Hz are indicated by triangles.

the enzyme-bound state. The analysis of the conformation of APDT bound to the enzyme is in progress using intramolecular NOEs.

ACKNOWLEDGMENTS

This work was supported in part by Grants-in-Aid for Scientific Research from the Ministry of Education, Science, and Culture of Japan.

REFERENCES

1. Y. Koaze, H. Goi, K. Ezawa, Y. Yamada, and T. Hara, Fungal proteolytic enzymes. Part I. Isolation of two kinds of acid-proteases excreted by *Aspergillus niger* var. *macrosporus*, *Agric. Biol. Chem.* **28**: 216 (1964).
2. W.-J. Chang, S. Horiuchi, K. Takahashi, M. Yamasaki, and Y. Yamada, The structure and function of acid proteases. VI. Effects of acid protease-specific inhibitors on the acid proteases from *Aspergillus niger* var. *macrosporus*, *J. Biochem. (Tokyo)* **80**: 975 (1976).
3. K. Iio, and M. Yamasaki, Specificity of acid proteinase A from *Aspergillus niger* var. *macrosporus* towards B-chain of performic acid oxidized bovine insulin, *Biochim. Biophys. Acta* **429**: 912 (1976).
4. E. Ido, T. Saito, and M. Yamasaki, Substrate specificity of acid proteinase A from *Aspergillus niger* var. *macrosporus*, *Agric. Biol. Chem.* **51**: 2855(1987).
5. K. Takahashi, H. Inoue, K. Sakai, T. Kohama, S. Kitahara, K. Takashima, M. Tanji, S. B. P. Athauda, T. Takahashi, H. Akanuma, G. Mamiya, and M. Yamasaki, The primary structure of *Aspergillus niger* acid proteinase A, *J. Biol. Chem.* **266**: 19480 (1991).
6. H. Inoue, T. Kimura, O. Makabe, and K. Takahashi, The gene and deduced protein sequences of the zymogen of *Aspergillus niger* acid proteinase A, *J. Biol. Chem.* **266**: 19484 (1991).

7. D. Marion, and K. Wüthrich, Application of phase sensitive two-dimensional correlated spectroscopy (COSY) for measurements of ^1H-^1H spin-spin coupling constants in proteins, *Biochem. Biophys. Res. Commun.* **113**: 967 (1983).
8. A. Bax, and D. G. Davis, MLEV-17-based two-dimensional homonuclear magnetization transfer spectroscopy, *J. Magn. Reson.* **65**: 355 (1985).
9. M. Rance, O. W. Sörensen, G. Bodenhausen, G. Wagner, R. R. Ernst, and K. Wüthrich, Improved spectral resolution in cosy ^1H NMR spectra of proteins via double quantum filtering, *Biochem. Biophys. Res. Commun.* **117**: 479 (1983).
10. G. W. Vuister, R. Boelens, and R. Kaptein, Nonslective three-dimensional NMR spectroscopy. The 3D NOE-HOHAHA experiment, *J. Magn. Reson.* **80**: 176 (1988).
11. E. R. P. Zuiderweg, A proton-detected heteronuclear chemical-shift correlation experiment with improved resolution and sensitivity, *J. Mag. Reson.* **86**: 346 (1990).
12. D. J. States, R. A. Haberkorn, and D. J. Ruben, A two-dimensional nuclear overhauser experiment with pure absorption phase in four quadrants, *J. Magn. Reson.* **48**: 286 (1982).
13. T. Nakai, A. Kidera, and H. Nakamura, Intrinsic nature of the three-dimensional structure of proteins as determined by distance geometry with good sampling properties, *J. Biomol. NMR* **3**: 19 (1993).
14. K. Takahashi, M. Tanokura, H. Inoue, M. Kojima, Y. Muto, M. Yamasaki, O. Makabe, T. Kimura, T. Takizawa, T. Hamaya, E. Suzuki, and H. Miyano, Structure and function of pepstatin-insensitive acid proteinase from *Aspergillus niger* var. *macrosporus*, *in*: Structure and Function of the Aspartic Proteinases: Genetics, Structures, and Mechanisms, B.Dunn, ed., pp. 203-211, Plenum Press, New York (1991).

AUTHOR INDEX

SUBJECT INDEX

www.ingramcontent.com/pod-product-compliance
Lightning Source LLC
Chambersburg PA
CBHW050936160926
51501CB00061B/163

9780306448300